ISBN 978-1-5285-0538-3
PIBN 10933593

DEPARTMENT OF THE INTERIOR

UNITED STATES GEOLOGICAL SURVEY

GEORGE OTIS SMITH, DIRECTOR

BULLETINS

Nos. 336-340

WASHINGTON

GOVERNMENT PRINTING OFFICE

1908

CONTENTS.

3

ABBREVIATIONS.

In describing fuels, especially as to size, use is made of the following abbreviations:

e. = egg.
f. c. = finely crushed.
f. scr. = finely screened.
l. = lump.
n. = nut.
p. = pea.
r. o. m. = run of mine.

s. = slack.
sc. = screenings.
std. = standard.
thr. = through.
w. = washed.
″ = inch or inches.

WASHING AND COKING TESTS OF COAL AND CUPOLA TESTS OF COKE CONDUCTED BY THE UNITED STATES FUEL-TESTING PLANT AT ST. LOUIS, JANUARY 1, 1905, TO JUNE 30, 1907.

By RICHARD MOLDENKE, A. W. BELDEN, and G. R. DELAMATER.

INTRODUCTION.

By JOSEPH A. HOLMES.

The tests of washing and coking coals and of the behavior of the resulting coke in cupola practice, as reported herein, were made during the fiscal years 1905 and 1906 at the St. Louis fuel-testing laboratory of the United States Geological Survey. These tests were carried on in connection with similar investigations of the steaming and gas-producing qualities of the same coals and of the possibility of improving such coals by briquetting. This work was a part of the general inquiry concerning the most economical manner of utilizing each type of coal tested.

Many coals as received from the mine were found to be too high in ash, in sulphur, or in phosphorus to make satisfactory metallurgical coke without prior treatment, and some coals possessed better coking qualities than others. It was found that the washing of some coals so reduced the percentage of ash and sulphur as to make available for the production of coke a coal which otherwise would have had no value for this purpose. In the following pages are reported the details of the washing of coal, the production of coke therefrom, and the behavior of the coke in the cupola when utilized · for the production of castings, the results of each test being tabulated in full. A study of these tables indicates many important facts as to the behavior and treatment of the coals mined in the various portions of the United States when prepared as metallurgical coke.

The washing tests of 1905 were not as satisfactory as the later tests because of inadequate storage facilities and the lack of certain equipment, but the latter was added in time for the tests of 1906. An important result of the washing tests is shown in the percentage of ash and sulphur actually removed. The reduction of these impurities by washing, of course, increases the percentages of fixed carbon and volatile matter over the amounts present in raw coal. These facts, the number of washings, and the methods of washing, are recorded, thus furnishing valuable data as a guide to the treatment necessary to render each coal tested most suitable for coking. Altogether there were 101 regular washing tests and 12 special tests.

The results of these tests show an increase in moisture of 10 to 30 per cent, a reduction in ash in the 1905 tests of 15 to 50 per cent and in the 1906 tests of 20 to 60 per cent, and a reduction in sulphur in the 1905 tests of 10 to 40 per cent and in the 1906 tests of 10 to 50 per cent. A few examples of the total amount of reduction may be mentioned. A raw coal containing 5.05 per cent of sulphur contained after washing 2.47 per cent, a total removal of 55 per cent. Proportionate reductions in sulphur were made in coals containing lesser amounts. The ash in a raw coal containing 42.56 per cent was reduced by washing to 29.67 per cent, a total removal of 65 per cent. In a similar manner ash in a raw coal containing 15.72 per cent was reduced to 10.16 per cent, a total removal of 41 per cent; and in a coal containing 9.81 per cent to 5.38 per cent, a total removal of 59 per cent. It is evident that coals which are in the raw state utterly unfit for steaming purposes can be made fairly good steaming coals by washing, and that coals unsuited for coking can be made available in the same way.

It is proposed to conduct during the next fiscal year washing tests with much improved apparatus at the fuel-testing plant recently established at Denver, Colo., where experiments in washing and coking will be made on the coals mined in the Rocky Mountain region, with a view to determining what can be done to make them available for the production of metallurgical coke.

The coking tests were made in ovens of the regular beehive pattern, two of standard size 7 feet high, and one of standard diameter 6 feet 4 inches high. Samples of coke were taken from five different parts of the oven in practically the same location for each test, so as to give a standard method of comparison for each coke. The present report covers 192 tests, made on 100 coals, the samples having been collected from 17 States and 1 Territory. One hundred of these tests were made on raw coal, 82 on washed coal, and 10 under miscellaneous conditions. In some of these tests it was found that the addition of pitch produced coke from coal which when tested raw gave either no coke or coke of an inferior quality. In other tests the addition of pitch did not improve the quality of the coke. The tabulated results of the coking tests should be studied in the light of the description of the resulting coke which accompanies the tables. The physical tests to determine the compressive strength of the coke—or, in other words, the, height of the' furnace burden which the coke will support—showed only the worthlessness of such determinations. The compressive strength of a given coke made with the same coal ranged from about 700 pounds ultimate strength per square inch to over 2,000 pounds. As a coke with compressive strength of 48 pounds will support the burden of any modern furnace, it is evident that this test is of little or no practical value, especially as the burden borne by the coke may be greatly modified by the action of heat, by attrition, and by other factors. The inquiries seem to indicate that the yield of coke is increased and the proportion of breeze reduced by preliminary crushing. Further experiments are necessary to verify these determinations, as well as to indicate the limit of fineness of such crushing. Fine crushing appears to increase the strength of the coke, which is apparently influenced also by the amount and distribution of ash.

More complete coking tests will be carried on with a view to procuring more conclusive data along the lines above indicated, also with a view to determining more accurately the loss of sulphur from coal to coke, which varies with the coals and the method of treatment. These coking tests are being continued at the new plant at Denver on beehive ovens with two heights of crown, in order to determine the treatment necessary to produce good metallurgical coke from the coals mined in the Rocky Mountain region.

The great need of the immediate future in connection with coking experiments is the conduct of such tests in by-product ovens, and it is hoped that funds may soon be had which will permit the erection and operation of such ovens.

The cupola tests of coke in 1905 and 1906 were carried on along lines fully described in Professional Paper No. 48. The results as set forth in the following tables give the details of 170 cupola tests. The data concerning record of melt, taken in connection with the indications of the source of the coals and the analyses of the corresponding coke, furnish interesting facts as to the melting ratio of iron to coke, the rate of melting per hour, and the amount of iron recovered. Equally interesting is the table giving the chemical effect on iron from cupola tests of cokes made from coals mined in various States. It is not contemplated that these cupola tests will be continued during the fiscal year, in view of the necessity of devoting the available funds to the study of the coking qualities of western coals.

WASHING TESTS.

By G. R. Delamater.

REPORT FOR 1905.

IMPROVEMENT IN EQUIPMENT.

The lack of adequate storage facilities and the constant demand on the weighing and conveying apparatus for delivering coal to the other sections of the fuel-testing plant, which greatly interfered with the washing tests made during 1904,[a] were again felt in 1905 and tended somewhat to vitiate the results, although much improvement was made in the equipment.

In order to eliminate these difficulties, important changes were made during the year in the arrangement of the washer equipment. The storage capacity available during 1904 was increased from 175 tons to 350 tons. Additional "shed bins" of 150 tons aggregate capacity were provided outside the washer plant for storing coal at times when the regular washer bins were filled.

The coal was shoveled from the cars direct to the crusher or to a hopper scale. By means of a combination elevator conveyor the coal could be transferred from the hopper scale or crusher to any one of the twelve regular storage bins, or from one bin to another; or could be transferred to belt conveyors for delivering the coal to the boiler section and other divisions of the plant. The elevator conveyor referred to was relieved of a large portion of the work of handling coals for the gas-producer and boiler sections by a 30-inch Jeffrey belt conveyor, which was installed to run from the car siding to the bins of these sections for the purpose of delivering coal to them direct from the cars.

All washing tests made during 1905 were made on the Stewart jig used during 1904, at a speed of 35 revolutions per minute and 6-inch stroke. The sludge-recovery system, with the customary perforated-bucket elevators, was used in reclaiming the washed coal and refuse

---

[a] Compare the following U. S. Geological Survey publications: Bull. No 261, 1905, p. 60; Prof. Paper No. 48, 1906, p. 1460.

8

after washing.   Owing, however, to the fact that there were only two sludge tanks—one for the washed coal and one for the refuse—only one jig could be operated at a time, since if two or more jigs were operated their output would become mixed in the sludge tanks.

With the existing arrangement of the washery it was necessary to use the same water over and over again.   The washed-coal sludge tank, supplied with water from the city mains, was used as a reservoir from which the water was delivered, principally beneath the screens of the jigs.   A considerable amount of fine coal was thus carried over from the sludge tank.   The bulk of this fine coal settled to the bottom of the jig body, where it became mixed with the refuse and was carried on to the refuse sludge tank.

All coals tested were passed through an 18 by 24 inch Cornish tooth-roll crusher, which breaks the coal down to a maximum size of about 2 inches, although, of course, a considerable proportion may be much smaller, depending on the nature of the coal.

The power for operating the plant was furnished by a 12 by 16 inch Frost steam engine, belted to a main shaft from which the jigs and other machinery of the plant were driven.   The steam for this engine was received from the boiler section.

PERSONNEL.

The 1905 tests were made under the direction of John D. Wick.

REPORT FROM JANUARY 1, 1906, TO JUNE 30, 1907.

EQUIPMENT AND OPERATION.

On February 22, 1906, the washery plant was almost entirely destroyed by fire, and with it a few samples of coal that were on hand in the storage bins.   The plant was immediately rebuilt, the former arrangement being followed throughout.

From January 1 to December 15, 1906, one Stewart jig was used in making all the washing tests.   During December, 1906, a special jig was installed.   This jig was of the center-plunger type, i. e., the plunger was directly beneath the screen, and the upstroke of the plunger caused the pulsation.   The plunger had no valves, but valves were arranged in the sides of the jig body to admit the supply water on the downstroke of the plunger.   Cams and springs were used in such a manner that the plunger had a slow downward and a quick upward stroke.   The screen of this jig was 4 feet wide by 5 feet long and was made of strips of No. 10 wire running lengthwise of the screen frame and set one-sixteenth inch apart.   The length of the stroke was adjustable up to 4 inches.   The depth of the coal bed was also adjustable.

Owing to the fact that the power for operating the washer plant was furnished by a 12 by 16 inch Frost steam engine, belted to a main shaft from which the jigs and other machinery were driven, it was impossible to change the speed of the jigs. Better results could have been obtained on some coals tested had it been possible to change the speed to suit the length of stroke used.

As the only crusher available for this work was an 18 by 24 inch Cornish tooth-roll crusher, it was impossible to crush some coals down as fine as they should have been crushed. However, an adjustable-mesh bumping screen was installed in January, 1907, in such a manner that the coal was first passed over this screen, and the tailings then passed on to the crusher, while the fuel which went through the screen dropped into the bin over the jig. The product of the crusher was then elevated again to the screen, and this cycle of operation was repeated until all the coal passed through the screen.

In December, 1906, a float and sink testing equipment was installed. Before each washing test was made, samples of the raw coal, quartered down to 2 kilograms each, were tested on four different specific-gravity solutions. In this manner it was possible to make a preliminary determination of the result of a separation under varying percentages of washed coal and refuse. The coal was then washed with the jig regulated to discharge as refuse a percentage about equal to the percentage found advisable from the float and sink tests. After a washing test was made, a sample of the refuse was taken and quartered down to four samples of 2 kilograms each, and these were also tested on the specific-gravity solutions. The test showing the highest percentage of float coal and having an analysis which agreed fairly with that of the washed coal was then used in determining the percentage of "loss of good coal in the refuse." In this manner the efficiency of the test was shown.

PERSONNEL.

John D. Wick, assisted by Edward Moore, was in charge of the washing tests from January 1 to June 30, 1906; J. II. Gould from July 16 to October 12, 1906; and G. R. Delamater from November 15, 1906, to June 30, 1907.

EXPLANATION OF TABLES.

"*Percentage of reduction*" and "*amount actually removed.*"—The "percentage of reduction" is the comparison made of the percentages of the impurities in the raw coal and in the washed coal. It will be readily understood that if the ash alone is reduced by washing, the fixed carbon and volatile matter will form a higher percentage of the washed coal than of the raw coal. In actual practice, however, it is impossible to make so perfect a separation that the washing process will not remove portions of some constituents other than the impurities, and therefore the percentage of each constituent in the washed coal is affected by the reduction of each of the other constituents. This is clearly indicated in test 192, on Alabama No. 6, and test 198, on Virginia No. 6. A comparison of the raw-coal and washed-coal analyses in these two tests shows that in the test on Alabama No. 6 the percentage of sulphur was the same in the washed coal as in the raw coal; and in the test on Virginia No. 6 the sulphur in the washed coal was higher than in the raw coal. It will therefore be seen that a simple comparison of the raw-coal and washed-coal analyses will not always show whether any of the sulphur in the raw coal was actually removed with the refuse in washing.

Formulas.—In order that these percentages might be determined, the following formulas were compiled and used in making up this report. It will be noted by referring to the test data (P. 15) that 10 per cent of the original sulphur in the raw coal was actually removed with the refuse in washing Alabama No. 6, and that 13 per cent was actually removed in washing Virginia No. 6:

Let X = the percentage of reduction of any constituent.

Y = the percentage of any constituent removed by washing.

M = the percentage that the amount of the constituent in the washed coal is of the raw coal.

a = the percentage that the washed coal is of the raw coal.

b = the percentage of the constituent in the washed coal.

c = the percentage of the constituent in the raw coal.

Then $X = \dfrac{c-b}{c}$, $M = ab$, and $Y = \dfrac{c-M}{c}$.

TESTS MADE.

Sixty-three domestic samples of fuel from fourteen States and Territories and two samples from·Argentina were washed during the period covered by this report. The detailed results of the tests are given in the following tables.

Details of washing tests of bituminous coals in 1905.

Washing test No.	Field No. of coal.a	Bed.	Date of test.	Size of coal. (See p. 4.)		Amount of coal.				
				As shipped.	As washed.	Raw (lbs.).	Washed.		Refuse.	
							Lbs.	Per cent.	Lbs.	Per cent
	Illinois:									
101	6	No. 6	May 15	r. o. m.	thr. 2".	14,710	13,586	92	1,124	8
104	7 C	..do..	June 26	s.	s.	15,809	11,238	71	4,571	29
108	7 D	..do..	July 8	r. o. m.	thr. 2'.	14,000	11,790	84	2,210	16
102	9 A	..do..	May 22	r. o. m.	thr. 2".	18,000	13,920	78	4,080	22
103	10	No. 7	May 26	s.	s.	14,710	12,795	87	1,915	13
106	.12	No. 6	July 6	r. o. m.	thr. 2".	18,000	15,900	88	2,100	12
107	13	No. 7	..do....	1½"-0".	thr. 2'.	29,950	27,550	92	2,400	8
105	14	No. 5	June 29	l.	thr. 2'.	18,000	15,955	89	2,045	11
110	15	No. 6	July 12	l., e.	thr. 2'.	18,000	13,035	72	4,965	28
111	16	No. 7	July 28	l., e.	thr. 2'.	14,000	12,500	89	1,500	11
109	18	No. 2	July 12	l.	thr. 2'.	18,000	14,400	80	3,600	20
	Indiana:									
115	3	No. 5	Aug. 7	n., s.	thr. 1½".	50,000	36,000	72	14,000	28
112	4	No. 6	July 25	sc.	thr. 1½".	32,000	24,000	75	8,000	25
113	6	No. 4	Aug. 4	r. o. m.	thr. 2".	24,000	19,100	80	4,900	20
114	7 A	No. 5	Aug. 7	l., e., n.	thr. 2".	14,000	12,240	87	1,760	13
118	8	No. 7	Aug. 31	l.	thr. 2".	16,000	15,080	94	920	6
117	9 B	..do..	Aug. 24	r. o. m.	thr. 2".	18,000	15,460	86	2,540	14
116	10	No. 6	Aug. 21	l.	thr. 2".	18,000	15,300	85	2,700	15
	Maryland:									
130	1	Lower Kittanning.	Oct. 20	r. o. m.	thr. 2'.	45,500	37,450	82	8,050	18
	Ohio·									
119	1	No. 4	Sept. 8	r. o. m.	thr. 2".	26,900	23,750	88	3,150	12
121	2	No. 5	Sept. 13	r. o. m.	thr. 2".	32,420	25,625	79	6,795	21
122	3	No. 6	Sept. 19	r. o. m.	thr. 2".	47,125	34,800	74	12,325	26
125	4	No. 8	Sept. 27	over ¾".	tnr. 2".	29,120	26,000	89	3,120	11
133	6	..do..	Nov. 8	r. o. m.	thr. 2".	24,000	20,400	85	3,600	15
138	7	No. 7	Dec. 16	over 1½".	thr. 2".	16,000	14,000	88	2,000	12
137	8	No. 6	Dec. 15	r. o. m.	thr. 2".	17,200	15,560	90	1,640	10
131	9 B	No. 4	Oct. 25	n., s.	thr. 2".	46,530	37,830	81	8,700	19
	Pennsylvania:									
120	5	Pittsburg	Sept. 13	over ¾."	thr. 2".	30,920	29,000	94	1,920	6
124	6	..do..	Sept. 26	r. o. m.	thr. 2".	50,000	43,300	87	6,700	13
123	7	..do..	Sept. 23	r. o. m.	thr. 2".	32,000	27,180	85	4,820	15
126	9	Lower Kittanning.	Oct. 5	r. o. m.	thr. 2".	12,000	9,700	81	2,300	19
	Virginia:									
134	2	McConnell	Nov. 13	r. o. m.	thr. 2".	28,000	24,550	88	3,450	12
	West Virginia:									
127	4 B	Upper Freeport.	Oct. 9	r. o. m.	thr. 2".	26,000	21,000	79	5,600	21
128	16 B	Pittsburg	Oct. 14	s.	s.	22,825	19,800	87	3,025	13
132	17	Bakerstown.	Oct. 27	r. o. m.	thr. 2".	29,530	24,765	84	4,765	16
135	20	Keystone	Nov. 29	r. o. m.	thr. 2".	49,150	24,590	87	6,560	13
136	21	Peerless	Dec. 11	r. o. m.	thr. 2".	24,000	22,000	92	2,000	8
	Wyoming:									
129	3	(?)	Oct. 19	r. o. m.	thr. 2".	24,120	20,060	85	4,060	15

a Detailed account of the field origin and collection of each sample of coal may be found in Bull. U. S. Geol. Survey No. 290, 1906.

Details of washing tests of bituminous coals in 1905—Continued.

Washing test No.	Chemical analyses (per cent).								Reduction (per cent).		Actually removed (per cent).	
	Raw coal.					Washed coal.						
	Moisture.	Volatile matter.	Fixed carbon.	Ash.	Sulphur.	Moisture.	Ash.	Sulphur.	Ash.	Sulphur.	Ash.	Sulphur.
101	14.43	29.48	42.81	13.28	4.01	15.23	8.64	3.30	35	18	40	24
104	10.69	33.08	36.14	20.09	4.06	16.64	8.59	3.25	57	20	70	43
108	10.83	36.24	39.75	13.18	4.53	12.45	9.30	3.65	29	19	41	32
102	13.54	35.69	40.03	10.74	4.03	15.65	7.57	3.38	30	16	45	34
103	9.50	31.98	47.08	11.44	1.45	11.86	6.67	1.38	42	5	48	17
106	8.20	32.26	46.59	12.95	3.48	13.30	8.91	2.48	31	29	39	37
107	8.31	31.65	49.56	10.48	1.55	11.15	7.49	1.27	29	18	34	25
105	12.77	34.68	40.77	11.78	4.16	16.32	9.37	3.29	20	21	29	29
110	9.95	34.76	42.06	13.23	3.87	11.81	8.41	3.00	36	23	54	44
111	8.43	30.08	51.89	9.60	1.14	10.14	8.06	1.02	16	11	25	20
109	12.39	36.89	41.80	8.92	3.92	14.99	5.77	2.98	35	24	48	39
115	13.18	31.92	39.27	15.63	4.79	15.02	8.61	3.25	45	32	60	51
112	13.99	29.40	42.29	14.32	2.31	16.49	7.25	1.94	49	16	62	37
113	10.80	36.09	40.49	12.62	4.39	11.65	9.83	3.49	22	21	38	36
114	8.90	38.52	43.37	9.21	3.74	10.16	7.89	3.24	14	13	25	25
118	9.55	36.19	43.65	10.61	3.72	11.76	9.52	3.18	10	15	16	20
117	13.53	34.80	40.91	10.76	3.15	14.55	8.14	2.56	24	19	35	30
116	10.72	39.29	41.42	8.57	3.83	10.67	6.15	3.34	28	13	39	26
130	2.33	16.11	68.43	13.13	1.49	3.67	10.61	1.09	19	27	34	47
119	7.71	38.32	42.02	11.95	4.61	9.25	8.57	3.72	28	19	37	29
121	9.01	35.85	43.80	11.34	4.02	10.77	7.42	2.95	35	27	48	42
122	9.90	33.66	44.86	11.58	1.81	9.96	7.74	1.36	33	25	51	44
125	3.53	37.45	49.90	9.12	3.47	3.33	7.48	3.27	17	6	27	15
133	5.31	36.72	49.45	8.52	3.33	6.16	6.38	2.94	25	12	36	25
138	6.65	33.94	48.86	10.55	3.13	7.47	6.37	2.16	40	31	47	39
137	7.55	38.00	46.08	8.37	2.84	11.77	6.03	2.07	28	27	35	35
131	6.87	36.87	43.10	11.93	3.35	9.49	7.45	2.88	38	14	49	31
120	2.46	34.48	57.01	6.05	.88	4.91	4.57	.90	24	......	29	3
124	3.24	31.78	52.46	12.52	1.94	4.31	7.26	1.47	42	24	50	35
123	4.09	20.62	62.82	12.47	2.08	5.67	10.08	1.55	19	25	31	37
126	3.09	17.29	68.29	11.33	2.04	4.58	8.75	1.24	22	39	30	51
134	3.35	35.13	55.94	5.58	.92	6.39	3.95	.88	29	4	38	16
127	3.91	26.68	59.30	10.11	1.07	4.47	7.76	.81	23	24	39	31
128	5.57	31.61	54.45	8.37	1.20	5.41	5.91	.92	29	23	39	33
132	3.46	27.29	61.13	8.12	1.45	5.33	5.50	1.14	32	21	43	34
135	2.82	32.20	56.95	8.03	1.38	5.70	4.64	1.07	42	22	50	33
136	3.57	36.38	55.20	4.85	1.32	6.35	3.47	1.00	28	24	34	30
129	15.12	34.36	33.82	16.70	6.66	19.16	6.52	4.16	61	38	67	47

Details of washing tests, January 1, 1906, to June 30, 1907.

Washing test No.	Field No. of fuel.[a]	Designation of bed.	Date of test.	Jig used.[b]	Size of fuel. (See p. 4.)		Weight of raw fuel (tons).	Amount washed fuel.	
					As shipped.	As used.		Weight (tons).	Per cent.
	Alabama:								
163	2 B	Jagger	May 26, 1906	Stewart..	r. o. m.	thr. 2".	7.93	6.85	86
161	3	Underwood or Thompson.	May 23, 1906	......do....	r. o. m.	thr. 2".	9.00	8.25	92
159	4	Youngblood	May 21, 1906	......do....	r. o. m.	thr. 2".	8.50	7.23	85
195	5c	Black Creek	Jan. 15, 1907						
192	6	Pratt	Jan. 12, 1907	Special...	r. o. m.	thr. 1".	12.00	10.75	90
	Argentina:								
187	1		Nov. 13, 1906	Stewart..	r. o. m.	thr. 2".	18.00	9.00	50
187a	1		Dec. 24, 1906	......do....	r. o. m.	thr. 2".	5.50	3.30	60
	Arkansas:								
139	1 B	Huntington	Dec. 30, 1905	......do....	s.	s.	30.59	23.00	75
141	7 B	Hartshorne	Jan. 15, 1906	......do....	s.	s.	25.00	19.00	76
144	8	(?)	Jan. 29, 1906	......do....	No. 4.	thr. 2".	11.50	9.78	85
140	9	Huntington	Jan. 4, 1906	......do....	s.	s.	38.65	28.67	74
	Illinois:								
142	20	No. 6	Jan. 18, 1906	......do....	sc.	sc.	31.64	28.50	90
160	21	....do....	May 21, 1906	......do....	l.	thr. 2".	8.50	7.32	86
151	22 A	....do....	Feb. 16, 1906	......do....	l.	sc.	9.70	8.50	88
150	22 B	....do....	Feb. 13, 1906	......do....	sc.	sc.	20.00	16.00	80
146	23 A	....do....	Jan. 31, 1906	......do....	5" l.	thr. 2".	14.00	12.00	86

a Detailed account of the field origin of each sample of fuel may be found in Bull. U. S. Geol. Survey No. 332.

b Stewart jig—speed 35 revolutions per minute, with 6-inch stroke; special jig—speed, 70 revolutions per minute, with 2½-inch stroke.

e Not enough coal for other than special float and sink tests.

Details of washing tests, January 1, 1906, to June 30, 1907—Continued.

Washing test No.	Field No. of Fuel.	Designation of bed.	Date of test.	Jig used.	Size of fuel. (See p. 4.) As shipped.	As used.	Weight of raw fuel (tons).	Amount washed fuel. Weight (tons).	Per cent.
	Illinois—Cont'd.								
147	23 B	No. 6	Feb. 1,1906	Stewart	s.	s.	40.00	31.50	79
169	24 A	...do	June 8,1906	...do	sc.	sc.	10.00	7.50	75
166	24 B	...do	June 1,1906	...do	l.	thr. 2".	9.37	8.33	89
162	25 A	...do	May 24,1906	...do	r.o.m.	thr. 2".	7.27	6.00	83
164	26	No. 5	May 26,1906	...do	r.o.m.	thr. 2".	9.00	8.00	89
165	27	No. 6	May 29,1906	...do	r.o.m.	thr. 2".	9.00	7.77	86
181	28 C	No. 7	Sept. 26,1906	...do	l.	thr. 2".	12.00	9.96	74
183	29 A	No. 5	Oct. 13,1906	...do	sc.	sc.	9.00	6.87	76
184	29 A	...do	Oct. 16,1906	...do	sc.	sc.	29.75	20.75	70
190	30	No. 7	Jan. 5,1907	...do	n.	thr.2½".	15.00	11.60	77
190a	30	...do	Feb. 11,1907	Special	n.	thr. 1".	12.45	10.10	81
196	34 A	No. 5	Feb. 15,1907	...do	sc.	thr. 1".	24.65	19.55	80
197	34 B	...do	Feb. 12,1907	...do	r.o.m.	thr. 1".	14.00	11.81	85
	Indiana:								
145	12	...do	Jan. 30,1906	Stewart	r.o.m.	thr. 2".	20.00	17.53	89
185	20	Brazil Black	Nov. 14,1906	...do	sc.	sc.	30.00	20.30	68
	Indian Territory:								
176	2 B	Hartshorne	July 13,1906	...do	s.	s.	19.00	14.53	78
175	8	(?)	July 1,1906	...do	s.	s.	18.80	16.15	86
	Kansas:								
191	2 B	Weir-Pittsburg	Jan. 10,1907	Special	s.	s.	23.00	18.10	79
191a	2 B	...do	Jan. 21,1907	...do	s.	s.	39.00	25.25	65
148	6	...do	Jan. 8,1906	Stewart	l.	thr. 2".	12.00	11.00	92
	Kentucky:								
143	2 B	(?)	Jan. 19,1906	...do	coke br.	coke br.			
182	9 A	No. 9	Oct. 1,1906	...do	n.	n.	9.56	7.84	82
	Missouri:								
149	5	(?)	Feb. 10,1906	...do	r.o.m.	thr. 2".	7.65	6.45	84
155	6a	(?)	Feb. 21,1906	...do	l.	thr. 2".			
152	7 A	(?)	Feb. 17,1906	...do	No.1 n.	No.1 n.	12.50	10.73	86
154	7 Aa	(?)	Feb. 22,1906						
153	7 B	(?)	Feb. 19,1906	...do	No.2 n.	No.2 n.	11.75	9.30	79
	New Mexico:								
168	3 C	Main Raton,or Lower Laramie.	June 6,1906	...do	s.	s.	21.50	19.00	88
174	4 A	...do	June 19,1906	...do	r.o.m.	thr. 2".	10.00	8.14	81
170	4 B	...do	June 9,1906	...do	s.	thr. 2".	12.00	10.50	88
167	5	...do	June 2,1906	...do	r.o.m.	thr. 2".	7.50	6.65	89
	Ohio:								
193	12	No. 8	Jan. 25,1907	...do	r.o.m.	thr. 2".	6.70	5.10	76
	Pennsylvania:								
179	12	Pittsburg	Sept. 20,1906	...do	r.o.m.	thr. 2".	10.60	8.45	80
188	15	B. or Miller	Feb. 4,1907	Special	r.o.m.	thr. 1".	20.37	15.25	76
189	17	Upper Freeport	Feb. 6,1907	...do	r.o.m.	thr. 1".	7.28	6.30	87
194	20	Lower Kittanning.	Jan. 29,1907	...do	r.o.m.	thr. 1".	22.21	17.25	78
	Tennessee:								
171	1	Mingo	June 12,1906	Stewart	r.o.m.	thr. 2".	10.80	9.30	86
172	5	Brushy Mountain.	June 13,1906	...do	r.o.m.	thr. 2".	9.22	8.00	87
156	7 B	Wilder	May 1,1906	...do	s.	s.	15.50	10.50	68
157	8 A,8 B	First above Sewanee.	May 15,1906	...do	r.o.m.	thr. 2".	49.00	43.00	88
158	9 B,9 C	Sewanee	May 19,1906	...do	s.	s.	9.69	7.21	75
173	10	Battle Creek	June 14,1906	...do	1"s.	1"s.	30.25	23.70	78
178	11	(?)	Sept. 11,1906	...do	s.	s.	21.00	13.75	65
	Virginia:								
198	6	No. 4	Feb. 1,1907	Special	r.o.m.	thr. 1".	8.31	6.75	81
	West Virginia:								
186	22 A	(?)	Oct. 25,1906	Stewart	n. & s.	n. & s.	19.25	16.25	85
180	23 B	Cedar Grove	Sept. 25,1906	...do	n. & s.	n. & s.	18.00	16.99	94
	Miscellaneous:								
177	10		Aug. 7,1906	...do	s.	s.	28.28	20.08	71

a Destroyed by fire when plant was burned.

Details of washing tests, January 1, 1906, to June 30, 1907—Continued.

Washing test No.	Amount of refuse.		Chemical analyses of fuel (per cent).								Reduction (per cent).		Actually removed (per cent).	
			Raw.					Washed.						
	Weight (tons).	Per cent.	Moisture.	Volatile matter.	Fixed carbon.	Ash.	Sulphur.	Moisture.	Ash.	Sulphur.	Ash.	Sulphur.	Ash.	Sulphur.
163	1.08	14	3.95	30.70	50.76	14.59	1.12	6.29	9.39	1.22	36	.....	45	1
161	.75	8	3.03	30.94	55.31	10.71	.49	5.82	10.01	.58	7	.....	14	...
159	1.26	15	6.43	28.56	52.09	12.92	1.08	6.82	38.1	1.03	71	5	74	19
195			5.69	53.28	25.05	16.08	1.40							
192	1.25	10	3.39	63.57	26.20	6.84	.59	6.69	4.76	.59	30		37	10
187	9.00	50	7.10	30.97	19.37	42.56	.82	17.29	29.67	.64	30	22	65	61
187a	2.20	40	7.10	30.97	19.37	42.56	.82	22.73	34.57	.55	19	33	51	60
139	7.59	25	7.49	15.16	59.38	17.97	1.06	6.32	8.62	1.12	52		64	21
141	6.00	24	6.89	15.23	62.88	15.00	2.24	6.45	7.19	1.89	52	16	64	36
144	1.71	15	5.19	10.49	70.31	14.01	2.05	5.03	7.85	2.03	44	1	52	17
140	9.97	26	5.26	14.71	55.22	24.81	1.00	7.78	14.30	.98	42	2	57	27
142	3.14	10	14.68	31.32	40.32	13.68	3.88	16.80	10.26	3.21	25	17	33	25
160	1.18	14	15.30	30.59	43.40	10.71	1.43	8.25	8.09	1.25	25	13	35	25
151	1.20	12	11.91	35.65	39.43	13.01	5.34	14.02	8.58	3.69	34	31	42	39
150	4.00	20	13.03	32.65	39.79	14.53	4.35	16.78	9.99	3.79	31	13	45	30
146	2.00	14	13.47	34.35	40.65	11.53	4.41	13.81	8.78	3.44	24	20	35	31
147	8.50	21	15.68	31.28	37.45	15.59	3.98	16.83	8.75	3.22	44	19	56	36
169	2.50	25	11.44	33.93	43.92	10.71	4.94	15.10	9.75	3.18	9	36	32	52
166	1.04	11	11.44	33.93	43.92	10.71	4.94	14.36	8.38	3.31	22	33	30	41
162	1.27	17	11.35	34.62	40.63	13.40	4.76	14.14	8.98	3.05	33	36	44	47
164	1.00	11	15.68	32.41	39.82	12.09	3.52	15.96	9.40	2.76	22	22	30	36
165	1.22	14	16.00	32.41	37.82	13.77	4.05	16.11	7.76	3.26	43	20	52	31
181	2.03	26	7.78	29.85	52.39	9.98	1.32	9.75	7.12	1.05	29	20	47	42
183	2.12	24	13.10	30.78	40.12	16.00	4.17	15.86	7.70	3.06	52	27	63	44
184	9.00	30	13.10	30.78	40.12	16.00	4.17	.15.86	7.70	3.06	52	27	66	49
190	3.40	23	11.69	39.42	35.70	13.19	4.38	12.36	9.44	3.26	28	26	45	40
190a	2.35	19	11.69	39.42	35.70	13.19	4.38	13.67	7.89	3.15	40	28	52	42
196	5.10	20	9.33	47.86	30.92	11.89	2.76	8.68	7.44	2.19	37	21	50	37
197	2.19	15	7.81	50.27	33.54	8.38	2.36	10.12	6.52	1.76	21	26	34	37
145	2.47	11	10.57	35.30	42.75	11.65	3.87	14.16	7.85	3.29	33	15	39	24
185	9.70	32	16.91	38.87	26.85	17.37	1.89	16.86	7.09	1.35	59	29	72	51
176	4.47	22	6.27	32.37	47.07	14.29	1.79	6.61	8.27	1.55	42	13	55	32
175	2.65	14	3.77	32.65	51.15	13.43	1.79	8.97	8.46	1.56	37	13	46	25
	4.90	21	8.01	45-22	26.39	20.38	4.70	7.00	8.88	3.72	56	21	62	37
191a	13.75	35	8.01	45.22	26.39	20.38	4.70	9.53	10.87	3.80	47	19	65	48
148	1.00	8	9.04	29.69	45.55	15.72	5.01	12.63	10.16	2.47	35	51	41	55
143						46.30		26.10			44			
182	1.71	18	8.70	35.00	47.34	8.96	3.14	9.09	7.22	2.61	19	17	34	32
149	1.20	16	12.92	33.64	39.82	13.62	5.03	13.93	9.08	3.62	33	28	44	40
155														
152	1.76	14	16.36	29.12	35.01	19.51	3.53	17.30	9.45	3.04	52	14	58	15
154														
153	2.45	21	16.39	29.01	34.42	20.18	3.12	19.70	11.05	3.07	45	2	56	22
168	2.50	12	4.36	32.21	47.51	15.92	.83	6.01	12.43	.71	22	15	25	25
174	1.86	19	2.78	34.31	48.34	14.57	.61	3.71	11.39	.58	22	5	37	23
170	1.50	12	3.38	34.63	48.45	13.54	.61	5.97	9.41	.65	31	.....	39	7
167	.85	11	2.72	31.85	50.86	14.57	.69	4.68	11.87	.91	25		28	...
193	1.60	24	4.14	47.18	39.30	9.38	3.96	6.85	6.19	3.60	33	9	50	31
179	2.15	20	1.96	30.55	58.24	9.25	2.19	4.63	6.40	1.39	31	37	45	49
188	5.12	25	3.13	69.45	17.61	9.81	3.77	6.45	5.38	1.53	45	59	59	62
189	.98	13	4.35	55.99	27.76	11.90	1.51	5.18	8.02	1.16	33	23	41	33
194	4.96	22	4.00	69.57	15.89	10.54	2.85	6.48	6.76	1.30	36	54	50	65
171	1.50	14	4.81	32.91	51.13	11.15	1.58	5.28	5.33	1.32	52	16	58	28
172	1.22	13	5.59	33.62	51.03	9.76	3.23	5.29	5.64	2.46	43	24	50	34
156	5.00	32	7.88	28.28	46.43	17.41	3.43	7.04	10.12	2.26	42	34	60	55
157	6.00	12	3.12	32.91	49.85	14.12	4.74	1.71	9.99	2.94	29	38	38	45
158	2.48	23	5.68	25.36	50.41	18.55	.74	4.02	9.91	.85	47	.....	60	14
173	6.55	22	2.92			22.74	.95	7.02	13.75	.98	40	.....	53	20
178	7.25	35	3.53	20.75	47.85	27.87	.90	5.60	13.47	.92	52	.....	69	33
198	1.56	19	5.62	61.52	23.07	9.79	1.21	6.36	4.38	1.30	57	.....	64	13
186	3.00	15	4.59	52.23	33.38	9.80	1.01	7.06	5.76	.97	41	4	50	19
180	1.01	6	3.25	34.61	54.56	7.58	1.22	4.24	4.87	.93	34	24	40	29
177	8.20	29	6.67	31.61	51.19	10.53	1.55	11.06	6.38	1.30	39	16	57	41

Details of special washing tests, January 5 to February 15, 1907.

Washing test No.	Field No. of fuel.	Date and duration of test.	Float and sink tests with finer crushing.								
			Special test No.	Size—through square hole (inch).	Specific gravity of solution used.	Float (per cent).	Sink (per cent).	Analysis of float (per cent)			
								Ash.		Sulphur.	
								Determined.	Reduction.	Determined.	Reduction.
	Alabama:										
195	5	Jan. 15	1		1.35	81	19	2.18	86	0.81	42
			2		1.40	85	15	2.63	84	.98	30
			3		1.45	87	13	2.66	84	1.05	25
			4		1.52	87	13	3.19	80	1.13	19
a 192	6	Jan. 12, 1¼ hours	1		1.36	87	13	2.81	59	.54	8.5
			2		1.42	90	10	3.51	49	.57	4
			3		1.48	91	9	3.43	50	.53	10
			4		1.56	94	6	3.75	45	.56	5
	Argentina:										
187	1		1		1.55	45	55	22.56	47	.73	12
			2		1.60	59	41	24.96	41	.78	12
			3		1.65	59	41	27.68	35	.72	12
			4		1.70	61	39	27.90	34	.70	12
	Illinois:										
190	30	Jan. 5, 2 hours.	1		1.36	73	27	7.10	46	3.29	23
			2		1.41	84	16	8.69	34	3.29	23
			3		1.47	88	12	8.98	31	3.33	23
			4		1.56	90	10	9.59	27	3.41	22
196	34 A	Feb. 15, 4¼ hours	1		1.36	84	16	6.07	49	1.90	31
			2		1.41	88	12	6.12	48	1.83	34
			3		1.45	90	10	7.06	41	2.18	21
			4		1.51	92	8	7.10	40	1.97	29
a 197	34 B	Feb. 12, 2 hours.	1	1	1.35	87	13	5.91	29	1.71	28
			2	1	1.41	90	10	6.15	27	1.64	31
			3	1	1.46	92	8	6.20	26	1.68	29
			4	1	1.51	92	8	7.23	14	2.17	8
	Kansas:										
151	2 B	Jan. 10, 2 hours.	1	Slack.	1.36	66	34	4.48	78	2.63	44
			2	Slack.	1.41	74	26	5.31	74	2.78	41
			3	Slack.	1.47	78	22	5.73	72	3.19	32
			4	Slack.	1.56	81	19	6.18	70	3.31	30
	Ohio:										
193	12	Jan. 25, ½ hour.	1		1.35	77	23	5.12	45	3.23	18
			2		1.40	89	11	6.43	32	3.63	8
			3		1.45	92	8	6.78	28	3.88	2
			4		1.52	94	6	7.31	22	3.98	......
	Pennsylvania:										
a 188	15	Feb. 4, 2 hours.	1		1.35	72	28	5.47	44	1.30	66
			2		1.41	78	22	5.27	46	1.45	62
			3		1.45	80	20	5.54	43	1.54	59
			4		1.52	81	19	6.26	36	1.71	55
189	17	Feb. 6, 1½ hours	1		1.35	86	14	5.14	57	1.00	34
			2		1.40	90	10	5.69	52	1.08	28
			3		1.45	91	9	6.20	48	1.26	17
			4		1.52	91	9	7.51	37	1.13	25
a 194	20	Jan. 29, 2¼ hours	1		1.35	83	17	4.95	53	.93	67
			2		1.42	88	12	5.66	46	1.24	57
			3		1.45	88	12	4.72	55	1.02	64
			4		1.52	89	11	6.07	42	1.09	62
	Virginia:										
a 198	6	Feb. 1, 1½ hours	1		1.35	84	16	2.60	54	.95	21
			2		1.41	85	15	2.98	48	.92	24
			3		1.45	85	15	3.44	42	.95	21
			4		1.53	87	13	3.53	41	.97	20

a Finer crushing advantageous.

Details of special washing tests, January 5 to February 15, 1907—Continued.

Washing test No.	Float and sink tests on refuse.						Loss of good coal in refuse (per cent).	Analysis of refuse (per cent).		
	Special test No.	Specific gravity of solution used.	Percentage of float—		Analysis of float (per cent).			Moisture.	Ash.	Sulphur.
			To refuse.	To total sample.	Ash.	Sulphur.				
a 192	1	1.35	18.40	1.91	2.81	0.89	1.00	8.21	34.92	2.20
	2	1.40	20.80	2.16	3.48	1.01				
	3	1.45	20.80	2.16	4.06	1.17				
	4	1.52	22.30	2.32	5.09	1.06				
190	1	1.36	15.80	2.98	8.00	3.12	2.98	11.22	46.50	9.59
	2	1.40	20.18	4.13	11.30	3.42				
	3	1.45	29.90	5.63	12.00	3.43				
	4	1.51	33.50	6.31	14.60	4.63				
196	1	1.35	12.10	2.52	6.95	2.05	2.52	9.29	58.43	11.91
	2	1.41	13.59	2.82	7.78	2.32				
	3	1.45	14.21	2.95	9.58	2.45				
	4	1.51	16.78	3.43	11.20	3.05				
a 197	1	1.35	50.00	7.80	6.27	2.36	1.75	15.35	61.00	15.90
	2	1.40	52.00	8.14	5.99	2.50				
	3	1.45	55.00	8.60	7.40	2.79				
191	1	1.35	9.00	1.94	4.30	2.58	2.16	....	76.50	11.32
	2	1.40	10.00	2.16	5.05	2.93				
	3	1.46	10.00	2.16	7.53	3.57				
	4	1.53	10.00	2.16	7.71	3.81				
193	1	1.35	39.00	9.35	5.75	3.67	9.00	5.78	19.91	6.62
	2	1.41	57.00	13.29	6.42	4.04				
	3	1.45	59.00	14.09	9.12	5.26				
	4	1.53	81.00	19.30	10.17	4.97				
a 188	1	1.35	11.80	2.95	4.95	1.71	2.00	5.78	47.18	19.78
	2	1.41	13.20	3.30	6.50	2.13				
	3	1.46	14.50	3.64	7.65	2.29				
	4	1.51	17.20	4.30	8.15	2.88				
189	1	1.35	13.00	1.70	5.39	1.28	1.50	4.58	41.50	8.85
	2	1.41	14.00	1.80	6.20	1.40				
	3	1.45	19.30	2.60	8.15	1.47				
	4	1.51	23.75	3.20	9.51	1.67				
a 194	1	1.35	17.20	3.91	5.42	1.69	2.00	10.21	46.25	17.40
	2	1.41	18.50	4.20	5.69	1.69				
	3	1.45	19.88	4.51	6.45	2.15				
	4	1.53	20.20	4.59	7.89	2.08				
a 198	1	1.35	15.30	2.90	4.80	1.39	2.20	3.64	63.98	6.15
	2	1.41	15.75	2.99	5.35	1.78				
	3	1.45	15.90	3.02	5.62	1.75				
	4	1.51	19.25	3.65	9.31	2.79				

a Finer crushing advantageous.

COKING TESTS.

By A. W. BELDEN.

EQUIPMENT.

The ovens in which the tests of the coking qualities of coals have been made are of the regular beehive pattern. Of the battery of three ovens two are of standard size, 12 feet in diameter and 7 feet high, the third is 12 feet in diameter and 6 feet 4 inches high. This change was made by raising the bottom of one of the standard ovens 8 inches with well-tamped loam and bottom tile of the usual size. The object of the change was to bring the charge nearer the dome of the oven and effect a more rapid penetration of heat.

For the first nineteen tests the small oven only was used. In the twentieth charge one of the 7-foot ovens was blown in, and two ovens were used continuously during the remainder of the work—one of each size. Owing to the small supply of coal it has not been possible to use more than two ovens, and they may, therefore, be considered as end ovens. Some suppose that end ovens yield results less favorable than those from ovens located between other heated ovens, but, even if this supposition is correct, the difference is fully balanced by the greater care bestowed on these experimental ovens as compared with ovens operated under normal conditions. As both of the ovens used are, in the sense indicated, end ovens, the results obtained in each are comparable one with the other.

In charging the ovens for the first nineteen tests the larry used held less than 1 ton. This necessitated the filling and emptying of the larry six to eight times before the charge was completed. Each portion thus became hot and began invariably to gas, and often to blaze before the next portion of the charge was added. This unfortunate state of affairs is believed to be responsible, at least in some measure, for cross lamination and cross breakage of the coke, layers of coal as charged showing plainly in many of these tests in each oven drawn. The average time of charging with this device was about one hour. After the nineteenth charge a standard-size larry was installed and the time of charging was reduced to an average of seven minutes. With this change the lamination and cross breakage referred to disappeared, showing that the whole charge should be put in at once.

18

PERSONNEL.

The writer took charge of this work in May, 1905, succeeding Fred W. Stammler, of Johnstown, Pa. He was assisted by W. E. Vickers, of Pocahontas, Va., to whom in large measure is due whatever success has been obtained during these investigations.

PROCEDURE OF TESTS.

All coal was finely crushed through a Williams mill unless otherwise tested for definite comparison of results, and these exceptions are noted in the subjoined detailed report (pp. 21–26). The coals not crushed were, when unloaded from the cars, put through rolls having an aperture of 1¼ inches. The coals put through the Williams mill will vary somewhat, depending on the nature of the coal, but will practically all pass through a 10-mesh sieve, as shown by the following report by the laboratory on an average sample: Amount remaining on 10-mesh, 15.08 per cent; on 20-mesh, 35.71 per cent; on 30-mesh, 12.89 per cent; on 40-mesh, 8.53 per cent; on 60-mesh, 9.33 per cent; on 100-mesh, 9.13 per cent; through 100-mesh, 9.33 per cent.

Both the door and the trunnel head of the oven were always closed directly after the oven was drawn and it was allowed to gather heat, the length of time varying as necessity demanded. The average time was one and one-half hours.

The sample of coal was taken at regular intervals as the charge was emptied from bin to larry, by means of a small shovel holding about one-fourth pound. The total weight of the sample averaged 45 pounds.

The sample of coke was taken from five different parts of the oven, as nearly as possible from the same location for each test, as follows: 2 feet from the oven door; 2 feet from each side, on a line drawn from the center of the oven; at the center; and 2 feet from the back wall, on a line with the point of selection of the pieces taken from the door and the center. The separate pieces of coke extended the whole height of the charge and were as nearly uniform in size as possible.

In beginning the series of tests, before the ovens were fully seasoned, the first charges showed a rather large percentage of breeze, and black butts due to cold bottom were produced. It was unfortunate that these first tests should have been made on coals that were supposedly noncoking, as the condition of the oven did not permit it to give as effective service as it would probably have given under other and more favorable conditions.

EXTENT OF TESTS.

In the scope of this report, covering the period from July 7, 1905, to February 20, 1907, are included results from 192 tests of 102 coals from seventeen States and one Territory, as shown in the accompanying table. Of these tests, 100 were made on raw coal, 82 on washed coal, 1 on raw coal with the addition of pitch, 6 on washed coal with the addition of pitch, 1 on washed coal with the addition of asphalt, and 2 on coals of widely varying composition. Of the 102 different coals, 8, viz, Arkansas No. 9, Illinois No. 19, Indiana Nos. 3 and 18, Ohio No. 3, Maryland No. 1, and Wyoming Nos. 3 and 5, produced no coke. Arkansas No. 9 and Maryland No. 1 were coked by the addition of pitch to washed coal. Four tests were made on Pennsylvania No. 9 (PP. 24, 32, 41); two tests with raw coal gave only a few pieces of coke; a third, with washed coal, produced coke of inferior quality; and the fourth, with the addition of 5 per cent pitch to raw coal, produced coke of no better quality than that from washed coal. Of Indiana No. 3, Ohio No. 3, and Wyoming Nos. 3 and 5, there was not enough for further tests.

TABULATION OF RESULTS.

The results of the coking tests will be found in the detailed report on each sample, presented below. For convenience of comparison data are given as to the yield of dry coke from dry coal as well as coke as received from coal as charged. The analyses of both coal and coke as received and on dry basis are also given. No distinction is made between breeze and ash, as it was found impossible to separate them with any degree of accuracy, and both are represented in this report in the item "breeze." This breeze includes everything that will pass through a fork with tines 1¼ inches apart, after thorough shaking, and its percentage is much higher than that from regular operations, but is comparable in all tests. It was not deemed necessary or advisable to size the coke, and under this caption is included everything over the 1¼-inch tine fork. Except in a few special cases the determination of phosphorus was not made on coke having over 2 per cent of sulphur, and when more than one test was made on a coal in the same condition this determination was also omitted.

Details of coking tests of coals, January 1, 1905, to June 30, 1907.

Test No.	Field No. of coal	Designation of bed.	At or near—	Duration of test (hours).	As shipped.	As used.	Real	Apparent	Dry	As received (wet)	Coke	Cells	1	2	3	4
		Origin of coal sample.a			Size of coal (see p. 4).		Specific gravity.		Pounds per cubic foot.		Percentage by volume.		6-foot drop test: Percentage over 2-inch mesh.			
							8	9	10	11	12	13	14	15	16	17
Alabama:																
82	2 B (w.)	Jagger	Carbon Hill	51	r. o. m.	f. c.	1.88	0.92	55.63	87.44	49.00	51.00	93.00	78.50	77.50	58.00
38	3	Underwood or Thompson	Garnsey	54	r. o. m.	f. c.	1.91	1.04	63.51	92.20	54.00	46.00	97.50	94.50	92.00	88.00
139	3 (w.)	...do...	Belle Ellen	44	r. o. m.	f. c.	1.99	.95	58.67	91.10	48.00	52.00	95.00	90.00	87.00	83.50
131	4 (w.)	...do...	...do...	47	r. o. m.	f. c.	1.95	.96	59.70	91.52	49.00	51.00	96.00	91.50	86.00	82.00
136	5	Youngblood	Lehigh	49	r. o. m.	f. c.	1.95	.87	54.06	88.35	45.00	55.00	92.00	88.00	83.50	81.00
171		Black Creek	Dolomite	42	r. o. m.	f. c.	1.99	.98	60.77	92.58	49.00	51.00	93.50	92.50	88.50	86.00
172	6	Pratt	...do...	40	r. o. m.	f. c.	1.97	.93	57.72	90.79	47.00	53.00	93.50	87.50	88.50	79.50
174	6 (w.)	...do...	...do...	50	r. o. m.	f. c.	1.95	.91	56.43	89.50	47.00	53.00	94.50	91.50	88.00	85.00
Arkansas:																
95	1 B (w.)	Huntington	Huntington	49	s.	s.	1.78	.58	35.13	76.92	33.00	67.00	97.00	94.50	90.50	88.00
96	1 B (w.)	...do...	...do...	71	s.	f. c.	1.90	.75	46.18	83.59	40.00	60.00	96.50	93.00	90.50	87.00
97	1 B (w.) c	...do...	...do...	48	s.		1.96	.83	50.37	86.53	42.00	58.00				
100	1 B (w.)	...do...	...do...	50	s.		1.95	.68	42.33	82.87	35.00	65.00				
104	7 B (w.) c	Hartshorne	Midland	67	s.	s.	2.01	.71	44.20	84.73	35.00	65.00	94.50	90.50	88.00	87.50
105	7 B (w.) c	...do...	...do...	57	s.	f. c.	1.97	.93	57.61	90.68	47.00	53.00				
98	9 (w.)	Huntington	Bonanza	28	1½" s.	s.										
99	9 (w.)	...do...	...do...	24	1½" s.											
101	9 (w.) d	...do...	...do...	39	1½" s.		2.04	1.00	58.37	90.18	49.00	51.00				
102	9 (w.) b	...do...	...do...	43	1½" s.		2.02	.94	58.45	91.52	47.00	53.00				
103	9 (w.) c	...do...	...do...	46	1½" s.											
Georgia:																
173	1	Little River	Menlo	58	"overly".	f. c.	2.01	.98	60.85	92.66	49.00	51.00	99.00	98.00	96.50	95.00
Illinois:																
1		No. 6	Collinsville	43	r. o. m.	f. c.	1.91	.85	51.82	86.11	45.00	55.00	92.50	86.50	78.00	73.50
4	7 D (w.)*	No. 7	...do...	65	r. o. m.	f. c.	1.85	.82	50.60	85.53	44.00	56.00	91.50	84.50	77.50	75.00
5	11 D (w.)*	...do...	Carterville	48	No. 3.	f. c.	1.87	.82	50.14	85.08	44.00	56.00	95.50	85.00	80.50	75.00
2	13 (w.)*	...do...	Benton	65	1½" by 6" e.	f. c.	1.85	.84	49.99	83.67	46.00	54.00	93.50	87.50	82.50	75.00
3	13*	...do...	...do...	65	1½" by 6" e.	f. c.	1.90	.90	54.60	87.67	47.00	53.00	95.00	89.50	85.50	81.00
7	16*	...do...	Herrin	65	l. e.	f. c.										
10	16 (w.)*	...do...	...do...	66	l. e.	f. c.	1.85	.87	53.11	86.18	47.00	53.00	93.50	84.50	80.50	76.00

a Additional details of origin of samples tested in 1905 (designated by * in column 2) can be found in Bull. U. S. Geol. Survey No. 290; of other samples in Bull. No. 332.
b With 10 per cent pitch.
c With 5 per cent pitch.
d With 3½ per cent asphalt.

Details of coking tests of coals, January 1, 1905, to June 30, 1907—Continued.

Test No.	Field No. of coal	Designation of bed.	At or near—	Duration of test (hours).	As shipped.	As used.	Real.	Apparent.	Dry.	As received (wet).	Coke.	Cells.	1.	2.	3.	4.
					Size of coal (see p. 4).		Specific gravity.		Pounds per cubic foot.		Percentage by volume.		6-foot drop test: Percentage over 2-inch mesh.			
1	2	3	4	5	6	7	8	9	10	11	12	13	14	15	16	17
Illinois—Con'td.:																
11	19 A*	No. 7	Zeigler	48	3″ s.	l.c.										
15	19 A*	do	do	24	1r. s.	l.c.										
19	19 B*	do	do	47	1r. s.	l.c.										
106	20 (w.)	No. 6	Staunton	47	sc.	l.c.	1.85	0.90	55.82	87.63	49.00	51.00	92.00	86.50	82.00	75.00
407	20 (w.)	do	do	36	sc.	l.c.	1.82	.84	52.12	85.80	46.00	54.00	94.00	87.50	82.00	75.50
126	21	do	Troy	44	L.	l.c.										
137	21 (w.)	do	do	45	sc.	l.c.	1.89	.98	60.50	90.45	52.00	48.00	95.50	88.50	83.00	75.50
117	22 B	do	Maryville	76	sc.	l.c.	1.84	.89	55.17	87.59	48.00	52.00	89.50	79.50	73.00	68.50
118	23 A (w.)	do	Donkville	39	L.	l.c.	1.85	.79	48.92	84.47	43.00	57.00	88.50	78.00	68.00	62.00
111	23 B (w.)	do	do	43	s.	l.c.	1.81	.79	48.81	83.74	44.00	56.00	89.00	78.50	78.00	62.00
114	23 B (w.)	do	do	72	sc.	l.c.	1.86	.85	52.27	85.95	46.00	54.00	88.00	78.00	66.50	58.50
119	24 A	No. 6	Nw (?)	24	sc.	l.c.										
155	24 A (w.)	do	do	78	L.	l.c.										
145	24 B (w.)	do	do	55	sc.	l.c.	1.80	.93	54.71	84.66	52.00	48.00				
120	24 B (w.)	do	Germantown	79	L.	r.o.m.										
140	25 (w.)	No. 5	do	62	r.o.m.	l.c.	1.89	.88	53.42	86.49	47.00	53.00				
143	36 (w.)	do	Lincoln	59	r.o.m.	l.c.	1.80	.89	52.58	84.39	46.00	51.00				
144	27 (w.)	No. 7	Auburn	53	r.o.m.	l.c.	1.82	.84	50.56	84.24	46.00	54.00				
166	28 C (w.)	No. 5	Herrin	60	L.	l.c.	1.83	.86	52.12	85.19	47.00	53.00				
169	29 A (w.)	d	Livingston	74	sc.	l.c.	1.83	.78	48.31	83.96	43.00	57.00	91.50	82.00	74.00	64.00
170	29 A (w.)	do	do	54	sc.	l.c.	1.83	.80	48.48	83.48	44.00	56.00	92.00	84.00	76.00	70.00
190	34 B*	do	Harrisburg	51	r.o.m.	l.c.	1.85	.78	48.58	84.73	42.00	58.00	92.00	84.50	78.00	74.00
Ind.:																
14	3 (y)*	d	Boonville	24	n., s.	l.c.	1.93	.89	52.01	85.69	46.00	54.00	96.50	90.50	85.50	80.00
6	4 *	No. 6	Star City	62	sc.	l.c.	1.86	.79	49.00	84.54	43.00	57.00	87.50	82.50	74.50	68.50
9	4 (w.)*	No. 5	do	58	sc.	l.c.	1.91	.82	50.37	85.92	57.00	57.00	96.00	89.50	85.50	80.50
8	5.	No. 5	Hymera	49	sc.	l.c.	1.89	.85	52.62	86.91	45.00	55.00	92.50	89.00	86.50	80.50
12	6 (w.)*	No. 4	do	112	L.e., n.	l.c.	1.85	.76	46.86	83.67	41.00	59.00	92.50	84.50	79.00	74.50
13	7 A (w.)*	No. 7	Littles	90	H.*,	l.c.	1.86	.95	58.75	89.31	50.00	49.00	94.50	87.50	82.50	78.00
16	9 A*	do	Macksville	112	r.o.m.	l.c.	1.84	.91	56.39	87.59	50.00	50.00	94.50	86.50	82.50	79.00
17	9 B*	do	do	90	r.o.m.	l.c.	1.84	.87	53.76	86.83	47.00	53.00	94.50	86.50	82.50	79.00
18	9 B (w.)*	do	do	50	r.o.m.	l.c.	1.83	.84	51.59	85.27	46.00	54.00	94.00	94.50	91.00	88.50
51	11 D*	No. 4	Dugger	50	r.o.m.	l.c.	1.92	.97	59.97	90.53	51.00	49.00	95.50	91.50	88.50	79.00
108	12	No. 5	Hartwell	37	r.o.m.	l.c.										

No.	Coal	Bed	Locality	Size	Coke	No.										
109	12 (w.)	do	do	r. o. m.	f. c.	35	1.82	.87	53.91	86.33	48.00	52.00	95.50	90.00	84.50	80.50
110	12 (w.)	do	do	r. o. m.	f. c.	44	1.87	.84	52.16	88.45	45.00	55.00	92.50	89.50	82.00	75.50
163	17 (t.)	do	do	r. o. m.	f. c.	50	1.92	.81	49.66	85.84	42.00	58.00	94.00	86.50	80.00	74.00
158	18 A (w.)	do	Bickmell	sc.	s.	16										
168	18 A (w.)	do	Ayrshire	sc.		24										
	Ind.:															
613	6 (w.)	Weir-Pittsburg	Jewett	L.	f. c.	47	1.97	.94	58.03	90.45	48.00	52.00	94.00	89.50	85.50	82.50
115	6 (w.)	do	do	L.	f. c.	49	1.90	.85	52.69	86.98	45.00	55.00	93.50	88.00	83.50	80.00
	Ky.:															
76	1 B*	Straight Creek	Straight Creek	1″ to 3″.	f. c.	51	1.94	.92	56.85	89.27	48.00	52.00	91.00	85.00	75.50	69.50
71	1 C*	do	do	1½″ s.	f. c.	67	1.88	.94	58.00	89.27	50.00	50.00	90.00	82.00	74.50	61.50
75	5*	High Splint	Big Black Mountain	r. o. m.	f. c.	49	1.84	.91	56.62	87.82	50.00	50.00	89.00	79.00	75.50	69.50
86	6*	No. 1	Paintsville	large l.	f. c.	56	1.79	.93	56.01	87.85	52.00	48.00	86.50	79.50	70.50	62.50
90	6*	do	Central City	large l.	f. c.	50	1.78	.93	55.21	85.95	51.00	49.00	91.50	84.00	78.50	75.00
85	7*	No. 9	Sturgis	r. o. m.	f. c.	49	1.83	.82	50.63	85.76	49.00	55.00	94.50	88.00	85.00	82.50
164	8	No. 1 B.	McHenry	r. o. m.	f. c.	49	1.90	.92	55.25	84.32	47.00	53.00	96.00	92.50	90.00	88.00
165	8	do	do	r. o. m.	f. c.	51	1.90	.91	56.36	88.32	43.00	52.00	96.00	93.00	89.50	86.50
167	9 A (w.)	No. 9	do	n.	f. c.		1.86	.80	49.42	88.93	43.00	57.00	93.00	87.50	83.50	79.00
	Maryland:															
50	1*	Lower Kittanning	Westernport	r. o. m.	r. o. m.	36	1.91	.90	61.61	91.55	52.00	48.00	96.50	94.00	91.00	90.00
54	1 (w.)*	do	do	r. o. m.	f. c.	48										
58	1 (w.)*	do	do	r. o. m.	f. c.	54										
	Missouri:															
116	3 B.	(?)	Higbee	r. o. m.	f. c.	33	1.88	.84	51.82	86.11	45.00	55.00	92.50	85.50	81.50	78.00
148	3 B.	Min. or Laramie	Van Houten	s.	f. c.	52	1.92	.99	61.30	91.25	52.00	48.00	95.00	89.00	84.00	81.00
149	3 B (w.)	do	do	s.	f. c.	48	1.90	.99	61.23	91.17	52.00	48.00	95.50	91.00	86.00	80.50
152	3 B, 4 B, 5 (w.)	do	Brilli at	s.	f. c.	49	1.91	1.01	59.47	90.08	50.00	50.00	96.50	91.00	86.00	90.00
150	4 B.	do	do	r. o. m.	f. c.	43	1.95	.96	62.29	92.24	48.00	50.00	96.00	88.00	81.50	75.00
151	4 B (w)	do	Bioseburg	r. o. m.	f. c.	56	1.92	.96	59.06	90.26	50.00	50.00	93.00	85.00	76.00	70.00
146	5 (w.)	do	do	n.	f. c.	50	1.88	.96	59.32	88.88	51.00	49.00	94.50	88.00	83.50	73.50
147					f. c.		1.91	.91	56.20	88.62	52.00	48.00	95.50	91.00	87.00	83.50
	Ohio:															
24	1 (w.)*	No. 4	Wellston	r. o. m.	f. c.	48	1.82	.89	54.70	86.60	49.00	51.00	96.50	92.50	90.00	86.50
27	2 (w.)*	No. 5	do	r. o. m.	f. c.	59	1.85	.89	59.55	88.24	54.00	46.00	95.00	90.50	84.50	78.50
31	3 (w.)*	No. 6	do	r. o. m.	f. c.	60										
28	4*	do	Bradley	s.	f. c.	55	1.85	.89	52.54	84.96	48.00	52.00	97.00	94.50	93.50	91.00
22	4*	do	Rush Run	r. o. m.	f. c.	46	1.84	.86	53.19	86.26	47.00	53.00	95.50	82.50	74.50	66.50
59	6 (w.)*	No. 7	Neffs	L.	f. c.	60	1.88	.84	51.97	86.25	45.00	55.00	95.00	90.50	86.50	85.00
66	7 (w.)*	do	Danford	L.	f. c.	45	1.82	.78	48.46	84.01	43.00	57.00	95.00	90.00	85.50	81.00
94	8 (w.)*	No. 6	do	r. o. m.	f. c.	60	1.82	.87	53.80	87.44	46.00	54.00	95.00	91.00	87.00	81.50
81	8 (w.)*	do	do	r. o. m.	f. c.	69	1.86	.83	51.47	85.76	45.00	55.00	89.50	83.00	75.50	73.00
93	9 A*	No. 4	do	1½″ l.	f. c.	71	1.89	.90	55.74	88.16	48.00	52.00	91.00	82.50	75.00	68.50
72	9 B*	do	do	n. s.	f. c.	53	1.81	.85	49.49	83.78	45.00	55.00	88.00	75.00	68.00	64.00
55	9 B (w.)*	do	do	n. s.	f. c.	57	1.86	.83	52.04	85.73	46.00	54.00	84.00	81.00	68.00	67.00
57	12 (w.)	No. 8	Bellaire	r. o. m.	f. c.	45	1.94	.83	57.57	89.99	46.00	54.00	88.00	81.00	74.50	70.50
							1.82	.83	51.44	85.12	48.00	52.00	92.00	81.00	80.00	74.50
180							1.90	.82	50.75	86.30	43.00	57.00	92.50	85.00	79.00	72.50

a With 10 per cent pitch.

Details of coking tests of coals, January 1, 1905, to June 30, 1907—Continued.

Test No.	Field No. of coal	Origin of coal sample: Designation of bed	At or near—	Duration of test (hours)	Size of coal (see p. 4): As shipped	As used	Specific gravity: Real	Apparent	Pounds per cubic foot: Dry	As received (wet)	Physical properties of coke: Percentage by volume: Coke	Cells	6-foot drop test: Percentage over 2-inch mesh: 1.	2.	3.	4.
1	2	3	4	5	6	7	8	9	10	11	12	13	14	15	16	17
Pennsylvania:																
25	5 *	Illg.	Illrth.	62	3″	f. c.	1.92	0.94	58.33	90.14	49.00	51.00	94.00	80.50	77.50	72.00
26	5 (w.) *	do.	do.	92	4″	f. c.	1.84	.83	51.66	85.95	45.00	55.00	91.00	81.50	74.50	67.00
32	6 *	do.	East Millsboro	46	r. o. m.	f. c.	1.96	1.09	67.74	95.17	56.00	44.00	91.50	84.50	88.00	84.50
34	6 (w.)	do.	do.	45	r. o. m.	f. c.	1.92	1.05	65.34	93.42	55.00	45.00	91.00	84.50	80.50	75.50
35	6 *	do.	do.	88	r. o. m.	f. c.	1.91	.97	68.88	95.10	55.00	42.00	97.50	95.50	94.00	92.00
38	6 (w.) *	do.	do.	72	r. o. m.	f. c.	1.84	1.11	59.97	89.27	53.00	47.00	91.50	91.50	88.00	74.50
41	6 *	do.	do.	90	r. o. m.	f. c.	1.91	1.18	73.42	97.12	62.00	38.00	94.50	88.50	88.00	80.00
30	7 *	do.	Ligonier	93	r. o. m.	f. c.	1.92	.94	58.10	89.92	49.00	51.00	94.50	88.50	84.00	78.00
33	7 (w.) *	Lower Kittanning	Ehrenfeld	51	r. o. m.	f. c.	1.83	.91	56.54	87.74	50.00	50.00	91.50	86.00	81.00	80.00
37	8 *	do.	Kimmelton	45	r. o. m.	f. c.	1.76	.65	40.16	79.44	37.00	63.00	94.50	88.50	84.00	80.00
39	9 *	do.	do.	66	r. o. m.	f. c.	1.99	.85	52.69	88.24	43.00	57.00	98.50	96.50	95.50	95.00
42	9 (w.) *	do.	do.	51	r. o. m.	f. c.	1.94	.87	53.87	88.16	45.00	55.00	97.00	94.50	93.00	92.00
56	9 *	Illg.	ai eoi	38	1½″	f. c.	1.85	.87	54.03	87.10	47.00	53.00	93.50	93.00	84.00	79.50
47	10 *	do.	am.	49	r. o. m.	f. c.	1.92	.84	52.20	87.13	44.00	56.00	93.50	90.00	85.00	82.00
159	10 *	do.	do.	47	r. o. m.	f. c.	1.97	1.00	62.03	91.90	52.00	48.00	96.00	90.50	83.50	80.00
161	11	do.	Welrum	41	r. o. m.	f. c.	1.92	.98	60.69	91.97	50.00	50.00	97.50	94.50	87.00	80.50
162	12 (w.)	B. or Miller.	do.	46	r. o. m.	f. c.	1.98	.72	57.07	90.14	47.00	53.00	95.00	94.50	92.00	90.00
185	15 (w.)	do.	White	61	r. o. m.	f. c.	1.96	.78	44.65	84.58	36.00	64.00	87.00	90.00	84.00	90.50
188	15 (w.)	Jr Freeport	do.	54	r. o. m.	f. c.	2.01	1.00	48.39	86.45	39.00	61.00	92.00	94.00	88.00	83.50
178	17	do.	Herm ie	52	r. o. m.	f. c.	1.96	.88	60.22	94.03	49.00	51.00	95.00	90.00	86.50	67.50
186	17 (w.)	Illg.	do.	44	r. o. m.	f. c.	1.95	1.10	54.75	89.04	56.00	44.00	94.50	92.00	90.50	70.00
176	19	do.	do.	36	r. o. m.	f. c.	1.97	1.08	67.28	95.90	56.00	44.00	91.50	80.00	73.50	92.00
179	19	Lower Kittanning	do.	40	r. o. m.	f. c.	2.04	.86	68.47	95.36	55.00	45.00	98.00	84.00	75.00	87.50
177	20	do.	Connellsville	68	r. o. m.	f. c.	1.95	.73	53.49	80.65	42.00	58.00	87.00	95.50	93.50	73.50
182	20 (w.)	Pittsburg	do.	78	r. o. m.	f. c.	1.94	.96	45.30	84.58	37.00	63.00	98.00	94.50	90.00	75.00
183	21	do.	do.	70	r. o. m.	f. c.	1.90	.92	50.47	90.68	50.00	50.00	96.50	94.00	90.00	87.50
187	21	do.	d	78	r. o. m.	thr. 1″	1.93	.95	57.04	89.46	48.00	52.00	92.50	84.50	77.50	73.50
189	21	do.	d	51	r. o. m.	thr. ¾″	1.97	.98	58.75	90.56	49.00	51.00	93.50	85.50	79.50	75.00
191	21	do.	do.	47	r. o. m.	thr. ¾″	1.98	.96	60.92	92.13	50.00	50.00	95.00	87.00	81.50	77.50
192	21	do.	do.	43	r. o. m.	f. c.	1.96	1.00	59.70	91.52	49.00	51.00	95.00	91.00	87.00	81.50
Tennessee:																
133	1	Ming o.	Fork Ridge	49	r. o. m.	f. c.	1.96	1.00	62.14	92.70	51.00	49.00	95.00	91.00	87.00	81.50
153	1 (w.)	do.	do.	43	r. o. m.	f. c.	1.87	.88	54.41	87.48	47.00	53.00	90.00	78.00	71.00	66.00

No.	Ref.	Coal bed	Locality	Lab. No.	Condition	t.c.	1	2	3	4	5	6	7	8	9	10
127	2	Log Mountain	Gatliff	54	r. o. m.	t.c.	49.50	59.00	72.00	87.00	50.00	50.00	87.33	56.12	.91	1.81
128	3	Regal Block	...do.	42	r. o. m.	t.c.	58.50	68.50	79.00	90.00	52.00	48.00	87.55	55.13	.89	1.87
125	4	Windrock, or Dean	Oil or Springs	60	r. o. m.	t.c.	63.00	69.00	79.00	90.00	52.00	48.00	90.11	57.68	.93	1.93
129	4 (w.)	...do.	Petros	53	r. o. m.	t.c.	56.50	63.00	78.00	87.00	50.00	50.00	89.61	58.41	.96	1.91
154	5 (w.)	Brushy Mountain	Wilder	49	r. o. m.	t.c.	77.00	81.50	87.00	93.00	54.00	46.00	88.28	54.60	.88	1.91
122	6	Lower Sewanee	...do.	43	r. o. m.	t.c.	90.00	92.00	94.50	97.00	52.00	48.00	90.94	58.52	.94	1.95
121	7 B (w.)	Wilder	...do.	50	s.	t.c.		92.50	96.00	98.00	49.05	51.00	93.92	63.36	1.02	2.01
123	8 B (w.)	First above Sewanee	Clifty	45	s.	t.c.	91.50	87.00	91.00	95.00	56.00	44.00	87.02	52.08	.84	1.93
134		...do.	...do.			t.c.	82.00	89.00	92.00	96.00	55.00	45.00	88.39	52.24	.84	1.98
124	9 (w.)	Sewanee	Mint	49	¾″ s.	t.c.	85.50	95.00	92.00	96.00	55.00	45.00	88.32	54.03	.87	1.93
156	10 (w.)	Battle Creek	Ozone	53	1″ s.	t.c.	93.00	95.00	97.00	98.00	56.00	44.00	92.43	65.90	1.06	1.90
160	11 (w.)	(?)	Ozone	40	s.	t.c.	90.00	90.00	95.00	95.00	51.00	49.00	92.20	61.65	1.00	1.97
Utah:																
130	1:	(?)	Huntington Creek	49	r. o. m.	t.c.	36.50	42.50	61.50	81.50	38.00	62.00	91.17	67.48	1.11	1.78
141	1 c:	(?)	...do.	60	r. o. m. and luken.	t.c.										
157	1 d	(?)	...do.	48	r. o. m. and broken.	t.c.					38.00	62.00	94.98	71.29	1.16	1.87
Virginia:																
64	1*	Wilson	Gab	71	r. o. m.	t.c.	80.50	85.50	90.50	95.50	48.00	52.00	87.06	57.11	.93	1.80
65	1*	...do.	...do.	36	r. o. m.	t.c.	74.50	78.50	85.00	92.50	49.00	51.00	87.21	56.65	.91	1.79
67	1*	...do.	...do.	67			67.00	71.00	80.00	89.00	48.00	52.00	89.08	59.13	.96	1.83
68	1*	...do.	...do.	65	r. o. m.	t.c.	67.00	72.00	80.00	91.00	48.00	52.00	88.01	58.06	.94	1.81
77	1*	...do.	...do.	51			78.00	82.00	81.50	93.00	55.00	45.00	86.56	56.56	.85	1.87
63	1*	Bonnell	...do.	41	r. o. m.	t.c.	73.00	78.50	81.00	93.00	55.00	45.00	87.17	52.88	.83	1.87
69	2*	...do.	...do.	48			68.50	83.50	89.50	93.50	57.00	43.00	88.43	57.23	.93	1.85
70	2*	...do.	...do.	45	r. o. m.	t.c.	61.50	83.00	85.00	92.50	50.00	50.00	85.23	49.68	.80	1.87
61	2 (w.)*	Banner	Tams Creek	50	1:		75.50	83.50	89.00	93.50	40.00	60.00	89.67	67.72	1.25	1.93
88	3*	...do.	...do.	36	1:		80.50	82.00	87.50	92.50	35.00	65.00	94.26	69.30	1.12	1.87
62	4*	Darby	Darby	50	r. o. m.	t.c.	77.00	82.50	87.50	93.50	56.00	44.00	85.99	51.70	.88	1.84
181		No. 4	Richlands	48	r. o. m.	t.c.	76.00	82.00	87.50	94.00	44.00	56.00	89.65	54.51	.88	2.00
184		...do.	...do.		r. o. m.	t.c.	87.00	89.50	92.50	96.50	41.00	59.00	87.82	51.02	.82	2.01
Washington:																
135	2:	(?)	Roslyn	36	1:	t.c.	54.50	63.00	72.00	84.50	50.00	50.00	89.84	58.64	.95	1.90
West Virginia:																
40	4 B*	Upper Freeport	Bretz	57	r. o. m.	t.c.	92.50	94.50	96.50	98.50	54.00	46.00	88.39	54.71	.88	1.92
44	4 B (w.)*	...do.	...do.	72	r. o. m.	t.c.	88.50	91.00	93.00	95.50	50.00	50.00	91.33	60.12	.97	1.96
46	4 B (w.)*	...do.	...do.	74	r. o. m.	t.c.	50.50	93.50	96.00	97.50	54.00	46.00	87.78	54.10	.87	1.88
21	13*	Austed	Page	84	r. o. m.	t.c.	84.50	90.50	90.50	96.00	55.00	45.00	91.21	63.13	1.02	1.84
23	13 and 14*	...do.	...do.	77	r. o. m.	t.c.	87.00	89.50	90.00	95.00	52.00	48.00	87.10	54.67	.88	1.85
30	14*	...do.	...do.	71	r. o. m.	t.c.	78.00	82.00	89.50	94.50	44.00	56.00	87.74	52.84	.85	1.92
36	15*	...do.	Clarksburg	46	r. o. m.	t.c.	87.00	89.50	90.50	96.00	56.00	44.00	93.88	66.45	1.07	1.90
43		Pittsburg	...do.	73	r. o. m.	t.c.	85.50	95.86	89.50	95.50	55.00	45.00	95.86	70.29	1.13	1.93
73	16 A*	...do.	...do.	60	¾″ s.	t.c.	79.50	80.00	89.00	94.50	45.00	55.00	88.20	53.91	.87	1.92
45	16 B*	...do.	Monongah	72	7″ s.	t.c.	91.00	93.00	95.50	97.50	45.00	55.00	88.32	54.03	.87	1.95

a With 5 per cent pitch.
b Over ¾-inch screen, with 18 per cent of slack returned to it.
c Mixed with one-third Rhode Island No. 1.
d Mixed with one-fourth Rhode Island No. 1.

Details of coking tests of coals, January 1, 1905, to June 30, 1907—Continued.

Test No.	Field No. of coal	Designation of bed	At or near—	Duration of test (hours).	As shipped.	As used.	Real.	Apparent.	Dry.	As received (wet).	Coke.	Cells.	1.	2.	3.	4.
		Origin of coal sample.			Size of coal (see p. 4).		Specific gravity.		Pounds per cubic foot.		Percentage by volume.		6-foot drop test: Percentage over 2-inch mesh.			
1	2	3	4	5	6	7	8	9	10	11	12	13	14	15	16	17
	Wet Va—Continued.															
48	16 B (w.)*	Pittsburg	...do	47	s.	f.c.	1.87	0.90	49.72	85.27	43.00	57.00	94.00	90.50	86.50	83.00
49	16 B (w.)*	do.	do.	25	s.	f.c.	1.84	.73	45.34	82.75	40.00	60.00	95.00	91.50	88.00	84.50
60	17 (w.)*	Bakerstown	Bretz	42	r.o.m.	f.c.	1.91	.93	57.95	89.76	49.00	51.00	96.50	93.50	90.50	88.00
74	18*	Glen ...	Glen Alum	43	r.o.m.	f.c.	1.93	.98	60.73	91.29	51.00	49.00	93.00	85.00	78.00	73.00
78	18*	...do.	...do.	48	r.o.m.	f.c.	1.87	.90	55.74	88.16	48.00	52.00	94.50	89.50	86.00	83.00
79	19*	Sewell	McDonald	49	r.o.m.	f.c.	1.98	.86	53.45	88.00	43.00	57.00	96.50	95.50	92.50	90.50
83	19*	do.	d.	53	r.o.m.	f.c.	1.94	.84	52.16	87.71	43.00	57.00	99.00	89.50	96.50	95.00
84	20*	Keystone	Acme	42	r.o.m.	f.c.	1.80	1.12	69.72	93.42	62.00	38.00	92.00	85.00	80.50	75.00
87	20 (w.)*	do.	do.	50	r.o.m.	f.c.	1.85	1.02	63.44	91.52	55.00	45.00	97.50	89.50	85.00	81.50
92	20 (w.)*	do.	do.	53	r.o.m.	f.c.	1.90	1.01	62.87	92.16	61.00	47.00	94.50	88.50	83.50	78.50
82	21*	Peerless	do.	46	r.o.m.	f.c.	1.92	1.18	73.49	97.19	62.00	38.00	92.50	87.50	82.00	77.00
91	21 (w.)*	do.	do.	52	r.o.m.	f.c.	1.96	.89	53.91	86.98	48.00	52.00	94.50	90.50	85.50	81.00
175	25.	Black Band	Charleston	44	L.	f.c.	1.87	.87	60.31	90.87	51.00	49.00	87.50	79.50	73.50	69.50
	Wyoming:															
52	3 (w.)*	Rock Springs	Aladdin	27	r.o.m.	f.c.										
132	5.	(?)	Rock Springs	39	r.o.m.	f.c.										

Details of coking tests of coals, January 1, 1905, to June 30, 1907—Continued.

Test No.	Field No. of coal	Condition[a]	Weight of coal (pounds)	Production (pounds) Coke	Production (pounds) Breeze	Production (per cent) Coke	Production (per cent) Breeze	Production (per cent) Total	Chemical analysis of coal Mois-ture	Chemical analysis of coal Volatile matter	Chemical analysis of coal Fixed carbon	Chemical analysis of coal Ash	Chemical analysis of coal Sul-phur	Chemical analysis of coke Mois-ture	Chemical analysis of coke Volatile matter	Chemical analysis of coke Fixed carbon	Chemical analysis of coke Ash	Chemical analysis of coke Sul-phur	Chemical analysis of coke Phos-phorus
1	2	18	19	20	21	22	23	24	25	26	27	28	29	30	31	32	33	34	35
Alabama:																			
142	2 B (w.)	1	10,530	6,197	684	58.85	6.50	65.35	6.26	31.99	52.66	9.09	1.36	3.04	1.06	82.15	13.75	1.16	0.070
		2	9,871	6,009	663	60.88	6.72	67.60		34.12	56.18	9.70	1.45		1.09	84.73	14.18	1.20	.0057
138	3	1	12,150	7,802	489	64.06	4.01	68.07	2.77	28.99	53.14	15.10	.62	2.03	1.80	74.89	21.28	.60	.0008
		2	11,841	7,644	479	64.54	4.04	68.58		29.82	54.65	15.53	.64		1.84	76.44	21.72	.61	.0008
139	3 (w.)	1	11,660	7,072	258	60.65	2.21	62.86	6.36	30.54	53.10	10.00	.62	.99	1.06	83.51	14.44	.58	.0126
		2	10,918	7,002	255	64.13	2.34	66.47		32.61	56.71	10.68	.66	1.07	1.07	84.34	14.59	.59	
131	4	1	12,000	7,706	281	64.22	2.34	66.56	4.17	31.69	54.50	10.96	1.18	.29	.84	83.21	15.56	1.08	
		2	11,500	7,684	280	66.82	2.44	69.26		30.37	56.87	11.44	1.23		.84	83.45	15.71	1.08	.0126
136	4 (w.)	1	12,000	6,809	239	56.74	1.99	58.73	7.28	30.46	58.38	3.88	1.00	.35	.42	92.99	6.24	.87	.008
		2	12,000	6,785	228	60.98	2.14	63.12		32.85	62.97	4.18	1.08		.42	93.32	6.26	.87	
171	5	1	11,126	7,560	390	65.65	3.22	68.87	3.98	26.55	56.92	12.55	1.44	.59	.89	81.10	17.42	1.16	.0077
		2	12,110	7,903	388	67.97	3.34	71.31		27.65	59.28	13.07	1.50		.90	81.58	17.52	1.17	.0512
172	6	1	11,628	8,350	316	69.01	2.61	71.62	3.28	25.30	64.50	6.92	.59	.46	.35	89.37	9.82	.59	.0377
		2	12,100	8,312	315	71.02	2.69	73.71		26.16	64.69	7.15	.61	.63	.35	89.78	9.87	.59	
174	6 (w.)	1	11,703	7,900	221	65.66	1.86	67.52	6.73	24.84	63.57	4.86	.59	.63	.27	92.36	6.74	.60	
		2	11,080	7,751	220	69.95	1.99	71.94		26.63	68.16	5.21	.63		.27	92.95	6.78	.60	.0377
Arkansas:																			
95	1 B (w.)	1	10,000	5,532	574	58.32	5.74	64.06	10.96	16.66	66.51	5.87	1.01	2.89	3.67	85.32	8.21	1.25	.006
		2	8,904	5,603	557	63.60	6.26	69.86		18.71	74.70	6.59	1.13		3.78	87.77	8.45	1.29	.0135
96	1 B (w.)	1	10,000	5,606	1,290	58.06	12.90	70.96	6.77	15.04	69.32	8.87	1.14	1.31	2.44	84.53	11.72	1.11	
		2	9,333	5,730	1,273	61.46	13.65	75.11		16.13	74.35	9.52	1.22		2.47	85.65	11.88	1.12	
97	1 B (w.)	1	10,000	6,055	214	60.55	2.14	62.69	7.80	18.93	64.92	8.35	1.08	2.74	1.29	85.81	10.16	1.05	
		2	9,290	5,889	208	63.87	2.26	66.13		20.53	70.41	8.30	1.17		1.33	88.23	10.44	1.07	
100	1 B (w.)	1	10,000	5,976	391	59.76	3.91	63.67	5.69	17.34	68.67	8.30	1.12	.18	2.03	87.26	10.55	1.70	.0116
		2	9,431	5,963	390	63.25	4.14	67.39		18.39	72.81	8.80	1.12		.53	87.42	10.55	1.71	
104	7 B (w.)	1	10,000	2,730	3,750	27.30	37.50	64.80	6.98	14.86	76.97	7.19	1.78	.13	.53	89.72	9.62	1.70	
		2	9,000	2,726	3,745	29.31	40.26	69.57		15.98	76.29	7.03	1.91		.55	89.84	9.63	1.71	.0062
105	7 B (w.)	1	10,000	4,868	1,604	48.68	16.04	64.72	7.52	16.66	68.83	6.97	1.65	.67	.85	89.14	9.34	1.60	
		2	9,246	4,835	1,593	52.28	17.23	69.51		18.01	74.45	7.54	1.78		.86	89.74	9.40	1.61	
98	9 (w.)	1	10,000						7.43	13.84	65.55	13.18	.96						
		2								14.95	70.81	14.24	1.04						
101	9 (w.)	1	8,000	6,252	458	62.52	4.58	67.10	6.30	14.74	65.01	13.95	.98	.30	.81	81.48	17.41	1.07	.0329
		2	10,000	6,233	457	60.03	4.84	64.87		17.22	64.03	13.15	1.01		.81	81.73	17.46	1.07	
102	9 (w.)	1	9,440	5,107	1,693	51.07	16.93	68.00	5.60	18.24	67.83	13.93	1.07	.30	.80	83.70	15.17	1.07	.0329
		2	10,000	5,090	1,667	54.01	17.90	71.91		18.24	66.63	12.77	1.02		.80	83.70	15.17	1.07	
103	9 (w.)	1	10,000	5,107	1,693	51.07	16.93	68.00	5.76	14.84	66.63	12.77	1.02	.33	.80	83.70	15.17	1.07	.0293
		2	9,454	5,090	1,667	54.01	17.90	71.91		15.75	70.70	13.55	1.08		.80	83.98	15.22	1.07	.0293

a Condition 1 means "as charged" with reference to weight of coal (column 19), and "as received" ("wet") with reference to other items; condition 2 means "on dry basis."

Details of coking tests of coals, January 1, 1905, to June 30, 1907—Continued.

Test No.	Field No. of coal	Condition	Weight of coal (pounds)	Production (pounds)			Production (per cent).			Chemical analysis of coal.					Chemical analysis of coke.					
				Coke.	Breeze.		Coke.	Breeze.	Total.	Moisture.	Volatile matter.	Fixed carbon.	Ash.	Sulphur.	Moisture.	Volatile matter.	Fixed carbon.	Ash.	Sulphur.	Phosphorus.
1	2	18	19	20	21	22	22	23	24	25	26	27	28	29	30	31	32	33	34	35
	Georgia:																			
173	1	1	12,180	8,100	549		66.50	4.51	71.01	3.35	16.54	66.07	14.04	1.20	0.45	0.35	81.69	17.51	1.00	0.0113
		2	11,772	8,004	547		68.50	4.65	73.15	....	17.11	68.36	14.53	1.23	....	.35	82.06	17.59	1.00	....
	Illinois:																			
1	7 D	1	8,000	3,907	452		48.84	5.65	54.49	10.88	35.27	38.44	15.41	4.53	2.26	1.86	72.66	23.20	3.95	
		2	7,120	3,819	442		53.57	6.20	59.77		39.58	43.13	17.29	5.08		1.90	74.36	23.74	4.04	
4	7 D (w.)	1	10,000	5,200	280		52.00	2.80	54.60	12.45	39.58	42.08	9.30	3.64	1.04	.61	82.10	16.25	3.64	
		2	8,755	5,145	257		58.78	2.94	61.72		41.32	48.06	10.62	4.16		.61	82.96	16.43	3.27	.0065
5	11 D (w.)	1	10,000	5,326	300		54.00	3.00	57.00	8.24	31.64	52.81	7.31	1.55	1.19	.93	85.97	11.91	1.46	
		2	9,176	4,600	296		58.15	3.23	61.38		34.48	52.55	7.97	1.69		.93	87.01	12.06	1.91	.007
2	13 (w.)	1	10,000	4,600	718		46.00	7.18	53.18	11.44	34.95	50.16	7.45	1.25	4.00	1.08	82.08	11.64	1.27	
		2	8,856	4,386	665		49.55	7.73	57.28		30.95	56.64	8.41	1.41		1.76	86.04	12.20	1.37	
3	13	1	12,000	4,386	1,327		44.65	11.06	55.71	10.56	30.96	49.06	10.28	1.71	2.73	1.36	82.08	16.61	1.33	.0162
		2	10,733	5,212	1,291		48.56	12.03	60.59		33.63	54.87	11.50	1.91		1.40	79.30	17.07	1.82	
10	16 (w.)	1	10,000	5,579	910		55.79	9.10	64.89	9.79	30.35	51.79	8.07	1.09	2.14	1.46	81.53	17.07	1.02	.0076
		2	9,021	5,400	891		60.53	9.88	70.41		33.64	57.41	8.96	1.21		1.49	83.96	12.71	1.04	
106	20 (w.)	1	10,000	4,285	628		42.55	6.28	48.83	17.04	32.59	40.77	9.60	3.23	.57	.66	85.80	16.28	3.01	
		2	8,296	4,231	524		51.00	6.32	57.32		39.28	49.15	11.57	3.89		.67	82.96	16.37	3.09	
107	20 (w.)	1	10,000	4,099	385		46.59	3.98	50.57	14.36	34.61	42.63	8.40	3.23	.53	.29	82.49	14.31	2.72	
		2	8,862	4,634	396		54.11	4.61	58.72		40.41	49.78	9.81	3.77		.29	84.87	14.39	2.73	
126	21	1	12,000	5,046	534		50.47	5.34	55.80	13.37	31.17	44.74	12.31	1.46						
137	21 (w.)	2	11,690	4,997	529		56.77	6.01	62.78	11.98	30.01	44.74	7.80	1.10	.96	.72	72.18	26.12	4.61	
117	22 B	1	8,892	5,616	574		56.80	5.67	61.68	16.19	33.87	37.72	16.43	4.74		.73	72.89	26.38	4.66	
		2	12,000	5,579	570		55.47	5.47	61.14		38.48	39.53	18.67	5.39	.65	1.00	80.76	26.99	4.05	
118	22 B (w.)	1	12,000	4,183	347		52.11	3.47	46.58	13.74	34.14	40.14	10.14	4.62	.66	1.61	81.29	17.10	3.67	
		2	10,037	4,407	345		48.40	4.00	52.40		36.47	47.17	12.09	3.57		.74	83.45	15.15	3.09	
111	23 A (w.)	1	8,026	4,365	389		54.07	3.89	47.96	15.85	36.47	41.01	8.78	3.14	.96	.74	84.01	15.15	3.11	
		2	10,000	4,365	386		51.87	4.58	56.45		42.28	41.54	10.18	3.27		1.14	82.66	15.24	2.87	
112	23 B (w.)	1	8,415	6,443	601		46.02	4.29	50.31	15.93	35.02	40.57	8.56	3.89		1.15	83.46	15.39	2.60	
114	23 B (w.)	2	14,000	6,355	593		53.99	5.04	59.03		35.98	48.21	10.17	3.25	1.36	1.19	83.46	14.62	2.84	
			11,770								42.68	40.16	8.03	3.87		1.21	83.97	14.82	2.88	
119	24 A	1	10,000	4,710	1,350		39.81	11.41	51.22	13.28	29.93	39.03	17.76	4.05		1.64	70.01	13.73	2.97	
155	24 A (w.)	2	11,410	4,445	1,274		41.26	11.83	53.09	15.18	32.13	39.55	9.23	3.41	5.62	1.74	83.71	14.55	3.15	
145	24 B (w.)	1	11,830					8.93			35.22	46.29	9.56	3.07						
		2	11,000								38.67	50.83	10.50	3.74						
120	25	1	10,773	5,355	710		45.27	6.00	51.27	12.96	33.01	43.66	14.43	4.04		4.67	77.69	14.93	2.32	
140	25 (w.)	2	11,830	5,210	691		50.85	6.74	57.59	13.40	39.06	50.42	10.52	3.45	2.71	4.90	79.85	15.35	2.38	

No.	Sample	n																	S
143	26 (w.)	1	11,730	5,880	1,250	49.79	10.04	60.43	15.18	33.46	41.53	9.83	2.73	5.27	3.35	75.98	15.40	2.80	
144	27 (w.)	2	9,966	6,542	1,184	51.88	11.88	67.38	16.39	39.45	41.57	11.88	3.22	3.51	3.54	80.20	16.26	2.95	
166	28 C (w.)	1	11,550	4,987	405	43.87	3.51	47.38	9.37	34.28	41.57	7.78	3.22	2.82	1.55	81.14	13.80	3.40	.0075
169	29 A (w.)	2	12,000	6,380	661	50.40	5.25	55.65	16.63	40.38	53.36	9.30	1.00	.72	1.60	86.12	10.46	3.52	
170	29 A (w.)	1	11,410	4,990	642	50.63	4.05	59.66	18.30	33.88	49.81	6.99	1.20	2.78	.62	86.62	10.76	.98	
190	34 B (w.)	2	12,010	4,914	684	41.22	6.70	55.20	9.91	40.16	50.74	7.68	3.13	.19	.90	84.62	13.86	2.57	.0205
			10,133	4,980	679	48.50	2.89	43.27		42.87	41.83	9.10	3.06		.74	84.35	13.29	2.59	
			11,090	4,962	338	42.38	3.43	53.53		40.28	50.65	7.21	3.75		.76	83.35	13.61	2.56	
			9,483	4,658	329	50.48	2.43	52.68		33.33	50.65	8.11	1.75		.35	85.73	13.55	1.49	
			11,094	6,511	316	50.25	2.69	58.37		37.00	56.22	6.78	1.94		.35	87.91	11.57		
					315	55.68										88.08			
Indiana:																			
14	3 (w.)	1	10,000	5,010	536	50.10	5.36	55.46	13.53	34.83	43.10	8.54	3.50	6.31	1.60	68.35	23.74	2.80	
6	4	2	10,000	4,664	502	54.77	5.86	60.63	14.30	39.45	40.96	15.26	2.60		1.71	72.95	25.34	2.99	
9	4 (w.)	1	8,570	5,153	332	51.53	3.32	54.85		34.40	47.79	17.81	3.03	.55	.38	85.91	13.16	2.06	
8	5 (w.)	2	10,000	5,125	330	61.25	3.94	65.19	16.33	40.38	44.19	9.07	2.08		.38	86.39	13.23	2.07	
12	6 (w.)	1	8,367	5,340	232	53.40	2.32	55.72	10.74	38.11	52.82	10.60	2.49	.76	.54	81.91	17.41	4.24	
13	7 A (w.)	2	10,000	5,299	230	59.37	2.58	61.95	12.39	37.13	47.00	11.87	4.61	.75	.55	81.90	17.54	3.16	
16	9 A	1	10,000	6,464	501	54.34	5.01	59.35	10.33	36.17	40.85	9.99	3.59		.73	82.52	16.62	3.18	
17	9 B	2	8,761	5,868	497	56.55	5.67	62.36	12.30	41.97	42.63	11.40	3.12	1.13	.73	81.90	16.75	2.69	.0424
18	9 B (w.)	1	12,240	6,903	633	57.13	5.23	62.36	12.51	38.35	42.93	8.39	3.48	.81	2.35	82.52	16.29	2.72	
51	11 D	2	10,976	6,914	455	62.99	3.77	68.76	14.14	42.77	41.62	9.36	3.41		2.38	84.18	18.45	3.41	
108	12	1	12,000	7,034	464	58.62	3.79	68.41	11.90	34.96	47.87	12.68	3.89	.68	2.00	78.74	18.60	3.44	
109	12 (w.)	2	12,000	7,005	451	66.30	4.29	70.59	11.77	39.86	41.62	11.36	3.26		2.02	79.36	18.03	3.50	
110	12 (w.)	1	10,524	6,977	442	58.38	3.87	62.25	13.79	35.65	47.46	12.98	2.61	.91	2.85	78.44	18.15	3.52	
163	17 (w.)	2	12,000	6,946	461	66.27	3.68	70.66	12.82	40.75	46.27	12.28	3.73		2.87	78.98	13.52	2.24	
158	18 A (w.)	1	12,000	6,915	438	57.63	4.25	70.75	10.57	36.70	46.90	8.86	3.04	1.55	.97	84.60	13.64	2.26	
168	18 A (w.)	2	10,303	6,882	460	66.50	3.83	75.74	15.09	32.50	47.64	9.62	1.53	.88	.98	84.09	12.70	1.34	
			10,572	6,496	453	61.46	4.28	75.95	13.97	36.89	53.28	9.83	1.74	.60	1.69	85.41	12.90	1.36	
			10,000	5,306	464	52.58	4.64	62.41		33.78	40.98	14.45	4.32	.42	.62	75.95	22.55	3.84	
109	12		8,823	5,212	460	59.07	5.21	64.28		38.29	45.33	16.38	4.90	1.65	.63	76.62	22.75	3.87	
			10,000	5,304	255	53.04	2.55	64.08		35.43	39.59	8.03	3.22		1.16	84.32	14.01	2.86	
8	5 (w.)		8,621	5,272	253	61.15	2.93	54.08		41.10	49.52	9.31	3.73		1.03	84.83	14.18	2.88	
			10,402	6,365	296	52.96	2.47	63.31		37.25	47.63	8.41	3.82		1.03	84.37	14.24	2.89	
			12,050	6,328	295	60.49	2.76	62.51		42.73	43.77	9.64	4.20		.67	84.73	16.53	2.90	
			10,776	7,900	332	59.75	2.39	68.74		35.65	48.95	10.01	1.48		.68	81.42		2.39	
			11,900	7,081	327	65.71	3.03			39.86	46.42	11.19	1.49			82.79		3.45	
			12,740							30.97	48.07	7.43							
												6.99							
Kansas:																			
113	6	1	10,000	5,443	341	54.43	3.41	57.84	8.58	30.27	45.92	15.23	3.47	1.01	.64	75.07	23.28	3.45	
115	6 (w.)	2	9,142	5,388	338	58.04	3.71	62.65	12.29	33.11	50.23	16.66	3.80	.59	.64	75.84	23.52	3.49	
		1	12,000	6,439	299	58.66	2.49	56.15		30.30	47.21	10.20	2.63		.56	82.78	16.07	2.49	
		2	10,525	6,401	297	60.82	2.82	63.64		34.55	53.82	11.63	3.00		.56	83.27	16.17	2.50	.0300
Kentucky:																			
76	1 B	1	12,000	8,026	257	66.88	2.14	69.02	3.55	35.49	55.88	5.08	1.17	1.01	1.25	91.40	6.42	1.05	
		2	11,574	7,951	255	68.70	2.20	70.90		30.79	57.94	5.27	1.21	.93	1.26	92.26	6.48	1.06	

Details of coking tests of coals. January 1, 1905, to June 30, 1907—Continued.

Test No.	Field No. of coal	Condition	Weight of coal (pounds)	Production (pounds) Coke	Production (pounds) Breeze	Production (per cent) Coke	Production (per cent) Breeze	Production (per cent) Total	Chemical analysis of coal — Moisture	Volatile matter	Fixed carbon	Ash	Sulphur	Chemical analysis of coke — Moisture	Volatile matter	Fixed carbon	Ash	Sulphur	Phosphorus
		(18)	(19)	(20)	(21)	(22)	(23)	(24)	(25)	(26)	(27)	(28)	(29)	(30)	(31)	(32)	(33)	(34)	(35)
Kentucky—Cont'd.																			
71	16 C	1	12,000	7,319	335	60.98	2.96	63.93	5.64	32.41	52.77	9.18	1.11	1.32	1.53	84.68	12.47	0.96	0.0435
		2	11,323	7,222	350	63.78	3.09	66.87	4.67	34.35	55.92	0.73	1.18	.30	1.55	85.81	12.64	.97	.0115
75	5	1	12,000	7,903	256	66.61	2.13	68.74		35.33	56.30	3.70	.45		1.21	93.26	5.25	.41	.0021
		2	11,440	7,960	255	69.66	2.23	71.89	5.21	37.06	59.06	3.88	.47	2.81	1.21	91.93	3.99	.40	
86	6	1	12,000	6,780	587	56.50	4.89	61.39		36.82	58.68	2.35	.51		1.27	92.26	4.41	.37	.0033
		2	11,375	6,589	571	57.93	5.02	62.95	5.42	36.84	58.58	2.48	.54	1.64	1.31	94.59	4.10	.41	
90	6	1	12,000	6,650	447	55.49	3.73	59.22		38.82	55.30	2.46	.48		1.09	93.90	4.48	.38	
		2	11,300	6,550	440	57.71	3.88	61.59	8.82	38.93	58.47	2.60	.51	.23	1.72	92.97	14.64	3.16	
85	7	1	11,380	6,965	342	57.21	2.85	60.06	4.97	33.42	46.78	10.28	3.56	.50	2.16	83.16	14.67	3.17	
		2	10,492	6,849	371	62.59	3.12	65.81		38.49	51.30	9.28	3.90		2.17	83.16	10.89	.93	.0091
164	8	1	11,831	7,200	369	57.83	2.98	60.81	5.49	32.49	56.66	7.50	1.23	.47	.65	87.96	10.95	.93	.0120
		2	12,940	7,164	361	61.10	3.12	63.67		30.36	59.02	7.89	1.27		.50	88.40	12.93	1.14	
165	8	1	12,135	7,845	369	60.55	2.81	63.91	5.49	30.51	55.49	8.66	1.27		.50	86.10	12.99	1.15	
		2	12,190	7,808	359	61.30	2.96	63.96		30.12	58.72	9.16	2.58		.69	86.51	11.84	1.96	
167	9 A (w.)	1	12,190	6,360	309	52.17	2.54	54.71	9.12	35.42	47.88	7.58	2.58	1.01	.69	86.46	11.84	1.96	
		2	11,078	6,296	306	56.83	2.76	59.59		39.97	52.08	8.35	2.84		.70	87.34	11.96	1.98	
Maryland:																			
50	1	1	8,000						2.24	13.77	71.04	12.95	1.39		.88	87.47	11.38	.95	.0278
54	1 (w.)	1	8,000	4,650	565	58.13	7.06	65.19	5.37	14.23	70.54	9.86	1.04	.27	.88	82.64	11.41	.95	
58	1 (w.)	2	7,652	4,616	561	60.32	7.33	67.65	4.35	17.81	68.86	8.98	1.09		.73	82.64	15.51	3.40	
Missouri:																			
116	5 (w.)	1	10,000	4,903	299	49.03	2.99	52.02	13.68	34.82	42.26	9.24	3.60	1.12	.73	82.64	15.51	3.40	.0000
		2	8,632	4,884	296	56.16	3.43	59.59		34.34	46.96	10.70	4.17		.74	83.58	15.68	3.44	
New Mexico:																			
148	3 B	1	12,120	7,655	410	63.16	3.38	66.54	3.68	33.06	48.94	14.32	.78	.76	.67	78.14	20.43	.71	.0042
		2	11,674	7,597	407	65.08	3.49	68.57		34.32	50.81	14.81	.81		.67	78.48	20.59	.71	
149	3 B (w.)	1	11,710	7,727	360	61.77	3.07	64.84	5.74	34.32	48.99	12.25	.72	.88	1.51	80.42	17.19	.65	.0348
		2	11,088	7,169	357	64.95	3.23	68.18		35.03	51.97	13.00	.77		1.52	81.14	17.34	.66	
152	3 B, 4 B, 5 (w.)	1	12,000	7,506	326	63.30	2.73	66.03	5.13	33.88	50.06	10.93	.99	.69	1.48	82.18	15.65	.63	.1004
		2	11,384	7,544	326	66.27	2.86	69.13		35.71	52.77	11.52	.73		1.49	82.75	15.76	.63	
150	4 B	1	11,700	7,610	362	64.55	3.07	67.62	3.69	34.02	47.83	13.86	.66	1.10	.94	78.48	19.48	.58	.0946
		2	11,365	7,526	358	66.28	3.15	69.43		35.95	40.66	14.39	.67		.95	79.35	19.70	.60	
151	4 B (w.)	1	11,430	6,986	307	63.79	2.80	66.59	5.52	35.95	49.87	9.32	.67	1.39	.85	83.66	14.10	.61	.0003
		2	10,700	6,980	303	63.51	2.71	66.22		31.96	52.78	9.87	.71		.86	84.84	14.30	.96	
146	5	1	11,810	7,500	320	63.79	2.77	67.38	3.05	31.91	49.11	15.28	.76	1.04	1.44	70.33	20.80	.87	.0001
		2	11,770	7,422	317	64.82	2.78	67.38		32.97	51.77	15.76	.78		.84	77.74	20.80	.76	
147	5	1	11,572	7,660	290	65.00	2.86		4.23	32.25	51.72	11.73	.80	.99	1.38	81.38	16.79	.76	
		2		7,574	277	67.19	2.46	69.65		33.67	54.08	12.25	.93		.85	82.19	16.96	.77	

Ohio:																					
24	1 (w.)	1	10,000	5,708	589	57.08	5.80	62.07	8.67	38.81	44.02		8.50	3.70	1.30	1.95	82.98	13.77	3.30		.0166
27	2 (w.)	2	10,133	5,634	581	61.69	6.36	08.05	10.38	42.49	48.20	3.55	9.31	4.05		1.98	84.07	13.95	3.34		
31	3 (w.)	1	8,902	4,980	690	49.80	6.90	56.70	12.68	37.30	45.04		7.28	3.07		2.27	82.52	11.66	2.84		
28	4	2	8,000	4,803	666	53.59	7.43	61.02	3.86	41.62	50.26	.33	8.12	3.43		2.35	85.56	12.00	2.94		.0087
22	5	1	10,000	6,770	332	67.70	3.32	71.02	4.49	32.55	40.38	5.33	6.79	3.60	.87	1.28	79.79	13.40	2.87		
59	6 (w.)	2	9,614	6,397	314	66.54	3.27	69.81	5.26	37.09	41.98		9.67	3.74		1.35	84.29	14.36	3.03		.0045
66	7	2	10,000	6,383	358	63.83	3.58	67.41	5.53	38.58	51.36	.77	10.06	1.77		.98	86.15	12.00	1.64		
89	7 (w.)	1	9,551	6,327	355	63.08	3.72	69.96	5.77	34.57	52.85		8.09	1.85	.42	.99	86.01	12.10	1.65		
94	8	2	11,300	7,500	370	66.00	3.08	63.08	7.37	36.20	51.30		8.47	3.49		1.26	83.71	14.31	2.96		.0058
81	8 (w.)	1	11,336	7,166	367	02.85	3.24	66.18	8.43	37.36	47.53	.75	9.85	3.68		1.21	84.36	14.37	2.51		
93	9 A	2	12,000	7,117	399	59.72	3.12	62.43		38.95	50.17		10.40	2.96		1.20	88.25	10.07	2.52		
72	9 B	2	12,308	7,064	397	62.94	2.53	65.75	5.58	35.13	50.36		6.56	3.13	.57	.90	83.35	12.04	2.34		
53	9 B (w.)	1	12,000		371	59.31	2.78	62.09	9.03	35.43	53.95		6.94	3.06		1.58	88.72	13.04	2.32		
57	12 (w.)	2	12,000		303	62.47	1.61	60.41	9.73	30.36	52.87	.66	8.98	2.23	.66	1.98	85.33	9.77	1.91		
180		1	12,000		301	57.48	3.77	64.41	6.19	37.25	49.82	2.46	9.53	2.90		.94	83.97	9.83	2.61		
		2	12,116		454	61.70	3.92	61.09	11.43	37.75	49.72		6.07	2.91		.93	88.58	15.60	1.93		
		1	12,000		451	54.38	4.94	60.35		40.68	44.56	1.80	9.76	2.10	.75	1.76	88.36	15.70	1.88		
		2	10,988		433	59.20	4.04	55.39	5.58	37.90	44.64		10.66	2.37		1.80	86.05	9.98	3.06		
		1	10,628		422	62.34	1.78	61.61		39.80	46.03		6.03	3.84	.68	1.08	84.43	12.69	3.12		
		2	11,330		452	57.64	1.61	58.74	9.03	42.15	49.72	.70	6.80	2.94		1.10	85.98	19.32	2.94		
		1	12,000		444	54.97	3.77	61.09	9.73	33.56	53.56		8.18	3.17		1.35	82.81	9.46	2.96		
		2	12,000		385	57.17	3.92	60.35		37.27	46.18	.68	8.66	2.37		.89	83.36	11.58	2.42		
		1	10,916		591	60.46	4.96	56.18		37.27	50.44		16.31	3.84		1.36	86.22	10.80	2.08		
		2	12,000		441	52.50	5.41	61.80		41.29	41.10	.73	6.98	3.15		.89	84.43	19.46	2.40		
		1	9,270		438	57.76	4.04	58.40		39.04	45.08		6.73	3.63	.51	.51	78.63	11.58	2.96		
		2	8,686		223	55.99	2.41	61.80		37.04	50.98		6.41	2.84		.89	86.93	10.80	2.42		
				5,152	221	59.25	2.54	61.79		41.62	51.55		6.68	3.87	.73	.51	87.96	11.58	3.08		
																	88.61	10.88	3.10		
Pennsylvania:																					
25	5	1	12,000	7,433	319	61.94	2.66	64.60	2.44	34.28	55.76	.49	6.52	.97		.90	88.03	10.58	.79		.0123
28	5 (w.)	2	11,707	7,307	317	63.18	2.71	65.89	4.73	35.14	58.18	.23	6.68	.99		.91	88.46	10.63	.70		.0053
32	6	1	13,000	8,596	242	66.14	1.95	68.00	3.22	34.59	58.07		4.71	.99		1.19	91.63	6.95	.81		.0241
34	6 (w.)	2	13,865	8,578	315	66.26	1.95	71.21		31.52	59.07	.34	13.39	2.10		1.19	91.84	6.97	1.59		.0162
35	6	1	10,000	7,350	314	73.30	3.15	76.45	6.58	33.00	51.87		13.83	1.46	.24	.71	82.31	16.64	1.60		.0229
38	6 (w.)	2	9,342	6,905	246	73.48	2.46	78.52		33.22	53.60		7.46	1.58		.66	82.59	10.49	1.21		.0153
41	6	1	12,000	6,888	295	69.05	2.62	71.51	3.64	33.59	52.96	.48	7.99	1.93		.66	88.61	10.52	1.70		.0187
30	7	2	11,563	8,646	294	73.73	2.54	76.35	4.21	32.01	56.69		13.32	2.00	.89	1.02	88.30	16.70	1.24		.0366
33	7 (w.)	1	13,000	8,604	373	72.05	2.87	76.54	2.53	32.96	51.03		7.18	1.51		1.63	81.59	16.78	1.25		.0369
29	8	2	13,000	8,927	364	68.67	2.97	74.02	3.75	33.40	54.59	.27	12.49	1.78		2.28	86.77	10.08	1.52		.0040
			12,000	8,848	363	71.05	3.03	75.89	3.98	34.87	55.21		12.17.	1.83	.46	2.25	87.55	10.17	1.87		
			11,696	8,743	325	72.86	3.10	77.65	3.32	32.99	53.21		12.49	2.33	.43	1.26	82.66	15.85	1.88		
			12,000	8,750	323	72.92	2.80	73.63		32.92	62.04		12.32	2.42		1.16	82.39	15.99	1.43		
			11,540	8,400	509	75.31	2.71	78.11		22.74	64.46		12.80	2.64		1.27	82.77	16.08	.91		
			11,522	8,364	507	70.00	4.24	74.24		23.32	60.75		11.95	1.71		1.84	83.40	14.33	.92		
			10,000	5,223	1,600	72.59	4.40	76.99		24.29	74.29		6.83	1.12		2.16	83.76	14.39			
			9,668	5,175	1,565	53.53	16.00	68.23		15.56	76.85		7.06	1.16		2.18	89.81	8.01			

Details of coking tests of coals, January 1, 1905, to June 30, 1907—Continued.

Test No.	Field No. of coal	Condition	Weight of coal (pounds)	Production (pounds) Coke	Production (pounds) Breeze	Production (percent) Coke	Production (percent) Breeze	Production (percent) Total	Chem. anal. of coal Moisture	Volatile matter	Fixed carbon	Ash	Sulphur	Chem. anal. of coke Moisture	Volatile matter	Fixed carbon	Ash	Sulphur	Phosphorus
1	2	18	19	20	21	22	23	24	25	26	27	28	29	30	31	32	33	34	35
	Pennsylvania—Con.																		
37	9.	1	7,000	5,000	801	70.00	10.01	80.01	3.26	16.18	69.44	11.12	1.90	0.54	1.21	86.84	11.41	1.06	0.0101
39	9.	1	8,000	5,500	797	72.94	10.44	83.38	2.86	16.24	68.60	12.30	2.14		1.22	87.31	11.47	1.07	.0070
42	9 (w.)	2	8,000	5,570	495	66.25	6.19	72.44	4.55	17.59	72.08	9.06	1.39	.72	.61	86.02	12.65	1.53	.0081
56	9.	1	8,000	5,300	491	67.53	6.30	73.83	2.60	17.92	69.36	9.69	1.80		.62	86.64	12.74	1.54	.0081
47	10.	2	7,792	7,400	340	61.67	2.83	64.50	2.56	18.40	71.21	10.39	1.85	.39	1.14	90.40	8.07	1.00	.0081
53	10.	1	12,000	8,371	339	63.04	2.90	65.94	2.73	34.80	56.81	5.83	1.25		1.14	90.75	8.10	1.00	.0307
159	11.	2	12,000	8,321	368	69.35	2.23	71.58	2.20	35.41	55.58	5.98	1.28	.36	1.15	89.93	8.57	1.05	.0184
161	12.	1	12,040	8,100	267	71.12	2.29	73.41	2.46	36.40	55.14	6.28	1.36		.97	90.25	8.60	1.19	.0087
162	12 (w.)	1	11,950	8,109	296	67.28	2.46	69.74	4.50	33.18	55.46	6.46	1.40	.52	.98	85.02	13.56	1.20	.0064
185	15 (w.)	2	11,656	7,831	294	68.43	2.50	70.93	7.19	33.33	56.70	9.16	1.36		.29	86.29	12.73	1.66	.0050
188	15 (w.)	1	12,140	8,100	286	65.98	2.44	68.37	4.53	31.28	56.71	9.36	1.39	.69	.29	86.80	12.82	1.67	.0547
178	17.	2	11,594	8,038	257	67.18	2.39	69.02	4.41	32.07	57.66	9.56	1.39		1.28	89.13	9.07	1.11	.0582
186	17 (w.)	1	9,750	5,779	256	66.72	2.12	68.84	6.30	31.35	56.70	9.80	2.03	.52	1.29	88.59	9.02	1.12	.0147
176	19.	2	9,040	5,147	261	69.50	2.21	71.71	3.01	32.83	58.13	6.49	2.08		.32	89.10	8.07	1.46	.0153
177	19.	1	12,460	8,144	332	59.27	2.69	61.96	3.57	17.98	60.38	6.79	1.40	.56	.32	91.61	8.02	1.47	.0092
179	20.	1	11,896	8,096	330	63.51	2.88	66.39	3.91	19.24	69.57	5.38	1.47		.55	90.23	8.66	1.54	.0083
182	20 (w.)	2	11,200	7,006	355	65.36	2.66	68.02	6.30	18.56	74.96	5.80	1.63	.57	.55	90.75	8.70	1.55	.0164
183	21.	1	11,662	7,923	306	68.00	2.92	70.94	5.37	19.44	70.63	6.28	1.76		.38	84.55	14.90	1.37	.0115
187	21.	2	11,920	7,501	315	67.70	3.04	67.98	5.53	28.83	73.96	6.58	1.85	.22	.36	84.74	11.19	1.37	.0099
190	21.	1	11,169	7,900	314	67.16	2.74	63.69	4.05	30.16	57.86	8.90	1.94		.56	87.96	11.21	1.00	

No.	State / sample	Test																	
191	21	1	11,940	7,983	66.99	368	3.08	69.77	4.28	20.92	58.22	7.58	.91	.33	.15	88.54	10.98	.87	.0104
	21	2	11,430	7,637	66.45	367	3.21	72.19		31.26	60.82	7.92	.95		.15	88.83	11.02	.87	.0065
	21	1	11,500	7,733	67.24	339	3.05	70.19	3.66	21.86	60.13	7.43	.91	.29	.69	88.54	10.48	.76	.0238
	21	2	11,088	7,711	66.44	338	3.05	72.50		22.93	62.36	7.71	.94		.69	88.30	10.51	.76	.0253
83	Tennessee: 1 (w.)	1	12,000	7,730	64.33	347	2.89	67.22	3.71	32.81	51.69	11.79	1.38	.37	.73	83.00	15.90	1.35	.0094
153	2	2	11,556	7,661	66.56	346	2.90	68.55	4.96	34.07	63.08	12.25	1.64	.93	.73	83.31	8.99	1.28	.0125
87	3	1	11,900	7,270	61.00	270	2.86	63.59	3.56	35.06	64.72	5.36	1.17	1.13	.77	89.61	8.77	.93	.0215
128	4	2	11,323	7,202	61.61	243	2.83	60.97	4.32	36.63	67.82	5.63	1.23	.67	.78	90.45	7.23	.94	.0233
85	4	1	11,573	7,213	61.03	383	2.93	64.55	3.82	37.88	57.36	4.76	1.05	.54	.64	92.04	7.31	.78	.0634
89	5 (w.)	2	11,280	7,131	61.63	394	3.63	66.75	3.82	33.90	56.85	6.22	.99	1.43	.65	87.44	11.08	.70	.0468
84	6	1	10,708	7,088	64.67	391	2.61	60.29	5.53	33.43	58.07	6.60	1.10	.56	.81	88.03	12.63	.88	.0068
92	7 B	2	12,000	7,671	64.38	31	2.89	64.87	2.06	33.17	63.17	9.84	.89	.22	1.56	84.27	13.70	.89	.0090
121	7 B (w.)	1	11,542	7,431	64.28	266	2.61	67.07	7.88	34.49	65.28	9.84	.93	.43	1.57	84.73	13.70	.72	.0053
123	8 B (w.)	2	11,270	7,500	68.55	282	2.36	68.91	8.37	34.06	55.28	10.23	.93	.57	.28	87.98	11.74	.72	
134	9 (w.)	1	10,639	7,363	68.21	314	2.20	70.63	3.09	35.06	46.40	10.13	2.44	.32	.28	90.81	8.56	.70	.0005
24	10 (w.)	2	10,980	6,897	58.55	253	2.32	66.89	4.02	37.11	56.28	6.49	.90	.39	.42	90.97	19.86	2.08	
56	11 (w.)	1	12,000	7,712	61.93	313	2.66	68.13	7.80	27.93	40.02	15.47	2.58	1.67	.91	90.46	19.90	2.09	
80		2	11,764	7,496	64.37	316	2.92	61.64	5.29	28.29	46.43	17.41	.92	1.14	.92	70.18	24.78	.99	.0005
130	Utah: 1	1	11,810	5,650	46.99	1,418	12.01	59.00	5.83	41.80	47.44	4.84	.56	2.53	1.37	88.06	8.04	.64	.0041
141	1	2	11,121	5,410	48.65	1,382	12.43	61.08	4.08	44.48	48.38	5.14	.59	1.50	1.40	90.35	8.25	.66	
157	1	1	12,000	3,564	29.20	3,206	26.78	55.93	4.60	28.43	53.07	9.97	.45		1.38	96.07	11.05	.57	.0041
64	Virginia: 1	1	14,000	9,079	64.85	415	2.96	67.81	5.70	32.52	56.15	5.63	.98	1.52	.99	89.20	8.29	.88	.0036
65	1	2	13,202	8,941	67.72	409	3.09	70.69	4.44	34.49	59.54	5.97	1.04	1.23	1.01	90.58	8.41	.89	
67	1	1	10,000	6,811	68.11	258	2.58	73.07	4.95	33.44	56.27	5.85	1.13		1.67	90.24	7.86	.94	.0041
68	1	2	9,656	6,727	70.40	255	2.67	63.03		34.99	58.89	6.12	1.18	.21	.89	90.35	7.96	.95	
77	1	1	10,000	5,867	61.73	424	4.24	66.18	4.82	35.99	55.86	4.98	1.12		1.16	90.99	7.91	1.01	.0046

a Laboratory sample showed air-drying gain and was thrown out, analysis of sample for test 125 being substituted.

Details of coking tests of coals, January 1, 1905, to June 30, 1907—Continued.

Test No.	Field No. of coal	Condition	Weight of coal (pounds)	Production (pounds) Coke	Production (pounds) Breeze	Production (per cent) Coke	Production (per cent) Breeze	Production (per cent) Total	Chemical analysis of coal Moisture	Volatile matter	Fixed carbon	Ash	Sulphur	Chemical analysis of coke Moisture	Volatile matter	Fixed carbon	Ash	Sulphur	Phosphorus
1	**2**	**18**	**19**	**20**	**21**	**22**	**23**	**24**	**25**	**26**	**27**	**28**	**29**	**30**	**31**	**32**	**33**	**34**	**35**
	Virginia—Cont'd.																		
63	2	1	12,000	7,518	291	62.65	2.43	65.08	3.88	34.11	57.01	5.00	1.02	0.25	1.08	91.25	7.42	0.68	0.0026
		2	11,634	7,409	290	65.02	2.51	67.53	3.86	35.49	59.31	5.20	1.06	.69	1.08	91.48	7.44	.68	
66	2	1	12,037	7,314	433	60.95	3.61	64.56		34.13	56.39	5.62	.79		.93	90.33	8.05	.65	
		2	11,337	7,264	430	62.97	3.73	66.70	5.96	34.50	56.65	5.85	.82	.45	.94	90.96	8.10	.65	
70	2 (w.)	1	10,404	6,099	294	60.96	2.94	63.90		34.17	56.03	3.84	.91		1.23	92.25	6.07	.69	.0011
		2	12,066	6,069	253	64.54	3.12	67.66	2.87	34.34	59.58	4.08	.97	.29	1.24	92.67	6.09	.69	
61	3	1	12,000	8,160	240	68.00	2.00	70.00		31.58	61.43	4.12	.50		1.21	92.60	5.92	.61	.0060
		2	11,666	8,136	239	69.80	2.05	71.85	2.49	32.51	63.25	4.24	.58	.16	1.21	92.87	5.73	.55	
88	3	1	12,000	7,907	336	65.89	2.80	68.69		31.90	61.16	4.45	.57		1.26	91.85	6.74	.42	.0046
		2	12,701	7,894	335	67.46	2.86	70.32	3.87	32.72	62.72	4.56	.58	.16	1.26	92.00	5.81	.42	
62	4	1	10,818	6,272	241	62.72	2.41	65.13		36.39	55.60	4.14	.39		1.14	92.90	5.81	.42	.0076
		2	9,813	6,262	241	65.14	2.51	67.65	3.87	37.85	57.84	4.31	.41	.38	1.14	93.05	5.81	1.44	
181	6	1	10,910	6,686	267	61.39	2.45	63.84	5.05	22.95	62.11	9.99	1.49		1.35	86.05	12.22	1.45	.0063
		2	11,189	6,675	287	64.42	2.57	66.99		24.77	65.41	10.42	1.57	.24	1.35	86.38	12.27	1.24	
184	6 (w.)	1	10,507	6,135	191	55.01	1.71	56.72	5.48	24.77	64.96	4.79	1.45		.33	93.73	5.71	1.24	
		2		6,135	191	58.06	1.81	59.87		26.21	68.73	5.06	1.53		.33	93.96	5.72		
	Washington:																		
135	2	1	10,000	5,477	444	54.77	4.44	59.21	3.07	37.42	47.35	12.16	.44	1.02	2.10	77.53	19.35	.44	.0847
		2	9,063	5,421	409	55.93	4.53	60.46		38.60	48.85	12.55	.46	.35	2.12	78.33	19.55	.44	
	West Virginia:																		
40	4 B	1	10,000	7,533	286	75.33	2.86	78.19	4.24	26.57	58.89	10.30	.97		.68	86.01	12.96	.82	.0141
		2	9,876	7,537	285	76.60	2.88	81.58		27.75	61.50	10.75	1.01	.60	1.01	86.31	13.00	.82	
44	4 B (w.)	1	10,000	8,159	337	67.74	2.81	70.55	5.40	26.83	59.90	7.87	.93		1.20	87.04	11.16	.72	.0186
		2	11,862	8,080	335	71.17	2.75	74.12		28.36	63.32	8.32	.98	.27	1.20	87.57	11.23	.78	
46	4 B (w.)	1	9,760	7,201	367	73.98	2.85	76.73	3.87	27.72	63.67	8.74	.87		1.37	86.95	11.38	.78	.0112
		2	9,382	7,555	413	66.75	3.44	79.00		28.84	62.07	9.09	.90	.75	1.37	87.22	11.41	.77	
21	13	1	11,613	7,886	410	67.99	3.53	69.73	3.23	31.12	61.98	3.79	.89		1.05	93.36	4.84	.78	.0046
		2	11,432	8,080	214	67.44	1.79	71.52		32.16	64.05	4.05	.88	.43	1.37	94.07	4.87	.84	
23	13 and 14	1	11,423	8,087	215	70.53	1.87	69.23	4.81	29.28	62.84	3.07	.93		1.06	94.20	4.00	.84	.0057
		2	10,896	7,127	364	65.47	3.31	68.35		30.76	66.80	3.22	1.04	.38	1.38	94.61	4.01	.94	
20	4	1	10,886	6,867	363	65.04	3.33	72.40	1.04	29.28	66.02	2.88	1.05		.96	95.55	3.48	.94	.0046
		2	10,631	7,127	275	65.47	2.76	68.80		29.59	67.50	2.91	1.05	.45	.97	88.70	10.07	2.09	
36	15	1	8,831	6,835	289	69.60	2.80	71.43	1.79	37.90	67.31	7.31	2.73		.78	88.10	10.12	2.09	.0153
		2	12,000	6,835	268	70.03	2.24	72.27	2.33	38.33	63.97	7.44	2.78	.26	1.89	89.10	10.55	2.27	
43	15	1	11,720	8,382	384	71.52	2.29	73.81		39.24	51.72	7.62	2.72		1.89	87.30	10.58	2.27	
		2		9,700		69.29	2.74	72.03		39.34	52.95	7.81	2.78		1.31	87.63	9.08	.81	
73	16 A	1	13,560	9,648	382	70.99	2.81	73.80	2.93	36.26	55.93	6.81	.95	.54	1.32	89.55	9.13	.81	.0226

No.	Sample	Tests																	
45	16 B	1	12,000	8,124	276	67.70	2.30	70.00	5.70	32.52	53.44	8.34	1.20	.40	1.55	93.61	12.44	1.23	.0218
48	16 B (w.)	2	11,214	8,092	275	71.51	2.30	73.94	6.31	34.49	54.80	8.84	1.27	.37	1.18	83.96	12.49	1.24	.0166
49	16 B (w.)	1	12,365	7,954	289	64.00	2.41	68.94	4.50	32.67	55.52	5.97	.97	.38	1.19	89.87	8.58	.91	.0133
60	17 (w.)	2	7,560	5,207	288	64.76	2.24	72.63	4.84	33.53	55.71	5.31	1.02	.07	1.19	89.20	8.61	.89	.0047
74	18	1	7,448	5,187	175	66.75	2.34	69.00	3.42	33.53	55.80	6.08	.95	.63	2.35	88.25	8.36	.59	.0058
78	18	2	10,816	5,676	174	66.98	2.08	72.04	4.18	28.33	60.49	6.04	1.00	.69	2.36	89.25	8.39	.54	.0037
79	19	1	9,515	6,670	293	70.37	2.58	73.17	2.82	30.82	67.02	6.34	1.25	.38	.63	91.36	7.94	.53	.0026
83	19	2	11,380	8,444	300	71.09	2.53	71.95	2.43	30.09	67.07	4.96	1.31	.45	.97	91.42	8.86	.77	.0030
80	20	1	11,360	8,391	307	72.40	2.63	75.06	3.79	32.98	57.91	6.74	.70	.17	1.19	89.54	8.92	.82	.0040
84	20 (w.)	2	11,456	8,422	340	70.18	2.04	75.68	5.34	33.46	89.04	4.96	.72	.29	1.20	89.11	8.65	.90	.0044
87	20 (w.)	1	11,682	8,396	338	64.10	2.16	66.25	4.47	31.63	60.21	6.31	.61	.17	1.68	89.47	8.71	.90	.0036
92	20	2	11,705	7,663	287	65.71	2.36	67.81	3.01	21.23	70.30	6.26	.64	.18	1.61	90.96	7.98	.86	
83	21	1	12,030	7,914	283	65.93	2.41	66.70	3.65	21.34	72.24	5.73	1.00	.24	.55	91.67	8.01	.85	.0027
91	21 (w.)	2	11,640	7,573	282	67.29	2.18	72.24	5.72	21.87	72.28	5.87	1.03	.55	.56	91.98	7.46	1.30	
75	25	1	12,359	8,371	208	72.51	2.27	74.78	4.56	32.24	55.72	7.77	.98	.30	.46	89.10	9.74	.90	.0062
			12,030	8,014	302	66.98	1.83	68.71		33.51	57.00	5.10	1.15		.81	91.73	7.82	.86	
			12,030	5,014	264	65.78	2.53	71.30		34.39	60.22	4.87	1.02		.57	92.00	6.73	.85	
			11,464	7,880	264	68.74	2.27	71.37		34.32	60.58	7.99	1.62		.81	92.45	6.74	1.30	
			11,639	8,277	242	71.11	2.02	73.38		36.32	68.60	8.23	1.67		.81	88.11	11.32	.98	
			11,574	8,106	241	67.55	2.08	71.95		37.66	55.18	4.95	1.23		1.03	91.45	7.74	.76	
			12,000	8,087	249	69.87	2.08	67.21		36.54	87.21	5.13	1.28		.63	93.16	5.26	.70	
			11,314	7,516	248	65.13	2.19	70.89		38.76	67.65	3.39	1.01		.49	93.08	12.50	.47	
			11,340	7,773	381	68.70	2.36	64.38		34.38	63.39	3.59	1.07		.49	86.81	12.54	.47	
			10,823	6,899	380	63.74	3.51	67.25		36.02	55.94	7.67	.64			86.97			

W'yoming:

No.	Sample	Tests																	
52	3 (w.)	1	10,000						19.20	37.99	36.13	6.68	4.09						
132	5	1	8,000						11.09	34.53	50.50	3.88	.84						

DESCRIPTIONS OF COKE AND REMARKS.

The following notes are given to supplement the information contained in the preceding table:

"Cell structure" refers to the general appearance as to size and not to the number of cells as given by percentage of cells by volume. In many tests the cell structure as determined from general appearance is small when the percentage by volume indicates quite the reverse. (See, for example, test 29, Pennsylvania No. 8 coal, p. 24.)

Alabama No. 2 B.—Test 142: Soft, dense coke; dull appearance; cell structure very small; breakage, lumps of irregular size; 1-inch black butts.

Alabama No. 3.—Test 138: Dark-gray color, some deposited carbon; cell structure good; breakage good; long, large pieces; good, hard, heavy coke, with exception of ⅜-inch black butts, which should be easily removed; ash high; washing would probably reduce ash and improve quality of coke.

Test 139: Light-gray color, some silvery deposit of carbon; good ring; cell structure good; breakage good; long, large pieces; good, strong, hard, heavy coke; improved very materially by washing.

Alabama No. 4.—Test 131: Light-gray and silvery color; metallic ring; cell structure good; breakage somewhat cross fractured, but pieces of good, large, uniform size; good, strong, heavy coke; ash high; probably could be reduced by washing.

Test 136: Light-gray and silvery color; metallic ring; cell structure rather large; breakage somewhat cross fractured, but pieces of good, uniform size; good, strong coke; ash very materially reduced by washing.

Alabama No. 5.—Test 171: Light-gray and silvery color; metallic ring; cell structure good; breakage good; uniform size; ash and sulphur high, both would probably be reduced by washing.

Alabama No. 6.—Test 172: Light-gray and silvery color; metallic ring; cell structure good; breakage somewhat cross fractured, but pieces of good, uniform size; good, heavy coke.

Test 174: Light-gray and silvery color; metallic ring; cell structure good; breakage somewhat cross fractured, but pieces of good, uniform size; good, heavy coke, somewhat better than coke from raw coal, but low ash and sulphur of this coal would not warrant washing.

Arkansas No. 1 B.—Test 95: Dull-gray color; soft, dense, punky coke; cell structure very small; breakage very bad and irregular; large and small chunks.

Test 96: Dull-gray color; soft, dense, punky coke, with no apparent cell structure; drawn from oven in large and small chunks, very easily crushed; test was run slowly and high enough heat was not obtained, which accounts for the large percentage of breeze.

Test 97: Dull-gray color; soft, dense coke; cell structure small; better than coke from washed coal.

Test 100: Dull-gray color; soft, dense, punky coke; possibly little better than coke from this coal, with addition of 10 per cent of pitch.

Arkansas No. 7 B.—Test 104: Dull, dark color; very soft, light-weight coke; no apparent cell structure; drawn from oven in large and small lumps; bottom 6 inches did not coke, burning to ash, all volatile being expelled, but did not stick together.

Test 105: Soft, dense, punky coke; drawn from oven in large and small chunks; somewhat better and heavier than coke from coal containing no pitch.

Arkansas No. 9.—Test 98: No coke produced; charge ashed over top and down about 5 inches.

Arkansas No. 9.—Test 99: No coke produced; ashed down about 4 inches.

Test 101: No coke produced; ashed down about 6 inches.

Test 102: Soft, dense, punky coke; drawn from oven in large and small chunks.

Test 103: Soft, dense, punky coke; drawn from oven in large and small chunks; high yield of breeze, due to large amount of coal whose volatile was expelled, not sticking together; 5 per cent of pitch not sufficient for this coal.

Georgia No. 1.—Test 173: Poor, dense coke; large pieces of irregular size; ash high; probably reduced ash and materially improved by washing.

Illinois No. 7 D.—Test 1: Good, hard coke with medium cell structure; breakage straight and long. This was the first charge after firing ovens. and results were not as good as might be expected.

Test 4: Light-gray and silvery color; metallic ring; cell structure good; breakage somewhat marred by cross fracture, but pieces of good size; good, strong coke, much improved by washing.

Illinois No. 11 D.—Test 5: Light-gray and silvery color; metallic ring; cell structure good; breakage, good long pieces; good, strong coke.

Illinois No. 13.—Test 2: Dull-gray color; cell structure good; breakage marred by cross fracture, probably due to successive charging of small portions.

Test 3: Dull-gray color; cell structure small; cross breakage more pronounced than from washed coal.

Illinois No. 16.—Test 7: Accident to charging larry necessitated discontinuing test. Coal burned to keep oven hot.

Test 10: Dull-gray color; cell structure small.

Illinois No. 19 A.—Tests 11, 15, and 19: No coke produced; the whole charge was burned and volatile was expelled, but the residue would not bind together.

Illinois No. 20.—Test 106: Dull-gray color; cell structure small; breakage bad; separate and distinct cross fracture all over oven, coking in layers; ash and sulphur high.

Test 107: Dull-gray color; some little deposit of carbon; metallic ring; cell structure small, but not dense; breakage somewhat marred by cross fracture; pieces of good size; great improvement over former test; ash and sulphur high.

Illinois No. 21.—Test 126: No coke produced.

Test 137: Burned very vigorously for 12 hours, afterwards falling off rapidly to small candles all over surface of charge; when pulled, after 45 hours, product was mixture of unburned coal and slightly coherent mass of coal of original size showing no trace of cell structure; all volatile apparently expelled.

Illinois No. 22 B.—Test 117: Dull-gray color; cell structure medium; breakage very irregular, probably owing to high amount of slate; poor coke; heavy clinker over whole surface; ash and sulphur high.

Test 118: Light-gray color; upper 12 inches fingered, two 6-inch sections below in chunks; upper 12 inches had metallic ring and good cell structure; the remaining coke poor. This oven was held 72 hours on account of accident. Under more favorable conditions, the whole charge would have probably been better coke. Ash and sulphur high.

Illinois No. 23 A.—Test 111: Light-gray color; some silvery deposit of carbon; cell structure a little large; breakage large-fingered pieces; metallic ring; ash and sulphur high.

Illinois No. 23 B.—Test 112: Light-gray color; a little silvery deposit of carbon; metallic ring; cell structure a little large; breakage, long, thin pieces; larger charges would probably make better coke; ash and sulphur high.

Test 114: Light-gray color; a little silvery deposit of carbon; metallic ring; cell structure good; breakage, long, thin pieces and large 6-inch chunks; bottom very hot; bottom 6 inches probably coked upward; ash and sulphur high.

Illinois No. 24 A.—Test 119: No coke produced; ashed down about 4 inches.

Test 155: No coke produced; all volatile driven off; high heat of by-product ovens quickly applied might produce coke.

Illinois No. 24 B.—Test 145: Dull-gray color; practically no cell structure; barely stuck together; very poor, dense coke, with high sulphur.

Illinois No. 25 A.—Test 120: No coke produced.

Test 140: Dull-gray color; cell structure small; soft, dense coke; breakage poor; two distinct layers of 16 inches and 8 inches, the lower coming out in chunks; high ash and sulphur.

Illinois No. 26.—Test 143: Dull-gray color, soft, dense coke; breakage poor; practically no cell structure; ash and sulphur high.

Illinois No. 27.—Test 144: Poor, soft, dense coke; breakage poor; sulphur high.

Illinois No. 28 C.—Test 166: Dark-gray color; cell structure small; breakage, good, uniform size.

Illinois No. 29 A.—Test 169: Dark-gray color; drawn from oven in three distinct layers; breakage poor; large chunks and small-fingered pieces; poor, dense coke; high sulphur.

Test 170: Dull-gray color; some silvery coloration; metallic ring; drawn from oven in 6-inch chunks of practically uniform size; cell structure good; more rapid burning and higher heat produced gave much better coke than former charge; sulphur high.

Illinois No. 34 B.—Test 190: Light-gray and silvery color; metallic ring; cell structure a little large; breakage good; uniform-sized pieces; yield low on account of burning, but could be easily increased on better acquaintance; good coke; sulphur high.

Indiana No. 3.—Test 14: No coke produced; ashed down about 10 inches and blaze lost.

Indiana No. 4.—Test 6: Light-gray color; cell structure a little large; breakage somewhat marred by cross fracture.

Test 9: Light-gray and silvery color; metallic ring; fine-fingered pieces; cell structure large; ash and sulphur reduced by washing.

Indiana No. 5.—Test 8: Light-gray and silvery color; metallic ring; cell structure large; breakage, good, long pieces; good coke.

Indiana No. 6.—Test 12: Light-gray and silvery color; cell structure good; breakage good; metallic ring; good coke, but ash and sulphur high.

Indiana No. 7 A.—Test 13: Light-gray and silvery color; metallic ring; cell structure large; breakage somewhat marred by cross fracture, somewhat brittle; ash and sulphur high.

Indiana No. 9 A.—Test 16: Light-gray and silvery color; cell structure small; breakage somewhat marred by cross fracture, brittle; ash and sulphur high.

Indiana No. 9 B.—Test 17: Light-gray and silvery color; cell structures mall; long-fingered, heavy coke; high ash and sulphur.

Test 18: Light-gray and silvery color; metallic ring; breakage somewhat brittle; cell structure good; ash and sulphur somewhat reduced by washing, but still high.

Indiana No. 11 D.—Test 51: Light-gray color; metallic ring; breakage, long, fine-fingered pieces; cell structure medium.

Indiana No. 12.—Test 108: Light gray, with a little silvery coloration; metallic ring; cell structure a little large; breakage, good-sized pieces; ash and sulphur high.

Test 109: Light-gray color; some silvery deposit of carbon; cell structure large; breakage, good-sized pieces; ash and sulphur reduced by washing, but still high.

Test 110: Light-gray color; some silvery deposit of carbon; breakage practically the same as in test 109; somewhat larger size; cell structure not quite so large; metallic ring; good weight; ash and sulphur high.

Indiana No. 17.—Test 163: Dark-gray color; breakage, large pieces of irregular size; cell structure large; ash and sulphur high.

Indiana No. 18 A.—Test 158: No coke produced; ashed down about 3 inches, and blaze lost.

Test 168: No coke produced.

Kansas No. 6.—Test 113: Light-gray color, some silvery coloration; cell structure good; breakage good; long, large, heavy pieces; heavy clinker over whole surface of coke; ash and sulphur high; washing would probably reduce ash very materially, and produce better grade of coke.

Test 115: Light-gray and silvery color; metallic ring; breakage good; long, large, heavy pieces; cell structure good; strong heavy coke; washing reduces ash and sulphur, but both still high.

Kentucky No. 1 B.—Test 76: Light-gray and silvery color; much deposited carbon; metallic ring; cell structure good; a fine-fingered coke; breakage bad, brittle.

Kentucky No. 1 C.—Test 71: Light-gray and silvery color; metallic ring; cell structure a little large; breakage, long, thin-fingered pieces; good coke, but very brittle.

Kentucky No. 5.—Test 75: Light-gray and silvery color; much deposited carbon; metallic ring; cell structure good; breakage bad, very brittle.

Kentucky No. 6.—Test 86: Light-gray and silvery color; metallic ring; cell structure small; breakage, long, fine-fingered pieces, very brittle.

Test 90: Light-gray and silvery color; metallic ring; cell structure small; breakage bad, brittle; fine-fingered coke.

Kentucky No. 7.—Test 85: Light-gray and silvery color; metallic ring; cell structure good; breakage good; long, large pieces; coke contains a large amount of hard clinker on top and through cracks; good weight coke; ash and sulphur high.

Kentucky No. 8.—Test 164: Dark-gray color, with some little silvery deposit of carbon; cell structure large; breakage good; regular-sized pieces.

Test 165: Light-gray color; breakage good; large pieces of regular size; cell structure a little large; some little improvement over test No. 164.

Kentucky No. 9 A.—Test 167: Light-gray color, with black top and some silvery deposit of carbon; cell structure good; breakage, long-fingered pieces; sulphur high.

Maryland No. 1.—Tests 50, 54: No coke produced.

Test 58 (with 10 per cent pitch): Dull-gray color; cell structure small; breakage, large and small chunks; poor, soft coke.

Missouri No. 5.—Test 116: Light-gray and silvery color; cell structure good; breakage somewhat cross fractured but pieces of good, large size; good weight coke; ash and sulphur high.

New Mexico No. 3 B.—Test 148: Light-gray color, some silvery deposit of carbon; metallic ring; cell structure medium; breakage good; long, large pieces; good, heavy coke, but ash high.

Test 149: Light-gray color; some silvery deposit of carbon; metallic ring; cell structure medium; breakage good; long, large pieces; good, heavy coke; ash reduced by washing, but still high.

New Mexico Nos. 3 B, 4 B, and 5.—Test 152: Light-gray and silvery color; metallic ring; cell structure good; breakage good; long, large pieces; good, strong, heavy coke.

New Mexico No. 4 B.—Test 150: Light-gray color; silvery deposit of carbon; metallic ring; cell structure good; breakage somewhat cross fractured, but pieces of good, large, uniform size; good, heavy coke; high ash.

Test 151: Light-gray and silvery color; large deposit of carbon; metallic ring; cell structure good; breakage good; lo. z, large pieces; good, strong, heavy coke; ash reduced by washing.

New Mexico No. 5.—Test 146: Light-gray color; cell structure a little large; breakage somewhat marred by cross fracture; good, heavy coke; ash high; blaze lost after 15 hours, and necessary heat not attained.

New Mexico No. 5.—Test 147: Light-gray color, some silvery deposit of carbon; metallic ring; cell structure a little large; breakage good; long, large, heavy pieces; ash reduced by washing, but still high.

Ohio No. 1.—Test 24: Light-gray and silvery color; metallic ring; breakage good; fine-fingered pieces; cell structure good; good weight coke; high sulphur.

Ohio No. 2.—Test 27: Dull-gray color; cell structure close; poor coke, soft and easily broken.

Ohio No. 3.—Test 31: Charge burned to ash.

Ohio No. 4.—Test 28: Light-gray and silvery color; metallic ring; breakage good; long, large pieces; cell structure good; very heavy; sulphur high.

Ohio No. 5.—Test 22: Light-gray and silvery color; metallic ring; cell structure good; long-fingered coke, brittle.

Ohio No. 6.—Test 59: Light-gray and silvery color; metallic ring; cell structure a little large; breakage good; large, long, heavy pieces; high sulphur.

Test 66: Light-gray and silvery color; metallic ring; cell structure large; breakage somewhat crosswise, but good-sized pieces; ash and sulphur reduced by washing, but sulphur still high.

Ohio No. 7.—Test 89: Light-gray and silvery color; metallic ring; cell structure a little large; breakage good; long pieces; large-fingered coke; high sulphur.

Test 94: Light-gray and silvery color; metallic ring; cell structure a little large; breakage fine-fingered; very brittle; ash and sulphur reduced by washing, but sulphur still high.

Ohio No. 8.—Test 81: Light-gray color; metallic ring; breakage long, thin pieces; cell structure small; fingered coke, very brittle; ash and sulphur high.

Test 93: Light-gray color, with black-fused bottom, not a butt; metallic ring; cell structure small. About three-fourths of oven coked up 8 inches, and the upper 16 inches coked down, showing clear demarcation; the lower 8 inches in chunks, the upper 16 inches fingered; very brittle; ash and sulphur reduced by washing, but sulphur high.

Ohio No. 9 A.—Test 72: Light-gray and silvery color; metallic ring; breakage long and thin pieces; fine-fingered coke, very brittle; sulphur high.

Ohio No. 9 B.—Test 55: Dull-gray color; cell structure small; breakage bad; very brittle; ash and sulphur high.

Test 57: Light-gray color; metallic ring; cell structure good; breakage bad; fine-fingered coke, very brittle; sulphur high; ash greatly reduced by washing.

Ohio No. 12.—Test 180: Light-gray color; some silvery deposit of carbon; metallic ring; cell structure large; breakage somewhat cross fractured; sulphur high.

Pennsylvania No. 5.—Test 25: Light-gray and silvery color; much deposited carbon; metallic ring; cell structure a little large; breakage good; long, large pieces; good, heavy coke.

Test 26: Light-gray and silvery color; metallic ring; cell structure good; breakage good; long, large pieces; good, heavy coke; ash and phosphorus reduced by washing, the phosphorus over 50 per cent.

Pennsylvania No. 6.—Test 32: Light-gray and silvery color; much deposited carbon; metallic ring; cell structure good; breakage good; long, large pieces; very heavy coke; sulphur and ash high.

Test 34: Light-gray and silvery color; metallic ring; cell structure good; breakage somewhat irregular, but not so good as from raw charge; very heavy coke; ash and sulphur reduced by washing.

Test 35: Light-gray and silvery color; metallic ring; cell structure good; breakage good; long, large pieces; very heavy; ash and sulphur high.

Test 38: Light-gray and silvery color; metallic ring; cell structure a little small; breakage good; ash and sulphur reduced by washing.

Pennsylvania No. 6.—Test 41: Light-gray and silvery color; much deposited carbon; metallic ring; cell structure small; breakage good; long, large pieces; very heavy; ash and sulphur high.

Pennsylvania No. 7.—Test 30: Light-gray color; cell structure small; breakage long and irregular, but in large pieces; very heavy; ash and sulphur high.

Test 33: Light-gray color; breakage, large and small lumps, very irregular; cell structure small; coke heavy; sulphur reduced by washing; ash not materially affected.

Pennsylvania No. 8.—Test 29: Dull-gray color; breakage bad; large and small chunks; cell structure small; soft, dense coke.

Pennsylvania No. 9.—Test 37: Some few pieces of coke obtained, but the amount was so small that it was not determined.

Test 39: Some few pieces of coke; mostly large lumps of closely adhering ash.

Test 42: Dull-gray color; cell structure small; poor, dense coke.

Test 56 (with 5 per cent pitch): Dull-gray color; cell structure medium; breakage very irregular; large and small lumps; poor, soft coke, scarcely any better than coke from washed coal.

Pennsylvania No. 10.—Test 47: Light-gray and silvery color; metallic ring; breakage poor, somewhat brittle; cell structure large.

Test 53: Light-gray and silvery color; metallic ring; cell structure good; breakage bad; increase in yield of coke and decrease in amount of breeze probably due to fine grinding.

Pennsylvania No. 11.—Test 159: Light-gray color; metallic ring; cell structure a little small; breakage good; long, large pieces.

Pennsylvania No. 12.—Test 161: Light-gray and silvery color; metallic ring; cell structure a little small; breakage good; large pieces; good, heavy coke; sulphur a little high.

Test 162: Light-gray color; some deposit of carbon; metallic ring; cell structure a little small; breakage good; uniform-sized pieces; ash reduced by washing; good, strong coke.

Pennsylvania No. 15.—Test 185: Dull-gray color; soft, dense coke; cell structure small; breakage badly cross fractured, and pieces of irregular size; sulphur high.

Test 188: Light-gray color, some silvery deposit of carbon; cell structure medium; breakage somewhat cross fractured, but pieces of good, uniform size; much improvement over coke from finely ground charge; sulphur high.

Pennsylvania No. 17.—Test 178: Light-gray and silvery color; cell structure a little small; breakage good; long, large pieces; good, heavy coke.

Test 186: Light-gray and silvery color; metallic ring; cell structure good; breakage good; large pieces of uniform size; good, strong, heavy coke; ash and sulphur reduced by washing.

Pennsylvania No. 19.—Test 176: Light-gray color; some silvery deposit of carbon; cell structure small; breakage marred by cross fracture, probably due in large measure to uncrushed slate; good, heavy coke, somewhat brittle.

Test 177: Light-gray and silvery color; metallic ring; cell structure a little small; breakage good; large, uniform pieces; crushing improves physical appearances and increases total yield.

Pennsylvania No. 20.—Test 179: Gray color, some silvery deposit; cell structure small; breakage irregular, but pieces of good size; soft, dense coke; high sulphur.

Test 182: Gray color; soft, dense coke; no evident physical improvement over raw charge; ash and sulphur reduced by washing.

Pennsylvania No. 21.—Test 183: Light-gray and silvery color; metallic ring; cell structure small but not dense; breakage somewhat marred by cross fracture, but pieces of good, uniform size; good, heavy coke.

Pennsylvania No. 21.—Test 187: Light-gray and silvery color; metallic ring; cell structure small, not dense; breakage good; long, large pieces; good, heavy coke.

Test 189: Light-gray and silvery color; metallic ring; cell structure small, not dense; breakage somewhat marred by cross fracture, but pieces of good, uniform size; good, heavy coke.

Test 191: Light-gray and silvery color; cell structure small, not dense; metallic ring; breakage good; uniform size; good, heavy coke.

Test 192: Light-gray and silvery color; metallic ring; cell structure a little small, not dense; breakage good; uniform size; good, heavy coke.

Tennessee No. 1.—Test 133: Light-gray and silvery color; metallic ring; cell structure a little large; breakage good; long, large pieces; good, strong, hard, heavy coke; ash a little high.

Test 153: Light-gray and silvery color; metallic ring; cell structure good; breakage somewhat marred by cross fracture, but pieces of good, uniform size; good, heavy coke; ash and sulphur reduced by washing.

Tennessee No. 2.—Test 127: Light-gray and silvery color; metallic ring; cell structure medium; breakage poor; very brittle, long-fingered pieces.

Tennessee No. 3.—Test 128: Light-gray and silvery color; metallic ring; cell structure good; breakage poor; very brittle, long-fingered pieces.

Tennessee No. 4.—Test 125: Light-gray and silvery color; metallic ring; cell structure good; breakage, long-fingered pieces; ½-inch black butt.

Test 129: Light-gray and silvery color; metallic ring; cell structure good; breakage, long-fingered pieces; black butt removed.

Tennessee No. 5.—Test 154: Light-gray and silvery color; metallic ring; cell structure a little large; breakage good; long, large pieces; good, heavy coke; sulphur high.

Tennessee No. 6.—Test 122: Light-gray and silvery color; metallic ring; cell structure good; breakage, good; long, large pieces; good, strong, heavy coke; ash high; probably reduced by washing.

Tennessee No. 7 B.—Test 121: Poor coke; soft, tough, and punky; drawn from oven in large chunks; ash and sulphur high.

Test 123: Light-gray and silvery color; metallic ring; cell structure a little large; breakage good; large, uniform-sized pieces; strong coke, great improvement over raw charge; ash and sulphur reduced by washing, but still high.

Tennessee No. 8 B.—Test 134: Light-gray and silvery color; metallic ring; cell structure large; breakage somewhat cross fractured, but pieces of good, large, uniform size; sulphur high.

Tennessee No. 9.—Test 124: Light-gray color; some silvery deposit of carbon; metallic ring; cell structure large; breaks in irregular pieces of good size.

Tennessee No. 10.—Test 156: Poor coke; drawn from oven in large, irregular lumps; very tough and dense; with black butt and high ash.

Tennessee No. 11.—Test 160: Poor coke; breakage, large pieces of irregular size; cell structure small; dense and punky; small amount not coked well at bottom; ash high.

Utah No. 1.—Test 130: Dull-gray color; practically no cell structure; soft, dense coke; very fine-fingered pieces, very brittle, and easily broken into small pieces.

Test 141: With R. I. No. 1. No coke produced; all volatile expelled and charge burned entirely to bottom.

Test 157: With R. I. No. 1. Very poor, dense coke; half the product did not cement together; the other half very finely fingered coke, very brittle and easily broken, similar to coke from Utah No. 1.

Virginia No. 1.—Test 64: Light-gray color; metallic ring; cell structure good; breakage somewhat marred by cross fracture, but pieces of good size; hard, heavy coke, not dense.

Test 65: Light-gray and silvery color; much deposited carbon; cell structure medium; metallic ring; breakage somewhat marred by cross fracture, but pieces of good size; good, hard coke.

Test 67: Light-gray and silvery color; much deposited carbon; cell structure medium; metallic ring; breakage somewhat marred by cross fracture, probably due to uncrushed slate; lower yield of coke and higher amount of breeze probably due to fact that coal was not crushed.

Test 68: Light-gray and silvery color; much deposited carbon; metallic ring; cell structure medium; breakage somewhat marred by cross fracture, but pieces of good size; large amount of breeze and lowered percentage yield probably due to fact that coal was not crushed.

Test 77: Light-gray and silvery color; much deposited carbon; metallic ring; cell structure good; breakage somewhat marred by cross fracture, but pieces of good size; good, hard, heavy coke; increased yield of coke and decreased amount of breeze probably due to fine grinding.

Virginia No. 2.—Test 63: Light-gray and silvery color; metallic ring; cell structure a little large; breakage somewhat marred by cross fracture, but pieces of good size.

Test 69: Light-gray and silvery color; much deposited carbon; metallic ring; cell structure medium; breakage good; long, large pieces, somewhat brittle; decreased yield of coke and increased amount of breeze probably due to fact that coal was not crushed.

Test 70: Light-gray and silvery color; metallic ring; cell structure large; breakage somewhat marred by cross fracture, but pieces of good size; good, hard coke, somewhat brittle; washing does not seem to benefit it materially.

Virginia No. 3.—Test 61: Light-gray and silvery color; much deposited carbon; metallic ring; cell structure small; breakage good; long, large, heavy pieces; very heavy coke.

Test 88: Light-gray and silvery color; much deposited carbon; metallic ring; cell structure small; breakage good; good, heavy coke; decreased yield of coke and increased amount of breeze probably due to fact that coal was not crushed.

Virginia No. 4.—Test 62: Light-gray and silvery color; much deposited carbon; metallic ring; cell structure good; breakage, long, thin pieces; light weight; fingered coke.

Virginia No. 6.—Test 181: Light-gray color; cell structure small; dense coke; breakage very irregular; pieces of various sizes; high sulphur.

Test 184: Light-gray and silvery color; much deposited carbon; cell structure small, but not dense; breakage, irregular pieces of various sizes; washing reduces ash and sulphur and improves quality of coke.

Washington No. 2.—Test 135: Light-gray color; some deposit of carbon; fair ring; cell structure small; breakage, long-fingered pieces, very brittle; dense coke; high ash.

West Virginia No. 4 B.—Test 40: Light-gray and silvery color; much deposited carbon; metallic ring; cell structure good.

Test 44: Light-gray and silvery color; much deposited carbon; metallic ring; cell structure good; breakage good; long, large pieces.

Test 46: Light-gray and silvery color; metallic ring; cell structure good; breakage somewhat marred by cross fracture, but pieces of good size; high yield of coke and decreased amount of breeze probably due to fine grinding.

West Virginia No. 13.—Test 21: Light-gray and silvery color; metallic ring; cell structure good; breakage good; long, large, heavy pieces. ·

West Virginia Nos. 13 and 14.—Test 23: Light-gray and silvery color; metallic ring; cell structure medium; breakage good; long, large, heavy pieces.

West Virginia No. 14.—Test 20: Light-gray and silvery color; metallic ring; cell structure a little large; breakage good; long, large, heavy pieces; good, hard coke.

West Virginia No. 15.—Test 36: Light-gray and silvery color; much deposited carbon; metallic ring; cell structure good; breakage good; long, large, pieces; good, heavy coke; high sulphur.

Test 43: Light-gray and silvery color; metallic ring; cell structure a little small; breakage good; long, large pieces; hard, heavy coke; sulphur high.

West Virginia No. 16 A.—Test 73: Light-gray and silvery color; much deposited carbon; metallic ring; cell structure a little large; breakage somewhat marred by cross fracture, but pieces of good size; good, hard, heavy coke.

West Virginia No. 16 B.—Test 45: Cell structure good; breakage good; long, large pieces; good, heavy coke.

Test 48: Light-gray and silvery color; much deposited carbon; metallic ring; cell structure a little large; breakage somewhat marred by cross fracture, but pieces of good size; somewhat brittle; washing does not appear to improve coke, on the contrary the coke from the raw charge is decidedly better.

Test 49: Light-gray and silvery color; metallic ring; cell structure large; breakage somewhat marred by cross fracture; coke brittle; washing does not appear to improve physical properties of coke; sulphur and ash somewhat lowered.

West Virginia No. 17.—Test 60: Light-gray and silvery color; metallic ring; cell structure good; breakage good.

West Virginia No. 18.—Test 74: Light-gray and silvery color; much deposited carbon; metallic ring; cell structure good; breakage somewhat crosswise; coke brittle; good, hard, heavy coke.

Test 78: Light-gray and silvery color; metallic ring; cell structure good; breakage somewhat crosswise; coke brittle; good, hard, heavy coke; no appreciable difference in yield between the crushed and uncrushed charges.

West Virginia No. 19.—Test 79: Dull-gray color; some silver; cell structure small, rather dense; breakage good; poor, light-weight coke.

Test 83: Dull-gray color; some silver; cell structure small; breakage good. This oven was burned with a smaller draft and coke was much heavier and better than that from test 79.

West Virginia No. 20.—Test 80: Light-gray and silvery color; much deposited carbon; metallic ring; cell structure small, not dense; breakage good; long, large, pieces, somewhat brittle; good, hard, heavy coke.

Test 84: Light-gray and silvery color; metallic ring; cell structure good; breakage marred by cross fracture, but pieces of good size; washing does not materially benefit; on the contrary, the coke is not as good as that from raw coal.

Test 87: Light-gray and silvery color; much deposited carbon; metallic ring; cell structure a little small; breakage very irregular, but pieces of good size; decreased percentage of coke and increased percentage of breeze probably due to fact that coke was not crushed.

Test 92: Light-gray and silvery color; metallic ring; cell structure small, not dense; breakage somewhat marred by cross fracture, but pieces of good size; good, heavy coke.

West Virginia No. 21.—Test 82: Light-gray and silvery color; much deposited carbon; metallic ring; cell structure good; breakage good; long, large pieces; good, hard, heavy coke.

Test 91: Light-gray and silvery color; much deposited carbon; metallic ring; cell structure large; breakage very irregular and brittle; washing does not materially improve coke, on the contrary the coke from raw charge is decidedly better.

West Virginia No. 25.—Test 175: Light-gray color, some silvery deposit of carbon; cell structure a little small; breakage, long, large, fingered pieces, very brittle.

Wyoming No. 3.—Test 52: Charge burned to ash down about 8 inches.

Wyoming No. 5.—Test 132: No coke produced.

CONCLUSIONS.

It is unfortunate that the necessary routine work in order to cover so many coals permitted so few tests on each, and that the supply of coal in many cases permitted only one test to be made on that particular coal. The data here presented show the results obtained under the best conditions possible to one not conversant with the burning of these coals, based on observations made from time to time as coking proceeded. These facts should be distinctly borne in mind when analyzing the results here presented. It is hoped that in future work it may be possible to vary conditions, make changes as they suggest themselves, and compare results on many different tests of the same coal and thus draw conclusions of a more definite nature. It is to be regretted that no comparisons can be made between beehive and by-product coke, but the nature of the work here recorded and the facilities provided confined operations to ovens of the beehive pattern exclusively.

No data are given in the detailed statement for compressive strength or height of furnace burden supported, as the results obtained show conclusively the worthlessness of these determinations. This conclusion was reached after careful attempts to obtain results on 1-inch cubes. Four cubes were selected from each coke made, care being taken to obtain pieces with no fracture and representing as nearly as possible the average of the coke. The cubes were cut by means of an emery wheel and guide, and although by no means perfect they were as nearly so as possible and always the two sides used in the machine were parallel. The machine used for breaking was a Tinius Olsen patent machine of 10,000 pounds capacity and gave direct readings of the ultimate strength.

Only a few of these results, taken at random, are given, and these only to show their great variation and the worthlessness of this method of drawing conclusions. Illinois No. 16, test 10, 910 pounds, 1,330 pounds, 2,190 pounds, and 2,270 pounds; Indiana No. 4, test 6, 640 pounds, 790 pounds, 1,060 pounds, and 1,245 pounds; Kentucky No. 1, test 76, 880 pounds, 1,065 pounds, 1,920 pounds, and 2,570 pounds; Ohio No. 9, test 94, 535 pounds, 890 pounds, 1,170 pounds, and 1,600 pounds; Virginia No. 1, test 68, 740 pounds, 1,120 pounds, 1,280 pounds, and 2,060 pounds; West Virginia No. 16, test 49, 520 pounds, 1,500 pounds, 1,780 pounds, and 2,100 pounds.

The difficulty of obtaining a cube, or any number of cubes, to represent anything more than the piece of coke from which it is taken is so apparent that results pretending to show compressive strength of any

amount of coke are worse than useless—in fact, misleading. Even
if coke is selected the whole height of the charge and tests are made
on cubes in number representing the number of inches the results
still show only the strength of the one piece of coke from some particular part of the oven and it is practically impossible to procure
even approximately similar results from other pieces taken from
different places. The condition of burning, the quenching either
inside or out, and any number of factors which it is not possible to
know, much less control, make different portions of the same oven
vary greatly.

A simple calculation will show that coke with a compressive
strength of 48 pounds will support the burden of any modern furnace;
consequently this test gives no data of practical value. Moreover,
there are so many other factors, such as action of heat and gases, attrition of coke against coke, against other ingredients of charge, and
against the side walls, etc., that any calculation to show the burden-
bearing capacity of the coke, even if it were possible to select cubes
representing the whole charge, would be inaccurate if based simply
on a compression test.

An endeavor was made to compare the different cokes by approximating the amount of breakage under conditions of present-day
handling, showing the percentage of coke over 2-inch size that may be
expected to reach the top of the charge in the blast furnace. Fifty
pounds of each coke were selected, as nearly as possible representing
the average size of the coke after handling at the ovens. This coke
was dropped a distance of 6 feet onto a rigid (1-inch) iron plate. All
pieces over 2 inches in size were weighed and again dropped, the
operation being repeated three times. The results of these drop tests
are shown in the detailed statement.

The yield of coke appears to be increased and the amount of breeze
reduced by preliminary crushing. Whether there is a limit to the
degree of fineness, or whether a point may be reached beyond which
finer crushing gives no appreciable improvement or has opposite
effects, can not be determined from the present results; but the data
available indicate that it would be economical to crush all coal before
charging into the ovens, even though a coke of good quality may be
obtained without this preliminary treatment. Fine crushing also
appears to increase the strength of the coke and make the fracture
less irregular, by the greater uniformity and distribution of the ash,
but the weight per cubic foot is reduced. The strength of the coke
is probably influenced by the amount, composition, and distribution
of the ash, but the results so far obtained show no definite relations
between these factors or their relative importance.

The matter of investigating the action of CO_2 on red-hot coke as
determining its value for furnace work was thoroughly considered.

The conclusion was reached that it was of no practical importance, as there are so many other factors in the blast furnace. In view of the fact that the gases in the furnace are mixtures of CO_2, CO, H, O, N, water vapor, and probably others, it appears that action of CO_2 is of little value unless the action of these other gases, either independently or in connection with CO_2, is known. An investigation of the action of CO_2 on red-hot coke, as a means of making comparison of hardness, is being made and gives evidence of yielding some positive results, but work along this line has not progressed far enough to draw any definite conclusions.

The loss of sulphur from coal to coke by volatilization varies with the different coals, depending on several factors, among which, in the order of their importance, are the condition in which sulphur exists in the coal, the heat of the oven, the rapidity of coking, and watering. The sulphur loss ranged from 20.79 per cent on Arkansas No. 1 (test 95) to 63.07 per cent on Illinois No. 29 (test 170), the average for all tests being 43.27 per cent.

CUPOLA TESTS OF COKE.

By RICHARD MOLDENKE.

EQUIPMENT.

Owing to the removal of one of the cupolas which served for the foundry tests during the Louisiana Purchase Exposition all the tests made since then have been conducted in the 36-inch foundry cupola loaned by the Whiting Foundry Equipment Company, of Chicago. The remaining apparatus was rearranged and the 36-inch shell of the cupola was relined to 26 inches internal diameter. There were four horizontal tuyeres measuring 4 by 6 inches on the outside and 3 by 13 inches on the inside of the cupola which were situated 11 inches above the sand bottom. The total tuyere area was 96 square inches, giving a ratio of 1 to 5.96 with the cupola area. A No. 6 Sturtevant fan run at 2,514 revolutions per minute furnished the blast, which was kept at about 7 ounces.

By proper training, the crew was able to run off two heats a day without interruption. The melted iron was poured into molds for sash weights, thus reducing to a minimum the amount of scrap made.

PERSONNEL.

The cupola tests were conducted by W. G. Ireland, under the direction of A. W. Belden, coke expert of the Geological Survey, and by the advice of Richard Moldenke, foundry expert in charge of the cupola tests of the fuel-testing plant.

METHOD OF TESTING.

The method of testing has been fully described in the report of the fuel-testing plant for 1904.[a] Toward the end of the tests it was sometimes necessary to vary the proportion of scrap to pig iron according to the supply, but the total amounts were kept correct as planned for the general series of tests.

After completing the tests on the available cokes in the regular way, so that the results might be comparable with the previous work

---

[a] Prof. Paper U. S. Geol. Survey No. 48, part 3, 1906, pp. 1367-1370.

of the division, a series of further tests was made on some of these cokes. In these tests the coke bed was not kept at a constant height above the tuyeres, but the carbon content was calculated from the analysis of the particular coke and an amount taken to make up 175 pounds of carbon regardless of the height above the tuyeres. The results show interesting features. Some cokes gave melting ratios and melting rates per hour which were better than with the ordinary test methods and others gave inferior results. The tests were made to show the advisability on the part of the manufacturer as well as of the foundryman of studying the conditions of cupola practice in order to determine those which give the best results.

DETAILED RESULTS.

The detailed results of the regular tests as well as of the special 175-pound carbon bed tests will be found in the following tables. Results of a typical test of Connellsville 72-hour coke are given at the head of the first table as a standard for comparison. All the tests here reported were made within the calendar year 1906 except test 190, on coke from Pennsylvania No. 21 coal, the date of which was February 13, 1907. Many of the coals tested, however, were received during 1905.

23975—Bull. 336—08——4

Cupola tests of coke from coals received in 1905.

1	2	3	4	5	6	7	8	9	10	11	12	13	15	16	17	18	19
Cupola test No.	Field No. of coal.	Coke test No.	Date.	Coke bed.	Pig Iron.	Scrap.	Coke.	Pig Iron.	Scrap.	Coke.				Scrap.	Coke.	Pig Iron.	Scrap.
19	Connellsville coke		1904 Nov. 30	220	660	220	53	398	133	53	398		397	132	52	397	132
	Illinois:		1906.														
54	11 D (w.)	5	Sept. 17	190	570	190	60	420	140	60	420		420	140	60	420	140
21	13 (w.)	2	July 10	200	600	200	58	413	137	58	413		413	138	57	412	138
93	13	3	Oct. 6	220	660	220	53	398	133	53	398		397	132	52	397	132
94	16	10	Nov. 8	220	660	220	53	398	133	53	398		397	132	52	397	132
	Indiana:																
82	4 (w.)	9	Sept. 27	180	540	180	63	428	143	63	428		427	142	62	427	142
83	5	8	Sept. 28	170	510	170	65	435	145	65	435		435	145	65	435	145
187	5	8	Dec. 7	180	720		80	570		80	570		570		80	570	
48	7 A (w.)	13	Sept. 12	170	510	170	65	402	146	65	435		411	146	65	435	146
175	7 A (w.)	13	Nov. 30	215	645	215	54	402	134	54	401		411	134	53	411	133
173	9	17	Nov. 28	200	800		44	550		44	550		550		43	550	
174	9	17	...do	210	840		73	540		73	540		540		72	540	
81	9 B (w.)	18	Sept. 26	180	540	190	63	428	143	63	428		427	142	62	427	142
185	9 B (w.)	18	Dec. 6	205	615	205	57	408	137	57	408		408	137	56	408	136
22	11	51	July 11	190	570	190	60	408	140	60	408		408	140	60	408	140
184	11	51	Dec. 5	205	615	205	57	408	137	57	408		408	136	56	408	136
	Kentucky:																
29	1 C	71	July 17	190	570	190	60	420	140	60	420	56	420	140	60	420	140
26	1 B	76	July 14	190	570	190	60	420	140	60	420	55	420	140	60	420	140
27	5	75	July 16	200	600	200	58	413	138	58	413	62	412	137	57	412	137
28	5	90	...do	205	615	205	58	405	136	58	405	60	405	136	56	405	136
89	6	90	Aug. 30	205	615	205	57	405	135	57	405		405	135	55	405	135
67	6	86	Aug. 16	180	630	210	55	405	135	55	405		405	135	55	405	135
47	6	85	Sept. 11	190	570		68	570		68	570		570		62	570	
55	7	85	Aug. 7	190	570	190	57	408	137	57	408		408	135	60	408	135
	Maryland:																
75	1 (w.) c	58	Sept. 22	200	600	200	58	413	138	58	413	57	412	137	57	412	137
	Ohio:																
23	6 A, 6 B (w.)	22	July 12	200	600	200	58	413	138	58	413	57	412	137	57	412	137
80	6 A, 6 B (w.)	66	Sept. 26	170	510	170	60	435	145	60	435	65	435	145	65	435	145
57	7 (w.)	94	Aug. 8	190	570	190	60	420	140	60	420	60	420	140	60	420	140
70	7	89	Aug. 18	190	570	190	58	420	140	58	420	60	420	140	60	420	140
56	8 A (w.)	93	Aug. 8	200	600	200	58	413	138	58	413	57	412	137	57	412	137
165	9	72	Nov. 22	190	760		47	560		46	560	46	560		46	560	

Pennsylvania:
5 B (w.)
5 B (w.)
6.
6.
6.
6 A, 6 B (w.)
6 A, 6 B (w.)
6 A, 6 B (w.)
6 A, 7 B
7 A, 7 B
8.
9 (w.)
9 (w.)
9 (w.)
10.

Virginia:
1 A
1 A
1 B
1 A
1 A
1 A
2 (w.)
2 (w.)
2 B
2 B
2 B (w.)
3.
3.
3.

West Virginia:
4 (w.)
4.
4 D
4 B (w.)
4 B (w.)

a Details of origin of coal samples can be found in Bull. U. S. Geol. Survey No. 290, 1906.
b Pig iron used from car 131943.
c Plus 10 per cent pitch.
d Pig iron used from car 27633.
e Plus 5 per cent pitch.

Cupola tests of coke from coals received in 1905—Continued.

Cupola test No.	Designation of coke.			Coke bed.	Charges (pounds).														Total.			Ratio iron to coke.
	Field No. of coal.	Coke test No.	Date.		Pig iron.	Scrap.	Coke.	Pig iron.	Scrap.	Coke.	Pig iron.	Scrap.	Coke.	Pig iron.	Scrap.	Coke.	Pig iron.	Scrap.	Coke.	Pig iron.	Scrap.	
1	2	3	4	5	6	7	8	9	10	11	12	13	14	15	16	17	18	19	20	21	22	23
	West Virginia—Con.																					
86	4 B (w.)	46	Oct. 1	190	570	190	60	420	140	60	420	140	60	420	140	60	420	140	430	2,250	750	7
20	13 and 14	23	July 7	190	570	190	60	420	140	60	420	140	60	420	140	60	420	140	430	2,250	750	7
51	15	36	Sept. 14	180	880		53	530		53	530		52	530		52	530		430	3,000		7
49	15	43	Sept. 12	220	920		53	530		55	530		52	530		52	530		375	3,000		8
67	15	36	Nov. 23	220	920		37	520		68	520		67	520		66	520		500	3,000		6
168	15	36	Nov. 24	200	900		58	413		49	413		57	412		57	412		430	2,250		7
82	16	45	Dec. 4	200	730		78	570		78	570		77	570		77	570		375	3,000		8
69	16	45	Nov. 24	180	760		60	560		60	560		60	560		60	560		430	3,000		(a)
70	16 B	45	Nov. 26	190	760		60	420		60	420		60	420		60	420		430	2,250		7
50	16 A	50	Sept. 14	190	570	190	60	435	140	60	435	140	65	435	140	65	435	140	430	2,250	750	7
36	16 B (w.)	36	July 24	190	510	190	65	416	145	65	416	145	65	416	145	65	416	145	430	2,250	750	7
37	16 B (w.)	49	do	195	385	170	59	405	139	59	405	139	55	398	135	58	398	138	430	2,250	750	7
181	16 B (w.)	49	Sept. 24	210	690	195	62	397	132	62	397	132	57	398	133	56	398	133	430	2,250	750	7
38	17 (w.)	60	Dec. 4	200	690	220	60	413	140	60	413	140	57	412	137	57	412	137	430	2,250	750	7
39	18	74	July 25	190	690	200	60	420	140	60	420	140	60	420	140	60	420	140	430	2,250	750	7
46	18	78	do	190	670	190	60	420	140	60	420	140	60	420	140	60	420	140	430	2,250	750	7
62	19	79	Sept. 11	190	570	190	63	386	133	63	386	133	52	397	132	53	397	132	430	2,250	750	7
43	19	83	Sept. 10	190	660	220	48	383	128	48	383	128	47	382	127	47	382	127	430	2,250	750	7
63	19	84	Aug. 13	240	720	240	50	390	130	50	390	130	57	413	130	57	413	130	430	2,250	750	7
90	20 A (w.)	84	Sept. 5	200	690	200	63	390	130	63	390	130	55	390	135	55	390	135	430	2,250	750	7
177	20 A	80	Aug. 6	210	640	210	55	405	135	55	405	135	62	405	135	62	405	135	430	2,250	750	7
71	20 A	87	Aug. 13	180	640	180	63	413	143	63	413	143	67	407	143	67	407	143	430	2,250	750	7
42	20 A (w.)	82	Oct. 1	190	600	190	60	420	138	60	420	138	60	412	137	60	412	137	430	2,250	750	7
44	20 A (w.)	91	Dec. 1																			
72	21 (w.)	91	Aug. 20		570			420			420			420			420		430	2,250	750	7
88	21 (w.)		Aug. 29																			

a Pig iron used from car 131943.

Cupola tests of coke from coals received in 1905—Continued.

Cupola test No.	Field No. of coal	Coke test No.	Moisture	Volatile matter	Fixed carbon	Ash	Sulphur In coke	Sulphur In ash	Phosphorus	Specific gravity	Fluidity strip (per cent), full	Maximum blast pressure (ounces)	Pounds of iron Poured	Pounds of iron Additional melted	Pounds of iron Total	Melting rate Per hour (pounds)	Melting rate Increase or decrease	Recovered Iron	Recovered Coke	Melting loss (per cent)	Iron to coke	Melting ratio Increase or decrease	Coke bed Increase(+) or decrease(−) (pounds)	Height above top of tuyeres (inches)
1	2	8	24	25	26	27	28	29	30	31	32	33	34	35	36	37	38	39	40	41	42	43	44	45
19	Connellsville coke		0.18	0.32	88.75	10.75	0.87	0.033	0.018	1.92	(b)	7½	2,470		2,470	5,489		283	20	8.2	6.02			45
	Illinois:																							
54	11 D (w.)	5	1.19	.93	85.97	11.91	1.44	.135	.0065	1.87	99.9	7	1,874	490	2,364	4,728		418	71	7.23	6.58			
21	13 (w.)	3	4.60	.68	82.08	11.64	1.27	.15	.007	1.85	99.9	7	822	860	1,682	2,925		1,064	110	8.46	6.26			
93	16.	2	2.73	1.36	79.30	16.61		.10	.0162	1.90	97.12	7	1,770	462	2,232	5,356		606	71	8.4+	6.22			
94	16.	10	2.14	1.46	83.96	12.44	1.02	.05	.0076	1.85	81.94	7	1,220	578	1,808	3,874		988	59	7.80	4.87			
	Indiana:																							
82	4 (w.)	9	.55	.38	85.91	13.16	2.06	.16	.044	1.86	99.9	7	1,568	347	1,915	3,990		921	57	5.47	5.13			
83	5.	8	.76	.54	81.29	17.41	4.21	.20	.03	1.91	99.9	7	1,562	605	2,167	4,642		611	62	7.4	5.89			
187		13	.76	.54	81.29	17.41	4.21	.20	.012	1.91	93.05	7	1,622	145	1,767	4,022	inc.	920	67	10.43	6.83	Inc.		
48	7 A (w.)	12	1.13	2.35	83.23	13.29	2.69	.07	.0033	1.85	90.28	7	1,761	719	2,480			288	65	6.55	3.90			
175	7 A (w.)	17	1.13	2.35	83.23	13.29	3.50	.07	.0033	1.84	94.44	7	2,247	403	2,650	5,483		22	59	10.93	7.26			
173	9.	17	.68	2.85	78.44	18.03	3.50	.210	.0033	1.84	94.44	7	782	235	1,017	2,653		1,636	44	4.90	3.22			
174	9 B (w.)	18	.91	.97	84.60	13.52	2.24	.210	.042	1.84	88.88	7	1,854	276	2,130	3,651		650	61	7.33	4.67			
81	9 B (w.)	13	.91	.97	84.00	13.52	2.24	.04	.042	1.84	99.9	7	1,834	279	2,113	4,372		717	51	5.67	5.80			
85	11.	15	1.55	1.66	84.00	12.70	1.34	.04	.042	1.83	94.44	7	1,827	363	2,190	4,106	Dec.	688	56	6.73	5.73			
22			1.55	1.66	84.09	12.70	1.34			1.83	99.9	7	1,364	357	1,721	1,912	(d) inc.	688	41	12.66	4.54			
84													2,163	253	2,416	4,832		233	51	11.7	6.34			
	Kentucky:																							
29	1 C.	71	1.32	1.53	84.08	12.47	.98	.07	.044	1.88	97.2	7	1,639	450	2,089	2,507		775	36	4.60	5.29	Inc.	e.+45	21.4
26	1 B.	76	.93	1.25	91.40	6.42	1.05	.04	.03	1.94	95.83	7	1,970	284	2,254	5,009		476	30	9.03	5.04			
75	5.	78	.30	1.21	93.26	5.23	.41	.025	.012	1.84	94.44	7	1,885	482	2,367	3,737		645	64	2.70	6.47			
28	6.	80	1.64	1.09	92.28	4.41	.37	.05	.0033	1.78	99.10	7	1,624	290	1,914	3,589		709	77	8.03	4.96			
89	6.	67	.64	1.27	92.28	4.41	.37	.05	.0033	1.79	98.61	7	1,872	247	2,119	4,384		208	88	5.73	6.00			
80	7.	47	2.81	1.93	91.93	3.99	.40	.02	.0033	1.83	99.30	7	1,885	314	2,199	4,123		425	68	6.00	5.82	Dec.	e.+25	17.9
67	7.	55	.23	2.16	82.97	14.64	3.16	.01	.0021	1.83	95.83	7	2,120	529	2,649	4,907			88	10.36	7.14			
47			.23		82.97	14.64	3.16	.01		1.83	97.91		1,849	377	2,225	5,136			58	4.73	6.05	Inc.	e.+15	16.2
55																				11.3	6.56			
75	Maryland: 1 (w.)[c]	58	.27	.88	87.47	11.38	.95	.09	.0278	1.91	99.9	7	1,748	326	2,074	3,969		782	114	6.47	6.56			

a For chemical analyses of coals from which these cokes were made, see pp. 27-35.

b Ran up well.

c Bed rearranged.

d Plus 10 per cent pitch.

d Trouble with iron notch accounts for long time between ladles, consequently melting rate is low.

Cupola tests of coke from coals received in 1905—Continued.

Cupola test No.	Field No. of coal.	Coke test No.	Moisture.	Volatile matter.	Fixed carbon.	Ash.	Sulphur In coke.	Sulphur In ash.	Phosphorus.	Specific gravity.	Fluidity str'gh (per cent), full.	Maximum blast pressure (ounces).	Poured.	Additional melted.	Total.	Per hour.	Increase or decrease.	Iron.	Coke.	Melting loss (per cent).	Iron to coke.	Increase or decrease.	Increase (+) or decrease (—) (pounds).	Height above top of tuyeres (inches).
Ohio:																								
23	5.	29	0.87	0.98	86.15	12.00	1.64	0.02	.0067	1.84	99.9	7	2,093	179	2,272	3,495		400	73	10.93	6.36			
90	6 A, 6 B (w).	66	.42	1.29	88.35	10.03	2.51	.08	.0045	1.82	98.61	7	1,888	431	2,319	5,797		385	74	9.86	6.51			
57	7 (w).	94	.57	1.88	88.08	9.77	1.90	.03	.0058	1.86	97.91	7	2,365	152	2,517	4,196		249	44	7.80	6.52			
70	8 A (w).	92	.75	1.76	85.33	9.73	2.32	.03		1.88	98.61	7	1,285	1,067	1,997	3,528		819	42	6.13	6.05			
165	9.	73	2.46	1.08	86.05	9.49	1.88	.05		1.81	92.36	7	1,666	301	2,352	3,725		346	19	10.00	6.58			
166	9.	72	1.80	1.08	84.43	12.69	3.06	.06		1.86	93.05	7	2,065	303	1,967	4,646		841	36	5.53	6.34			
Pennsylvania:																								
68	5 B (w).	26	.22	1.19	91.63	6.95	.81	.015	.0063	1.84	99.9	7	2,247	362	2,600	5,218	Dec.	347	89	7.87	7.65	Dec. +10		15.5
180	5 B (w).	95	.34	1.71	91.63	6.95	.81	.015	.0063	1.96	92.36	7	1,067	326	1,747	3,176		135	89	9.80	5.12			
60		32	.41	1.71	82.66	16.64	1.69	.11	.0241	1.91	97.22	7	890	493	1,392	2,784	Dec.	454	154	5.13	5.04			11.4
85		41	.54	1.26	82.61	15.81	1.52	.11	.0187	1.92	99.9	7	1,096	329	2,040	4,946		1,028	114	4.97	4.51			
87	6 A, 6 B (w).	34	.24	.66	88.61	10.49	1.21	.035	.0162	1.84	94.44	7	1,710	139	2,362	5,311	Inc.	1,028	96	4.13	5.28	Inc. 6.—20		
84	6 A, 6 B (w).	36	.39	2.26	86.77	10.08	1.24	.045	.015	1.84	99.9	7	639	163	2,470	4,903		218	100	8.70	3.24			
176	6 A, 6 B (w).	35	.46	1.84	86.64	10.08	1.24	.045	.0160	1.83	99.22	7	2,283	257	2,464	5,244	Inc.	985	82	10.43	7.47			
25	6 A, 6 B (w).	38	.16	1.16	83.46	14.33	1.42	.04	.0366	1.83	97.22	7	1,913	410	2,353	4,671		71	100	10.43	7.88			
61	7 A, 7 B (w).	33	.91	2.16	83.83	15.99	1.87	.03	.0366	1.76	93.05	7	1,941	552	2,404	5,183	Inc.	223	102	9.07	7.20	Dec. a+15		16.33
66	7 A, 7 B.	30	.91	1.92	82.39	7.94	.91	.10	.0049	1.76	94.61	7	1,989	317	2,258	4,471		480	81	9.60	6.34			
79		29	.72	.61	89.99	7.94	1.53	.015	.007	1.94	93.05	7	2,256	565	2,313	4,166	Inc.	228	49	5.83	6.19	Inc. a+10		15.5
188		96	.54	1.21	86.02	12.66	1.06	.04	.0101	1.99	97.22	7	1,886	257	2,313	4,160		199	132	6.37	6.54			
64	9 (w). b	55	.34	1.21	86.84	11.41	1.06	.04	.0101	1.99	98.61	7	1,903	272	2,343	4,016		934	94	10.30	6.73			
76	9 (w).	42	.39	1.14	86.84	11.41	1.05	.03	.008	1.85	97.91	7	1,578	750	2,358	4,016		729	82	9.06	6.92			
189	9 (w).	47	.39	.99	89.93	8.57	1.01	.08	.0081	1.92	98.61	7	1,950	808	2,300	5,013		347	74	7.37				
30	10.	53																400	104				
73																		521	59				
Virginia:																								
24	1 A	64	1.52	.99	89.20	8.29	.88	.07	.0036	1.80	95.83	7	1,418	402	2,220	3,096	Inc.	485	46	9.83	5.78	Inc. a—10		12.4
78	1 A	77	1.23	1.67	89.24	7.86	.94	.025	.0041	1.79	97.22	7	1,783	400	2,183	4,366		625	87	6.40	6.36			
31	1 B	75	.20	.80	91.52	7.48	1.02	.04	.005	1.87	98.60	7	2,253	200	2,453	4,460		322	58	7.50	6.57			
69	1 A	67	.21	.89	90.99	7.91	1.01	.08	.004	1.83	97.22	7	1,373	454	1,827	3,780		965	90	6.93	4.94			
186	1 A	67	.21	.89	90.99	7.91	1.05	.08	.004	1.83	98.9	7	2,007	293	2,300	4,182		261	59	11.30	6.20			

59	1 A	68	.30	1.16	90.85	7.69	.97	.04	.004	1.81	90.9	7	1,816	706	2,522	4,881		261	55	7.23	6.72			
183	1 A	68	.30	1.16	90.85	7.69	.97	.04	.004	1.81	94.44	7	2,019	295	2,314	4,207	Dec.	391	38	9.83	5.90	Dec.	a—10	12.4
158	2	69	.69	.93	90.33	8.05	.65	.07	.0026	1.85	97.22	7	2,153	337	2,490	4,819		318	48	6.40	7.50			
171	2 (w.)	70	.45	1.23	92.25	6.07	.69	.04	.0026	1.87	91.67	7	2,256	286	2,542	5,101		255	58	6.76	8.02			
172	2 (w.)	70	.45	1.23	92.25	6.07	.69	.04	.0026	1.87	94.44	7	2,475	205	2,680	5,956		122	107	6.60	6.82			
92	2 B	63	.25	1.08	91.25	7.42	.68	.008	.0026	1.87	94.44	7	1,882	130	2,012	3,658		786	30	6.73	5.90			
178	2 B	63	.25	1.08	91.25	7.42	.68	.008	.0026	1.87	97.22	7	1,900	514	2,414	5,003	Inc.	348	84	11.37	6.98	Inc.	a+15	16.2
52	2 B	69	.69	.93	90.33	8.05	.65	.07	.0026	1.85	99.9	7	1,989	485	2,474	4,948		301	52	7.50	6.54			
32	2 B (w.)	70	.45	1.23	92.25	6.07	.69	.04	.0026	1.87	99.3	7	2,104	126	2,230	4,181		578	56	6.40	5.96			
33	3	61	.39	1.21	92.60	5.90	.61	.085	.0011	1.93	95.10	7	1,584	807	2,391	4,947		255	98	11.80	7.09			
91	3	61	.39	1.21	92.60	5.90	.61	.085	.0011	1.93	99.9	7	2,422	176	2,598	5,028		288	70	3.80	7.22			
58	3	66	.16	1.26	91.85	6.73	.55	.07	.006	1.87	94.44	7	1,710	490	2,200	4,551		686	104	5.46	6.75			
40	3	66	.16	1.26	91.85	6.73	.55	.07	.006	1.87	99.9	7	1,761	310	2,071	3,858		663	108	8.20	5.42			
34	4	62	.16	1.14	92.90	5.80	.42	.07	.0046	1.84	99.3	7	2,356	122	2,478	4,130		287	33	7.88	6.24			
	West Virginia:																							
159	4 (w.)	44	.60	1.20	87.04	11.16	.72	.010	.0186	1.96	95.83	7	1,433	282	1,715	3,118		988	74	9.90	5.70			
160	4	44	.60	1.20	87.04	11.16	.72	.010	.0186	1.96	99.09	7	2,188	254	2,442	4,726		316	66	6.07	5.68			
35	4 B	40	.35	.68	87.01	12.96	.82	.010	.014	1.92	99.4	7	2,107	305	2,412	4,256		323	88	8.83	7.05			
53	4 B (w.)	44	.60	1.20	87.04	11.16	.72	.10	.0186	1.96	99.9	7	2,106	332	2,438	5,851		422	72	4.67	6.53			
74	4 B (w.)	46	.27	1.37	86.98	11.38	.78	.010	.0112	1.88	99.9	7	1,954	419	2,373	4,815		360	119	8.90	7.69			
86	4 B (w.)	46	.27	1.37	86.98	11.38	.78	.010	.0112	1.88	99.9	7	1,286	279	1,565	3,854		1,280	147	5.87	5.63			
20	13 and 14	28	.43	1.37	94.20	4.00	.84	.02	.0018	1.85	99.4	7	2,143	200	2,343	3,905		353	56	9.13	6.26			
51	15	36	.45	.78	88.70	10.07	2.09	.07	......	1.90	99.9	7	2,380	180	2,560	5,815		214	88	7.23	7.51			
49	15	43	.26	1.89	87.30	10.55	2.26	.24	.0153	1.93	98.61	7	1,645	335	1,980	2,829		891	106	4.63	6.11			
167	15	36	.45	.78	88.70	10.07	2.09	.07	......	1.90	95.14	7	1,495	183	1,678	3,248		1,252	66	2.33	5.43			
168	15	36	.45	.78	88.70	10.07	2.09	.07	......	1.90	96.52	7	2,196	155	2,351	4,596		474	87	5.83	5.69			
182	15	43	.26	1.89	87.30	10.55	2.26	.24	.0153	1.93	97.22	7	1,825	317	2,142	5,588	Inc.	612	127	8.20	7.07	Inc.	a—20	11.4
169	16	45	.40	1.55	85.61	12.44	1.23	.51	.0218	1.95	90.97	7	1,206	397	1,603	2,539		1,174	42	4.43	5.08			
170	16	45	.40	1.55	85.61	12.44	1.23	.55	.0218	1.95	......	7	2,379	179	2,558	5,292		270	55	5.73	5.75			
50	16 B	45	.40	1.55	85.61	12.44	1.23	.51	.0218	1.95	95.83	7	1,579	512	2,091	5,018		590	103	10.63	6.39			
36	16 A	73	.54	1.31	89.07	9.08	.81	.06	.0226	1.92	95.83	7	1,937	481	2,418	4,396		256	84	10.86	6.99			
37	16 B (w.)	48	.27	1.18	89.87	8.58	.91	.04	.0166	1.87	95.1	8	1,591	554	2,145	2,798		539	71	10.53	5.97			
77	16 B (w.)	49	.38	2.35	88.91	8.36	.89	.010	.0133	1.84	97.22	7	2,081	371	2,452	5,073		344	83	6.80	7.06			
181	16 B (w.)	49	.38	2.35	88.91	8.36	.89	.010	.0133	1.84	95.14	7	2,089	239	2,328	4,656	Dec.	485	49	6.23	6.11	Dec.	a+25	18.1
38	17 (w.)	60	.07	.83	91.36	7.94	1.04	.015	.0047	1.91	93.8	7½	1,706	404	2,110	3,332		611	106	9.30	6.51			
39	18	74	.63	.97	89.54	8.86	.54	.03	.0058	1.93	99.9	7	2,163	260	2,423	4,276		306	70	9.08	6.73			
46	18	78	.69	1.19	89.47	8.65	.53	.02	.0037	1.87	98.61	7	2,135	310	2,445	3,965		315	103	8.00	7.48			
45	19	79	.38	1.68	89.96	7.98	.77	.02	.0026	1.98	98.61	7	2,187	298	2,485	4,119		202	64	10.40	6.79			
62	19	83	.45	.55	91.57	7.43	.82	.01	.0030	1.94	99.30	7	2,312	132	2,444	6,110		184	86	12.40	7.10			
41	19	83	.45	.55	91.57	7.43	.82	.01	.0030	1.94	91.67	7	1,850	508	2,358	4,287		409	83	7.77	6.79			
43	20 A (w.)	84	.29	.46	91.73	7.52	.90	.04	.0044	1.85	99.9	7	2,211	165	2,376	4,193		453	82	5.70	6.82			
63	20 A (w.)	84	.29	.46	91.73	7.52	.90	.04	.0044	1.85	96.53	7	1,984	238	2,222	4,444		503	106	9.16	6.86			
90	20 A	80	.17	1.08	89.01	9.74	1.10	.05	.004	1.80	99.9	7	2,136	375	2,511	5,022		204	85	9.50	7.28			
177	20 A	80	.17	1.08	89.01	9.74	1.10	.05	.004	1.80	80.55	7	905	352	1,257	2,285	Dec.	1,540	131	6.77	4.20	Dec.	a—40	8.5
71	20 A (w.)	87	.17	.81	92.29	6.73	.85	.03	.0036	1.90	99.9	7	1,912	144	2,056	3,628		727	72	7.23	5.74			
42	20 A (w.)	87	.17	.81	92.29	6.73	.85	.03	.0036	1.90	91.67	7	1,729	345	2,074	4,610		657	113	8.97	6.54			
44	21	82	.24	.81	91.23	7.72	.98	.03	.0027	1.86	99.9	7	1,654	557	2,211	4,574		557	93	7.73	6.56			
72	21 (w.)	91	.55	1.08	93.16	5.26	.76	......	.0027	1.87	95.83	7	1,989	182	2,171	4,491		579	77	8.33	6.15			
88	21 (w.)	91	.55	1.08	93.16	5.26	.76	......	.0027	1.87	98.61	6½	1,881	432	2,313	4,477		503	68	6.13	6.39			

a Bed rearranged.

b Plus 5 per cent pitch.

Cupola tests of coke from coals received in 1905—Continued.

Cupola test No.	Designation of coke.		Record of melt—Continued.		Weight and time of each ladle of melted iron.																
	Field No. of coal.	Coke test No.	Blast on at—	Iron running.	1.		2.		3.		4.		5.		6.		7.		8.		9.
					Lbs.	At—	Lbs.	At—	Lbs.	At—	Lbs.	At—	Lbs.	At—	Lbs.	At—	Lbs.	At—	Lbs.	At—	At—
1	2	3	46	47	48	49	50	51	52	53	54	55	56	57	58	59	60	61	62	63	65
19	Connellsville coke.	......	10.57 a. m.	11.03	175	11.07	115	11.11	185	11.13	150	11.14	220	11.16	150	11.18	225	11.19	155	11.22	11.2
	Illinois:																				
54	11 D (w.)	5	11.15 a. m.	11 26	39	11.29	72	11.31	103	11.37	96	11.37½	72	11.41	92	11.41½	89	11.42	65	11.43	11.43½
21	13 (w.)	2	9.40 a. m.	9.55	79	10.02	29	10.03	70	10.07	57	10.09	75	10.11	63	10.12	61	10.13	53	10.15	10.16
93	13	3	10.25 a. m.	10.35	90	10.40	54	10.40½	111	10.43	91	10.43½	31	10.44	118	10.46	93	10.46½	70	10.47	10.49
94	16	10	10.31 a. m.	10.47	88	10.55	107	10.55½	86	10.57	93	10.58	93	10.59	85	11.00	83	11.01	94	11.01½	11.03
	Indiana:																				
82	4 (w.)	9	2.25 p. m.	2.33	51	2.36	78	2.38	93	2.41	112	2.45	116	2.45½	117	2.46	106	2.50	115	2.50½	2.51
83	5	8	2.34 p. m.	2.45	145	2.51	121	2.56	146	2.56½	100	2.57	91	3.00	102	3.00½	120	3.01	89	3.05	3.05½
187	5	8	10.46 a. m.	10.58	28	11.01	102	11.04	96	11.08	37	11.08½	78	11.12	92	11.12½	62	11.13	82	11.16	11.16½
48	7 A (w.)	13	10.15 a. m.	10.25	101	10.32	28	10.32½	70	10.38	76	10.38½	88	10.39	64	10.41	66	10.41½	82	10.42	10.44
175	7 A (w.)	13	10.47 a. m.	10.52	68	10.56	109	10.59	85	11.01	87	11.01½	43	11.02	85	11.04	95	11.04½	88	11.05	11.07
173	9	17	11.07 a. m.	11.18	97	11.25	96	11.30	96	11.30½	85	11.31	79	11.36	91	11.36½	88	11.37	63	11.40	11.40½
174	9	17	3.45 p. m.	3.56	90	4.05	88	4.05½	91	4.06	94	4.09	110	4.09½	90	4.10	82	4.14	93	4.14½	4.15
81	9 B (w.)	18	2.55 p. m.	3.06	82	3.10	122	3.13	41	3.13½	102	3.16	62	3.16½	145	3.20	111	3.20½	102	3.21	3.24
185	9 B (w.)	18	10.52 a. m.	11.00	24	11.03	87	11.07	99	11.11	75	11.11½	44	11.12	94	11.16	69	11.16½	82	11.17	11.19
22	11	51	2.44 p. m.	2.58	77	3.01	99	3.09	72	3.13	106	3.27	68	3.29	96	3.34	103	3.36	92	3.37	3.38
184	1	51	3.47 p. m.	3.55	91	4.00	59	4.00½	92	4.03	60	4.04	106	4.05	78	4.08	80	4.08½	96	4.09	4.10
	Kentucky:																				
29	1 C	71	9.26 a. m.	9.33	68	9.40	66	9.42	55	9.43	92	9.44	91	9.45	104	9.47	78	9.49	81	9.50	9.52
26	1 B	76	2.18 p. m.	2.31	85	2.35	79	2.37	93	2.38	113	2.39	63	2.40	95	2.41	108	2.41½	108	2.42	2.44
27	5	75	9.09 a. m.	9.19	53	9.24	180	9.26	56	9.28	101	9.29	107	9.30	93	9.31	97	9.33	105	9.34	9.35
28	6	90	3.09 p. m.	3.18	77	3.25	85	3.27	88	3.28	91	3.29	112	3.31	91	3.32	84	3.33	105	3.34	3.35
89	6	90	10.41 a. m.	10.51	101	10.55	100	10.57	98	11.00	89	11.01	74	11.01½	97	11.03	97	11.03½	66	11.04	11.06
67	6	86	9.52 a. m.	9.56	47	10.02	62	10.08	87	10.08½	78	10.09½	93	10.10	97	10.11	61	10.12	100	10.12½	10.13
47	7	85	3.20 p. m.	3.34	56	3.41	82	3.41½	64	3.42	96	3.42½	117	3.43	50	3.46	88	3.46½	80	3.47	3.49
55	7	85	4.24 p. m.	4.39	93	4.47	109	4.47½	103	4.48	64	4.50½	92	4.51	89	4.51½	102	4.52	87	4.54	4.54½
	Maryland:																				
75	1 (w.)[a]	58	10.35 a. m.	10.45	73	10.50	108	10.53	102	10.55	132	10.58	113	10.58½	105	10.59	114	11.01	40	11.01½	11.05
	Ohio:																				
23	5	22	1.51 p. m.	1.57	117	2.04	97	2.05½	108	2.07	83	2.09	89	2.10	101	2.11	92	2.12½	82	2.14	2.15
80	6 A, 6 B (w.)	66	10.8 a. m.	10.20	81	10.24	43	10.24½	132	10.27	50	10.27½	119	10.30	53	10.30½	108	10.31	121	10.33	10.33½
57	7 (w.)	94	3.57 p. m.	4.05	93	4.11	31	4.11½	112	4.14½	89	4.15	18	4.15½	98	4.20	85	4.20½	97	4.21	4.22
70	7	89	8.05 a. m.	8.12	91	8.20	107	8.20½	109	8.23½	90	8.24	94	8.24½	103	8.28	96	8.28½	56	8.29	8.29½
56	8 A (w.)	93	11.03 a. m.	11.16	81	11.35	92	11.35½	98	11.36	27	1.36½	78	11.44	69	11.45	46	11.45½	44	11.52	11.52½
165	9	72	3.31 p. m.	3.40	100	3.44	85	3.48	94	3.48½	46	3.49	73	3.52	101	3.52½	80	3.53	76	3.55	3.55½
166	9	72	11.18 a. m.	11.29	54	11.33	85	11.37	86	11.37½	76	11.38	70	11.42	85	11.42½	77	11.43	63	11.45	11.45½

	Pennsylvania:																						
68	5 B (w.)	26	2.31 p. m.	2.37	82	2.42	78	2.44	93	2.47	103	2.47½	54	2.48	96	2.50	111	2.50½	38	2.51	92	2.52
180	5 B (w.)	26	4.16 p. m.	4.26	21	4.30	69	4.33	59	4.33½	65	4.36	78	4.36½	86	4.37	59	4.41	77	4 4.41½	117	4.42
60	6	32	10.25 a. m.	10.40	46	10.45	85	10.50	94	10.50½	66	10.53	78	18.53½	77	10.56	98	10.56½	50	11.02	74	11.02½
85	6	41	1.10 p. m.																			
87	6	41	2.07 p. m.	2.17	55	2.20	69	2.23	29	2.26	90	2.26½	68	2.27	74	2.33	80	2.33½	74	2.34	66	2.36
84	6 A, 6 B (w.)	34	10.00 a. m.	10.09	91	10.14	115	10.17	106	10.18	98	10.21	130	10.21½	130	10.22	63	10.24	127	10.26	130	10.30
176	6 A, 6 B (w.)	34	3.25 p. m.	3.33	92	3.37	32	3.38	77	3.40	97	3.40½	82	3.42	89	3.42½	54	3.43	83	3.45	102	3.45½
25	6 A, 6 B (w.)	38	7.49 a. m.	8.07	8	8.18	52	8.22½	95	8.29	80	8.31	86	8.33	91	8.35	75	8.37	72	8.39	46	8.41
61	6 A, 6 B (w.)	38	2.54 p. m.	3.04	50	3.07	105	3.10	76	3.10½	83	3.14	122	3.14½	84	3.15	92	3.17	112	3.17½	116	3.18
66	7 A, 7 B (w.)	33	9.50 a. m.	9.59	38	10.03	77	10.05	72	10.05½	85	10.07	85	10.07½	82	10.08	67	10.10	80	10.10½	91	10.11
79	7 A. 7 B	30	3.04 p. m.	3.13	35	3.17	135	3.19	141	3.21	25	3.21½	115	3.24	148	3.24½	116	3.25	104	3.27	133	3.27½
65	8	29	11.35 a. m.	11.42	22	11.47	89	11.50	74	11.50½	79	11.52	96	11.52½	91	11.53	66	11.55	82	11.55½	80	1.56
188	8	29	3.34 p. m.	3.42	88	3.47	94	3.47½	30	3.48	78	3.50	86	3.50½	58	3.51	86	3.52	88	3.52½	71	3.53
64	9 (w.)[b]	56	9.02 a. m.	9.15	35	9.18	90	9.21	43	9.21½	80	9.25	82	9 25½	93	9.26	69	9.32	71	9.32½	87	9.33
76	9 (w.)	42	10.12 a. m.	10.25	47	10.28	95	10.31	41	10.31½	120	10.37	105	10.37½	130	10.38	118	10.41	116	10.41½	143	10.42
189	9 (w.)	42	3.40 p. m.	3.48	93	3.53	61	3.57	63	3.57½	34	3.58	68	4.01	80	4.01½	67	4.02	79	4.04	87	4.04½
30	10	47	9.19 a. m.	9.29	42	9.36	97	9.42	86	9.43	28	9.45	65	9.47	73	9.48	80	9 49	60	9.50	62	9.51
73	10	53	2.00 p. m.	2.09	74	2.12	102	2.14	110	2.17	105	2.17½	83	2.18	110	2.21	116	2.21½	108	2.23	110	2.23½
	Virginia:																						
24	1 A	64	8.53 a.m.	9.01	97	9.12	73	9.15	77	9.18	95	9.19	93	9.20	95	9.21	91	9.22	89	9.23	93	9.24
78	1 A	65	10.01 a.m.	10.09	68	10.13	131	10.15	106	10.17	87	10.22	130	10.22½	136	10.23	88	10.26	136	10.26½	151	10.27
31	1 B	77	3.31 p.m.	3.37	83	3.45	75	3.48	81	3.49	102	3.50	114	3.51	79	3.52	91	3.54	104	3.55	81	3.56
69	1 A	67	10.22 a.m.	10.33	69	10.38	106	10.41	47	10.41½	81	10.42	47	10.42½	101	10.46	77	10.46½	56	10.47	103	10.49
186	1 A	67	3.49 p.m.	3.59	56	4.02	108	4.05	89	4.08	92	4.08½	77	4.09	86	4.13	108	4.13½	76	4.14	88	4.16
59	1 A	68	3.00 p.m.	3.10	69	3.14	63	3.17	98	3.20	78	3.20½	66	3.23	101	3.23½	86	3.25	106	3.25½	54	3.26
183	1 A	68	11.04 a.m.	11.10	80	11.19	89	11.23	89	11.23½	66	11.24	82	11.25	108	11.25½	66	11.26	68	11.28	87	11.28½
158	2	69	11.00 a.m.	11.06	85	11.13	103	11.13½	55	11.14	88	11.17	116	11.17½	112	11.18	93	11.20	100	11.20½	116	11.21
171	2 (w.)	70	10.48 a.m.	10.57	97	11.01	95	11.05	108	11.05½	102	11.06	96	11.07	118	11.07½	106	11.08	90	11.09	111	11.09½
172	2 (w.)	70	3.45 p.m.	3.55	40	3.59	62	4.02	116	4.02½	72	4.03	91	4.05	101	4.05½	128	4.06	76	4.08	99	4.08½
92	2 B	63	8.12 a.m.	8.22	85	8.27	52	8.27½	129	8.29	76	8.29½	123	[illegible]	135	8.34½	91	8.35	121	8.29	141	8.39½
178	2 B	63	3.52 p.m.	3.59	101	4.03	100	4.07	114	4.09	97	4.09½	57	4.10	88	4.12	96	4.12½	95	4.13	87	4.16
52	2 B	69	7.42 a.m.	7.54	47	7.57	65	7.57	88	8.04	83	8.04½	66	8.05	83	8.09	77	8.09½	74	8.10	74	8.12
32	2 B (w.)	70	10.19 a.m.	10.29	95	10.34	76	10.34½	90	10.35	80	10.36	108	10.37	97	10.39	98	10.40	110	10.41	94	10.42
33	3	61	4.03 p.m.	4.12	85	4.15	103	4.19	98	4.20	94	4.21	88	4.22	84	4.23	73	4.24	71	4.25	81	4.25½
91	3	61	3.30 p.m.	3.37	132	3.45	104	3.45½	111	3.47	135	3.47½	92	3.51	110	3.51½	151	3.52	94	3.54	112	3.54½
58	3	88	10.23 a.m.	10.36	102	10.45	107	10.45½	77	10.50½	74	10.51	84	2. 51½	64	10.54	64	10.54½	83	10.55	62	10.56
40	3	88	11.14 a.m.	11.23	64	11.27	62	11.29	80	11.31	62	11.32	72	11.38	75	11.38½	95	11.39	55	11.41	92	11.41½
34	4	62	8.56 a.m.	9.01	104	9.08	101	9.10	105	9.11	101	9.13	100	9.14	113	9.16	89	9.18	92	9.18½	101	9.19
	West Virginia:																						
159	4 (w.)	44	3.21 p.m.	3.30	36	3.35	92	3.38	140	3.45	74	3.45½	98	3.46	130	3.48	73	3.49½	96	3.50	115	3.51
160	4	44	10.51 a.m.	11.01	36	11.06	52	11.08	29	11.08½	120	11.14	80	11.14½	126	11.16	112	11.16½	122	11.19	112	11.19½
35	4 B	40	2.31 p.m.	2.38	104	2.48	102	2.50	106	2.50½	81	2.51	73	2.54	105	2.54½	99	2.55	91	2.55½	71	2.56½
53	4 B (w.)	44	9.54 a.m.	10.09	74	10.12	100	10.14	55	10.14½	92	10.16	99	10.16½	42	10.17	97	10.19	99	10.19½	119	10.20
74	4 B (w.)	46	8.18 a.m.	8.27	87	8.31	81	8.34	29	8.34½	70	8.40	54	8.41	88	8.42	65	8.42½	78	8.43	100	8.45
86	4 B (w.)	46	9.33 a.m.	9.45	48	9.50	100	9.52	138	9.54	36	9.54½	85	[illegible]	104	9.58	108	10.00	89	[illegible]	132	10.03
20	13 and 14	23	2.24 p.m.	2.38	48	2.41	105	2.44	110	2.46	65	2.47	100	2.48	78	2.49	93	2.49½	79	[illegible]	57	2.50½
51	15	36	1.29 p.m.	1.43	95	1.48	94	1.51	107	1.51½	33	1.52	94	1.54	108	1.54½	87	1.55	100	1.57	113	1.57½
49	15	43	3.29 p.m.	3.44	30	3.49	99	3.54	112	3.54½	32	3.55	79	3.59	96	3.59½	69	4.00	74	4.04	96	4.04½
167	15	36	3.55 p.m.	4.02	35	4.06	93	4.09	97	4.09½	108	4.13	91	43½	112	4.14	146	4.23	97	4.23½	161	4.24
168	15	36	11.01 a.m.	11.11	88	11.15	90	11.18	67	11.18½	72	11.22	99	122½	89	11.23	64	11.28	99	1. 28½	93	11.29

[a] Plus 10 per cent pitch.

[b] Plus 5 per cent pitch.

Cupola tests of coke from coals received in 1905—Continued.

Cupola test No.	Designation of coke.		Record of melt—Continued.		Weight and time of each ladle of melted iron.																	
	Field No. of coal.	Coke test No.	Blast on at—	Iron running.	1.		2.		3.		4.		5.		6.		7.		8.		9.	
					Lbs.	At—	Lbs.	At—	Lbs.	At—	Lbs.	At—	Lbs.	At—	Lbs.	At—	Lbs.	At—	Lbs.	At—	Lbs.	At—
1	2	3	46	47	48	49	50	51	52	53	54	55	56	57	58	59	60	61	62	63	64	65
	West Virginia—Continued.																					
182	15.	43	4.10 p. m.	4.20	35	4.24	93	4.28	71	4.28½	25	4.29	75	4.32	81	4.32½	76	4.33	68	4.34	78	4.34½
169	16.	45	2.43 p. m.	2.53	38	2.56	104	2.59	99	3.04	95	3.04½	94	3.05	70	3.11	71	3.11½	87	3.12	58	3.19
170	16.	45	10.57 a. m.	11.07	40	11.10	96	11.15	112	11.15½	79	11.16	95	11.17	103	11.17½	83	11.18	96	11.19	108	11.19½
50	16 B.	45	8.23 a. m.	8.37	91	8.42	72	8.42½	100	8.44	57	8.44½	84	8.45	95	8.51	90	8.51½	78	8.53	87	8.53½
36	16 A.	73	9.18 a. m.	9.26	98	9.32	61	9.32½	104	9.35	89	9.36	119	9.36½	109	9.37	85	9.41	95	9.42	90	9.44
37	16 B (w.).	48	2.45 p.m.	2.50	29	2.55	96	2.57	73	2.59	25	2.59½	92	3.02	71	3.03	64	3.03½	67	3.04	68	3.06
77	16 B (w.).	49	2.31 p.m.	2.38	101	2.42	120	2.44	117	2.46	92	2.50	113	2.50½	116	2.51	102	2.53	111	2.53½	119	2.54
181	16 B (w.).	49	11.08 a.m.	11.15	101	11.19	83	11.22	54	11.23	65	11.26	97	11.26½	33	11.27	112	11.30	107	11.30½	134	11.31
38	17 (w.).	60	9.32 a.m.	9.38	121	9.47	111	9.47½	83	9.50	79	9.51	76	9.52	88	9.53	63	9.54	63	9.56	58	9.58
39	18.	74	3.15 p.m.	3.23	93	3.29	65	3.30	98	3.32	80	3.33	73	3.33½	82	3.36	84	3.37	73	3.38	97	3.39
46	18.	78	10.20 a.m.	10.32	29	10.35	51	10.37	100	10.39	77	10.44	94	10.44½	79	10.45	78	10.46	83	10.46½	85	10.47
45	19.	79	3.07 p.m.	3.17	30	3.23	90	3.25	13	3.25	75	3.27	71	3.27½	80	3.28	67	3.31	75	3.31½	79	3.32
62	19.	83	9.51 a.m.	10.09	88	10.15	111	10.15½	107	10.16	85	10.17½	123	10.18	100	10.18½	77	10.20	101	10.20½	103	10.21
41	19.	83	3.23 p.m.	3.34	90	3.39	84	3.43	105	3.43½	49	3.44	86	3.45	106	3.45½	46	3.46	71	3.49	102	3.49½
43	20 A (w.).	84	3.28 p.m.	3.37	92	3.43	99	3.43½	34	3.44	82	3.46	96	3.46½	112	3.47	77	3.49	98	3.49½	81	3.50
63	20 A (w.).	84	3.30 p.m.	3.40	69	3.47	102	3.52	102	3.52½	98	3.53	59	3.53½	87	3.54	110	3.54½	100	3.55	83	3.57
90	20 A.	80	10.27 a.m.	10.34	46	10.37	128	10.40	97	10.40½	89	10.44	120	10.44½	96	10.45	74	10.48	137	10.48½	99	10.49
177	20 A.	80	10.54 a.m.	11.05	20	11.09	56	11.14	81	11.20	51	11.20½	71	11.22	81	11.22½	60	11.23	64	11.26	74	11.26½
71	20 A (w.).	87	10.56 a.m.	11.04	80	11.12	110	11.15½	81	11.16	76	11.17	74	11.20	99	11.21	90	11.22	92	11.22½	90	11.23½
42	20 A (w.).	87	10.32 a.m.	10.44	45	10.47	97	10.50	103	10.53	73	10.53½	73	10.59	86	10.59½	91	11.00	81	11.02	73	11.02½
44	21.	82	11.00 a.m.	11.09	58	11.13	90	11.17	57	11.17½	66	11.22	78	11.22½	97	11.23	63	11.25	80	11.25½	88	11.26
72	21 (w.).	91	10.32 a.m.	10.40	91	10.49	115	10.50	106	10.50½	88	10.51	90	10.54	100	10.54½	88	10.55	92	10.55½	98	10.57
88	21 (w.).	91	3.39 p.m.	3.47	77	3.51	76	3.52	102	3.55½	100	3.56	66	3.59	98	3.59½	93	4.00	97	4.02	97	4.02½

Cupola tests of coke from coals received in 1905—Continued.

Cupola test No.	Designation of coke.		Record of melt—Continued. Weight and time of each ladle of melted iron—Continued.																					
	Field No. of coal.	Coke test No.	10.		11.		12.		13.		14.		15.		16.		17.		18.		19.		20.	
			Lbs.	At—	Lbs.	At—	Lbs.	At—	Lbs.	At—	Lbs.	At—	Lbs.	At—	Lbs.	At—	Lbs.	At—	Lbs.	At—	Lbs.	At—	Lbs.	At—
1	2	3	66	67	68	69	70	71	72	73	74	75	76	77	78	79	80	81	82	83	84	85	86	87
19	Connellsville coke	......	155	11.26	250	11.27	125	11.28	170	11.29	140	11.30	......	......	......	......	......	......	......	......	......	......	......	......
	Illinois:																							
54	11 D (w.)	5	74	11.44	108	11.46	107	11.46½	87	11.47	93	11.48	101	11.48½	79	11.49	95	11.51	101	11.51½	76	11.52	75	11.55
21	13 (w.)	2	90	10.17	53	10.18	31	10.20	63	10.21	29	10.24	13	10.28½	3	10.29½	......	......	......	......	......	......	......	......
93	13	3	90	10.49½	58	10.50	132	10.51	93	10.51½	60	10.52	125	10.54	90	10.54½	95	10.55	118	10.58	77	10.58½	50	11.00
94	16	10	73	11.03½	89	11.04	60	11.07	88	11.07½	76	11.08	44	11.15	......	......	......	......	......	......	......	......	......	......
	Indiana:																							
82	4 (w.)	9	89	2.54	109	2.54½	120	2.55	78	2.59	105	2.59½	114	3.00	43	3.03	......	......	......	......	......	......	......	......
83	5	8	111	3.06	88	3.08	118	3.08½	92	3.09	64	3.13	76	3.14	......	......	......	......	......	......	......	......	......	......
187	5	8	65	11.17	79	11.20	96	11.20½	67	11.21	77	11.24	91	11.24½	67	11.25	71	11.29	90	11.29½	53	11.30	61	11.32
48	7 A (w.)	13	63	10.44½	92	10.45	87	10.47	56	10.48	92	10.50	77	10.50½	48	10.51	92	10.54	72	10.54½	77	10.55	52	10.56
175	7 A (w.)	13	90	11.07½	80	11.08	92	11.09	94	11.10	93	11.10½	93	11.11	84	11.12	96	11.12½	87	11.13	81	11.14	92	11.14½
173	9	17	13	11.41	......	......	......	......	......	......	......	......	......	......	......	......	......	......	......	......	......	......	......	......
174	9	17	91	4.16	94	4.16½	102	4.17	92	4.19	104	4.19½	98	4.20	73	4.24	92	4.24½	90	4.25	62	4.30	80	4.30½
81	9 B (w.)	18	114	3.24½	97	3.25	116	3.29	113	3.29½	118	3.30	132	3.34	112	3.34½	127	3.35	......	......	......	......	......	......
185	9 B (w.)	18	131	11.19½	121	11.20	127	11.23	130	11.23½	137	11.24	63	11.25	81	11.28	74	11.28½	68	11.29	100	11.30	73	11.31
22	11	51	89	3.40	76	3.41	80	3.42	53	3.43	56	3.44	57	3.46	83	3.48	50	3.51	20	3.52	......	......	......	......
184	11	51	97	4.10½	38	4.11	89	4.12	90	4.12½	57	4.13	83	4.15	94	4.15½	120	4.16	78	4.18	94	4.18½	122	4.19
	Kentucky:																							
29	1 C	71	56	9.55	73	9.55½	66	9.58	49	10.00	67	10.01	51	10.02	73	10.04	40	10.07	47	10.10	65	10.12	40	10.13
26	1 B	76	113	2.45	86	2.46	93	2.47	112	2.48	88	2.49	82	2.50	97	2.51	88	2.53	88	2.53½	102	2.54	82	2.55
27	5	75	94	9.36	111	9.37	84	9.39	84	9.40	97	9.41	72	9.41	79	9.44	75	9.46	63	9.47	68	9.48	81	9.50
28	6	90	88	3.37	99	3.39	67	3.40	72	3.41	76	3.42	64	3.43	68	3.44	105	3.45	61	3.47	53	3.49	59	3.50
89	6	90	94	11.06½	93	11.08	87	11.08½	109	11.10	85	11.10½	107	11.12	45	11.12½	96	11.14	64	11.14½	105	11.15	89	11.17
67	6	56	98	10.16	56	10.16½	107	10.17	94	10.18	102	10.18½	104	10.19	87	10.21½	96	10.23	69	10.23	106	10.25	14	10.25½
47	7	85	88	3.49½	67	3.50	95	3.51	58	3.51½	89	3.53	46	3.53½	91	3.54	90	3.55	41	3.55½	96	3.56	39	3.58
55	7	85	81	4.55	101	4.55½	75	4.57	88	4.57½	93	4.58	73	4.59	83	5.01½	87	5.02	88	5.02½	82	5.03	86	5.04
	Maryland:																							
75	1 (w.) [a]	58	83	11.05½	117	11.07	79	11.07½	117	11.10	79	11.10½	82	11.12	113	11.14½	109	11.15	68	11.16	......	......	......	......
	Ohio:																							
23	5	22	90	2.17	97	2.19	106	2.20	89	2.22	79	2.23	107	2.24	97	2.26	73	2.27½	100	2.28	74	2.31	75	2.33
80	6 A, 6 B (w.)	66	114	10.34½	141	10.35	120	10.36½	108	10.37	122	10.38½	60	10.39	108	10.40	59	10.40½	81	10.42	97	10.43	86	10.44
57	7 (w.)	94	93	4.22½	87	4.23	87	4.26	81	4.26½	83	4.27	88	4.28	84	4.28½	88	4.29	92	4.30	84	4.30½	89	4.31
70	7	89	89	8.33	107	8.33½	87	8.34	104	8.35	88	8.35½	89	8.36	91	8.37	93	8.42	107	8.43	91	8.44	128	8.45
56	8 A (w.)	93	53	11.53	43	11.57	53	11.57½	53	11.58	52	11.58½	34	11.59	53	11.59½	64	12.00	29	12.00½	47	12.01	33	12.01½
165	9	72	89	3.56	92	3.59	97	3.59½	84	4.00	85	4.03	98	4.03½	84	4.04	78	4.09	96	4.09½	79	4.10	46	4.12
166	9	72	65	11.46	82	11.48	79	11.48½	99	11.49	88	11.50	83	11.50½	114	11.52	64	11.52½	105	11.53	113	11.56	85	11.56½

[a] Plus 10 per cent pitch.

Cupola tests of coke from coals received in 1905—Continued.

	Designation of coke.			Record of melt—Continued.																		
				Weight and time of each ladle of melted iron—Continued.																		
Cupola test No.	Field No. of coal.	Coke test No.		10.		11.		12.		13.		14.		15.		16.		17.		18.		19.		20.
			Lbs.	At—	Lbs.	At—	Lbs.	At—	Lbs.	At—	Lbs.	At—	Lbs.	At—	Lbs.	At—	Lbs.	At—	Lbs.	At—	Lbs.	At—	Lbs.	At—
1	2	3	66	67	68	69	70	71	72	73	74	75	76	77	78	79	80	81	82	83	84	85	86	87

Pennsylvania:

Virginia:

	3.......
West Virginia:	
	4 (w.).......
	4 ....
	4 B (w.).......
	4 B (w.).......
	4 B (w.).......
	4 B (w.).......
	13 and 14.......
	15.......
	15.......
	15.......
	15.......
	16.......
	16 B.......
	16 A (w.).......
	16 B (w.).......
	16 B (w.).......
	16 B (w.).......
	17 (w.).......
	18.......
	19.......
	19.......
	19.......
	20 A (w.).......
	20 A (w.).......
	20 A ....
	20 A ....
	20 A (w.).......
	21 (w.).......
	21 (w.).......

a Plus 5 per cent pitch.

b Blast off at 11.11 a. m.; belt on fan broke while pouring sixteenth ladle.

Cupola tests of coke from coals received in 1905—Continued.

Record of melt—Continued.

Cupola test No.	Designation of coke. Field No. of coal.	Coke test No.	21. Lbs.	21. At—	22. Lbs.	22. At—	23. Lbs.	23. At—	24. Lbs.	24. At—	25. Lbs.	25. At—	26. Lbs.	26. At—	Melting time (minutes).	Remarks.
1	2	8	88	89	90	91	92	93	94	95	96	97	98	99	100	101
19	Connellsville coke														27	Iron hot.
	Illinois:															
54	11 D (w.)	5	80	11.55½											30	Do.
21	13 (w.)	2			72	11.56									34½	Iron very hot and fluid, but chilled at bottom; bed burned out.
93	13	3													25	Temperature of iron medium.
94	16	10													28	Iron cold.
	Indiana:															
82	4 (w.)	9													30	Temperature of iron ...
83	5	8	90	11.32½	44	11.43	49	10.58	54	11.01	46	11.02			28	Temperature of iron medium; blast off 1 minute.
187	5	13	02	10.56½	79	10.57	101	11.17½	91	11.18	65	11.20	93	11.21	45	Iron cold. ...iron medium.
48	5 A (w.)	13	91	11.15	79	11.17									37	Iron do.
175	7 A (w.)	17													29	Temperature of first 7 ladles ... diam; balance hot.
173	9	17	46	4.31											23	Iron hot.
81	9 B (w.)	18													35	Temperature of iron medium.
185	9 B (w.)	18	30	11.32											29	Iron very hot and fluid; coke used for further trial.
22	11	51													32	Iron hot.
184	11	51	69	4.22	90	4.22½	130	4.23	71	4.24	88	4.25			54	Do.
	Kentucky:															
29	1 C	71	42	10.16	96	10.17	60	10.18	41	10.20	23	10.21	33	10.23	30	Iron very hot.
26	1 B	76	71	2.57	31	2.58									50	Iron hot and fluid.
27	5	75	52	9.52	45	9.57									27	Iron hot.
89	6	90													38	Iron hot and fluid.
88	6	86	81	11.20	98	10.27	95	10.28	77	4.01½	106	4.02	41	4.03	32	Iron hot; all pig iron used to determine effect of sulphur; 27th ladle—80 pounds at 4.03½; 28th ladle—67 pounds at 4.04; 29th ladle—35 pounds at 4.06.
67	6	88	45	10.26	97	3.59	47	4.01							32	Iron hot and fluid.
47	7	85	83	3.58½											32	Iron hot.
55	7 (w.) a	85	75	5.05											28	Iron very hot and fluid.
	Maryland:															
75	1 (w.) a	58													31	Iron hot.
	Ohio:															
23	7	23	113	2.34	87	2.34½	31	2.36							39	Iron very hot and fluid.
90	6 A, 6 B (w.)	22													24	Iron hot.
57	7 (w.)	94	100	4.32	83	4.32½	86	4.33	99	4.37	89	4.37½	91	4.38	36	Iron very hot and fluid: 27th ladle—76 pounds at 4.39; 28th ladle—65 pounds at 4.41.

											Temp.	Remarks
7.	80	42		12.02		12.04	38	12.05		12.01	33	Iron very hot and fluid.
8 A (w.)	93										40	Iron hot and fluid; blast off 9 minutes on account of hard iron notch.
9.	72	79		11.57		11.56		12.04½		12.01	32	Temperature of iron medium.
	72										32	
Pennsylvania:												
5 B (w.)	26	80	104	3.01		3.02	80	3.02½	53	3.05½	30	Iron hot.
5 B (w.)	26		102				97	3.05	90	12.01	33	Iron dd.
6.	32									73	3.07	30	Iron dd; coke in bed very hard to ignite; bed turned very slowly.
6.	41												Bottom dropped at 1.40 p. m.; blast on 30 minutes; no iron melted.
6 A, 6 B (w.)	34	101		3.54		3.54½	77	3.55	92	3.59	29	Iron cold
6 A, 6 B (w.)	34		101			3.31½		3.56			33	Iron hot; blast off 1 in.
6 A, 6 B (w.)	33										26	Iron very hot.
6 A, 6 B (w.)	38										40	Iron hot; charges hung in cupola and bottom had to be dropped; blast off 2 minutes.
7 A, 7 B	38	102	41	3.31	41	3.34	88	3.35	40	3.36	32	Iron hot.
7 A, 7 B	33	41	80	10.22	80	10.26	43	10.26½	58	10.27	28	Iron hot. Nature of iron medium.
8.	30		79	12.06	79	12.08	95	12.10½	26	12.11	29	Do.
	29	50	57	4.01	57	4.02	57	4.04	92	4.04	29	Iron hot; 27th ladle—79 pounds at 4.07; 28th ladle—58 pounds at 0l.
9 (w.)	56	66								4.06	32	Iron dd.
9 (w.)	42		84	4.22	54	4.22½	70	4.23		89	30	Iron hot.
10.	42		55	10.06	55	10.12	60	10.14			35	Tempert. nature of iron ütn.
10.	47	46									39	Iron hot and fluid; last off 6 minutes.
Virginia:	53	55									27	Iron hot.
1 A	64		92	4.11	92	4.12	94	4.12½	67	4.14		Iron hot after the th. dd.
1 B	65										29	Nature of iron
1 A	77	81	123	4.28½	123	4.31	77	4.32		4.13	33	Iron very hot and fluid; blast off 4 minutes.
1 A	67	61	33	3.37	33	3.37½	77	3.40	80		33	Nature of iron medium.
1 A	67	72	86	11.40½	86	11.41	65	11.42	26	3.41	31	km hot.
2 (w.)	68	88	46	11.32	46	11.36	82	11.36½	51	11.37	25	Tmpe nature of tm rdd.
2 (w.)	68	83	104	11.17½	104	11.18	82	11.20		11.22	33	Iron hot.
2 B	69	101	121	4.17½	121	4.18	99	4.20	116	4.20½	27	Do.
2 B	70	95									33	Do.
2 B	70										33	Do.
2 B	63	33	95	4.28	95	8.22½	104	8.23	92	8.24	30	Temperature of iron medium.
2 B (w.)	69	74	149	8.22	149							Iron hot.
2 B			37		37							rdn hot; all pig tm used to determine effect of shr.
3.	70	14		10.56½		11.01	66				32	Iron hot.
3.	61	27		4.42		4.44		4.45			29	Iron very hot and ufd.
3.	61		121	4.06	121	4.08	64	11.09½	28		31	Iron hot and lufd; melting too fast to handle; blast off 4 minutes.
3.	88	98	103	10.05½	103	11.05	77	11.54½	44		24	km hot.
4.	40	46	60	11.54	60	11.58	78	9.35	97	9.37	37	rdn hot; blast off 5 tks. Temperature of iron medl m.
	34	51	74	9.30½	74	9.31		9.36	66		36	Iron very hot and fluid.

Cupola tests of coke from coals received in 1905—Continued.

Record of melt—Continued.

Cupola test No.	Designation of coke. Field No. of coal.	Coke test No.	21. Lbs.	At—	22. Lbs.	At—	23. Lbs.	At—	24. Lbs.	At—	25. Lbs.	At—	26. Lbs.	At—	Melting time (minutes).	Remarks.
1	2	3	88	89	90	91	92	93	94	95	96	97	98	99	100	101
	West Virginia:															
159	4 (w.)	44	119	11.30½	97	11.31	65	11.32							33	Temperature of iron medium.
160	4	44		3.04	83	3.05	43	3.05	100	3.12					31	Iron hot.
35	4 B	40	101	10.28	92	10.32	35	10.34							34	Iron very hot and fluid.
63	4 B (w.)	44	91	8.57	61	8.59	157	8.59½	47	9.00					25	Iron hot; all pig iron used to determine effect of sulphur.
74	4 B (w.)	46	72												33	Iron hot.
86	4 B (w.)	46	78	3.03	107	3.04	72	3.07	64	3.10	92	3.12	64	3.13	28	Temperature of iron medium.
20	13 and 14	23	71	2.04½	85	2.07	102	2.07½	85	2.08	79	2.10	69	2.12	36	Iron very hot and fluid.
51	15	36													29	Iron hot; all pig iron used to determine effect of sulphur.
49	15	43	30	4.26											42	Do.
167	15	36													31	Iron hot.
168	15	36	96	11.39	102	11.39½	83	11.40	97	11.42	100	11.41			32	Temperature of iron medium.
182	16	38	148	4.42½	140	4.43									23	Iron very cold.
169	16	45													40	Iron hot.
50	16 B	45	102	11.29½	92	11.30	83	11.34	100	11.34½	88	11.35	45	11.36	25	Temperature of iron medium; all pig iron used to determine effect of sulphur.
36	16 A	73	63	9.59	82	10.00	118	10.02	89	10.03	49				33	Iron hot; blast off 4 minutes.
37	16 B (w.)	49	44	3.19	43	3.20	82	3.21	42	3.21		3.22	31	3.23	40	Iron fairly hot but sluggish.
77	16 B (w.)	49	77	11.44	49	11.45	70	10.14	54	10.15	43	10.16	68	3.56	30	Iron hot.
181	16 B (w.)	60	42	10.11	62	10.13	77	3.51	118	3.53	27	3.54	78	11.00	30	Do.
38	17 (w)	74	67	11.02	91	11.04	93	11.04½	66	11.05	70	11.06	75	3.50	34	Iron hot but sluggish.
39	18	78	97	3.44½	87	3.45	58	3.46	100	3.47	72	3.46			37	Iron very hot and fluid.
45	18	79	104	10.29	76	10.29½	45	10.32	92	10.32½	82	10.33			36	Iron hot.
62	19	83	82	4.05	110	4.05	98	4.07							24	Do.
41	19	84	82	4.06	90	4.10	85	4.10½	68	4.11					33	Iron hot and fluid.
63	20 A (w.)	84	94	11.00	52	11.03	88	11.04							30	Do.
90	20 A (w.)	80													33	Iron cold.
177	20 A	87	78	11.37	66	11.38		4.18							33	Iron hot and fluid.
71	20 A (w.)	82	83	11.37	54	11.38									34	Iron hot and fluid.
42	20 A (w.)	82	50	11.00	69	4.17	120								27	Temperature of iron medium.
44	21 (w.)	91	107												29	Iron hot.
72	21 (w.)	91	70	4.16½											29	Do.
88															31	Do.

Cupola tests of coke from coals received from January 1, 1906, to June 30, 1907.

Cupola test No.	Designation of coke.		Date.	Charges (pounds).																Total.			Ratio of iron to coke.
	Field No. of coal.[a]	Coke test No.		Coke bed.	Pig iron.	Scrap.	Coke.	Pig iron.	Scrap.	Coke.	Pig iron.	Scrap.	Coke.	Pig iron.	Scrap.	Coke.	Pig iron.	Scrap.	Coke.	Pig iron.	Scrap.	
1	2	3	4	5	6	7	8	9	10	11	12	13	14	15	16	17	18	19	20	21	22	23
Alabama:																						
107	2 B (w.)	142	Aug. 1	200	800	200	58	413	138	58	413	138	57	412	137	57	412	137	430	2,250	750	7
131	2 B (w.)	142	Aug. 25	200	800	200	58	413	138	58	413	138	57	412	137	57	412	137	430	2,250	750	7
132	...do	138	...do	210	680	210	55	405	135	55	405	135	55	405	135	55	405	135	430	2,250	750	7
101	3 (w.)	130	Sept. 7	210	680	210	55	405	135	55	405	135	55	405	135	55	405	135	430	2,250	750	7
106	3 (w.)	130	Aug. 2	200	680	200	58	405	138	58	405	135	57	405	135	55	405	137	430	2,250	750	7
103	4	131	Sept. 8	200	660	200	53	413	138	58	413	135	57	412	135	52	405	135	430	2,250	750	7
124	4 (w.)	131	Sept. 21	190	660	230	60	398	133	60	398	133	60	397	132	52	397	132	430	2,250	750	7
109	4 (w.)	136	Aug. 21	190	570	190	53	420	140	53	420	140	52	420	140	60	420	140	430	2,250	750	7
133	4 (w.)	136	Aug. 27	200	570	200	58	413	138	58	413	137	57	412	137	57	412	132	430	2,250	750	7
Arkansas:																						
96	1 B (w.)[b]	97	Sept. 4	190	570	190	60	420	140	60	420	140	60	420	140	60	420	140	430	2,250	750	7
115	1 B (w.)[b]	97	Aug. 9	190	570	190	60	420	140	60	420	140	60	420	140	60	420	140	430	2,250	750	7
116	7 B (w.)[c]	105	Aug. 10	220	660	220	53	398	133	53	398	133	53	397	132	52	397	132	430	2,250	750	7
142	7 B (w.)[c]	105	Sept. 1	210	630	210	55	405	135	55	405	135	55	405	135	55	405	135	430	2,250	750	6
95	9 (w.)[b]	102	Sept. 4	220	660	220	53	398	133	53	405	135	55	405	135	55	405	135	430	2,250	750	6
117	9 (w.)[b]	102	Aug. 11	220	660	220	53	398	133	53	398	133	52	397	132	52	397	132	430	2,250	750	8
Illinois:																						
125	22 B (w.)	118	Aug. 21	210	630	210	55	405	135	55	405	135	55	405	135	55	405	135	430	2,250	750	7
150	29 (w.)	170	Nov. 13	190	780		78	550		78	550		77	550		77	560		430	3,000	(d)	7
157	29 (w.)	170	Nov. 16	190	780		44	560		44	560		44	560		43	560		500	3,000	(d)	6
164	29 (w.)	170	Nov. 16	190	800		63	550		63	550		62	550		62	560		375	3,000	(e)	8
Indiana:																						
121	12 (w.)	110	Aug. 17	180	540	180	63	428	143	63	428	143	63	427	142	63	427	142	430	2,250	750	7
148	17 (w.)	163	Nov. 12	180	700		60	560		60	560		60	560		60	560		430	3,000	(d)	7
Kansas:																						
122	6 (w.)	115	Aug. 23	180	540	180	63	428	143	63	428	143	62	427	142	62	427	142	430	2,250	750	7
Kentucky:																						
147	8	164	Nov. 10	200	800		58	550		58	550		57	550		57	550		430	3,000	(e)	7
155	8	164	Nov. 15	210	840		73	540		73	540		72	540		72	540		500	3,000	(e)	6
156	8	164	Nov. 16	200	800		44	560		44	560		44	560		43	560		375	3,000	(e)	8
149	9 A (w.)	167	Nov. 12	170	680		65	560		65	560		65	560		65	580		430	3,000	(e)	7
162	9 A (w.)	167	Nov. 21	170	680		51	580		51	580		51	580		51	580		375	3,000	(e)	8
163	9 A (w.)	167	...do	190	690		78	560		78	560		77	560		77	560		500	3,000	(e)	6

[a] Details of origin of coal samples can be found in Bull. U. S. Geol. Survey No. 332.
[b] Plus 10 per cent pitch.
[c] Plus 5 per cent pitch.
[d] Pig iron used from car 27633.
[e] Pig iron used from car 131945.

Cupola tests of coke from coals received from January 1, 1906, to June 30, 1907—Continued.

	Designation of coke.							Charges (pounds).												Total.			Ratio of iron to coke.
Cupola test No.	Field No. of coal.	Coke test No.	Date.	Coke bed.	Pig iron.	Scrap.	Coke.	Pig iron.	Scrap.	Coke.	Pig iron.	Scrap.	Coke.	Pig iron.	Scrap.	Coke.	Pig iron.	Scrap.	Coke.	Pig iron.	Scrap.	
1	2	3	4	5	6	7	8	9	10	11	12	13	14	15	16	17	18	19	20	21	22	23
	Missouri:																					
123	5 (w.)	116	Aug. 20	190	570	190	60	420	140	60	420	140	60	420	140	60	420	140	430	2,250	750	7
	New Mexico:																					
119	4 B (w.)	151	Aug. 15	200	600	200	58	413	138	58	413	138	57	412	137	57	412	137	430	2,250	750	7
120	5 (w.)	147	...do...	220	660	220	53	398	133	53	398	133	52	397	132	53	397	132	430	2,250	750	7
98	4 B (w.)	152	July 27	210	630	210	55	405	135	58	405	135	55	405	135	57	405	135	430	2,250	750	7
130	5.		Aug. 24	200	600	200	58	413	138	58	413	138	58	412	138	57	412	137	430	2,250	750	7
	Pennsylvania:																					
143	11	159	Nov. 8	200	800		58	550		58	550		57	550		57	550		430	3,000		7
161	11	159	Nov. 20	220	840		70	530		70	530		70	530		70	530		500	3,000		6
145	12.	161	Nov. 9	210	840		55	540		55	540		55	540		55	540		430	3,000		7
146	12 (w.)	162	Nov. 10	210	800		55	550		55	550		55	550		55	550		430	3,000		6
151	12.	161	Nov. 13	200	840		44	530		44	530		44	530		43	530		375	3,000		8
152	12.	161	Nov. 14	200	840		73	540		73	540		72	540		72	540		500	3,000		6
153	12 (w.)	162	...do...	220	880		55	530		55	530		55	530		55	530		375	3,000		8
154	12 (w.)	162	Nov. 15	210	840		55	540		55	540		55	540		55	540		500	3,000		7
190	21.	187	Feb. 13	210	630	210	55	405	135	55	405	135	55	405	135	55	405	135	430	2,250	750	7
	Tennessee:																					
126	1 (w.)	133	Aug. 22	200	600	190	58	430	140	58	430	140	58	430	140	58	430	140	430	2,250	750	7
99	1 (w.)	133	July 30	200	600	200	57	413	137	57	413	137	57	413	138	58	413	138	430	2,250	750	7
128	1 (w.)	133	Aug. 23	180	570	200	55	412	137	55	412	138	57	412	137	57	413	137	430	2,250	750	7
140	2.	127	July 31	180	570	190	58	430	140	58	430	140	58	430	140	58	430	140	430	2,250	750	7
104	2.	127	Aug. 31	190	570	190	58	430	140	58	430	140	58	430	140	58	430	140	430	2,250	750	7
139	3.	126	July 31	190	570	190	58	430	140	58	430	140	58	430	140	58	430	140	430	2,250	750	7
102	3.	125	Sept. 7	190	570	190	58	430	140	58	430	140	58	430	140	58	430	140	430	2,250	750	7
105	4.	139	July 31	190	570	190	58	430	140	58	430	140	58	430	140	58	430	140	430	2,250	750	7
138	4.	139	Aug. 30	190	570	190	58	430	140	58	430	140	58	430	140	58	430	140	430	2,250	750	7
106	4 (w.)	154	Aug. 1	190	570	190	58	430	140	58	430	140	58	430	140	58	430	140	430	2,250	750	7
159	5 (w.)	132	Aug. 24	220	660	220	53	398	133	53	398	133	53	397	132	53	397	132	430	2,250	750	7
112	5 (w.)	132	Sept. 1	190	570	190	58	430	140	58	430	140	58	430	140	58	430	140	430	2,250	750	7
113	6.	123	Sept. 6	190	570	190	58	430	140	58	430	140	58	430	140	58	430	140	430	2,250	750	7
137	7 B (w.)	128	Aug. 29	190	540	180	63	428	143	63	428	143	63	428	143	62	427	142	430	2,250	750	7
14	7 B (w.)	131	Aug. 28	180	570	190	58	430	140	58	430	140	58	430	140	58	430	140	430	2,250	750	7
135	8 B (w.)	134	Aug. 3	180	570	190	58	420	140	58	420	140	58	420	140	58	420	140	430	2,250	750	7
10	9 (w.)	124	Aug. 28	200	600	200	58	413	138	58	413	138	58	412	137	57	412	137	430	2,250	750	7
136	9 (w.)	124																				

111	10 (w.)............	156	Aug. 4	230		230	48	510/398	130	48	510/398	47/52	510/397	130	300	50	300	130	430	2,280	750	7
127	10 (w.)............	156	14	250			53			53		52								430	3,000	750	7
144	11 (w.)............	160	Nov. 9	240	140	220																	7
170	11 (w.)............	90	Dec. 3	220	600	300	53	510/398	133	53	510/398	47/52	510/397	132	510/397	57	510/397	132	430	2,250	750	7
	Utah:																					
118	1................	130	Aug. 14	200	600	300	58	413	138	58	413	57	412	137	412	57	412	137	430	2,250	750	7
	Washington:																					
97	2................	135	July 26	190	570	190	58	420/413	140	58	420/413	60/57	420/412	140	420/412	60/57	420/412	140	430	2,250	750	7
134	2................	135	Aug. 27	200	600	300	58		138	58				137				137	430	2,250	750	7

a Pig iron used from car 27633. b Pig iron used from car 131943. c 1907.

Cupola tests of coke from coals received from January 1, 1906, to June 30, 1907—Continued.

Cupola test No.	Field No. of coal.	Coke test No.	Moisture.	Volatile matter.	Fixed carbon.	Ash.	Sulphur in coke.	Sulphur in ash.	Phosphorus.	Specific gravity.	Fluidity at rip, (per cent), full.	Maximum blast pressure (ounces).	Poured.	Additional melted.	Total.	Melting rate per hour (pounds).	Iron.	Coke.	Melting loss (per cent).	Melting ratio, iron to coke.
1	2	8	24	25	26	27	28	29	30	31	32	33	34	35	36	37	39	40	41	42
Alabama:																				
107	2 B (w.)	142	3.04	1.06	82.15	13.75	1.16		.0700	1.88	99.9	7	2,257	322	2,579	4,991	214	37	6.9	6.56
131	2 B (w.)	142	3.04	1.06	82.15	13.75	1.16		.0700	1.88	99.9	7	1,852	365	2,217	4,587	413	48	12.33	5.80
132	3 (w.)	38	2.03	1.80	74.89	21.28	.60		.0057	1.91	97.22	7	740	214	954	2,201	1,911	132	4.46	3.21
101	3 (w.)	39	.99	1.06	83.51	14.44	.58		.0008	1.99	94.44	7	1,695	305	2,000	3,429	807	75	6.43	3.63
108	3 (w.)	39	.99	1.06	83.51	14.44	.58		.0008	1.99	94.44	7	1,822	213	2,035	3,700	774	99	6.36	6.15
103	4	31	.29	.84	83.21	15.66	1.08	.05	.0126	1.95	98.61	7	1,451	217	1,668	3,033	1,185	126	4.90	5.49
124	4 (w.)	31	.29	.84	83.21	15.66	1.08	.05	.0126	1.96	80.55	7	1,368	542	1,910	4,244	1,770	99	10.66	5.76
109	4 (w.)	36	.35	.42	92.99	6.24	.87		.008	1.95	99.9	7	2,328	153	2,481	4,115	334	72	6.16	6.93
133	4 (w.)	36	.35	.42	92.99	6.24	.87		.008	1.95	98.61	7	1,952	178	2,130	4,280	697	124	5.76	6.96
Arkansas:																				
96	1 B (w.) b	97	2.74	1.29	85.81	10.16	1.02	.05	.0135	1.96	93.06	7	1,663	319	1,982	3,303	730	58	9.60	5.33
115	1 B (w.) b	97	2.74	1.29	85.81	10.16	1.02	.05	.0135	1.96	94.44	7	1,648	404	2,052	4,925	627	68	10.70	5.58
116	7 B (w.) c	105	.67	.83	89.14	9.34	1.60	.07	.0082	1.97	99.9	7	772	318	1,090	1,982	1,816	95	3.13	3.25
142	7 B (w.) c	142	.67	.67	89.14	9.34	1.60	.07	.0082	1.97	93.05	7	1,637	399	2,036	3,460	1,736	74	7.60	5.72
95	9 (w.) b	95	.30	.81	81.48	17.41	1.07	.17	.0329	2.04	98.89	7	679	354	1,033	2,214	1,815	140	5.07	3.56
117	9 (w.) b	102	.30	.81	81.48	17.41	1.07	.17	.0329	2.04	98.61	7	1,005	288	1,293	3,103	1,464	108	8.10	4.01
Illinois:																				
125	22 B (w.) b	118	.65	.65	80.76	16.99	3.65	.05		1.84	98.61	7	1,191	325	1,516	2,675	1,280	43	6.80	3.92
150	29 (w.)	170	2.78	.74	83.35	13.13	2.49			1.83	94.44	7	2,095	325	2,424	4,278	410	34	5.53	6.02
157	29 (w.)	170	2.78	.74	83.35	13.13	2.49			1.83	97.22	7	2,195	466	2,661	5,150	178	58	5.37	6.12
164	29 (w.)	170	2.78	.74	83.35	13.13	2.49			1.83	94.44	7	1,331	295	1,626	3,252	1,171	40	6.77	4.85
Indiana:																				
121	12 (w.)	110	.42	1.03	84.37	14.18	2.89	.06		1.87	96.53	7	1,752	199	1,951	4,682	852	74	6.57	5.48
148	17 (w.)	163	1.65	.67	81.42	16.26	3.39			1.92	94.44	7	1,548	274	1,822	3,416	1,043	53	4.50	4.82
Kansas:																				
122	6 (w.)	115	.59	.56	82.78	16.07	2.49	.02		1.90	97.22	7	1,468	230	1,698	3,087	1,098	132	6.80	5.70
Kentucky:																				
147	8 (w.)	164	.50	.65	87.96	10.89	.93		.0091	1.90	97.22	7	1,556	403	1,959	3,791	900	76	4.70	5.53
155	8 (w.)	164	.50	.65	87.96	10.89	.93		.0091	1.90	93.05	7	2,101	361	2,462	3,887	360	79	5.93	5.85
156	8 (w.)	164	.50	.65	87.96	10.89	.93		.0091	1.90	93.06	7	2,124	203	2,327	4,651	499	82	5.80	7.94
149	9 A (w.)	167	1.01	.69	86.46	11.84	1.96			1.88	95.83	7	2,282	228	2,510	5,378	250	53	8.00	6.66
162	9 A (w.)	167	1.01	.69	86.46	11.84	1.96			1.88	94.44	7	2,161	356	2,517	5,719	319	43	5.47	6.70
163	9 A (w.)	167	1.01	.69	86.46	11.84	1.96			1.88	98.61	7	2,230	123	2,353	3,361	558	29	2.97	5.00

a Analysis of coke (per cent).

No.	Locality																			
	Missouri:																			
123	5 (w.)	116	1.12	.73	82.04	15.51	3.40		.0946	1.88	92.36	7	1,904	423	2,387	4,212	341	48	9.06	6.25
	New Mexico:																			
119	4 B (w.)	151	1.39	.85	83.66	14.10	.60		.0001	1.92	98.61	7	2,210	351	2,561	4,957	162	79	9.23	7.30
120	5 (w.)	147	.99	.84	81.38	16.79	.76			1.91	99.9	7	1,354	124	1,478	3,167	1,300	107	7.40	4.58
98	3 (w.)	152	.69	1.48	82.18	15.65	.63		.0348	1.91	99.9	7	1,162	325	1,487	3,076	1,288	85	7.50	4.31
130	4 B (w.)		.69	1.48	82.18	15.65	.63		.0348	1.91	99.9	7	1,778	252	2,030	4,511	759	118	7.03	6.51
	5.																			
	Pennsylvania:																			
143	11.	159	.52	.97	85.02	13.49	1.19		.0307	1.92	97.22	7	1,636	278	1,914	3,278	947	71	4.63	5.33
161	11.	159	.52	.97	85.02	13.49	1.19		.0307	1.92	93.75	7	2,054	542	2,866	5,563	272	76	4.40	6.12
146	12.	161	.69	.29	89.13	12.73	1.66		.0184	1.97	97.22	7	1,439	1,109	2,578	7,931	47	82	9.17	7.41
151	12 (w.)	162	.52	1.28	90.46	9.07	1.11		.0067	1.95	99.9	7	1,831	206	2,038	3,550	812	95	4.07	6.08
152	12.	161	.69	.29	86.29	12.73	1.66		.0184	1.97	96.53	7	1,841	256	2,291	4,522	211	72	5.03	6.26
153	12.	162	.69	.29	86.29	12.73	1.66		.0184	1.97	98.61	7	2,382	450	2,638	4,106	586	74	4.10	7.61
154	12 (w.)	162	.52	1.28	89.13	9.07	1.11		.0067	1.95	99.9	7	2,352	120	2,472	3,917	388	85	4.66	5.96
190	12 (w.)	187	.63	.66	87.78	10.93	.82		.0087	1.95	97.22	7	2,162	355	2,517	873	309	79	5.80	8.50
	21.							.0115	1.90			1,793	691	2,644	5,982	211	90	10.16	7.31
	Tennessee:																			
126	1 (w.)	133	.37	.73	83.00	15.90	1.35		.0238	1.96	97.22	7	1,240	244	1,464	2,966	1,342	112	5.80	4.67
99	1 (w.)	153	.93	.77	89.61	8.69	.93		.0253	1.87	92.00	7	1,443	563	2,095	3,540	646	106	11.00	6.19
128	2.	153	.93	.77	89.61	8.69	.93		.0253	1.87	99.9	7	1,833	210	2,043	3,606	769	119	6.26	6.57
100	2.	127	1.13	.64	91.00	7.23	.78		.0094	1.81	99.9	7	1,551	556	2,107	3,830	699	66	6.47	5.79
140	3.	127	1.13	.64	91.00	7.23	.78		.0094	1.81	94.44	7¾	2,080	270	2,380	4,147	477	36	5.77	5.97
104	3.	128	.67	.81	87.44	11.08	.88		.0125	1.87	95.83	7½	1,477	429	1,905	3,267	856	84	7.93	5.51
139	4.	128	.67	.81	87.44	11.06	.88	.03	.0125	1.87	97.81	7	1,913	239	2,132	4,056	650	50	6.60	5.55
102	4.	125	.54	1.56	84.27	13.63	.72		.0215	1.93	98.61	7	1,731	131	1,980	3,466	852	79	6.70	5.57
105	4 (w.)	129	1.43	.29	86.73	11.56	.69		.0233	1.91	990	7½	1,915	252	2,044	3,960	753	78	6.16	5.81
138	5 (w.)	129	.56	.42	86.73	11.56	.69		.0233	1.91	94.44	7	1,875	144	2,127	3,191	688	81	8.30	6.09
106	5 (w.)	154	.22	.42	90.46	8.56	2.08			1.91	98.51	7	2,328	418	2,472	3,454	279	73	6.13	6.92
129	6.	154	.22	.91	90.46	8.56	2.08			1.95	93.05	7	1,825		2,243	3,568	573	74		6.30
112	7 B (w.)	122	.57	.91	79.01	19.86	.69	.03	.0834	1.95	99.9	7	749	296	1,045	2,060	1,408	197	4.90	4.48
141	7 B (w.)	122	.57	.87	79.01	19.86	1.77	.03	.0834	1.93	98.61	6¼	2,384	432	2,516	3,336	122	141	12.06	8.70
113	8 B (w.)	123	.32	.87	83.59	14.97	1.77	.04	.0488	1.93	95.83	7	1,446	444	1,880	3,918	942	128	5.60	6.26
137	8 B (w.)	123	.33	.11	83.59	14.97	2.45		.0238	1.98	93.14	6½	2,301	217	2,518	5,956	231	66	8.36	6.92
114	9 (w.)	134	.38	.39	85.66	13.91	.61		.0238	1.93	97.22	7	1,547	1,034	2,581	3,635	169	38	8.33	6.59
135	10 (w.)	124	.39	.39	85.78	13.44	.95	.04	.0968	1.90	99.9	7	1,950	110	2,060	5,437	735	102	6.83	6.28
110	10 (w.)	124	.67	.81	85.78	13.44	.95	.04	.090	1.90	93.05	7	2,106	250	2,356		615	85	1.13	6.83
136	11 (w.)	156	1.67	1.60	77.81	19.71	.69		.0390	1.97										
111	11 (w.)	160	1.14	1.60	80.14	17.12	.69													
127		160	1.14		80.14	17.12														
	Utah:																			
118	1.	130	2.53	1.37	88.06	8.04	.86		.0005	1.78	86.11	6	424	317	741	2,223	2,189	143	2.03	2.58
	Washington:																			
97	2.	135	1.02	2.10	77.53	19.35	.44		.0447	1.90	91.60	7	903	299	1,202	2,487	1,529	33	8.96	3.03
134	2.	135	1.02	2.10	77.53	19.35	.44		.0447	1.90	97.22	7	1,411	365	1,776	3,532	1,122	32	3.40	4.46

a For chemical analyses of coals from which these cokes were made, see pp. 27-35.   *b* Plus 10 per cent pitch.   *c* Plus 5 per cent pitch.

Cupola tests of coke from coals received from January 1, 1906, to June 30, 1907—Continued.

Cupola test No.	Designation of coke. Field No. of coal.	Designation of coke. Coke test No.	Record of melt—Continued. Blast on at—	Iron running.	Weight and time of each ladle of melted iron. 1. Lbs.	1. At—	2. Lbs.	2. At—	3. Lbs.	3. At—	4. Lbs.	4. At—	5. Lbs.	5. At—	6. Lbs.	6. At—	7. Lbs.	7. At—	8. Lbs.	8. At—	9. Lbs.	9. At—
1	2	3	46	47	48	49	50	51	52	53	54	55	56	57	58	59	60	61	62	63	64	65
	Alabama:																					
107	2 B (w.)	142	1.59 p. m.	2.04	69	2.11	69	2.11½	89	2.16	98	2.16½	82	2.19	101	2.19½	70	2.20	81	2.22	83	2.22½
131	2 B (w.)	142	10.42 a. m.	10.49	33	10.51	79	10.53	77	10.56	30	10.57	74	10.58½	73	10.59	68	11.03	72	11.04	76	11.04½
132	3	138	2.49 p. m.	3.01	92	3.10	100	3.10½	61	3.11	77	3.13	77	3.13½	70	3.16	34	3.16½	65	3.18	34	3.19
101	3 (w.)	139	11.17 a. m.	11.27	77	11.32	91	11.36	97	11.36½	72	11.43	72	11.43½	70	11.44	62	11.47	76	11.47½	73	11.48
108	3 (w.)	139	3.39 p. m.	3.44	103	3.55	119	3.55½	51	3.56	69	4.00	116	4.00½	102	4.01	110	4.01½	86	4.03½	1.07	4.04
103	4	131	8.19 a. m.	8.30	66	8.35	95	8.38	75	8.41	36	8.42	96	8.43	61	8.45	101	8.45½	70	8.51	88	8.51½
124	4	131	10.04 a. m.	10.14	91	10.19	34	10.19½	90	10.26	76	10.26½	83	10.27	69	10.30	72	10.31	92	10.32	57	10.33
109	4 (w.)	136	10.30 a. m.	10.37	54	10.41	112	10.45	125	10.45½	102	10.46	112	10.47	110	10.47½	69	10.48	108	10.51½	114	10.52
133	4 (w.)	136	10.53 a. m.	11.01	51	11.07	84	11.10	94	11.11	69	11.13	82	11.15	105	11.15½	80	11.16	72	11.18½	116	11.19
	Arkansas:																					
96	1 B (w.)[a]	97	3.50 p. m.	3.58	64	4.02	78	4.07	85	4.07½	65	4.12	73	4.12½	41	4.13	79	4.13½	70	4.14	70	4.19
115	1 B (w.)[a]	97	3.33 p. m.	3.47	106	3.57	107	3.57½	104	3.58	110	4.05	112	4.06	111	4.06½	129	4.07	107	4.07½	128	4.08
116	7 B (w.)[b]	105	10.41 a. m.	10.53	83	11.05	94	11.06	74	11.11	90	11.12	97	11.12½	23	11.13	65	11.17	72	11.17½	89	11.18
142	7 B (w.)[b]	105	10.55 a. m.	11.01	81	11.07	87	11.10	75	11.10½	73	11.12	70	11.12½	80	11.17	75	11.17½	80	11.18	84	11.19
95	9 (w.)[a]	102	10.50 a. m.	11.02	82	11.16	50	11.19	92	11.19½	53	11.20	58	11.26	72	11.26½	91	11.27	61	11.29	59	11.29½
117	9 (w.)[a]	102	8.02 a. m.	8.18	78	8.26	52	8.31	53	8.3[illegible]½	70	8.32	67	8.32½	37	8.34	60	8.34½	66	8.35	83	8.35½
	Illinois:																					
125	22 B (w.)	118	3.04 p. m.	3.12	80	3.20	74	3.20½	96	3.23	81	3.24	94	3.26	62	3.26½	62	3.31	74	3.32	78	3.32½
150	29 (w.)	170	10.59 a. m.	11.11	55	11.14	96	11.18	110	11.19	69	11.23	103	11.23½	103	11.24	75	11.25	98	11.25½	103	11.26
157	29 (w.)	170	3.15 p. m.	3.25	19	3.28	95	3.32	107	3.32½	86	3.33	87	3.36	107	3.36½	85	3.37	81	3.40	100	3.40½
164	29 (w.)	170	11.10 a. m.	11.19	86	11.25	32	11.25½	93	11.28	88	11.28½	52	11.29	90	11.30	69	11.31	65	11.38	83	11.38½
	Indiana:																					
121	12 (w.)	110	10.42 a. m.	10.54	80	10.57	77	10.58	110	10.59	60	11.02	83	11.02½	58	11.03	106	11.05	95	11.05½	57	11.06
148	17 (w.)	163	11.17 a. m.	11.28	24	11.30	95	11.34	101	11.34½	79	11.39	94	11.39½	108	11.40	76	11.41	98	11.42	78	11.46
	Kansas:																					
122	6 (w.)	115	8.57 a. m.	9.04	133	9.16	111	9.16½	102	9.19½	83	9.20	100	9.22	98	9.22½	89	9.23	103	9.23½	80	9.24
	Kentucky:																					
147	8	164	3.30 p. m.	3.41	86	3.47	86	3.47½	40	3.48	72	3.52	91	3.52½	94	3.53	67	3.56	102	3.56½	98	3.57
155	8	164	3.40 p. m.	3.48	67	3.54	97	3.57	39	3.57½	81	4.00	113	4.01	85	4.03	87	4.03½	103	4.04	90	4.07
156	8	164	10.56 a. m.	11.07	32	11.10	94	11.13	82	11.13½	70	11.18	102	11.18½	100	11.19	70	11.22	101	11.22½	102	11.23
149	9 A (w.)	167	3.08 p. m.	3.17	90	3.23	50	3.23½	78	3.25	96	3.25½	69	3.26	88	3.28	102	3.28½	118	3.29	91	3.31
162	9 A (w.)	167	10.45 a. m.	10.55	84	11.00	96	11.03	105	11.03½	96	11.04	80	11.07	101	11.07½	89	11.08	82	11.09	97	11.09½
163	9 A (w.)	167	3.35 p. m.	3.45	66	3.49	114	3.51	110	3.54	103	3.55	30	3.55½	128	4.02	96	4.02½	97	4.03	100	4.07
	Missouri:																					
123	5 (w.)	116	3.16 p. m.	3.21	67	3.29	85	3.33	81	3.34	54	3.37	53	3.37½	92	3.38	62	3.39	66	3.40	69	3.41

	New Mexico:																					
119	4 B (w.)	151	8.57 a. m.	9.03	95	9.12	53	9.12½	96	9.16	94	9.16½	28	9.17	88	9.19	102	9.20	70	9.20½	89	9.21
120	5 (w.)	147	2.17 p. m.	2.27	93	2.34	50	2.34½	57	9.39	107	2.39½	99	2.40	43	2.40½	83	2.41	92	2.42	95	2.43
98	4 B (w.)	152	1.48 p. m.	1.59	83	9.07	86	2.07½	39	2.08	78	9.15	79	2.15½	86	2.16	58	2.18	66	2.18½	70	2.19
130	5		3.18 p. m.	3.28	94	2.34	52	2.34½	97	2.38	88	2.38½	79	2.39	109	2.39½	83	3.40½	82	3.41	91	2.42
	Pennsylvania:																					
143	11	159	3.08 p. m.	3.18	68	3.22	73	3.24	104	3.26	90	3.27	90	3.32	101	3.32½	87	3.33	93	3.35	99	3.35½
161	11	159	1.58 p. m.	2.08	93	2.12	110	2.13	80	2.16	82	2.16½	125	2.17	68	2.22	73	2.22½	59	2.23	109	2.23½
145	12	161	3.13 p. m.	3.24	106	3.32	112	3.32½	108	3.35	91	3.35½	100	3.36	97	3.37	89	3.38	87	3.38½	106	3.39
146	12 (w.)	162	11.14 a. m.	11.26	100	11.30	97	11.32	100	11.33	101	11.34	88	11.38	110	11.38½	69	11.39	151	11.42	111	11.42½
151	12	161	3.29 p. m.	3.40	55	3.43	98	3.46	55	3.46½	94	3.49	103	3.49½	96	3.50	90	3.53	101	3.53½	105	3.54
152	12	161	10.39 a. m.	10.47	97	10.53	96	10.59	110	10.59½	91	11.00	95	11.04	101	11.04½	112	11.05	118	11.05½	100	11.07
153	12 (w.)	163	2.56 p. m.	3.07	95	3.11	29	3.11½	97	3.14	95	3.14½	108	3.18	108	3.18½	108	3.19	102	3.23	108	3.23½
154	12 (w.)	163	10.57 a. m.	11.06	92	11.15	133	11.15½	78	11.16	122	11.16½	93	11.19	93	11.19½	93	11.20	135	11.22	95	11.22½
190	21	187	11.00 a. m.	11.06	61	11.10	132	11.13	80	11.15	62	11.17	147	11.17½	55	11.18	49	11.20	136	11.20½	73	11.21
	Tennessee:																					
126	1	133	9.58 a. m.	10.08	85	10.13	95	10.19	98	10.19½	82	10.20	82	10.22	81	10.22½	76	10.23	68	10.28	95	10.28½
99	1 (w.)	158	8.47 a. m.	8.54	66	9.05	59	9.06	68	9.09	61	9.10	62	9.11	51	9.12	51	9.13	61	9.14	43	9.15
128	1 (w.)	158	3.11 p. m.	3.16	65	3.29	110	3.29½	100	3.30	75	3.32	104	3.32½	117	3.33	41	3.35½	91	3.36	108	3.36½
100	2	127	1.20 p. m.	1.27	52	1.35	34	1.35½	64	1.40	70	1.40½	52	1.41	94	1.42	79	1.42½	73	1.43	80	1.45½
140	2	127	2.47 p. m.	2.53	28	2.59	110	3.01	44	3.01½	90	3.04	78	3.07	106	3.07½	102	3.08	80	3.10	101	3.10½
104	3	128	9.08 a. m.	9.16	50	9.21	64	9.21½	62	9.22	77	9.24	82	9.25	20	9.25½	57	9.31	68	9.31½	64	9.32
139	3	128	10.43 a. m.	10.52	93	10.56	93	10.59	100	10.59½	81	11.02	62	11.02½	89	11.06	81	11.06½	101	11.07	87	11.09
102	4	125	3.15 p. m.	3.23	104	3.32	62	3.32½	95	3.35	79	3.35½	86	3.38	22	3.38½	79	3.40	91	3.41½	76	3.42
105	4	129	2.25 p. m.	2.32	94	2.38	98	2.38½	48	2.39	89	2.42	111	2.43	113	2.43½	87	2.46	104	2.46½	97	2.47
138	4	129	3.51 p. m.	4.00	87	4.03	73	4.06	84	4.06½	46	4.07	89	4.09	79	4.11	106	4.11½	60	4.12	98	4.14
106	5 (w.)	154	9.13 a. m.	9.20	103	9.25	107	9.25½	80	9.26	86	9.29	109	9.30	116	9.31	89	9.32	108	9.32½	89	9.35
129	5 (w.)	154	9.44 a. m.	9.50	80	9.58	65	9.59	91	10.05	64	10.06	87	10.09	88	10.09½	64	10.10	76	10.11½	84	10.12
112	6	122																				
141	6	122	8.05 a. m.	8.10	36	8.18	78	8.23	88	8.24	78	8.28	69	8.28½	40	8.29	57	8.32	57	8.32½	45	8.33
113	7 B (w.)	123	3.30 p. m.	3.38	93	3.47	56	3.47½	102	3.51½	93	3.52	78	3.56	102	3.56½	83	3.57	71	3.59	106	3.59½
137	7 B (w.)	123	10.30 a. m.	10.39	88	10.48	96	10.48½	61	10.49	76	10.53	88	10.53½	93	10.54	67	10.55	70	10.55½	76	10.56½
114	8 B (w.)	134	11.05 a. m.	11.12	61	11.18	92	11.24	105	11.24½	64	11.25	74	11.27	98	11.27½	87	11.28	78	11.28½	61	11.30
135	8 B (w.)	134	9.47 a. m.	9.54	30	9.56	72	9.58	102	10.00	75	10.01	87	10.04	87	10.04½	84	10.06½	78	10.07	21	10.07½
110	9 (w.)	124	3.27 p. m.	3.33	46	3.37	86	3.42	103	3.42½	24	3.43	95	3.45½	100	3.46	101	3.47	94	3.49	103	3.49½
136	9 (w.)	124	3.41 p. m.	3.48	96	3.57	110	3.57½	43	3.58	77	3.59	103	4.00	111	4.00½	30	4.01	88	4.01½	101	4.03
111	10 (w.)	156	10.51 a. m.																			
127	10 (w.)	156																				
144	11 (w.)	160	10.50 a. m.																			
179	11 (w.)	160	11.28 a. m.	11.44	84	11.51	41	11.52	75	11.56	92	11.57	55	12.03	77	12.04						
	Utah:																					
118	1	130	10.41 a. m.																			
	Washington:																					
97	2	135	9.41 a. m.	9.50	24	9.53	36	9.56	83	10.00	91	10.04	96	10.04½	17	10.05	58	10.08	76	10.09	64	10.10
134	2	135	4.27 p. m.	4.34	78	4.43	30	4.43½	80	4.46	44	4.46½	78	4.48	41	4.48½	73	4.49	98	4.50	26	4.50½

a Plus 10 per cent pitch.

b Plus 5 per cent pitch.

Cupola tests of coke from coals received from January 1, 1906, to June 30, 1907—Continued.

Cupola test No.	Designation of coke.		Record of melt—Continued. Weight and time of each ladle of melted iron—Continued.																					
	Field No. of coal.	Coke test No.	10.		11.		12.		13.		14.		15.		16.		17.		18.		19.		20.	
			Lbs.	At—	Lbs.	At—	Lbs.	At—	Lbs.	At—	Lbs.	At—	Lbs.	At—	Lbs.	At—	Lbs.	At—	Lbs.	At—	Lbs.	At—	Lbs.	At—
1	2	8	66	67	68	69	70	71	72	73	74	75	76	77	78	79	80	81	82	83	84	85	86	87
	Alabama:																							
107	2 B (w.)	142	89	2.23	87	2.23½	95	2.24	82	2.25	99	2.25½	75	2.26	79	2.26½	81	2.27	89	2.28	69	2.29	85	2.30
131	2 B (w.)	142	46	11.05	71	11.07	60	11.07½	94	11.08	47	11.08½	90	11.11	62	11.11½	102	11.12	94	11.13	60	11.13½	86	11.14
132	3	138	64	3.23	2	3.23½	64	3.27																
101	3 (w.)	139	91	11.49	74	11.49½	74	11.50	85	11.53	86	11.53½	67	11.54	78	11.56	95	11.56½	71	11.57	73	11.59	97	11.59½
108	3 (w.)	139	91	4.04½	106	4.05	79	4.07	101	4.07½	102	4.08	58	4.08½	61	4.12	85	4.12½	95	4.13	81	4.16	100	4.17
103	4	131	99	8.52	94	8.56	97	8.56½	90	8.57	104	8.59	96	8.59½	88	9.00	95	9.03						
124	4	131	66	10.33½	90	10.34	64	10.34½	53	10.35	80	10.37	53	10.37½	99	10.38	66	10.38½	72	10.40	61	10.41		
109	4 (w.)	136	92	10.52½	113	10.53	104	10.55	105	10.55½	108	10.56	100	10.56½	97	10.57	118	10.57½	110	10.58	80	10.59	110	11.02
133	4 (w.)	136	92	11.19½	83	11.20	107	11.22	102	11.22½	80	11.23	106	11.23½	68	11.24	101	11.26	104	11.27	77	11.27½	84	11.28
	Arkansas:																							
96	1 B (w.)[a]	97	75	4.19½	61	4.20	108	4.22	115	4.22½	61	4.23	97	4.24	112	4.24½	58	4.25	77	4.28	98	4.28½	55	4.29
115	1 B (w.)[a]	97	111	4.09	108	4.10	100	4.10½	126	4.11	95	4.11½	94	4.12										
116	7 B (w.)[b]	105	61	11.25	24	11.26																		
142	7 B (w.)[b]	105	61	11.23	99	11.23½	83	11.24	64	11.25	104	11.25½	77	11.26	55	11.28	98	11.28½	72	11.29	58	11.33	63	11.33½
95	9 (w.)[a]	102	61	11.30																				
117	9 (w.)[a]	102	77	8.38	54	8.38½	55	8.39	64	8.39½	70	8.41½	48	8.42	45	8.42½	26	8.43						
	Illinois:																							
125	22 B (w.)	118	79	3.33	60	3.38	68	3.38½	105	3.39	29	3.39½	92	3.44	24	3.45	33	3.46						
150	29 (w.)	170	70	11.27	102	11.27½	103	11.28	66	11.30	99	11.30½	97	11.31	90	11.34	98	11.34½	100	11.35	97	11.40	94	11.40½
157	29 (w.)	170	80	3.41	101	3.43	101	3.43½	88	3.44	107	3.45	97	3.45½	98	3.46	106	3.48	102	3.48½	90	3.49	88	3.52
164	29 (w.)	70	82	11.39	60	11.43	83	11.43½	84	11.44	57	11.46	78	11.46½	136	11.47	53	11.48	40	11.49				
	Indiana:																							
121	12 (w.)	110	99	11.08	95	11.08½	76	11.09	106	11.10	100	11.10½	32	11.11	102	11.13	96	11.13½	49	11.14	24	11.15	27	11.16
148	17 (w.)	163	68	11.46½	96	11.47	80	11.49	73	11.49½	96	11.50	65	11.54	68	11.54½	85	11.55	56	11.59	108	12.00		
	Kansas:																							
122	6 (w.)	115	80	9.26	99	9.27	82	9.28	61	9.34	75	9.35	71	9.36	47	9.36½	54	9.37						
	Kentucky:																							
147	8	164	88	4.03	84	4.03½	94	4.04	168	4.04½	86	4.08	90	4.08½	84	4.09	82	4.11	44	4.12				
155	8	164	79	4.07½	108	4.08	84	4.11	75	4.11½	106	4.12	85	4.14	81	4.14½	105	4.15	76	4.17	76	4.17½	104	4.18
156	8	164	59	11.24	104	11.24½	92	11.27	77	11.27½	107	11.28	94	11.29	91	11.29½	93	11.30	89	11.32	102	11.32½	100	11.33
149	9 A (w.)	167	100	3.31½	115	3.32	79	3.33	98	3.33½	106	3.34	83	3.35	96	3.35½	110	3.36	83	3.37	104	3.37½	102	3.39
162	9 A (w.)	167	82	11.11	83	11.11½	93	11.12	88	11.14	84	11.14½	100	11.15	80	11.18	86	11.18½	96	11.19	126	11.23	69	11.23½
163	9 A (w.)	167	91	4.07½	126	4.08	100	4.11	100	4.11½	136	4.12	86	4.17	95	4.17½	130	4.18	90	4.24	148	4.24½	98	4.25
	Missouri:																							
123	5 (w.)	116	51	3.42	63	3.43	60	3.43½	79	3.44	91	3.45	62	3.45½	58	3.46	70	3.46	94	3.47	59	3.48	45	3.48½

New Mexico:
  4 B (w.)
  5 (w.)
  3.
  4 B (w.)
  5.
Pennsylvania:
  11.
  11.
  12.
  12. (w.)
  12.
  12 (w.)
  12 (w.)
  21.
Tennessee:
  1. (w.)
  1 (w.)
  1.
  2.
  3.
  4.
  4.
  5 (w.)
  5 (w.)
  6.
  7 B (w.)
  7 B (w.)
  8 B (w.)
  8 B (w.)
  9 (w.)
  9 (w.)
  10 (w.)
  11 (w.)
Utah:
  1.
Washington:
  1.
  2.

a Plus 10 per cent pitch.

b Plus 5 per cent pitch.

Cupola tests of coke from coals received from January 1, 1906, to June 30, 1907—Continued.

Cupola test No.	Field No. of coal	Coke test No.	21 Lbs.	21 At—	22 Lbs.	22 At—	23 Lbs.	23 At—	24 Lbs.	24 At—	25 Lbs.	25 At—	26 Lbs.	26 At—	Melting time (minutes)	Remarks
1	2	3	88	89	90	91	92	93	94	95	96	97	98	99	100	101
	Alabama:															
107	2 B (w.)	142	102	2.30½	79	2.31	69	2.31½	83	2.33	103	2.33½	52	2.34	31	Iron very hot and fluid; 27th ladle—97 pounds at 2.35.
131	2 B (w.)	142	20	11.14½	86	11.15	53	11.15½	86	11.16	82	11.17	131	11.18	29	
132	3 (w.)	138	65	12.00	49	12.02									26	Iron hot.
101	3 (w.)	139													35	Iron hot; slag filled up tuyeres after 20th ladle and bottom had to be dropped.
108	3 (w.)	139													33	Temperature of iron medium.
103	4	131													33	Iron sluggish.
131	4 (w.)	131	96	11.03	80	11.03½	109	11.04							27	Iron very hot and fluid.
136	4 (w.)	136	52	11.29	103	11.32	40	11.33							27	Iron hot.
133	4 (w.)	136													32	
	Arkansas:															
97	1 B (w.)[a]	97	80	4.31	41	4.34									36	Iron hot. Bed burned out and charges hung; bottom had to be dropped after 15th ladle.
115	1 B (w.)[a]	97													25	Iron hot. Charges hung and bottom dropped after 11th ladle.
106	7 B (w.)[b]	106	58	11.34	40	11.36									33	Temperature of iron medium.
105	7 B (w.)[b]	105													35	Iron cold and dull.
95	9 (w.)[a]	105													28	Iron hot and fluid.
102	9 (w.)[a]	102													25	
	Illinois:															
125	22 B (w.)	118	88	11.41	92	11.43	67	11.44	20	11.45					34	Iron hot.
150	20 (w.)	170	109	3.52½	85	3.53	83	3.55	93	3.56					34	Temperature of iron medium.
157	20 (w.)	157													31	Iron hot.
164	20 (w.)	164													30	Iron cold.
	Indiana:															
121	12 (w.)	110	94	11.18	126	11.19									25	Iron dull.
148	17 (w.)	148													33	Temperature of iron medium.
	Kansas:															
122	6 (w.)	116													33	Temperature of iron medium.
	Kentucky:															
147	8	144	66	4.22	64	4.23½	85	4.24	65	4.25	55	4.25½	25	4.26	31	Temperature of iron medium.
155	8	144	85	11.36	100	11.36½	103	11.37	75	11.39	85	3.45			38	Do.
156	8	144	75	3.39½	97	3.40	97	3.41	67	3.43					32	Do.
149	9 A (w.)	167	70	11.24	66	11.25	130	11.26	69	11.27					28	Iron hot.
163	9 A (w.)	167	88	4.26	98	4.27									32	Temperature of iron medium.
163	9 A (w.)	167													42	Do.
	Missouri:															
123	5 (w.)	116	90	3.49	70	3.50	56	3.51	90	3.52	45	3.52½	69	3.53	34	Temperature of iron medium; 27th ladle—92 pounds at 3.54; 28th ladle—52 pounds at 3.54; 29th ladle—39 pounds at 3.55.

		70	9.29½	101	9.31	82	9.32	65	9.32½	69	9.33	142	9.34		Remarks
New Mexico:															
119	4 B (w.)													31	Iron hot and fluid.
120	5 (w.)													29	Iron very hot and fluid. Large quantities of slag closed up tuyeres after 16th ladle.
98	3. (w.)														
130	4 B (w.)	84				57	3.55					9.34		28	Iron very hot and fluid.
	5.														
Pennsylvania:															
143	11.													34	Iron hot.
161	11.	83	2.33½	99	2.34	37	2.36							28	Iron cold.
146	12.													41	Blast off 19 minutes; melting too fast to handle.
151	12.	109	4.10	71	4.10½	106	4.11	109	4.11½	66	4.15			24	Iron hot.
152	12 (w.)													35	Iron hot.
153	12 (w.)	98	3.40	102	3.49½	103	3.41	89	3.43					36	Do.
154	12 (w.)	70	11.35½	88	11.36	17	11.37							31	Temperature of iron medium.
190	21.	100	11.88	23	11.38½	87	11.39	17	11.39					25	Temperature of iron medium. Blast off 8 minutes.
Tennessee:															
126	1 (w.)	111	9.26	63	9.27	30	9.27½	100	9.28					33	Iron very hot.
99	1 (w.)													34	Blast off 1 minute.
128	2.	82	1.55	40	1.55½	40	1.56	60	1.59		2.00			33	Iron very hot and fluid.
100	2.	71	3.22	77	3.23	101	3.23½	57	3.24	58	3.25½	43	3.27	34	Temperature of iron medium.
104	3.	33	9.44	74	9.44½	24	9.47	29	9.48	63	9.48½	27	9.49	35	Iron hot and fluid; 27th ladle—62 pounds at 9.51.
139	3.													32	Iron hot.
102	4.													32	Do.
105	4.	127	3.03	74	3.03	27	4.35	40	4.40					31	Iron very hot and fluid.
138	4. (w.)	25	4.32½	105	4.35	124	4.35½	50	4.40					40	Iron hot.
106	5 (w.)	74	9.46	75	9.46½		9.47		9.51					26	Iron very hot and fluid.
112	6.	90	10.23	75	10.24									34	
141	6.	83	4.07½	90	4.08	106	4.08½	84	4.09	96	4.09½	79	4.11	30	Blast on 30 minutes; no iron melted; coke very high in ash and dirty.
113	7 B (w.)													30	Iron cold.
137	7 B (w.)	92	11.38½	74	11.39	83	11.39½	104	11.40	57	11.46½	67	11.47	34	Blast off 3 minutes. Iron very hot and fld.
114	8 B (w.)													36	Iron cold.
135	8 B (w.)													27	Iron very hot; 27th last off 1 minute; tap hole cut out and iron run over teeth. Iron hot.
110	9 (w.)	91	4.06	98	4.07	57	4.13½	65	4.14					34	Iron very hot and fld.
136	9 (w.)	92	4.11½	108	4.13									26	
111	10 (w.)														Blast on 30 minutes; no iron melted.
127	10 (w.)														fire bed put in and blast put on for 20 minutes; coke would not burn; bottom dropped without doing fan.
144	11 (w.)													20	Blast on 32 minutes; no iron melted.
179	11 (w.)														Iron cold and sluggish.
U'tah:															
118	1.	88	5.04												Blast on 20 minutes; no iron melted; bed burned out.
Washington:															
197	2.													29	Iron cold.
134	2.													30	

a P us 0 per cent pitch.   b Plus 5 per cent pitch.

Chemical effect on iron in cupola tests of coke from coals received from January 1, 1905, to June 30, 1907.

Cupola test No.	Field No. of coal.	Coke test No.	Silicon.			Manganese.			Sulphur.			Total in coke taken up by iron.	Analysis of coke on page—
			In pig iron.	In melted iron.	Lost by oxidation.	In pig iron.	In melted iron.	Lost by oxidation.	In pig iron.	In melted iron.	Increase.		
1	2	3	102	103	104	105	106	107	108	109	110	111	112
	Illinois:												
150	29 (w.)	170	2.12	1.91	9.90	0.178	0.155	12.93	0.059	0.086	0.027	6.59	29
157	29 (w.)	170	2.12	1.84	13.21	.178	.133	25.29	.059	.108	.049	11.69	29
164	29 (w.)	170	2.10	1.68	20.00	.163	.111	31.90	.098	.133	.035	6.74	29
	Indiana:												
187	5	8	2.10	1.66	20.96	.163	.115	29.44	.098	.165	.067	6.13	29
173	9	17	2.10	1.58	28.16	.163	.112	31.30	.098	.148	.050	4.51	29
174	9	17	2.10	1.71	18.58	.163	.124	23.93	.098	.156	.058	7.22	29
148	17 (w.)	163	2.12	1.75	17.47	.178	.126	29.20	.059	.108	.049	6.88	29
	Kentucky:												
47	7	85	1.74	1.39	20.11	.178	.133	25.28	.051	.085	.034	7.56	30
156	8	164	2.12	1.82	14.15	.178	.111	37.64	.050	.083	.024	19.95	30
147	8	164	2.12	1.80	15.12	.178	.096	46.06	.059	.067	.008	4.65	30
155	8	164	2.12	1.72	18.85	.178	.123	30.90	.059	.079	.020	12.37	30
162	9 A (w.)	167	2.10	1.76	16.19	.163	.120	26.39	.098	.118	.020	7.73	30
149	9 A (w.)	167	2.12	1.83	13.68	.178	.133	25.29	.059	.079	.020	6.69	30
163	9 A (w.)	167	2.10	1.73	17.61	.163	.096	41.09	.098	.135	.037	8.29	30
	Ohio:												
165	9	72	2.10	1.78	15.24	.163	.110	34.33	.098	.143	.045	8.16	31
166	9	72	2.10	1.74	19.06	.163	.111	31.89	.098	.151	.053	9.17	31
	Pennsylvania:												
143	11	159	2.12	1.85	12.74	.178	.111	37.66	.059	.070	.011	4.83	32
161	11	159	2.10	1.74	17.15	.163	.113	30.68	.098	.113	.015	7.57	32
152	12	161	2.12	1.86	12.28	.178	.130	26.97	.059	.070	.011	4.92	32
145	12	161	2.12	1.89	10.85	.178	.133	25.28	.050	.069	.010	4.35	32
151	12	161	2.12	1.91	9.90	.178	.123	30.90	.050	.078	.019	5.94	32
154	12 (w.)	162	2.12	1.81	14.62	.178	.136	23.50	.059	.074	.016	11.13	32
146	12 (w.)	162	2.12	1.84	13.21	.178	.141	20.78	.059	.080	.021	11.20	32
153	12 (w.)	162	2.12	1.78	16.05	.178	.128	28.09	.059	.088	.029	15.27	32
	Virginia:												
158	2	69	2.12	1.77	16.52	.178	.144	19.09	.059	.074	.015	17.10	34
171	2 (w.)	70	2.10	1.69	19.52	.163	.107	34.35	.098	.116	.018	25.01	34
172	2 (w.)	70	2.10	1.76	17.61	.163	.124	23.92	.098	.111	.013	12.49	34
52	2 B	69	1.89	1.55	17.77	.163	.133	18.42	.048	.042		None.	34
	West Virginia:												
159	4 (w.)	44	2.12	1.94	8.50	.178	.126	29.22	.059	.077	.018	9.41	34
53	4 B (w.)	44	1.89	1.54	18.53	.163	.126	22.70	.048	.042		None.	34
160	4 (w.)	44	2.12	1.90	10.38	.178	.126	29.21	.050	.085	.026	19.99	34
167	15	36	2.10	1.65	21.42	.163	.111	35.54	.098	.137	.039	9.92	34
51	15	36	1.89	1.45	23.27	.163	.104	36.18	.048	.047		None.	34
168	15	36	2.10	1.71	18.57	.163	.104	36.19	.098	.146	.048	12.79	34
49	15	43	1.74	1.35	22.41	.178	.111	37.63	.051	.060	.009	2.35	34
169	16	45	2.10	1.68	20.00	.163	.106	34.96	.098	.138	.040	16.33	34
170	16	45	2.10	1.67	20.47	.163	.111	31.90	.098	.126	.028	12.94	34
50	16 B	45	1.74	1.26	27.57	.178	.111	37.64	.051	.052	.001	.26	35

O

Bulletin No. **337**

DEPARTMENT OF THE INTERIOR

UNITED STATES GEOLOGICAL SURVEY

GEORGE OTIS SMITH, DIRECTOR

THE

FAIRBANKS AND RAMPART QUADRANGLES

YUKON-TANANA REGION, ALASKA

BY

L. M. PRINDLE

WITH A SECTION ON THE RAMPART PLACERS

BY

F. L. HESS

AND A PAPER ON THE WATER SUPPLY OF THE FAIRBANKS REGION

BY

C. C. COVERT

WASHINGTON

GOVERNMENT PRINTING OFFICE

1908

CONTENTS.

ILLUSTRATIONS.

5

By Alfred H. Brooks.

In planning the surveys and investigations of Alaska, the attempt was made to cover first those regions which were of the greatest economic importance. As a result of this, many of the mapped areas are very irregular in outline, and it now seems desirable to introduce greater uniformity into the published maps as rapidly as the data available for their preparation will permit. With this end in view a system of maps has been projected covering quadrangular areas outlined by parallels of latitude and meridians of longitude, this being in conformity with practice in surveys made within the United States proper. But as the Alaska surveys are for the most part of a reconnaissance character and the region is very thinly populated, it has seemed best to adopt a map unit larger than that used in the States. This unit will include 4 degrees of longitude and 2 degrees of latitude, making a map about as large as can be conveniently handled. It is hoped that eventually these published reconnaissance topographic maps can be accompanied by sheets showing the geology and the economic resources, but in view of the great demand for the topographic maps it has been deemed advisable to publish them immediately with such accounts of the geology and mineral resources as may be available. Nor is it deemed desirable to delay the issuing of maps until the areas have been completely covered.

The following report, with its accompanying maps (Pls. I, IV, and V), is the second of this series to be issued, and it will be followed by others as fast as the accumulation of the field notes will permit. The topographic surveys on which the maps are based were made under the direction of T. G. Gerdine in 1903 and D. C. Witherspoon in 1904, 1905, and 1906; the geology is by L. M. Prindle, who has worked in this general field in 1903, 1904, 1905, and 1906. Mr. Prindle was assisted in his geologic work in 1904 by Frank L. Hess and in 1905 by Adolph Knopf. The present writer, who crossed this

7

region in 1902, has added some features to the geologic map, and also material regarding developments in the Fairbanks and Rampart regions during 1907. Mr. Prindle has presented the salient features of the geology, so far as known, in simple language devoid of technicalities. A more elaborate discussion of the scientific results is in preparation. A section by F. L. Hess on the placers of the Rampart region, which was originally printed in Bulletin No. 280, "The Rampart gold placer region, Alaska," is here reprinted, in order that all the placers in the region indicated by the maps may be covered in the written description.

As the publication goes to press, it has been possible to add to it the preliminary statement of C. C. Covert on the water resources of the Fairbanks region, based on surveys made in 1907.

THE FAIRBANKS AND RAMPART QUADRANGLES, YUKON-TANANA REGION, ALASKA.

By L. M. PRINDLE.

INTRODUCTION.

The Yukon-Tanana region is the area extending westward from the International boundary between Yukon and Tanana rivers to their confluence. The greatest east-west dimension of this area is about 300 miles, the greatest north-south dimension about 175 miles. The gold placers of Fortymile, Birch Creek, Rampart, and Fairbanks are situated in this region, and their economic importance has led to several years' work by the Geological Survey in making topographic and geologic maps and in studying conditions in the gold-producing regions.

A topographic map of the Fortymile quadrangle was made by E. C. Barnard in 1898 and was included in a preliminary report on the gold-producing regions.[a]

In 1903 a comprehensive scheme for mapping the entire remaining area was planned by A. H. Brooks, geologist in charge. In view of the facts that the Alaska surveys are for the most part of a reconnaissance character and that the region is thinly populated, the unit of publication adopted (see fig. 1) is a quadrangular area embracing 4 degrees of longitude and 2 degrees of latitude. The first of these maps, that of the Circle quadrangle, which was made by parties under the direction of T. G. Gerdine and D. C. Witherspoon, was published in 1906 in a bulletin of the Survey.[b] The maps of the Fairbanks and Rampart quadrangles (Pls. IV and V, in pocket), with topography by D. C. Witherspoon and R. B. Oliver, are published with this report. The area mapped on the Fairbanks sheet extends westward from the western edge of the Circle quadrangle to the 150th meridian, and the mapping has been continued northward to the

---

[a] Prindle, L. M., Gold placers of the Fortymile, Birch Creek, and Fairbanks regions, Alaska: Bull. U. S. Geol. Survey No. 251, 1905.
[b] Prindle, L. M., The Yukon-Tanana region, Alaska; description of Circle quadrangle: Bull U. S. Geol. Survey No. 295, 1906.

edge of the Yukon Flats. The mapping of the areas marked with broken contour lines immediately north of Tanana River, in the southeastern part of the quadrangle, is to be completed during 1907. The town of Fairbanks is the main supply point for the creeks of the Fairbanks placers, and its name has been given to the quadrangle. In the Rampart quadrangle the mapping of the area between the western boundary of the Fairbanks quadrangle and the confluence of

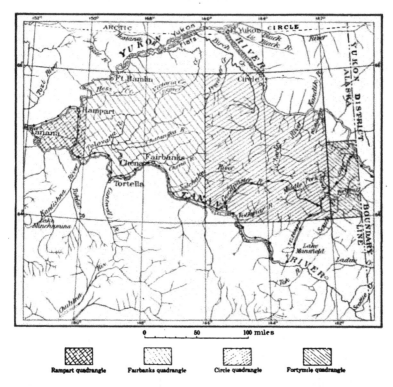

Fig. 1.—Index map showing location of quadrangles in the Yukon-Tanana region.

the two rivers has been completed. This includes all the areas of economic interest in the Rampart quadrangle and finishes the western portion of the interstream area. In the preparation of the report which has been written to accompany these maps free use has been made of the other publications of the Survey bearing upon the areas under consideration. The material for the account of the Fairbanks region has been taken largely from the progress reports[a] for 1905

[a] Brooks, Alfred H., and others, Report on progress of investigations of mineral resources of Alaska in 1905: Bull. U. S. Geol. Survey No. 284, 1906, pp. 109-123, and Bull. No. 314, 1907, pp. 36-37.

and for 1906. Additional material obtained by Brooks and Covert in 1907 has been added. The description of the Rampart region is taken largely from a Survey bulletin,[a] and as that report is now out of print the account of the placers, by Frank L. Hess, is here reprinted. This account is supplemented by data obtained by Mr. Brooks in the Rampart region.

GEOGRAPHY.

The area under consideration in the two quadrangles, about 14,000 square miles, is bounded by the 146th meridian and Yukon and Tanana rivers and has about the shape of an equilateral triangle, whose eastern side is the 146th meridian and whose western angle is at the confluence of the two rivers. The area is made up of ridges and valleys bordered on the north by a part of the Yukon Flats, deeply embayed on the south by the Tolovana Flats, and terminated on the west by the low area between the two rivers.

The dominant type of country is one of ridges more or less uniform in height, with altitudes of 2,000 to 3,000 feet, separated by valleys of equally uniform depth, but certain areas, like the White Mountains and, 50 miles farther west, the group of hills near Rampart, attain altitudes of 4,000 to 5,000 feet or more, and these areas are accentuated on the maps by the greater number and more closely crowded character of the contour lines that represent their slopes. The country between these two main features comprises many ridges, which are in general higher toward the northern part of the area and break off abruptly to the Yukon Flats or Yukon River. The southern limit of the hill country is less abrupt, and the ridges merge more gradually into the valleys of the Tanana and its tributaries. The valleys exhibit much variety; some of them are deep, steep walled, and narrow, with but little floor; others have well-developed floors and gentler slopes, and others become extensive flats. Some of them are complexly interwoven and follow most tortuous courses before leaving the hills and joining the main lines of drainage. The important tributaries of the Yukon are Beaver, Hess, and Minook creeks; those tributary to the Tanana are the Salcha, Chena, and Tolovana rivers and Baker Creek.

The Yukon Flats extend far northward from the base of the ridges to the plateau country fronting the Rocky Mountains. The sparsely timbered surface is somewhat uneven and broken where minor ridges run out into it from the base of the hills. It is dotted with a few small lakes, and the small streams that furrow the northern slopes of the hills cross it in rather ill-defined valleys. Distant shimmering

---

[a] Prindle, L. M., and Hess, F. L., The Rampart gold placer region, Alaska: Bull. U. S. Geol. Survey No. 280, 1906.

lines and crescent-shaped areas of water indicate the many inter-
lacing channels of the Yukon which are spread widely over this great
flat. The lowlands of the southern border, the Tolovana and Baker
flats, are embayments from a similar great flat that forms a large part
of Tanana Valley.

The main features of the surface can be learned from the topographic maps and need no further comment in the text. The relief,
the drainage, the relations of the various drainage systems to each
other, the grades of the streams, the comparative elevations and
grades of different valleys are all brought out on the maps, and many
of their economic relations are thus made clear to those who read the
maps with care. If, for example, the digging of a ditch or the construction of roads or railways is under consideration, the most feasible routes can be approximately determined by a study of the maps.

The explanation of the uniformities in altitudes is to be found in
the history of the region. The ridge level is a remnant of a former
continuous more or less uniform surface lying at a lower altitude
near sea level. Subsequent elevation has resulted in the cutting of
the present valleys. The explanation of local differences of relief,
like that of Pedro Dome with its surroundings or the White Mountains in the Fairbanks region, or Wolverine Mountain in the Rampart
region, is to be found largely in differences in the character of the
bed rock that are accentuated under the conditions of the present
downcutting. The local granitic intrusives in Pedro Dome and
Wolverine Mountain have withstood erosion and held up these
areas, and the limestone of the White Mountains performs the same
function.

Stream adjustments are also clearly shown on the map. An example may be taken from the Beaver Creek system. One of the
small tributaries draining the southern part of the White Mountains
has been called, for purposes of description, Fossil Creek, from the
occurrence of fossiliferous limestone pebbles in the gravels. East of
the single narrow limestone ridge that terminates the White Mountains is a parallel ridge of shorter extent. Between the two lies a
low-grade valley, 2 miles wide, extending northeastward for nearly
10 miles and drained by a stream that flows along its western side
close to the base of the limestone ridge. From the base of the ridge
on the east, broad, flat spurs, separated by open depressions so shallow
as to be hardly noticeable, extend westward to the stream. On close
examination the apparently continuous valley is seen to be composed
of two parts—a lower, drained by a stream which seems dispropor-
tionately small for the size of the valley and which has its rise in a
few small stagnant ponds strung longitudinally along its course
about 5 miles from Beaver Creek, and an upper, drained by a stream
of about the same length, which flows at first southwestward, in line

with the stream of the lower portion of the valley, and then turns abruptly westward to Beaver Creek through a narrow canyon that interrupts inconspicuously the continuity of the limestone ridge. This upper part of the valley has been reduced somewhat below the level of the lower part, but not to an extent appreciable in a general view. It is an example of stream diversion, a minor tributary of the Beaver having cut through the limestone ridge and diverted to itself the waters of the upper portion of a valley that formerly drained southwestward along the entire eastern base of the limestone ridge. The diversion of the drainage from the upper valley has weakened the stream that occupies the lower portion, and its forceless character is indicated by the string of small ponds along its present headwaters.

The streams, therefore, are by no means permanent, independent units of drainage, but are most delicately adjusted and may in their development encroach upon one another's drainage areas.

CLIMATE AND VEGETATION.

Owing to the high latitude of the area, 65° to 66° north, there are great differences in the characteristics of winter and summer. The annual range of temperature is great. At Tanana, for the period from August, 1901, to December, 1902, the temperature varied from 76° F. below zero in January to 79° F. above zero in August. The intense cold of winter is not accompanied by excessive snowfall, but the water circulation is reduced to a minimum and the long continuance of such conditions has resulted in freezing to great depths a large part of the superficial deposits. But even during the winter, when the larger streams are covered with ice up to 6 feet thick, a considerable amount of water is in circulation, and it frequently breaks through the ice, causing overflows. Many small streams thus form thick accumulations of ice that may remain throughout much of the summer. The ice begins to go out of the Yukon at Tanana at dates varying in different years from about May 10 to May 15, and a few days later the river is clear. Mush ice begins to run at dates varying from October 15 to 25, and in from one to two weeks later the river is generally closed to navigation. The Tanana at the mouth of Baker Creek froze October 20, 1905, and the ice went out May 6, 1906. The summers are characterized essentially by conditions like those in temperate latitudes, except that the sun is above the horizon a much larger portion of a summer day than in more temperate regions and that the season itself is short. The summer advances rapidly from the time of the break-up of the ice, the days become hot to a degree comparable with those of regions much farther south, and generally no killing frosts occur till about the first of September. The mean annual precipitation in the interior of Alaska, including both

melted snow and rain, is not great. Observations taken at Tanana
at intervals from 1882 to 1886 gave an annual average of 15.45 inches.
The rainfall varies greatly in the different seasons; sometimes long
continued drought lowers the quantity of water in the streams below
the economic limit, while in other seasons water is almost constantly
in excess of the amount required. A detailed statement of the water
supply and rainfall, by C. C. Covert, is given on pages 51–59 of this
report.

The distribution of vegetation is shown in fig. 2. The high ridges
are mostly bare of trees and are covered only with the low vegetation

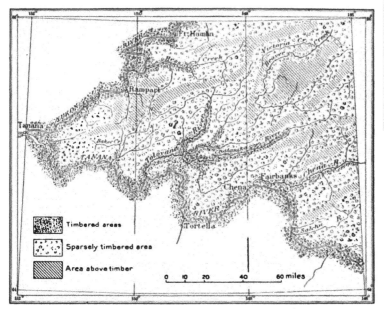

FIG. 2.—Map showing distribution of timber in the Fairbanks and Rampart quadrangles.

of the tundra. The lower ridges are largely covered with a dense
growth of small spruce, accompanied by birch and poplar and in the
larger valleys by a small proportion of tamarack. Willows and scrub
alders grow abundantly near the streams. Good timber is confined to
the larger valleys, where it is most thickly concentrated in scattering
patches along the streams. Spruce is the most important tree and attains in many places a diameter of 2 or more feet.

The valley sides and lower ridges of the Fairbanks region are covered with a light growth of spruce and poplar, and fine patches of
birch are common on the hillsides facing the Tanana. Good timber
is reported from the lower valleys of the Chatanika and Tolovana.

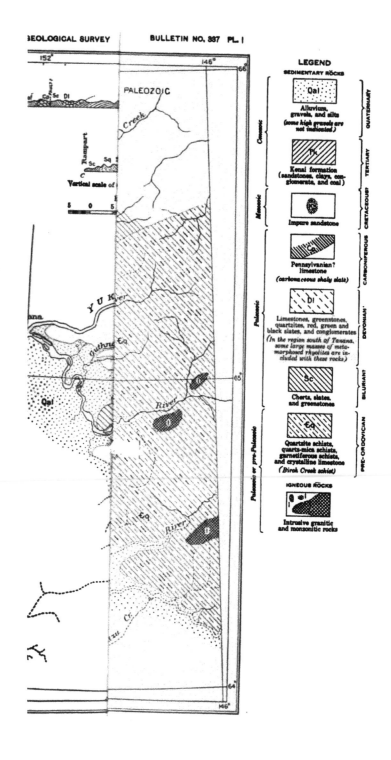

The low ridges forming the eastern side of Beaver Valley east of the White Mountains present in places the appearance of a well-timbered country, but on closer view are found to be covered with only a thick growth of small spruce. The upper valley of Hess Creek and the valleys of its tributaries contain also abundant small spruce, with some larger timber near the streams. The valley of Minook Creek has contained considerable timber of sufficient size for mining purposes, but only a small amount of such timber is left. The valleys of the southern slope facing the Baker flats contain but little timber of large size, but some good timber is reported to be present in the valleys of Hutlina and Baker creeks. On the hillsides near the Baker Hot Springs is a luxurious growth of birch and poplar. Timber for fuel purposes is abundant in most of the valleys and lower slopes throughout the region.

Feed for horses can be found on the sunward-facing slopes of most of the main valleys and there are large grass-covered areas in the Tolovana flats. Blueberries are almost everywhere to be found in abundance on the open hillsides and valleys, and locally currants, cranberries, and red raspberries are sufficiently abundant to be of practical importance. Most of the hardy vegetables can be grown successfully, and at Hot Springs they are produced in large quantities and sold to the miners of the Fairbanks and Rampart regions.

DESCRIPTIVE GEOLOGY.

GENERAL STATEMENT.

The area covered by the maps is composed essentially of two rock groups, for the most part widely different in characteristics. (See Pl. I.) The members of the one are highly schistose and are predominantly quartzite schists and quartz-mica schists, with some crystalline limestone and greenstone. The members of the other are predominantly carbonaceous slate and coarser fragmentals, chert, greenstone, and limestone. The older group is the schists, which are characteristically developed in the Fairbanks, Birch Creek, and Fortymile regions and are the oldest rocks known in the Yukon-Tanana region. Their age has not been definitely determined, but they are probably older Paleozoic or pre-Paleozoic. The younger group is characteristically developed in the Rampart region and in areas farther east, including the White Mountains. Rocks provisionally correlated with them occur also south of the Tanana. Their age is Paleozoic, for fossils determined as possibly Silurian, as Devonian, and as Carboniferous have been found in them. The main line of demarcation between the two groups extends northeast and southwest in the area between the White Mountains and the ridge limiting on the northwest the valley of the Chatanika. A small area of the older rocks occurs also in the Rampart region. Although the two groups have not been found in unconformable contact there was probably a long time interval between the deposition of the two, and the difference in the degree of their metamorphism is sufficiently great to constitute a metamorphic unconformity of stratigraphic value. The general strike is northeast and southwest, and the rocks wherever observed are closely folded. There are small areas of later rocks, probably not older than the Cretaceous; slightly consolidated sediments of Tertiary age; and unconsolidated Pleistocene and Recent alluvial deposits. Granitic and more basic intrusives are common in both of the older groups, and greenstones are especially abundant in them. Fresh volcanic rocks occur to a slight extent in the vicinity of Rampart and metamorphosed rhyolites are present in the area south of the Tanana.

The following tabular statement is based only on reconnaissance trips and is therefore of a preliminary nature. The variety of rocks, the complexity of structure, and the paucity of paleontologic material in the areas studied render possible only the barest outlines of the stratigraphy. The areal distribution of the rocks is shown on the geologic map (Pl. I). The predominant lines of structure are well exhibited by the areal development of the rocks, and the formation occurring in the White Mountains finds its continuation most probably in the rocks outcropping along the Tanana between the mouth of the Kantishna and that of Baker Creek.

Provisional tabular statement of stratigraphy of Fairbanks and Rampart quadrangles.

	System	Series	Period or formation.	Lithologic character.
Cenozoic	Quaternary.......	{Recent ... ...{Pleistocene ...	....	Stream gravel. Bench gravel.
	Tertiary.........	Eocene	Kenai	Slightly consolidated conglomerate, sandstone, consolidated clay.
Mesozoic	Cretaceous			Black carbonaceous stone and slate.
Paleozoic......	{Carboniferous....	Permian or Pennsylvanian.	....	Gray, greenish, and shale with thin silic beds.
	Devonian and possibly older rocks.	....		Gray and blue, partly sili limestone, greenst quartzite, red, black green slate, conglom and finer fragmenta interbedded limestor
	{Silurian (")....	....	....	Chert, slate, greenstor limestone.
Metamorphics (Paleozoic or pre-Paleozo-	....	....	Birch Creek schist.	Quartzite schist, qua and garnetiferous greenstone schist, s talline limestone.

STRATIGRAPHY.

METAMORPHICS (PALEOZOIC OR PRE-PALEOZOIC).

The rocks characterized as metamorphics were, for the most part, originally sediments, including sandstone, shale, and a small proportion of limestone. This material, through various processes included under the term metamorphism, has all been more or less changed until it now exhibits characteristics of structure and composition that differ greatly from those of the original rocks. Of the new minerals formed mica is the most generally distributed, and its extensive development in nearly all the rocks under consideration has imparted a schistose character to them. They split readily along the planes parallel to the cleavage of the micas, and are easily recognized as schists by the miners who are working in the areas where this older group forms the bed rock. These rocks in places have been but partly metamorphosed and retain many characteristics of their sedimentary origin. Some of the quartzitic rocks, for example, contain but a small proportion of mica and are but slightly schistose. In a more advanced stage of metamorphism, however, there is a more or less complete rearrangement of material by solution and recrystallization, until the rock that owed its characteristics originally to sedimentation becomes one with most of its characteristics due to entirely different processes. In the same way igneous rock, originally formed by consolidation of molten material, may lose the structure, texture, and mineral composition characteristic of its mode of origin and take on those of metamorphism.

In a region of metamorphic rocks, therefore, rocks of both sedimentary and igneous origin may be present as schists. In the eastern part of the Yukon-Tanana region, metamorphosed igneous intrusives form an essential part of the metamorphic assemblage. In the western part of the area, however, including the Birch Creek and Fairbanks regions, the schists are predominantly of sedimentary origin and include quartz schists, quartz-mica schists, feldspathic, carbonaceous, garnetiferous, and hornblendic schists, and crystalline limestone.

The structure of the rocks is complex. The trend or strike, where the bedding is inclined at a considerable angle or is vertical, is in most places northeast and southwest. The rocks are so closely folded in some places that the sides of the folds are approximately parallel, and the planes of their axes are commonly so nearly horizontal as to make the whole assemblage appear level or horizontal. Quartz veins are common throughout the schist areas. They are generally small, occurring as stringers from a fraction of an inch up to a few feet in thickness, but are so numerous as to form probably a considerable proportion of the entire mass.

Although the proportion of metamorphosed igneous material is small, fresh granitic intrusives are rather common and occur as irregular masses up to a few miles in diameter, fringed with dikes that complicate the structure of the schists. The elaborate system of repeated intrusion, so extensively developed in the Fortymile region, with its far-reaching metamorphic effects, has not extended to the area under consideration.

The western limit of the metamorphics is in the ridge northwest of the valley of the Chatanika. This ridge is formed of quartzitic schists, with some carbonaceous schists and crystalline limestone. A few miles nearer Beaver Creek, at the ends of spurs extending laterally from this ridge toward Beaver Creek, there are outcrops of less-metamorphosed feldspathic quartzite that is included in the later group. A small area of garnetiferous schist with associated crystalline limestone occurs in the Rampart region on Minook Creek and is included in the Birch Creek schist.

PALEOZOIC ROCKS.

SILURIAN AND DEVONIAN.

Between the two areas of schists above described, one the westward extension of the schists of the Fairbanks region, the other a small area whose limits are not known, lie rocks most of which are regarded provisionally as Silurian and Devonian. The bulk of the formation, as already mentioned, is made up of cherts, slates with alternating beds of conglomerate, finer fragmentals, limestones, and greenstones. These rocks may be roughly divided into two groups according as cherts, slates, and greenstones, or conglomeratic rocks with massive limestones, greenstones, slates, and quartzites predominate. It is believed that those forming the first division are older. The term "Rampart series" has been used by Spurr[a] in grouping similar rocks occurring in other parts of the Yukon-Tanana country. In those areas, however, diabase, tuffs, and green slates are most abundantly developed, and carbonaceous slates and limestones, though frequently present, are of minor importance. In the areas under consideration, while greenstones are found throughout the assemblage, black slates, cherts, and massive dark and light-gray limestones are more common and more strikingly characteristic. A section of these rocks is crossed in traveling from Chatanika River northward to the southern edge of Yukon Flats, a distance of about 60 miles, and this section, though incompletely studied, has afforded some information in regard to the structure and age of the rocks composing it.

[a] Spurr, J. S., Eighteenth Ann. Rept. U. S. Geol. Survey, pt. 3, 1897, pp. 155–169.

The general strike in the southern part of the section is about N. 50° E., and it is instructive to note that the direction of structure so strongly emphasized topographically in the Alaska Range is here repeated. In the northern part of the section the strike is more nearly east and west. The rocks are closely folded and in most cases their attitude is nearly vertical. There is a distinct symmetrical arrangement with reference to a northeast-southwest axis, and, so far as our present knowledge extends, this symmetrical disposition of the rocks seems to be a fact of importance in regard to their structure and succession. This repetition is most noticeable in the occurrence of limestone. The limestones of the White Mountains, with associated greenstones, flanked on the northwest by red and black slates and prominent masses of impure quarzite, are repeated 15 miles to the north in another limestone belt, less conspicuous topographically than that of the White Mountains, with a similar association of rocks, and flanked on the northwest by similar slates and quartzites. The rocks of the middle portion of the section are slates, greenstones, and cherts, and, although the relations are not clear, there are two main areas of cherts about 3 miles apart, with black, purple, and greenish slates and some greenstones in the intervening space. North of the northern belt of cherts, toward the northern belt of limestones, there are conglomerates containing abundant chert pebbles. Farther south, toward the limestones of the White Mountains, there are also chert conglomerates with pebbles an inch or more in diameter, associated with finer rocks containing grains of chert and fragments of slate. Black slates are also common and thin beds of conglomerate are interbedded with them. The slates are closely succeeded by the limestones on either side, and although the direct relation of the two kinds of rock was not observed, it is believed that the limestones are younger than the rocks above described. The quartzite flanking the limestone on the north, which is very similar to other quartzites apparently interbedded with the limestones, contains occasional fragments of chert, and it seems best until further knowledge is available to consider the flanking slates and quartzites as partly of the same age as the limestones and partly younger.

Fossils found in the limestones of the White Mountains have been determined as characteristic of early to middle Devonian, and some of them are possibly of Silurian age; but the stratigraphic association makes it probable that all the remains belong approximately to the same horizon, and the determinations point more definitely to the Devonian. No fossils were found in the limestones of the northern belt, but the rocks and their associations are very similar, and no reasons were found for assigning them to a different position.

The contact relations of these rocks with the Carboniferous rocks north of them and the metamorphics south of them are not clear.

At the north end of the section, fossils determined as Pennsylvanian or Permian were found in a low, outlying ridge overlooking the flats. South of these are black slates and cherts at the base of a high ridge, the topmost points of which are quartzites containing occasional chert pebbles interbedded with red and black slates and some dark limestones. The relations with the metamorphics on the south were not observed.

The evidence at hand seems sufficient to justify only the statement that in the section from Chatanika River to Yukon Flats there is a large area of closely folded rocks, in part of Devonian age and in part probably older, flanked on the south by highly metamorphosed schists and on the north by shaly slates containing Carboniferous fossils.

In the Rampart region a variety of rocks like those already described is found. These rocks have undergone greater disturbance and have been more generally intruded by igneous material. They have been closely folded; many of them have been greatly sheared and in some places brecciated. The strike of the rocks is nearly east and west. The garnetiferous mica schists and marbles of Ruby Creek, considered pre-Ordovician, are followed on the north by cherts and greenstones and on the south by rocks including slate, chert, sheared chert conglomerate, fine-grained rocks having the same composition as the chert conglomerate and also sheared until they have been rendered schistose, massive quartzites, and siliceous limestones, in places much brecciated. Here, also, the succession seems to be from chert and slates through chert conglomerate to fine slate and limestone.

A partial succession has been observed in other localities in the Rampart region. The main divide east of Lynx Mountain is composed mostly of chert flanked by chert conglomerate, and at one locality near the southeastern base of Lynx Mountain there are fine exposures showing at the base a conglomerate containing sheared chert pebbles several inches in diameter, changing gradually to alternating beds of finer material. Gray slates, highly folded and cleaved and pitching eastward, were observed not far from this locality. These slates contain thin beds of quartzite, a few inches to a foot or more thick, which contain grains of chert.

In Troublesome Valley the succession seems to be the same. A section southward between Troublesome and Hunter creeks shows greenstones, cherts, sheared chert conglomerate, and slates, with disconnected limestone masses in which fossils determined as Devonian were found. The chert conglomerates are at many places found in close association with the limestones, and at the locality where Devonian corals were collected conglomerate with chert and quartz pebbles is found in close association with the limestone, which also contains quartz pebbles.

Most of the rocks encountered in the Rampart region form a continuation along the strike of those occurring in the section from the White Mountains nearly to Yukon Flats. The same lithologic types and the same associations are found, and, so far as the meager fossil evidence is available, it corroborates the relationship of the rocks in the two areas. Similar rocks are developed south of Tanana River, where extensive masses of altered rhyolite occur with them. This correlation is based on lithologic and stratigraphic evidence alone.

The fossils collected by the Survey in 1904 are mentioned in the following list. The determinations were made by E. M. Kindle, of the United States Geological Survey.

Fossils from the White Mountains.

4 AP 240. This lot contains two corals, a *Michelina* and a *Zaphrentis*, neither of which is specifically determinable. Horizon probably Devonian.

4 AP 241. *Favosites* near *epidermatus* occurs in this collection indicating a horizon of Middle Devonian age.

4 AP 242. Specimens of a coral comparable with *Favosites winchelli* comprise the collection from this locality. Horizon probably Middle Devonian.

4 AP 243, 245, and 246. These lots represent the same horizon. The fauna represented in the collection comprises a single species of brachiopod *Gypidula* cf. *pseudogaleatus*. The horizon represented is either late Silurian or early Devonian, probably the latter.

4 AH 186. Includes only poorly preserved specimens of *Cladopora*. Probably of Middle Devonian age.

4 AH 193. The fossils represented are *Cytherella* sp., *Cladopora* sp., and *Ptilodictya ?* cf. *frondosa*. The horizon is probably Silurian.

4 AH 194. *Favosites* near *limitaris* occurs in this lot, indicating a horizon probably near Middle Devonian.

4 AH 195. Contains an undetermined *Stromatopora*. Age probably Devonian.

Fossils from the Rampart region.

4 AP 303. Two poorly preserved specimens of *Aulacophyllum* comprise this lot. The horizon represented may be either Devonian or Silurian so far as the evidence from this material indicates.

4 AP 317. Minute fragments of small corals in a breccia comprise this lot. Fossils too fragmentary for determination of the age.

Additional material from the Rampart region collected during 1907 was also referred to Mr. Kindle, and the following is quoted from his report:

7 AP 277 (Quail Creek). This lot contains several species of corals and fragmentary representatives of a large lamellibranch and a gasteropod. They are referred provisionally to the following genera:

Cladopora ?	Diphyphyllum.
Syringopora.	Megalomus??
Amplexus ?	Pleurotomaria ?
Streptelasma ?	

The minute characters used for specific limitation in corals are not well enough preserved to justify any attempt at the specific determination of the corals. Since the genera noted above are all common to both Devonian and Silurian horizons, they afford no definite evidence as to which of the two the corals represent. The chief interest of the collection lies in the lamelli-branch fragments, which represent a very large, thick-shelled form that appears almost certainly to be specifically identical with a shell occurring in the limestones of Glacier Bay and similar beds at Freshwater Bay in southeastern Alaska. This southeastern Alaska shell has been referred to the genus *Mega-lomus* and considered to belong to a late Silurian fauna.

7 AP 318 and 7 AP 320 (Head of Little Minook Creek). This material has yielded four or five species of brachiopods, represented by very poor fragments, so that even approximate determination is difficult. They may be tentatively referred to the following genera:

Chonetes?   Amphigenia?
Stropheodonta.   Rhipidomella?
Delthyris?

The determination of *Stropheodonta* is based on a fragment of a single valve, but the distinctly denticulated hinge line leaves but little doubt of the correctness of this generic determination. This genus is unknown in the Carboniferous "and is emphatically characteristic of the Devonian." *Amphi-genia*, which is believed to be represented by a fragmentary mold of a pedicle valve, is also limited to the Devonian. The specimen referred to *Delthyris?* is unique in its ornamentation and may possibly belong to another genus. There are also two other doubtfully determined genera of brachiopods, represented by fragments. The small gasteropods, which are represented by molds belonging to two or three species, contribute no evidence as to the age of the fauna. Although more and better material is needed to determine the horizon with certainty, it is believed that the forms determined as *Stropheodonta* and *Amphigenia?* necessitate placing the fauna in the Devonian, at least provisionally.

The occurrence of this fauna in conglomerates interbedded with black shales leads me to believe that it belongs near the top of the Devonian of the Yukon section, in beds corresponding to the black shales of the Calico Bluff section.

The fauna is quite unlike any other Alaskan fauna which has come under my observation, but for the reasons already stated it should, in my judgment, be placed provisionally in the Devonian.

CARBONIFEROUS.

At the extreme northern limit of the hill country, in a minor ridge bordering Yukon Flats, there are greenish, grayish, and black slates, with siliceous material, scattered fragments of which were found to contain fossils indicating a Pennsylvanian or Permian age. At another locality, about 15 miles farther southwest, just south of Hess Creek, fossils also regarded as Carboniferous were found in soft, black carbonaceous shales. At both localities the rocks are in close association on the south with cherts and slates. It is possible that there is a fault between the two formations, and the relation is so represented in the section. All that can be affirmed at present is that in the northern part of the area there occur Carboniferous rocks, whose extent and relations to the rocks of Devonian age are not yet determined.

The following fossils, determined by G. H. Girty, were collected at the two localities, and the discussion of them is quoted from his report:

Fossils from near Yukon Flats and from Hess Creek, Alaska.

4 AH 213.

Stenopora 2 sp.	Rhombopora sp.
Fenestella sp.	Productus? sp.
Rhombopora sp.	Lima? sp.

4 AP 270.

Fistulipora sp.	Rhombopora sp.
Stromatopora? sp.	Rhombopora sp.
Coral sp.	Spirifer n. sp.?
Fistulipora sp.	Hustedia cf. H. compressa Meek.
Fistulipora? sp.	

4 AP 277.

Coral? sp.	Polypora? sp.
Lithostrotion? sp.	Archimedes? sp.
Fistulipora 3 sp.	Productus sp.
Rhombopora sp.	Euomphalus sp.

The presence of the form identified as *Hustedia compressa* seems to show that lot 270 belongs in the Pennsylvanian, perhaps in the "Permo-Carboniferous." The ages of the other lots, although without much doubt being Carboniferous, are less certain. While probably no species is common to all three collections, yet in a general way the facies is much the same, and it is quite possible that all represent the same fauna.

It will be observed that only in one case have the forms collected been identified specifically. In many instances the material is too imperfectly preserved to admit of more than the genus being determined. In others the species are distinct from those of the Mississippi Valley sections, and entirely new unless some of them have been described in European and Asiatic publications not included in my bibliography and therefore difficult of reference.

I have consulted freely with Mr. Bassler wherever the Bryozoa were concerned.

MESOZOIC ROCKS.

CRETACEOUS.

On the flanks of Wolverine and Lynx mountains there are black, rather massive, carbonaceous sandy shales. In those of Wolverine Mountain, which form great rock piles along the upper parts of the spurs at an altitude of over 1,000 feet above the base of the mountain, there were found in 1905 fragments of dicotyledonous leaves and a part of an indeterminable bivalve, and on this basis the rocks were assigned to the Cretaceous. The Lynx Mountain rocks are correlated on only lithologic grounds.

During 1907 additional material was obtained and referred to Dr. T. W. Stanton for determination. The following is quoted from his report:

> While the fossils are fairly well preserved, they have been considerably distorted, so that it is not practicable to make specific determination. The better preserved forms appear to be undescribed. The following list will show the forms recognized in each lot:

4278. 7 AP 271 (Spur of Wolverine Mountain).

Hemiaster? sp.	Lucina sp.
Pecten sp.	Pleuromya sp.
Inoceramus cf. labiatus Schloth.	Pachydiscus sp.
Cucullæa sp.	Pachydiscus? sp.

4279. 7 AP 278 (Ridge on left limit south fork of Quail Creek).

Hemiaster? sp.	Pachydiscus sp.
Cucullæa sp.	Pachydiscus? sp.

4280. 7 AP 279 (Right limit south fork of Quail Creek).

Pachydiscus sp.

> These fossils evidently all belong to practically a single horizon which is confidently referred to the Upper Cretaceous. * * * The species of *Inoceramus* is very likely one that has been previously found on the Yukon, but the specimens in the present collection are too imperfect to serve as the basis for a positive identification. The most important forms are ammonites, which make up the bulk of the collection and which I have referred, in some cases doubtfully, to the genus *Pachydiscus*. These are unquestionably Upper Cretaceous types.

These Cretaceous rocks have been abundantly intruded by dikes of granular igneous rocks of intermediate composition that are described below and, furthermore, are much seamed with small quartz veins, many of which are ferruginous.

CENOZOIC ROCKS.

TERTIARY.

Rocks determined as Kenai (Eocene) occur along the Yukon above Rampart and are found also for a short distance up the valley of Minook Creek, where they contain a small amount of coal. They are conglomerates, sandstones, and clays resting unconformably upon the older rocks. In some places they are hardly consolidated sufficiently to withstand the pick, and prospect holes have been sunk into this formation under the impression that the material belonged to the stream deposits. These rocks are tilted but otherwise are little changed. The same formation is found in considerable areas south of Tanana River (see Pl. I), where it carries some good lignitic coal seams.

The changes in elevation with reference to sea level which the Yukon-Tanana country has undergone have left benches at various altitudes, some of them of considerable extent, which stand generally in a definite relation to the present drainage lines. Gravel was deposited on some of the benches during their formation. Part of this gravel is regarded as of Pleistocene age.

Such deposits have been found in the Fairbanks region in the valley of Fairbanks Creek, and in the Rampart region along Hess Creek and its tributaries, along the Minook, and along the tributaries of Baker Creek. The deposits of the high bench of the Minook, approximately 500 feet above the present stream, are of interest with reference to the occurrence of gold in the tributaries of Minook Creek. The bench gravels of the Baker drainage have proved in some places to be of great economic importance. The description of these gravels and the deposits of the present streams is given elsewhere in this report in the account of the gold placers.

Silts also have accumulated in great quantities in the larger valleys throughout the interior of Alaska, and as gravels have been repeatedly deposited on successively lower benches, so, too, silts have been deposited at different levels down to that of the present flood plains. These were probably laid down under lacustrine conditions and the interaction of lacustrine and fluviatile conditions, and the age of deposition dates from the Pleistocene, or earlier, to the present time.

IGNEOUS ROCKS.

In the geologic history of the Yukon-Tanana region igneous rocks have played a very important part, and though not so widely distributed in this western part of the area they form nevertheless an essential part of the bed rock.

The igneous rocks include material that in a more or less molten condition has penetrated the other rocks in various forms at various times and at various depths below the contemporary surface and also material that has been poured out upon the surface as lava flows or has been ejected as fragmental material. The rocks formed by the consolidation of the molten material have characteristics so indicative of their origin that miners generally distinguish with little difficulty the igneous from the sedimentary rocks with which they are associated, or the igneous material present in the gravels from the sedimentary material with which it is mixed.

The differences in the composition of the igneous rocks are not so readily observed. Coarse-grained, fine-grained, and glassy rocks entirely different in appearance may result from the same material solidifying under different conditions, and furthermore, the igneous material from which the rocks are derived has a wide range of chemical composition, and the rocks that result from its solidification present a great variety of mineral composition and many gradational types. Therefore, no hard and fast lines can be drawn between the kinds of igneous rocks resulting from solidification through more or less complete crystallization. Furthermore, some of the igneous rocks, since their solidification, have been greatly changed by metamorphism along with the sedimentary rocks with which they are associated, and their original characteristics have been more or less obscured by the processes that have altered them. For the purposes of this brief report the igneous rocks of this area need to be considered only in a very general way. For sake of convenience they are divided into granitic rocks, monzonitic rocks, greenstones, and basalts.

GRANITIC ROCKS.

Intrusive biotite granite and hornblende granite occur in parts of the Fairbanks region, notably in the ridge south of Gilmore Creek, on Twin Creek, Pedro Dome, and at the head of Chatham Creek. The rock of some of these localities is porphyritic with feldspar crystals an inch or more in diameter, and that at other localities is fine and even grained. The rocks are fresh and have not undergone the metamorphism that has altered the schists. Intrusive gneisses, such as are common in the Fortymile region, were not observed, nor are the injection zones that are so common in the Fortymile country present in this western part of the Yukon-Tanana region. There are a few areas of coarse-grained biotite granite in the northern part of the region, near the head of Beaver Creek and between Beaver and Victoria creeks. Another area is located on the southern side of the divide, near the head of the Tolovana drainage. These intrusive masses were probably injected during Mesozoic time.

Granitic rocks are not of common occurrence in the Rampart region. The most extensive mass observed forms a part of the summit of Wolverine Mountain, where it occupies an area about 1,000 feet wide. It is a porphyritic, massive, gray rock composed chiefly of quartz, phenocrysts of orthoclase a half inch or more in diameter, considerable plagioclase, some biotite and hornblende, and a little pyroxene. The rock is finer grained toward the margin; the slates in contact with it have been indurated, and their fracture surfaces are flecked with the products of metamorphism, chiefly andalusite.

A similar rock occurs west of the mountain near the saddle where the trail passes through the ridge to descend toward the Hutlina. This is also a gray porphyritic rock, but the porphyritic feldspars, some of them an inch or more in diameter, have a tabular development. The proportion of pyroxene is greater, there is less quartz, and the composition of the rock is transitional to that of the monzonitic rocks. Rocks similar in composition to those of Wolverine Mountain occur also in Lynx Mountain along with monzonitic rocks, but their outcrops were not observed.

A porphyritic biotite granite is found in the vicinity of Baker Hot Springs, where it is intrusive, in black carbonaceous slates.

MONZONITIC ROCKS.

The most common intrusive in the Rampart region is a monzonitic rock that varies in color from dark, brown to nearly black. It is a medium to fine-grained rock, and the coarser varieties show abundant plates of reddish-brown mica, the most striking mineral present. All the minerals of this rock are fresh and include about equal proportions of an orthoclase feldspar and of plagioclase, which is embedded in the irregular limpid mottled grains of potash feldspar, abundant pale-green monoclinic pyroxene, biotite, and a small proportion of olivine. Hypersthene occurs frequently, its small prisms often fringing the grains of olivine, at the expense of which it has probably been formed. There is some apatite and often much magnetite.

This rock occurs in Lynx Mountain and in the ridge at the heads of Glenn, Rhode Island, and Omega creeks, where it forms a mass of considerable extent. The numerous small dikes of minette-like rock in the slates, from 1 foot to 3 feet thick, containing prominent plates of bleached biotite and a large proportion of nearly colorless prismatic crystals of monoclinic pyroxene, are probably to be referred to this type.

A portion at least of the granular rocks of intermediate composition in the Rampart region are intrusive in Upper Cretaceous rocks and it is probable that all of them are of the same general period.

GREENSTONES.

The greenstones include serpentine, altered gabbro, diabase, basalt, and much tuffaceous material, and have at many places been intruded by fresh diabasic rocks. Some show clearly their mode of origin. others are indefinite aphanitic chertlike rocks. They occur abundantly in the area lying between the White Mountains and Rampart. Their dark color contrasts strongly with the associated limestones of the White Mountains. They form the prominent ridges across Beaver Valley west of these mountains and occur in the area between this ridge and Yukon Flats. Farther west they become prominent in the ridge north of Hess Creek. In the Rampart region they form the bed rock in the lower part of Troublesome Valley and are the most widely distributed rocks in the lower valley of the Minook below the mouth of Florida Creek.

The greenstones are partly intrusive and partly extrusive in the rocks in which they occur. Those in association with the limestone are, so far as has been observed, parallel to the structure, and furthermore, some of them are altered basalts containing numerous amygdules filled with calcite. Diabasic intrusives cut the serpentine and in the Rampart region intrude the Rampart slates. The volcanic activity that resulted in the production of this material took place, principally at least, during the deposition of the rocks regarded as Silurian and Devonian.

BASALT.

A fresh olivine basalt occurs on Minook Creek about 1 mile above its mouth. On Hunter Creek, a short distance above its mouth, and apparently related to the basalt in their occurrence, are volcanic glasses containing basic feldspar phenocrysts. These volcanics are probably subsequent to the Kenai. A small mass of basalt was also observed in the Fairbanks region in the valley of Fairbanks Creek.

METAMORPHOSED RHYOLITES.

In the region south of Tanana River there are interbedded with the cherts and slates (Devonian?) large masses of augen gneisses which on microscopic examination proved to be altered rhyolites. These rocks range from a coarsely porphyritic gneiss with feldspars up to 2 inches in diameter to an evenly fine-grained sericite schist with no grains visible to the eye. The most common type is composed essentially of quartz and feldspar grains in a groundmass of quartz and feldspar largely sericitized. Remnants of original igneous textures are preserved, but the present structures and textures are due mostly to cataclastic action and recrystallization.

SUMMARY.

The greatest part of the area is formed of closely folded rocks striking northeast and southwest, separable into two large groups, one including highly metamorphosed schists regarded as early Paleozoic or pre-Paleozoic and the other being made up of rocks greatly variable in kind but characterized in general by a less degree of metamorphism and regarded as predominantly Silurian and Devonian. Igneous rocks are an essential constituent of the area. Some of them have been highly metamorphosed along with the sedimentary rocks that are now schists; others are fresh granitic or more basic intrusives that were probably injected at the end of Mesozoic time; and still others, which are present to a minor extent, are comparatively recent extrusives. Igneous rocks occur in all the areas that produce gold, but are not confined to these areas, and the gold occurrences are probably to be referred partly, at least, to after-effects of intrusion.

ECONOMIC GEOLOGY.

FAIRBANKS REGION.

GENERAL STATEMENT.

The Fairbanks region includes the gold placers of the Tanana Valley about 260 miles above the mouth of the river. They comprise parts of the valleys of small streams, most of which are within 25 miles of the navigable water of the Tanana. The location of a trading post in this part of the Tanana Valley in 1901 was followed in 1902 by the discovery of gold. The region began to attract attention and by the end of 1903, with a production of at least $40,000, had become of prospective importance. The work of 1904 resulted in a production of about $600,000, which was increased in 1905 to about $6,000,000, and in 1906 to over $9,000,000.

The Fairbanks region was visited by A. H. Brooks[a] in 1906 and 1907, and the condition of the creeks is thus described by him:

Cleary, Fairbanks, Dome, Vault, Esther, Goldstream, and Pedro creeks and their tributaries are the chief producing creeks of the district. Cleary continues to stand first in production, with Fairbanks in second place. The finding of values on Cripple and Treasure creeks definitely extends the producing area to the southwest, and reported discoveries of gold in the upper Chena may show a northeasterly extension of the same belt, though this is not yet established.

The facts in hand are, however, sufficient to determine that there is a gold-bearing zone, at least 10 miles wide, running northeast and southwest, which has been traced for about 30 miles. Its northeastern extension would intersect the upper Chena basin, while to the southwest it runs out into Tolovana flats. A logical deduction from these facts would suggest that the prospector could turn his attention to the Chena basin and to the streams draining the upland which bounds the Tolovana flats on the east. It should be remembered, however, that the investigations so far made indicate that the conditions which bring about mineralization are local, and hence the formation of placers probably does not persist over any great distance.

Worthy of special note are the rich placers found last year on Vault Creek, which had previously been unproductive. On nearly all the producing streams which are tributary to the Chatanika the pay streak has been traced well down to the main river. In fact, the origin of the rich gravels found in various places at 100 to 200 feet depth under the valley floor of the Chatanika is among the most puzzling of the phenomena connected with the placers of this district. Interest in Goldstream Creek was revived during last summer by some rich placer discoveries, and as a result, though the creek was almost abandoned in the early part of the summer, later it was studded with operators for several miles.

Of the outlying districts tributary to Fairbanks the Tenderfoot probably made the largest production, estimated at-$300,000. The gravels on Tenderfoot Creek are deep, but in the smaller creeks are said not to exceed 8 to 10 feet in depth. It would appear that these deposits lie in a different zone from those of Fairbanks.

---

[a] Brooks, A. H., report on progress of investigations of mineral resources of Alaska, 1906: Bull. U. S. Geol. Survey No. 314, 1907, pp. 36–37.

Some work was done on the streams tributary to the upper Chatanika, where probably 30 men were at work. Some gold has been found on Faith, Hope, and Homestake creeks. The pay streak is thin and the values are said to be regularly distributed.

Up to the present time, 1907, transportation to the Fairbanks region from outside points has been by way of Dawson, St. Michael, or Valdez. The Valdez route is convenient for many who desire to reach Fairbanks before the opening of navigation and is traveled extensively during the winter season. Supplies shipped in the spring to Fairbanks by way of Dawson reach their destination earlier in the season than if shipped by way of St. Michael, as the upper Yukon is first open to navigation. Freight rates on supplies shipped from Seattle to Fairbanks vary greatly according to their position in the freight classifications and according as they are sent by way of Dawson or St. Michael. During 1905 the rates ranged from $55 to $75 per ton on ordinary supplies. First-class passenger rates have ranged from $125 to $150.

The town of Fairbanks is situated on a slough of the Tanana, near the head of easy navigation, and in 1905 had a population of about 2,500. During dry seasons the quantity of water in the slough is so small that some of the steamers have difficulty in reaching the town. The larger boats that ply occasionally on the Tanana are unable to reach Fairbanks except at high water, and their supplies are left at Chena. Wages for miners are generally $5 and board, in some cases $6 and board, per day.

Fairbanks has newspapers, a school system, and three banks—one of them a national bank with currency of its own in circulation—and is the seat of the court which has jurisdiction over the whole of the interior of Alaska.

The town of Chena is situated about 9 miles from Fairbanks, at the entrance of the slough into the main river. It is accessible for the largest boats, but has the disadvantage of being several miles farther from most of the gold-producing creeks, and thus far its development has not kept pace with that of Fairbanks. The construction of the railroad has, however, favorably affected the growth of the town.

The transportation of supplies from the towns to the creeks has been a source of much trouble and expense, but cheaper and better transportation has been gained by constructing a narrow-gage railroad between Fairbanks and Chena, by extending a branch road from an intermediate point up the valley of Goldstream Creek to the junction of Pedro and Gilmore creeks, and also (1907) up Fox Gulch and along the divide between Vault and Dome creeks to claim 15 below on Cleary Creek, and by constructing Government wagon roads from the main supply points and from the terminus of the railroad to the various creeks.

The region, though dependent on the outside for the greatest part of its supplies, is in the matter of lumber and fuel mostly independent. The spruce timber along the sloughs of the Tanana and the lower parts of the valleys of its largest tributaries is of good quality and much of it exceeds 2 feet in diameter at the butt. The small spruce and birch, so abundant on the hillsides, furnish a supply of fuel which has not up to the present time been heavily taxed, but which is becoming year by year a more important item in the cost of production.

BED ROCK.

The bed rock is predominantly schist, closely folded and striking in general northeast and southwest. The main structural planes vary from nearly horizontal to nearly vertical. The alternating beds of blocky quartzite schist and schists with a large proportion of mica. give rise to a bed-rock surface of varying influence on the distribution of the gold. The gold sinks along the structural planes of the blocky bed rock to a depth of several feet in some places, but the compact clayey surface of the softer beds is practically impermeable to the gold. Quartz veins, some of them being 2 feet or more thick, are common in the schists, but are not so abundant that quartz becomes a conspicuous constituent of the gravels.

Granitic intrusions are present at several localities in the main ridge. They form the bed rock at the upper ends of some of the valleys and have furnished material for a small proportion of the gravels in many of the valleys. Greenstone schist forms a large part of the ridge west of Cleary Creek, and this rock is generally conspicuous by the quantity of garnets it contains. A small amount of black basaltic rock outcrops at the point of the ridge that forms the northern limit of the valley of Fairbanks Creek.

ALLUVIAL DEPOSITS.

The loose material formed by the weathering of the bed rock varies in character according to the amount of transportation it has undergone. Outcrops of bed rock in the Fairbanks region are confined mostly to the summits of the ridges and to the steeper slopes of the valleys, while on the gentler slopes and in the bottoms of the valleys the bed-rock surface is covered with a mantle of material ranging in thickness from a few feet to over 300 feet. This mantle is composed partly of heterogeneous talus, which is continually working down the sides of the valleys, and partly of the material in the valley floors, which has been worked over so many times by running water that it has a fairly uniform structure throughout. All of these deposits are, for the most part, frozen throughout the year.

As the streams generally flow close to one side of their valleys the deposits are mostly on but one side. Their upper surface slopes gradually toward the base of the hills. The bed-rock surface, so far as known (1905), is in general nearly flat, or has a very gentle grade hillward from the creek. The deposits are in most places separable into three divisions, which, from surface to bed rock, are referred to by the miners as muck, barren gravels, and pay gravel.

The muck ranges in thickness from a few feet to a maximum of about 70 feet, the line of separation between it and the underlying gravels being fairly sharp. It is a black deposit containing a large amount of material derived from the decomposition of moss and other vegetation, a considerable percentage of clay and sand being either intermingled with the organic matter or distributed as layers and thin lenses irregularly through the mass. Horizontal, and occasionally vertical, sheets of ice several feet thick occur in this deposit.

The underlying gravels, ranging in thickness from 10 to over 60 feet, are derived from the rock occurring within the areas drained by each particular stream. As quartzite schist is the most common bed rock and also the most resistant to the processes of wearing and weathering, the largest proportion of the coarse material in the gravels is composed of it. The gravels also include quartz-mica and carbonaceous schist, vein quartz, and some igneous material, mostly granite. Teeth of the mammoth and bones of other animals now extinct are occasionally found. The coarse material being mostly schistose occurs as more or less flattened angular pieces, only slightly waterworn. Few of them exceed a foot in diameter and the proportion of bowlders is therefore small. The fine material is composed partly of smaller pieces of the more resistant rocks and partly of clay derived from the decomposition of the micaceous and carbonaceous schists. These gravels contain also a small percentage of individual minerals released by the process of weathering. The proportion of clay in the barren gravels is small, but that in the underlying pay gravels is large. All the material, both coarse and fine, is irregularly intermingled, the larger pieces being usually nearly horizontal.

The pay gravels resemble those above them, but contain a considerable amount of clay, which adheres tightly to the gravel and to the surface of the blocky fragments of bed rock. This clay is prevailingly of a yellowish color in the more shallow diggings and of a bluish color in the deeper gravels. The proportion of clay varies, but in most places there is enough present to render the pay gravels easily distinguishable in the drifts from the barren ground above them. The thickness of the pay gravels ranges from a few inches to 12 or more feet, a range which is rather uniformly maintained. The under surface of the gravels not only rests upon the bed rock, but where the latter is blocky the fine material is found within it to a depth of from 1 to 3 or more feet. The pay gravels vary in width in different creeks and in different parts of the same creek, but make up only a small part of the width of most of the valleys. Pay streaks ranging in width from 30 feet or less to 450 feet, and in one place reaching a width of 800 feet, have been reported. The average width of gravels carrying values sufficient to pay for working under conditions in 1905 is probably about 150 to 200 feet, and this, like the thickness, is fairly constant. The pay streaks in the valley floors bear no uniform relation to the present stream beds.

The gold is either evenly distributed throughout the pay gravels or lies mostly near the bed rock, or, occasionally, is found chiefly within the bed rock. The great bulk of the gold is composed of flattish pieces of various sizes up to one-fourth inch in diameter and of granular pieces, some of which are minute. The proportion of very fine gold, however, is apparently small, and there is but little flaky gold. Nuggets form an inconsiderable part of the clean ups; those worth a few dollars are common, however, and a few of considerable value have been found. Some of the largest were worth approximately $145, $160, $190, $233, and $529. Many of the nuggets contain quartz. Most of the gold found near the heads of the creeks is angular. Downstream there is in general a gradual decrease in the average size of the pieces and an increase in the amount of wear they have sustained. Nuggets, too, are less common in the lower parts of the valleys. In some places coarse and fine gold occur together; in others the coarse gold is found mostly on one side of the pay streak. At a few localities there appears to be an abrupt change from gravels carrying a large percentage of coarse gold to others—immediately below on the same stream—whose gold content is chiefly fine. The values in the pay gravels exploited in 1905 range from about 2 cents to 20 cents or more to the pan, and a large part of the ground will average about 8 cents to the pan, or about $10 to the cubic yard or $2 to the square foot of bed rock. Some of this ground averages $3 to $3.50 to the square foot and some carries even better values. Assay values were reported ranging from $16.16 to $18.25 of gold per ounce, and the gold from one locality was said to assay as high as $19.25.

The minerals most commonly associated with the gold, aside from the quartz with which it is often intergrown, are garnet, rutile, and black sand. The black sand occurs in but small proportion and is composed mostly of magnetite. Cassiterite is rather common, and there is some stibnite. Bismuth is occasionally intergrown with the gold.

The frozen gravels are tough, in distinction from the muck. They can not be broken with the pick and are with difficulty rent by explosives. A sudden caving in of the ground undermined in drifting is rare, the sinking usually being so gradual as to permit the removal of mining apparatus. In such cases a parting often takes place between the gravels and the overlying muck leaving the latter as a roof. The solidly frozen gravels are practically impermeable to the surface waters and to any underground water that may be present and the underground mining operations are comparatively dry. Unfrozen areas are encountered at many places, and where they occur in the deeper ground the presence of " live water " adds to the expense of mining. In other places, notably near the heads of some creeks where the gravels are shallow, unblanketed by muck, and well drained, the greater part of the ground thaws during the summer.

Bench gravels are not common in the Fairbanks region.' A deposit of gravels composed essentially of quartz-mica schist, carbonaceous schist, and vein quartz has, however, been found on the northern valley slope of Fairbanks Creek, 600 feet above the valley floor. The gravel is rather well rounded and contains bowlders up to 1 foot in diameter. These gravels have been somewhat prospected, but so far as known, without favorable results.

FORMATION OF PLACERS.

It appears from a cursory examination that the pay gravels were deposited under conditions somewhat different from those which now prevail. Though the details can not be discussed here, some of the facts bearing on this matter deserve mention. The general uniformity in the altitude of the ridges has been noted. This uniformity is the result of the erosion of a surface which formerly stood at a lower level than at present. Its topography was then undulating, dotted with rounded hills, and broken by isolated groups of hills and ridges of greater prominence. The valleys, furthermore, were open and of low grade. It is probable that the stream deposits were deep and that the interstream areas bore much weathered bed rock, awaiting transportation. Elevation of the region enabled the streams to cut the present valleys and thus form the avenues, or sluice boxes as they might be called, through which passed the products of long-continued weathering as well as the deposits of the former streams. The bench deposits above described form a remnant of these old deposits.

In the constant, slow, and often interrupted progress of the unsorted coarse and fine material down the valleys the particles of gold, because of their high specific weight, tend to lag behind the particles of other materials and to find a lower position in the mass or a lodgment in the crevices of the bed rock. They offer a passive resistance to onward motion and an active assistance to vertical downward motion. The accumulating deposit of gold is mixed with unsorted material which, for the most part, was probably not originally associated with the gold but was derived from some source farther up the valley. This deposit closely follows the cutting action of the stream and is the first to cover the bare surface of the bed rock when opportunity is offered.

Active erosion and an abundance of previously accumulated auriferous material appear to be favorable to the formation of placers. The so-called "wandering" placers which have been noted in Australia,[a] where the pay dirt is shifted at times of melting snows to claims lower down the valley, appear to represent an early stage in the development of placers. With the lessening of the stream's activity, accompanied often by the exhaustion of the great part of the auriferous material, the mobility of the deposits is diminished. An increasing amount of barren material is then deposited over the pay gravels, the stream may abandon the part of the valley in which it has hitherto worked and the pay streak may become practically a stationary deposit. In the interior of Alaska the pay streak has become not only permanently stationary, but also, through the cementing agency of ice, for the most part permanently consolidated.

Few facts are known regarding the amount, distribution, or circulation of the underground waters and the consequent extent of the permanently or only temporarily unconsolidated gravels. In the Yukon-Tanana country there are valleys whose deposits are so "spotted," as it were, with live water that it is practically impossible to work them by drifting. The presence of large amounts of live water in many valleys during the winter is shown by the repeated overflows to which streams are subject and by the unexpected filling of prospect holes with water from below. It is possible, therefore, that the extent of the unfrozen ground is greater than is generally supposed. The extent of consolidation, while dependent primarily on the climate, is probably greatly modified by local conditions. The slope of the valley, the character and thickness of the deposits, and the quantity of water are factors which together may become of dominating importance, counteracting successfully the tendency of the climatic conditions to cause consolidation of deposits to great depths. As a result, a part of the deposits of a valley, whether talus or stream gravels, where not too deep, may be further differentiated. This differentiation, under the mobility imparted by the contained water and by the stream action to which they may be subjected, may bring about the gradual accumulation of the gold on or near the bed rock. In the Fairbanks region this process would be most active in the shallow deposits which are generally confined to the headwaters of the valleys. Although the quantity of weathered material now at hand is not so large as formerly, when the product of long-continued weathering had accumulated, and although the proportion

---

[a] Schmeisser, Karl, Die Goldfelder Australiens, p. 100.

of gold may now be different from what it was formerly, nevertheless it is reasonable to suppose—and the occurrence of gold near the
headwaters renders such a supposition entirely justifiable—that the
deposition of auriferous material is there in progress. At present
the streams come in closest relation in the vertical section to the bed
rock near the heads of the valleys, and there, if anywhere within
the valleys, downward cutting of the bed rock is in progress. The
lower parts of the valleys have been areas of abundant deposition.
Near the heads deposition closely follows cutting and there the
deeply buried, more or less permanently frozen pay streaks of the
lower valleys merge into the deposits within the zone of the present
streams' activity.

The gravels in the valleys of the Fairbanks region are composed
of materials derived from the bed rock in which the valleys have
been cut and were deposited through stream action, uninfluenced by
any general glaciation, yet under conditions somewhat different from
those of the present. The position of the pay streak in the valley
marks the position of an earlier drainage way as well as that part
of the cross section of the valley which was at the time of its deposition probably the deepest.

The successive stages of development may have been somewhat as
follows: (1) Elevation of the region, the surface being laden with
much unassorted weathered material and older stream deposits; (2) a
period of active erosion by the streams during which there was little
opportunity for the formation of permanent deposits; (3) a period
of deposition when the streams were nearly down to grade and when
the pay streaks were, for the most part, laid down with their clay
content, which may have been derived in part from the abundant
clay of the weathered material and in part directly from the bed
rock; and (4) a period of stream shifting, valley widening, and further deposition, with the gradual development of the unsymmetrical
type of valley of the present day. This unsymmetrical shape—one
side steep and the other a more or less gradual slope—is a characteristic feature of many valleys in Alaska and results probably from
several causes, among which are local elevation or depression, lithologic character and structure, and insolation. It suffices here to
emphasize only one of these, often observed by miners—that the sunny
side of valleys is subject to more rapid wear than the shady side,
which remains locked in frost for a much greater part of the open
season. Slides are of frequent occurrence on the sunny sides of the
valleys, even on very gentle slopes, and in the course of time these
produce an accumulation of waste which forces the stream toward
the opposite slope of the valley.

The greater mobility of the material was due probably in part
to the greater activity of the streams, which were at that time just

becoming graded; in part to more abundant precipitation, as is suggested by the much greater extension of the glaciers of the Alaska Range, and in part, perhaps, to a higher average temperature, though it would seem that with the other factors present no essential difference in the temperature would be required. Whatever the conditions of formation—and these are only imperfectly known—the dominant facts of economic importance are that in general but one pay streak has been laid down; that this is next to bed rock beneath a considerable thickness of other deposits, and that its formation is, for the most part, a closed incident.

SOURCE OF THE GOLD.

The origin of the gold in the placers, although not definitely determined, is suggested by the character of the gold itself and by its association. Most of that found near the heads of the creeks is rough and practically unworn; much of it is flat, as if derived from small seams; most of the coarse pieces are intimately intergrown with quartz and many of them are flat, like the small fragments of thin quartz seams which are common in the schists. That mineralization has not been confined to gold is shown by the occurrence of native bismuth intergrown with gold, of veins of stibnite, and of the cassiterite often found in the gravels. The most acidic igneous rocks observed in the Fairbanks region are intrusive porphyritic biotite granite. The acidic dikes so common in the Fortymile region are absent and the gold of the placers has probably been derived from small quartz seams in the schists.

It is often a subject of surprise to the miners that when gold is abundant in the placers it should be found so rarely in the bed rock. It might be said that if gold were commonly encountered in the bed rock the proportion of it in the placers, considering the amount of bed rock that has been removed, ought to be much greater. There is the possibility also that the veins in the country rock which contributed the material for the first deposits of the valleys were richer in gold than those now exposed. Be that as it may, it is certain that through long-continued weathering and sorting of the rock material a concentration of the heavier indestructible contents, including the gold, takes place, yielding auriferous detrital deposits which are made richer in gold than the parent rock. Furthermore, much of the gold may have been contributed by the bed rock that has been removed in the formation of the present valleys.

PLACERS.

The productive areas of the Fairbanks region at present (1907) include the valleys of streams draining the ridge that extends northeast and southwest between Chatanika and Little Chena rivers. The distance between these two streams is about 25 miles and the gold-bearing creeks are located at intervals along a distance of about 30 miles.

FAIRBANKS CREEK.

Fairbanks Creek is about 10 miles long and flows in an easterly direction to Fish Creek, a tributary of Little Chena River. The floor of the valley is 200 to 450 feet broad but widens rapidly about 3 miles from the mouth. The productive area of Fairbanks Creek comprises about 4 miles of the valley, starting from a point about 2 miles below the source. The gravels in general range from 12 to over 30 feet in thickness but in the lower part of the valley are much thicker. The pay streak ranges from 4 to 8 feet in thickness, averaging about 5½ feet, while in some places 2 to 3 feet additional of bed rock are mined. The pay streak ranges from 40 to 200 feet in width and in the upper part of the valley lies close to the present stream bed but lower down the valley it diverges toward the north valley slope. Ground has been worked containing values in the pay streak ranging from 5 to 10 cents to the pan. As this streak has not been traced through the lower part of the valley it is uncertain whether it continues as a well-defined pay streak or becomes disseminated or distributed over a considerable area. Fairbanks Creek has been a good producer but is somewhat spotted, and ground should be carefully prospected to determine values before the introduction of expensive machinery. The following table of depths to bed rock, etc., is based on data obtained by Mr. Covert in 1907.

Depth to bed rock, thickness of muck, and width of valley floor along Fairbanks Creek.

Claim No.	Depth to bed rock.	Thickness of muck.	Width of valley floor.
	Feet.	*Feet.*	*Feet.*
8 above	14	4	250
7 above	13	6	250
6 above	16	8	300
5 above	30	20	300
4 above	30	20	300
3 above	26	6	400
2 above	25	5	400
1 above	16	7	400
Discovery	17	5	400
1 below	19	5	450
2 below	18	6	450
3 below	23	6	450
4 below	23	6	450
5 below	16	4	400
6 below	26	8	450
7 below	32	11	450
10 below			

GOLDSTREAM BASIN.[a]

Goldstream Creek is 50 or 60 miles in length and the upper 7 miles of it have been found to be auriferous. The floor of the auriferous

[a] The information regarding the Goldstream basin was obtained during the field season of 1907 by Alfred H. Brooks and C. C. Covert.

part of the valley is from one-half to three-fourths of a mile in width and from this floor the walls rise with gentle slopes. Gold has been found on Pedro and Gilmore, Engineer, Big Eldorado, and O'Connor creeks and on Fox Gulch, all tributary to Goldstream Creek. The bed rock throughout most of the basin is probably mica schist, but granite has been found on several of the tributaries, notably on Pedro and Gilmore creeks. Pedro Creek was the scene of the first discovery of gold (1902) and this creek and the main stream have been large producers in the last two years. The alluvium of the basin ranges in depth from a few feet in the headwater region to more than 200 feet near the lower limit of discovery on Goldstream Creek. The gravels consist predominantly of mica schist, with considerable quartz and some granite.

Most of the work on Goldstream Creek during 1907 was confined to the portion of the valley from Discovery claim, near the confluence of Pedro and Gilmore creeks, to about 17 below Discovery. The pay streak from Discovery to about 7 below lies on the north side of the creek; below this point it is on the south side. The limit of profitable mining, as determined by present methods and costs, is probably $1.50 to $2 a cubic yard. If a cheaper method of handling the gravels could be devised it would make available for mining an enormous body of gravel. The following table shows the thickness of the deposits in the upper part of the valley so far as known.

Depth of bed rock, thickness of muck, and width of valley floor along Goldstream Creek.

Claim No.	Depth to bed rock.	Thickness of muck.	Width of valley floor.
	Feet.	*Feet.*	*Feet.*
Discovery to 8 below	15-20		1,000
8 below	22	9	
9 below	25	12	
10 below	20		1,500
12 below	45-60	30	2,500
14 below	70		3,000
15 below	110		3,000
16 below	80		3,000
17 below	65		3,000
27 below	60		3,500
29 below		90	
32 below	160	130	2,500
37 below	158	150	

The Pedro Valley is similar in general character to the Fairbanks Valley, already described. The productive area includes about 2 miles of the stream between the mouth of Twin Creek and the point where it is joined by Gilmore Creek. The deposits have a thickness ranging from about 8 to 30 feet. Values have been found in some places through 8 feet of gravel and 4 feet of bed rock; in other places they are confined mostly to the bed rock.

Gilmore Creek, which joins Pedro to form Goldstream, flows in a general northwesterly course and has a length of about 5 miles. The valley floor at its mouth is about 1,000 feet wide, narrowing rapidly upstream. The bed rock of the main creek is chiefly quartz-mica schist, much of it carbonaceous, but a granite mass is found to the south and also in the valley of Hill Creek, a small tributary. The alluvium, which is 50 to 60 feet deep near the mouth of Gilmore Creek, shoals rapidly upstream and 2 miles above is only about 10 feet deep. Most of this is muck, which ranges from 45 feet near the mouth of the stream to less than 5 feet 2 miles above. The gravels are chiefly schist of several varieties, with considerable quartz, which occurs to some extent in large bowlders, and much granite. An important commercial feature of the creek is the presence of an ill-defined bench on the north slope of the valley. This is in some places separated from the main valley floor by a rock rim with an escarpment of a few feet, but usually slopes directly to it. The bench gravels are auriferous and have been mined at three localities. Much of the material is angular and ill stratified and in many places it resembles a talus slope rather than a water-laid deposit. These bench placers are favorably located for economic mining and have yielded some gold but in dry times there is not sufficient water for hydraulicking. The pay streak in the Gilmore Creek gravels is said to be 30 to 40 feet wide but in places attains a width of 100 feet. On the hill-slope terraces it appears to embrace the entire bench, from 100 to 300 feet wide. The placer gravel of this creek is usually found down to or in the bed rock. In places 3 to 4 feet of the bed rock is mined, with but a foot or two of the overlying gravels. The gold ranges from medium coarse to fine and that of the benches appears to be more angular than that of the creek. Nuggets up to a value of $20 are reported. The gold is said to assay $19.25 to the ounce. Some bismuth has been found with it.

Tom Creek, a tributary to Gilmore Creek 3 miles from the mouth, lies in the extension of the main valley. No values have thus far been reported on it. Hill Creek, a southerly tributary of Gilmore Creek, has its source well back in a high granite dome and the bed rock for much of its length appears to be granite. Near its head is a small basin 200 or 300 yards long. Below this basin the valley floor falls away with steep descent for about a quarter of a mile. The bed rock is a coarse porphyritic biotite granite, which in the basin is deeply weathered and stained with iron. Here some open-cut mining has been done. No one was at this place when visited by the geologists but the pay streak appears to be very narrow. The alluvium consists of granitic sand about 8 feet deep. This occurrence is of interest because it shows that the gold has been derived from the granite. This deposit is near the contact with the schists and indicates that the granite intrusives may bear a genetic relation to the occurrence of the gold in bed rock.

Fox Gulch is a small northerly tributary of Goldstream Creek. Little mining has been done on the lower part of the creek where the alluvium is probably 30 to 40 feet deep. Several claims have been worked about 3 or 4 miles from the mouth, at a point where the gravels are 8 to 10 feet deep, overlain by muck about 10 feet thick. The pay streak here is rather narrow and the reported values are not high.

Some gold has been found on Big Eldorado and O'Connor creeks but as neither has been visited by a geologist details regarding them are lacking. Both these creeks are tributary to Goldstream from the north, and each is about 5 miles in length. The pay streak on Big Eldorado Creek is said to be narrow, but some mining has been done. Work on O'Connor Creek is said not to have gone beyond the prospecting stage (1907). Ground is reported to be from 100 to 130 feet deep.

Engineer Creek, about 5 miles in length, a southerly tributary of Goldstream, has also become a producing creek during the last two years. The placers of this stream have not been studied by any geologist.

The gravels on some other small tributaries of Goldstream Creek have been found to be auriferous, but no values have been discovered. This does not signify, however, that they are not worthy of careful prospecting.

CLEARY CREEK.

Cleary Creek has been the best producer of the region. Workable deposits have been found along about 7 miles of the stream and far out into the Chatanika flats. The limit of their extension into the flats has not been determined. Chatham Creek, only about a mile long, has been a good producer. Considerable work was done on Wolf Creek in 1903, but since that time little gold has been found there.

The deposits of the main valley range in thickness from a few feet to more than 120 feet, averaging about 60 feet. The pay streak has a maximum thickness of about 14 feet and an average for the creek of about 5 feet. The width of the pay streak, under present mining costs, ranges from 30 feet or less to several hundred feet. The average value in the pay streak for much of the valley appears to be about $10 to the cubic yard. The pay streak is rather uniformly developed, but the valley is so wide that its location requires much prospecting. The position of the pay streak in the valley is at variance with the course of the present stream. The creek makes one large bend in its course, above which the pay streak is altogether on the west side, several hundred feet from the creek, except at the head. It crosses the valley at the bend, and throughout the lower part is found on the right side, 1,000 feet from the stream. As the valley of Cleary Creek opens into the Chatanika flats the pay streak swerves back to the left side and has been found there within a short distance of the creek. The pay streak occupies, in general, the center of the valley, being about equidistant, both above and below the bend, from the ridges on the sides. Gold was discovered at the point where the pay streak crosses the valley, where good surface prospects were found. The following table of depths to bed rock, etc., is based on data obtained by Mr. Covert in 1907:

Depth to bed rock, thickness of muck, and width of valley floor along Cleary Creek.

Claim No.	Depth to bed rock.	Thick- ness of muck.	Width of val- ley floor.
	Feet.	*Feet.*	*Feet.*
8 to 1 above	40-50		300-500
Discovery	14-20		800
2 below	40		500-600
6 below	90		800
7 below	25-27	7	800
10 below	80		800
13 below	110-120	40	1,000
15 below	100 (left limit)		
	60 (creek claim)		1,000
	90 (first tier, right limit)		

The important characteristics of the Cleary deposits are their thickness, the only shallow diggings being on Chatham and Wolf creeks and at the very head of Cleary Creek, the relation of the pay streak to the present course of the creek, and the extension of the pay throughout the lower part of the valley.

ELDORADO CREEK.[a]

Eldorado Creek is a southerly tributary of the Chatanika, its valley lying between that of Cleary Creek on the northeast and Dome Creek on the southwest. Its length is about 4½ miles, and the valley floor is from 100 yards to half a mile in width. The bed rock of the basin is, so far as determined, a quartz-mica schist, and the presence of diorite bowlders in the gravels indicates the presence of intrusives in the basin. The creek falls about 150 feet to the mile, but the slope of the bed-rock floor has not been determined. It appears, however, to be somewhat irregular.

Depths to bed rock range from 60 to 120 feet in the lower half of the creek, which is the only part visited by the writer. The muck is 20 to 40 feet thick. The gravels are well rounded and are made up largely of mica schist, with some dioritic rock and some large bowlders of white quartz. The alluvium appears to be generally frozen.

---

[a] These notes are based on a hurried examination made by Alfred H. Brooks in 1906.

Though staked some years before, the first systematic prospecting was done in 1906 and met with only moderate success. The gravels were found to be auriferous, but no pay streak had been discovered at the time of the writer's visit. It is reported that in 1907 values were found on one claim and some gold taken out.

The valley of Eldorado Creek, lying as it does between two of the important gold producers, certainly deserves careful prospecting. There appears to be good reason to believe that it will yet become a producer.

DOME CREEK.[a]

Dome Creek, also a tributary of the Chatanika, heads in the west side of Pedro Dome. The bed rock and gravels are similar to those of Cleary Creek. The depth to bed rock where the ground is being worked ranges from 30 to 200 feet. The slope of the bed rock in the lower part of the valley is about 15 feet to the mile. The surface of the bed rock is uneven, variations of 15 to 20 feet being common. The pay streak has been found to be from 130 to 165 feet wide. The material includes 2 to 3 feet of bed rock and 2 to 3 feet of gravel. Good reports were coming from nearly the entire length of the creek in September, 1907.

VAULT CREEK.[b]

Vault Creek, about 6 miles in length, flows northward into the Chatanika, and is adjacent on the west to Dome Creek. Bed rock is probably schist throughout the basin, but there are no exposures except on the round ridges.

The east wall of the valley rises by a gentle slope; the west wall is more abrupt. This gentle slope appears to be underlain by a heavy deposit of muck having a maximum thickness between 50 and 75 feet and thinning out toward the creek as well as toward the ridge. Below this muck there is a deposit of sand and fine gravel with some clay—30 to 60 feet thick—overlying coarse gravel which is auriferous. The depth to bed rock varies greatly, being determined by the position of the shaft, whether on the creek bed, where the alluvium is said to be not more than 50 feet deep, or on the talus slope, where it may be 200 feet.

The pay streak is reported to be 100 feet or more wide, with a thickness of 3 to 6 feet. It has been rather definitely traced for a couple of miles along Vault Creek below the mouth of Treasure Creek, and for half a mile up the latter stream. Little prospecting has been done on the lower part of Vault Creek, where the alluvium is said to be very deep. Near its mouth, in the Chatanika flats, the

---

[a] By Alfred H. Brooks and C. C. Covert.
[b] The information on Vault Creek was obtained by Alfred H. Brooks in 1907.

alluvium is 319 feet deep, of which 60 feet is muck. The pay streak was struck at 160 feet, on a false bed rock. Prospecting has been done on Vault Creek for several years, but it was not until 1906–7 that good values were found. In 1907 there was a considerable production. The creek has been rendered easily accessible since the railway was extended down the side of its valley in 1907.

OUR CREEK.[a]

Our Creek is tributary to the Chatanika west of Vault Creek. It has a length of about 7 miles and a general northerly course. This creek has not been visited by a geologist, but it is reported to be very similar to Vault. In 1907 good prospects were reported to have been found. If values are obtained in this basin it will show that the gold-bearing area extends to the west.

ESTHER AND CRIPPLE CREEKS.[a]

Esther Creek adjoins Alder Creek to form Cripple Creek, which is tributary to Chena Slough. The lower part of Cripple Creek flows through a broad flat, but about a mile from the slough the valley becomes well defined. Its floor here is about half a mile wide and is bounded by ridges 500 to 1,000 feet high. The Esther and Alder creek valleys are about one-half mile wide at their mouths and gradually narrow upstream. So far as known the bed rock of the entire basin is principally mica schist, but some granite is known to occur on Esther Creek.

Esther Creek has a length of about 5 miles, and gold has been found about 4 miles from the mouth. The depth to bed rock ranges from between 120 and 135 feet near the mouth to 15 feet 4 miles above. The following table shows some of the important features of the alluvial deposits.

Depth to bed rock, thickness of muck, and width of valley floor along Esther Creek.

Claim No.	Depth to bed rock.	Thickness of muck.	Width of valley floor.
	Feet.	Feet.	Feet.
7 and 8 below	90	40–60	1,000
2 to 4 below	65–75	25–30	
1 below	60	20	700
Discovery	55	15	
3 to 5 above	25	9–12	
7 above	15	3±	200

The above depths to bed rock are only approximate, as they vary greatly in any given cross section of the valley. For example, on

[a] Based on information obtained by Alfred H. Brooks in 1907.

claim No. 4 the depth to bed rock near the east side of the valley was found to be nearly 100 feet, while on the claim above, a hole on the west side of the valley encountered bed rock at 40 feet. The matter is further complicated by a terrace (20 to 40 feet) of fine silt, with a little sand and gravel, remnants of which are preserved at various places along both slopes of the valley. The bed rock underneath this terrace appears to lie at about the same altitude as that under the creek bed, but holes sunk through the terrace to bed rock are necessarily much deeper than those in the valley floor. A still higher terrace has been found along the east side of the valley near a tributary called Ready Bullion Creek. Here a heavy deposit of silt 40 to 60 feet deep was found resting on some well-washed gravels which overlie a bed-rock floor standing 100 to 200 feet above that of the creek bed. These high gravels are known to carry colors and are worthy of further investigation. The gravels are iron stained, like those of the bench placers on Cripple Creek.

The gravels of Esther Creek are in the main well rounded and are made up principally of mica and quartz-mica schist. Some large bowlders of quartz, the larger of which reach 2 feet in diameter, are also present. Granite pebbles are not uncommon, and in some places, notably on No. 4 below, they form a large percentage of the gravels. The bed rock, as stated, is generally a schist, though in some places granite has been found. The schist ranges from a fine-grained mica variety, which is as a rule deeply weathered, to a harder quartzose phase which approaches a quartzite and is commonly blocky. The ground is usually frozen, but one belt of thawed ground has been found. The pay streak as mined averages between 5 and 6 feet, of which from half to two-thirds is gravel and the balance weathered bed rock. The width of the pay streak, as defined by the present cost of mining, is about 40 to 50 feet for the upper part of the creek. The values in the material hoisted probably average $4 to $6 to the cubic yard.

Esther Creek was first staked in 1903, and some careful prospecting was done on it during the following year, but its output did not become important until 1906. It now has an established position as a large gold producer, and there is much ground on the creek which has not been prospected.

In 1907 several mines were in operation on the east side of Cripple Creek and are reported to be working on bench gravels, but these deposits have not been studied. These gravels are red like the bench gravels on Esther Creek. Some gold has been taken out, but little mining except this has been done on Cripple Creek.

Alder Creek has yielded no gold, though the gravels are said to be auriferous, and it appears probable that this creek also lies in the gold belt. Some extensive prospecting was done on this stream with a churn drill during the winter of 1907 but the results are not available for publication.

All these creeks are readily accessible from the Tanana Valley Railway. A good road leads up Esther Creek and several feeders have been built.

SMALLWOOD CREEK.[a]

Smallwood Creek, a westerly tributary of the Little Chena, is about 9 miles in length. The floor of its valley where work is in progress is about half a mile wide, and from this floor the valley walls rise with gentle slope to the bounding ridges, which stand at 1,000 to 1,600 feet above the sea. Smallwood Creek has not been gaged, but appears to carry a large amount of water. The gradient of the creek is low, probably not over 60 feet to the mile in the upper half of its course and somewhat less in the lower half. There is little clue to the character of the bed rock except the constitution of the gravels. It is known, however, that the ridge lying northwest of Smallwood Creek is made up largely of granite and that granite forms the bed rock of much of Nugget Creek, a westerly tributary of Smallwood. To judge from the alluvium, mica schist is probably the country rock of much the larger part of this basin. Among other rocks noted in the gravels is a fine-grained rock carrying porphyritic crystals of feldspar.

Comparatively little prospecting has been done on Smallwood Creek, and it is therefore possible to make but few statements as regards the character and depth of the alluvium. It appears that the bed-rock floor has in general about the same slope as the present stream valley, but locally some irregularities occur. The depths to bed rock as reported by miners are as follows:

Depth to bed rock and condition of ground along Smallwood Creek.

Claim number.	Depth to bed rock.	Condition of ground.
	Feet.	
1 above (mouth of Nugget Creek)	50	10 feet of muck; ground thawed.
2 above	50	Ground frozen.
3 below	108	Do.
4 below	130	25 feet of muck; ground frozen.
5 below	135	Ground frozen.
7 below	145	45 feet of muck; ground frozen.
17 below	317	60-100 (?) feet of muck; ground frozen.

This table indicates that there is a body of gravel ranging in thickness from 40 feet near the head of the creek to 200 feet 5 miles below. As the creek valley is from 100 yards to half a mile or more

[a] This description, by Alfred H. Brooks, is based on one day's examination of the upper part of the creek and on compiled information.

in width, it is evident that there is here an enormous body of gravel. The overburden, usually termed muck by the miners, is a fine silt or clay, in general of a dark-gray color. This overlies sands and gravels which appear to increase in coarseness toward bed rock. The gravels are mostly well rounded except on bed rock, where they are angular. They are made up in the main of mica schist, with much vein quartz, and also carry granite, which at the mouth of Nugget Creek predominates over the other material. The pebbles of gray porphyritic rock have already been referred to.

Claims were first staked on Smallwood Creek in November, 1904, but no considerable work appears to have been done until 1906. During 1906 and 1907 probably a dozen claims were worked. This prospecting has been sufficient to establish the fact that the gravels of at least the upper 2 miles of Smallwood Creek and some of its tributaries are auriferous, and at some localities values have been found. Perhaps the most significant feature is the reported finding of good prospects at No. 17 below, at a depth of 317 feet. The writer was unable to visit this locality, but it appears that the discovery is sufficiently encouraging to warrant further developments. If values have been found in this lower part of the creek, as they are known to occur in the upper part, it augurs well for the occurrence of gold between.

So little ground has been opened up that it is impossible to make any generalizations as to the width of the pay streak. It is reported that on one claim values have been found for a width of 120 feet, and that these occur in the lower 3 to 4 feet of gravels and in the decomposed bed rock to a depth of 2 to 2½ feet.

The gold seen by the writer is medium fine and occurs in small scales. Nuggets are relatively rare, the two largest reported being valued at $2.75 and $11.50. The gold is said to run about $18.11 to the ounce.

The ground in the upper mile of the creek is said to be thawed and is shallow enough to be worked by dredges or open-cut methods.

The meager information here set forth clearly indicates that this creek and its tributaries are well worth careful prospecting. The lower valley is so wide that it will be expensive to prospect it carefully, and it would appear that this can best be done by grouping the claims and systematically crosscutting the entire body of gravel.

MINING METHODS.

The methods of mining in the Fairbanks region (1905) are the same as those used extensively for similar types of deposits in the Klondike region. The methods have been necessarily determined by the grade of the valleys, the thickness and character of the deposits, and the available water supply. Only a small part of the ground has a grade of over 100 feet to the mile and most of it has considerably less. The deposits worked range from a few feet to over 120 feet in thickness. The creeks are small, carrying ordinarily 200 to 400 miner's inches of water. In dry seasons the present sources of water supply would be inadequate, and, while thus far only short ditches have been in use, a project is under way to supply some of the areas with water from the upper part of Chatanika Valley by means of a ditch about 75 miles long.

PROSPECTING.

As the pay streak, if present, lies almost everywhere on bed rock, the chief work of prospecting consists of sinking holes to bed rock. It is usually necessary to thaw the ground, and while crude methods requiring wood fires or hot water are still in use, the most approved method, and the one most commonly employed, is that carried on by means of steam. Small, portable, knockdown steam-thawing outfits that can be packed on horses are now obtainable, thus permitting prospecting in remote areas. After the ice has been melted the material to be excavated is loosened with a pick, shoveled into a bucket, and hoisted to the surface, usually by hand windlasses. If the ground is deep the prospect shaft is generally timbered to the depth of the overlying muck. The most formidable difficulty encountered in sinking is live water, which often necessitates the abandonment of shafts. Great depth of ground also increases the difficulty of sinking holes, and consequently makes the work of locating the pay streak slow.

OPEN-CUT MINING.

The ground is generally stripped first of all by sluicing off the overlying muck. A bed-rock drain is then constructed, and an open cut of sufficient width for one or two sets of boxes is carried gradually up the valley. In some cases the gravel is hoisted by steam power entirely out of the cut to boxes set above the surface and to one side of the workings. By this method a frequent resetting of the boxes is avoided and there is a better disposal of tailings. Gravel is hoisted by derrick, by automatic trolley, or by a rock pump. Where the last method is used a set of boxes is placed on the bottom of the cut, the coarsest pieces are forked out, and all the rest of the material is elevated through the pump to the boxes on the surface. Owing to the depth of the gravels the open-cut method and its modifications are of limited application.

DRIFTING.

The methods of working the deep gravels of this region are similar to those employed in the deep gravels of other fields, with the modifications rendered necessary by the frozen character of the ground. These methods have gradually developed in the Yukon Territory and in Alaska, and from year to year have become more efficient in solving the problems that are met. In the Fairbanks region in 1903 thawing was accomplished by the cruder methods mentioned, and equipments for thawing by steam, which had been found so effective in the Klondike region, were not plentiful. Since then extensive steam plants have been introduced, capable of thawing and handling daily large quantities of gravel.

The process in general includes the following operations: (1) The sinking of a shaft to bed rock, ranging in depth from 20 to 300 or more feet; (2) the timbering of the shaft and the portion of the drifts near the shaft; (3) the opening up of the ground by drifts which are run either parallel to or across the pay streak and from which crosscuts are driven; (4) the extraction of the gravel from the crosscuts, beginning at the farther limits of the drifts and working toward the shaft; (5) the hoisting of the pay gravel with as little waste as possible to the surface; and (6) the recovery of the gold by ordinary sluicing. The main drift is usually carried to a maximum distance of about 200 feet in each direction from the shaft, and the ground is blocked off by crosscuts having a variable length up to about 100 feet. Fortunately but little timbering is generally required. Where the ground is weak, pillars are left at intervals of about 25 feet when working back the faces toward the shaft. Ordinarily, as mining commences at the extreme limit of the area to be worked, the ground from which the pay dirt has been removed is allowed to settle if it will. Experience has shown that settling is generally so gradual that the work can be carried away from the settling ground with sufficient speed to avoid trouble.

The steam-point method of thawing is the one most commonly in use. The steam point is a piece of one-half or three-eighths inch hydraulic pipe, 5 to 8 feet or more in length, with a blunt, hollow point of tool steel for piercing the ground and a solid head of tool steel or machine steel, sufficiently strong to withstand the impact of a maul or sledge. Steam is admitted through a pipe fitted laterally in a small aperture near the head. The points are placed about $2\frac{1}{2}$ feet apart, and from a dozen to twenty or more are used in a plant of average size. The power needed is 1 to 2 horsepower per point and the duty of a point is 3 to 4 or more cubic yards per day of ten hours. In use the point is driven in gradually as the ground becomes thawed. It is customary in most places to use either hot water at a temperature of about 140° F. or a mixture of hot water and steam while driving the points, and then to complete the thawing by means of steam alone, since by employing hot water in a part of the operation the atmosphere of the mine does not become so vitiated through the condensation of the steam and the conditions for working are consequently better.

Hot-water hydraulicking by means of the pulsometer or other steam pump has been very successful in some places. Pulsometers in use in 1905 were reported to do the work of 20 points, and as by this method a jet of hot water is thrown forcefully against the frozen face, the gold particles are more easily released from adhesive material in which they may be embedded than by the use of points. Pulsometers are generally suspended in a sump at the bottom of the shaft, and the hot water is supplied by siphon from the boiler. Surplus water is generally removed by centrifugal pumps. It seems probable that hot-water hydraulicking will be more generally employed.

After thawing, the gravel is removed with pick and shovel and carried by wheelbarrows to the shaft, whence it is hoisted to the surface by buckets attached generally to an automatic trolley. In summer it is conveyed directly to the sluice boxes, or, when the water for sluicing is available for only part of the shift, to a hopper connected with the set of boxes. In winter the gravel is conveyed to a dump under which sets of boxes have been arranged and later, in the spring, it is passed through the sluices. Ground which stands well without timbering is worked both winter and summer, but summer work is cheaper. Ground having a tendency to cave is often left for winter exploitation, as it is found that the expense of rehandling in the spring is more than counterbalanced by the greater facility with which the gravel can be extracted.

The ordinary sluice boxes with pole riffles are universally employed, usually 12 by 14 inches in cross section and 12 feet long. An average size dump box or rock box is 20 to 22 feet in length and 36 to 40 inches or more in width. This catches from 60 to 90 per cent of all the gold saved, and most of the remainder is caught in the next three boxes, which have grades generally ranging from 9 to 12 inches to the box. Ordinarily two clean-ups a week are made. The concentrates are dried in mining pans on stoves or blacksmiths' forges, and as a rule are cleaned by dry panning and blowing.

COSTS.

The cost of mining under conditions in 1905 was so great that most of the ground worked had to carry in the pay streak values of at least 2 cents to the pan, or approximately $2.75 to the cubic yard. Most of the claims are 1,500 feet in length, measured parallel with the courses of the creeks on which they are located, and there are generally two or three outfits working on a single claim. At many claims the ground is worked by laymen, who give from a third to a half of the output to the owners.

The prevailing wage for miners is $5 a day and board, but in some places it reaches $6 a day and board. The duty per man per day of ten hours is from 75 to 100 wheelbarrows of dirt broken down with the pick, shoveled into a wheelbarrow or cars, and delivered to the shaft bucket; the average is probably about 9 cubic yards a day, but under very favorable conditions for short periods of time this quantity may be nearly doubled. The conditions under which work in the drifts is carried on vary with the character and form of the deposit. Where the pay streak is thin the drifts are made as low as possible to avoid removing more waste than is absolutely necessary, from which it is seen that the most favorable conditions occur when the pay streak is of such a thickness, 6 feet or more, that on its removal there is space for perfect freedom of movement and sufficient ventilation.

SUMMARY.

Although up to 1905 the producing creeks were few and comparatively short and most of the deposits were so deep and so consolidated by ice that machinery and much time were required for their development, the returns were for the most part satisfactory. The introduction of much machinery met with a quick response in a greatly increased production. With the lower cost of mining resulting from increased facilities in transportation, there is the opportunity every season of working ground containing lower values; there are, further, the potentialities of the undeveloped creeks which have just become producers, and the possibilities of new discoveries.

The problem of water supply is becoming more important every year, and has led to extensive plans, to which reference has already been made, for bringing water from the upper valley of the Chatanika. An inspection of the map shows a considerable difference between the level of Beaver Creek at the great bend and that of streams to the south and southeast, tributary to the Tanana, and this has been suggested by R. B. Oliver as an important possible source of water supply.

WATER SUPPLY OF THE FAIRBANKS REGION, 1907.

By C. C. COVERT.

The future development of the Fairbanks mining district depends more or less on the economical development of its water resources. Most of the producing creeks have small drainage areas and will furnish but a scanty water supply, especially during the dry season.

During July and part of August, 1907, the operators were obliged to resort to various schemes to procure sufficient water for sluicing. In some places the water was returned for a second and third time to the sluice box by means of the steam pump, entailing extra expense both in fuel and equipment, and on a number of the creeks only about half of the mines were in operation.

In the early days of the camp, when but a few operators were at work on each stream and its watershed was well protected by timber, little thought was given to the supply of water for the sluice box, but as the camp developed from year to year and the demand for water was greatly increased it became evident that a larger supply must be procured. Consequently, as with other and older camps, numerous ditch lines were planned to bring water into the district.

The general topography of the country is such that ditch lines from the larger drainage areas are not practical. (See map, Pl. I.) Most of the producing creeks rise in a high, rocky ridge, of which Pedro Dome, with an elevation of nearly 2,500 feet above sea level, is the center. At least 50 per cent of the mining is done at an elevation of over 800 feet and 25 per cent above 1,000 feet. The drainage basins of sufficient area and elevation to supply water to the upper reaches of these producing creeks lie at a distance of more than 50 miles in a direct line and over 100 miles by ditch line. The cost of building and maintaining such ditches, especially as they could furnish but a moderate supply of water, would be excessive.

In the older mining camps of Alaska, especially those of Seward Peninsula, many hydraulic enterprises have failed owing to the lack of reliable information concerning the available water supply. In order that like failures may be avoided in the Tanana Valley, the United States Geological Survey, during the summer of 1907, extended to the Fairbanks district the stream-gaging work started in the Nome region in 1906 and continued there this year.

The field work in the Fairbanks district was carried on from June 20 to September 15 and the region covered includes the drainage basins of Little Chena River, Goldstream Creek, Chatanika River, Beaver Creek, and Washington Creek, comprising an area of approximately 2,200 square miles. Owing to the lack of adequate funds the work was largely of a reconnaissance character. However, the keeping of systematic records on some of the more important streams was made possible through the hearty cooperation of the people interested. Among the many who rendered valuable assistance in procuring the data given in the accompanying tables are Mr. John Zug, superintendent of the good roads commission; Mr. A. D. Gassaway, general manager of the Chatanika Ditch Company; Mr. Falcon Joslin, president of the Tanana Mines Railroad; Mr. Herman Wobber, Fairbanks Creek; Mr. C. D. Hutchinson, electrical engineer, Tanana Electric Company; and Mr. Martin Harris, Chena.

After making a careful study of the general topographic conditions of the mining district proper and its surrounding country it was decided to establish a few regular stations at the most convenient points in the larger drainage areas and study the daily run-off during the open season from records thus obtained. This plan afforded greater opportunities for procuring comparative data than that of covering a larger territory in a less definite way. In this country without storage, daily records are an important factor, and such records could not have been obtained over an extended area. Outside of the producing creeks the country is almost a wilderness, and it is practically impossible to get observations other than those which would be made on the occasional visits of the engineer. No daily or even weekly records could have been assured, and the results obtained from scattering measurements would have furnished no comprehensive idea as to what the daily run-off of the streams really was throughout the open season.

On account of the location of the stations the results published in the following tables have a more direct bearing on the development of water power for electric transmission than on that of a water supply to ditch lines for hydraulicking, though a properly constructed ditch may furnish water for either or both.

The records kept on the upper Chatanika establish the fact that the volume of water is more nearly what would be required for a ditch supply than that of any other drainage area within a practicable distance of the Fairbanks district, except that of Beaver Creek. While the upper Chatanika may thus be considered for furnishing water to the Fairbanks district, the supply would have to be conveyed for more than 100 miles through a ditch line difficult to construct and maintain before it would be available for use, and then on account of the low head but a small number of the producing creeks would be benefited. The Beaver Creek basin would furnish a greater supply at perhaps a higher elevation, but its greater distance from the seat of operations makes it a less practical source of water than the Chatanika.

From the data at hand it appears that hydro-electric development is the most practical solution for the various industrial problems of the camp. Electric power could be readily transmitted to the various creeks and easily supplied to the individual miner as a cheap and practical power for pumping water to the sluice box, for running the hoist, for elevating the tailings, for pumping water out of the mines, for lighting the underground work, and in some localities for supplying power to the dredge.

The following tables indicate in a general way the work done in the Fairbanks district during the past season. A more detailed report of the work done in this territory will be published in a water-supply paper of the United States Geological Survey.

Table 1 gives a list of discharge measurements made at the several gaging stations together with the approximate elevation above sea level, the drainage area in square miles above the gaging station, and the discharge in cubic feet per second.

Table 2 gives the daily discharge in second-feet at the regular gaging stations. Second-feet is an abbreviation for cubic feet per second. A second-foot is the rate of discharge of water flowing in a stream 1 foot wide, 1 foot deep, at a velocity of 1 foot per second.

The "miner's inch" expresses the rate of discharge of water that passes through an orifice 1 inch square under a head which varies locally, that commonly used in the Fairbanks district being 6 inches. To obtain the discharge in miner's inches multiply the cubic feet per second by 40.

Table 3 gives the drainage area in square miles above the gaging station and the mean run-off in second-feet per square mile; $\frac{(22-31)}{0.577}$ signifies that the records covered the period from the 22d to the 31st of the month and that the mean run-off was 0.577 second-foot per square mile for that period.

TABLE 1.—*Discharge measurements made in the Fairbanks district in 1907.*

Date.	Stream.	Locality.	Elevation above sea level.	Drainage area.	Gage height.	Discharge.
			Feet.	Sq. miles.	Feet.	Sec.-ft.
July 21	Fish Creek	Above Fairbanks Creek.	925	39	1.00	23.7
July 25	----do	----do			.99	24.2
August 3	----do	-----do			1.55	47.8
August 4	----do	----do			1.35	37.6
August 19	----do	----do			1.00	20.8
June 24	Fairbanks Creek	Near Crain Creek				1.4
Do	----do	Near claim 2 above				2.2
July 5	----do	Near claim 9 above				.72
July 20	----do	Near Claim 10 above				1.30
Do	Bear Creek	Near Tecumseh	900	12		8.4
August 22	----do	----do				7.0
July 6	Miller Creek	Near mouth	750	15		7.0
July 24	----do	----do				7.6
August 20	----do	----do				8.0
August 6	----do	Just below Helm Creek	790	10		8.0
August 7	----do	----do				8.0
Do	----do	Just above Helm Creek	800	6		4.9
July 23	Elliott Creek	Near mouth of Sorrel Creek.	800	13.8	1.6	5.1
August 5	----do	----do			1.85	13.8
August 20	----do	----do			1.615	7.1
July 23	Sorrels Creek	Near mouth	800	21	1.0	10.3
August 5	----do	----do			1.4	28.2
August 20	----do	----do			1.02	12.0
July 22	Little Chena River	Above mouth of Elliott Creek.	800	79	.60	44.2
July 24	----do	----do			.565	39.7
August 4	----do	----do			1.10	113.0
August 5	----do	----do			1.05	103.0
August 20	----do	----do			.73	56.7
June 21 a	Goldstream Creek	Lower line of claim 6 below.	886	28.6		12.3
June 28 a	----do	----do				22.4
June 28	Goldstream ditch	Below intake				10.8
Do	Fox Creek	----do	900			2.0
June 27	Ditch on Dome Creek	Near claim 2 below		7		.84
June 26	Little Eldorado			4		.45
July 4	Cleary Creek	Near Cleary		10.5		2.9
July 10	McManus Creek	At mouth	1,375	80		15.6
Do	----do	----do				16.4
July 12	----do	500 feet above mouth of Smith Creek.	1,400	42		10.2
Do	----do	At mouth	1,375	80		15.6
July 13	----do	¾ mile above Montana Creek.	2,000	8		1.8

a Includes flow through small sluice box that diverts water from creek above gaging station.

TABLE 1.—*Discharge measurements, etc.*—Continued.

Date.	Stream.	Locality.	Elevation above sea level.	Drainage area.	Gage height.	Discharge.
			Feet.	*Sq. miles.*	*Feet.*	*Sec.ft.*
July 13	McManus Creek	Just below Montana Creek.	1,975	10		3.8
Do	do	1½ miles below Idaho Creek.	1,800	26		6.5
Do	do	⅜ mile above mouth	1,380			21.4
July 14	do	500 feet above mouth of Smith Creek.	1,400	42		12.4
Do	do	⅜ mile above mouth	1,380			19.4
July 12	Smith Creek	Near mouth	1,400	34		7.8
July 14	do	do				8.7
Do	do	Above mouth of Pool Creek.	1,450	17.0		5.4
Do	Pool Creek	Above mouth		14.0		2.4
Do	Charity Creek	1 mile above Hope Creek.				5.7
Do	Hope Creek	Near Zephyr				7.7
July 16	Chatanika River	Below Faith and McManus creeks.	1,350	132	1.58	51.9
July 26	do	do			1.805	80.5
August 3	do	do			1.80	96.5
August 7	do	do			2.255	188
August 15	Boston Creek	1 mile above mouth	800	6.5		3.9
Do	McKay Creek	do	800	6.2		3.7
Do	Belle Creek	do	800	11		10.0
July 9	Kokomo Creek	do	750	26		13.8
August 14	do	do				22.7
July 27	Poker Creek	⅜ mile above mouth		40	1.09	22.3
July 30	do	do			1.10	22.6
August 9	do	do		40	1.32	36.6
August 10	do	do			1.33	37.8
Do	do	1 mile above Caribou				21.1
Do	Little Poker Creek	Near mouth				3.9
Do	Caribou Creek	Above Little Poker Creek.				10.4
June 22	Chatanika River	Below mouth of Poker Creek.	700	456	1.08	246
July 4	do	do			.83	178
August 9	do	do			1.98	669
August 27	Trail Creek	About 3 miles above mouth.	1,700	27		39.9
Do	Brigham Creek	1 mile above mouth	1,500	15		16
August 28	Fossil Creek	Near mouth				19.2
August 29	Bryan Creek	5 miles above mouth	1,800	48		75.3
August 30	Beaver Creek, R. Branch.		1,800	122		267
Do	Beaver Creek, L. Branch.		1,800	67		124
Do	Nome Creek	¾ mile above mouth	1,700	120		135

TABLE 2.—*Daily discharge in second-feet of various streams in Fairbanks district, 1907.*

JUNE.

Day.	Goldstream Creek.	Faith Creek.	McManus Creek.	Chatanika River below mouth of Poker Creek.	Day.	Goldstream Creek.	Faith Creek.	McManus Creek.	Chatanika River below mouth of Poker Creek.
20	10.8	44.7	34.8	250	27	6.4	34.4	24.3	192
21	10.8	44.7	34.8	250	28	20.7	45.9	31.1	216
22	9.3	42.8	31.2	250	29	30.2	43.6	26.0	250
23	4.9	39.3	34.8	250	30	26.3	36.8	23.2	216
24	7.8	38.8	25.0	232					
25	7.8	35.3	21.7	216	Mean	13.4	40.5	28.5	228
26	12.3	36.5	25.0	192					

TABLE 2.—*Daily discharge in second-feet of various streams, etc.*—Continued.

JULY.

Day.	Fish Creek.	Elliott Creek.	Sorrels Creek.	Little Chena River.	Goldstream Creek.	Faith Creek.	McManus Creek.	Chatanika River, near Faith Creek.	Kokomo Creek.	Chatanika River, below mouth of Poker Creek.
1					20.7	32.6	21.6			192
2					12.3	28.5	20.1			192
3					10.8	26.4	19.0			192
4					9.3	24.8	18.5			167
5					4.9	22.1	17.8			167
6					3.6	21.6	16.1			167
7					15.4	22.0	17.5			167
8					12.3	20.8	17.8			204
9					10.8	20.1	15.8		13.9	180
10					6.4	19.2	15.0		10.9	167
11					32.2	21.0	16.1		25.8	192
12					30.2	20.5	15.0		19.8	192
13					17.1	20.1	15.4		19.8	204
14					13.8	21.0	17.8		16.8	192
15					34.4	20.9	18.5		16.8	216
16					28.2	21.7	19.0		16.8	282
17					18.9	35.3	21.6	60	19.8	250
18					15.4	35.0	34.7	90	19.8	250
19					12.3	62.5	40.0	80	13.9	283
20					12.3	43.9	31.6	73	13.9	266
21					10.8	38.6	26.0	66	13.9	250
22	24	5.8	10.3	42	9.3	31.4	21.2	57	10.9	250
23	24	5.8	10.3	42	7.8	25.5	17.8	54	7.9	204
24	24	5.8	10.3	42	12.3	28.8	21.4	60	13.9	192
25	24	5.8	10.3	42	9.3	26.4	19.1	73	10.9	192
26	27	9.0	14.7	80	9.3	61.0	38.6	88	13.9	232
27	24	9.0	14.7	66	6.4	42.0	29.1	78	13.9	250
28	21	5.8	10.3	53	6.4	28.4	23.9	63	13.9	250
29	21	5.8	10.3	42	4.9	30.6	21.8	60	10.9	250
30	18	4.1	8.2	42	4.9	26.7	18.8	54	7.9	216
31	18	2.5	6.0	42	2.2	25.0	16.7	54	7.9	192
Mean	22.5	5.94	10.5	49.3	13.1	29.2	21.4	67	14.2	211

AUGUST.

Day.	Fish Creek.	Elliott Creek.	Sorrels Creek.	Little Chena River.	Goldstream Creek.	Faith Creek.	McManus Creek.	Chatanika River, near Faith Creek.	Kokomo Creek.	Chatanika River, below mouth of Poker Creek.
1	155	9.0	14.7	53	30.2	36.4	81.2	80	12	752
2	100	23	27.8	157	32.2	41.1	80.8	122	68	1,160
3	39	23	27.8	113	20.7	35.9	56.1	92	43.8	680
4	35	17.2	32.1	113	15.4	34.7	51.2	101	37.9	530
5	37	12.3	25.6	113	13.8	42.5	63.4	106	31.8	530
6	39	12.3	23.4	95	15.4	40.6	60.6	106	31.8	480
7	47	12.3	23.4	104	24.4	87.4	98.6	186	34.8	405
8	35	12.3	23.4	113	26.3	62.7	84.3	147	31.8	530
9	50	15.6	27.8	134	22.5	52.4	75.6	128	43.8	620
10	50	12.3	23.4	113	32.2	44.2	77.8	122	37.9	590
11	39	12.3	23.4	95	26.3	39.0	62.2	101	31.8	455
12	31	9.0	19.0	80	15.4	35.0	49.8	88	28.9	405
13	27	12.3	19.0	95	13.8	42.8	45.5	88	25.8	342
14	27	12.3	19.0	66	12.3	35.6	40.0	80	22.7	363
15	27	9.0	14.7	80	12.3	33.6	37.2	80		300
16	24	9.0	14.7	66	13.8	34.4	42.4	80		283
17	24	9.0	14.7	66	10.8	30.8	39.0	80		250
18	24	7.4	12.5	60	10.8	30.6	37.4	78		250
19	24	7.4	10.3	60	10.8	28.5	34.7	73		216
20	24	5.8	10.3	53	13.8	27.8	33.6	73		250
21	24	5.8	10.3	53	18.9	26.9	32.2	73		250
22	24	5.8	10.3	53	20.7	44.2	68.7	111		216
23	24	5.8	10.3	53	20.7	39.4	50.3	101		266
24	24	5.8	10.3	53	20.7	49.8	67.1	122		283
25	27	7.4	12.5	73	18.9	62.8	81.2	177		342
26	27	9.0	14.7	95	22.5	82.6	102	186		430
27	24	9.0	14.7	80	20.7	69.3	92.6	154		321
28	24	12.3	19.0	80	22.5	62.6	91.2	147		363
29	27	12.3	19.0	88	26.3	70.5	92.5	114		430
30	31	12.3	19.0	95	28.2	72.5	112	186		505
31	27	12.3	19.0	95	28.2	67.8	94.1	186		455
Mean	36.8	11.0	18.2	85.4	20.0	47.5	66.4	117	41.6	428

TABLE 2.—*Daily discharge in second-feet of various streams, etc.*—Continued.

SEPTEMBER.

Day.	Fish Creek.	Elliott Creek.	Sorrels Creek.	Little Chena River.	Goldstream Creek.	Faith Creek.	McManus Creek.	Chatanika River, near Faith Creek.	Kokomo Creek.	Chatanika River, below mouth of Poker Creek.
1	24	12.3	19	95	18.9	59	71.5			384
2	24	12.3	19	80	17.1	52.5	62.8			363
3	24	9.0	14.7	80	17.1	50.2	57.8			321
4	24	9.0	14.7	66	18.9	66.4	57.2			321
5	27	9.0	14.7	80	20.7					300
6	27	9.0	14.7	88	17.1					321
7	27	9.0	14.7	88	17.1					321
8	27	9.0	14.7	95	15.4					321
9	27	10	14.7	95	17.1					384
10	35	12	19	95	15.4					342
11	(a)	(a)	(a)	(a)	22.5					342
12					36.6					2,160
13					36.6					3,160
14					28.2					1,780
15					28.2					1,390
16					41.0					2,620
17					30.2					2,980
18					26.3					942
19					20.7					1,060
20					32.2					901
21					26.3					942
22					24.4					901
23					30.2					860
24					26.3					788
25					24.4					680
26					24.4					680
27					20.7					680
28					18.9					680
29					22.5					788
30					24.4					942
Mean	26.6	10.0	16.0	86.2	24.0					954

OCTOBER.

Day.	Goldstream Creek.	Chatanika River below mouth of Poker Creek.	Day.	Goldstream Creek.	Chatanika River below mouth of Poker Creek.
1	20.7	860	10		232
2	20.7	680	11		384
3	24.4	590	12		590
4	20.7	530	13		560
5	20.7	505	14		530
6	20.7	480	15		(b)
7	17.1	455			
8	(b)	384	Mean	20.7	506
9		300			

a No records for remainder of season.   b Stream frozen over.

TABLE 3.—*Mean run-off at various gaging stations in the Fairbanks district, 1907.*

Stations.	Elevation (feet).	Drainage area (square miles).	June.	July.	Aug.	Sept.	Oct.
Fish Creek above Fairbanks Creek........	925	39		(22–31) 0.577	0.944	(1–10) 0.682	....
Elliott Creek near mouth of Sorrels Creek.	800	13.8	....	(22–31) .430	.797	(1–10) .724	....
Sorrels Creek near mouth..................	800	21	....	(22–31) .500	.867	(1–10) .762	....
Little Chena River above mouth of Elliott Creek.............................	800	79	....	(22–31) .624	1.08	(1–10) 1.09	....
Goldstream Creek, lower line of claim 6 below..............................	880	28.6	(20–30) 0.469	.542	.700	.826	(1–7) .724
Faith Creek near mouth..................	1,375	51	(20–30) .794	.572	.932		
McManus Creek near mouth	1,375	80	(20–30) .356	.268	.830		
Chatanika River near Faith Creek........	1,350	132	....	(17–31) .508	.886		
Kokomo Creek near mouth................	750	26	....	(9–31) .546	1.60	.	
Chatanika River below mouth of Poker Creek..........................	700	456	(20–30) .500	.463	.938	2.09	(1–14) 1.11

In connection with these investigations the following rainfall stations were established:

Summit Road House near Pedro Summit, elevation 2,310 feet.
Cleary, elevation 1,000 feet.
Chatanika River near mouth of Poker Creek, elevation 730 feet.
Chatanika River near mouth of Faith Creek, elevation 1,400 feet.

The results of the observations taken at these stations, together with other records kept in the Tanana and Yukon basins in 1907, are as follows:

TABLE 4.—*Daily rainfall, in inches, at stations in Fairbanks district, 1907.*

Day.	May.	June.	July.				August.					September.		
	Fairbanks village.	Fairbanks village.	Fairbanks village.	Summit Road House.	Cleary.	Chatanika River near Faith Creek.	Fairbanks village.	Summit Road House.	Cleary.	Chatanika River near Poker Creek.	Chatanika River near Faith Creek.	Fairbanks village.	Cleary.	Chatanika River near Poker Creek.
1							0.72	1.27	1.17	(a)	0.49			
2						0.02	.01	.06	.12	(a)	.19			Tr.
3		0.04		0.09						(a)				
4									.09	(a)	.20	0.18	0.08	0.10
5		0.15				.04	.13	.27	.04	0.05	.08	.03	.14	
6		.09		0.30					.11					
7		.11	.35	.06	.30	.14		.07	.22	.24	.15	.12		.01
8		.07		.01		.09		.01	.42		.16	.18	.11	.02
9		0.15							.11	.46	.33	.10	.02	.22
10		.15		.50	.47	.03	.25		.08	.05	.02	.05		Tr.
11		.05	.02	.12	.09	.14							.23	.63
12		.05		.22	.32							.71	.21	.88
13	.06				.03	.05			.01					.80
14				.30		.05								
15		.02	.01	.05	.19	.28	.09				.07	.22	.85	.70
16		.18	.19	.24	.20	.11					.01	.04	.27	.10
17		.10		.03	.01	.01	.09						.01	.01

a No record.

TABLE 4.—*Daily rainfall, in inches, at stations in Fairbanks district, 1907*—Con.

Day.	May. Fairbanks village	June. Fairbanks village	July. airbanks village	July. Summit Road House	July. Cleary	July. Chatanika River near Faith Creek	August. Fairbanks village	August. Summit Road House	August. Cleary	August. Chatanika River near Poker Creek	August. Chatanika River near Faith Creek	September. Fairbanks village	September. Cleary	September. Chatanika River near Poker Creek
18		0.02	0.09	0.24	0.15	0.35	0.12		0.05	0.07	0.01			
19				.13										
20				.14		.13						0.15	0.15	0.40
21							0.19			.13		.15		.13
22				.01			.05	.04	.10	.02	.13	.16	.52	.27
23							.20		.11	.13		.37	.23	.15
24				.25	.22	.27	.23	.18	.13	.09	.04	.15		
25				.18	.15			.12	.03	.13	.15	.36		
26				.05	.02	.12	.31		.03					
27						.07								
28		.23							.26	.02	.09			
29	.13	.30		.13	.06		.13		.13	.22	.15	.54	.39	.30
30	.01			.12		.12					.13			
Total	.35	1.47	1.51	2.71	2.55	1.87	1.81	3.27	2.88		3.00		3.82	3.70

TABLE 5.—*Monthly precipitation, in inches, at stations in drainage basin of Tanana River, 1907.*

	Jan.a	Feb.a	Mar.a	Apr.a	May.	June.	July.	Aug.	Sept.
Cleary							2.55	2.88	3.82
Chatanika River, near Poker Creek									3.70
Chatanika River, near Faith Creek							1.87	3.00	
Fairbanks	3.30	0.86	2.42	0.03	0.35	1.47	1.51	1.81	
Summit Road House							2.71	3.27	
Summit (headwaters Tanana)	1.8	.10		.40	.80	2.15			

* Records for these months are practically 10 per cent of snowfall.

TABLE 6.—*Monthly precipitation, in inches, at stations in drainage basin of Yukon River, 1907.*

	Jan.a	Feb.a	Mar.a	Apr.a	May.	June.	July.	Aug.	Sept.
Central House	1.04	0.42	2.57	0.93	0.57	2.21	1.40		
Circle	1.02	.57	.28	.15	2.29		1.86	2.79	1.73
Fort Egbert (Eagle)	1.45	.20	0.0	.15	.40	1.89	1.48	1.98	
Fort Gibbon	1.26	.21	.53	0.0	.30		2.56	2.31	2.32
Holy Cross Mission	2.08	.55	4.49			2.95	3.73	5.39	
Ketchumstock	.12	.20	.27	Tr.	1.3	2.03	1.60	2.14	
North Fork	.69	.28	.27	Tr.	1.34	1.92	1.57	3.19	2.0
Rampart	1.17	.44	1.17	.02	.44	1.64	2.29	3.38	

* Records for these months are practically 10 per cent of snowfall.

SALCHA REGION.

The area drained by streams in the southeastern part of the Fairbanks quadrangle has been under investigation by prospectors for several years. The bed rock throughout this region consists essentially of schists, gneiss, 'crystalline limestone, some greenstone, serpentine, and intrusive hornblende granite.

Tenderfoot is the only creek that up to the present time (1907) has proved productive. Steamers run occasionally from Fairbanks to the mouth of Banner Creek and supplies are thence carried by pack train a distance of about 3 miles to Tenderfoot Creek. The creek is only about 6 miles long and carries probably not more than 3 to 4 sluice heads of water. It flows for a part of its course in a narrow channel in the muck 15 to 20 feet below the valley floor, which is a quarter of a mile or more wide and has a grade of about 100 feet to the mile. There are remnants of a bench of soft deposits in parts of the valley just to the west of the creek and about 40 feet above it. The valley is filled with deposits ranging, in the prospect holes that were being sunk in 1905, from 48 to 120 feet in thickness. The overlying muck is from 36 to 80 feet thick. The gravels are similar to those of the Fairbanks region and comprise quartzite schists, mica schist, carbonaceous schist, feldspathic schist, and granite.

RAMPART REGION.

GENERAL STATEMENTS.

The Rampart region (see Pl. II) is about 80 miles northwest of the Fairbanks region, and all the creeks of present economic importance are within 30 miles of the Yukon and belong to the drainage systems of both the Yukon and the Tanana.

The region has passed through many stages characteristic of the life of a placer camp. Some of the creeks were prospected as early as 1893 and were active producers by 1896, when the region became of equal prominence with the Fortymile and Birch Creek regions. Many were attracted by the favorable results, and during the winter of 1898–99 the town of Rampart, the suply point of the camp, contained about 1,500 people. After the preliminary stage of prospecting and the subsequent excitement of the boom days, with their excess of hopes and population, the camp settled down to the laborious existence of an average producer, influenced from time to time by the discoveries of gold in other portions of Alaska and rewarded occasionally by discoveries in its own territory. Discoveries have recently been made which have contributed to the permanence of the camp and illustrated the possibilities still existing in a region which has already been under investigation for several years. The introduction of hydraulic methods, too, entailing the expenditure of considerable capital, has given further importance to this region.

The town of Rampart, with a population of a few hundred, is on the south bank of the Yukon, 170 miles below Circle and about 70 miles above the junction with the Tanana, at a point where the

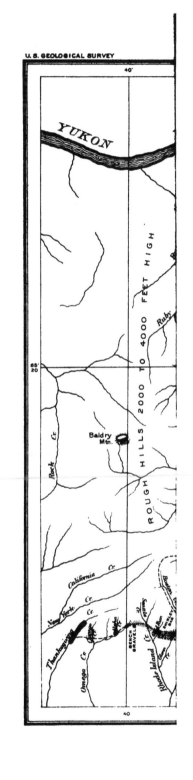

Yukon, after pursuing for a few miles a southerly course, bends squarely to the west and sweeps in a deep channel past the hills which bound the south side of the rather open valley. A narrow-terraced slope between these hills and the river is picturesquely occupied, along the water front and the hillside in the background, by the irregular collection of buildings that forms the town. There is an air of importance about the place, and it possesses, also, a kind of dignity which the pervasive majesty of the great river and the vast loneliness of the country through which it flows have conferred upon every one of these small isolated outposts of civilization.

Conditions in the Rampart region during 1906 [a] were as follows:

The total gold output of the Rampart district for 1906 is estimated to have a value of $270,000. The writer is indebted for valuable information to Messrs. H. F. Thumm and E. H. Chapman, of Rampart. Mr. Thumm states that about 33 claims were worked during the winter of 1906 and 17 during the summer, giving employment to about 100 men in winter and about twice as many in summer. New creeks not producing last year are Boothby and Skookum.

Three hydraulic plants were operated during part of the summer, one each on Hoosier, Ruby, and Hunter creeks. The Alaska road commission has begun the construction of a highway from Rampart up Big Minook. This when completed will materially reduce the cost of all mining operations.

Another road has been built from Baker Hot Springs to Glenn Creek, a distance of 24 miles, by Thomas Manley, a large owner of mining property. This road affords a natural outlet to Tanana River for the Glenn Creek region. Mr. Manley has also surveyed a ditch line from Hutlinana Creek to Thanksgiving Creek, a distance of 15 miles. If the scheme is carried out and there is sufficient water it will lead to extensive mining developments in the Glenn Creek region. It is of interest to note that the same operator has imported a churn drill for prospecting, the first in the district.

Baker Hot Springs, on a slough about 6 miles from the Tanana, was rapidly developing during 1906 as a supply point for the creeks. Since all the common vegetables can be grown there it has become a productive center of supplies for both the Fairbanks and the Rampart regions. Twelve acres were under cultivation in the vicinity of the hot springs and about 50 acres had been prepared for cultivation. Although the most favorable conditions for cultivation prevail in the vicinity of the hot springs, on what is called the warm ground, where a large variety of vegetables can be grown, there are large areas of cold ground where the common vegetables like cabbage, beets, radishes, and potatoes can be grown in abundance.

There is a station of the Government telegraph line at Rampart and another at Hot Springs, and these afford communication with other portions of Alaska and the outside at rates which are low in comparison with the advantages which may thus be secured.

The total production of the Rampart region, including that of 1906, is approximately $1,582,000.

---

[a] Brooks, A. H., Bull. U. S. Geol. Survey No. 314, 1907, pp. 37–38.

During 1907 there was much activity on the southern side of the divide and the discoveries on Patterson Creek indicate the possibility of another productive area.

BED ROCK.

Most of the gold-producing creeks tributary to Minook Creek from the east head in areas composed of slates, quartzites, feldspathic quartzites, chert, and sheared chert derivatives, and flow in the lower parts of their valleys through areas of greenstone, which are largely tuffaceous. The schistose, fine-grained fragmentals, alternating with the slates and quartzites, form the greater part of the bed rock in the valleys of the streams tributary to Baker Creek. The same rocks strike northeastward and occupy large areas in the valleys of the headwaters of the Tolovana, which were traversed by the Brooks party in 1902, and still farther in the same direction are found in the White Mountain section. There is no essential difference in the bed rock of the northern and southern sides of the divide in the Rampart region, except that the greenstones are confined mostly to the northern side in the lower part of Minook Valley below Florida Creek.

ALLUVIAL DEPOSITS.

The alluvial deposits of the Rampart region are in general much shallower than those of the Fairbanks region. They include muck and underlying gravels. The gravels reflect the variety of the bed rock and are present as both stream gravels and bench gravels. The stream gravels are composed of both angular and subangular material from the bed rock and well-rounded material from the benches. Bowlders at some localities are rather common. Bench gravels are of common occurrence, and much of the mining at present is confined to them. In this respect the region differs from the Fairbanks region, where bench gravels are the exception. The bench gravels lie at various levels up to 600 feet or more above the streams, and the greatest thickness that has been determined is about 100 feet. The material is both fine and coarse and includes a large proportion of quartzite pebbles. The pay streak of the pay gravels is next to bed rock, or partly within it when this is blocky, or in the lowermost few feet of gravel. Prospects have been found at several localities in the high bench gravels of Minook Creek, but the distribution of gold in these gravels has not been determined. In the Baker area the pay of a part of the bench gravels at least occurs as a streak similar to that of the creeks. The gold is mostly well worn and on some creeks is associated with nuggets of silver.

SOURCE OF THE GOLD. ·

A large part of the gold in the stream gravels has been concentrated from the bench gravels. The occurrence of gold has not been directly traced to a definite relation to any particular bed rock or to the quartz seams, which are rather common in the slates. Many of the dikes are more or less mineralized, and some of them are reported to carry values. Light-colored acidic dikes like those of the Fortymile region, with associated quartz veins, were not observed in the Rampart region. The slates contain generally a large amount of carbonaceous matter, and anthracitic material is common in some of the small quartz seams. Pyrite is often found in both the slates and the quartz seams. On creeks where the conditions are apparently least complex the only rocks observed were carbonaceous slates and grits with quartz seams, which occasionally are a foot or more in thickness, and the monzonitic intrusives in the ridge about the headwaters. Many of the nuggets have a considerable quantity of quartz attached, and it seems probable that the gold has been derived from the small quartz seams. The only general fact which seems to emphasize itself is that the occurrence of gold in quantities of economic importance is limited to an area where deformation of the rocks has been intense and where there has been much igneous activity.

PLACERS OF THE RAMPART REGION.

By Frank L. Hess.

GENERAL STATEMENT.

The placers of the Rampart region were studied by the Survey party for ten days during the first part of September, 1904. Every working claim was visited except those on Gunnison Creek, but in a number of cases where claims are worked only during the winter the operators could not be seen. Foot traverses were carried over the region and a sketch map was made (Pl. II). The time allowed only a hasty reconnaissance, though if it had been possible to know the dates of arrival and departure of the boats more time could have been put upon the study of region. The miners met were universally generous, hearty, and hospitable, ready to help whenever possible with information or otherwise, and the work was thus made much more effective and pleasant.

The placer diggings near Rampart may be grouped according to the drainage systems to which they belong. The three general groups, Minook Creek, Baker Creek, and Troublesome Creek, are separated by a divide having the general shape of a Y whose stem runs northeastward between the Minook and Baker Creek drainage, whose left arm runs nearly northward from Wolverine Mountain, about 13 miles southeast of Rampart (Pl. II), and whose right arm runs nearly eastward, from the eastern base of the mountain. Between the arms, extending northward, is the "Troublesome country," as the region surrounding the creek of that name is known from its steep, rocky ridges and deep, narrow valleys. Each group embraces only the diggings located on the creek that gives the group its name, or upon its tributaries.

In the Minook Creek group most of the gold-bearing creeks are on the east side, nearer the left-hand arm of the Y, only a few diggings being on the west side of the creek. In the Baker Creek group the diggings now known are on the side nearer the Y, and in the Troublesome Creek group the only diggings known are on the west side of the creek, on branches flowing from the left arm of the Y. The extreme length of the area containing known gold-bearing localities is about 30 miles and its greatest width is about 12 miles, the total area being probably less than 350 square miles.

Winter prospecting is being done on Squaw Creek, a tributary of the Yukon about the size of Minook Creek, which enters the river nearly opposite Rampart.

The first placer claim in the Rampart region was located and worked in 1896 on Little Minook Creek by F. S. Langford, though gold had been previously discovered by John Minook, a Russian half-breed, who seems to have sluiced out a small amount of gold, and for whom the creek was named. Some prospecting had probably been done along Minook Creek a number of years before. Since the first systematic work in 1896 the region has been a constantly productive one. Though the amounts taken out have not been so large as those mined at places in the Klondike district or at a few of the claims near Nome, yet a number of creeks in this region produce a fair amount of gold. At first Little Minook and Hunter creeks were the only producers, and during 1897 no new ground seems to have been found, but in 1898 a small amount was taken out of Quail Creek. Afterwards gold was discovered upon Little Minook Junior, Hoosier, Ruby, and Slate creeks of Minook Creek Valley. In the meantime prospecting was carried on over the divide on the south, and deposits along Baker Flats were discovered. In fact, each year has shown some new source of production, and it seems likely that more may still be found. The output to the fall of 1904, from the best available data, was $1,112,000, and that for the year ending at the same time was about $232,900.

As Rampart lies only about 1 degree south of the Arctic Circle, the cold of winter is severe and the open season is comparatively short. During the early part of June thawing is generally so far advanced that some preliminary work and sluicing can be done. Cold snaps are likely to make the work intermittent at first, but the latter part of June and all of July and August can be depended upon for outside operations. Frosts are likely to occur the first part of September, though mining can sometimes be carried on during practically the whole month. In 1904 the sluice boxes froze up on the 5th of September, and after that date there were only a few days on which sluicing could be done.

The surficial deposits are always frozen, and the limit of the frozen ground has not yet been reached, but there are channels in the frozen gravels through which water circulates freely at all seasons. Large masses of ground ice often occur in the muck, though none are found in the gravels. The depth of the alluvial deposits sometimes exceeds 100 feet, but it is generally less than one-fifth of that amount.

The larger part of the mining has been carried on by drifting and open cuts, depending on the season and the local conditions, but during the season of 1904 two hydraulic plants began active operations, and two more were under construction. Ordinarily, wherever the gold-bearing alluvials are of sufficient depth they are mined by drifting during the winter and the dirt taken out is washed in the spring. In some cases the presence of water interferes very seriously with the drift mining and renders gravels otherwise workable comparatively valueless. Drifting can not ordinarily be carried on in the summer time, because the warm air melts the ground and causes it to cave. In thawing the ground for drift mining steam points have generally superseded wood fires, though the latter are still sometimes used.

During 1904 wages were $5 and board for a 10-hour day. This is equivalent to $6.50 to $9 a day, varying with the locality. The men who work for wages are generally strong and healthy and render a full equivalent for their pay.

The currency of the country, as in the early stages of most placer camps, is gold dust. The different values of the gold from the different creeks makes the fixing of the price at which it should pass rather difficult, and the result is that, while some gold passes considerably below its value, some passes at more than it is actually worth. The gold assays from $14.88 to over $19 per ounce, and passes at $15.50 to $18 per ounce.

MINOOK CREEK GROUP.

This group includes the placers of Minook Creek and its tributaries within limits of 5 to 13 miles from Rampart. Most of the diggings, and much the richest so far discovered in the group, are upon the east side of the valley, and none have been found in the main valley above the mouth of Slate Creek, 11 miles from Rampart.

The hills are generally rounded or flat-topped. The valleys are canyon-like, with steep walls 500 feet or more high, and benches are prominent features of the topography. The larger streams have cut their valleys down to a grade varying from 40 to 80 feet to the mile. The watershed of Minook Valley is narrow on the west, sometimes not over a half mile or a mile wide, and is probably at no place over 4 miles wide. On the east it is 5 to 7 miles wide through the greater part of the length of the creek.

The total production of the Minook Creek group has been about $702,600, of which $75,500 was produced during the winter of 1903-4 and $10,900 during the summer of 1904, making a total for the year of $86,400.

The surficial deposits are derived from the country rocks, mostly slate, quartzite, and greenstone, and reach occasionally a depth of over 100 feet, though usually much less than that, and there is generally a large proportion of muck.

MINOOK CREEK.

General description.—Minook Creek empties into the Yukon just east of Rampart, and is about 25 miles long. Near its mouth it is a shallow stream 50 or 60 feet wide, with a flow of possibly 200 second-feet or 8,000 miner's inches. It flows in a northerly direction through a deep valley whose width varies from a few hundred feet to about a half mile. The creek receives a number of large tributaries from the east—Hunter, Little Minook, Little Minook Junior, Hoosier, Florida, Chapman—and a number of creeks whose names are unknown. From the west it receives Montana, Ruby, Slate, and Granite creeks and a few small tributaries. Granite Creek, about 17 miles from the Yukon, is the largest western tributary, carrying probably 30 to 40 second-feet; Minook Creek carries perhaps 40 to 50 second-feet at the junction. These approximate estimates are given to convey some idea of the comparative sizes of the streams.

Aneroid barometer readings by Arthur J. Collier [a] showed a descent of about 760 feet from the "106 road house," about 1½ miles above Granite Creek, to the Yukon. As the distance is about 18 miles, these readings indicate a gradient of about 42 feet to the mile. In the next 3½ miles above he noted a rise of 240 feet, showing a gradient of about 68 feet to the mile. According to M. E. Koonce,[b] of Rampart, the creek has a fall of about 40 feet in the vicinity of the mouths of Ruby and Slate creeks. Aneroid barometer readings of L. M. Prindle and the writer showed a somewhat higher grade for the central portion of the creek. It seems likely that Minook Creek has an average gradient of somewhat over 40 feet per mile from the Yukon to Slate Creek and a somewhat steeper gradient above Slate Creek.

Just below the mouth of Slate Creek the Minook spreads into a number of branches in a wide gravel flat. This flat, which is typical of many Alaskan streams, is probably due to a change in the grade of the creek. The stream here is unable to carry the gravels of the swifter water above, and so spreads them upon the flat. Here are found the so-called "winter glaciers," which sometimes last through the short summers. In 1904 a quarter or half acre of ice still remained when the September frosts occurred. This ice owes its origin to the fact that the channel which carries the water is greatly con-

---

[a] Personal communication. In giving aneroid barometer readings their lack of reliability is recognized in all cases.
[b] Personal communication.

tracted by freezing in the fall. The resulting hydrostatic pressure cracks the ice and the water overflows and freezes. This process is repeated until a considerable thickness of ice is accumulated.

The valley is V-shaped in cross section, and the eastern slope is often benched, while the western is more abrupt and has remnants of benches at but few places. Five well-marked benches rise at irregular intervals above the floor between Little Minook Junior Creek and Hoosier Creek (fig. 3), the highest of which is about 500 feet. These benches are features of much importance in both the physiography and economic geology of the region. Important gravels cover the highest one, which lies on the east side of the Minook and extends from Hunter Creek to about a mile above Florida Creek. It will be described later. A small remnant of the same bench is found on the north side of the mouth of Montana Creek and another on the north side of the mouth of Ruby Creek. Other remnants are found on the north side of the mouth of Chapman Creek and at a point about

Fig. 3.—Diagrammatic sketch of Minook Valley.

4½ miles above the Chapman on the same side of Minook Creek. The last two benches show no gravel. On the west side of the creek but few remnants of benches are found. One, about 50 feet high, extends to a little above the mouth of Hunter Creek, and is probably an extension of a corresponding bench on the south side of the Yukon. It seems probable that all of the benches of Minook Creek may be more or less closely correlated with the benches of the Yukon. In the vicinity of the mouth of Slate Creek is a bench cut in the upturned slates and thin-bedded quartzites to a depth of 12 to 16 feet and covered by 4 or 5 feet of gravel and a foot or more of muck. No gravel has yet been found upon the benches of intermediate height, but further investigation may show its presence.

In its upper course the creek flows somewhat north of east for about 2 miles, and here the topography of its valley is altogether different from that of the lower part. The north side is a long, gentle slope with a greater rise in the upper part, while the south side is steep and the stream flows near its base. The asymmetry of this part of the valley is repeated in Eureka, Pioneer, Hutlina, Omega, New

York, California, and many other creeks of the region whose valleys lie in parallel or nearly parallel directions.

The rocks in the upper part of the valley are mostly closely folded slates and limestones. Garnetiferous schists occur at Ruby Creek, and greenstones form the bed rock of the lower valley except near the mouth, where they are partly covered by the Kenai rocks.

The alluvials of the valley are said to be 10 to 12 feet thick and consist of the usual muck (soil mixed with much vegetal matter), peaty soil, and gravel, with much angular débris at the foot of many of the hillsides. In the middle part of the valley they consist of about 5 to 6 feet of muck and the same thickness of gravel. The muck thickens toward the sides while the bed rock remains about level. The gravel deposits are derived from local bed rock and contain large numbers of smoothly rounded quartzite bowlders from a few inches to 3 feet in diameter, whose source has been a mystery to many. Some of these bowlders have undoubtedly descended to the present creek bed from the high benches already referred to, in whose gravels they are abundant.

The outcrop of quartzite near the " 72 road house " would in itself seem sufficient explanation for the bowlders below, but above this point the thinner quartzite beds have added many more to the stream. The quartzites are so hard and their abrasion is so slow that while the other rocks wear into sand and small pebbles, or decompose and are swept away, the quartzite bowlders remain and make up a continually larger proportion of the gravels.

Mining.—Minook Creek has not produced a large amount of gold. The wide valley, large stream, and heavy gravels have made mining difficult, so that men with the limited means of the ordinary prospector have found it more advantageous to work the smaller streams. The total production to 1904 is placed by miners of the region at $9,900. The gold produced is said to have been taken from the central portion of the valley, partly from bar diggings and partly by drifting, but in general the gravels do not seem to be rich enough for working by pick and shovel methods.

Nothing was learned of the occurrence of gold in the gravels of Minook Creek above the mouth of Slate Creek, except that colors have been found throughout its length. Below the mouth of Ruby Creek colors of gold are said to have been found in the gravels of a bench on the west side of Minook Creek, a few feet above the present stream, but not in paying quantity. The débris here is largely a carbonaceous slate somewhat schistose and highly impregnated with pyrites. An assay [a] of some of the material gave a trace of silver.

Two small areas worked in the gravels between Ruby and Slate Creeks are said to have given values of about $3 per square yard of

---

[a] Burlingame, E. E., & Co., Denver, Colo.

bed rock; another small area is said to have given $4 per square yard, and nuggets of values up to $90 are reported to have been found. The gold is stated to be practically all upon bed rock. The width of the gravel in which gold is found is not known, but it is supposed to occur throughout the gravels which floor the valley for a width of half a mile.

A company has been formed to hydraulic this portion of the creek; considerable preliminary work has been done, some pipe, lumber, etc., were on the ground September 20, 1904, and a large amount of pipe and other supplies for the company were brought to Rampart by the steamer *Susie* on her last trip up the Yukon for the season.

Several schemes were on foot for working the gravels in the lower part of the valley near the mouth of Hoosier Creek. One proposition was to work them with a dredger, and another with power scrapers. Little was learned of either plan, but from the roughness of the bed rock dredging would seem a difficult undertaking, except in the limited area in which the bed rock seems to be the Kenai sediments.

HIGH BENCH.

The high bench mentioned on the east side of Minook Creek, the most prominent feature of Minook Valley, needs to be treated here, as on its gravels depends probably in large measure the richness of most of the placers of the Minook region.

This bench, starting at a point about a mile above the mouth of Ruby Creek and about 9 miles in a straight line above the mouth of Minook Creek, continues to Hunter Creek, to a point within about 3 miles of the Yukon. The eastern line bounding the bench runs about N. 60° E., so that between Hunter Creek and Little Minook Creek the bench has a width of between 2½ and 3 miles. At its extreme eastern side the bench has a height of about 800 feet above Minook Creek, and it slopes toward the west until the height above the stream is only about 500 feet. The surface of the bench is remarkably smooth and continuous between the various streams that have cut across it, and resembles a plain through which deep ditches have been cut. It seems to narrow somewhat and crosses Hunter Creek at the mouth of " 47 Pup " continuing in a northeast direction toward the Yukon. Although the writer was unable to follow the bench farther than Hunter Creek, miners assured him that they were able to trace it beyond in a northeast direction by gravels on the surface.

The gravels contain chert, diabasic and metamorphic rocks, vein quartz, and some other pebbles, with also many very large quartzite bowlders. They are exposed on the sides of the valleys of the different creeks cutting the bench and have rolled down into the present stream beds where the great number of large quartzite bowlders make considerable trouble for the miner.

The origin of these gravels has been puzzling to the miners and prospectors. Their great width and depth, their position so far above the present gravels of Minook Creek, and the presence of the great quantity of heavy quartzite bowlders, where the bed rock would afford no such material, have made it seem to many miners necessary to assume that some larger stream, possibly the Yukon itself, once flowed across the country. This view received some support from the apparent course of an old channel either toward or from the northeast, while the present stream flows somewhat west of north from the mouth of Hunter Creek.

The data at hand suggest that Minook Creek while flowing toward the Yukon to the northeast of its present course, when the land stood at a lower altitude, had formed a flood plain of approximately the dimensions of the present high bench. With the elevation of the land along the Yukon, the effects of which are to be seen over hundreds of miles, the mouth of Minook Creek may have been raised through local variations, its grade may have been lessened, and the former flood plain may have had the gravels under discussion deposited over it. As the elevation went on, the creek was forced to the west and finally found a new outlet to the Yukon. The elevation continued, and Minook Creek cut downward, leaving its gravels on a bench above it. The elevation did not, perhaps, proceed steadily but periodically, and thus intermediate benches were formed.

The smaller creeks, Hunter, Little Minook, and Hoosier, all give some support to this hypothesis. By reference to the map (Pl. II, p. 60) it will be noticed that each of them, upon reaching the edge of the bench gravels, sharply changes its course and flows westward through the high bench. In the case of Hunter Creek and Little Minook Creek the change in direction amounts to about a right angle, while with Hoosier Creek the angle is less acute but still noticeable. The eastern limit of the bench gravels probably marks the mouths of the various streams when this line represented the course of Minook Creek. As the course of Minook Creek was shifted to the west the tributary creeks followed under the influence of the same force that shifted the larger stream. The age of the bench is probably Pleistocene, as is shown by vertebrate fossils found in the gravels of Little Minook Junior Creek, which seem to be the oldest gravels of the streams cutting the bench.

Gold has been found in the bench at many places, and between Little Minook Creek and Hunter Creek a large amount of prospecting has been done. This portion of the bench is known as " Idaho Bar."

One shaft near the middle of the divide toward their eastern edge is said to have shown the gravels to be over 100 feet thick. Many other prospect holes have been sunk in them at various places, and tunnels were run above Little Minook Creek at their eastern edge. Three claims upon this portion of the bench have been patented. From the bottom of one prospect hole, between Little Minook Junior Creek and Hoosier Creek, $27 was reported to have been taken, but drifting failed to show pay. Above Florida Creek, in the small area of high gravels known as "Macdonald Bar," prospect holes gave colors but no pay. Apparently the gravels of the bench are nowhere rich enough to pay for drifting, although if it were possible to get hydraulic water to them cheaply they might, perhaps, be worked at a profit. The aneroid barometer readings, though not very reliable, suggest the possibility of bringing water from a point 3 or 4 miles above the mouth of Granite Creek under sufficient head to work at least a part of these gravels, if prospecting should show them to be rich enough to warrant the expense.

CREEKS CUTTING THE HIGH BENCH.

Hunter, Little Minook, Little Minook Junior, Hoosier, and Florida creeks cut through the high bench just described. Of these, Little Minook Junior and Florida creeks have their channels in large part or wholly within this area, while, as already noted, the other creeks lie partly outside and change their courses noticeably upon reaching it. The three longer creeks head close together in the hills which extend northward from Wolverine Mountain and divide the Minook drainage from that of Troublesome Creek. Their valleys, even at the heads, are so steep that the trails leading out of them are exceedingly difficult to travel.

Hunter and Hoosier creeks not only have had a sufficiently large flow to cut their canyons, but they did it quickly enough to have since had opportunity to widen them, while Little Minook Creek with its smaller volume has not yet graded its valley sufficiently to do so much side cutting, and Little Minook Junior and Florida creeks lack much of having cut their beds down to grade.

HUNTER CREEK.

General description.—Hunter Creek is the first tributary of any size above the mouth of Minook Creek. It is between 12 and 15 miles long, carries probably a little over 40 second-feet, and flows in a steeply walled canyon-like valley through its whole length. In its upper 7 or 8 miles it flows almost north until it comes to the line of the high bench, when it turns at a right angle and flows west to Minook Creek.

Through the upper part of its course it is a crooked stream with a narrow V-shaped valley, probably indicating a rejuvenated drainage, while at its turn into the high bench the course becomes almost straight, showing a young, rapidly cut valley. It has but one tributary below the bend—Dawson Creek—entering from the south about 4 miles above the Minook. In the lower part of the valley of Hunter Creek the two sides are unlike. On the south side the upper 300 or 400 feet of the valley wall is very steep, almost precipitous. The descent then becomes gentler and forms a broad bench which slopes easily to the creek where it ends abruptly with a face 15 to 40 feet high. This bench is probably to be correlated with the lowest one on Minook Creek. It is covered with gravel, varying in thickness from 5 or 6 feet to 15 feet, and with muck varying in thickness from 1 foot near the creek to 40 feet or more near the hillside.

The creek flows tortuously through its bench, retaining the meanders it had before the bench was formed, and generally is close to the north side of the valley, but occasionally, as about 4 miles above the mouth, it wanders toward the south side, cutting away most of the bench. The valley has a grade in its lower part of 75 to 80 feet per mile.

Gold was discovered in Hunter Creek Valley by William Hunter (for whom the creek is named) in 1896, at a point about 1¼ miles above the mouth. Few definite data were obtainable concerning the gold production of the creek, but it is believed to have been approximately $24,000, of which $3,000 was produced during the winter of 1903–4 and $3,000 during the summer of 1904, a total for the year of $6,000. Hunter Creek has so far not proved to be a rich creek, though gold has been found in the gravels of both the bench and the present stream bed.

At the head of the creek the bed rock is mostly Rampart slate and quartzite; tuffaceous greenstones which predominate in the lower part of the valley are overlain near the mouth by Kenai sandstones and conglomerates. The tuffs contain some rounded pebbles, and a hole 228 feet deep was sunk in them under the impression that they belonged to the frozen muck and gravels of the creek. The rocks are much jointed and contain many small veins of quartz and calcite. Pyrite occurs at many places.

The gravels of the creek are 2 to 12 feet thick and are mostly diabase, slate, and chert pebbles from the bed rock, with many heavy bowlders of quartzite, occasionally reaching 3 feet in diameter. These larger bowlders are residuals from the gravels of the old bench through which the valley is cut. Much of the diabase gravel is angular or subangular. The muck over the gravel varies in thickness from 1 foot in places along the stream to 40 feet or more where the small streams pour their débris upon the valley floor.

Mining.—During the summer of 1904 work was being done on Discovery claim, about 1¼ miles above the mouth of the creek; and on No. 17, a claim about 4 miles above the mouth. Several claims, between these were being prospected.

On Discovery claim a flume 2,000 feet long, 30 inches wide, and 20 inches deep carried water to a bench about 16 feet above the creek. The bench was covered by 5 to 6 feet of gravel and over this was 1 to 4 feet of muck. The muck and gravel were practically all sluiced off, and the loose bed rock, in which the gold is found to a depth of about 18 inches, was shoveled into sluice boxes. The bed rock is partly a diabase and partly a much-folded brown, cherty shale standing on edge. The large bowlders were moved by hand. Three men ground-sluiced an area 75 by 150 feet in thirty days. The gold is mostly bright, smooth "pumpkin seed," with a few small rough pieces (Pl. III). A considerable amount of small barite pebbles and some hematite occur in the concentrates.

On claim No. 17-A a hydraulic plant was installed during the season of 1904. About a mile of combined ditch and flume had been put in, about 3,000 feet of which was flume, 32 by 18 inches, delivering 300 inches of water under a head of about 75 feet. A No. 1 Hendy Giant with a 3-inch nozzle was being used.

In working the ground the niggerheads and moss are torn up with a team and harrow and washed off with the nozzle. The ground is then left for a week, during which time the muck will thaw 1 to 2 feet. This is then washed off. The process is repeated until the top is removed. The remaining gravel thaws much more rapidly than the muck. It is found that the gravel can thus be thawed and made ready for sluicing much faster than a hydraulic giant working steadily can wash the gravel into the sluice boxes. In one instance an area of 125 by 250 feet was worked out in 40 days. The gravels are 2 to 12 feet thick, averaging about 6 feet, and are covered by 1 to 40 feet of muck. The maximum thickness was found at the mouth of a small tributary gulch where the gravel is mixed with angular fragments of rock. The muck contains much ground ice, which thaws readily when hydraulicked. The ice occurs occasionally as "dikes." One such was encountered over 400 feet long and 2 feet thick, intersecting the surface layer of muck and a flat lenticular mass of ground ice down to the gravel, making a depth of 12 to 15 feet.

The gold is found through the lower 3 feet of gravel and in the rough broken bed rock, which is made up of diabase and thin-bedded quartzite. It is bright and smooth, and nuggets up to 10 ounces in weight have been found. There is a small amount of rougher gold. Colors of gold are said to occur throughout the length of the creek, but no workable deposits have been found above the eastern limit of old gravels on the high bench already described. The larger part

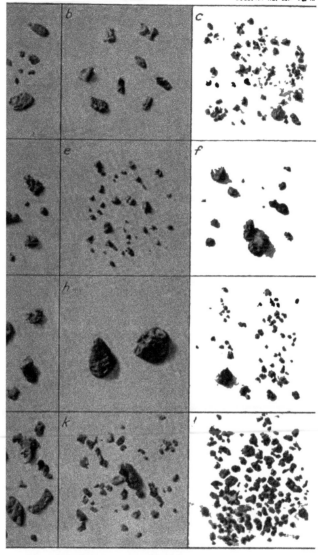

GOLD SPECIMENS FROM THE RAMPART REGION.

ery claim, Hunter Creek.
Claim No. 10 above, Slate Creek.
Bar.
No. 1 above, Thanksgiving Creek.
No. 3 above, Thanksgiving Creek.
No. 11 above, Omega Creek.

g. Gold from Seattle Bar.
h. Gold from Discovery claim, Doric Creek.
i. Gold from What Cheer Bar.
j. Gold from Claim No. 8 above, Little Minook Creek.
k. Gold from Claim No. 3 below, Little Minook Creek.
l. Gold from Claim No. 4 below, Little Minook Creek.

All the specimens are natural size.

of the gold has probably been reconcentrated from this bench. The smaller portion of rough gold has probably a local source in the rocks of the creek valley. Drifting is done in the winter at a number of places on the creek, but little information could be obtained as to results.

LITTLE MINOOK CREEK.

General description.—Little Minook Creek empties into Minook Creek about 5 miles from the Yukon and about $1\frac{1}{2}$ miles above the mouth of Hunter Creek, and has so far been the largest producer of the region. In drier years it carries scarcely a sluice head of water. It has a grade of 100 feet or less per mile in the lower 3-mile section to which all the mining has been confined, and its course is remarkable in being nearly parallel to Hunter Creek, though considerably shorter, as it has a length of only about 8 miles. Like Hunter Creek it makes a sharp bend upon entering the high bench about 3 miles from Minook Creek. Above this bend Little Minook Creek has a maturer character, as shown by the more crooked valley and greater number of tributaries, while in its course through the bench it has a straight sharply V-shaped valley which has been cut to a depth of 500 to 700 feet, and is so narrow that for over three months of the winter the sun can not be seen from the bottom of the valley.[a]

The creek follows closely the southern side of the valley through its lower 3 miles, and mostly the western side above this. It seems likely that the greater accumulation of talus on the north side of the creek is due to the greater amount of sunshine it receives, resulting in a greater amount of breaking down of bed rock by alternate freezing and thawing.

Gold was first discovered upon the creek in the early nineties by John Minook, who is reported to have taken out some gold near the mouth of the creek. The first claim, however, was located and worked by F. S. Langford in 1896, since which time the creek has been worked continuously. The total production of the creek is calculated, from the best ascertainable figures, to be $486,100, of which $40,000 was taken out during the winter of 1903–4 and $2,900 during the summer of 1904, making the output for the season of 1904 $42,900.

Little Minook Creek heads among slates and quartzites cut by small decomposed acid dikes. A little over a mile below the head, the creek is crossed by a belt of clayey, nonfossiliferous limestone, accompanied, as is often the case with the Rampart rocks, by green fine-grained slates. Below this there is an indistinct series of interbedded quartzites, cherts, siliceous shales, and some sandstones, all greatly contorted and accompanied by large masses of greenstones which form probably the larger part of the bed rock of the lower valley. At the foot of the

[a] Peck, C. W., and Laboskie, Wallie, personal communication.

valley walls the exposure of igneous rocks seems to be greater than in
their upper parts; that is, erosion seems to have exposed larger masses
of igneous rocks. Small veins of quartz and calcite occur in the rocks,
but none of great extent. The rocks of the valley have a considerable
impregnation of iron pyrites, the oxidation of which has stained them
the familiar rusty brown of iron oxide.

The placer deposits are all in the stream bed. The valley has been
cut down so quickly that no bench deposits have formed. The alluvial
deposits of Little Minook Creek vary in thickness from 7 to 25 feet,
of which gravel forms 3 to 12 feet, and muck, though occasionally
absent, generally 3 to 16 feet. The deposits are shallowest in the
lower part of the creek. The gravels contain fragments of many
rocks, of which diabase is probably most abundant, but slate, grit, and
much vein quartz also occur, and there are many large quartzite
bowlders from the bench above. Much of the gravel, as would be
expected in a weak stream, is subangular.

In the gravels mammalian bones are said to be found, although
none were seen by the writer. In places clear ice is uncovered in
digging, the structure of the alluvium showing how sudden floods had
drifted detritus over the ice in the spring, and had thus preserved it.
Locally there is much wood in the muck.

Occurrence and character of the gold.—Values are found in the
lower part of the gravels through a thickness of 1 to 3 feet, and a
width of 50 to 200 feet. The gold frequently occurs in the bed rock,
particularly the broken diabase, to a depth of 1 to 2 feet. The pay
streaks extend up the creek only as far as the creek has cut through
the high bench gravels, a distance of about 3 miles. There are sixteen
1,000-foot claims within these limits.

The creek has been well prospected throughout its length, and
although colors are found there is no pay above the line of the high
bench. A small amount of gold, in which were some large nuggets,
has been found in the gulches leading from the high bench. The
amount of gold carried by the gravels varies greatly, but in the pay
streak probably runs from $2 to $10 per square yard. The gold is
generally smooth, chunky, and bright (Pl. III, j, k), and shows a
large amount of wear. In the upper part many nuggets are found
weighing 1 to 12 ounces apiece, but the gold gets finer downstream
until near the mouth it is nearly all flat, smooth, bright, and even in
size, looking like golden bran when seen in quantity. There is a very
small amount of rough gold, probably of local origin, but the larger
part is probably reconcentrated from the old bench gravels of Minook
Creek. The gold of the creeks cutting this bench is said to assay over
$19 per ounce. This would make it of about the same value per
ounce as the Koyukuk gold. The gold is taken in trade by the stores

at $18 per ounce. Some small nuggets of copper and a small amount
of silver have been found with the gold.

Mining.—Most of the claims are worked by drifting in the winter,
though the three lower ones are worked during the summer by open
cuts. On the latter the muck and gravel are first ground-sluiced off
within 1 or 2 feet of bed rock by means of a dam provided with an
automatic gate, and the remaining gravels are afterwards shoveled
into the sluice boxes. The drifted pay gravels have often been " coy-
oted " or " gophered ; " that is, holes have been sunk here and there
without system until, although there is probably much pay still left,
the ground is frequently almost unworkable on account of the ice
in the old holes which floods the new workings when thawed by a
steam point. When workings are filled with water, the mass is said
to freeze on the top, sides, and bottom, while the central part remains
unfrozen through several years. Much of the ground is worked on
" lays " or leases, the lessees paying from 25 to 55 per cent of the
gross output, an amount that is apt to leave the worker little for his
labor if things do not run very smoothly. Freight rates are 2 cents
per pound in winter and 4 cents per pound in summer.

The remaining gold in Little Minook Creek would seem to be best
recovered by working the claims in cooperation as one company, for
it is certain that some of the richer claims can no longer be profitably
worked by drifting. The quickest, but an initially expensive, mode
of working would be to hydraulic the gravels by bringing water from
Minook Creek. A ditch 10 miles in length above the mouth of Little
Minook Creek would probably give a head of over 100 feet and plenty
of water at the upper limit of the pay gravels. A way requiring
less capital, but much slower, and the one that will likely be carried out in the end, is the ground-sluicing of the claims, successively,
from the mouth of the creek upward, by means of dams and automatic gates, but as the claims belong to different parties, some of
whom are unwilling to sell, there will probably be only a small
amount of work carried on along the creek for a number of years to
come.

LITTLE MINOOK JUNIOR CREEK.

General description.—Little Minook Junior Creek, between Little
Minook and Hoosier creeks, is about 2½ miles long. Its valley lies
wholly within the high bench of Minook Creek. It is a weak stream,
generally dry during the summer, and rarely carries a sluice head of
water. With a valley of hard rocks it has not been able to cut its
bed down to the depth reached by the larger tributaries of Minook
Creek. In the lower half the grade of the creek is torrential and the
valley is narrow with steep sides. In the upper half the grade is

much easier and the valley is wider with gentler slopes, especially on the north side. The rocks of the valley are the same as along Little Minook Creek. The lower part is entirely in diabase.

The total output of the creek was estimated by Donald McLean at about $150,000, and the output for the year 1904 at about $17,000.

The steep grade of the lower part of the creek has allowed little accumulation of alluvium, but in the upper part the deposits have reached a depth of 12 to 30 feet, of which gravel forms the lower 4 or 6 feet. The gravels are angular and largely composed of diabase with well-washed quartzite bowlders from the bench gravels through which the stream has cut.

In the gravels are many bones of bison, musk ox, mammoth, and horse. A very fine specimen of the skull of *Bison alleni*, with the shell upon the horns, was taken out of Donald McLean's claim, No. 25, near the head of the creek, by Mr. McLean and Thomas Evans. This is the only specimen of this species that has been reported from Alaska. It was carefully removed and is now in the National Museum at Washington. Some teeth obtained by C. W. Peck from gravel next to bed rock on the same claim and referred to Dr. T. W. Stanton for identification were called by him " horse teeth of Pleistocene or more recent age."

Mining.—There are 29 500-foot claims upon the creek, numbered from the mouth upward, the upper 9 or 10 of which are said to have paid wages or more upon working. The pay streak is 30 to 60 feet wide and 1 to 6 feet thick, averaging probably 3 feet, but gold is sometimes found through the whole thickness of the gravel. The gravels were reported to carry $10 per square yard on one claim, which is probably the highest value on the creek, the values on other claims running down to amounts too small to pay for working.

The gold is similar to that of Little Minook Creek, mostly smooth and bright with a little that is rough. It is generally coarse and chunky, nuggets sometimes reaching 3 ounces in weight. The larger part of the gold is undoubtedly reconcentrated from the high bench of Minook Creek. The small amount of rough gold has probably had its origin in the bed rock.

The gravels have been mined by drifting with steam points, but advantage was taken of the wet season of 1904 and some ground sluicing was done in gravel and muck 16 feet thick. Trees and brush in the lower part of the creek were cleared away in preparation for further ground sluicing. The cost of mining by drifting is 50 per cent or more of the output, but as there is so little water it has been the only feasible mode of work. The creek is considered to be nearly worked out.

HOOSIER CREEK.

Hoosier Creek flows into Minook Creek from the east side between 5 and 6 miles from the Yukon. It is a stream of about the same volume as Hunter Creek and has a valley of about the same gradient and general section, but it shows no sign of the bench that appears along Hunter Creek. Like Hunter and Little Minook creeks, its course bends to the left upon entering the area of the high bench of Minook Creek, although in a less degree.

The production of Hoosier Creek is unknown. There is assigned to it but $500 in a previous report, $227 of which was in one nugget. Other small amounts have been taken out, but the production has not been large, and it has been almost impossible to thoroughly prospect the creek on account of live water in the gravels.

The bed rock is similar to that of the other creeks cutting the bench. Quartz veins up to 18 inches in width occur in the diabase, and there is some pyrite distributed through the rocks. The alluvial deposits vary in thickness from 6 to 15 feet, of which 1 to 9 feet is gravel, averaging probably about 6 feet, and 1 to 10 feet is muck, averaging perhaps 6 or 7 feet. There is thought to be a pay streak about 100 feet wide whose length coincides with the distance the creek flows through the high bench, but the gravels of the valley are broader than those of the other creeks described, and with the live water the pay is hard to locate.

Two miles above the mouth of the creek a hydraulic plant has been installed and had just gotten in shape to begin work at the end of the season of 1904. A combined ditch and flume 4.300 feet in length delivered 500 miner's inches of water under a head of about 80 feet. A hydraulic elevator is used to dispose of the tailings.

FLORIDA CREEK.

Florida Creek is only about 2 miles long, lying in the high bench of Minook Creek about 2 miles south of Hoosier Creek. Ordinarily it is dry during the summer and fall. The gradient of the stream is high and the valley narrow. The bed rock is almost entirely of diabase. The alluvial deposits are narrow, but in places reach a depth of 15 or 20 feet. Nuggets up to $33 in value have been taken from the creek, but so far as known not more than a total of $2,000 has been obtained, though the stream has been well prospected. The first prospect showed up so well that miners at once located the whole of the creek, and a number of good cabins were erected on the different claims. Some ground sluicing was done on the lower part of the creek during the season of 1904, but no other work was done.

ORIGIN OF THE GOLD.

The great difference in the richness of the several creeks flowing through the high bench of Minook Creek, and the variation in the richness of claims and size of nuggets on the same creek within the limits of the bench, show that the gold is not evenly distributed through the gravels of the bench. Thus Hunter Creek has so far shown no rich claims, while Little Minook Creek has been very rich in places, and along the latter the gold is very coarse on the upper claims but grows much finer toward the mouth (Pl. III, j, k, and l), showing that probably the larger part of the gold in the lower portions of the stream has been washed down from the upper claims. The gold in the bench gravels was probably concentrated from local gold-bearing zones in the rocks worn away above the level of the high bench. How great a thickness of these rocks was disintegrated and carried away can not be told, but there may have been many hundred feet. The rocks were probably the same as those now forming the bed rock. The gold in the bench gravels is said to be well worn, but gold found in the gravels of a stream as large as Minook Creek is generally well worn, and in this case we have no clue as to the length of time through which wearing may have continued.

OTHER TRIBUTARIES OF MINOOK CREEK.

RUBY CREEK.

Ruby Creek flows into Minook from the west side about 9 miles from the Yukon. It is a stream carrying 300 to 500 miners' inches (7.5 to 12.5 second-feet) of water, with a grade of about 150 feet per mile in the lower part. In this part the valley is broadly V-shaped, with steeply sloping sides. The upper part was not seen.

The first pay was taken out of the creek in 1901, and the total product is said to have been $13,000 or $14,000, although this estimate may be a little high. About $5,000 was reported during 1904. No pay has been found above 1½ miles from the mouth of the creek, but it is claimed that no holes have been sunk to bed rock on account of the live water in the gravel.

The bed rock is the calcareous schist, garnetiferous mica-schist, carbonaceous slate, chert, and grit, intruded by greenstones (diabase?). The bedded rocks strike almost north and south across the creek with the dip downstream (east). The alluvial deposits are 6 to 10 feet thick and 300 to 500 feet wide. In some places there is almost no muck and nowhere is its depth more than about 4 feet. The gravels are 5 to 7 feet in thickness and the total thickness of muck and gravel is 6 to 10 feet, averaging nearer the lower figure. No large chert or quartzite bowlders are seen, as in the creeks, cutting the high bench. There are some gneiss pebbles, which indicate the probable presence of gneiss on the creek. The gravel is comparatively fine but contains a few bowlders a foot or more in diameter.

The gold is all on bed rock and is distributed through the whole width of the gravels. The only gold seen came from a point about one-half mile above Minook Creek. It was somewhat iron stained and in general rougher than the gold of the creeks cutting the high bench. The larger pieces were very smooth, but the smaller pieces were rough and most of the gold is rather flat. The gold is said to be rougher in the claims below. Nuggets up to about 2 ounces in weight are obtained. In the concentrates with the gold are large quantities of garnets that sometimes reach 1 inch in diameter. A handful of garnets was obtained from a pan of dirt. There are so many of them that they give considerable trouble by filling up the spaces in the riffles and must be cleaned out once or twice a day. Some barite is said to be present, and an occasional silver nugget appears, one weighing 2 ounces having been reported. The silver nuggets are very rough.

It seems likely that the origin of the gold is in the local bed rock, which along this part of the creek is a carbonaceous slate of irregular cleavage. In places much pyrite is distributed through it. The creek has been worked during the summer by open cuts and in winter by drifting, but it has probably paid little, if anything, more than wages. Preparations were being made to install a hydraulic plant, and a mile of steel pipe, consisting of 720 feet each of 20, 19, 18, 17, 16, 15, and 14 inch pipe with branches of 11-inch pipe for an elevator, and 7-inch pipe for a giant, was to be put in. It was said that it would deliver the water under a head of 154 feet.

SLATE CREEK.

Slate Creek, a western tributary of Minook Creek, about 12 miles from Yukon River, is about 4 miles long and is said to always carry at least a sluice head of water. It has a grade in the lower portion of about 150 feet to the mile, and the valley is narrowly V-shaped.

The creek has been worked only since 1902. Freights from Rampart are 8 cents per pound in summer and 4 cents per pound in winter.

The bed rock in the lower part is much-folded shaly limestone, green and purple slates, and cherty beds, with a northeast strike. The main rock of the valley is a dark carbonaceous schist which breaks into pencil-like fragments and contains many quartz seams. Most of the work has been done nearly 2 miles above the mouth by drifting in the winter. The deposits here are 26 feet thick.

Gold is found in as much as 3 feet of gravel and to a depth of 1½ feet in bed rock and over a width of 50 feet. An $8 piece is the coarsest thus far taken out. Silver is a common associate (Pl. III, b), and an 8-ounce nugget has been found. Copper is also said to occur in the gravels. The absence of garnets indicates that the schists of Ruby Creek do not extend into the valley. The gold in this case has probably been derived from the small stringers common in the bed rock.

THE BAKER CREEK GROUP.

GENERAL DESCRIPTION.

The Baker Creek diggings are situated from 28 to 32 miles by trail almost south from Rampart, and occupy a narrow belt with a northeast-southwest extension of about 9 miles. Along Baker Creek itself there have been no placers discovered so far, all at present known being on the tributaries flowing from the divide separating the Minook and Troublesome drainages from that of Baker Creek. The principal diggings are located along Pioneer, Eureka, Glenn, Gold Run, Omega, and Thanksgiving creeks.

The topography is strikingly different from that of the other two areas. Baker Creek flows along the southwestern side of a large flat, 7 to 9 miles broad in its widest part, and perhaps 10 miles long, its longer extension being northeast-southwest in the line of flow of Eureka and Hutlina creeks. Instead of sharp canyon-like valleys the streams flow through open valleys, and where they flow in general parallel to the Baker-Minook divide—that is, approaching a northeast-southwest or an east-west direction—the southern bank is steep, while the northern one is gently sloping, the creeks flowing close to the steeper side. Even along the broad Baker Flats this feature is still prominent. The north side is a long gentle slope toward the divide, rising more sharply in its upper part, while across the flats the southern side may be seen rising abruptly from the valley floor.

The main streams of the Baker Creek gold area are: Eureka Creek, lying next to the Baker-Minook divide, flowing southwesterly for about 5 miles, then turning to the south; Pioneer Creek, flowing parallel to Eureka between 1 and 2 miles to the southeast and joining it on Baker Flats; Rhode Island Creek, flowing in a southerly direction, about 1½ miles west of Eureka Creek; and Omega Creek, in the western part of the gold area. Into these creeks flow all of the smaller creeks of the area along the Baker-Minook divide. The streams are all small, many of the smaller ones being ordinarily dry during the summer and fall. The gradient of the larger streams is comparatively low and it is with difficulty that water is carried to the benches.

The only practical trail to and from the Baker Creek area is from Rampart along Minook Creek, a trail that in most parts of "the States" would be considered practically impassable during the summer time. Most of the way it is soft and miry. The pack horses sometimes sink to their girths, floundering and wallowing their way through. As a choice, there is the bed of Minook Creek along which, if the creek is not too high, the horses can make their way on the bars and through the icy water; but at best this trail is hard on the animals. The foot traveler can not take the creek bed, and if he carries a pack, as he often does, he must make his way along the mucky trail. Over this trail all provisions for the camps are carried. Freight rates, until the summer of 1904, were 25 cents a pound in summer and 6 cents in winter, and it is said that one man had to pay freight on 47 pounds of "grub" and 53 pounds of box and packing at the higher rate. During the summer of 1904 freight rates came down to 15 cents a pound, but the packers declared they could make nothing at that rate, and this is probably true, as hay and oats when cheapest are $100 a ton. There is another trail to the mouth of Baker Creek, but it is said to be bad and is not used.[a]

Lumber is high, most of it being shipped from the States of the Pacific slope. Some is whip-sawed along Baker Creek in the winter, which costs about 20 cents a foot, board measure. It would seem that a small sawmill operated through a portion of the year would be a paying investment, as there is said to be plenty of timber along Baker Creek for local needs.

The total production of the region is estimated to have been about $406,100, of which $84,700 was produced during the winter of 1903–4 and $61,300 during the summer of 1904. These figures are probably under rather than over the actual amount.

Gold was discovered in the Baker Creek area on Eureka Creek, where mining was begun during the winter of 1898–99 and a small amount was taken out. On Glenn Creek gold was discovered in July, 1901, on the benches along Pioneer Creek in 1902, on Thanksgiving Creek in February, 1903, and other discoveries were made during the summer of 1904. Prospecting is in active progress in other valleys of the vicinity, and it is altogether possible that new discoveries may be made.

The rocks of the Baker Creek group show less variety than those of the Minook Creek group. In the gold-bearing region the rocks are schistose grits with interbedded slates and with quartzites, the latter in thin strata, generally 1 to 3 feet thick. Both slates and schistose beds are generally carbonaceous. The strike is northwest and the dip about vertical. No igneous rocks occur except along the top of

* During 1907 and 1908 a wagon road was being built by the Government from Rampart up the Minook Valley to connect with the Baker Creek region.

the 'divide, where there are some dikes and masses of a monzonitic rock. Quartz occurs generally only in small veins, and these are not prominent. In places there is a considerable amount of pyrite in the rocks in small crystals and grains, but no large masses or veins have been seen.

THE CREEKS AND BENCHES.

EUREKA CREEK.

General description.—Eureka Creek, on which gold was first discovered in this area (in February, 1899), flows southwestward along the foot of the Baker-Minook divide. It runs in a straight southwest course for about 4½ miles, then turns and runs south 2½ miles to its junction with Pioneer Creek. It has a number of small tributaries from the northwest side, but none from the southeast. The largest is Boston Creek, about 2 miles long, which joins Eureka Creek at its bend. The other tributaries are mere rills. Eureka is a small creek carrying barely a sluice head of water above the mouth of Boston Creek during the ordinary seasons. From aneroid barometer readings the gradient of the stream is about 100 feet per mile. The valley slopes gently to the divide on the northwest side, but on the southeast side the slope is almost precipitous, rising 400 to 600 feet above the valley. The creek flows close to the foot of the steeper side.

The gravels of the creek are not much worn, as is characteristic in weak streams, and have been left for a considerable distance, in places at least 500 feet, up the slope of the hill as the stream bed has moved to the southeast. The bench gravels, like those of the present stream bed, are made up entirely of the country rocks. The deposit varies in thickness from 5 to 18 feet, and the overlying muck varies from nothing to 8 feet, the distribution being rather irregular. The total thickness varies from 5 to 20 feet. The gravel contains a considerable amount of very sticky clay, which makes sluicing difficult. The clay seems to come from the decomposition of both the grit and the slates.

Mining.—Only one claim above and one below the mouth of Boston Creek have so far been made to pay, but prospectors on the bench gravels about 2 miles above the mouth of Boston Creek reported that they had found gold in sufficient quantities to pay for ground sluicing, if not for drifting. On this part of the bench it is 8 feet to bed rock near the creek, and 450 or 500 feet back from the stream it is 20 feet. The elevation above the creek at this distance, as shown by the aneroid barometer, is 70 feet.

The gold is said to be in the lower 18 inches of gravel and in a foot of bed rock. Along the creek the bed rock is largely blocky, and in it gold is found to a depth of 3 feet; but it is not found at such depths where the bed rock decomposes into clay. The gold may be distributed through the gravels to a depth of 4 or 5 feet, but it is generally close to bed rock, which must be scraped.

The larger part of the mining has been done by drifting, but on Discovery claim, just below the mouth of Boston Creek, an open cut is being worked. The muck and upper gravel are ground sluiced through sluice boxes, so as to save any fine gold that may be in them, and the lower gravel is shoveled in. Fifty-seven 12-foot boxes are used, 37 of which contain pole riffles and 2 contain Hungarian riffles. The lower boxes are lined with sheet iron to facilitate the movement of the gravel. Some gold is probably carried off by the sticky clay in spite of the length of the sluice box.

PIONEER CREEK.

General description.—Pioneer Creek heads against the Baker-Minook divide, flows around the head of the Eureka, and then, at a distance of 1 to 2 miles, flows parallel to the main course of that creek. After traversing 7 or 8 miles it joins Eureka Creek and they are said to lose themselves on Baker Flats. Pioneer Creek is larger than Eureka Creek; probably it never carries less than three or four sluice heads of water, and its gradient along its lower course is about 60 feet per mile. The valley is similar to that of Eureka Creek. Its northwest side is a gentle slope running back for about a mile, and the southeast side is of almost precipitous steepness, but not so high.

On the gentle slope of the northwest side there are perceptibly flatter places or benches, but only one of these is persistent. This bench is traceable along Pioneer Creek for over 4 miles. Its northeast end is but little above the present level of the creek while its southwest end is about 250 feet above the creek. Over this bench and covering much of the slope below is a deposit of auriferous gravel left by the creek as it moved to the southeast. The different diggings upon it are known as " bars."

Five small tributaries, Doric, Boothby, Seattle Junior, Skookum, and Joe Bush, flow across this bench at right angles to the course of Pioneer Creek. Near the upper end of the bench at Joe Bush Creek prospect holes showed a well-defined old stream channel. Upstream the bench rises so that a ditch supplying water to What Cheer Bar is below the workings at Seattle Bar, but crosses the bench and is on the upper side when it reaches What Cheer Bar. There can be no doubt that the bench is of stream origin.

Like many other Alaskan creeks Pioneer Creek was staked and then each man waited for his neighbor to do the hard work necessary to locate the pay streak, if there was one. Meanwhile the claims lapsed and were then restaked by other parties, and pay was discovered on What Cheer Bar in 1902. After this discovery pay was found on Doric Creek and at several other points along the bench.

The production of Pioneer Creek Valley to the end of the summer of 1904 was about $35,800.

The bed rock is the same as on Eureka Creek, schistose grit, with interbedded slates and thin beds of quartzite. The grits sometimes become very carbonaceous, particularly on Doric Creek. The general strike of the rocks is N. 70° or 75° E., with a steep northerly dip. There is some quartz in small veins and stringers, and on Doric Creek at places there is considerable pyrite distributed through the rocks. The pyrite is often oxidized, so that only small holes lined with iron rust indicate its former presence. On Doric Creek inclusions of a carbonaceous substance the size of a walnut occur with small quartz seams. Little is known of the alluvial deposits along the creek bed. The deposits on the gentle slope already referred to are 3 to 12 feet thick. They consist of the usual muck and gravel, and extend over 2,000 feet back from the creek.

What Cheer Bar.—What Cheer Bar is located in the lower part of Pioneer Creek Valley, about a mile from Eureka Creek, 2,000 feet from Pioneer Creek, and 250 feet above the latter. The season of 1903 was spent in putting in about 4 miles of ditch, with the necessary flumes. This ditch carries about three sluice heads of water to the upper edge of the workings. The ground is excellent for ditching, compared with other Alaska localities, for there is little ground ice and the soil is tenacious enough to make good banks. The bed rock is much jointed and broken and exhibits fine examples of creep, the rock leaning downhill and gradually blending with the gravels.

The average depth to bed rock is about 12 feet. The overlying material is composed of 1 to 1½ feet of muck, 3 feet of rather fine flat wash, 5 feet of medium-sized yellowish gravel, and 3 to 4 feet of rather heavy wash, including some bowlders of vein quartz 2 feet or more in diameter. There are some bowlders of conglomerate similar to that found in Quail Creek, in Troublesome Valley, and it is probable that beds of it outcrop on the headwaters of Pioneer Creek. Most of the gold is found in the lower part of the gravels and the upper 1 or 2 feet of bed rock. It is well worn and bright (Pl. III, i), and probably is derived from the bed rock in the vicinity. The largest nugget found weighed somewhat less than 2 ounces and was worth $28. It contained considerable quartz. The gold is taken in trade at $15.50 per ounce.

The muck and upper gravel are ground sluiced and the lower gravel and upper bed rock shoveled in. The water could not be used until August 15, and only fifteen days were available for washing. Fifteen men were employed.

Seattle Bar.—Seattle Bar is about 2½ miles farther northeast on the same bench, on the northeast side of Seattle Junior Creek, and about the same distance back from Pioneer Creek as What Cheer Bar. Pay was discovered here in the spring of 1904. The depth to bed rock is about 9 feet and the bed rock and gravel are similar to those of What Cheer Bar. The gold occurs in the lower foot of gravel and the upper foot or more of bed rock. It is bright, chunky, and well worn. Some of it is rather flat, but all is easily saved. The largest nugget obtained was worth $9.40. Water is obtained for sluicing by a ditch and hose from Skookum Creek, which in a dry year will furnish but a scant supply.

Doric Creek.—Doric Creek is a small tributary of Pioneer Creek about three-fourths mile above What Cheer Bar, and is dry most of the summer and fall. It has an open valley, at its greatest depth probably not over 50 feet below the level of the bench. Gold was discovered here in 1902, and in the winter of 1903–4 a portion of the valley about one-fourth mile from Pioneer Creek was found to be very rich. As with other weak streams its wash shows almost no wear, but there is also a large amount of more rounded gravel from the bench through which it has cut.

Only one claim has produced much gold. Some pay was found on the lower part of the next claim above, but none in the upper part. The richness of the deposit is probably due to the reconcentration of the gold from the gravels of the bench. The ground is worked by drifting during the winter, and the largest bowlders are left in the drifts. It is worthy of note that a large degree of the success obtained in locating this claim was attributed to the remarks upon concentration in the Survey report on the Nome region,[a] which apply with much force to many of the deposits of the Rampart region.

Other bench gravels.—A mantle of gravels similar to that which covers the gentle slope on the northwest side of Pioneer Creek bends around a spur from the divide on the west side of Eureka Creek and continues to Omega Creek, a distance of about 2½ miles. Beyond this point it has not been traced. In the space described the gravels are cut by Glenn Creek, Gold Run, Rhode Island Creek, and Seattle Creek.

SHIRLEY BAR.

The bench gravels have been prospected at many places and shown to carry gold, but at only one point outside of the creeks crossing them have they proved sufficiently rich to pay for working. This place, known as Shirley Bar, is located between Glenn Creek and Gold Run. It is at an elevation of about 200 feet above the lower workings on

[a] Brooks, A. H., and others, Reconnaissances in the Cape Nome and Norton Bay regions, Alaska, in 1900; a special publication of U. S. Geol. Survey, 1901, pp. 149–151.

Glenn Creek, and was first worked in the fall of 1901. The bed rock is the same schistose grit, slate, and quartzite. The wash is small and subangular, with a few quartzite bowlders and some monzonitic rocks from the divide above. The gravel varies in thickness from 2 feet at the lower side of the claim to 7 or 9 feet in the middle and 5 feet at the upper end. The gold is bright, rounded, and " shotty," well distributed through the gravel, and, though it seems strange, the nuggets come from near the surface. There are few large pieces of gold, the largest nugget taken out weighing a little over 1¼ ounces.

A ditch from Rhode Island Creek, 1 mile long and capable of carrying 2 sluice heads of water (about 100 miner's inches), has been dug; but water is so scarce that it is collected in a pool after being used and pumped back to the sluice boxes. For this purpose a 30-horsepower boiler, two twinned 4-horsepower upright engines, and a 4-inch centrifugal pump are used. Seven men have been employed on the claim during the season.

GLENN CREEK.

General description.—Glenn Creek is located about 1 mile west of Eureka Creek, and flows across the bench gravels. It is about 3 miles long, running almost south down the slope from the Baker-Minook divide. The valley is shallow and open and probably stands not more than 50 feet below the bench. It is practically dry during ordinary summers, but during the wet summer of 1904 it carried water sufficient for sluicing.

Gold was discovered on Glenn Creek by Messrs. Beardsley, Belsea, and Dillon in July, 1901. The total production up to the fall of 1904, according to the most reliable information obtainable, had been about $277,500, not including the output of one claim known to have produced a considerable amount. During the year ending with the fall of 1904 $50,500 is known to have been produced, and this again does not include some smaller outputs. Of the 1904 output $11,000 was obtained by drifting during the winter of 1903–4 and $39,500 during the summer of 1904.

The bed rock is similar to that on the other creeks. The alluvial deposits are 7 to 9 feet thick and are in large part composed of the angular fragments usually found in so weak a stream, with rounded material from the bench through which it flows. There are occasional small bowlders of monzonitic rock from the divide above.

The quartzite interbedded in the softer slates has given rise to a peculiar condition in the gravels of the lower part of the creek. On claim No. 1A a section of the gravels shows about a foot of muck, under which is a discontinuous layer of angular quartzite blocks 8 to 10 inches thick and 2 feet or more broad, showing no water wearing. Under these is a thickness of about 2 feet of washed gravel, and fine broken slate lies below this upon vertical strata of slate which strike about northeast. These angular blocks of quartzite on top of the gravel have been very puzzling. It is likely that they are to be explained by the supposition that the creep, which acts very strongly here, has broken down a thin bed of quartzite that can be seen on the side of the claim, and as the creep has moved the gravels the blocks of quartzite have been broken off and have crept with the gravels.

Mining.—The pay seems to have been mostly, if not wholly, in that part of the creek which cuts the mantle of gravel covering the hillside. The pay streak is said to vary in width from 50 to 100 feet. In places it was very rich; one pan taken by the writer gave about $3.75. Many pans of $10 and upward were said to have been taken. At one place fine gold could be seen all through the broken slate. On this claim the pay was in the lower 3 feet of gravel and 2 feet of the bed rock. A plat 20 by 48 feet yielded $4,000 to 4 men working three and one-half days.

The gold is bright, clean, generally worn, and fine, but " shotty " and easily saved. Such nuggets as are found generally contain considerable quartz. The largest nugget found weighed nearly 6 ounces. It was bright, clean, beautiful gold, and showed the impression of large quartz crystals. It is said to assay a little over $16 per ounce.

A small ditch 1 mile long brings about a sluice head of water from Rhode Island Creek. Another ditch dug to bring a sluice head or more from Boston Creek was just ready to use when freezing began in the fall of 1904. An average number of 24 men are said to have been employed during the year. Some drifting is to be done this winter (1904-5), but it is said that most of the ground fit for drifting has been worked out, the remaining pay gravels being too shallow to give a good roof. The creek is probably more than half worked out.

GOLD RUN.

A creek, about 1½ miles long, flowing into Rhode Island Creek 1 mile west of Glenn Creek, is called Gold Run. It carries little water at any time and is practically dry during the summer and fall. The valley is shallow and open and the lower part is cut through the gold-bearing gravels covering the extension of What Cheer Bar. The creek was staked in the spring of 1899 and the first work was done during the winter of 1900-1901. There are six claims, each 500 or 1,000 feet in length, upon the creek, but only the lower four have so far been producers. The total production during the winter of 1903-4, as reported by the miners, was about $16,000, not including the output of one claim, which was probably small. The production of former years is estimated at about $9,000.

The bed rock of the lower part of the creek is a carbonaceous schistose grit, which in places becomes slaty. The rather well-rounded gravels are of slate, quartzite, and grit, 16 to 18 feet deep, with a covering of about 2 feet of muck. The creek is difficult to work on account of the live water in the gravels. Through part of their length the gravels are frozen on the bottom and thawed for several feet above, so that drifts must be timbered throughout. The pay frequently goes down into the bed rock 3 or 4 feet.

The gold is generally bright, fine, and somewhat worn. One nugget weighing nearly 4 ounces has been taken out. When poured out of a sack [a] or pan part of it rolls almost like shot, owing to its rounded form. This characteristic is common to a large part of the gold of the Rampart region, but particularly so of this area, and is due to its crystalline form. The crystal faces are often observable on the pieces. The placers of the creek are probably derived largely from reconcentration of the gold from the gravels of the bench through which the creek has cut its course, and in part from the local bed rock. The creek is probably more than half worked out.

RHODE ISLAND CREEK.

Rhode Island Creek is somewhat larger than Gold Run and heads nearer the top of the ridge. Its general conditions of bed rock, gravel, etc., are similar to those of Gold Run, below the mouth of which Rhode Island Creek flows close against a bluff on its western side, while its eastern side rises more gently.

Considerable work has been done on the creek, but during the summer of 1904 no claims were worked. The output is unknown. Miners on other creeks are of the opinion that the gravels would pay for working if water for hydraulicking could be obtained.

SEATTLE CREEK.

Seattle Creek, although the longer stream, is called a tributary of the Rhode Island. It probably carries less than a sluice head of water during an ordinary season. The bed rock in the lower part is carbonaceous schistose grit. The gravels contain bed-rock fragments, quartzite, vein quartz, and carbonaceous slate, and are rather fine. They are said to be 8 to 30 feet thick and covered with 1 to 3 feet of muck. They are well frozen and have no live water. About $100 was taken out in the course of prospecting during the winter of 1903-4. The gold is said to be bright, fine, and shotty. Prospecting was to be continued during the winter of 1904-5.

----

[a] Known among miners as a "poke."

A spur on the west side of Rhode Island Creek, similar to the one on the west side of Eureka Creek, has a well-defined bench cut upon it, extending about one-half mile to Omega Creek Valley. The bench is about 300 feet above the bed of Rhode Island Creek and is covered with subangular gravel, through which gold is said to be found. It is about on a level with Shirley Bar.

OMEGA CREEK.

General description.—Omega Creek, another small stream, heads in a ridge about 2 miles southwest of the head of Minook Creek and about one-half mile west of Seattle Creek, flows almost south for one-half mile or more, and then swings gradually to the west. As soon as it takes a westerly course the shape of its valley becomes similar to that of Eureka Creek, having a steep hill on the south, against which the stream flows, and on the north side a gentle slope to the ridge above, rising more steeply in its upper portion.

Gold was discovered in Omega Creek in 1899, but the first pay was found in 1901. The creek has been worked in only a small way, and the production has been small. The bed rock is a black, fissile, much-broken slate, and a yellowish, somewhat schistose grit. It has a strike of N. 70° E., with a high northerly dip. The gravel is about 7 feet deep, very angular and fine, and is made up of the country rock with a small amount of quartzite. There is little or no muck over the gravel, but there is a sticky clay through it which probably carries off some of the fine gold.

Values and mining.—The pay is known to extend for about 1 mile down the creek from a point due west of the mouth of Seattle Creek. The width of the pay is unknown. One cut 30 feet wide has been taken out, and it is known that the pay extends to both sides, rising on a low bench on the right (northwest). This cut is at the upper end of the pay streak. The gold is distributed through the gravel both top and bottom. It is "shotty" and coarse, and much of it is very rough. Many pieces show crystal faces, and all the larger pieces and many of the smaller pieces contain quartz. In color the gold is more brassy than most of the gold of the region. A great many small crystals of pyrites occur in the concentrates with the gold. So far the claim has been worked only by an open cut, but some of the gravel was thought to be deep enough to be workable by drifting and this method was to be tried during the winter of 1904–5. The water supply is small and a dam has been put in to collect the water so that sluicing can be carried on about half the time during an ordinary season.

CHICAGO CREEK.

Chicago Creek is a small rivulet flowing down the northern slope to Omega Creek about 2½ miles west of the mouth of Seattle Creek. Pay was reported to have been discovered near its mouth during the summer of 1904, and it was the intention to work it during the winter.

THANKSGIVING CREEK.

Thanksgiving Creek is a small tributary of Omega Creek, between 4½ and 5 miles west of Eureka Creek. It occupies a shallow, open depression in the southern slope of the ridge, on the north side of Baker Flats, and can hardly be said to have a valley in its lower part. It is almost dry in the summer and fall. Gold was discovered on it . in February, 1903. The combined output of Omega and Thanksgiving creeks has been about $18,200.

The bed rock is exposed only in the diggings, but where seen was a yellowish, somewhat schistose grit. The gravel varies in depth from 6 to 18 feet where the creek is worked, though it is said to be deeper farther downstream. It is composed of subangular pieces of quartzite, schistose grit, vein quartz, slate, and a small amount of monzonitic rock. The overlying muck is 1 to 4 feet in thickness.

The gravel is peculiarly mixed with a sticky yellow clay, which in places seems to be half ice. In some of the deeper holes there is 10 feet of this mixed clay and ice. It can not be worked with wood fires, for when melted it runs down upon the fires and quenches them. In open cuts the sides when melted move together like a mass of yellow tar. In some of the holes the section is said to show 10 to 12 feet of finely mixed yellow clay and ice, of which 5 feet is fully half ice and below this there is 6 feet of subangular gravel. The pay streak varies in width from 25 to 45 feet, and is 1½ to 9 feet thick. Gold is sometimes distributed through the yellow clay and colors always occur through the mixture of clay and ice. At one place where the pay is found through 7 feet of the ice-clay mixture, when the mass is thawed the pay sinks to the lower 4 feet. If the clay is dried it is difficult to part the gold from it, and at one claim, on which open-cut work was progressing, angular pieces of sheet iron like saw teeth were driven into the poles used in the sluice boxes to break up the clay. The iron pieces were left projecting about three-fourths of an inch, and 25 were used to a 6½-foot pole. The device is said to work well.

The gold is generally rough and somewhat iron stained, but some of it is smooth, bright, and "shotty." Some "black sand" is said to be with it in the concentrates. R. H. Wright picked out 8.48 ounces of the smooth, bright gold, and the United States assay office at Seattle gave it a value of $15.64 per ounce. In it there were 1.68 ounces of silver and 0.4 ounce of impurities; 32.03 ounces of the gold as it came from the sluice boxes contained 6.38 ounces of silver and 2.41 ounces of impurities, and had a value of $15.17 per ounce. Each assay gives about 20 per cent as the silver content of the gold.

Water for sluicing is brought from Eureka and Chicago creeks, but the supply is scanty. The probabilities are that the production of gold upon Thanksgiving Creek will increase considerably, but as in most of the diggings of the Baker Creek group more water is needed.

HUTLINA CREEK.

Hutlina Creek is a large tributary of Baker Creek, several miles southeast of, and having a generally parallel course to Pioneer Creek. As seen from Glenn Creek its valley and that of its principal tributary are shaped similarly to those of Pioneer and Eureka creeks, and prospectors confirm this impression. A stampede to the Hutlina occurred in 1902, and it is reported that colors and occasionally good prospects were found, but live water in the gravels prevented their being worked without machinery. The bed rock is said to be similar to that of Pioneer Creek. At the time the Geological Survey party left Rampart, September 20, 1904, several prospectors were going into the valley with tools and provisions to prospect the benches during the winter.

PATTERSON CREEK.[a]

The following notes, collected from various sources, are considered reliable, though the writer was not able to verify them by personal observation.

Patterson Creek is a tributary of the lower Tanana, some of whose tributaries rise in Rough Top Mountain and others in a broad, flat divide between the Tanana and Left Fork of Baker Creek. Gold was discovered on Patterson Creek at the mouth of Easy Money and Sullivan creeks, which join to form the main valley. The scene of the discovery lies close to the trail from Hot Springs to the mouth of the Tanana. The gravels are from 35 to 70 feet deep, and the bed rock is chiefly slate. The gold is similar in character to that found on Glenn Creek. Although there was no actual mining in this district in 1907, sufficient gold had been found to warrant careful prospecting. This occurrence appears to be similar to the placers of the Glenn Creek region. Its position and character suggest that the Patterson Creek deposit may represent a southwestern extension of the Glenn Creek district.

WATER FOR HYDRAULICKING.

A large part of the gravels of the creeks and benches of the Baker Creek area, while they will not pay for shoveling in, would probably pay for working if water for hydraulicking could be obtained at a

---

[a] By Alfred H. Brooks.

reasonable cost. But, as has been said, the creeks of the region are small and furnish hardly enough water for ordinary sluicing operations.

Miners say that Hutlina Creek would furnish plenty of water for hydraulicking, but the distance it would have to be carried is variously estimated at 8 to 15 miles. Were water brought from this creek it would have to be piped through a large part of the distance to retain the head. In connection with hydraulic mining in this region the writer can do no better than quote the remarks of L. M. Prindle [a] upon the subject:

Outlook for hydraulic mining.—The installation of a hydraulic plant in any of the placer regions of the Yukon-Tanana country involves the expenditure of an amount of money several times in excess of that required for similar work in the States and should be preceded by much careful preliminary study of all the conditions. The transformation of available water supply into a powerful tool of excavation and transportation and the use of this tool in the most skillful and efficient manner are among the most important problems of mining. Lack of knowledge and skill may be covered by the results where the ground is very rich, but with ground like that under consideration the possession of these qualities or the lack of them may make all the difference between success and failure. Directors and stockholders of companies planning such work should insist upon and be constantly ready to bear the expense of the intelligent study of conditions and careful management of operations.

GENERAL CONCLUSIONS.

The rocks of the Baker Creek area are interbedded schistose grits, slates, and quartzites, with the grits forming the larger part. The grits and slates are often carbonaceous. Igneous rocks were found only along the crest of the divide, and not in large quantity. When compared with the Klondike or the Nome gold-producing regions, the small amount of metamorphism and mineralization and the scarcity of quartz veins and stringers are very noticeable.

The source of the gold is probably local, and the richer placers are generally in the vicinity of carbonaceous phases of the rocks. There is frequently reconcentration from older gravels where the streams cut across gravel-covered benches and hillsides. The gold, though generally close to bed rock, is sometimes distributed through a considerable thickness of gravel and muck. It often occurs in small crystals, and is thus shotty and chunky and easy to save. Large nuggets are rare. It contains a large amount of silver, so that its value per ounce is much lower than that of the gold of the Minook Creek area, running from $14.88 to a little over $16. There are few minerals accompanying the gold, a little pyrite, magnetite, and hematite being the only ones noticed.

---

[a] Prindle, L. M., and Hess, F. L., The Rampart placer region, in Report on progress of investigations of mineral resources of Alaska: Bull. U. S. Geol. Survey No. 259, 1905, pp. 104–119.

The creeks are all small and some have been half worked out or more, but new deposits have been discovered each year and more will probably be found. Water for working the claims is scarce, and, although some ground which will not pay for shoveling in would probably pay for hydraulicking under favorable conditions, water in adequate quantity and under a sufficient head can not be obtained without considerable expense. One of the greatest needs is a good road from Rampart, and until that is made supplies must continue excessively high.

TROUBLESOME CREEK GROUP.

GENERAL DESCRIPTION.

The Troublesome Creek group is situated between the arms of the Y formed by the divides separating the drainage basins of Minook, Baker, and Troublesome creeks. It is 18 or 20 miles southeast of Rampart. Troublesome Creek, rising among the hills east of Wolverine Mountain, flows northeast to Hess Creek, a tributary of the Yukon. The tributary valleys are often narrow and shut in by hills with steep sides and ridges, closely resembling each other, and making traveling so difficult that the country has come by its name honestly.

So far pay dirt has been found upon two creeks, Quail and Gunnison, though colors are found through the gravels over a wide area.

The rocks include all the varieties present in the Rampart formation, but slates are characteristic of the upper valley and greenstones of the lower. The slates have been intruded by a variety of igneous dikes. The creeks have cut benches upon the hills, but to a less degree than in Minook Valley.

CREEKS PROSPECTED.

QUAIL CREEK. ·

Quail Creek heads opposite Hoosier Creek and flows eastward into Troublesome Creek, having a length of between 5 and 6 miles. A large branch of this creek, known as South Fork, joins Quail Creek about a mile above Troublesome Creek. Between the two branches is a gravel-covered bench 400 feet high, upon which colors have been found, and which is being prospected. Parts of this bench occur at various places farther up Quail Creek. On the north side of Quail Creek is another bench about 50 feet above the creek, and this, too, is being prospected. In one hole bed rock was reached at 29 feet. There were 19 feet of muck and 10 feet of well-washed gravel. Colors were found all through the gravel, but no pay.

A number of igneous dikes which cross the lower part of Quail Creek show considerable mineralization by metallic sulphides. An assay of a porphyry gave no gold, but 0.52 ounce of silver [a] per ton.

The creek was located in 1898, and it is said that it was desired to call the stream " Ptarmigan " Creek, but as no one in the party could spell ptarmigan it was named " Quail," the spelling of which was easier. Some gold is said to have been taken out in that year, and a little was taken out during the summer of 1904. The total is thought to be about $3,300.

There seems to be a considerable accumulation of gravels at some places, while at others the bed rock rises to the surface. The gravels are of the country rock, with many bowlders of porphyritic granitic rocks.

A number of miners were fixing up old cabins and building new ones, and getting ready to prospect the creek during the winter.

GUNNISON CREEK.

Gunnison Creek is located a few miles farther down Troublesome Creek on the same side as Quail Creek. Miners are said to have worked upon it during the summer of 1904, and to have taken out some gold, but no further particulars were learned. The creek was not visited by the Geological Survey party.

GENERAL SUMMARY.

The alluvial deposits formed from the rocks of the valleys in which the deposits occur are found both in stream channels and on benches, and are probably all of stream origin. They are of Recent and Pleistocene age. and their thickness is generally near 5 feet, but varies from 5 to 100 feet.

The gold is generally found in the lower 2 or 3 feet of the gravel and upper 1 or 2 feet of the bed rock. but on Shirley Bar and Omega Creek it is in places distributed through the whole depth of the gravel, 5 to 7 feet, and on Omega Creek the gold is found not only in the gravel but through several feet of intimately mixed ice and clay.

The placers are of two general types as regards their origin, placers of ordinary concentration from the disintegration and wearing down of the bed rock. and placers formed through reconcentration of the gold in older gold-bearing gravels by the cutting of streams. The bench gravels of the region and the placers of Ruby and Slate creeks belong to the first class. To the second class belong the placers of the creeks cutting the high bench of Minook Creek and the placers of Doric, Glenn. and Seattle creeks and Gold Run. The other

---

[a] Burlingame, E. E., & Co., Denver, Colo.

placers of the region probably belong to the first class, although there may be some reconcentrated gold in Thanksgiving and Quail creeks.

The gold of the reconcentrated placers is generally smoother and brighter than that from the others, contains less quartz and iron, owing to abrasion and oxidation, and is thus higher in value per ounce, though the higher value of the gold of the Minook group is principally due to its containing less silver than the gold of other creeks. There is much crystallization in the gold, particularly of the Baker Creek group, where the gold contains a large percentage of silver. It is notable that along Minook Creek, where the gold contains so little silver, native silver nuggets are found in the placers, while in the Baker Creek group, where the placer gold contains about 20 per cent of combined silver, there are no silver nuggets. The only other minerals known in the concentrates with the gold are hematite, a small amount of magnetite on Thanksgiving Creek, pyrite, garnets on Ruby Creek, barite on a few other creeks, and copper on Hunter, Little Minook, and Slate creeks.

In all cases the origin of the gold has probably been in the immediate neighborhood of the placers, though it may be the result of the concentration of many hundreds of feet of bed rock. There seems to be no indication that the gold has been derived from any great "mother lode," and it has probably come from comparatively small veins distributed through the country rock.

All of the creeks at present known to be gold bearing to a paying extent, except Slate and Ruby creeks, take their rise in the Minook-Baker-Troublesome divide. Along this divide are dikes of monzonitic, dioritic, and acid igneous rocks, and it may be that these were associated with causes that introduced gold-bearing solutions into the rocks. The large mass of diabasic rocks in the Minook group may have been related in some way with the mineralization of that area.

As to the origin of the silver and copper nuggets with the gold in the creeks of the Minook group, little can be said. On all the creeks in which they are found, except Slate Creek, both limestones and diabases occur. On Slate Creek diabase was not seen, although there is much of it on Ruby Creek. It is likely that the silver and copper nuggets come from the oxidation of contact minerals resulting from the diabasic intrusions.

The average depth of gravel worked is probably between 10 and 20 feet. The gravels are mostly frozen, but much trouble in working them is sometimes had on account of live water. Hydraulicking has been introduced and apparently works well in the frozen gravels. This form of mining will probably become of considerable importance in the Minook Creek group wherever plenty of water and head a obtainable. In the Baker Creek group the expense of obtainir water for hydraulicking seems to be very much greater, though da are lacking. There is much gravel in this group, which can probab be worked at a profit only by this method. In the Troubleson country there is plenty of water, with sufficient fall for hydraulickir in that valley.

The following tables give the most important statistical data (the Rampart region and include the total gold production, so far : ascertainable, up to the fall of 1904:

Distances, men employed, and freight rates of Rampart region, 1904.

Name of diggings.	Distance from Rampart.	Number of men em- ployed.	Freight rates.	
			Winter.	Summe
	Miles.		*Cents per pound.*	*Cents p pound*
Hunter Creek	3	15	2	
Little Minook Creek	4½	30	2	
Little Minook Junior Creek..................................	5	10	2	
Hoosier Creek..	6	7	2	
Florida Creek...	8	....	....	....
Ruby Creek...	9	10	3	
Slate Creek...	11	5	4	
Eureka Creek...	28		6	
Bench bars..	30	40	6	
Doric Creek..	30		6	
Glenn Creek..	30	24	6	
Gold Run...	31	10	6	
Seattle Creek...	31	....	6	
Omega Creek.	32	3	6	
Thanksgiving Creek	34	10	6	
Quail Creek	20	5	6	15-
Total	....	169	....	....

Gold production of Rampart region.

Name of diggings.	Winter of 1903-4.	Summer of 1904.	Total, 1904.	Previous to 1904.	Total t fall, 19
Minook Creek	....	....	....	$10,000	$10,0
Hunter Creek....................................	$3,000	$3,000	$6,000	18,000	24,0
Little Minook Creek...........................	40,000	2,900	42,900	443,200	486,1
Little Minook Junior Creek	17,000	....	17,000	133,000	150,0
Hoosier Creek	500	....	500	1,500	2,0
Florida Creek	....	....	....	2,000	2,0
Ruby Creek	3,000	2,000	5,000	8,500	13,5
Slate Creek..	12,000	3,000	15,000	....	15,0
Eureka Creek......................................					
Bench bars..	45,500	16,500	62,000	23,300	85,3
Doric Creek.......................................					
Glenn Creek.......................................	11,000	39,500	50,500	227,000	277,5
Gold Run	16,000	....	16,000	9,000	25,0
Seattle Creek......................................	100	....	100	....	.
Omega Creek......................................	12,100	5,300	17,400	800	18,2
Thanksgiving Creek...............................					
Quail Creek	....	500	500	2,800	3,3
Total	160,200	72,700	232,900	879,100	1,112,0

Total production previous to 1905... 1,112,0
Total production for 1905... 200,0
Total production for 1906... 270,0

Total production of Rampart region previous to 1907.............................. 1,582,0

INDEX.

102 INDEX.

RECENT SURVEY PUBLICATIONS ON ALASKA.

[Arranged geographically. A complete list can be had on application.]

All of these publications can be obtained or consulted in the following ways:

1. A limited number are delivered to the Director of the Survey, from whom they can be obtained, free of charge (except certain maps), on application.

2. A certain number are delivered to Senators and Representatives in Congress for distribution.

3. Other copies are deposited with the Superintendent of Documents, Washington, D. C., from whom they can be had at prices slightly above cost.

4. Copies of all Government publications are furnished to the principal public libraries throughout the United States, where they can be consulted by those interested.

GENERAL.

The geography and geology of Alaska, a summary of existing knowledge, by A. H. Brooks, with a section on climate by Cleveland Abbe, jr., and a topographic map and description thereof, by R. U. Goode. Professional Paper No. 45, 1906, 327 pp.

Placer mining in Alaska in 1904, by A. H. Brooks. In Bulletin No. 259, 1905, pp. 18-31.

The mining industry in 1905, by A. H. Brooks. In Bulletin No. 284, 1906, pp. 4-9.

The mining industry in 1906, by A. H. Brooks. In Bulletin No. 314, 1907, pp. 19-39.

Railway routes, by A. H. Brooks. In Bulletin No. 284, 1906, pp. 10-17.

Administrative report, by A. H. Brooks. In Report on progress of investigations of mineral resources of Alaska in 1904: Bulletin No. 259, 1905, pp. 13-17.

Administrative report, by A. H. Brooks. In Report on progress of investigations of mineral resources of Alaska in 1905: Bulletin No. 284, 1906, pp. 1-3.

Administrative report, by A. H. Brooks. In Report on progress of investigations of mineral resources of Alaska in 1906: Bulletin No. 314, 1907, pp. 11-18.

Notes on the petroleum fields of Alaska, by G. C. Martin. In Bulletin No. 259, 1905, pp. 128-139.

The petroleum fields of the Pacific coast of Alaska, with an account of the Bering River coal deposits, by G. C. Martin. Bulletin No. 250, 1905, 64 pp.

Markets for Alaska coal, by G. C. Martin. In Bulletin No. 284, 1906, pp. 18-29.

The Alaska coal fields, by G. C. Martin. In Bulletin No. 314, 1907, pp. 40-46.

Methods and costs of gravel and placer mining in Alaska, by C. W. Purington. Bulletin No. 263, 1905, 362 pp. (Out of stock; can be purchased from Superintendent of Documents, Washington, D. C., for 35 cents.) Abstract in Bulletin No. 259, 1905, pp. 32-46.

Geographic dictionary of Alaska, by Marcus Baker, second edition by J. C. McCormick. Bulletin No. 299, 1906, 690 pp.

Administrative report, by A. H. Brooks. In Report on progress of investigations of mineral resources of Alaska in 1907. Bulletin No. 345, pp. 5-17.

The distribution of mineral resources in Alaska, by A. H. Brooks. In Report on progress of investigations of mineral resources of Alaska in 1907. Bulletin No. 345, pp. 18-29.

The mining industry in 1907, by A. H. Brooks. In Report on progress of investigations of mineral resources of Alaska in 1907. Bulletin 345, pp. 30-53.

Prospecting and mining gold placers in Alaska, by J. P. Hutchins. In Bulletin No. 345, 1908, pp. 54-77.

Water-supply investigations in Alaska in 1906-7, by F. F. Henshaw and C. C. Covert. Water-Supply Paper No. 218, 1908, 156 pp.

Topographic maps.

Alaska, topographic map of; scale, 1: 2500000. Preliminary edition by R. U. Goode. Contained in Professional Paper No. 45. Not published separately.

Map of Alaska showing distribution of mineral resources; scale, 1: 5000000; by A. H Brooks. Contained in Bulletin 345 (in pocket).

Map of Alaska; scale, 1: 5000000; by Alfred H. Brooks.

In preparation.

Methods and costs of gravel and placer mining in Alaska, by C. W. Purington.   Second edition.   -

SOUTHEASTERN ALASKA.

Preliminary report on the Ketchikan mining district, Alaska, with an introductory sketch of the geology of southeastern Alaska, by Alfred H. Brooks.   Professional Paper No. 1, 1902, 120 pp.   .
The Porcupine placer district, Alaska, by C. W. Wright.   Bulletin No. 236, 1904, 35 pp.
The Treadwell ore deposits, by A. C. Spencer.   In Bulletin No. 259, 1905, pp. 69–87.
Economic developments in southeastern Alaska, by F. E. and C. W. Wright.   In Bulletin No. 259, 1905, pp. 47–68.
The Juneau gold belt, Alaska, by A. C. Spencer, pp. 1–137, and A reconnaissance of Admiralty Island, Alaska, by C. W. Wright, pp. 138–154.   Bulletin No. 287, 1906, 161 pp.
Lode mining in southeastern Alaska, by F. E. and C. W. Wright.   In Bulletin No. 284, 1906, pp. 30–53.
Nonmetallic deposits of southeastern Alaska, by C. W. Wright.   In Bulletin No. 284, 1906, pp. 54–60.
The Yakutat Bay region, by R. S. Tarr.   In Bulletin No. 284, 1906, pp. 61–64.
Lode mining in southeastern Alaska, by C. W. Wright.   In Bulletin No. 314, 1907, pp. 47–72.
Nonmetalliferous mineral resources of southeastern Alaska, by C. W. Wright.   In Bulletin No. 314, 1907, pp. 73–81.
Reconnaissance on the Pacific coast from Yakutat to Alsek River, by Eliot Blackwelder.   In Bulletin No. 314, 1907, pp. 82–88.
Lode mining in southeastern Alaska in 1907, by C. W. Wright.   In Bulletin No. 345, 1908, pp. 78–97.'
The building stones and materials of southeastern Alaska, by C. W. Wright.   In Bulletin No. 345, 1908, pp. 116–126.
Copper deposits on Kasaan Peninsula, Prince of Wales Island, by C. W. Wright and Sidney Paige.   In Bulletin No. 345, 1908, pp. 98–115.

Topographic maps.   '

Juneau Special quadrangle; scale, 1:62500; by W. J. Peters.   For sale at 5 cents each or $3 per hundred.
Topographic map of the Juneau gold belt, Alaska.   Contained in Bulletin 287, Plate XXXVI, 1906.   Not issued separately.

In preparation.

Physiography and glacial geology of the Yakutat Bay region, Alaska, by R. S. Tarr, with a chapter on the bed-rock geology by R. S. Tarr and B. S. Butler.
The Ketchikan and Wrangell mining districts, Alaska, by F. E. and C. W. Wright.
Berners Bay Special map; scale, 1:62500; by R. B. Oliver.   (In press.)
Kasaan Peninsula Special map; scale, 1:62500; by D. C. Witherspoon and J. W. Bagley.

CONTROLLER BAY, PRINCE WILLIAM SOUND, AND COPPER RIVER REGIONS.

The mineral resources of the Mount Wrangell district, Alaska, by W. C. Mendenhall.   Professional Paper No. 15, 1903, 71 pp.   Contains general map of Prince William Sound and Copper River region; scale, 12 miles = 1 inch.   (Out of stock; can be purchased from Superintendent of Documents for 30 cents.)
Bering River coal field, by G. C. Martin.   In Bulletin No. 259, 1905, pp. 140–150.
Cape Yaktag placers, by G. C. Martin.   In Bulletin No. 259, 1905, pp. 88–89'
Notes on the petroleum fields of Alaska, by G. C. Martin.   In Bulletin No. 259, 1905, pp. 128–139.   Abstract from Bulletin No. 250.
The petroleum fields of the Pacific coast of Alaska, with an account of the Bering River coal deposits, by G. C. Martin.   Bulletin No. 250, 1905, 64 pp.
Geology of the central Copper River region, Alaska, by W. C. Mendenhall.   Professional Paper No. 41, 1905, 133 pp.
Copper and other mineral resources of Prince William Sound, by U. S. Grant.   In Bulletin No. 284, 1906, pp. 78–87.
Distribution and character of the Bering River coal, by G. C. Martin.   In Bulletin No. 284, 1906, pp. 65–76.   -

Petroleum at Controller Bay, by G. C. Martin. In Bulletin No. 314, 1907, pp. 89–103.
Geology and mineral resources of Controller Bay region, by G. C. Martin. Bulletin
No. 335, 1908, 141 pp.
Notes on copper prospects of Prince William Sound, by F. H. Moffit. In Bulletin
No. 345, 1908, pp. 176–178.
Mineral resources of the Kotsina and Chitina valleys, Copper River region, by F. H.
Moffit and A. G. Maddren. In Bulletin No. 345, 1908, pp. 127–175.

Topographic maps.

Map of Mount Wrangell; scale, 12 miles = 1 inch. Contained in Professional Paper
No. 15. Not issued separately.
Copper and upper Chistochina rivers; scale, 1:250000; by T. G. Gerdine. Contained
in Professional Paper No. 41. Not issued separately.
Copper, Nabesna, and Chisana rivers, headwaters of; scale, 1:250000. D. C. Wither-
spoon. Contained in Professional Paper No. 41. Not issued separately.
Controller Bay region Special map; scale, 1:62500; by E. G. Hamilton. For sale at 35
cents a copy or $21.00 per hundred.
General map of Alaska coast region from Yakutat Bay to Prince William Sound; scale,
1:1200000; compiled by G. C. Martin. Contained in Bulletin No. 335.

In preparation.

The Kotsina-Chitina copper region, by F. H. Moffit.
Chitina quadrangle map; scale, 1:250000; by T. G. Gerdine and D. C. Witherspoon.

COOK INLET AND SUSITNA REGION.

The petroleum fields of the Pacific coast of Alaska, with an account of the Bering River
coal deposits, by G. C. Martin. Bulletin No. 250, 1905, 64 pp.
Coal resources of southwestern Alaska, by R. W. Stone. In Bulletin No. 259, 1905,
pp. 151–171.
Gold placers of Turnagain Arm, Cook Inlet, by F. H. Moffit. In Bulletin No. 259,
1905, pp. 90–99.
Mineral resources of the Kenai Peninsula; Gold fields of the Turnagain Arm region, by
F. H. Moffit, pp. 1–52; Coal fields of the Kachemak Bay region, by R. W. Stone,
pp. 53–73. Bulletin No. 277, 1906, 80 pp.
Preliminary statement on the Matanuska coal field, by G. C. Martin. In Bulletin No.
284, 1906, pp. 88–100.
A reconnaissance of the Matanuska coal field, Alaska, in 1905, by G. C. Martin. Bulle-
tin No. 289, 1906, 36 pp. (Out of stock; can be purchased of Superintendent of
Documents for 25 cents.)
Reconnaissance in the Matanuska and Talkeetna basins, by S. Paige and A. Knopf.
In Bulletin No. 314, 1907, pp. 104–125.
Geologic reconnaissance in the Matanuska and Talkeetna basins, Alaska, by S. Paige
and A. Knopf. Bulletin No. 327, 1907, 71 pp.

Topographic maps.

Kenai Peninsula, northern portion; scale, 1:250000; by E. G. Hamilton. Contained
in Bulletin No. 277. Not published separately.
Reconnaissance map of Matanuska and Talkeetna region; scale, 1:250000; by T. G.
Gerdine and R. H. Sargent. Contained in Bulletin No. 327. Not published
separately.
Mount McKinley region; scale, 1:625000; by D. L. Reaburn. Contained in Professional Paper No. 45. Not published separately.

ALASKA PENINSULA AND ALEUTIAN ISLANDS.

Gold mine on Unalaska Island, by A. J. Collier. In Bulletin No. 259, 1905, pp. 102–103.
Gold deposits of the Shumagin Islands, by G. C. Martin. In Bulletin No. 259, 1905,
pp. 100–101.
Notes on the petroleum fields of Alaska, by G. C. Martin. In Bulletin No. 259, 1905,
pp. 128–139. Abstract from Bulletin No. 250.
The petroleum fields of the Pacific coast of Alaska, with an account of the Bering River
coal deposits, by G. C. Martin. In Bulletin No. 250, 1905, 64 pp.
Coal resources of southwestern Alaska, by R. W. Stone. In Bulletin No. 259, 1905,
pp. 151–171.
The Herendeen Bay coal field, by Sidney Paige. In Bulletin No. 284, 1906, pp.
101–108.

YUKON BASIN.

The coal resources of the Yukon, Alaska, by A. J. Collier. Bulletin No. 218, 1903, 71 pp.

The gold placers of the Fortymile, Birch Creek, and Fairbanks regions, by L. M. Prindle. Bulletin No. 251, 1905, 89 pp.

Yukon placer fields, by L. M. Prindle. In Bulletin No. 284, 1906, pp. 109–131.

Reconnaissance from Circle to Fort Hamlin, by R. W. Stone. In Bulletin No. 284, 1906, pp. 128–131.

The Yukon-Tanana region, Alaska; description of the Circle quadrangle, by L. M. Prindle. Bulletin No. 295, 1906, 27 pp.

The Bonnifield and Kantishna regions, by L. M. Prindle. In Bulletin No. 314, 1907, pp. 205–226.

The Circle Precinct, Alaska, by Alfred H. Brooks. In Bulletin No. 314, 1907, pp. 187–204.

The Yukon-Tanana region, Alaska; description of the Fairbanks and Rampart quadrangles, by L. M. Prindle, F. L. Hess, and C. C. Covert. Bulletin No. 337, 1908, 102 pp.

Occurrence of gold in the Yukon-Tanana region, by L. M. Prindle. In Bulletin No. 345, 1908, pp. 179–186.

The Fortymile gold placer district, by L. M. Prindle. In Bulletin No. 345, 1908, pp. 187–197.

Water supply of the Fairbanks district in 1907, by C. C. Covert. In Bulletin No. 345, 1908, pp. 198–205.

Topographic maps.

Fortymile quadrangle; scale, 1:250000; by E. C. Barnard. For sale at 5 cents a copy or $3 per hundred.

Yukon-Tanana region, reconnaissance map of; scale, 1:625000; by T. G. Gerdine. Contained in Bulletin No. 251, 1905. Not published separately.

Fairbanks and Birch Creek districts, reconnaissance maps of; scale, 1:250000; by T. G. Gerdine. Contained in Bulletin No. 251, 1905. Not issued separately.

Circle quadrangle, Yukon-Tanana region; scale, 1:250000; by D. C. Witherspoon. Contained in Bulletin No. 295. Not issued separately.

In preparation.

Water-supply investigations in Alaska, 1906 and 1907, by F. F. Henshaw and C. C. Covert. Water-Supply Paper No. 218, 1908, 156 pp.

Fairbanks quadrangle map; scale, 1:250000; by D. C. Witherspoon. Contained in Bulletin No. 337, 1908.

Rampart quadrangle map; scale, 1:250000; by D. C. Witherspoon. Contained in Bulletin No. 337, 1908.

Fairbanks Special map; scale, 1:62500; by T. G. Gerdine and R. H. Sargent.

SEWARD PENINSULA.

A reconnaissance of the Cape Nome and adjacent gold fields of Seward Peninsula, Alaska, in 1900, by A. H. Brooks, G. B. Richardson, and A. J. Collier. In a special publication entitled "Reconnaissances in the Cape Nome and Norton Bay regions, Alaska, in 1900," 1901, 180 pp.

A reconnaissance in the Norton Bay region, Alaska, in 1900, by W. C. Mendenhall. In a special publication entitled "Reconnaissances in the Cape Nome and Norton Bay regions, Alaska, in 1900."

A reconnaissance of the northwestern portion of Seward Peninsula, Alaska, by A. J. Collier. Professional Paper No. 2. 1902, 70 pp.

The tin deposits of the York region, Alaska, by A. J. Collier. Bulletin No. 229, 1904, 61 pp.

Recent developments of Alaskan tin deposits, by A. J. Collier. In Bulletin No. 259, 1905, pp. 120–127.

The Fairhaven gold placers of Seward Peninsula, by F. H. Moffit. Bulletin No. 247, 1905, 85 pp.

The York tin region, by F. L. Hess. In Bulletin No. 284, 1906, pp. 145–157.

Gold mining on Seward Peninsula, by F. H. Moffit. In Bulletin No. 284, 1906, pp. 132–141.

The Kougarok region, by A. H. Brooks. In Bulletin No. 314, 1907, pp. 164–181.

Water supply of Nome region, Seward Peninsula, Alaska, 1906, by J. C. Hoyt and F. F. Henshaw. Water-Supply Paper No. 196, 1907, 52 pp. (Out of stock; can be purchased of Superintendent of Documents for 15 cents.)

Water supply of the Nome region, Seward Peninsula, 1906, by J. C. Hoyt and F. F. Henshaw. In Bulletin No. 314, 1907, pp. 182–186.

The Nome region, by F. H. Moffit. In Bulletin No. 314, 1907, pp. 126–145.

Gold fields of the Solomon and Niukluk river basins, by P. S. Smith. In Bulletin No. 314, 1907, pp. 146–156.

Geology and mineral resources of Iron Creek, by P. S. Smith. In Bulletin No. 314, 1907, pp. 157–163.

The gold placers of parts of Seward Peninsula, Alaska, including the Nome, Council, Kougarok, Port Clarence, and Goodhope precincts, by A. J. Collier, F. L. Hess, P. S. Smith, and A. H. Brooks. Bulletin No. 328, 1908, 343 pp.

Investigation of the mineral deposits of Seward Peninsula, by P. S. Smith. In Bulletin No. 345, 1908, pp. 206–250.

The Seward Peninsula tin deposits, by Adolph Knopf. In Bulletin No. 345, 1908, pp. 251–267.

Mineral deposits of the Lost River and Brooks Mountain regions, Seward Peninsula, by Adolph Knopf. In Bulletin No. 345, 1908, pp. 268–271.

Water supply of the Nome and Kougarok regions, Seward Peninsula, in 1906–7, by F. F. Henshaw. In Bulletin No. 345, 1908, pp. 272–285.

Topographic maps.

The following maps are for sale at *5 cents a copy*, or $3 per hundred:

Casadepaga Quadrangle, Seward Peninsula; scale, 1:62500; by T. G. Gerdine.

Grand Central Special, Seward Peninsula; scale, 1:62500; by T. G. Gerdine.

Nome Special, Seward Peninsula; scale, 1:62500; by T. G. Gerdine.

Solomon Quadrangle, Seward Peninsula; scale, 1:62500; by T. G. Gerdine.

The following maps are for sale at *25 cents a copy*, or $15 per hundred:

Seward Peninsula, northeastern portion of, topographic reconnaissance of; scale, 1:250000; by T. G. Gerdine.

Seward Peninsula, northwestern portion of, topographic reconnaissance of; scale, 1:250000; by T. G. Gerdine.

Seward Peninsula, southern portion of, topographic reconnaissance of; scale, 1:250000; by T. G. Gerdine.

In preparation.

Water-supply investigations in Alaska, 1906 and 1907, by F. F. Henshaw and C. C. Covert. Water-Supply Paper No. 218, 1908, 156 pp.

Geology of the area represented on the Nome and Grand Central Special maps, by F. H. Moffit, F. L. Hess, and P. S. Smith.

Geology of the area represented on the Solomon and Casadepaga Special maps, by P. S. Smith.

The Seward Peninsula tin deposits, by A. Knopf.

NORTHERN ALASKA.

A reconnaissance from Fort Hamlin to Kotzebue Sound, Alaska, by way of Dall, Kanuti, Allen, and Kowak rivers, by W. C. Mendenhall. Professional Paper No. 10, 1902, 68 pp.

A reconnaissance in northern Alaska across the Rocky Mountains, along the Koyukuk, John, Anaktuvuk, and Colville rivers, and the Arctic coast to Cape Lisburne, in 1901, by F. C. Schrader and W. J. Peters. Professional Paper No. 20, 1904, 139 pp. (Out of stock; can be purchased of Superintendent of Documents for 40 cents.)

Coal fields of the Cape Lisburne region, by A. J. Collier. In Bulletin No. 259, 1905, pp. 172–185.

Geology and coal resources of Cape Lisburne region, Alaska, by A. J. Collier. Bulletin No. 278, 1906, 54 pp.

Topographic maps.

Fort Yukon to Kotzebue Sound, reconnaissance map of; scale, 1:1200000; by D. L. Reaburn. Contained in Professional Paper No. 10. Not published separately.

Koyukuk River to mouth of Colville River, including John River; scale, 1:1200000; by W. J. Peters. Contained in Professional Paper No. 20. (Out of stock.) Not published separately.

O

DEPARTMENT OF THE INTERIOR
UNITED STATES GEOLOGICAL SURVEY
GEORGE OTIS SMITH, Director

BULLETIN 338

THE IRON ORES

OF THE

IRON SPRINGS DISTRICT

SOUTHERN UTAH

BY

C. K. LEITH AND E. C. HARDER

WASHINGTON
GOVERNMENT PRINTING OFFICE
1908

CONTENTS.

3

ILLUSTRATIONS.

5

6 ILLUSTRATIONS.

OUTLINE OF PRINCIPAL RESULTS.

In the Iron Springs district of southern Utah, described in this paper, sediments of Carboniferous, Cretaceous, and Tertiary ages have been intruded by large masses of andesite taken to be laccoliths. Erosion following the intrusion developed mountains out of the laccoliths, with surrounding rings of outward-dipping sediments on the lower slopes. Later extrusion of lavas covered the eroded laccoliths and sediments. Subsequent erosion has reexhumed the laccoliths and part of the sediments, leaving the lavas in a surrounding zone. The lower slopes and flats adjacent to the laccoliths are covered by the usual unconsolidated deposits of the Great Basin country. Faulting, principally of the tension type, is prevalent.

The iron ores constitute (1) fissure veins in the andesite, (2) fissure and replacement deposits along the contact of the andesite with the Carboniferous limestone, and (3) breccia cement in the Cretaceous quartzite. The larger deposits are the ones along the limestone-andesite contact. Their longer diameters in general follow the contact with rough lens shape, but there are many irregularities due to faulting and other causes.

About 1,600 exploration pits have been sunk, but the deepest goes down only 130 feet, and has not reached water level. The vertical dimensions of the deposits are therefore unknown. Their total area is 5,430,000 square feet, and their total tonnage, so far as can be measured by area and pits, is about 40,000,000 tons. Probably this estimate is much too small, for the pits do not go to the bottom of the deposits.

The ore is magnetite and hematite, containing a small amount of limonite. Much of it runs above 60 per cent in iron, but the average is about 56 per cent. Phosphorus is uniformly high. Sulphur, copper, and titanium are not present in prohibitive amounts. There is no evidence of increase in sulphur with depth, but water level has not been reached by the pits. The texture is hard and crystalline at the surface; beneath the surface the ores are relatively soft. The gangue is principally quartz or chalcedony near the surface; with depth calcite increases. Garnet, diopside, apatite, mica, hornblende, and other silicates constitute minor constituents of the ore.

The ores are closely related in origin to the intrusion of the andesite laccoliths. The first effect of the intrusion was the contact metamorphism of the adjacent limestone, produced principally by driving out its lime and magnesia carbonates and recrystallizing the remaining constituents, but also by the introduction of silica, iron, and soda. The net result was diminution of volume. Limestone was actually fused, for residues of glass still remain. The net density of the contact minerals is somewhat higher than the density of the unaltered limestone, but the density of the rock as a whole is less than that of the fresh limestone, owing to the content of glass and the pore space. It is suggested that the fusion of the limestone contact phase denied access to the andesite solutions, thus explaining the lack of abundant introduction of minerals from the intrusives along the contact. Had the temperature not reached the fusion point of the limestone (probably about 1,000° C.), other conditions remaining unchanged, the contact phase of the limestone might have been left porous, affording openings for the entrance of the solutions.

The ore-bearing solutions followed shortly after the contact metamorphism of the limestone, and after the outer part, at least, of the andesite laccoliths had crystallized, as shown by the occurrence of ore in fissures in the andesite and limestone contact phase. The ore-bearing solutions carried silicates similar to those previously formed in the limestone contact. The conspicuous feature was the introduction of soda by the ore-bearing solutions into both limestone and andesite along the contact. A comparison of the alteration of the andesite next to the ore with its alteration by weathering brings out clearly the effect of the ore-bearing solutions.

A graphic method of comparison of the fresh and altered rocks brings out various possibilities in the interpretation of analyses.

A consideration of the temperatures of crystallization of some of the silicates deposited with the ores and possible pressures resisting the solutions seems to indicate a probability that the solutions may have been vapors and that they were pneumatolytic after-effects of the andesite intrusions.

It is suggested that the iron was carried as ferrous chloride, which became oxidized at a temperature above 500° C. by breaking up associated water and simultaneously developing free hydrogen. Precipitation came about partly by the neutralization of the acid when it came into contact with the limestone, and partly by the evaporation of the acid with the water which would accompany the diminution of temperature and pressure as the solutions left the andesite. The conspicuous introduction of soda into wall rocks suggests further that the same solution carried salt. This would be decomposed by silicic acid and the soda be deposited as sodium silicate, largely albite, the form in which it is now found.

The ore-filled fissures in the andesite and limestone resulted from the cooling of the andesite mass, which developed stretch fractures within the mass and pulled it away from the limestone wall as a casting away from the mold. Where the ore-bearing solutions reached the unaltered limestone rather than the silicated contact phase of the limestone the ore replaced the limestone. Where they reached the overlying quartzite they filled fractures.

A considerable secondary concentration may be traced to the Tertiary lavas which once covered the ore deposits. The ores were coarsely recrystallized and chalcedony was introduced as a gangue material in place of calcite. No heavy silicates were introduced. The concentration under the influence of the lavas is a relatively shallow one, which can be fairly separated from the deep primary concentration.

The entire effect of weathering is not known, because the ores have been observed only above water level. Above this they are porous, suggesting that there may have been a considerable amount of solution of the gangue materials down to this point. Calcite has obviously been leached near the surface and deposited farther down; apatite has been altered to osteolite; small amounts of sulphides have been oxidized and partly removed; limonite has developed in thin films along the margins of cavities and along fissures. In general, however, the obvious relations of the characteristics of the ore to conditions other than weathering seem to indicate that the iron oxides themselves have been little affected by weathering.

Inferences as to the extension and shape of the ore deposits in depth are drawn from a consideration of origin. The ores may have a considerable depth. The ores are not likely to be so thick along the contact on upper slopes, because gravity tended to close the openings formed by the cooling of the andesite mass. There seems to be no reason why ore should not be found at the limestone-andesite contact at places where the contact is now overlain by Cretaceous quartzite.

The Iron Springs deposits are typical of many other western iron-ore deposits, and it is believed from personal observation that their origin has essential features in common. Description is made of the structure of ore of similar occurrence in the adjoining Bull Valley district, and reference is made to deposits regarded as belonging to the same class in other parts of the West.

THE IRON ORES OF THE IRON SPRINGS DISTRICT OF SOUTHERN UTAH.

By CHARLES KENNETH LEITH and EDMUND CECIL HARDER.

CHAPTER I.

INTRODUCTION:

PURPOSE OF THE INVESTIGATION.

The enormous and rapidly increasing consumption of iron ore in the United States during the last few years has led to careful inventories of available supplies, both in producing districts and in relatively unknown fields, with the result in general of emphasizing the limitation of the deposits now worked and of pointing out the necessity for the early exploitation of new ones. Up to the present time less than 2 per cent of the iron ore mined annually has come from west of Mississippi River, but there have long been known in that region large deposits of iron ore which have not been exploited because of lack of transportation facilities or distance from consuming centers. The part these deposits are to play in meeting the future demand for iron ore, the possibilities or probabilities of finding other large deposits in the West, and the geological features of significance in connection with these questions remain largely to be ascertained.

Mr. Leith began a general geological reconnaissance of western iron ores in 1903 and continued it during 1904, 1905, and 1906, visiting some of the better known ore deposits of Utah, Colorado, Wyoming, Washington, British Columbia, California, and Missouri. It early became apparent to him, as it had been apparent to others, that the iron-ore deposits of the West are prevailingly of a distinct and uniform type—an irregular replacement of limestone near the contact with igneous rocks, or a vein filling in both limestone and igneous rock—a type fundamentally different from that of the important producing districts east of Mississippi River, and probably of different origin. The deposits of the Iron Springs district of southern Utah were selected as typical of this class of deposits. well located for study, and of such size and quality as to warrant the belief that their exploitation would follow in a short time; accordingly, detailed geological work was begun in this district in 1905 by Mr. Leith, with E. C. Harder and Frank J. Katz as field assistants, and was continued during the early part of the field season of 1906 by E. C. Harder and Freeman Ward, the senior author being with the party for three weeks. The results of this work are presented in the following pages. It is believed that they will be found to apply in their essential features to many other deposits of the same general class in western United States.

PREVIOUS WORK.

Lying near the northwest side of the High Plateau region and in the southeast part of the Basin Range province, the Iron Springs district has received passing attention from Powell (1875 and 1879),[a] and Dutton (1880 and 1882)[b] and associates in their general surveys of the High Plateaus and Colorado River, and from the geologists of the Wheeler Survey—Howell (1875),[c] Gilbert (1875),[d] and Marvine (1875).[e] Their maps and descriptions are necessarily so generalized that they are of little value in detailed study of the district under discussion. They are important, however, in establishing a basis for the general correlation of formations.

The Powell Survey prepared a reconnaissance topographic map of southern Utah, part of which, covering the Iron Springs district, is included in the St. George sheet of the United States Geological Survey. While this is too much generalized to be of use in detailed mapping, it expresses the general topographic and geographic relations of the Iron Springs district to the surrounding territory.

Howell (1875)[f] was the first of the early geologists to make anything like a specific description of the Iron Springs district and its iron-ore deposits. He includes analyses of the ores by C. E. Dutton.

In 1880 the deposits received attention from J. S. Newberry,[g] then America's leading authority on iron ore. He describes the deposits as constituting "perhaps the most remarkable deposit of iron ore yet developed on this continent," and refers the sediments of the district

a Powell, J. W., Exploration of the Colorado River of the West and its tributaries; explored in 1869, 1870, 1871, and 1872, under the direction of the Secretary of the Interior; U. S. Geog. and Geol. Surv. Rocky Mountain Region, Washington, 1875, 291 pp.; Report on the lands of the arid region of the United States, with a more detailed account of the lands of Utah; U. S. Geog. and Geol. Surv. Rocky Mountain Region (2d ed.), 1879, 195 pp.

b Dutton, C. E., Report on the geology of the High Plateaus of Utah; U. S. Geog. and Geol. Surv. Rocky Mt. Region, 1880, 307 pp.; Tertiary history of the Grand Canyon district; Mon. U. S. Geol. Survey, vol. 2, 1882, 264 pp.

c Howell, E. E., Report on the geology of portions of Utah, Nevada, Arizona, and New Mexico, examined in the years 1872 and 1873; U. S. Geog. Surv. W. 100th Mer., vol. 3, Geology, 1875, pt. 3, pp. 227–301

d Gilbert, G. K., Report on the geology of portions of Nevada, Utah, California, and Arizona examined in the years 1871 and 1872; U. S. Geog. Surv. W. 100th Mer., vol. 3, Geology, pt. 1, 1875, pp. 17–187.

e Marvine, A. R., Report on the geology of route from St. George, Utah, to Gila River, Arizona, examined in 1871; U. S. Geog. Surv. W 100th Mer., vol. 3 Geology, pt. 2, 1875, pp. 189–225.

f Op. cit., pp. 261–263.

g Newberry, J. S., The genesis of the ores of iron: School of Mines Quart., Nov., 1880, pp. 9–12.

doubtfully to the Lower Silurian. Both the ores and the andesite (called granite) he regarded as metamorphosed sediments.

The Iron Springs deposits figure in the Tenth Census report (1886),[a] where they were mapped and described by Putnam from an economic standpoint. He referred to them as "probably the largest mass of iron ore in the whole West."

The district lies close to the southeastern shore of Lake Bonneville, and accordingly it appears on the maps accompanying Gilbert's Bonneville report (1890).[b]

The physiographic record of the region to the east and south of the Iron Springs district was the subject of careful study in 1904 by Huntington and Goldthwait[c] under the direction of W. M. Davis.

The senior author[d] of the present report has published two short statements of the results of his examination of the district (1903 and 1906).

A general economic account of the district with special reference to the chemistry of the ores was made by Fred Lerch (1904).[e]

A considerable amount of exploration of the district has been done in recent years for commercial interests, the results of which have not been published, but which are incorporated in the present report.

ACKNOWLEDGMENTS.

The authors are indebted to Mr. Fred Lerch, of Biwabik, Minn., to the late Col. S. B. Milner, and to Mr. Archibald Milner, of Salt Lake City, Utah, for many chemical analyses, blueprints of claims, and much general information concerning the district, and to Mr. R. N. Dickman, of Chicago, for chemical analyses. With a single exception, calls for information from the property holders were met fully and promptly. The authors record here their appreciation of the many courtesies afforded them by the people of Cedar City and adjacent territory, which contributed not a little to the comfort and effectiveness of the field work.

---

a Putnam, B. T., Notes on the samples of iron ore collected west of the one hundredth meridian: Tenth Census U. S., vol. 15, Mining Industries, 1886, pp. 469–505.

b Gilbert, G. K., Lake Bonneville: Mon. U. S. Geol. Survey, vol. 1, 1890, pp. 438.

c Huntington, Ellsworth, and Goldthwait, J. W., The Hurricane fault in the Toquerville district, Utah: Bull. Mus. Comp. Zool., Harvard Coll., No. 42 (Geol. ser., vol. 6), 1904, pp. 199–259.

d Leith, C. K., Iron ores in southern Utah: Bull. U. S. Geol. Survey No. 225, 1904, pp. 229–237; Iron ores of the western United States and British Columbia: Bull. U. S. Geol. Survey No. 285, 1906, pp. 194–200.

e Lerch, Fred, The iron-ore deposits in southern Utah: Iron Trade Review, May 19, 1904, pp. 49–50.

CHAPTER II.

GEOGRAPHY, TOPOGRAPHY, AND GENERAL GEOLOGY.

GEOGRAPHY AND TOPOGRAPHY.

The Iron Springs district lies between longitudes 113° 10′ and 113° 26′ 30″ and latitudes 37° 35′ and 37° 47′ 30″, in Iron County, southern Utah, about 250 miles south of Salt Lake City and 550 miles from the harbor of San Pedro, Cal., on the Pacific Ocean. The San Pedro, Los Angeles and Salt Lake Railroad runs within 22 miles of the district on the west, Lund being the nearest station. The district can also be reached by way of a spur of the Denver and Rio Grande Railroad running down to Marysvale, 80 miles northeast of the district, thence on by stage.

The area (see fig. 1) lies near the eastern margin of the Basin Range province and includes several basin ranges and hills, principal among which are the Harmony Mountains, Iron Mountain, Antelope Range, Granite Mountain, The Three Peaks, and Swett Hills (Pl. I, pocket). To the south are the Pine Valley Mountains. Immediately west, north, and east of the district lies the desert, beyond which on the west and north are other basin ranges, and on the east, 12 miles away, is the Hurricane fault scarp of the High Plateaus. On the southwest the district is continuous with a series of ranges and hills extending west of the Pine Valley Mountains well into Nevada.

The elevation ranges from 5,300 to 8,000 feet.

The drainage is through small creeks leaving the mountains and hills and soon losing themselves in the desert.

The tops of the Harmony Mountains retain snow until the middle of summer, and consequently have an abundance of vegetation, such as yellow pine, fir, cottonwood, quaking aspen, and mountain mahogany. The tops and slopes of the other mountains are dry and are covered with a growth of scrub cedar and piñon. Shrubs, sagebrush, and several species of cacti are also abundant, but grasses are lacking. The surrounding desert presents the variety of sagebrush, rabbit brush, greasewood, and shad scale characteristic of the desert elsewhere in the Great Basin.

14

The location of the Iron Springs district affords unusual opportunities for the study of Basin Range structure, the relation of the Basin Ranges to the High Plateaus, and the associated problems. From a scenic standpoint also the district is difficult to surpass, affording a view of the massive Pine Valley Mountains, the desert

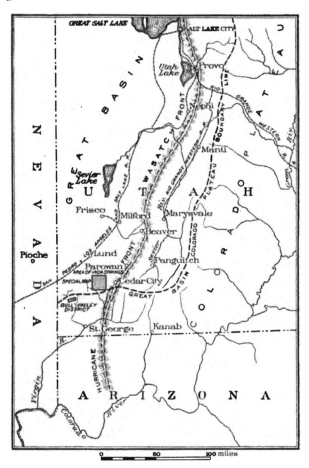

Fig. 1.—Sketch map of parts of Utah, Arizona, and Nevada, showing geographic relation of Iron Springs district to physiographic regions.

in its typical development, and the still more striking, brilliant-colored escarpments of the High Plateaus, notched immediately opposite the Iron Springs district by Cedar Canyon (Coal Creek Valley), which rivals in form and beauty of coloring the canyon of the Virgin, 50 miles to the south, and even the Grand Canyon, 100 miles to the south.

A. IRON SPRINGS VALLEY AND THE THREE PEAKS, FROM LINDSAY HILL.

B. CRETACEOUS SEDIMENTS DIPPING AWAY FROM THE ANDESITE ON SOUTHWEST CORNER
OF THE THREE PEAKS LACCOLITH.

GENERAL GEOLOGY.

The dominating geological features of the district are 3 large andesite laccoliths, constituting The Three Peaks (Pl. VII, *A*), Granite Mountain, and Iron Mountain, lying in a northeast-southwest line across the area mapped. Three unconformable sedimentary series, aggregating 4,000 feet in thickness, outcrop in successive rings around these laccoliths (Pl. VII, *B*), dipping outward asymmetrically, very steeply at the contact, less steeply farther away. Still farther from the laccoliths lava flows 2,000 feet thick rest in nearly horizontal attitude on the tilted sedimentary rocks. These general relations are modified by faulting. All of the rock formations of the district are more or less covered on the middle and lower slopes by unconsolidated and partly consolidated erosion débris, both aqueous and subaerial, which spreads out on the lower ground to make the deserts. The detailed succession is shown in fig. 2.

The laccoliths are exposed only in their upper parts—at no place has erosion shown their bottoms in section. The rock is an andesite of remarkably uniform texture and composition. Within the area of the laccoliths are a few veins of iron ore and fault blocks of ore and Carboniferous and Cretaceous sediments.

In contact with the laccoliths for the most part is Carboniferous limestone, a pure, dense, blue limestone, with.a few feet of sandy material appearing locally at the base. The contacts are at most localities nearly vertical, this being due partly to faulting and partly to the fact that erosion has cut down to the sides of the laccoliths, exposing the vertical part of their contact. Locally, where erosion has not gone down so far, the limestone dips distinctly away from the andesite at an angle as low as 10°. The limestone is altered at the contact with the laccolith for a maximum of 1,000 feet, as measured on the erosion slopes, by loss of carbonates, development of anhydrous silicates, and replacement by ore.

Cretaceous sediments outcrop in a zone outside of the Carboniferous limestone, except where locally they are faulted down against the laccolith, or where the laccolith has penetrated the Cretaceous and erosion has not yet cut down to the Carboniferous limestone, as on the west side of Iron Mountain. The Cretaceous rocks are principally sandstone, with layers or lenses of shale, conglomerate, and limestone breccia. At the contact with the laccoliths the sandstone has been indurated and amphibole has developed.

The relations of the Cretaceous sandstone to the underlying Carboniferous limestone are those of apparent conformity, but the contact is rendered somewhat obscure by the presence of shale at the base of the Cretaceous sandstone. The lower portion may be in part Jurassic, but separation could not be made. The greatly varying

A. IRON SPRINGS VALLEY AND THE THREE PEAKS. FROM LINDSAY HILL.

B. CRETACEOUS SEDIMENTS DIPPING AWAY FROM THE ANDESITE ON SOUTHWEST CORNER
OF THE THREE PEAKS LACCOLITH.

thickness of the Carboniferous limestone may be due partly to erosion and partly to intrusion of andesite at different horizons. If it represents erosion, this is evidence of unconformity. Notwithstanding the conformity of structure, it is not unlikely that there may be a hiatus between the two systems, for the correlation of

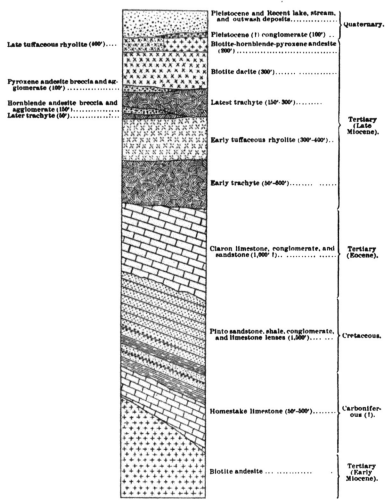

FIG. 2.—Geologic column of the Iron Springs district.

the two is fairly well based, and the Permian, Triassic, and some Jurassic sediments are lacking between them.

The Eocene series outcrops in a zone still farther away from the laccoliths. It consists of limestone and conglomerate, characterized principally by pink and bright-red colors. A basal conglomerate separates it from the Cretaceous sediments below. The dip of the Cretaceous averages about the same as that of the Tertiary, but there is a marked erosion unconformity between them.

Miocene lavas and tuffs rest nearly horizontally upon the eroded edge of the Eocene sediments and to a less extent upon the Cretaceous.

In the northwestern part of the area, near Antelope Springs, a Pliocene or Pleistocene fluviatile deposit occupies an embayment in the lavas. It is principally a conglomerate containing fragments of the lavas, of the earlier andesite laccolith, and of the sediments.

Pleistocene and Recent lake, stream, and outwash deposits, consisting of gravels, sands, and clays derived from the erosion of all of the rocks of the district, occupy nearly half of the area and mask the rock formations of the lower ground. As a whole this material is a result of disintegration to a larger extent than of decomposition. It is accumulating with extreme rapidity. The hillsides are covered with it and during heavy rain storms a great mass of débris, including fragments many feet in diameter, creeps down the slopes. The finer material is carried down by the torrents and distributed in broad, low fans on the deficient slopes below. The apparently flat "desert" is made up principally of overlapping fans. An hour's heavy rain brings about a conspicuous modification of alluvial fans or other deposits at these places. The extreme rapidity of erosion and transportation in an arid region has often been described, but the amount of material moved in a few hours can scarcely be realized without direct observation.

FAULTS, JOINTS, AND FISSURES.

That faulting complicates the elementary relations above sketched is apparent from an inspection of the general map (Pl. II, pocket). Fault scarps are common. Streams or canyons follow faults and joints, especially the former, so prevailingly that in the mapping faults were looked for whenever a canyon was encountered. Faults have been mapped only where they could be actually proved to exist by the relations of the rock formations, otherwise they are not shown on the map, even where their absence or their abrupt termination looks structurally improbable. It is certain that many have been missed.

The larger and more conspicuous faults have a prevailing north-south trend, but many have other directions, as summarized in fig. 3. At several localities, especially near the ore deposits, a tendency in the faults to follow about the periphery of the andesite laccoliths is observable.

The fault planes are vertical, or nearly so. The displacements vary from vertical to horizontal—principally the former in the faults

mapped.   The maximum vertical displacement known is 2,000 feet;
the horizontal displacement is unknown.   Hinge faults, the displacement at one end of which is opposite to that at the other, were found
locally in the lava area.   It is probable that many of the faults with
a displacement parallel to the dip of the beds have not been mapped
because not detected.   Allowing for a considerable unknown horizontal component of displacement, there remains a probable dominance of vertical movement sufficient to indicate that the net result

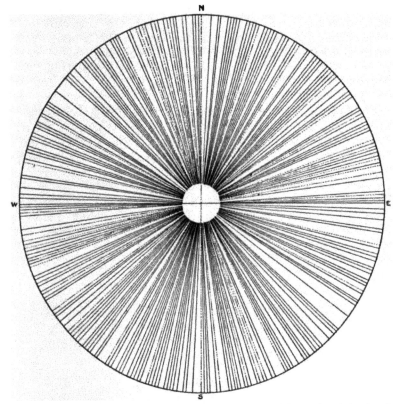

Fig. 3.—Plat of faults and iron-ore veins in the Iron Springs district.   Solid lines show faults; dotted
lines, iron-ore veins; and solid and dotted lines, faults and iron-ore veins.

of the faulting has been the extension of the area; in other words, the
faults are primarily tension phenomena.

Conspicuous in the andesite are fissures, some of them filled with
ore, which curve, taper, branch, and parallel themselves in such a
manner as to suggest that they are stretch fractures.   In the Three
Peaks andesite area there is a conspicuous parallel fracturing spaced
at intervals of a fraction of an inch to a number of feet.   In the
neighborhood of ore veins they have the same trend as the veins.

The origin óf the faults and other fractures is believed to have been principally due to the successive extension, shortening, and settling accompanying the intrusion and the cooling of the laccoliths—the origin cited by Spurr for some of the faults of the Tonopah district. The faults are tension phenomena, and are so numerous and so intricate in their intersections that it is easier to explain them by somewhat local strains accompanying cooling than by strain on the district as a whole. The faults outlining the periphery of the andesite and the stretch fractures within the andesite may be ascribed definitely to this cause. A rough calculation of the cubic shrinkage of a mass of andesite the size of the Iron Mountain laccolith in crystallizing from a glass to andesite indicates a horizontal radial shortening of between 200 and 500 feet, depending upon the depth assumed for the mass—in any case a large enough factor to be important in the development of fractures.

So far as the faulting was due to the intrusion and cooling of the andesite, it followed the intrusion closely and is of Tertiary age.

Other faults cut the lava flows and are therefore considerably later than the laccolith intrusions. These faults are, from their nature, probably in part due to the cooling of the lavas.

Certain of the larger faults, and especially those having north-south directions, showing great extent and continuity across both igneous and sedimentary formations, may be caused otherwise than by cooling either of laccoliths or of lavas. They seem to belong more to the order of deformation producing the Hurricane fault to the east than to the deformation associated with local igneous action. The whole Iron Springs district represents a downthrow on the west side of the Hurricane fault. The stresses have affected large areas. There can be little doubt that these great faults are essentially tensional in nature, but the cause of the tensional stresses is not clear from the facts available in this area, nor, judging from the literature, from the facts available elsewhere in the Great Basin. King [a] first emphasized the parallelism of faults with folds in the Great Basin region, and Gilbert [b] suggested that the Great Basin faults are but the surface expressions of folds similar to those of the Appalachian Mountains. The parallelism of the faults with axial planes of folds is shown also by the work of Huntington and Goldthwait [c] on the folded and faulted district separating the Great Basin from the high plateaus southeast of the Iron Springs district, where the fault planes and displacements are nearly vertical and the phenomena in general are those of tension.

The correlation of the tensile nature of the strike faults with the shortening of the lithosphere shown by the folds seems possible only

---

a King, Clarence, Systematic geology: U. S. Geol. Explor. 40th Par., vol. 1, 1878, p. 735.

b Gilbert, G. K., Report on the geology of portions of Nevada, Utah, California, and Arizona, examined in the years 1871 and 1872: U. S. Geog. Surv. W. 100th Mer., vol. 3, Geology, pt. 1, 1875, p. 62.

c Huntington, Ellsworth, and Goldthwait, J. W., The Hurricane fault in the Toquerville district, Utah; Bull. Mus. Comp. Zool., Harvard Coll., No. 42 (Geol. Ser., vol. 6), 1904, pp. 199–259.

by assuming tension along the crests of anticlines, resulting in strike joints and faults. It may be pointed out that no considerable uplift of the arch necessarily precedes faulting of this kind, for the faulting may follow the tendency to uplift so closely that the two movements are nearly simultaneous. It is of interest in this connection to note that Huntington and Goldthwait have developed the fact that the relative vertical movements produced by the folds whose axes are parallel to the Hurricane fault were in opposite directions from those produced by subsequent faults. In the former case the west or Great Basin side was uplifted with reference to the east or High Plateau side, while in the subsequent faulting the Great Basin became the downthrow side, suggesting a drop of the anticline—that is, normal faulting consequent on the folding.

SEQUENCE OF GEOLOGIC EVENTS.

The principal features in the development of the present geological and physiographic features of the district have been in order as follows:

(1) Deposition of Carboniferous limestone, with thin fragmental base, on basement unknown in this district.

(2) A period of nondeposition, perhaps erosion, followed by deposition of Cretaceous and perhaps some Jurassic sandstone with layers of shale, conglomerate, and limestone.

(3) An erosion interval, followed by deposition of Eocene limestone and conglomerate in an inclosed basin. The conditions are principally those of shallow water, strong currents, and rapid changes through both Cretaceous and Tertiary times.

(4) In early Miocene time intrusion of andesite laccoliths, principally into the Carboniferous limestone, but also into the overlying Cretaceous, accompanied by tilting of all the formations away from the laccoliths, steeply near the laccoliths, less steeply farther away. Limestone and sandstone were metamorphosed near the contact.

(5) Fissuring and faulting caused by cooling of laccoliths.

(6) Immediate advent of ore-depositing solutions through fissures in the andesite, depositing iron ore in the andesite fissures and in the adjacent limestone and affecting other alterations.

(7) Erosion, exposing the laccoliths and rings of sediments and ores around them.

(8) Extrusion of late Miocene lavas over the entire area, except possibly some of the higher peaks of the exposed laccoliths, effecting a secondary concentration of the ores and further altering the underlying rocks.

(9) Further faulting.

(10) Vigorous erosion, reexhuming the andesite cores and developing the Pleistocene conglomerate and the Pleistocene and Recent mantle of stream, lake, and other alluvial deposits.

DEVELOPMENT OF TOPOGRAPHY.

The outlines of the andesite cores at their contacts with the sediments are determined largely by their original forms. Their surface forms are in general those of mature topography. While nearly all the surfaces are smooth and rounded, differential erosion along vertical and concentric fissures has developed most irregular and fantastic forms. (See Pl. VIII, *A, B*.)

Erosion of the sediments has developed monoclinal ridges with their steeper edges toward the andesite and dip slopes away from it. Especially conspicuous are the ridges of pink and red conglomerates, sandstones, and limestones of the Eocene, which may be seen for many miles about from points on the laccoliths. In general the relief and slopes of the sediments are low.

The horizontal lava flows, overlying the sedimentary deposits and fringing the outer edge of the sedimentary area, form conspicuous foothills or mountains, only a little lower than the laccoliths themselves. Erosion has worked down along joints and faults, developing a mesa type of topography, and in detail often presenting the same erosion forms as the laccolithic andesite (Pl. VIII, *C*).

In a broad way the Carboniferous, Cretaceous, and Eocene sediments occupy a valley between mountains formed by the andesite laccoliths and the flat-topped lava hills.

Faults considerably modify the general relations thus outlined. Fault scarps are common, some of them ranging up to 1,000 feet in height, but as a rule erosion has partially or completely masked them. Erosion has frequently worked down along fault planes, with the result that many depressions mark faults. Instances of drainage along fault lines are seen in both branches of the Queatchupah and in several minor valleys to the southeast, in Stoddard Canyon, southeast and northeast of Joel Springs Canyon, around Antelope Range, Antelope Springs, and westward, in the Eightmile Hills, and west of the northern part of The Three Peaks. Instances of canyons following joints were seen abundantly in the western part of The Three Peaks, where there are a number of short parallel canyons following a nearly east-west slicing of the andesite.

The lower slopes in general are grade slopes in the heavy mantle of débris resulting from rapid erosion and inadequate transportation. On the desert this material is spread out in broad, low fans. The grade slopes are notched by sharp ravines and canyons.

The principal events in the development of the present topography have been: (1) Intrusion of laccoliths in early Miocene time, followed by faulting and erosion, exposing the laccoliths and surrounding belts of sediments and developing in the laccolith cores a mature type of topography; (2) extrusion of late Miocene lavas, followed by fault-

A EROSION FORMS OF THE LATEST TRACHYTE, SWETT
HILLS.

B. EROSION FORMS OF LACCOLITHIC ANDESITE
THE THREE PEAKS.

C. CONCRETIONARY EROSION OF THE LACCO-
LITHIC ANDESITE. THE THREE PEAKS

ing and by erosional exhumation of the laccolith cores and surrounding sediments, so far as these have been covered by the lavas, and developing a mature type of topography in the lavas themselves; (3) Hurricane faulting to the east, causing a slight renewal of activity, exhibited in the sharp canyons and steep sides of some of the lava hills.

There have thus apparently been three partial erosion cycles, the second one now shown chiefly by the flat tops of mesas like the Harmony Mountains and the third by the sharp ravines and canyons. But it is doubtful if the term cycle should be used here, for it implies uplift and renewal of activity of streams, the evidence for which is not satisfactory. There has been an interruption of erosion by the extrusion of the lavas, but the only evidence which can be cited for uplift and renewal are the sharp ravines and canyons; and it may be observed that both these and the grade slopes are being developed at the present time. The rapid development and modification of the grade slopes has already been referred to. But the rapid cutting of the ravines and canyons in these same grade slopes is no less striking. There is an alternate leveling up of the deep ravines to the grade slopes and the development of new ravines in periods of alternating high and low precipitation.

Huntington and Goldthwait, in their excellent paper on the Toquerville area, cut by the Hurricane fault to the southeast of the Iron Springs district, work out two cycles of erosion similar to the second and third here presented. Their first cycle occurred after the extrusion of trachyte, rhyolite, and andesite and subsequent faulting, and corresponds to the second cycle in the Iron Springs district. There appears to be no evidence in the Toquerville district for the prelava and postintrusion cycle of the Iron Springs district. The second cycle of Huntington and Goldthwait occurred after the basalt flows and the later Hurricane faulting. For this cycle there is scant evidence in the Iron Springs district. They cite the Pine Valley Mountains and similar structures of the Iron Springs district and southwest as an expression of the mature topography of their first cycle, and correlate it with certain remnants on the High Plateaus, which were then very much lower than at present, the Pine Valley Mountains being the dominant topographic feature. During the Hurricane faulting the High Plateaus were raised, and renewed erosion cut deep vertical-walled canyons into them, which are the present expression of the second cycle. This faulting produced very little difference of elevation west of the present Hurricane scarp, hence this region, including the Iron Springs district, shows mainly the old mature topography. There are, however, small canyons in the Iron Springs district which may show a renewal of activity and may be comparable to the steep-sided canyons of the High Plateaus.

CHAPTER III.

SEDIMENTARY FORMATIONS.

CARBONIFEROUS SYSTEM.

HOMESTAKE LIMESTONE.

DISTRIBUTION.

The Homestake limestone outcrops in or around the andesite laccolith areas and immediately in contact with the andesite.

West of The Three Peaks laccolith it is exposed in a band extending from the northern boundary of the district southwestward for about 3 miles and then southeastward to a point northeast of Iron Springs, except where, by faulting, the Cretaceous formations are brought into contact with the andesite for a distance of about a mile.

Northeast of the Granite Mountain laccolith a band of Homestake limestone is exposed for about 2 miles. Both ends of this band are overlapped by the Pinto sandstone, which here again comes in contact with the andesite. At the Desert Mound, southwest of Granite Mountain, the Homestake limestone again appears. It is cut off by a fault on the west and disappears under the lake and stream deposits on the east.

The Iron Mountain laccolith is bordered by the Homestake limestone on its northeast, east, south, and southwest sides, with a few interruptions due to faulting or covering by surface deposits. Southwest of the laccolith the limestone has considerable width, owing to the fact that the surface of the supporting laccolith is nearly horizontal here, as shown by the tongue of andesite extending westward from the main mass and by the few andesite outliers in the Homestake limestone. On the west and north sides of the Iron Mountain laccolith, as in Granite Mountain, the Homestake limestone is not exposed, the Cretaceous sandstones lapping against the laccolith.

A patch of Homestake limestone is present in the Comstock iron deposit in the Cretaceous area southwest of the Homestake mine; others are scattered irregularly within the area of the Iron Mountain laccolith.

24

CHARACTER.

The Homestake limestone is a dark bluish-gray limestone of a dense texture, with uniform characteristics throughout its entire extent, except near the laccolith contact. Under the microscope the limestone appears to be made up of exceedingly minute grains of calcite with scattered grains of pyrite, magnetite, and chert.

The bedding of the Homestake limestone is very indefinite and is easily confused with secondary fracturing. Where it is well defined the limestone is generally thin bedded.

CONTACT METAMORPHISM OF LIMESTONE BY ANDESITE LACCOLITHS.

Phases of alteration.—The limestone adjacent to the andesite has been locally replaced by iron ore and has been generally vitrified, silicated, and kaolinized in a band usually not more than 60 feet wide along the erosion surface, although locally it may be a few hundred yards wide where the erosion surface is nearly parallel to the limestone-andesite contact. Locally either or both contact phases are absent.

The altered limestone is a grayish, yellow, or greenish, fine-grained, argillaceous-looking rock. Near the contact it is soft, and farther away it is hard and fractured into small irregular blocks. The principal minerals are albite, kaolin, actinolite, diopside, quartz, orthoclase, serpentine, phlogopite, andradite, iron ores, osteolite, andalusite, wollastonite, calcite, etc., varying greatly in proportion in different places, but usually occurring in quantity in the order named. They are found in veins, in breccias, and disseminated through the rock. In addition there are local residues of a glassy base. The albite is probably not as abundant as here indicated, but very likely includes other sodium silicates which have not been detected. The glass can be distinguished only with difficulty from opal and other isotropic minerals. Its index of refraction was determined as 1.56 by means of the Becke method used in conjunction with liquids of known index of refraction. This distinguishes it from opal (1.45) and other isotropic minerals which might be found in contact-metamorphic limestones.

Another phase is coarsely crystallized limonite-stained marble, in some places found in a narrow belt between the andesite and the normal silicated contact phase and elsewhere outside of the normal phase or associated with the ore. It is thought possible that this limonitic marble is a later vein material, filling openings along the contact left by the cooling and crystallization of the intrusive and intruded masses.

Analyses of various phases of the Homestake limestone are given on the following pages.

Analyses of Homestake limestone.

UNALTERED HOMESTAKE LIMESTONE.

	A.a	B.	C.	D.	E.	F.	G.
SiO₂	8.08	6.90	3.78			16.48	3.12
Al₂O₃	1.95	1.03	1.14			1.78	.71
Fe₂O₃	.87	1.04	1.80				
Fe				0.28	0.46	.95	1.12
FeO	.06		.75				
MgO	2.86	4.52	.64	1.78	2.67		
CaO	46.67	47.15	50.39	50.77	49.46		
Na₂O	.13	.13					
K₂O	.77	.93					
H₂O+	1.01	1.04					
P₂O₅	.05	.04	.042	.03	.03	.037	.023
CO₂	37.00	36.42	39.92				
BaO	None.	None.					
	100.05	99.95	97.71				

a These letters are used throughout the discussion of the Homestake limestone in referring to the specimens and their analyses.

ALTERED HOMESTAKE LIMESTONE NEAR ANDESITE CONTACT, AND DIOPSIDE.

	Normal contact phase.			Exceptional contact phases.					Diopside.
	H.	I.	J.	K.	L.	M.	N.	O.	P.
SiO₂	50.73	52.00	57.05	11.31	2.38	4.08	45.03	75.16	51.72
Al₂O₃	14.63	9.32	9.86	1.64	2.02	1.05	1.95	2.39	1.99
Fe₂O₃	11.51	5.08	3.10	4.75	6.20		18.22		3.30
Fe						2.41		1.97	
FeO	1.13	2.41	1.80	2.55	None.		2.38		4.95
MgO	6.36	9.40	8.16	.41	45		.35		15.08
CaO	1.24	14.47	8.61	42.57	48.80		14.99		23.22
Na₂O	2.02	1.94	4.30	.12			1.65		.05
K₂O	4.24	1.41	1.56	.25			.72		.06
H₂O+	7.03	3.30	3.89	1.43	2.07		4.09		.50
P₂O₅	.32	.12	.13	2.30	.03	.02	.04	.074	
CO₂	.21	.03	1.74	32.30			9.34		
BaO	.01	None.	None.	.02	None.		.10		
MnO									.04
	99.43	100.08	100.20	99.71			99.49		100.91

A. Specimen No. 46319. Unaltered blue limestone west of Three Peaks. Analysis by R. D. Hall, University of Wisconsin.

B. Specimen No. 46375. Unaltered blue limestone from Desert Mound. Analysis by R. D. Hall, University of Wisconsin.

C. Unaltered limestone from Iron Mountain. Analysis by Fred Lerch, Biwabik, Minn.

D. Specimen No. 46123. Unaltered blue limestone. Partial analysis by George Steiger, U. S. Geol. Survey.

E. Specimen No. 46121. Unaltered blue limestone. Partial analysis by George Steiger, U. S. Geol. Survey.

F. Specimen No. 46319A. White limestone west of Three Peaks. Analysis by Fred Lerch, Biwabik, Minn.

G. Specimen No. 46325. White limestone west of Three Peaks. Analysis by Fred Lerch, Biwabik Minn.

H. Specimen No. 46338. Altered limestone between ore and andesite on Lindsay Hill. Analysis by R. D. Hall, University of Wisconsin.

I. Specimen No. 46349. Altered limestone between ore and unaltered blue limestone at Desert Mound. Analysis by R. D. Hall, University of Wisconsin.

J. Specimen No. 46376. Altered limestone between ore and andesite at Desert Mound. Analysis by R. D. Hall, University of Wisconsin.

K. Specimen No. 46437. Recrystallized limestone in vein a few feet thick at andesite contact, showing partial replacement by iron and silica. From Boston claim. Analysis by R. D. Hall, University of Wisconsin.

L. Specimen No. 46320. Recrystallized limestone impregnated with limonite, in vein near andesite-limestone contact west of Three Peaks. Analysis by R. D. Hall, University of Wisconsin.

M. Specimen No. 46321. Same as L. Analysis by Fred Lerch, Biwabik, Minn.

N. Specimen No. 46438. Same as K, showing a further stage in replacement by iron and silica. Analysis by R. D. Hall, University of Wisconsin.

O. Specimen No. 46326. Siliceous platy alteration phase of limestone west of Three Peaks. Analysis by Fred Lerch, Biwabik, Minn.

P. Specimen No. 46478. Diopside from long tunnel on Dear claim. Analysis by R. D. Hall, University of Wisconsin.

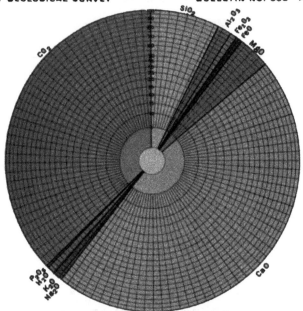

GRAPHIC REPRESENTATION OF COMPOSITION OF FRESH LIMESTONE

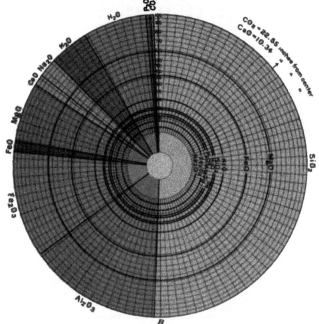

GRAPHIC COMPARISON OF COMPOSITION OF ALTERED LIMESTONE, ANALYSIS "H,"
WITH FRESH LIMESTONE (AVERAGE OF ANALYSES "A" AND "B")

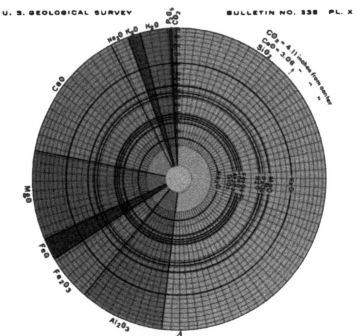

GRAPHIC COMPARISON OF COMPOSITION OF ALTERED LIMESTONE, ANALYSIS "I,"
WITH FRESH LIMESTONE

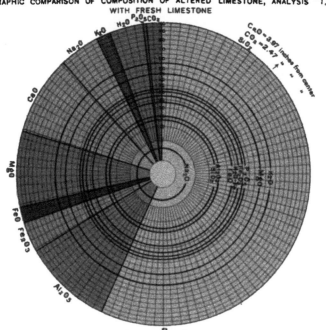

GRAPHIC COMPARISON OF COMPOSITION OF ALTERED LIMESTONE, ANALYSIS "J,"
WITH FRESH LIMESTONE

Mineralogical composition of fresh and altered Homestake limestones, calculated from chemical composition.

Mineral.	Fresh lime-stones.		Altered limestones.					Average of common type of altered limestone (H, I, J).
	A.	B.	H.a	I.	J.	K.	N.	
Calcite	78.30	72.40	0.50	1.40	3.90	70.80	21.10	1.93
Kaolin	4.90	2.58	24.77	11.61	3.09	3.10	....	13.16
Chert	2.64	....	18.78	1.50	6.30	7.92	34.14	8.86
Actinolite	....	....	....	15.55	19.00	....	....	11.52
Diopside	....	....	2.16	20.52	4.32	.86	1.94	9.02
Phlogopite	....	....	15.57	....	....	....	....	5.19
Serpentine	....	....	5.24	4.97	6.76	....	....	5.66
Andradite	....	....	....	11.20	3.14	....	....	4.78
Magnesite	5.96	8.65	....	....	....	.50	....	....
Wollastonite	6.03	9.86	....	4.76	3.13	....	5.10	2.63
Albite	....	....	16.77	16.24	36.15	1.05	6.81	23.02
Orthoclase	....	....	....	8.34	8.89	1.11	3.89	5.74
Limonite	1.12	.56	3.74	1.31	2.62	2.62	3.93	2.55
Magnetite	....	.69	3.48	....	....	3.48	7.66	1.16
Hematite	....	....	5.92	....	....	....	9.60	2.03
Enstatite	....	1.00	....	....	....	....	....	....
Apatite	....	....	.62	.31	.31	4.96	1.24	.41
Pyrite	.12	.84	....	....	....	....	....	....
Water	.16	.59	1.54	.81	2.21	.59	3.53	1.52
Siderite	....	....	....	....	....	2.32	....	....
Witherite	....	....	....	....	....	....	.20	....
	99.23	97.17	99.09	98.52	99.82	99.31	99.14	99.18

a Contains glass which has been calculated in terms of minerals.

Introduction of ore.—The introduction of ore took place after the development of the silicated contact phase, as is demonstrated by its occurrence in fissures that intersect this phase. The silicated contact phase is found also along parts of the contact where ore is absent. The introduction of ore-bearing solutions effected further metamorphism of the limestone of approximately the same sort, nearly all of the minerals found at the barren contacts being duplicated within and adjacent to the ore itself. (See analyses above.) Apatite, amphibole, biotite, pyrite, and garnet are more abundant in association with the ores than elsewhere in the contact phase, while albite and orthoclase appear in the contact phase and not in the ores. Beyond this it has not been found possible to separate the metamorphic effect of the ore-bearing solutions, aside from its deposition of ore, from the earlier contact effect of the andesite, although it is thought likely that additional careful field work with this object in view might lead to the discovery of further criteria for their separation. The replacement of the limestone by ore is discussed in connection with the origin of the ore (pp. 75–79).

Normal contact phase.—The following paragraphs are devoted to the silicated contact phase, which is described as a unit, without regard to the extent to which it has been developed under the influence of the first contact of the andesite or under the influence of later ore-bearing solutions, although the former is unquestionably dominant.

Analyses of fresh and altered rocks show clearly the net results of the alteration, but they tell little of the extent to which the transfers of materials have been additions or subtractions, unless it is assumed that one substance has remained constant during the change. A variety of inferences are therefore possible, depending upon which substance is selected as a basis against which to measure the changes of the other constituents of the rock. Some substances are easily eliminated as possible constants; for instance, calcium carbonate, in the alteration of a limestone to a silicated rock; for if this material were assumed to remain constant in amount during the alteration, it would require so large an addition of other substances to develop the proportions in the altered rock that the volume would be increased beyond the possibilities of the situation, even allowing for increased density. But other constituents are not so satisfactorily eliminated by such reasoning, and there arises uncertainty and error in the selection from among the various possibilities of the particular substance to serve as a basis of comparison of two analyses. If each substance in turn is considered as constant, the work of calculation becomes tedious and the probabilities of the situation are not easily discerned in the mass of figures and tables obtained.

To meet these difficulties in inferring from analyses the real nature of the rock alterations, the writers present graphic comparisons of the analyses (Pls. IX–XII, A), from which it is possible, by assumptions of constancy of any constituent during alteration, to read the percentage gains and losses of other constituents as a result of alteration, and, what is more important, to see at a glance what the probable constants are, without losing sight of the other possibilities.

In the accompanying diagrams a circle with $1\frac{3}{4}$-inch radius is divided by radii into 100 sectors, each sector therefore representing 1 per cent of the total area of the circle. It is also divided into a series of annuli by concentric circles so spaced [a] that the area of each annulus is equal to 5 per cent of the total area of the circle. The subdivisions of the area, bounded by the sectorial radii and the concentric circles, are therefore equivalent areas and each represents 0.05 per cent of the total area of the circle.

On the base diagram thus constructed, in Pl. IX, B, for example, the areas of sectors shown in different colors represent percentage weights of constituents of the altered limestone. Thus silica (analysis H) constitutes about 50 per cent of the altered rock. Of the fresh rock silica (average of analyses A and B) makes up about 7.5 per cent. The ratio of the two is 0.15, and hence the silica of the fresh rock is 15 per cent of the silica of the altered rock. This is indicated on the diagram by the area of the silica sector cut off by the black circle

---

a The area varies as the square of the radius.

marked "SiO₂" (which may be called the "silica circle"). This circle is drawn with such a radius that its area is 15 per cent of the base circle, and it follows that the portion of the silica sector included is 15 per cent of the whole silica sector.

In the same way the alumina (analysis H) makes up nearly 15 per cent of the altered rock and (average of analyses A and B) about 1.5 per cent of the fresh rock; that is, the alumina in the fresh rock is 10 per cent of that in the altered rock. In the diagram, therefore, the alumina in the altered rock is represented by a sector occupying 15 per cent of the area of the base circle, and to show the relations of that in the fresh rock a circle is drawn so as to include 10 per cent of the total area and consequently 10 per cent of the area of the alumina sector. In a similar manner the relations of the various other constituents in the altered and fresh rock are shown.

The important feature of the diagram is the fact that starting from the silica circle, or the circle representing any other constituent assumed to have remained constant, the actual gains or losses of other constituents may be obtained from the diagram by a simple subtraction.

If silica be assumed constant, the amount of magnesia, for instance, necessary to make up the percentage of magnesia in the altered rock is measured by the area of the magnesia sector included in the silica circle. But on the magnesia sector the magnesia in the original rock is represented by the area of the magnesia sector within the magnesia circle. There are indicated, therefore, both the amount of magnesia required, on the assumption that the silica is constant, and the amount of magnesia actually present in the original rock, which in this case is more than is necessary to meet the required proportion in the altered rock. Magnesia has therefore been lost by an amount measured by the area of the annulus between the silica and the magnesia circles as they cross the magnesia sector. If soda be assumed constant, all constituents other than soda have been lost in amounts measured by the difference between the area of their sectors inside their circles and that inside of the soda circle, which may be taken from radial scale. From the numbers thus obtained the percentage change is readily calculated. If, at the other extreme, calcium carbonate has remained constant, there has been gain of all other constituents, their circles all being smaller than the carbonate circle. Complete or nearly complete removal of a constituent is represented by an arrow.

In general it will be noted that the constituents whose circles lie outside of the circle of the constituent assumed as constant have suffered loss, while those whose circles lie inside have gained. The amount of gain or loss is proportional to the difference between the areas controlled by the superimposed polar coordinates of the several constituents. ¯In practice a comparison of size of circles gives the gains or losses of any constituent against any other constituent, without elaborate calculation.

Where several circles are of nearly the same size, this means that the constituents represented have maintained their mutual proportions during the alteration, and such maintenance of mutual proportions becomes presumptive evidence that the group has remained essentially constant during the alteration and that the change in composition has been effected by gain or loss of other constituents. It may be sufficient, therefore, instead of regarding each of the constituents as possible constants, to consider only those whose ratio circles lie near together. Thus in the present illustration the close grouping of the circles representing silica, alumina, and iron suggests that the constant is to be looked for in this group, and the conclusions then drawn as to the transfers of materials are probably on safer ground than they would be from assumption of constancy of one constituent, without taking into account the additional evidence afforded by the persistence of mutual relations within certain groups. The diagrams have been found also to have unlooked-for value in the recording and comparisons of long series of analyses, affording means for comparison of composites of series of similar alterations, which would involve tedious calculations by ordinary mathematics.

To avoid confusion, it may be stated that the diagrams represent weight alone, and that the gains and losses of weights mean nothing as to change in volume, unless the density is taken into account, as it may be on the same diagrams by a simple graphic device.

The attempt has been made to make the above explanation fairly brief and nontechnical, and not to emphasize the mathematical steps of the process, in order to bring to the reader the real simplicity and usefulness of the diagrams. They are easily grasped and understood by the student when explained empirically, but when the various steps in the construction of the diagrams are mathematically demonstrated the explanation becomes formidable to the nonmathematical reader. For the sake of those requiring more technical explanation the following supplementary account is given:

On the base figure constructed as stated above the percentage composition of any two rocks may be readily represented. For example, in Pl. IX, B, if the area of the whole circle be taken to represent a 100-unit mass of the altered rock (analysis H), then the proportional masses or percentages of the several constituents are shown by the areas of the variously colored sectors. Since the area of any sector varies as the arc which subtends it, the size of the sector of any constituent is readily ascertained by laying off an arc proportionate to the percentage of that constituent. The proportional masses or percentages of the several constituents in a 100-unit

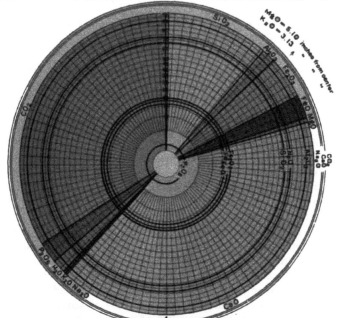

GRAPHIC COMPARISON OF COMPOSITION OF ALTERED LIMESTONE, ANALYSIS "K,"
WITH FRESH LIMESTONE

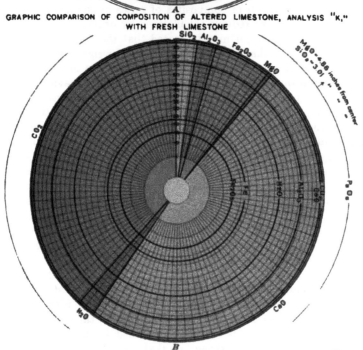

GRAPHIC COMPARISON OF COMPOSITION OF ALTERED LIMESTONE, ANALYSIS "L,"
WITH FRESH LIMESTONE

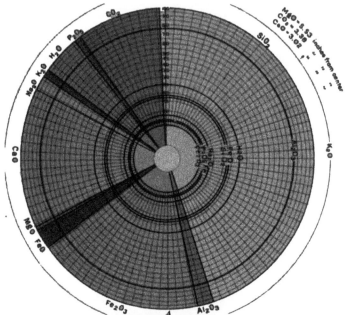

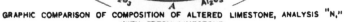

GRAPHIC COMPARISON OF COMPOSITION OF ALTERED LIMESTONE, ANALYSIS "N,"
WITH FRESH LIMESTONE

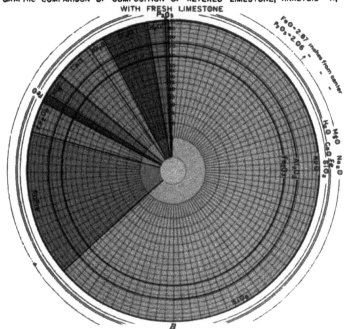

GRAPHIC COMPARISON OF COMPOSITION OF WEATHERED ANDESITE, ANALYSIS "B,"
WITH FRESH ANDESITE

mass of the original rock (average of analyses A and B) are shown by appropriate areas on the sectors already determined from the analysis of the altered rock. The construction for any constituent is effected by using the ratio of the percentages of this constituent in the original and in the altered rock, as follows: Lay off the determined ratio on the radial percentage scale of the base figure, and through the point thus formed draw a circle concentric with the base circle. The portion of the sector of the chosen constituent lying within the circle thus drawn represents the amount of this constituent in a 100-unit mass of the original rock. In order to make this relation clear, let us consider magnesia as an example. Since the altered rock (analysis H) contains 6.3 per cent of magnesia, take an arc of the base circle equal to 6.3 of the unit divisions. The area of the sector subtended by this arc is 6.3 per cent of the area of the base circle, so that the sector represents the number of unit masses of magnesia present in a 100-unit mass of the altered rock, or, what is the same thing, it represents the percentage of magnesia in the rock. Now, the original rock (analyses A and B) contains 3.7 per cent magnesia and the altered rock (analysis H) 6.3 per cent. The ratio of these numbers, 0.59, expresses the fact that the magnesia in a 100-unit mass of the original rock is only 59 per cent of that in an equal mass of the altered rock. Now, 59 per cent of the base circle and also 59 per cent of any sector lies within the circle drawn through point 59 on the radial scale, so that the area common to the magnesia sector and the 59 per cent circle (which may be called the magnesia circle) is 59 per cent of 6.3, or 3.7, which is the percentage of magnesia which it is desired to represent.

The proportional masses or percentages of the various constituents in both rocks being represented in the manner indicated, the completed diagram furnishes all the factors necessary for making any desired comparisons when it is assumed that the amount of any chosen constituent has remained unchanged.

An example will serve to indicate the principle on which the comparisons are made. The rock of analysis H has been derived by alteration from the rock of analyses A and B. Both analyses are represented on Pl. IX, B. Assume silica to have remained constant, what has been the gain or loss of magnesia during alteration? Apart from the diagram the change in magnesia may be calculated from the data given in the analyses in the following manner: A 100-unit mass of the original rock contains 7.5 units of silica, while a 100-unit mass of the altered rock contains 50 units of the same constituent. Therefore 15 units of the altered rock contain the same amount of silica as a 100-unit mass of the original rock. Now, 15 units of the altered rock contain $0.15 \times 6.3 = 0.95$ unit of magnesia. Evidently there has been a loss, since in 100 units of the original rock there are available 3.7 units of magnesia. Therefore, $3.7-0.95 = 2.7$ represents the number of units of magnesia lost during the alteration, if silica has remained unchanged, and $\frac{2.7}{3.7} = 0.75$, or 75 per cent, the ratio of loss. The number of necessary operations in this calculation may be reduced by using in the place of 3.7 the equivalent expression 0.59×6.3. Then the formula for the ratio of loss becomes $\frac{(0.59 \times 6.3) - (0.15 \times 6.3)}{0.59 \times 6.3}$, or, eliminating common factors, $\frac{0.59 - 0.15}{0.59} = \frac{0.44}{0.59}$, or 0.75, as before.

Now, by construction, the diagram gives the numbers 59 and 15, which by substraction gives 44; then $\frac{44}{59}$ gives 0.75 as above; or, converting this decimal fraction into percentage gives 75 per cent as the loss of magnesia during the alteration.

From the above the following rule is derived: To determine the percentage gain or loss of any constituent during alteration, any other constituent being assumed as constant, read on the radial scale the difference between the ratios expressed by the constant-constituent circle and the variable-constituent circle, and divide this by the ratio expressed by the variable-constituent circle. The result will be the percentage gain or loss of the variable constituent. Evidently gains will be shown by ratios greater than unity and losses by ratios less than unity.

We may take up now the specific inferences which may be drawn from the diagrams under discussion.

If calcium has remained constant there has been a great increase of all the other constituents and an increase of weight of the rock. If potassa be assumed constant there has been a considerable gain of all other constituents, except lime, in analyses I and J, and a gain of all other constituents, except magnesia and lime, in analysis H. If magnesia has remained constant there has been a loss in potassa and lime and a gain in other constituents in I and J, and a gain of all constituents except lime in H.

If silica and alumina remain constant this has involved a loss of all constituents except soda. The silica and alumina constants are nearly the same throughout. Soda is the only substance which, considered constant, shows a loss of all other constituents.

Of these various assumptions the constancy of alumina and silica seems to be the most probable. The conspicuous feature in the alteration is the *relative* increase of silica and alumina at the expense of calcium carbonate. This means either (1) that silica and alumina have been introduced from without to replace the calcite, in which case there may not have been any considerable change in the volume of the rock during alteration; or (2) that calcite has been taken out of the fresh limestone, leaving the silica and alumina in larger proportion, in which case the weight and, as will be shown later, the volume has considerably decreased; or (3) some combination of these two methods has accomplished the same result. Alternative 2 is favored by the fact that, while the alumina has gained slightly in percentage on the silica, the ratio of the two substances in the fresh and altered rocks remains substantially the same, in the fresh rocks average alumina standing to silica as 19 to 100, and in the altered rocks as 21 to 100. (See analyses, pp. 26–27, and Pl. X.) The near accordance of the alumina and silica circles in Pl. X brings out well the uniformity of the ratio. If silica and alumina have been added from without, it is a remarkable coincidence that they should have been added in so nearly the same proportion as in the fresh rock, especially when it is remembered that their proportion in the fresh limestone is not determined by silicate ratios but rather by the abundance of chert.

On the assumption that alumina remains constant, it will be noted that the greatest percentage lost is calcium, next magnesia, then potassa, and then silica, which is the order of solubility of these substances in weathering. The only exception is soda, which without much question has been introduced into the contact phase. It will be noted further that the composition of the altered rock is essentially that of a calcareous residual clay.

Comparison of average fresh Homestake limestone with normal contact phases, with alumina (Al_2O_3) assumed to be constant.

	A, B.	H.		I.		J.		H, I, J.	
	Average, fresh.	Altered.	Variation from A, B.	Altered.	Variation from A, B.	Altered.	Variation from A, B.	Average.	Variation from A, B.
SiO_2	7.49	5.168	− 2.322	8.313	+ 0.823	8.62	+ 1.13	7.367	− 0.123
Al_2O_3	1.49	1.49	0.00	1.49	0.00	1.49	0.00	1.49	0.00
Fe_2O_3	.95	1.172	+ .222	.812	− .138	.468	− .482	.817	− .133
FeO	.41	.115	− .285	.385	− .025	.271	− .139	.290	− .120
MgO	3.69	.648	− 3.042	1.502	− 2.088	1.233	− 2.467	1.127	− 2.563
CaO	46.91	.126	− 46.784	2.313	− 44.597	1.301	− 45.609	1.243	− 45.665
Na_2O	.13	.206	+ .076	.310	+ .180	.649	+ .519	.388	+ .258
K_2O	.85	.432	− .418	.225	− .625	.236	− .614	.298	− .414
H_2O+	1.03	.716	− .314	.527	− .503	.588	− .442	.610	− .419
P_2O_5	.04	.032	− .008	.019	− .101	.019	− .021	.023	− .043
CO_2	37.01	.021	− 36.989	.100	− 36.910	.263	− 36.747	.128	− 36.882
	100.00	10.126	· 89.864	15.996	83.984	15.138	84.872	86.104	− 13.791

Exceptional contact phases.—These phases represent introduction of calcite, iron, and silica. Analysis L is of an iron-stained marble in a vein. Probably all of the constituents are secondary, but if the alumina or silica is the same as in the fresh limestone, there has been a considerable introduction of iron and a slight loss of calcite and soda, or if alumina has remained constant there has been also a gain of silica. Sample N is a contact phase of the limestone in which there has been conspicuous introduction of soda, silica, water, and iron and loss of calcite and potassa. K is an intermediate stage between the fresh limestone and N.

General result of alteration.—It is concluded that in general there has been large loss of material from the limestones at the contact without conspicuous introduction of new material except soda, but that exceptional phases show clearly the introduction of silica and iron. Both normal and exceptional phases of the contact rocks are cut by later veins of calcite and iron ore, with associated minerals similar to those of the limestone contact. (See p. 85.)

Changes in density and volume during contact metamorphism.— Specific-gravity determinations of the fresh and contact phases of the Homestake limestone are as follows:

Specific gravities of fresh and contact phases of Homestake limestone.

	Fresh limestone.		Contact phase.				Average of fresh limestones.	Average of all contact phases.
	A.	B.	H.	I.	J.	Average of 5 specimens from contact, including H, I, J, and L.		
Specific gravity determined from mineral calculations	2.739	2.775	2.74	2.93	2.73	....	2.76	2.86
Specific gravity determined from hand specimens	2.705	2.706	2.22	2.59	2.57	2.56	2.70	2.56
Specific gravity determined from powdered specimens	......	......	2.20	2.92	2.68	....	....	2.60
Pore space calculated from specific gravities of hand specimens and powdered specimens	......	......	......	10.2	4.1	....	2.2	7.1

The difference between the gravity determined in the hand specimen and that in the crushed specimen measures the pore space. The fresh limestone when powdered is slightly soluble in water, vitiating gravity determinations. In this case the difference between calculated and measured gravities measures the pore space, for there is little chance for error in the computed gravity. The only altered rock not having pore space, H, is made up largely of glass. Taking the glass into account, there has been little average increase of pore space in the contact rocks as a group, if the analyses represent proper proportions of contact phases. Leaving this out of consideration, the pore space at the contact has increased slightly, from 4 per cent to 10 per cent.

The calculated specific gravities correspond fairly well with those determined from the powdered specimens, affording a check on the computation of the mineral composition from the chemical analyses. Sample H is again an exception because the contained glass has a specific gravity considerably lower than that computed from the possible mineralogical composition based on chemical analysis. By eliminating from consideration the mica and iron oxide observed in the glass, it is easy to determine from the measured density of the rock that the density of the glass is not far above 2. Thus the theoretical complete crystallization of glass or fused equivalent in the contact phase of the limestone has involved a diminution in volume of perhaps 19 per cent, or about 15 per cent if pore space be taken into account.

It may be concluded in general that the actual density of the altered rocks is less than that of the fresh limestone, largely because of the development of pore space and glass, but that the densities of the minerals of the altered rocks are on the whole slightly higher than those of the fresh limestone.

Exceptional contact phases, consisting almost entirely of silica or garnet or other constituents, give other density results.

If the principal chemical change in the development of the contact phase has consisted in the elimination of calcite and to a less extent of magnesia, iron, and potassa, leaving alumina and silica substantially unchanged in their ratios, as held to be possible in an earlier paragraph (P. 32), this has involved a very considerable loss of weight (see analyses, p. 27), and, as the densities of the fresh and altered rocks differ so little, the loss in volume also has been large. Kemp,[a] Lindgren,[b] and others have cited lack of structural evidence of diminnution in volume at limestone contacts as favoring the view that materials must have been introduced from without to take the place of the calcium carbonate. In the Iron Springs district the field evidence does not positively prove or disprove important volume change, but there is no apparent field evidence to contradict the evidence for diminution of volume here calculated. The limestone, though tilted away from the andesite laccoliths, nowhere shows evidence of crumpling or crowding where the bedding can be observed. In the altered phase the bedding has been destroyed, and it is easy to conceive that this structurally amorphous zone may represent only a part of the volume of the original rock, the calcium carbonate having been driven off and the other constituents concentrated. The change in volume of the limestone would scarcely be expected to stand out conspicuously in the field relations, for it has occurred, if at all, in the band which now does not show original textures or structures, by which change of volume can be measured in crumpling or folding.

In general it appears that there may have been important diminution of volume, accomplished essentially by loss of materials and not by change of density of minerals.

a Kemp, J. F., Ore deposits at the contacts of intrusive rocks and limestone and their significance as regards the general formation of veins; Econ. Geol., vol. 2, 1907, pp. 1-13.

b Lindgren, Waldemar, The copper deposits of the Clifton-Morenci district, Arizona; Prof. Paper U.S. Geol. Survey No. 43, 1905, 375 pp.

Physical conditions of contact metamorphism.—The andesite in its molten state may be supposed to have had a temperature common to acidic silicate solutions under similar conditions anywhere from 500° C. up; probably above 1,000° C. After sufficient heat had been communicated to the limestone to raise the temperature to 550° C., expulsion of carbon dioxide began, and continued during the subsequent increase in temperature. For any given temperature above 550° C. expelled carbon dioxide has a definite pressure which must find relief, if liberation of the carbon dioxide is to continue. The nearly complete elimination of carbon dioxide along the andesite contact indicates that the pressure was not sufficient to restrain its exit; that the pressure was therefore small; and that continuous openings for escape to the surface probably existed.

When the temperature reached 1,100° C. there probably began the incipient fusion of the alumina, silica, lime and magnesia, which remained in the limestone in about the proportions of a calcareous clay. The presence of lime in clay permits fusion at about this temperature.[a] The pore space in the limestones, possibly increased by the expulsion of carbon dioxide, was eliminated, probably in part by pressure, but certainly and completely by fusion. Had the temperature not reached 1,100° it is possible that a porous texture would have resulted, such as that sometimes seen along the limestone contacts with igneous rocks,[b] which would have favored the introduction of new substances in the contact phase to a larger extent than is here observed.

After the limestone reached its maximum temperature, probably somewhat below that of the andesite magma, the temperature slowly fell, allowing fairly coarse crystallization of both andesite and limestone. Local chilling, however, developed glass in parts of the fused limestone mass, and the glass remains, at least in part, to-day. The introduction of the ore-bearing solutions accompanied or followed the crystallization, and still later came the mineral-bearing waters from the late lavas. The minerals of these different periods are listed on page 85.

SANDY PHASE OF LIMESTONE.

The altered contact phases of limestone are often hard to distinguish from a much-fractured quartzite or clayey sandstone which is locally exposed below the limestone and constitutes a part of the same formation. The sandstone lies between the ore deposits and the andesite, or between the ore and the limestone. It is conspicuous also in the limestone patches faulted into the andesite. In

---

a Ries, Heinrich, and Kümmel, H. B., The clays and clay industry of New Jersey: Final Rept. State Geologist New Jersey, vol. 6, 1904, p. 103.
b Barrell, Joseph, The physical effects of contact metamorphism: Am. Jour. Sci., 4th ser., vol. 13, 1902, pp. 279-296.

the large-scale detail maps of the ore deposits (Pls. III–VI, in pocket) the lower sandy phase and the metamorphosed phases of the blue limestone have been separated from the typical blue limestone but not from each other.

Unless the structural relations are clear, both the sandy and silicated contact phases of the limestone are likely to be confused with the Cretaceous sandstone which occurs in isolated patches in the andesite.

FOSSILS.

The fossils found in the Homestake limestone are few and poorly preserved. Only one genus, *Aviculopecten*, ranging from Silurian to Triassic, was identified by Prof. Eliot Blackwelder, of the University of Wisconsin. There are fragments, however, which appear to be referable to *Edmondia*, a Carboniferous genus.

STRUCTURAL RELATIONS AND THICKNESS.

The Homestake limestone is intruded by the laccoliths. The contacts are usually nearly vertical, sometimes because of later faulting. In a few localities, as at the Desert Mound, in the southwestern part of the Iron Mountain laccolith, and south of The Three Peaks, the limestone rests against and upon the laccolith with a gently dipping plane of contact, though locally even here the contacts are steep because of faulting. In the circular ridge east and southeast of the Iron Mountain laccolith the dip of the Pinto sandstone and the Claron formation is steeply toward the andesite, suggesting an overturned fold in which the Homestake limestone probably participates.

The Homestake limestone is overlain conformably by the Pinto sandstone, with a shale at the base.

The thickness of the Homestake limestone, as shown by exposures, ranges from 500 to 50 feet. The average thickness has been taken to be about 200 feet. The variation is probably due to intrusion of laccoliths at different horizons. If the intrusion has followed the base the variation in thickness may indicate unconformity with the overlying Pinto sandstone.

CRETACEOUS SYSTEM.[a]

PINTO SANDSTONE.

DISTRIBUTION.

The Pinto sandstone for the most part borders the Homestake limestone on the side away from the laccoliths. On the west side of The Three Peaks it extends several miles westward from the band of Homestake limestone and disappears under the desert clays and

---

[a] Jurassic in part, perhaps.

gravels. On the southwest end of The Three Peaks laccolith the Pinto sandstone is in contact with the andesite in several places, but farther east it again appears in its normal place outside of the Homestake limestone, and continues until covered by the later clays. The formation borders the north, east, and south sides of the Homestake limestone and the andesite of Granite Mountain, and a tongue extends northward into the desert. South and west of the Desert Mound there are also considerable exposures. Patches are exposed south and about three-fourths of a mile northwest of Iron Springs.

In the southwest quarter of the district a belt of Pinto sandstone extends around the Iron Mountain laccolith, widening out to the west and southwest. A small area is found southwest of the Homestake mine, within the area of the Iron Mountain laccolith. North of Iron Mountain there are several small areas of sandstone exposed outside of the Claron limestone surrounding the Antelope Range lavas. A belt surrounds the Stoddard Mountain intrusive, which is just on the southern edge of the area mapped.

CHARACTER.

The Pinto formation is composed mainly of brown, yellow, gray, maroon, and spotted sandstone, both coarse and fine grained. Interstratified with these in the lower part of the formation are maroon, purple, and green shales, a few beds of conglomerate, and a few of gray, sandy, brecciated limestone. None of these interstratified beds are continuous throughout the area. In the Granite Mountain and Three Peaks areas the lower part of the Pinto formation has a characteristic succession, only a few members of which are present in the area about Iron Mountain. A generalized section for the first two areas is as follows:

Generalized section of Pinto formation in Granite Mountain and Three Peaks areas.

		Feet.
7.	Yellowish brown and gray sandstone	1,000+
6.	Conglomerate with interbedded sandstone	20–40
5.	Variegated sandstone and shale	40–75
4.	Conglomerate	8
3.	Cherty limestone breccia	10–20
2.	Maroon and spotted sandstone	40–60
1.	Purple and green shale	30–50

The shales (1) at the bottom, in contact with the Homestake limestone, are fairly continuous, being found throughout the northeast quarter of the district, and at one locality east of Iron Mountain, though here only the green shale is present. They are soft, sandy, and much fractured.

The maroon and spotted sandstones (2) are abundantly exposed in the northeast quarter of the area, but appear locally at other points, as east of the Homestake mine, south of Crystal Springs, and around the southwest corner of the Iron Mountain laccolith. They are comparatively soft and friable and are generally of a deep maroon color; at the base are beds of white sandstone having deep red spots. The grains are mostly quartz and the different colors are due to iron oxide in the cement.

The cherty limestone breccia (3), although thin, is another characteristic bed in the lower part of the Pinto formation. It is present around The Three Peaks laccolith, north of the Granite Mountain laccolith, and east of the Desert Mound, at all of which places it occupies the same part of the series. It is a dark grayish-blue brecciated limestone. The fragments vary in dimensions from a fraction of an inch to 6 inches, and are separated by narrow bands or veinlets of chert, which on weathered surfaces may project as much as a quarter of an inch above the rest of the rock. Under the microscope the fragments appear to be made up of exceedingly fine grains of calcite, with here and there veins of coarser calcite.

The lower conglomerate (4) is associated, almost without exception, with the brecciated limestone, but it is also found in parts of the district where the latter is not present. Wherever identified it has been mapped with the conglomerate symbol. It is exceedingly well cemented, so that on breaking, the fractures occur for the most part through the pebbles instead of around them. This is true especially in the northeast quarter of the area. The pebbles are chiefly quartz, black chert, quartzite, and dark limestone, the latter probably being obtained from the Homestake formation. The matrix is composed of rather fine-grained sand, the whole being well cemented by silica.

Between the lower and upper conglomerates there are a number of sandstone and shale beds (5), none of which are continuous over any considerable area. West of The Three Peaks the lower part is characterized by light-green shale about 20 feet in thickness, overlain by a purple shale about half as thick. Above this are fine-grained flaggy sandstones, white, gray, and mottled in color. West of Granite Mountain and south of The Three Peaks the shales are entirely absent, being replaced by coarse- and fine-grained sandstone.

The upper conglomerate (6) appears throughout most of the northeast quarter of the area. It consists of heavy conglomerate layers alternating with thin layers of sandstone, the pebbles composing the former being of the same kind as those of the lower conglomerate but as a rule larger, especially where the formation is thick, and less firmly cemented. Around the Iron Mountain laccolith only one conglomerate is present, presumably the lower, since at several places a maroon sandstone is found between it and the Homestake formation. In the northeast quarter of the district this conglomerate also has been mapped.

Above the conglomerate beds throughout the district is a great thickness of gray, yellowish-brown, or red sandstones (7), forming by far the largest part of the Pinto formation. These are normally coarse and friable, but where parts of the Pinto formation come in contact with the laccolithic andesite they are fine grained and quartzitic. The sandstone becomes coarser in its upper part, and near the top contains a layer of coarse conglomerate. During the erosion interval at the end of the Cretaceous this conglomerate was so largely removed that at present it is found only locally near the overlying Eocene deposits.

Limestone lenses appear in the Pinto sandstone along the southwestern edge of The Three Peaks, immediately southwest of Iron Springs, and in the middle of the valley halfway between these areas. Other lenses are exposed in Oak Springs Flat, east of Iron Mountain, where they are partly obscured by late unconsolidated deposits. These lenses are in the lower part of the Pinto formation, but their exact position in the preceding succession is not known.

The limestones are much fractured and brecciated. Their color varies, in different parts of the district, from dark gray to light gray and even pink. The latter is very much like the Claron limestone and is distinguished from it only by its brecciated character.

CONTACT METAMORPHISM BY LACCOLITHS.

At or near the contact of the Pinto sandstone with the laccolithic andesite the sandstone is quartzitic, and the yellow, brown, and red colors are replaced by gray and white. In places the rocks are mottled or have concentric dark-colored rings (Pl. XXI, *A*, p. 74), which, under the microscope, appear to be irregular aggregates of small elongated crystals of colorless amphibole. The metamorphism is commonly noticeable for a few hundred yards from the contact, and west of Iron Mountain it is seen as far as a quarter of a mile.

FOSSILS.

No fossils were found in the Pinto sandstone. This absence of fossils and the association of red sandstones and shales suggest terrestrial deposition.

STRUCTURAL RELATIONS AND THICKNESS.

The relations of the Pinto sandstone to the underlying Homestake limestone have already been discussed. The relations with the overlying Claron formation are unconformable, as evidenced by a basal conglomerate and by the partial erosion of certain well-recognizable beds near the top of the formation. There seems to be little unconformity of dip, however, indicating that the period of erosion between Cretaceous and Tertiary was not characterized by folding.

In general the Pinto sandstone dips steeply away from the laccoliths, but south of The Three Peaks, at the Desert Mound, and west of Iron Mountain the dip is gentle. East of Iron Mountain the dip is steeply toward the laccolith; to the southeast it gradually becomes vertical; and south of Iron Mountain it is steeply away from the laccolith. East of Upper Point, south of Joel Springs Canyon, and elsewhere the beds are nearly horizontal.

It is hard to determine the thickness of the Pinto formation because of the abundance of faults. The greatest thickness exposed in cliffs is about 500 feet, but thicknesses across the formation southeast of Iron Mountain, calculated from the dip and width of exposure, are as high as 3,500 feet. The average thickness is considerably over 1,000 feet.

TERTIARY SYSTEM (EOCENE).

CLARON LIMESTONE.

DISTRIBUTION.

The Claron formation surrounds the laccoliths outside of the Pinto sandstone. In the northeast quarter it occupies an area just inside of the north boundary of the area shown on the map. It is much faulted, giving an irregular contact between it and the Pinto sandstone to the south.

East of Granite Mountains the Claron limestone occupies an area which, on account of faulting, is broad at the north and narrow at the south and west. Narrow strips of the formation follow the northwestern base of the Swett Hills.

Southwest of the Antelope Range the Claron limestone comes out from underneath the lavas and extends eastward to where it is covered by the Pleistocene lake deposits and southward to where it has been eroded away from the underlying Pinto sandstone. In Chloride Canyon, southwest of the Antelope Range, and northwest of Iron Mountain it appears again, owing to the erosion of the overlying lavas.

The largest area of Claron limestone in the district borders the eastern and southern sides of Iron Mountain, then goes outside of the area, but reappears west of Iron Mountain, on the border of the area shown on the map. Both by its distribution and by its structure it brings out well the laccolithic shape of the andesite mass.

On the northeast side of the Harmony Mountains little areas of Claron limestone are brought up by faulting. It appears also west of the Stoddard Mountain intrusive.

CHARACTER.

The Claron formation is mainly limestone with numerous thin layers of conglomerate and a few heavy beds of sandstone. It is separated from the Cretaceous sandstones by a basal conglomerate 2 to 25 feet thick, made up of coarse quartzite pebbles where it is thick and of finer pebbles where it is thin. Below it is generally a soft, pink, calcareous sandstone or soft limestone, which probably represents residual material from the weathering of the Pinto formation formed before the deposition of the Eocene. It is only a few feet thick and under it are the red and yellow sandstones of the Cretaceous. Locally the underlying Cretaceous bed is a coarse conglomerate difficult to distinguish from the basal conglomerate of the Claron formation. In most places it was eroded away before the deposition of the Claron. Above the basal conglomerate is another conglomerate, thin and sandy, with small pebbles, and discontinuous. It is like numerous other conglomerates interbedded with the limestone higher up in the formation. These as a rule are thin and any single bed is not continuous over a large area. The pebbles are small and are separated by a matrix of coarse sandstone. They are mainly quartzite, limestone, quartz, and chert, in varying proportions in the different beds. The conglomerate beds may grade laterally into coarse sandstone beds.

The principal part of the formation is a sandy limestone varying from white and gray to pink, red, and purple. On weathering, the sand grains, being less easily dissolved, protrude above the surface and give it a rough pitted appearance. Some of the layers have spherical and irregular cherty concretions which vary from less than half an inch to an inch or more in diameter. The limestones are more resistant to erosion than the sandstones and conglomerates of the Eocene and Cretaceous, and hence form cliffs and ridges above them. On weathering they acquire bright-red colors, making the exposures conspicuous for many miles.

Heavy yellowish-brown sandstone beds containing a number of discontinuous layers of conglomerate are interbedded with the limestones. Owing to faulting, it is hard to tell the thickness of these sandstones, but it may be several hundred feet. In some places they are distinguished with difficulty from the Cretaceous sandstone. Criteria for their identification are their occurrence in the midst of Tertiary rocks, and the absence of a basal conglomerate between them and the overlying limestone and conglomerate beds. The first of these might be caused by faulting, but the invariable absence of the basal conglomerate in such cases was taken as evidence that they were Tertiary.

FOSSILS.

No fossils were found in the Eocene limestone, but a number were found in the limestone and chert pebbles of the conglomerate beds. The following forms, which are mainly Carboniferous, were identified by Prof. Eliot Blackwelder:

Fossils from the Claron limestone.

Productus sp.	Spirifer sp.
Fenestella sp.	Zaphrentis sp.
Lithostrotion sp.	Crinoid stems and bryozoans.
Rhynchonella sp.	

They show only that the Claron sediments are post-Carboniferous.

STRUCTURAL RELATIONS AND THICKNESS.

The relations of the Claron limestone to the underlying Pinto sandstone have been discussed. On the upturned and eroded edges of the Claron limestone lie the Tertiary lavas in nearly horizontal beds.

The dips of the Claron limestone are slightly less than those of the Pinto sandstone because of their greater distance from the laccoliths. In a few places the beds are nearly horizontal, as south of Joel Springs Canyon and at Mount Claron.

A large amount of faulting has taken place in Eocene areas, some of which has undoubtedly remained undetected because of the similarity of different parts of the formation. Most of it dated after the laccolithic intrusion and before the later lavas were poured over the area, as shown by the fact that certain of the limestone areas are much faulted while the adjacent lava beds are undisturbed. A few faults traverse both formations indiscriminately, thus showing a later period of faulting.

The thickness of the Claron formation varies in different parts of the quadrangle, owing to erosion both before and after the laccolithic intrusion. The average thickness is about 1,000 feet, but all estimates are largely vitiated by faulting.

CONTACT METAMORPHISM BY EFFUSIVES.

Adjacent to the lavas the limestone has a layer of white, gray, or red chert, chalcedony, and moss agate or jasper, which is sometimes 10 to 15 feet in thickness. The red moss agates and jaspers are colored with iron. With these cryptocrystalline varieties of quartz is often associated a white powdery calcium carbonate, apparently deposited by hot springs. This is especially abundant north of the Eightmile Hills. The same powdery carbonate is frequently associated with the ore deposits.

At Chloride Canyon and elsewhere the limestone contains mineral veins near the contact with the lavas. The minerals are mainly calcite, barite, and quartz, with subordinate amounts of galena, pyrite, chalcopyrite, siderite, limonite, magnetite, and copper carbonates. Gold and silver are reported, the latter probably being present in the galena. Similar mineral veins are found in the Homestake limestone, both associated with and away from iron-ore deposits.

QUATERNARY SYSTEM.

PLEISTOCENE CONGLOMERATE.

The Pleistocene conglomerate is exposed only east of Antelope Springs, in the northwest corner of the area, where it occupies a little embayment in the lava area. To the north it is covered by later lake and outwash deposits. In the Pleistocene conglomerate area a few exposures of Tertiary limestone and conglomerate were noted, but it was not possible to tell whether these were outcrops of actually underlying formations or merely huge bowlders partly buried. There being no definite proof of the former, these exposures were not mapped. The formation is probably present west of The Three Peaks, overlying the Cretaceous and overlain by surface deposits, but only one outcrop was found and this was too small for the scale of the map.

The conglomerate is composed of both rounded and irregularly shaped fragments of limestone, chert, sandstone, and lavas, derived from immediately adjacent rock formations and cemented by a reddish sandy calcareous material, forming in places a fairly well-consolidated rock. It is rather thinly bedded, some of the beds being made up of coarse conglomerate, and others of finer sandy materials. From the thinly bedded character and rapidly varying materials of the different beds it seems clear that the formation is of fluviatile or terrestrial origin, deposited by streams as they emerged from the mountainous areas.

The Pleistocene conglomerate is unconformable upon the Tertiary lava series, in horizontal beds which retain no evidence of faulting. Its thickness is not known, but it is likely not more than 200 feet at the most.

PLEISTOCENE AND RECENT GRAVEL, SAND, AND CLAY.

Gravel, sand, and clay, derived from the neighboring rock hills, cover the lower slopes of the hills and the desert areas as lake, stream, and outwash deposits. To the east are the deposits of Rush Lake Valley, which run northeastward along the Colob front. To the west and north are the deposits of the Escalante Desert, which extend 35 or 40 miles westward and far to the north.

In the Escalante Desert clay is found mainly along the washes and in the low areas bordering them, while the gravels and sands are found in the areas between the washes. These coarser materials are fairly well rounded and appear to have been sorted.

The area around Shirts Lake is composed of fine clay with a white coating of alkaline carbonates. Around the entrance to Leach Canyon and in Rush Lake Valley east of Iron Springs are rolling hills of coarse subangular and rounded gravels, similar to those in the Escalante Desert.

Outwash fans of fine and coarse angular material border all the hilly areas and extend out into the desert for half a mile or more. Some of these are perfectly fan shaped, but in general the separate fans encroach upon one another laterally and form continuous outwash aprons along the borders of the hills. They have a relatively steep slope near the hills and gradually become less steep until they merge imperceptibly into the flat deserts.

These are the latest deposits in the area, and they are still accumulating. Their thickness is not known.

CHAPTER IV.

IGNEOUS ROCKS.

The igneous rocks of the district are both intrusive and effusive. The intrusives are biotite andesite laccoliths intruded into the Paleozoic and to a less extent into the Mesozoic rocks after the deposition of the Tertiary sediments. The effusives are later and form a bedded series varying in thickness from 1,000 to 2,000 feet and consisting of rhyolitic, trachytic, and andesitic flows, tuffs, and breccias.

The lavas rest on the eroded and upturned edges of the Eocene and Cretaceous sediments, indicating a considerable period of erosion between the intrusion of the biotite andesite and the outpouring of the effusives. The latter have been correlated with the Miocene lavas of the Wasatch Mountains, hence the andesite intrusion is post-Eocene and probably early Miocene.

The Quaternary basalts, present on all sides of the area, are not present within the district.

LACCOLITHS (EARLY MIOCENE).

DISTRIBUTION AND STRUCTURE.

The laccoliths are exposed in three main areas forming the cores of the principal mountains of the district—The Three Peaks and Granite Mountain in the northeastern and Iron Mountain in the southwestern part. A fourth, Stoddard Mountain, lies mainly beyond the southwest side of the area mapped. The laccolith areas are circular, with local irregularities due to faulting and other causes. Northwest of the The Three Peaks a small area of biotite andesite is brought up in contact with the Claron limestone by a great fault. Northwest of Iron Mountain a wide dike of andesite breaks through the Pinto sandstone a short distance from the main laccolithic mass, while to the southwest, surrounded by sediments, are several small andesite areas, parts of the large andesitic mass exposed through erosion of the low-dipping overlying limestone. In The Three Peaks and Iron Mountain laccolithic areas there are small patches of sediments faulted down into the andesite or left as erosion remnants.

In general the sediments dip away from the andesite, although locally, as east of Iron Mountain, the dip is steeply toward it, suggesting overturned strata. The strike of the beds is always parallel, or nearly so, to the andesite contact, except where faulting has

46

caused irregularities of structure. Individual strata of sediments may be followed almost entirely around the andesite areas. The contact with the sediments is commonly steep and the dip of the sediments less steep. Less commonly both the dip of the contact and of beds is low. The dip of beds and contact varies with depth of erosion. Nowhere has erosion exposed the base of the laccoliths, though the tops are well stripped. The maximum thickness exposed is 1,600 feet.

The circular outline of the intrusions, the manner in which the sediments encircle them and rest against and upon them, and the texture of the andesite favor the view that they are laccoliths (Pl. II, pocket). With their relations to the sedimentary rocks, the only alternative explanation would be that of a batholitic intrusion, and against this stands their texture and the lack, in the surrounding sediments, of the schistosity that often characterizes a plutonic contact. It is with some confidence, therefore, that the andesite masses are concluded to be laccoliths, notwithstanding the lack of direct observation of shape of the lower parts.

According to Gilbert's study of the laccoliths of the Henry Mountains,[a] the pressure of injection remaining constant, the limital area of a laccolith is a direct function of its depth beneath the surface. The limital area is greater when the depth is greater, and less when the depth is less. A laccolith with the diameter of the Iron Mountain laccolith would require a minimum of 7,000 feet of covering. A possible covering of about 4,500 feet can be measured on the eroded edges of the surrounding sediments. If Gilbert's conclusion be a sound one, it may be inferred that in the Iron Springs district certain sediments have been completely removed by erosion. His calculation, however, is based on the assumption that the sediments have been elastic and free to slide one over the other during the intrusion. If, on the other hand, the overlying sediments have any considerable strength, developing resistance beyond that afforded by the weight, the size of the laccolith would be proportionally increased. Such resistance may well be possessed by the Homestake limestone, which in all but a few places immediately overlies the laccolith, and the large size of the Iron Springs laccoliths may be due to this cause rather than to any greater depth of covering than can be measured on the sediments now present in the area.

A similar conclusion as to the factors determining the horizon of the laccoliths is reached from a consideration of densities. The density of the andesite is 2.65. The mean density of the rocks known to overlie it is about 2.54. Gilbert argues for the Henry Mountains that the laccoliths came to rest in such a position that their density

[a] Gilbert, G. K., Report on the geology of the Henry Mountains: U. S. Geog. and Geol. Surv. Rocky Mtn. Region, 2d ed., 1880, p. 84.

is slightly above the mean density of any combination of the overlying sediments. The density of the Iron Springs laccoliths, now
2.65, was considerably lower before solidification, presumably lower
than 2.54, the mean density of the overlying sediments, suggesting
that the Iron Springs laccoliths came to rest not in places determined by the law of hydrostatic equilibrium but in places determined
by the competency of the restraining limestone and sandstone strata;
in other words, the Iron Springs laccoliths, if their density at the
time of the intrusion was that common to acidic intrusives of this type,
would have been intruded more nearly at the surface had it not
been for the strength of the restraining limestone.

PETROGRAPHY.

The intrusive andesite is the rock with which the iron-ore deposits
are associated. Deposits are found on the borders of all the andesite
areas except Stoddard Mountain, where its absence is probably due
to the absence of limestone.

The rock is a light-gray biotite andesite with porphyritic texture.
The phenocrysts, consisting of feldspar, biotite, hornblende, and
diopside, are numerous, occupying more than half of the rock mass,
and varying in size up to an eighth of an inch in diameter. The
most abundant phenocryst is plagioclase of the variety labradorite,
but orthoclase is occasionally present. A large number of feldspars
show zonal growth. They are comparatively fresh, showing alteration only along cracks, along lines of zonal growth, and on the surface. The alteration products are calcite, kaolin, quartz, and sericite, typical katamorphic products. The next most abundant
phenocryst is biotite, in shiny black hexagonal plates, often altered
to phlogopite with a golden luster and frequently having reaction
rims of magnetite. In the Stoddard Mountain area and parts of the
other areas, the biotite is almost entirely decomposed to ferrite. In
a few cases it has altered to green chlorite. The biotite has abundant inclusions, mainly apatite but sometimes quartz, magnetite
and zircon. The hornblende is generally in dark-green prismatic
crystals, with inclusions of magnetite, biotite, and quartz. Like
the biotite, the hornblende in the andesite of the Stoddard Mountain
area is almost entirely decomposed to ferrite. The diopside is of a
light-green color, and is noticeably associated with magnetite, which
is present as inclusions and around the borders. It is generally
more or less altered to uralite along cracks and around the border.
Fragments of magnetite are abundant throughout the rock. Ferrite
is present as an alteration product of the ferrous silicates.

The groundmass is cloudy from alteration, but seems to be composed mainly of fine crystalline quartz and feldspar, both orthoclase
and plagioclase. Biotite, hornblende, pyroxine, and magnetite are
also represented, but less abundantly.

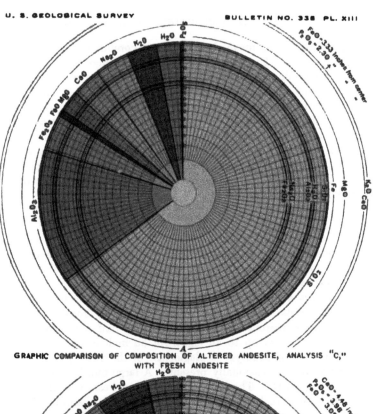

GRAPHIC COMPARISON OF COMPOSITION OF ALTERED ANDESITE, ANALYSIS "C,"
WITH FRESH ANDESITE

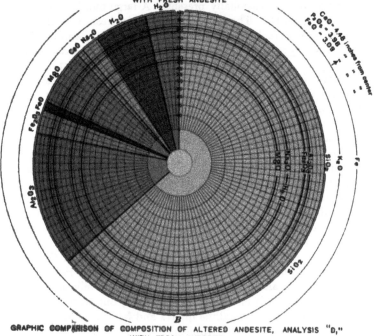

GRAPHIC COMPARISON OF COMPOSITION OF ALTERED ANDESITE, ANALYSIS "D,"
WITH FRESH ANDESITE

The biotite andesite has the same texture and mineral composition throughout the different areas. The absence of marginal facies (as a basic edge), dikes, and pegmatic veins is especially noticeable.

METAMORPHISM OF ANDESITE BY THE ORE-BEARING SOLUTIONS.

Near its contact with the ores for a few feet the andesite takes on a greenish-yellow color and dense texture, very similar to that of the silicated limestone at the andesite contact. The two are with great difficulty discriminated. The grain of the contact phase is much finer than that of the main mass of the andesite. The groundmass consists of finely crystalline feldspar and quartz, principally the former, cloudy with dark clayey-looking material, much of it not recognizable, but some of it in the larger masses clearly muscovite and rarely chlorite. The contact between the altered phase and the fresh andesite is sharp.

A comparison of the analyses of the fresh and contact phases by the circular-diagram method (Pls. XIII and XIV) brings out clearly the fact that at the contact the conspicuous change has been the introduction of soda. All other constituents show a possible loss, though ferric iron has developed at the expense of ferrous iron. In the present state of development of the district it is not possible to secure specimens from horizons surely below the influence of weathering. It may be that the oxidation of the ferrous iron here indicated is really a superposed weathering effect.

When the normal weathered andesite from the surface is compared with the fresh andesite, as in the following table (see also Pl. XII, B, p. 30), the change is found to be of quite a different character from that along the ore contacts. All constituents, including soda, show a loss relative to the alumina, as in Pl. XII, B.

Analyses of fresh and altered andesites.

[Analyst, R. D. Hall, University of Wisconsin.]

	A.	B.	C.	D.	E.
SiO$_2$	65 29	63 63	65 80	63.82	63.76
Al$_2$O$_3$	11.57	15 64	14.48	14.28	16.05
Fe$_2$O$_3$	2.10	3 59	3.96	2.72	1.91
FeO	2.67	93	.70	.81	.58
MgO	2 87	2 32	2.35	5.98	2.46
CaO	4.85	4. 46	3.19	.70	4.25
Na$_2$O	2.10	1.70	3.78	3.62	6.26
K$_2$O	5 18	5.22	3.32	4.24	2.84
H$_2$O −	.50	.40	12	2.30	1.22
H$_2$O +	1.82	1.70	2.71	1.68	.93
P$_2$O$_5$	.22	.15	.12	.04	.28
BaO	.17	.05	.06	.04	None.
	99.34	99.79	100.59	100 23	100.54

A. Specimen 46612. Fresh andesite east of Granite Mountains.
B. Specimen 46377. Weathered andesite from Desert Mound.
C. Specimen 46433. Altered andesite near iron-ore contact from Blowout, south of Iron Mountain.
D. Same nearer iron ore contact.
E. Specimen 46481. Altered andesite near iron-ore contact from Emma claim on east slope of Iron Mountain.

EFFUSIVES (LATE MIOCENE).

DISTRIBUTION AND STRUCTURE.

The effusives occupy two main areas in the district—the Antelope Range area and the Swett Hills, Eightmile Hills, and Harmony Mountains area. There are small exposures at Upper Point, in the northeastern quarter, and northwest of Iron Mountain.

There is a succession of 9 flows in the following order:

Succession of lava flows in Iron Springs district.

		Feet.
9.	Biotite-hornblende-pyroxene andesite	200
8.	Late tuffaceous rhyolite (Antelope Range)	400
7.	Biotite dacite	300
6.	Pyroxene andesite agglomerate and breccia	100
5.	Latest trachyte	150–300
4.	Hornblende andesite breccia and agglomerate	150
3.	Later trachyte	50
2.	Early tuffaceous rhyolite	300–400
1.	Early trachyte	50–600

Of these Nos. 1, 2, 5, and 7, extend throughout the area and for a number of miles to the south and west, while the rest are present only in parts of the district. There is little evidence of erosion between the successive flows, unless the absence of certain beds in different places be taken as such. The oldest formation forms the border of the lava series and inside of this successively younger flows outcrop, the latest in the center. Local faulting and sheets of outwash deposits have somewhat obscured these relations in places.

The Antelope Range lavas occupy a broad syncline pitching to the northeast, while those of the Swett Hills and Harmony Mountains have a general eastward dip and are not folded.

PETROGRAPHY.

Early trachyte.—The rocks of the early trachyte bed differ considerably in texture in different parts of the district as well as in different parts of the series. The color and mineralogical character, however, remain the same throughout the district. In general the rocks are dark and dense with few phenocrysts, principally feldspar, subordinately biotite.

In the Antelope Range, where it is typically exposed, the main part of the formation is a dense dark-red or purple porphyritic trachyte with very few phenocrysts. The phenocrysts are mainly orthoclase, and to a less extent plagioclase, of the variety andesine-labradorite. Both feldspars show Carlsbad twinning. They are altered along cracks and on the borders to quartz and calcite. The groundmass is dense, and in some layers contains amygdules, which are partly or

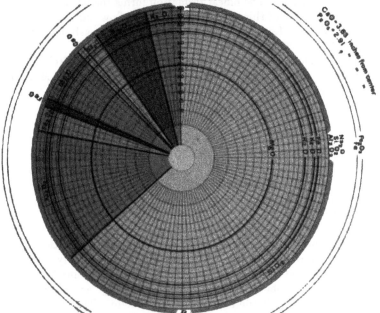

GRAPHIC COMPARISON OF COMPOSITION OF ALTERED ANDESITE, ANALYSIS "E,"
WITH FRESH ANDESITE

GRAPHIC COMPARISON OF COMPOSITION OF ALTERED ANDESITE, ANALYSIS "D,"
WITH ALTERED ANDESITE, ANALYSIS "C"

entirely filled with quartz and chalcedony. Near its base the trachyte in places becomes ferruginous, yielding from 6 to 7 per cent of iron.

The upper 30 feet of the bed has a somewhat varying character in the Antelope Range area, being composed of a main layer of lavender-gray trachyte, with thin layers of tuffaceous, scoriaceous, and dense trachytes, and has been mapped separately as far as it could be distinctly traced.

In the Swett Hills the early trachyte bed is dark gray and the phenocrysts are few. In the Eightmile Hills it is red, with numerous phenocrysts of sanidine and plagioclase and fewer of biotite. East of the Eightmile Hills and in the Harmony Mountains parts of the bed are porous and even scoriaceous, and have a predominance of lime-soda feldspar over orthoclase. In Cottonwood Canyon and westward it is complicated by additional beds of rhyolite, trachyte, and andesite similar to some of the overlying lavas. South of Iron Mountain the bed is amygdaloidal. The amygdules are spherical or oblong in shape, varying from 5 to 50 millimeters in diameter, and are filled with quartz and calcite.

Early tuffaceous rhyolite.—In the northeast and in the southern part of the Antelope Range area the early tuffaceous rhyolite bed is of uniform character throughout its vertical extent. It is a light-gray, pink, or white porphyritic rock with numerous phenocrysts, and small irregular fragments of other volcanic rocks. The phenocrysts are mainly quartz, orthoclase, and plagioclase, with less abundant biotite and hornblende. Zonal structure, secondary enlargement, and inclusions of long narrow crystals of apatite are common. The biotite is largely altered to phlogopite and is frequently associated with magnetite. Pyroxene appears in the included rock fragments, but not in the rhyolite itself. The groundmass is amorphous, except for small crystals of quartz and feldspar and cavities which are partly or wholly filled with chalcedony or calcite.

The early rhyolite bed in and near the Antelope Range in the western part of the Antelope Range area differs somewhat from the above. The succession, beginning at the bottom, is (1) a few feet of gray or black rhyolitic pitchstone, with numerous phenocrysts of quartz, feldspar, and mica; (2) hard red rhyolite which grades up into (3) a light-red rhyolite with numerous fragments of other volcanic rocks. (2) and (3) grade laterally into the tuffaceous phase of the eastern part of the area, but (1) does not change its character and is found in the same position, viz, at the bottom of the early rhyolite series, throughout this area as well as in other parts of the quadrangle. The minerals in these 3 phases are very much the same as in the tuffaceous phase, except that little or no hornblende is present. The feldspars are mainly orthoclase, but plagioclase is fairly abundant. They are altered to calcite, kaolin, and quartz, the plagioclase more so than orthoclase. Quartz shows strong resorption, but is otherwise fresh and limpid. Under the high power, it shows inclusions. The biotite is black and shiny and has inclusions of apatite. Magnetite and ferrite are present in small quantities, the latter as an alteration product of the ferrous silicates. Zircon is rare. The groundmass is partly crystalline and partly eutaxitic and amorphous. The crystalline part appears to be mostly quartz and feldspar. There are numerous chalcedony-filled cavities throughout the rock.

The band of early rhyolite along the northwestern front of the Swett Hills is bright pink with phenocrysts of quartz and sanidine, the latter with Carlsbad twinning. Both quartz and feldspar are fresh and glassy. The former frequently shows resorption. The groundmass is pink and dense, but with numerous little cavities. It appears to be amorphous with lenses of lighter colored crypto-crystalline material.

The areas of lower rhyolite in the Eightmile Hills and the Harmony Mountains, like those first described, are more or less tuffaceous in character, but as the groundmass becomes denser they grade into a fragmental rhyolite. The little neck of the early rhyolite bed west of Stoddard Canyon has a number of areas of trachyte and dacite very much like those in the purple trachyte east of Birch Canyon. These are very irregular in extent and distribution and could not be structurally so separated as to correlate them with the overlying trachyte and dacite.

Later trachyte.—Wherever the later trachyte is present it is composed of 2 distinct layers—at the base a black trachytic pitchstone averaging 5 to 10 feet in thickness and above this a dense red trachyte.

The pitchstone contains abundant phenocrysts of feldspar and biotite, and fewer of diopside, hornblende, magnetite, and apatite. The feldspars are largely sanidine, showing Carlsbad twinning. A little plagioclase is present. The feldspars are all fresh and unaltered, and have numerous inclusions of apatite and zircon. The biotite is dark brown and strongly pleochroic. It has few inclusions which are mainly zircon. Magnetite fragments are abundant, some of them showing alteration to hematite. Hornblende is rare.

The groundmass is black and glassy and under the high power appears to be made up of innumerable crystallites with simple and branching hairlike forms grouped along lines of flow. In it are red spherulites, some of which in addition to the spherulitic structure show flow structure like the rest of the groundmass.

The trachyte above the pitchstone is a reddish-brown porphyritic rock with phenocrysts of feldspar and biotite, about equally abundant, and together making up perhaps one-tenth of the mass of the rock. Both orthoclase and plagioclase are present, the latter being probably a little more abundant. They are very much altered to a white clayey material, probably kaolin. The biotite is fresh, dark brown, and strongly pleochroic. It has a few inclusions of apatite. Magnetite particles are numerous. The groundmass is cloudy from specks of ferrite and shows flow structure, especially around the phenocrysts. It appears to be partly amorphous with tiny specks of feldspar arranged along flow lines. The phenocrysts are also arranged parallel to the lines of flow.

Hornblende andesite breccia and agglomerate.—In the western and northern parts of the Antelope Range area the hornblende andesite is a greenish-gray breccia, but toward the south it grades into a coarse agglomerate, with dark-gray and black fragments. In the breccia the fragments are of dark andesite and rather small, the largest being about 10 inches in diameter, and by far the larger part being smaller. The material between the fragments is of the same character as the fragments but somewhat softer. In the agglomerate the fragments are larger, ranging up to perhaps 2 feet in diameter. Most of them are composed of the same material as the fragments of the breccia, but other fragments are red and trachytic. The intervening material in the agglomerate is light gray and tuffaceous and weathers readily, leaving the fragments strewn around over the surface like bowlders. These soon become black and shiny from desert varnish.

The hornblende andesite composing the fragments is a dark-gray porphyritic rock stained green in many places. The main phenocrysts are narrow crystals of hornblende, sometimes one-fourth inch long. A few phenocrysts of feldspar are present. The groundmass is dark gray and finely crystalline and makes up about nineteen-twentieths of the rock mass. The hornblende has a dark-brown color and shows pleochroism. Fragments of magnetite are found bordering it and included in it, but generally separated from it by an alteration rim of ferrite. The hornblende is altered to ferrite on the surface and sometimes well into the interior.

The ground mass is composed largely of small crystals of orthoclase and plagioclase of the variety labradorite, the latter being a little more abundant. Besides the feldspar, pyroxene, hornblende, and magnetite occur in the groundmass. The small crystals are separated by areas of cryptocrystalline material.

Latest trachyte.—The latest trachyte bed is easily recognized throughout its extent by a characteristic banded appearance. It is composed of several different layers; the full section, which is present only in the eastern part of the Antelope Range area, is as follows, beginning at the base: (1) a few feet of black trachytic pitchstone with phenocrysts mainly sanidine, but with plagioclase, biotite, and diopside present; (2) red trachyte with grayish-white bands; (3) gray and red trachytes, slightly banded in places; (4) gray tuffaceous trachyte; (5) hard, dense red trachyte. These phases differ only in texture, the phenocrysts (mainly sanidine, subordinately plagioclase, biotite, and magnetite) being the same throughout the series with the exception that layer 5 contains a few pyroxenes, and biotite and magnetite are slightly more abundant than in the other layers. There is also a slight variation in the amount of plagioclase in different parts of.the latest trachyte bed.

Outside of the eastern Antelope Range area only layers 1 and 2 and perhaps 3 are present. In the eastern part of the Antelope Range area layer 2 is comparatively thin, but throughout the rest of the quadrangle it forms almost the entire thickness of the bed, although the banding is not always as conspicuous as it is here. The black pitchstone is nearly always present at its base. At the top of the bed in other parts of the quadrangle there is often a layer of less distinctly banded material which may represent layer 3. The absence of the upper layers is probably due to nondeposition, since erosion could hardly have taken place so uniformly over the region as to leave practically the same thickness of trachyte throughout.

The latest trachyte of the Antelope Range (layer 2) is a reddish porphyritic rock, with parallel bands of white and gray material averaging generally less than one-eighth inch in thickness. The phenocrysts, averaging about one-sixteenth of an inch in diameter, form about one-tenth of the rock and are largely sanidine. At the surface these have been entirely leached out, leaving only the cavities. Biotite is also present, being generally altered to phlogopite and often entirely decomposed to ferrite. Phenocrysts appear to be present in both the gray and red bands. Nearly all the sanidine shows Carlsbad twinning. In general, they appear to be quite fresh, showing alteration only on the surface and along fractures. The alteration products appear to be calcite, kaolin, and quartz. The sanidine frequently has inclusions of zircon and apatite. A little plagioclase, variety andesine, is present. Magnetite occurs as fragments.

The red part of the groundmass shows flow structure and is more cloudy than the gray part, owing to numerous little specks of ferrite. It is amorphous for the most part, with here and there specks of feldspar. The gray part appears to be impregnated with quartz. It contains elongated quartz-filled cavities which grade into the more cloudy quartz-impregnated groundmass. In the center of some of the larger cavities there are brownish aggregates of fine cryptocrystalline material. These silicious fillings are probably later than the rest of the rock, having been deposited in the porous parts. This view is strengthened by the fact that the banded appearance of this formation in other parts of the district is caused by a difference in the character of the ground mass, the lighter bands being porous and the darker ones more dense.

The rocks of the latest trachyte bed over other parts of the district are very similar to the one described above with regard to mineral composition and texture, except for the lesser abundance of quartz in the lighter colored bands.

Pyroxene andesite agglomerate and breccia.— In the western part of the Antelope Range area the pyroxene andesite bed is a greenish-gray copper-stained breccia, with little or no difference between fragments and groundmass. Toward the east it grades into a coarse agglomerate with huge dark bowlder-like fragments, cemented by gray tuffaceous material. This agglomerate is very much like the hornblende andesite agglomerate, but as a general rule it has larger fragments.

The rock composing the fragments in the breccia and agglomerate is a pyroxene andesite. It is dark gray in color and has porphyritic texture, the phenocrysts making up perhaps one-tenth of the rock, and averaging one-eighth inch in diameter. The phenocrysts are pyroxene and plagioclase, variety labradorite, the former being the more abundant. The groundmass is dense and appears to be composed of minute crystals. The pyroxene is light green in color, shows a slight brownish pleochroism, and has included magnetite fragments. Often it occurs in aggregates of 2 or 3 and sometimes shows twinning parallel to the orthopinacoid (100). Plagioclase is not very abundant, and has suffered considerable alteration to calcite and kaolin. It shows zonal structure and, like the pyroxene, occurs in aggregates.

The groundmass is composed of numerous crystals of plagioclase and fewer of orthoclase, pyroxene, and magnetite, separated by amorphous areas.

Biotite dacite.—The biotite dacite has the same texture and mineral composition throughout both its vertical and horizontal extent. The color, however, varies slightly. In the Antelope Range area the main part has a reddish-brown color, while at the base there are 50 feet or less of a pinkish-gray color. Outside of the Antelope Range area the latter makes up the entire formation.

The biotite dacite is a porphyritic rock with phenocrysts ranging up to one-eighth inch in diameter and making up fully half of the rock mass. They are mainly feldspar and biotite, subordinately diopside, quartz, and brown hornblende. The feldspars are mainly plagioclase, ranging from basic andesine through labradorite to acidic bytownite. Orthoclase is also present, but not as abundant. A large number of the feldspars show well-developed zonal structure, the successive zones of growth being very thin, owing to rapidly alternating conditions in the formation of the crystal. There is considerable alteration along fracture lines, the most important decomposition product being calcite. The feldspars have inclusions of apatite and biotite. The biotite is dark brown and greenish brown, highly pleochroic, and has inclusions of feldspar. The diopside is light green in color and has numerous inclusions of magnetite. It is often very much decomposed, the products being calcite and ferrite. Fragments of quartz, which frequently show resorption, are present. The hornblende is of the brown basaltic variety. Fragments of magnetite are abundant.

The groundmass is amorphous and frequently shows flow structure. Through it are scattered tiny specks of the same minerals as compose the phenocrysts.

Late tuffaceous rhyolite.—The late tuffaceous rhyolite has the same character throughout, except for an agglomerate at the base. The latter consists of big bowlders and smaller fragments of dark igneous material, both acidic and basic. About halfway up in the tuff formation there is, in a few places, a thin layer of coarse sandy material, which appears to be water deposited. Frequently the upper part of the formation contains numerous concretion-like spheres, ranging from 1 inch to 5 inches in diameter, which are very much harder and darker than the rest of the rock. These have the same minerals as the rest of the formation, but the groundmass is denser and the cavities have all been filled with chalcedony.

The late tuffaceous rhyolite has a porphyritic texture with phenocrysts of feldspar, quartz, biotite, and hornblende, and a porous glassy groundmass. The main phenocryst is quartz, which is very fresh and generally shows crystal outlines. Orthoclase and plagioclase are present, but not abundant. Most of them have suffered surface alteration. The biotite is frequently twinned and has inclusions of apatite. Fragments of hornblende and pyroxene are present, the latter being in included rock fragments. Magnetite particles are scattered through the rock.

The groundmass is amorphous with numerous cavities filled or lined with calcite. For this reason calcite forms quite a large percentage of the rock.

Biotite-hornblende-pyroxene andesite.—In the Swett Hills the biotite-hornblende-pyroxene andesite bed consists of several types of rock. In the western part there is at the base a dark hornblende andesite breccia and above this several layers of rhyolite and trachyte interlayered with limestone and conglomerate. In the eastern part all but the breccia disappear, and a biotite-hornblende-pyroxene andesite, which to the west only caps the lower series, makes up almost the entire thickness. In the Harmony Mountains the formation consists of rhyolite, trachyte, and conglomerate, while in the lava area southwest of Iron Mountain only the hornblende andesite breccia is present, with a little limestone and conglomerate.

The biotite-hornblende-pyroxene andesite is a light pinkish-gray porphyritic rock, with feldspar as the main phenocryst. Biotite, hornblende, and diopside are of equal importance. Quartz and magnetite are present. The feldspars are mainly labradorite, but a little orthoclase is present. They show zonal growth and have inclusions of apatite. A few fragments of quartz are present. The diopside is light green in color, while the hornblende and biotite are dark brown and show alteration to ferrite. The groundmass is amorphous with finely crystalline feldspar and specks of ferrite.

· The hornblende andesite breccia is a dark purplish-gray porphyritic rock, with phenocrysts of brown hornblende and a groundmass of small lath-shaped crystals of labradorite, orthoclase, monoclinic and rhombic pyroxene, separated by irregular cavities. The rhyolite and trachyte layers have the same texture as the underlying tuffaceous rhyolite and trachyte formations. The limestones and conglomerate are like those of the Tertiary, but thinner and less consolidated. Their presence in the uppermost lava formation shows that all the lavas were submerged after its deposition.

CHEMICAL AND MINERAL COMPOSITION OF THE IGNEOUS ROCKS.

Analyses have been made of the intrusive andesite and of the four principal flows, namely, the early trachyte (1), early rhyolite (2), latest trachyte (5), and dacite (7). Of the rest of the flows, some, while they have considerable horizontal distribution, are very thin, while the others are local in their distribution.

The following table gives the chemical composition, the approximate mineral composition of the same specimens determined from thin sections, and the mineral composition of the same specimens · calculated as modes from the chemical composition for each of these formations, in order of age:

Analyses of igneous rocks from Iron Springs district.

CHEMICAL COMPOSITION.

	A.	B.	C.	D.	E.	E'.	F.	G.
SiO_2	65.29	63.63	64.83	58.04	66.38	70.03	73.17	61.05
Al_2O_3	11.57	15.64	16.68	18.96	13.72	14.47	13.34	16.03
Fe_2O_3	2.10	3.59	3.74	5.88	2.23	2.35	1.35	5.42
FeO	2.67	.93	1.22	1.33	.80	.84	.76	.98
MgO	2.87	2.32	.79	1.11	.54	.57	.81	3.03
CaO	4.85	4.46	2.85	6.12	5.49	2.42	1.32	5.40
Na_2O	2.10	1.70	.86	2.26	2.50	2.64	1.80	1.43
K_2O	5.18	5.22	7.56	4.08	5.20	5.48	7.10	5.58
H_2O+	2.32	2.10	.92	2.05	.92	.97	.54	.81
P_2O_5	.22	.15	.35	.34	.08	.08	.07	.30
CO_2					2.52			
BaO	.17	.05	.11	.04	.11	.11	.10	.08
	99.34	99.79	99.91	100.21	100.49	99.96	100.36	100.11

MINERAL COMPOSITION DETERMINED FROM THIN SECTIONS.

	A.	B.	C.	D.	E.	F.	G.
Quartz	Some	Some			Much	Some	Some.
Orthoclase	Some	Some	Some	Some	Much	Some	Much.
Plagioclase	Much	Much	Little	Much	Little	Little	Much.
Biotite	Some	Some	Little	Little	Some	Some	Some.
Hornblende	Some	Little					Some.
Diopside	Little	Little		Little			Some.
Magnetite	Little	Little	Little	Little	Little	Little	Little.

MINERAL COMPOSITION CALCULATED FROM CHEMICAL COMPOSITION.

	A.	B.	C.	D.	E.	F.	G.
Quartz	23.52	22.26	22.02	15.06	27.60	33.12	19.38
Orthoclase	28.35	27.80	43.37	22.79	28.35	39.47	28.35
Albite	14.67	13.10	7.34	18.86	20.96	15.19	12.05
Anorthite	5.84	15.29	11.12	25.85	10.56	5.56	15.84
Biotite [a]	5.08	4.63	3.81	3.71	3.70		2.70
Phlogopite [a]						1.62	4.16
Hornblende [a]	5.89	1.99					
Diopside	6.03	4.32	.86	2.16			7.12
Magnetite	1.39	1.16	1.85	2.78	1.62	.92	1.85
Apatite	.31	.31	.62	.62	.31	.31	.62
Limonite		2.24	.75	4.11	.56	.75	.75
Hematite			1.28				3.52
Calcite					5.70		
Kaolin		3.78	5.16	3.09			3.61
Serpentine	2.48	1.84					
Water	1.90	.90		.97	.76	.34	
Sillimanite			1.45				
Wollastonite	3.48						
	98.94	99.62	99.63	100.00	100.12	99.98	99.95

[a] Composition based on average of analyses in Dana's Manual of Mineralogy.

A. Specimen 46612. Fresh intrusive andesite east of Granite Mountain. Analysis by R. D. Hall, University of Wisconsin.

B. Specimen 46377. Slightly weathered intrusive andesite from Desert Mound. Analysis by R. D. Hall, University of Wisconsin.

C. Specimen 46533. Fresh early trachyte from Antelope Range (No. 1 of flows). Analysis by R. D. Hall, University of Wisconsin.

D. Specimen 46584. Andesite from same formation as specimen C north of Stoddard Mountain. Analysis by R. D. Hall, University of Wisconsin.

E. Specimen 46521. Early rhyolite from Eightmile Hills (No. 2 of flows). Analysis by R. D. Hall, University of Wisconsin.

E'. Specimen 46521. Early rhyolite. Recalculated on the basis of 100 per cent after removing CaO and CO_2 of the infiltrated calcite.

F. Specimen 46557. Latest trachyte from Antelope Hills (No. 5 of flows). Analysis by R. D. Hall, University of Wisconsin.

G. Specimen 46586A. Dacite from Swett Hills (No. 7 of flows). Analysis by R. D. Hall, University of Wisconsin.

Partial chemical analyses of igneous rocks from Iron Springs district.

[Analyst, Fred Larch, Biwabik, Minnesota.]

	H.	I.	J.	K.	L.	M.	N.
SiO₂	53.65	66.20					
Al₂O₃	1.59	2.72					
P₂O₃	.118	.117	.076	.138	.192	.090	.087
Fe	5.65	3.71	4.56	5.58	6.49	4.91	5.16

	O.	P.	Q.	R.	S.	T.	U.
P₂O₃	.133	.060	.074	.040	.032	.124	.127
Fe	5.79	2.43	2.29	2.78	1.74	4.03	4.24

H. Specimen 46317. Intrusive andesite. Three Peaks.
I. Specimen 46318. Same.
J. Specimen 46500. Early trachyte. Eightmile Hills.
K. Specimen 46533. Same. Antelope Range.
L. Specimen 46540. Same.
M. Specimen 46541. Same.
N. Specimen 46547. Same.
O. Specimen 46584. Same. North of Stoddard Mountain.

P. Specimen 46502. Early tuffaceous rhyolite. Eightmile Hills.
Q. Specimen 46503. Same.
R. Specimen 46476. Latest trachyte. Swett Hills.
S. Specimen 46477. Same (pitchstone at base).
T. Specimen 46504. Biotite dacite. Eightmile Hills.
U. Specimen 46380. Same.

Analyses A and B are average andesites, although A is a little too high in potash and low in alumina. C is a typical trachyte, while D is an andesite forming a subordinate part of the early trachyte bed.

The early rhyolite (E) is too high in lime, owing to secondary calcite occurring in amygdules. This has been subtracted from the analysis and the latter recalculated to 100 per cent. This recalculated composition (E') is that of a typical rhyolite. The composition of F, as given in the table, is too high in silica and too low in alumina for a trachyte. The excess silica is accounted for by secondary infiltration in amygdules, and if this be removed and the composition recalculated to 100 per cent the alumina percentage will be brought up within the limits. The lack of quartz phenocrysts determines its name.

The dacite (G) is too low in silica for a typical dacite, but the presence of quartz phenocrysts determines its name.

According to Professor Iddings the presence of trachyte and andesite in the same group of lavas is a rare occurrence, nevertheless chemical and mineral compositions indicate that both probably occur in the Iron Springs area.

The analyses represent the general acidic character of the series and show a slight range in chemical composition, especially in the silica and alkalies. Silica ranges from 58 to 70 per cent, the increase to 73 per cent in analysis F being due largely to the later cavity fillings. The alkalies range from 6½ to 9 per cent, the potash being in excess of the soda in all cases. This excess reaches a maximum in the trachytes C and F, a fact well illustrated in the mineral composition tables by the excess of potash feldspar over lime-soda feldspar in these rocks.

The alkalies vary approximately with the amount of silica, as follows:

Variations in alkali and silica in igneous rocks of Iron Springs district.

	F.	E.	A.	C.	B.	G.	D.
SiO$_2$	73.17	70.03	65.29	64.83	63.63	61.05	58.04
Alkalies	8.90	8.12	7.28	8.42	6.92	7.01	6.34

The only striking exception to this rule is "C," which is high in alkalies and comparatively low in silica.

The total alkalies vary inversely as the lime with minor exceptions.

Variations in alkali and lime in igneous rocks of Iron Springs district.

	F.	C.	E.	A.	G.	B.	D.
Alkalies	8.90	8.42	8.12	7.28	7.01	6.92	6.34
CaO	1.32	2.85	2.42	4.85	5.40	4.46	6.12

An increase in the amount of silica and alumina is attended by a decrease in iron, magnesia, and lime, illustrated in the following table:

Variations in silica and alumina corresponding to variations in iron, magnesia, and lime in igneous rocks of Iron Springs district.

	F.	E.	C.	B.	G.	D.	A.
SiO$_2$ and Al$_2$O$_3$	86.51	84.50	81.51	79.27	77.08	77.00	76.86
Fe$_2$O$_3$, FeO, MgO, and CaO	4.24	6.18	8.60	12.30	14.83	14.94	12.49

The relation between the potash and soda is similar to that shown by Professor Pirsson to exist in the rocks of the Highwood Mountains,[a] but the relation between the alkalies and the lime is reversed. The definite relation existing between potash, soda, and lime in the Highwood rocks, according to Professor Pirsson, must have been characteristic of the parent magma, and separates these rocks from rocks of other areas and groups them into a clan. The same argument may be applied to the rocks of the Iron Springs district.

The chemical character of the rocks is clearly expressed by the mineral composition. The predominance of potash is indicated by an abundance of orthoclase, while the basic feldspar varies in amount with the lime, except in the early rhyolite (E), in which the high

---

[a] Pirsson, L. V., Petrography and geology of the igneous rocks of the Highwood Mountains, Montana: Bull U. S. Geol. Survey No. 237, 1905, pp. 172-174.

lime percentage is accounted for by the presence of infiltrated calcite. In analyses C, D, E, and F the low magnesia explains the absence of hornblende and diopside and the lowering of the biotite percentage.

The mineral compositions calculated from the analyses (P. 58) are in general similar to those observed, yet on account of fine-grained or amorphous groundmass the mineral composition as determined from thin sections could only be approximate and in some instances does not correspond well with the calculated modes, the greatest discrepancy being in the relative amounts of quartz, ortho-clase, and lime-soda feldspar. Where the groundmass is crypto-crystalline, the mineral composition is based on the phenocrysts. In the dacite and andesites these form a large percentage of the mass of the rock, while in the early trachytes, the early rhyolite, and the latest trachyte they are subordinate in amount. Sillimanite and wollastonite were not seen in the rocks, but were added to take up surplus alumina and lime.

RELATIONS OF LACCOLITHS AND EFFUSIVES IN GENESIS.

The question naturally arises whether or not the intrusive andesite and the effusives came from the same reservoir. The chemical compositions of the laccolithic rocks and the different flows show but a small range. With very little differentiation all the different phases may have originated from the same parent magma. It is certain, however, that the laccoliths did not act as vents through which the outpouring of the effusives took place. The andesite had been intruded, solidified, and eroded when the lavas were poured out over the eroded edges of the uplifted sediments. That the lavas came to the surface through the laccoliths after their solidification and erosion does not seem likely, since the andesite areas show no dikes or stocks, unless the ore veins be so called. Neither are there stocks or dikes else-where in the district through which the eruptions might have occurred; hence we are driven outside of the area for the source of the effusives.

The following table shows an approximation of the average chemical composition of the lavas and andesite, obtained by averaging the preceding analyses. The general similarity of composition indicates that both may have come from the same reservoir.

Average chemical composition of laccoliths and effusives.

	Laccoliths.	Effusives.		Laccoliths.	Effusives.
SiO_2	64.46	66.42	K_2O	5.20	5.99
Al_2O_3	13.60	15.41	H_2O+	2.21	.95
Fe_2O_3	2.85	3.48	P_2O_3	.18	.19
FeO	1.80	.96	BaO	.11	.09
MgO	2.59	1.34			
CaO	4.65	3.40		99.55	100.09
Na_2O	1.90	1.86			

CHAPTER V.

CORRELATION.

SEDIMENTARY ROCKS.

Several excursions were taken into the surrounding country for the purpose of correlating the formations of the Iron Springs district with those of the Colob Plateau, one of the High Plateaus of Utah, and with the Pine Valley Mountains.

The rocks of these areas were first studied in 1871, 1872, and 1873 by G. K. Gilbert, A. R. Marvine, and E. E. Howell for the Wheeler Survey.[a] In 1875, 1876, and 1877, they were studied by Maj. C. E. Dutton[b] for the Rocky Mountain Survey and again in 1880 for the United States Geological Survey in connection with the work in Grand Canyon. In the summer of 1902 a small area east of the Pine Valley Mountains along the west margin of the High Plateaus was mapped by E. Huntington and J. W. Goldthwait,[c] under the direction of Prof. W. M. Davis, of Harvard University.

Dutton[d] classified the rocks of this region into systems from Carboniferous to Quaternary. Huntington and Goldthwait[e] subdivided the larger division of Dutton to some extent, and applied geographic names. The following tables show the correlation of the rocks of the Iron Springs district with those given by Huntington and Goldthwait and by Dutton:

---

[a] Gilbert, G. K., Report on the geology of portions of Nevada, Utah, California, and Arizona, examined in the years 1871 and 1872: U. S. Geog. Surv. W. 100th Mer., vol. 3, Geology, pt. 1, 1875, pp. 17-187. Marvine, A. R., Report on the geology of route from St. George, Utah, to Gila River, Arizona, examined in 1871: U. S. Geog. Surv. W. 100th Mer., vol. 3, Geology, pt. 2, 1875, pp. 189-225. Howell, E. E., Report on the geology of portions of Utah, Nevada, Arizona, and New Mexico, examined in the years 1872 and 1873: U. S. Geog. Surv. W. 100th Mer., vol. 3, Geology, pt. 3, 1875, pp. 227-301.

[b] Dutton, C. E., Report on the geology of the High Plateaus of Utah. U.S. Geog. and Geol. Surv. Rocky Mtn. Region, 1880, pp. 307; Tertiary history of the Grand Canyon district· Mon. U. S. Geol. Survey, vol. 2, 1882, pp. 264.

[c] Huntington, Ellsworth, and Goldthwait, J. W., The Hurricane fault in the Toquerville district, Utah: Bull. Mus. Comp. Zool., Harvard Coll., No. 42 (Geol. Ser., vol. 6), 1904, pp. 199-259.

[d] Dutton, C. E., op. cit.

[e] Op. cit., pp. 202-208.

Correlation of rocks of the Iron Springs district with rocks of the Colob Plateau.

Colob Plateau.		Iron Springs district.
Huntington and Goldthwait.	Dutton.	
Basalt.	Quaternary sand, gravel, and clay. Basalt.	Quaternary sand, gravel, and clay.
Pleistocene gravels.		Pleistocene conglomerate.
Trachyte, andesite.	Andesite, trachyte, rhyolite.	Andesite, trachyte, rhyolite, etc.
Tertiary limestone, shale, and conglomerate.	Tertiary limestone, shale, and conglomerate.	Claron limestone, including some sandstone and conglomerate.
Cretaceous sandstone, shale, and limestone.	Cretaceous sandstone, shale, and limestone.	Pinto sandstone, including some shale, conglomerate, and limestone lenses.
Colob sandstone.	Jurassic shale. Jurassic sandstone.	(?)
Kanab sandstone. Painted Desert sandstone and shale. Shinarump conglomerate.	Triassic sandstone, shale, and conglomerate.	(Missing.)
Moencopie shale and sandstone.	Permian shale and sandstone.	(Missing.)
Super-Aubrey shale and limestone. Aubrey limestone.	Carboniferous limestone.	Homestake limestone.

The fossils found in the Homestake limestone are few and poorly preserved. They were referred for determination to Prof. Eliot Blackwelder, of the University of Wisconsin, and Dr. George H. Girty, of the United States Geological Survey. Professor Blackwelder determined some of the better preserved forms as *Aviculopecten*, a genus ranging from Silurian to Triassic, but here probably Carboniferous. Doctor Girty says:

> The specific determination of these forms is hardly possible from the imperfect condition of the material. . . . I believe that should you obtain more complete collections they would prove that the beds from which they were obtained should be correlated with that portion of the Wasatch Mountains section which the geologists of the Fortieth Parallel Survey designated the Permo-Carboniferous. This is likely to be the highest Paleozoic horizon found in your region.

The overlying Pinto formation is satisfactorily determined as Cretaceous. The Homestake limestone could not be Jurassic, according to its fossils, and since the limestone is dissimilar to the Triassic and Permian sediments of the adjacent High Plateaus, which are masses of sandstone and shale, it is referred to the next preceding period, the Carboniferous. Specific correlation with the Aubrey (Carboniferous) limestone naturally suggests itself, but it differs much from the latter in appearance and lithologic character, in that it is more sandy, shows bedding plainly, and is yellow, while the Homestake limestone is a pure, massive, dark-blue-gray limestone, lacking a conspicuous bedding.

So far as known, the Pinto sandstone of the Iron Springs quadrangle bears no fossils, but in the southern part it contains a few layers of carbonaceous shales, which were followed southward and connected with the anthracite beds of Stoddard Mountain (mapped by the Wheeler Survey as Cretaceous and correlated by that survey with the Cretaceous rocks on the Colob Plateau), which contain numerous workable seams of bituminous coal, associated with oyster beds often several feet in thickness. The lower part of the Pinto formation may be Jurassic, but no evidence for the separation could be found.

The Claron formation contains only a few fossils and these are Carboniferous forms in pebbles in the conglomerates. They are of value only in pointing to a Carboniferous limestone outside of the district as the source of the conglomerates. Nevertheless, the Claron formation may be satisfactorily correlated with the Tertiary of Dutton,[a] which is the Uinta formation of Smith,[b] on the Markagunt Plateau. At the head of Cedar Canyon (Coal Creek valley), cutting the Markagunt Plateau, there are 3,000 feet of many-colored Tertiary limestones, sandstones, and conglomerates. The limestone of the Cedar Canyon Tertiary and of the Claron were found to be much the same both in regard to general appearance and lithology. Both are many colored, although red predominates, and both are sandy. The cherty concretions described above are also characteristic of many layers in both areas. The Claron formation of · the Iron Springs district contains more fragmental material in the form of beds of conglomerate, but this is to be expected, because it is nearer the shore line of the Eocene lake supposed by Smith to lie west of the High Plateau region. The Uinta formation is supposed to have been deposited in a fresh-water lake basin, which during the early stages of deposition extended into southwestern Utah nearly to the Nevada and Arizona boundaries but gradually retreated northeastward into central Utah. The lower part of the Cedar Canyon Tertiary was not seen. It may contain more beds of conglomerate than the middle and upper parts, and hence this part might be more specifically correlated with the Claron formation of the Iron Springs district.

In general, then, the correlation of the Cretaceous and Tertiary rocks of the Iron Springs district is based on similarity in lithology, thickness, and succession of the Pinto and Claron formations with the Cretaceous and Tertiary series, respectively, of the High Plateaus beginning 15 miles to the east, and on the direct connection of the Pinto formation with the Cretaceous coal-bearing beds of Stoddard Mountain. The assignment of the underlying Homestake formation to the Carboniferous is based on obscure fossil forms and

a Op. cit.   b Smith, J. H., The Eocene of North America: Jour. Geol., vol. 8, 1900, p. 452.

on the fact that, though differing lithologically from the Aubrey limestone (Carboniferous) of the High Plateaus, it is more like it lithologically than any other formation underlying the Cretaceous of this region. A possible hiatus between the Pinto formation (Cretaceous) and the Homestake limestone (see p. 17) may account for the absence of Permian, Triassic, and perhaps Jurassic sediments— if the last named, indeed, be absent (see p. 16)—between the Cretaceous and Carboniferous of this district. The weak features of the correlation are the absence of sufficient fossil evidence, the dissimilarity of the Homestake and Aubrey formation, and the absence of Permian, Triassic, and Jurassic sediments with so slight a structural discordance.

IGNEOUS ROCKS.

Only the earlier trachyte and the tuffaceous rhyolite have been correlated with rocks previously studied outside of the district, namely, the "rhyolite" of Dutton's High Plateaus survey.[a] These rocks were examined on Brian Head, near the south end of the area mapped by Dutton as rhyolite, and were found to be composed of a considerable thickness of rhyolite underlain by a rather thin layer of dark scoriaceous material like some of the scoriaceous phases of the earlier trachyte of the Iron Springs district. Below this, as in the Iron Springs district, were found Tertiary limestones and conglomerates. Thus a general correlation with the lavas may be made on the basis of their superposition above the Tertiary limestones and conglomerates in the two districts.

[a] Dutton. C. E., Report on the geology of the High Plateaus of Utah: U. S. Geog. and Geol. Surv. Rocky Mtn. Region, 1880, pp. 61 et seq.

CHAPTER VI.

DESCRIPTION OF THE IRON ORES.

DISTRIBUTION, EXPOSURES, AND TOPOGRAPHY.

The iron ores occur in disconnected masses within a general area about 1½ miles wide by 20 miles long, running northeast and southwest through the district mapped. (See Pls. II, XV, XVI.) They lie for the most part on eastern and southern slopes or foothills of The Three Peaks, Granite Mountain, and Iron Mountain, between elevations of 5,600 and 6,700 feet, but some of them, as on Iron Mountain, appear at or near the tops of the mountains at elevations between 7,000 and 8,000 feet.

Some of the iron-ore exposures stand out as much as 200 feet above the surrounding country as black, jagged ridges (Pls. XVII, B, to XIX, A). Others, including several of the larger deposits on the lower slopes, do not stand above the surrounding rocks (Pl. XIX, B), but are known by isolated exposures and black iron-formation fragments disseminated in the loose detrital material at the surface. Some of the ore does not appear at the surface at all, being covered by andesite detritus washed from the upper slopes, though, even here, fragments of ore are likely to appear in the detritus farther down the slopes. In such places the exact shape and distribution of the deposits can not be determined without trenching or pitting. Fortunately such work will suffice fairly well throughout the possible ore-bearing areas, though there are places where areal extensions of iron-ore belts may be found by underground exploration, or where belts, mapped as continuous on the basis of the surface fragments, may really be discontinuous. The deepest pits in the district, 130 feet, have not yet reached water level.

GEOLOGICAL AND STRUCTURAL RELATIONS OF THE ORE DEPOSITS.

The ore deposits for the most part lie at or near the contact of the andesite laccoliths and the Homestake limestone. Some of them occur entirely within the andesite well up the slopes, and others entirely within the limestone, but seldom far from the contact. (See Pls. III–V, pocket.)

66

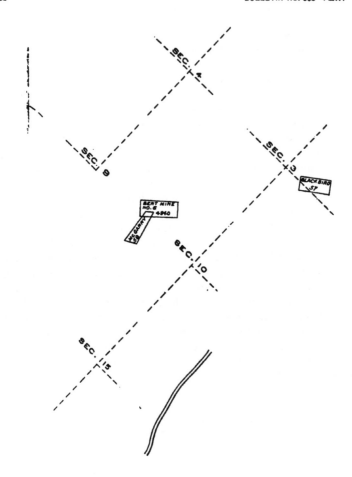

½   2 miles

ORE DEPOSITS IN ANDESITE.

The deposits within the andesite appear at the surface in long, narrow bands, ranging from 20 feet down to less than 1 foot in width, and usually standing from a few feet to 20 or 30 feet above the adjacent andesite. (Pl. XVIII, A.) These are true veins or fissure deposits. The fissures which they fill are usually somewhat curved, tapered at one or both ends, are almost invariably branching, and are accompanied by subsidiary parallel fissures. (See Pl. III.) Their orientation is diverse; in general they follow the directions of the adjacent jointing and faulting.

ORE DEPOSITS AT ANDESITE-LIMESTONE CONTACT.

The larger and more numerous deposits are along the andesite-limestone contact. (Pls. III, V, and XVII, A.) As exposed in the erosion surfaces they are commonly lens shaped, with their longer diameters parallel to the contours of the hills, but from this there are important variations toward irregular polygonal shapes, due partly to faulting and partly to the variation in the angle between the erosion surface and the plane of the andesite-limestone contact which the ores follow. The deposits at the contact have as a hanging wall either the fresh limestone or the silicated phase characteristic of the contact with the andesite. The ore protrudes irregularly into the limestone in large and small masses and veins. Small masses of the ore, measuring from a few inches to a few feet, may be seen entirely within the limestone, and in turn fragments of limestone are found in the ore. Along fault planes the limestone is brecciated and cemented by ore. Notwithstanding this local irregularity, measured by inches and a few feet, the contacts on a large scale are usually even and continuous. The dip of the contact of the ore and the hanging wall is almost invariably steeper than the dip of the bedding of limestone; that is, almost vertical but with a slight dip away from the andesite, so far as can be determined from the sections exposed.

The foot wall of the ore is principally andesite, but at many localities the ore is separated from the andesite by a thin layer of the silicated contact phase of the limestone or the sandy basal phase of the limestone. The contact of the ore with the foot-wall andesite as a whole is considerably more regular than that with the hanging-wall limestone. There is less interpenetration of the two masses, yet occasional fragments of andesite protrude into the ore or are entirely surrounded by it, and andesite breccias with ore cement are not uncommon along faults. Andesite dikes or offshoots are rare in the ores and limestone near the contacts, but are known in one locality east of Iron Mountain. The andesite near the contact is altered to

a soft clay retaining andesite texture.   The contacts may be vertical
or inclined, but are commonly somewhat steeply inclined away from
the andesite.

These simple relations of ore to wall rock, are complicated to a considerable extent by faulting, probably to a larger extent than has

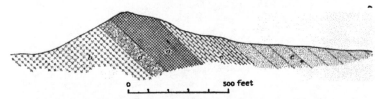

FIG. 4.—Cross section of Desert Mound contact deposit. a, Iron ore; b, laccolithic andesite; c, Homestake limestone; d, altered Homestake limestone; e, Pinto sandstone.

been proved.   Because of the faulting the ore may be nearly or quite
surrounded by andesite or by limestone or by any combination of
these rocks.   The map (Pl. II) indicates the effect of the faulting on
the surface distribution.   Its effect is probably equally marked on
the third dimension.

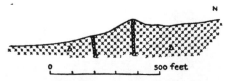

FIG. 5.—Cross section of Great Western fissure veins. a, Iron ore; b, laccolithic andesite.

The faulting is in considerable part earlier than the ore deposition,
as shown by the fact that the fault breccias are cemented by ore.
The Desert Mound and the Marshall claim in The Three Peaks area
afford good illustrations.   Other faults are distinctly later than the
deposition of the ore, as on the Chesapeake claim and others on the

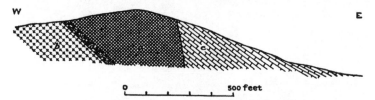

FIG. 6.—Cross section of Lindsay Hill contact deposit. a, Iron ore; b, laccolithic andesite; c, Homestake limestone; d, altered Homestake limestone.

slope and on the top of Iron Mountain.   The earlier and the later
faulting are not certainly to be distinguished in all places in the present state of development of the deposits, for the structural relations
developed are in part similar in the two cases.   The age of the late

. SOUTHERN CROSS IRON-ORE DEPOSIT, EAST OF GRANITE MOUNTAINS, LOOKING NORTH.

he ore forms the summit and the dark area on the slope to the right and foreground. The light area to the
left is foot-wall andesite. Hanging-wall sediments occupy the lower half of the slope to the right.

B. CHESAPEAKE IRON-ORE FISSURE VEIN, IRON MOUNTAIN. (BOUNDED BY ANDESITE.)

A. GREAT WESTERN IRON-ORE FISSURE VEIN, SOUTHERN PART OF THE THREE PEAKS
LACCOLITH, LOOKING NORTH.

B. IRON-ORE BLOWOUT SOUTH OF IRON MOUNTAIN LACCOLITH, LOOKING SOUTHEAST.

Crystalline magnetite constitutes the hill in the middle foreground. In the distance are the Harmony
Mountains.

faulting of the ore deposits is provisionally assigned to the postlava period, because this has been a period of considerable faulting throughout the district.

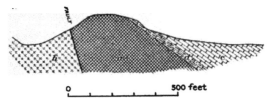

FIG. 7.—Cross section of Blowout contact deposit. *a*, Iron ore; *b*, laccolithic andesite; *c*, Homestake limestone; *d*, altered Homestake limestone.

The shape of the deposits in vertical cross section is incompletely known because exploration has been shallow. Available informa-

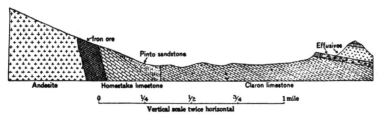

FIG. 8.—Cross section southeastward through Lindsay Hill, showing structural relations of ores.

tion is summarized in figs. 4, 5, 6, 7, 8, and 9. Inferences drawn from the manner of development of the ore are discussed on pages 76 and 85.

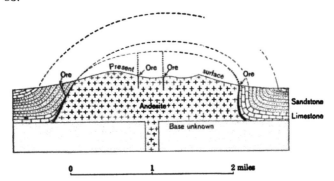

FIG. 9.—Ideal cross section through Iron Mountain laccolith, showing structural relations of ores.

ORE DEPOSITS IN BRECCIAS.

The ore constitutes cements or minute veins in fault breccias formed by the Homestake limestone, Pinto quartzite, or andesite at several localities, as follows: Limestone and quartzite at the Milner mine, Dear, Excelsior, Duluth No. 2, Desert Mound; andesite at the

Marshall, Blowout, Dexter, and Pot Metal claims. At the Desert Mound a fault breccia crosses andesite, ore, and limestone, and magnetite constitutes breccia fragments.

The iron ores lack the associated "iron formation" of ferruginous chert or jasper so characteristic of the Lake Superior iron ranges. The nearest approach to jasper is in ores banded parallel to the walls in the fissure veins in the andesite, and in contact ores with a banding representing original limestone bedding.

The ore deposits nowhere come in contact with the later effusives. However, it is a significant fact, to which attention will be directed in discussion of origin of the ores, that the principal ore deposits are approximately on the level of the general surface upon which the lavas rest, sometimes above it but never below, and that the effusive rocks before erosion must have rested upon the present deposits. (See fig. 9.)

KINDS AND GRADES OF ORE.

The following description applies to the ores as they appear above water level. Pits have not yet been sunk below this depth.

· The ore is mainly magnetite and hematite, usually intimately intermixed, but locally segregated. So far as present information goes (and it does not go far below the surface) the magnetite constitutes about 70 per cent and the hematite 30 per cent of the whole. As hematite appears more abundantly below the surface, it is thought likely that deeper exploration will develop a higher percentage of hematite. At the surface the ore is ordinarily hard crystalline magnetite and hematite in porous, gnarled, and contorted masses, with coarsely crystallized quartz and fibrous chalcedony as the principal gangue mineral, filling, wholly or partly, cavities in the ore. Other gangue minerals occurring in small and practically negligible amounts are apatite, mica, siderite, diopside, garnet, pyrite, chlorite, calcite, barite, galena, amphibole, copper carbonates, limonite, and amethyst. Of these minerals barite and galena are more closely associated with the limestone than with the ore. Melanterite, associated with pyrite, was found in process of formation in the long tunnel on the Duncan claim. Beneath the surface the ore is usually softer and contains a larger proportion of soft, bluish, reddish, brownish, grayish, and greenish banded hematite, limonite, and magnetite in greatly varying proportions and relations. The gangue materials are more abundant than near the surface, and calcite is in relatively increased proportion as compared with the quartz. The banding in the contact ores partly represents the bedding of the limestone, which, as will be shown later, the ore replaces. Banding in the dike or vein ores in the andesite is of unknown origin, possibly the result of original deposition. Some of the softer ore at lower levels entirely lacks *this* banding. Locally, as on the west side of Lindsay Hill, the

A. WAR EAGLE IRON-ORE DEPOSIT, NORTH OF IRON SPRINGS.

B. IRON-ORE CONTACT DEPOSIT AT DESERT MOUND, LOOKING WEST.

Iron ore forms the main hill. To the right is foot-wall andesite; to the left Homestake limestone. The lower slope in the foreground is composed of Homestake limestone and Pinto sandstone.

contact ore contains parallel streaks of a yellow clayey-looking material. On examination this resolves itself into a mixture of iron carbonate, iron sulphate, and glass, and probably some residual clay. Some of the narrow ore veins in the andesite possess a comb structure formed by the meeting and interlocking of apatite crystals projected from the walls (see fig. 10), sometimes not entirely closing the vein.

In the ore breccias the cements are magnetite, limonite, calcite, and quartz. At the Milner mine and elsewhere the magnetite has been deposited first about the fragments, here consisting of quartzite, then hematite, then limonite, but exceptionally in the same locality the reverse order appears.

The texture of the ore as a whole is good for furnace use. The harder ores will need crushing.

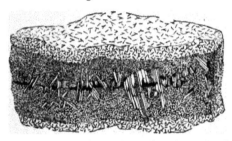

FIG. 10.—Vein of magnetite in andesite. The ore, indicated by the fine dots, does not quite fill the opening. Crystals of apatite penetrate the ore and interlock across the opening. One-half natural size.

For the following information concerning the composition of the ores the writers are indebted to Mr. Fred Lerch, of Biwabik, Minn., and to Mr. R. N. Dickman, of Chicago, Ill., both of whom have exhaustively sampled the ores of the district for commercial purposes. Corroborative figures were obtained from other commercial sources, and in a few cases analyses were made for the writers. In all, about 200 analyses from 400 samples of ores have been available, about two-thirds of them containing determinations only of iron, silica, and phosphorus, and one-third showing the percentages of all the common elements.

The average composition of the ores of the Iron Springs district, determined by combining all available analyses of the ores of the district from surface and pits, is as follows:

Average composition of iron ores of the Iron Springs district.

	Per cent.		Per cent.
Iron	56	Copper	0.027
Silica	7	Sulphur	.057
Phosphorus	.200	Manganese	.196
Lime and magnesia	4	Soda	1.19
Alumina	1	Potassa	.80
Water	3		

The samples run as low as 45 per cent in iron and as high as 69 per cent. The hard ore in the andesite runs higher on the average than the ore in the limestone. The ore at the surface, with few exceptions, has a higher percentage of iron than that below, the difference ranging from 3 to 12 per cent.

Phosphorus may diminish slightly in the deepest explorations, that is, below about 100 feet, but the distribution is so irregular and capricious that this generalization is doubtful. There are common variations within a short distance, both vertical and horizontal, of 0.050 to 3.18 per cent. A few 10-foot samples of ore run below the Bessemer limit in phosphorus, but practically all the ore as mined will be non-Bessemer ore.

Silica, averaging about 7 per cent, varies between 2 and 28 per cent, the lower figures being more common in the ore in the andesite. There is a distinct loss of silica at the surface, amounting to about 4 per cent as compared with that immediately below the surface.

Lime and magnesia range from 1.5 to 11 per cent. The hard ores in the andesite carry slightly less than the soft ores in the limestone. In both types the deeper ores carry the higher percentage. While both silica and lime are in greater quantity below the surface, corresponding to a lower percentage of iron, the lime and magnesia increase relatively faster than the silica with depth. Whereas at the surface a common ratio of silica to calcium and magnesium oxides is 2 to 1 by weight; below the surface it is more nearly 1 to 1.

Combined water varies from less than 1 per cent in the magnetite to 4 per cent in the soft ores, averaging about 3 per cent. One determination of moisture in crystallized magnetite[a] gives 0.45 per cent after heating to 110° C.

Sulphur is present in variable amounts, averaging .057 per cent, but in the deep exploration of one of the deposits, the Duncan, this figure is exceeded and the presence of sulphur becomes a serious consideration. There seems to be no general evidence of increase with depth, so far as exploration has yet gone, but water level has not yet been reached in the explorations.

Copper, titanium, and manganese are present, but not in injurious amounts.

Soda and potassa are determined in a single specimen.[b] Their significance is discussed on page 77.

Pl. XX summarizes the available information in regard to the variation of composition with depth.

A comparison of average analyses of Iron Springs iron ores with Lake Superior hematites and with the Clinton hematites of Alabama is made in the following table:

---

[a] Specimen 46113, Lindsay Hill. Analysis by R. D. Hall, University of Wisconsin.
[b] Specimen 46333A, Lindsay Hill. Analysis by R. D. Hall, University of Wisconsin.

72

ce
th
ha
fr

is
ca
v
0.
B
b

tl
T
4

Comparison of average analyses of Iron Springs iron ores with ores of Lake Superior and Alabama.

	Iron Springs ores.	Lake Superior ores.[a]	Alabama hematites.[b]
Iron (metallic)	56	59.6	37
Silica	7	7.5	13.44
Phosphorus	.200	.067	.37
Lime and magnesia	4	1.3	16.2
Alumina	1	1.5	3.18
Water, above 220°	3	4.0	.50
Copper	.027		
Sulphur	.057	.019	.07
Manganese	.196		
Carbonic acid			12.24

[a] Average cargo analyses for 1905.

[b] Birkinbine, John, The iron ores of Alabama (average analysis by Dr. William B. Phillips): Nineteenth Ann. Rept. U. S. Geol. Survey, pt. 6, 1898, p. 62.

It will be noted that the Iron Springs ores are intermediate in composition between the two other great classes of ores.

SIZE AND QUANTITY OF ORE DEPOSITS.

The iron-ore deposits vary from mere stringers to those having an area of 1,670,000 square feet. The aggregate surface of all the ore deposits of the district is 5,430,000 square feet or 0.2 square miles.

The aggregate tonnage of all grades of ore in the district, determined by multiplying the known area by the best available information as to depth in pits, drill holes, and erosion sections is 40,000,000 tons. The largest single deposit, figured on the same basis, has 15,600,000 tons. It is altogether likely that the figures are much too small rather than too large, because the depths used in the calculation have been those actually observed, and observation has not yet gone to the bottom. Inferences concerning the extension of the ore beneath the present workings, based on the manner of origin of the ore, are discussed on pages 87–89.

DEVELOPMENT.

During the years 1874 to 1876 a small furnace, with a daily capacity of 5 tons, was built and operated at Iron City, 5 miles southwest of Iron Mountain. The product was taken to the then prosperous silver mining camp at Pioche, Nev., and to Salt Lake, Utah. Later the old stack was torn down and a new one, projected to take its place, never rose higher than the foundation. The coal was derived from the Harmony Mountains, 5 miles to the southeast. The ore used in this furnace was taken out of the Duncan claim, one of the southernmost exposures of ore in the Pinto groups of claims, from shallow pits and short tunnels near the surface.

From time to time since the discovery of the deposits, pits and tunnels have been sunk in the ore, principally to meet assessment requirements, but partly to show up the ore bodies. Some of the more vigorous exploration was conducted during the years 1902 and 1903. The total number of pits sunk to date has been approximately 1,600, of which 30 have gone to a depth greater than 50 feet. The maximum depth has been 130 feet.

Diamond drills have been used only at the Desert Mound, where 5 holes have been sunk. The detailed results of this drilling have been withheld from the United States Geological Survey, though some of the cores have been examined.

A. CONTACT PHASE OF PINTO SANDSTONE WEST OF IRON MOUNTAIN.

B. RETICULATED IRON-ORE VEINS IN ANDESITE.

The dark bands are iron ore protruding on the weathered surface.

CHAPTER VII.

ORIGIN OF THE IRON ORES.

GENERAL.

The principal ore deposits, viz, those near the contact of the andesite and limestone, are partly replacements of limestone. The original bedding of the limestone has been preserved in the ore in a number of places, and there is gradation between the ore and the limestone. These deposits are also in part fillings of fissures in limestone or between limestone and andesite. Where ore occurs within the andesite it fills fissures. The source of the iron-bearing solutions is the same for the limestone replacements and for the vein fillings in the limestone and in the andesite, for their mineralogical and textural characters are the same and in a few cases they are actually connected. Several hypotheses as to this source have suggested themselves: (1) That the ore-bearing solutions were associated with the intrusion of the andesite as "igneous after-effects;" (2) that they were meteoric waters, cold, or heated by contact with the laccolith, acting after the laccolithic intrusions and before the eruption of the surface flows; (3) that they were hot solutions, magmatic or meteoric or both, connected with the late eruptives of the district, deriving the ores from the effusives or from the underlying rocks; (4) that they were cold meteoric waters later than the effusives; (5) that they were due to some combination of these sources. The source of the ore is best explained by the first hypothesis, but later concentrations of the ore have occurred in the order named.

CONCENTRATION AND ALTERATION.

DEPOSITION OF ORE FOLLOWING LACCOLITH INTRUSIONS AS IGNEOUS AFTER-EFFECTS.

INTRODUCTION OF THE ORE.

The general association of the ores with the andesite and their specific association with fissures and faults in the andesite and the immediately adjacent limestones, the nature of the ores and gangue materials, especially the primary association of magnetite with garnet, amphibole, pyroxene, mica, apatite, iron sulphide, and glass, seem to allow of but one general interpretation, and that is that the ore-bearing solutions were hot, rising from a deep-seated source through fissures in the andesite now filled with ore, at a period closely following the crystallization of at least the outer part of the laccolithic mass.

That the ore was introduced after the hardening and crystallization of the andesite and not before is shown not only by its occurrence in clear-cut fissures in the andesite and by the metamorphism of the andesite near the ore, but also by the lack of anything in the nature of a basic edge in the andesite, by the lack of irregularity in its composition, by the absence of ore for considerable intervals along the andesite contact, and by the fact that the intrusion of the andesite metamorphosed the limestone in a clearly recognizable manner, recrystallizing it, decarbonating it, rendering it more siliceous, and indurated the Pinto sandstone to a quartzite spotted by the segregation of ferrous iron in the form of amphibole—all before iron was introduced.

The ore-bearing fissures within the andesite (see p. 67) are of a kind which results from cooling and stretching. They are curved, tapered, branching, and in parallel sets, some of them minutely parallel and anastomosing (see Pl. XXI, B). The contraction accompanying the crystallization of the lava may also account for the ore-bearing fissures and faults following the periphery of the andesite, and affords a good explanation for the occurrence of ores at the contact in lens-shaped masses with their longer diameters parallel to the contact. The crystallization of a laccolith the size of the Iron Mountain laccolith from a viscous or glassy condition would yield a radial shortening of 200 to 500 feet, depending on the depth assigned to the laccolith. The parting of the andesite from the limestone during cooling finds its analogue in the parting of a casting from the mold.

The association of the ores with heavy anhydrous silicates, characteristic of slow-cooling intrusions, eliminates the possibility of development from the later effusives, which develop at their contacts minerals of a different kind and association, listed on page 85.

It is thought likely that the ore, at least the part within the zone of observation, was originally deposited as magnetite rather than as sulphide. Had it originally been sulphide, the subsequent alteration to magnetite would scarcely have left unchanged the closely associated silicates. The magnetites are nowhere observed to pass down into sulphides, although the ground-water level has not been reached. Until it has been reached, statements regarding the sulphides must be regarded as tentative.

Assuming, then, in the absence of negative evidence, that the iron was deposited primarily as magnetite, the iron may be supposed to have been carried as a ferrous compound of some kind, possibly chloride or carbonate or sulphate, and its deposition as magnetite required partial oxidation. At the time of the intrusion the covering of the laccolith was probably about 4,000 feet thick, which would give little opportunity for the entrance of oxygen from the air. The expulsion of the carbon dioxide from the limestone which preceded the deposition of the ore may also have aided in keeping out oxygen. The probable alternative source of the oxygen is that derived by the breaking up of the water. Ferrous chloride reacts with water at temperatures above 500° C., with the simultaneous development of magnetite, hydrochloric acid, and free hydrogen, as follows:[a]

$$3FeCl_2 + 4H_2O = Fe_3O_4 + 6HCl + H_2 + 77Cal._g$$

As magnetite is soluble in hydrochloric acid, there could obviously be no precipitation as long as this acid remained, but it is neutralized by the limestones which the solutions meet and the ore precipitated. Magnetite has abundantly replaced limestone in this district, as shown by the retention of the bedding structure of the limestone. Interchange of the ferrous iron with the lime would also precipitate the iron as iron carbonate, which is found associated with the magnetite. Had the iron been carried in ferric solution, limestone would have precipitated it as hematite. Possibly the hematite found in small amount in the ore may be so explained. Magnetite would also be precipitated by the evaporation and dispersion of the solution, which would result from the lowering temperature and pressure as the solutions left the andesite. Hydrochloric acid is easily volatilized with water, making it difficult to dehydrate a chloride without losing chlorine. So far as hydrochloric acid was lost, magnetite would be deposited. Kahlenberg[b] regards this process as entirely adequate to accomplish abundant precipitation of magnetite under the stated conditions. The existence of magnetite in dikes in the andesite or as cement in quartzite breccias, where limestone, as an alternative precipitating agent, is not present, is perhaps to be explained by this method of precipitation.

One of the conspicuous features of the contact metamorphism of the limestone and the metamorphism of the andesite adjacent to the ores is the introduction of soda. The ore of the first concentration itself shows a dominance of soda over potassa. These facts suggest that soda has been introduced not only into the limestone during its metamorphism but also later with the iron. It would be easy to explain the transportation of this salt in the chloride

a Moissan. Henri: Traité de chimie minérale, vol. 4, 1905. p. 330.

b Personal communication, Louis Kahlenberg, Univ. Wisconsin, 1907.

solution above hypothesized for the carrying of the iron. Lindgren finds salt in inclusions in vein quartz with a probable igneous origin. This most volatile of all the common mineral compounds is an abundant emanation product of volcanoes. It may be noted that sodium chloride would be decomposed by reaction with silicic acid undoubtedly met in the limestone contact, precipitating the sodium in its present silicate combination.

The introduction of apatite, garnet, amphibole, and pyroxene, all of them containing lime, and the greater abundance of these minerals nearer the limestone than elsewhere, suggests reaction between the hot solution and the limestone. It is not certain, however, that the limestone was necessary as a precipitating agent for the reason that these minerals are also found in fissures in the andesite entirely away from the limestone. The deposition of silicates may equally well have been a function of the change of temperature of the solutions. The order of deposition of the principal minerals of this first concentration seems to have been: (1) Magnetite (and other oxides), pyrites, and amphibole, apatite, and garnet; (2) diopside.

Interlayered with the magnetite and closely associated with glass is amorphous iron carbonate, the genetic relations of which to the other minerals are not known.

SOURCE AND CONDITIONS OF THE ORE-BEARING SOLUTIONS.

Judging from records available elsewhere as to flow of meteoric waters with depth, the 4,000 feet of rock covering the laccolith probably prevented ready access of abundant meteoric waters. Such as were present in the rocks may have reached high temperatures adjacent to the laccolith, and may have aided in the ore-depositing processes. But that the principal source of the solutions was the hot andesite magma seems to be implied by the nature of the minerals deposited and their association, and manner of association, with the laccolith. These solutions were ejected after at least the outer parts of the laccolith had crystallized.

That the temperature of the solutions was high is clearly shown by the nature of the materials deposited. The temperature was above the critical temperature of water—365°—judging from experimentally determined temperatures necessary for the crystallization of certain silicates similar to those here found.[a] A temperature higher than the critical temperature is further indicated by the presence of glass in the ores locally. From the fact that the fusion of the contact phase of limestone probably took place at temperatures ranging upward from 1,000° (see p. 36) and that the ore is in veins in this

---

[a] Allen, E. T., Wright, F. E., and Clement, J. K., Minerals of the composition $MgSiO_3$; a case of tetramorphism: Am. Jour. Sci., 4th ser., vol. 22, 1906, p. 399.

phase and does not replace it, it may be inferred that the ore-bearing solutions were introduced into the limestone after its temperature had fallen below 1,000°. There is nothing to show that the temperature of the wall rock may not have been much lower, but the introduction of the ore followed so closely that the presumption is that the temperature may not have been much below 1,000° C. The rock pressure above the laccolith, amounting at a maximum to the weight of 4,500 feet of rock, was sufficient to hold the solution in a liquid state at temperatures of less than 365°. The hydrostatic pressure, which was probably the only effective one, was sufficient to hold the solution in a liquid state below temperatures of 325° C.

It is concluded that the solutions were probably after-effects of the andesite laccolith intrusions.

It is perhaps better to refer the solutions vaguely to "after-action" than to attempt more specifically to indicate whether they have come from the andesite itself or from some deep reservoir common to the andesite and the solutions. The abundant presence of magnetite in the andesite—3½ to 4 per cent—as so-called "reaction rims" about biotite, and to a less extent about hornblende, suggests a possible connection between the andesite and the ore-bearing solutions. The occurrence of magnetite in these relations has been held by Washington[a] to indicate the instability of the biotite and hornblende under surface conditions and their breaking down into paramorphic magnetite and augite, the latter being less abundant about the biotite than about the hornblende, and thus unimportant or altogether lacking in the Iron Springs rocks. By others the magnetite rims have been explained as due to absorption (solution) by the surrounding magma. The abundance of hydrogen gas (see p. 81) suggests that the development of the magnetite has resulted from the breaking down of water. But whatever the origin of the rims, it is apparent that the separation of the magnetite has occurred late in the cooling of the andesite. It would simplify matters greatly if this segregation of magnetite could be shown to be definitely related in some way to the ore-bearing solutions coming out of the andesite, but the writers know of no way to do this.

CONCENTRATION OF ORE BY WEATHERING PRIOR TO ERUPTION OF TERTIARY LAVAS.

When erosion had uncovered the ore deposits, it is reasonable to infer that the same processes of concentration that may be observed to-day were effective. These are described on pages 82–84.

---

[a] Washington, H. S., The magmatic alteration of hornblende and biotite: Jour. Geol., vol. 4, 1896, pp 257–282.

Inference of the modification of the ores by contact of the late lavas is to be drawn from the topographic relations of the ores and lavas. The lavas at one time rested directly against the andesitic cores of the mountains, upon the eroded edges of limestones and other sediments, and upon the ores. The principal ore deposits on the lower slopes of the hills are exposed at about the same elevation as the lower part of the lavas. Erosion has cut the lavas back, so that they now fringe the ore-bearing areas. (See Pl. II.) The heated waters associated with the lavas flowed along the andesite slopes where the ores occur.

Under the influence of the lavas and accompanying solutions, limonite was dehydrated, limonite and hematite were deoxidized (a process easily brought about in the laboratory by passing steam over such ores) and all of the minerals were coarsely recrystallized, making the upper parts of the deposits nearest the lavas consist principally of coarsely crystallized magnetite and some hematite. Hot solutions from the lavas introduced new minerals into cavities in the ore then and previously formed. These minerals consist principally of quartz and chalcedony and to a subordinate extent of hematite, magnetite, siderite, limonite, chlorite, calcite, barite, galena, and the copper carbonates. The heavy anhydrous minerals of the early deep-seated concentration were not developed.

Immediately beneath the lavas at observed contacts with the Claron limestone, there has been introduced, obviously from the lava, an abundance of more or less iron-stained chalcedony, with a jaspery appearance, identical with that observed in the upper parts of the ore deposits. The lavas themselves contain abundant chalcedonic quartz, filling large and small cavities and also minutely disseminated through the rock so intimately as to suggest hot solutions accompanying and immediately following the cooling of the lavas.

To test the conclusion that the chalcedony, taken to be the chief contribution of the extrusions, was not really deposited from meteoric waters, determination was made of the gas content of the chalcedony associated with the ore, for which the writers are indebted to Rollin T. Chamberlin of the University of Chicago. This analysis is compared in the following table with analyses of gas content of the andesite of the Iron Springs district by Mr. Chamberlin, one of volcanic emanations by Fouqué, and one of crystalline rocks by Tilden.

Analyses of gas content of chalcedony associated with ore, of andesite, of volcanic emanations, and of the average of five crystalline rocks.

	A.	B.	C.	D.
H₂S	0.00	0.12	......	......
CO₂	13.93	13.93	0.22	34.104
CO	11.26	18.18	......	......
O₂	.00	.00	21.11	8.422
CH₄	4.00	3.63	.07	3.224
H₂	64.40	58.71	56.70	52.134
N₂	6.44	5.43	21.90	2.072
	100.03	100.00	100.00	99.958
Volume of gas per volume of rock	.82	.82	....	4.5

A. Specimen 46613. Chalcedony in ore. From Crystal claim, Iron Mountain, Iron County, Utah. Analysis by Rollin T. Chamberlin. 0° C. and 760 mm. pressure.

B. Specimen 46612. Andesite from Granite Mountain, Iron County, Utah. Analysis by Rollin T. Chamberlin. Same conditions.

C. Gaseous emanations from Santorin. Analysis by Fouqué. (Santorin et ses éruptions, p. 225.) When studied spectroscopically these gases were found to have traces of chlorine, soda, and copper.

D. Average of five crystalline rocks. Analysis by Tilden. (Chem. News, November 9, 1897.) Standard conditions.

The gases in the chalcedony associated with the ore are roughly similar in their proportions to those of igneous rocks and volcanic conditions and differ generally in their proportions from the gases of the atmosphere. The gas analyses therefore furnish corroborative evidence of the introduction of the chalcedony directly from the late lavas.

That iron was introduced into the underlying rocks by the hot lava solutions is shown by its intimate relations with the chalcedony in the jaspery phases at observed contacts of the lava and by its occurrence in crystalline masses with chalcedony, partly or wholly filling cavities in the ore. The lava masses themselves are locally rich in iron. In the Antelope Springs area samples of the lavas (not veins) run as high as 6.5 per cent metallic iron. A few veins of magnetite and hematite are found in the lavas, which may have come from late lava solutions or from meteoric waters acting subsequently. The microscope discloses reaction rims about the biotite (as in the andesite, p. 48) formed during and following the flow of the lavas. The segregation of iron salts in the magma is further demonstrated by schlieren in the groundmass of the lavas, some of them light colored and some of them darker colored owing to a higher content of iron, and still further by the occasional segregation of magnetite or siderite about amygdaloidal cavities, both as fillings and as segregations in the adjacent groundmass. It is easy to conceive that iron salts got into the aqueous solutions which were in the lava at this stage of the cooling.

The introduction of calcite, quartz, siderite, limonite, galena, barite, and copper carbonates from the lavas is well shown at Chloride Canyon, where immediately below the contact these minerals appear in veins or disseminated through the rock. Chlorite is not found here, but is present elsewhere in the same associations and was probably also introduced by the lava waters.

So obvious and striking are the effects of the extrusion of the lavas upon the principal ore deposits of the lower slopes of the laccoliths that the question was raised in the field study whether this concentration could not have been the principal and perhaps the only one, but the evidence seems to be conclusive that the concentration under the influence of the Tertiary lavas was after all relatively slight. Limestones in contact with the lava, but away from the andesite, do not carry iron-ore deposits, although reddened and silicified by solutions from the overlying lava. The minerals (barite, galena, copper carbonates, etc.) deposited in the ore by these solutions are distinctly later than the main mass of the ore, which is intimately associated with heavy anhydrous silicates characteristic of deep-seated intrusions. The coarse recrystallization of the ore under the lava influence seems to be shallow and the recrystallized ore to be superposed upon ore of finer and softer texture.

It seems a reasonably safe conclusion that during and following the lava extrusions both hot meteoric waters and waters contributed by the lavas flowed down the andesite slopes, that they were intermingled, and that the results of their work are not to be closely discriminated. The contribution of minerals from the lavas, however, would seem to require emphasis on the effectiveness of the waters from them.

ALTERATIONS SUBSEQUENT TO TERTIARY ERUPTIONS.

Observation has not yet gone below water level, so that a comparison of the ores of the weathered and unweathered zones can not be made. In general the characteristics of the ore are determined by conditions other than weathering. However, the ores as a whole are more porous near the erosion surface than below in pits. There is less calcite above than there is below. It is inferred that there has been leaching of the calcite above and perhaps redeposition below. The calcite is partly in fine granular form, incrusting ore and rock surfaces, and is similar in appearance to carbonate seen about some of the old vents of hot springs in this district. The possibility is therefore suggested that the solution and redistribution of the calcium carbonate went on partly through the agency of warm waters during the period of the cooling of the lava. However, redistribution is yet going on, through the agency of cold meteoric water, with sufficient rapidity to incrust rapidly changing erosion surfaces.

The magnetite commonly alters to limonite and hematite, in thin incrusting films, not directly at the surface, but in the solution cavities and in fissures near the surface. This alteration seems to be at a maximum near the contact of ores and the adjacent rocks at the surface. Iron sulphide is changed to limonite and melanterite, some of which is removed in solution. Iron carbonate alters to limonite.

Apatite has been altered to osteolite and leached in the upper parts of the ores to a very considerable extent. Frequently the entire crystal of apatite has disappeared, its former presence being indicated only by the shape of the cavity. Presumably this apatite is redeposited below. This does not necessarily mean that the ores beneath the surface will run higher in phosphorus than at the surface, for the reason that there may be concentrated at the present surface a large part of the phosphorus which has come from the erosion of the overlying materials—a relatively larger amount of phosphorus than has been leached from the present surface and carried farther down.

Erosion has cut down the ore deposits many feet, as shown by abundant ore débris on the slopes. The ore is fissured. Extreme temperature changes spall off considerable blocks bounded at the sides by fissure planes. The ultimate product is magnetic sand, abundantly found on the lower slopes. For most of the district this erosion has gone on much less rapidly in the ore than in the adjacent andesite and limestone, with the result that the ore stands up in conspicuous black masses above the surrounding rocks. In a number of places, however, as on the Vermilion, Lindsay, Enterprise, part of the Mount Lion group, Comstock, Sunbeam, Wellington, Queen of the West, Black Hawk, Pinto mine, Pinto Nos. 3, 5, and 6, Burke No. 5, Red Clouds, and Duncan No. 1 claims, the ore has been cut down flush with the surrounding rocks. In these places the ore is uniformly softer than where standing up in conspicuous crags. It may be that the hard crystalline surface ores, serving more or less as protecting caps, have been locally undermined and cut off, leaving the underlying softer ore unprotected, with the result that it is cut down fully as rapidly as the adjacent rocks.

There is no evidence that weathering contributed to the deposits any considerable amount of ore from adjacent rocks. During erosion the magnetic iron of the andesite is concentrated into magnetic sands, as is the magnetite derived from the disintegration of the ore deposits. If alumina be assumed to remain constant during weathering it will be apparent from the analyses of the andesite that a small percentage of iron has been lost. With this percentage of loss it would require the weathering of a mass 100 feet thick and 0.2 square mile in area to yield a million tons of 56 per cent ore.

Iron is only slightly leached from the limestone during weathering, all but a minute part remaining in the residual clay. The residual soil from the Homestake limestone has been removed and the ore deposits lack clay characteristic of residual deposits. If the slight amount carried off during weathering be assumed to be available for the development of ore deposits, there seems to be no reason why limestone of this low percentage of iron should yield sufficient ore for ore deposits in the particular localities where they now occur and not in others where the limestone contains as much iron.

SUMMARY OF MINERAL ASSOCIATIONS OF ORES IN RELATION TO ORIGIN.

The first effect of the andesite laccolith intrusions was the development of the silicated contact phase of limestone containing the minerals of column 1 in the following table. Slightly later, veins containing much the same association of minerals cut andesite and limestone (column 3). About the same time came the introduction of ore-bearing solutions in veins both in andesite and limestone and also replacing limestone (column 2). The minerals deposited are much the same as those previously developed in the contact limestone, though albite and orthoclase, present in the contact phase, have not been found in the ore, while apatite and garnet are more abundant in association with the ores than elsewhere in the contact phase. So similar are the groups of minerals developed up to this point and so close their association that they can not be sharply separated. There can be little doubt that they are developed under no greatly varying conditions with insignificant time intervals. According to Lindgren's classification,[a] the minerals of these groups are characteristic both of products of aqueo-igneous solutions, like pegmatites, and of products of aqueous solutions, in the lower contact zone. Criteria do not seem to be available in the Iron Springs district for clearly separating the two classes. The evidence appears to indicate that they all result from the andesite intrusion, principally through the transfer of pneumatolytic vapors, but that in the limestone contact they are also developed by simple elimination of lime and magnesia, and the recrystallization of the residue.

Later, solutions from the lavas introduced another and clearly distinguishable group of minerals, listed in column 5. These are found principally near the contact of the lavas with the prelava erosion surface, and were developed, therefore, under essentially surface conditions. They correspond closely to the minerals listed by Lindgren[a] as characteristic of these conditions.

a Lindgren, Waldemar, The relation of ore deposition to physical conditions, Econ. Geol., vol. 2, 1907, pp. 122-125.

Mineral associations of ores.

1.	2.	3.	4.	5.	6.	7.
Minerals developed in limestone at contact of andesite intrusive.	Minerals of first ore concentration under influence of andesite intrusive.	Minerals in veins in contact limestone and andesite.a	Minerals of late ore concentration under influence of effusives.	Minerals of same age and origin as 4 but occurring in late veins in limestones.	Minerals in veins and amygdules in lavas.	Weathering minerals.
Quartz......	Quartz......	Quartz......	Quartz, chalcedony, etc.	Quartz, chalcedony, etc.	Quartz, chalcedony, opal, etc.	
Calcite......	Calcite.......	Calcite......	Calcite.......	Calcite.......	Calcite......	
Diopside.....	Diopside.....	Diopside.				
Andradite (?).	Andradite...	Andradite.				
Apatite (?)..	Apatite......	Apatite......	....	....	....	Osteolite.
Actinolite....	Actinolite....	Actinolite...				
Magnetite ...	Magnetite ...	Magnetite ...	Magnetite ...	Magnetite.		
Hematite....	Hematite....	....	Hematite....	....	Hematite....	Hematite.
Limonite....	....	....	Limonite....	Limonite....	....	Limonite.
Albite.						
Orthoclase.						
Wollastonite (?).						
Phlogopite...	Biotite......	Phlogopite.				
Serpentine.						
Kaolin.......	Kaolin.					Kaolin.
Andalusite.		Epidote.				
	Pyrite......	....	....	Pyrite.		
	Titanite.					
			Chlorite.b	Copper carbonates.		
			Copper carbonates.	Siderite.		
			Siderite.	Barite.		
				Galena.		
				Chalcopyrite.		

a Slightly later than those of column 1 and of about the same age as ores, but containing no ore.
b Found in one locality to depth of 80 feet in porous ore, but probably still due to action of effusives.

CONCLUSIONS AS TO THE ORIGIN OF THE ORES.

If the foregoing reasoning is correct, the events leading up to the completion of the ore deposits in their present form are in outline as follows:

(1) Intrusion of andesite laccoliths in Paleozoic and Mesozoic sediments, with consequent tilting of the strata in quaquaversal manner about the laccolith and contact metamorphism of the zone adjacent to it, accompanied and followed by fissuring, jointing, and faulting.

(2) Entrance of hot ore-bearing solutions through fissures in the andesite into the adjacent sediments, depositing ore as dikelike masses in fissures in the andesite, as fissure fillings and replacements in the limestone, and as cements in breccias of andesite, limestone, and quartzite. The solutions introduced also garnet, diopside, amphibole, phlogopite, apatite, calcite, quartz, and pyrite. Most of these minerals had also been developed in the limestone by the preceding contact metamorphism. Soda was conspicuously increased in the wall rocks. It is thought that the solutions were pneumatolytic after-effects of the andesite intrusion.

(3) Erosion, developing mountains with andesite cores, encircled by belts of sediments at uniform elevations on the slopes, except where displaced by faults, or where cut back by differential erosion. The areas between the mountains were left with low relief. The ores were exposed and partly eroded, calcite, apatite and perhaps other gangue materials were leached and redeposited below. There was more or less oxidation and hydration of the ores along fissures beneath the surface.

(4) Extrusion of the Tertiary lavas over the entire area, furnishing hot magmatic waters and heat to meteoric waters, and thereby developing coarsely crystalline magnetite and hematite in the ore deposits and especially at the surface, leaching the gangue materials so far as they were left by weathering near the surface, and depositing in the cavities chalcedony and to a slight extent magnetite, hematite, limonite, siderite, chlorite, barite, calcite, galena, and the copper carbonates.

(5) Erosion, reexhuming the andesite mountains from under the lavas and bringing to light the sediments and ores on the slopes; this was accompanied by local surface oxidation and hydration of the ores and leaching of the gangue materials, chiefly calcite, but also apatite. Differential erosion caused the ore in some cases to stand above the adjacent rocks, and in others, where the hard cap was cut off, brought the softer underlying ores well down to the level of adjacent rocks.

(6) Preceding and accompanying (5) occurred faulting of the ore deposits, developing structural relations not in all cases to be distinguished from those determined by faulting before deposition of the ores.

BEARING OF THE THEORY OF ORIGIN UPON FURTHER EXPLORATION.

Exploration has not yet demonstrated the depth of the ore deposits. The greatest known depth is at the Pinto group of pits, south of Iron Mountain, where 130 feet is known from the pits and topography. At the Desert Mound drilling may have gone deeper, but the records are not available to the writers.

If the theory of the writers as to the origin of the ore is correct, there are certain fairly well-based inferences to be drawn as to depth and shape of the ore deposits.

If the ore-bearing fissures in the andesite are true stretch fissures determined by the crystallization of the lavas, it may be inferred that they will show the same features with depth as at the surface; that is, curving, pinching out, branching and occurrence in parallel sets. There seems to be no reason why these vein deposits should not go to a very considerable depth, this being determined by the depth to which fissuring had occurred at the time of the extrusion of the ore-bearing solutions.

The ores at and near the andesite contact are determined in their shape and distribution partly by replacement of limestone, partly by peripheral tension fissures and other joints and faults, partly by faults later than the ore deposition, and finally by erosion. A calculation of the radial shrinkage of the andesite mass in cooling makes it between 200 and 500 feet, depending on the thickness assigned to it. A part of this shrinkage is probably accounted for in the stretch fissures and faults in the andesite. A part of it has with equal probability furnished space for ore deposition at the periphery. At the sides of the andesite these spaces would remain open during and following their development, whereas on the slopes and on top gravity would tend to close them as fast as developed. The deposits therefore should have their maximum width where the contacts are vertical or nearly so and should thin out where the slopes become flatter. The conception of the authors as to the relations of the fissures to the andesite before erosion took place is illustrated in fig. 8. This figure does not take account of modifications of shape of the ore bodies due to replacement of the limestone or due to fissures normal to the andesite periphery, which are believed to be subordinate. Erosion has now cut down sufficiently far on the eastern and southeastern sides of the several laccoliths to expose vertical or nearly vertical contacts. On the southwestern side of the Granite Mountain and Iron Mountain laccoliths the erosion plane has exposed a very gently dipping contact plane between andesite and limestone. The wide extent of the deposits at these places, expecially at the southwestern side of Iron Mountain, is largely due to the fact that the erosion plane is so nearly parallel to the plane of contact. It does not indicate that the deposits are necessarily wider when measured in a direction normal to the contact than they are elsewhere in the district; indeed, if fig. 9 represents approximately the shape of the openings which determined the ore deposition, the deposits here may well have been thinner than elsewhere. Under these conditions vertical faulting has greater effect on the surface distribution.

It follows from the above considerations that the maximum depth of the ore deposits near the contacts may not be greater than the depth to which the andesite-limestone contact extends, and this is determined by the thickness of the andesite laccolith and whether it breaks across the limestone layers or has tilted them up in such a manner that the contact is approximately parallel to the bedding. In view of the fact that the limestone is tilted steeply where erosion has exposed the sides of the laccolith, it may be assumed that the limestone may be in contact with the andesite as far down as the laccolith goes. Therefore the thickness of the laccolith becomes the determining factor. Erosion has allowed no direct means of measuring this thickness in this district, for the bottom is nowhere exposed. A comparison of the horizontal dimensions of these laccoliths with those described by Gilbert at the Henry Mountains, where the vertical depth also is known, leads to the conclusion that the depth of the "bell" of the Iron Mountain laccolith (the one with outlines best determined) may not exceed 2,500 feet.   The other laccoliths are so covered with desert deposits that their horizontal dimensions are not known, and hence even this means of determining their depth is not available.   Such comparison affords a very uncertain means of determining the depth of laccoliths in view of the fact that their shape and size depend so much on the nature and structure of the rocks intruded, as well as on the relative densities of the intrusive and intruded rocks.   (See pp. 47–48.)

For the deposits resting on gently-dipping slopes, such as those southwest of Iron Mountain, the maximum depth is likely to be found in the direction of dip.   Vertically they rest directly upon andesite or upon the contact phase of the limestone, except where they are above ore-filled fissures in the andesite, in which case the deposit is likely to narrow considerably in a vertical direction when it reaches the fissures.

It will be noted that on the western side of Iron Mountain the Cretaceous quartzite comes directly into contact with the andesite at an elevation considerably higher than the limestone-andesite contact on the southern and eastern sides of the laccolith where the ore is exposed Ores are found cementing brecciated quartzite, but do not constitute important deposits.   The lack of important ore deposits along this contact may be explained by the fact that replacement is not possible in the quartzite and that any fissure veins developed at the contact through the contraction of the andesite would be immediately closed by gravity, for the quartzite rests against the andesite with a dip of 35°. There is nothing in the supposed origin of the ores to preclude the possibility that ore deposits may be found beneath the quartzite along the limestone-andesite contacts, especially farther down on the steeper slopes of the andesite.   If deposits are there, it may be pointed out that their size has not been diminished by erosion.

The uniform association of ore with andesite laccoliths in this local-ity, as well as in the district extending southwest to the Bull Valley and beyond, outlines the first rule of exploration—that the andesite laccoliths be found and their boundaries determined.   The effect of the laccoliths upon topography is so marked that this may be done at a distance of many miles.   (See p. 16.)   From high points in the Iron Springs district it is easy to determine the southwestward extension of the laccolithic area, and therefore of the possible ore-bearing area.

The laccolith determined, its contact with the adjacent sediments should be carefully followed.

The location of iron-ore deposits usually makes itself evident on the weathered surface, but not infrequently the mass of débris which has shifted down the slopes makes it impossible to determine the true outline of the deposit without sinking pits. Where iron-ore fragments are found in the wash at the surface it may be assumed that they came from farther up the slope. Following the wash up the slope, a point will be reached where no iron fragments are to be seen, usually indicating that the upper edge of the deposit has been passed; but it has been found in a few localities that the overwash from the upper slopes has been so thick and heavy that it completely obscures the upper side of the ore deposits, and the first iron-ore fragments appear some feet or yards farther down the slope. The presence of a few fragments, or even of a single fragment, of iron ore in the wash is often sufficient to indicate the existence of an ore deposit several hundred yards away, which may be found by following up the wash and carefully watching its change in iron content.

Where the laccolith and sedimentary rocks are found in direct contact it does not necessarily follow that ore will not be found deeper down along the contact. Replacements are usually irregular, and the source of the solutions is not from above. So far as the ores are fillings of fissures developed around the periphery of the andesite by its crystallization, it is easily conceivable that these openings would be irregular, both horizontally and vertically; and that where the dip of the laccolith contact is low, gravity might close them altogether, so that the appearance or nonappearance of ore on the erosion surface would be determined merely by the extent to which the erosion surface had cut down.

In underground exploration the fact should not be lost sight of that the deposit may show the same range of irregularity with depth that it does on the erosion surface; in other words, that all the factors known to have determined the peculiarities and shapes of deposits on the erosion surface, which are discussed in this paper, should be taken account of in the underground exploration.

CHAPTER VIII.

COMPARISON OF IRON SPRINGS IRON ORES WITH OTHER IRON ORES.

BULL VALLEY DISTRICT.

The Bull Valley district lies about 25 miles to the southwest of the Iron Springs district, extending from Garden Springs on the northeast southwestward to Bull Mountain and 40 or 50 miles beyond. (See fig. 11.) The district, which has only recently been explored

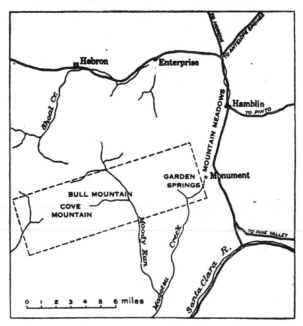

Fig. 11.—Sketch map of Bull Valley district, Utah.

and staked, is much more difficult of access than the Iron Springs district and is consequently less well known. It may be reached from the Iron Springs district by team by way of Pinto, Hamblin, and Enterprise, or on horseback by way of Pinto, Hamblin, Mountain

Meadows Valley, Magotsu Creek, and Moody Run. The nearest railway station is Modena, 28 miles distant by way of Enterprise. The area is covered by the St. George topographic sheet of the Powell Survey (1891), but the map is so imperfect and generalized that it is of little use.

The principal ore deposits lie several miles below the headwaters of Moody Run, which empties into Magotsu Creek 10 miles below the Mountain Meadows. From this point they extend eastward about 3 miles to Garden Springs, located a short distance west of the Mountain Meadows monument, the site of the famous Mountain Meadows massacre. To the southwest, deposits occur on Bull Mountain 2 or 3 miles distant and on Cove Mountain an equal distance beyond.

The essential geological features of the district are the same as those in the Iron Springs district—a series of laccoliths with sediments dipping quaquaversally away from them, surrounded and overlain by flat-lying lavas, the whole being bounded on north and west by later flows of basalt. The contour of the district is rougher than that of the Iron Springs district and the evidences of volcanism are more conspicuous on account of the presence of basalt flows and cinder cones. The general aspect is barren and forbidding. The same topographic and geologic conditions are said to extend for about 40 miles to the southwest into Nevada, and ores are reported from this area.

The ore deposits were examined in their discontinuous occurrence between the headwaters of Moody Run on the west and Garden Springs on the east, and were found to be similar in almost every feature to those of the Iron Springs district. The principal deposits lie within the andesite associated with limestone fault blocks, and subordinate ones follow the main contact of andesite and limestone, which crosses Moody Run in a northeast-southwest direction, dipping to the southeast. Flat-lying flows fringe the ore-bearing areas. On the south are acidic flows and tuffs, on the north acidic flows and tuffs and basalts. A white band near the base of the acidic lavas is very conspicuous and for most of the district is sufficiently near the limestone-andesite contacts to make it useful to explorers as a guide in locating the ores.

The greatest width of ore observed at the surface was 115 feet. However, it was not sufficiently well exposed to make it certain that this 115 feet was continuous ore. This particular deposit has a length of approximately 700 feet.

The iron is both magnetite and hematite, as in the Iron Springs district, but the hematite on the lower slopes takes on a fine granular texture and a steel-blue color which is not seen in the Iron Springs district.

The composition of the ore at the surface, as sampled by Lerch Brothers, is as follows:

Composition of the Bull Valley district ores.

[Analyst, Fred Lerch, Biwabik, Minn.]

	Iron.	Phosphorus.
Across 160 feet, Pilot No. 9	58.98	0.195
Across 250 feet, Pilot No. 8	62.38	.217
Across 69 feet, Pilot No. 7	62.06	.163
Pilot No. 12	66.40	.072
Across 40 feet, Pilot No. 7	64.13	.434

There has been a small amount of pitting, trenching, and tunneling, but these amount only to scratches in the upper parts of the deposits.

DEPOSITS OF SIMILAR ORIGIN IN OTHER DISTRICTS.

Ores similar to those of the Iron Springs district, consisting principally of magnetite in veins (not magmatic segregations in situ), in igneous rocks or near the contact of igneous rocks and limestone or in both combinations, are known at many localities in western North America, as follows:

In Mexico, where nearly all the iron ores are of this character, their distribution and occurrence being summarized by Hill[a] and by Aguilera.[b] The Durango deposit is described in a number of papers.

In several localities in San Bernardino County, Cal.,[c] and in the Redding quadrangle of northern California.[d]

In Lyon County, Nev., and in several other localities in the Great Basin.[e]

In Stevens County, northeastern Washington,[f] Texada and Vancouver islands, British Columbia,[f] and Kamloops, British Columbia.[f]

At Fierro[g] and Chupadera Mesa, New Mexico.[h]

In the Taylor Peak, White Pine, and Cebolla districts in Pitkin and Gunnison counties, Colo.[i]

At Iron Mountain, Missouri.[j]

This list of localities is far from complete and no attempt is made to cite all the publications concerning the districts named.   It is the

a Hill, R. T., The occurrence of hematite and martite iron ores in Mexico: Am. Jour. Sci., vol. 45, 1893, p. 112.

b Aguilera, José G., The geographical and geological distribution of the mineral deposits of Mexico: Trans. Am. Inst. Min. Eng., vol. 32, 1902, pp. 503-505.

c Leith, C. K., Iron ores of the western United States and British Columbia: Bull. U. S. Geol. Survey No. 285, 1906, pp. 194-200.

d Diller J. S., Iron ores of the Redding quadrangle, California: Bull. U. S. Geol. Survey No. 213, 1903, pp. 219-220.

e Manuscript notes furnished authors.

f Leith, C. K., op. cit., pp. 195-196.   Report on the iron ores of the coast of British Columbia, by the Provincial Mineralogist: Dept. of Mines, Victoria, B. C., 1903, 30 pp.

g Manuscript furnished the authors by R. W. Hills.

h Keyes, Charles R., Iron deposits of the Chupadera Mesa: Eng. and Min. Jour., vol. 78, 1904, p. 632.

i Manuscript notes, summers of 1905 and 1906, Van Hise, Leith, Harder, and Ward.

j Nason. Frank L., A report on the iron ores of Missouri: Missouri Geol. Survey, vol. 2, 1892, pp. 1-366.

purpose simply to emphasize the widespread distribution of deposits of these types and to cite some of the descriptions.

From the fact that the ores of these districts have certain essential features in common, it does not necessarily follow that the origin of the ores has been the same for all. Certainly there has not been enough detailed work on the ores of many of the localities named to warrant final conclusions as to their origin. Nevertheless, the similarities in these deposits are such as to suggest similarity of origin as a guiding hypothesis for study, and such results as have been obtained to the present time tend strongly to support this hypothesis rather than to break it down.

Several deposits which have been examined personally by the writers or associates are unhesitatingly assigned an origin similar to that here presented for the Iron Springs district of Utah. These include the Taylor Peak and White Pine ores of Colorado, studied by Van Hise and Leith and mapped in detail by Harder and Ward; the Cebolla district of Colorado, examined by Van Hise and Leith; the Bull Valley district of Utah, examined by Leith and Harder; and certain of the deposits in northeastern Washington, Texada and Vancouver islands, the Kamloops district, British Columbia, and San Bernardino County, California, examined in reconnaissance by Leith. All of these ores, in hand specimen and slide, show intimate association with anhydrous silicates and have structural relations to wall rock essentially similar to those of the Iron Springs district, though differing from these in structural and lithologic details.

Among the men who have given particular attention to the economic geology of the west emphasis has been uniformly placed on the genetic association of ores and igneous rocks along contacts, but there has been lack of agreement as to the real significance of the association—as to whether the igneous rocks have contributed both solutions and ores, or only hot water which has leached the ores from the adjacent rocks, or only heat which has enabled meteoric waters to leach ores from the adjacent rocks and redeposit them; whether the solutions have been liquid or gaseous; what their direction of movement has been, etc. In recent years there has been a rapidly growing tendency to emphasize the importance of liquid and gaseous solutions coming directly from the igneous rocks and bringing the ores with them. In discussions of genetic classifications of ores by Lindgren,[a] Weed,[b] Spurr,[c] and others, ore deposits developed along igneous contacts by "pneumatolytic after-action" find con-

[a] Lindgren, Waldemar, The character and genesis of certain contact deposits: Trans. Am. Inst. Min. Eng., vol. 31, 1902, pp. 226–244.

[b] Weed, W. H., Ore deposits near igneous contacts: Trans. Am. Inst. Min. Eng., vol. 33, 1903, pp. 715–746.

[c] Spurr, J. E., A consideration of igneous rocks and their segregation or differentiation as related to the occurrence of ores: Trans. Am. Inst. Min. Eng., vol. 33, 1903, pp. 288–340.

spicuous place. Iron-ore deposits have been included in this class largely because of the classic work of Vogt on the Christiania deposits. Individual deposits of western contact iron ores have not been specifically referred to this class because they have not been sufficiently known. The pneumatolytic origin of the Iron Springs ores now seems to be sufficiently well based and their similarity to other western contact iron ores sufficiently close to make it possible to assign a pneumatolytic origin to this general class of ores with some confidence.

The direct development of magnetite in pegmatitic veins is described by Spurr in the Georgetown quadrangle of Colorado. The case is conclusive, for magnetite crystals of considerable size are found in the interior of unaltered pegmatite veins.

Spencer[a] reached the conclusion that certain of the New Jersey magnetites are essentially pegmatitic in their origin. In addition to the Christiana deposits, Vogt[b] cites other European deposits[c] which he regards as of the same origin. Beck[d] does the same. Of interest also in this connection is the recent work of Stutzer[e] on the ores, which he regards as dikes, of the Kirunavaara and Luossavaara districts in northern Sweden, and the still more recent work of Sjögren[f] on the Scandinavian iron ores.

The prevalence of an iron cap or gossan above sulphide deposits in fissure veins or along the contact of igneous rocks and limestones is responsible for a widespread view that western iron ores of this occurrence will be found to grade down into sulphides; that the ores are the oxidized portions of sulphide veins which may have originated in the manner here outlined for the Iron Springs district; that the ores therefore have an ultimate igneous source to the same extent as the sulphide deposits, whatever this may be. Reasons are given on another page for the belief that the Iron Springs magnetite was deposited directly from hot solutions and not as an alteration product of sulphide, but that simultaneously there were deposited small and variable amounts of pyrite. So far as the writers know, there is no evidence of increase in sulphur content in depth in any of the ores of this class in the localities cited, beyond perhaps the first few inches or few feet, from which the sulphide, originally deposited with the

a Spencer, A. C., Genesis of the magnetite deposits in Sussex county, N. J.: Min. Mag., vol. 10, 1904, pp. 377–381.

b Vogt, J. H. L., Problems in the geology of ore deposits: Trans. Am. Inst. Min. Eng., vol. 31, 1902, pp. 125–169.

c Northern Sweden, Kristiania district, southern Hungary, Island of Elba and Dieletti, France.

d Beck, Richard, The nature of ore deposits, translated by W. H. Weed, 2 vols., New York, 1905, 685 pp.

e Stutzer, O., Die Eisenerzlagerstätte Gellivare in Nordschweden: Zeitschr. für prakt. Geol., bd. 14, No. 5, May, 1906, pp. 137–140; Die Eisenerzlagerstätten bei Kiruna: Zeitschr. prakt Geol., bd. 14, No. 3, March, 1906, pp. 65–71.

f Sjögren, Hjalmar, The geological relations of the Scandinavian iron ores: Trans. Am. Inst. Min. Eng., vol. 38, 1906, pp. 877–946.

ore, has been leached by weathering. It must be remembered, however, that no one of these deposits has been opened up sufficiently to demonstrate their character in depth.

The theory that iron ores at the contacts of igneous rocks and limestone have developed entirely by the action of meteoric waters from above, leaching ores from the adjacent rocks, has not been without supporters. This is the view held by Hill for the Durango deposit. The senior author presented this theory as possibly applicable to the Iron Springs deposits of Utah in an earlier publication. It would be entirely premature to exclude this hypothesis for all iron ores in fissure veins in igneous rocks or along their contacts with limestone, especially where these ores are largely limonite and apparently lack heavy anhydrous minerals, as in some of the deposits of northeastern Washington. It is usually where the ores are hard crystalline magnetites and hematites, intimately associated with the anhydrous silicates and adjacent to or within igneous rocks, that there is reasonable probability that the origin of the ores may be ascribed to the hot solutions coming from the igneous rocks.

STRUCTURAL DISSIMILARITIES OF THE IRON SPRINGS DEPOSITS TO OTHER CLASSES OF IRON-ORE DEPOSITS.

The structural features in which the iron-ore deposits of the Iron Springs district differ from ores of sedimentary origin, like the Lake Superior and Clinton ores, are obvious to all familiar with the great classes of iron-ore deposits, but for those who are not so familiar an elementary comparison may be of interest.

The Clinton hematites of Alabama and elsewhere are bedded deposits with all the stratigraphic and structural characteristics of sedimentary rocks. Given a bed, it may be expected to extend in strike and dip, thicken and thin, with about the same degree of uniformity as may be observed in other sedimentary layers, such as limestone or shale adjacent. Its structure also, as shown by strike and dip, is governed by the same laws as other deformed beds. It is thus frequently possible to determine with some certainty extensions of deposits for many thousands of feet. The Iron Springs deposits can not be followed or extensions calculated with any such certainty.

The Lake Superior iron-ore deposits are more or less irregular concentrations in a sedimentary "iron formation." The iron ores make up a very small per cent—less than 2—of the mass of the iron formation. The iron formation originally consisted of iron carbonate or iron silicate, and was altered to iron ore, ferruginous chert, or jaspilite, but retained its bedding. The structure and stratigraphy of the iron formation may be worked out as fully as for limestone or quartzite. Exploration is limited to the iron formation, and the presence of an iron formation is regarded as a prerequisite for exploration for iron ore. The Iron Springs deposits lack an associated sedimentary iron formation to serve as a guide for exploration, though there are locally present lean phases of ores for which the term "iron formation," without genetic significance, might be appropriate.   The deposits themselves in the Lake Superior region rest in structural basins which determine for each district certain uniformities and peculiarities of shape for the ore deposits for that district.   For the Lake Superior region as a whole the variety of shapes of ore deposits is probably greater than that in the Iron Springs district, but the Iron Springs deposits probably have a shape and size which can be less safely predicted in advance of exploration than can those of the deposits of any one of the Lake Superior districts.   They have, however, one considerable advantage in that they are largely exposed on the erosion surface, whereas the Lake Superior deposits are in great part covered by glacial drift.

BIBLIOGRAPHY.

BARRELL, JOSEPH. The physical effects of contact metamorphism: Am. Jour. Sci., 4th ser., vol. 13, 1902, pp. 279–296.

BECK, RICHARD. The nature of ore deposits. Translated by Walter Harvey Weed. 2 vols., 685 pp., New York, 1905.

BOUTWELL, J. M. Iron ores of the Uinta Mountains: Bull. U. S. Geol. Survey No. 225, pp. 221–228.

DILLER, J. S. Iron ores of the Redding quadrangle, California: Bull. U. S. Geol. Survey No. 213, 1903, pp. 219–220.

DUTTON, C. E. Report on the geology of the High Plateaus of Utah: U. S. Geog. and Geol. Survey Rocky Mountain Region, J. W. Powell in charge, 1880, pp. 307.

DUTTON, C. E. Tertiary history of the Grand Canyon district: Mon. U. S. Geol. Survey, vol. 2, 1882, pp. 264.

GILBERT, G. K. Report on the geology of portions of Nevada, Utah, California, and Arizona examined in the years 1871 and 1872: U. S. Geog. and Geol. Surveys W. 100th Mer., G. M. Wheeler in charge, vol. 3, Geology, 1875, pp. 17–187.

GILBERT, G. K. Lake Bonneville: Mon. U. S. Geol. Survey, vol. 1, 1890, pp. 438.

HOWELL, E. E. Report on the geology of portions of California, Nevada, Utah, Colorado, New Mexico, and Arizona examined in the years 1871, 1872, and 1873: U. S. Geog. and Geol. Survey W. 100th Mer., G. M. Wheeler in charge, vol. 3, Geology. 1875, pp. 227–301.

KEMP, J. F. Ore deposits at the contacts of intrusive rocks and limestone and their significance as regards the general formation of veins: Econ. Geol., vol. 2, 1907, pp. 1–13.

. LEITH, C. K. Iron ores in southern Utah: Bull. U. S. Geol. Survey No. 225, 1904, pp. 229–237.

LEITH, C. K. Iron ores of the western United States and British Columbia: Bull. U. S. Geol. Survey No. 285, 1906, pp. 194–200.

LINDGREN, WALDEMAR. The character and genesis of certain contact deposits. Trans. Am. Inst. Min. Eng., vol. 31, 1902, pp. 226–244.

LINDGREN, WALDEMAR. The relation of ore deposition to physical conditions: Econ. Geol., vol. 2, 1907, p. 123.

MARVINE, A. R. Report on the geology of route from St. George. Utah. to Gila River, Arizona, examined in 1871: U. S. Geog. and Geol. Survey W. 100th Mer., G. M. Wheeler in charge, vol. 3, Geology, 1875, pp. 189–225.

POWELL, J. W. Exploration of the Colorado River of the West and its tributaries, explored in 1869, 1870, 1871, and 1872, under the direction of the Secretary of the Interior, Washington. 1875, pp. 291.

POWELL, J. W. Report on the lands of the arid region of the United States, with a more detailed account of the lands of Utah: Report on survey of the Rocky Mountain region (2d ed.), 1879, pp. 116.

SPENCER, A. C. Genesis of the magnetite deposits in Sussex County, N. J.: Min. Mag., vol. 10, 1904, pp. 377–381.

SPURR, J. E. A consideration of igneous rocks and their segregation or differentiation as related to the occurrence of ores: Trans. Am. Inst. Min. Eng., vol. 33, 1903, pp. 340.

WEED, W. H. Ore deposits near igneous contacts: Trans. Am. Inst. Min. Eng., vol. 33, 1903, pp. 715–746.

INDEX.

O

102

II

'AIN

A HOEN & CO BALTIMORE MD

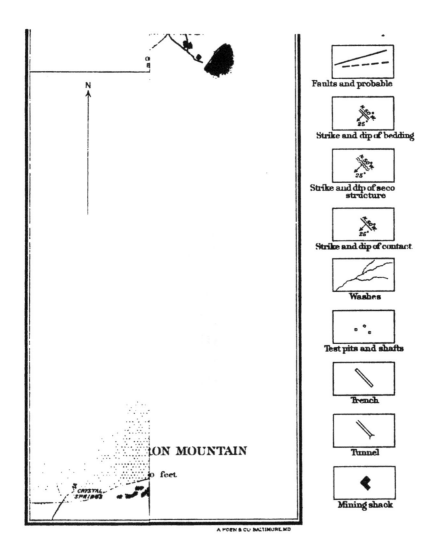

N

ON MOUNTAIN

o feet

CRYSTAL
SPRINGS

Faults and probable

Strike and dip of bedding

Strike and dip of seco
structure

Strike and dip of contact.

Washes

Test pits and shafts

Trench

Tunnel

Mining shack

A POEN & CO BALTIMORE, MD

IRON ORE D

I

CONTENTS.

3

THE PURCHASE OF COAL UNDER GOVERNMENT AND COMMERCIAL SPECIFICATIONS ON THE BASIS OF ITS HEATING VALUE.

By D. T. RANDALL.

INTRODUCTION.

The people of this country have been purchasing coal for years on the statement of the selling agent as to its quality, or on the reputation of the mine or district from which it was obtained. The farmers of the country show more business judgment in the purchase of fertilizer than do many manufacturers in buying coal, for the farmer demands a chemical analysis of the fertilizer before he accepts it.

Until recently there has been but little reliable information regarding the character of the coal supply of the United States. This fact was recognized in the establishment of the fuel-testing plant of the United States Geological Survey at St. Louis in 1904. Extended general investigations have been conducted at this plant and its successors and in the field, and in addition the technologic branch of the Survey has been charged with the duty of analyzing and testing the coals used by the Government.

The Government is a large purchaser of coal, reports from the various Departments indicating that nearly $6,300,000 is expended each year for fuel. Some time ago the necessity for a more uniform standard in the purchase of coal became apparent and the plan of buying it on the basis of its heating value was introduced in a few Departments. Many of the large commercial consumers in the United States have been purchasing their coal on contracts of this nature for some time.

The present paper summarizes the information on this subject obtained by the Survey, including examples of specifications that are now in use and analyses of some of the coal purchased by the Government during the winter of 1906–7.

ADVANTAGES OF DEFINITE SPECIFICATIONS IN THE PURCHASE OF COAL.

Under the old plan of purchasing coal, when the consumer had cause or thought he had cause to find fault with the quality of the fuel he received, he was in many cases assured that it must be good because, like all the other coal sent him, it came from a mine with an established reputation. Such a state of affairs made it difficult to take advantage of the competition which usually results when a considerable number of bidders are asked to submit prices. The purchaser was afraid to buy from any but such dealers as he knew and trusted, because, although each dealer claimed that his coal was equal in quality to that of the others, yet if it did not prove to be satisfactory there was no standard for settlement or for cancellation of the contract. Many thousands of dollars' worth of coal are bought each year in this manner, but the purchasers would consider it ridiculous if they were asked to contract for a building with no specifications and simply on the agreement that it should be of a certain size and well constructed. Neither would they buy gold, silver, or even copper and iron ores on the mere information that they were mined at certain localities. All products of mines are now purchased to a great extent on the basis of their value as shown by chemical analysis. This is true of coal in only a small degree, but the number of contracts made on this basis is increasing every year.

The purchase of coal on a specification is as advantageous as a definite understanding regarding the quality and other features of any other product, or of a building operation or engineering project. The man who buys under a specification gets what he pays for and pays for what he gets.

When the bidder is allowed to specify the quality of the coal he proposes to furnish as determined by a chemical analysis, he is placed on a strictly competitive basis with other bidders. Such a procedure broadens the field for both the bidder and the purchaser. It makes the bidder's proposal, when accepted, a contract that specifies an established standard of quality. This furnishes a basis for settling disputes regarding the quality of the coal delivered and the price to be paid if the fuel is either better or poorer than has been guaranteed. If other coal must be substituted, as often happens, there is a standard for settlement. If the coal is uniformly poorer than the standard as specified there is a basis for cancellation of the contract.

The quality of coal from a given mine may vary from time to time through the failure of the miners to reject impurities. Sometimes different beds of coal are mined at the same time and the output is mixed. When there is need of further preparation, such as picking slate and other impurities, or jigging or washing, a great deal depends on the care used in these processes. The mining companies are responsible in a large measure for variations in the grade of prepared coal. The purchase of coal under a contract on the basis of quality stimulates the operator to make a better preparation of the coal before it is shipped to market. An example of fluctuations in quality is furnished by the tables on pages 24–26, which show variations in the ash and British thermal units in coal delivered to some of the Government Departments at Washington.

It evidently will not be satisfactory to either the buyer or the seller to establish a standard for the coal unless this liability to variation is recognized and provision made for settlement when the coal is better or poorer than the standard. Experience with any method of buying coal shows that it will seldom be rejected when of poor quality, because of the difficulty, delay, and cost of removing it from the bins. The buyer is often confronted with the alternative of burning the coal delivered or going without fuel until more can be procured. Uuless the coal is very bad it is usually expedient to use it and pay a smaller price. This is also more favorable to the contractor, as to remove the coal would be costly and it would not be satisfactory as fuel to any other customer.

VALUE OF COAL AS A FUEL.

The purpose of burning coal under boilers is to abstract the heat for use in developing power, in drying various materials, or in warming buildings. The most valuable coal, therefore, is that which gives up the most heat to the boiler for a given weight burned.

Coal is now burned for power purposes in gas producers and boiler furnaces. For coals and lignites high in moisture or high in ash, the gas producer, used in connection with a gas engine, is best adapted to develop power. But for the generation of steam, which can be used for heating as well as for power purposes, a more convenient method is to burn the coal in a specially constructed furnace under a boiler.

The aim in the purchase of coal for any power plant should be to obtain a fuel which will produce a horsepower for the least cost, all things being considered, such as the equipment, the price of coal, and the cost of labor and repairs. Experiments have been made which seem to indicate that almost any fuel may be burned with reasonable efficiency in a properly designed apparatus. The recognized requirements are as follows:

A supply of fuel fed to the furnace as uniformly and continuously as possible.

An air supply slightly in excess of the theoretical amount required for complete combustion.

A sufficiently high temperature to ignite the gases which are driven off from the fuel.

A complete mixture of these gases with the air supplied before they reach a cooling surface, such as the shell or tubes of a boiler.

Some of the factors which may influence the commercial results obtained in a boiler are the cost of the coal, as determined by price and heating value; care in firing; design of the furnace and boiler setting, size of grate, etc.; formation of excessive amounts of clinker and ash; draft available; size of the coal (uniformity of size is desirable).

The value of a coal is indicated by the number of heat units it contains. This heating value is expressed in terms of British thermal units ª (abbreviated B. t. u.) per pound of coal, and is determined by means of a special apparatus called a calorimeter.

When coal is mined it contains moisture to a greater or less extent. It is exposed to the air in shipment and may either dry out or be drenched by rain. The moisture in the coal delivered is worthless to the purchaser, and really costs him a considerable amount in freight and cartage, and in the loss of the heat absorbed during its evaporation in the furnace. If all coal had the same proportion of moisture, or if the moisture in coal delivered by a given dealer was constant in amount, the purchaser's problem, so far as this factor is concerned, would be simplified. Under present conditions the moisture is an important element in the valuation of a ton of coal. It is evidently necessary to consider the coal just as it is received in order to determine its value to the consumer, but chemical reports should be made on the basis of both the " dry coal " and the " coal as received." The dry-coal basis is convenient for comparing several coals in regard to the relation of each element to the others; this is important because the moisture in the same coal varies from day to day. The dry-coal basis is also convenient for comparing the performance of boilers when burning the same or similar coals. Of several coals having a similar composition, the one which has the least moisture and the least ash will generate the most steam when burned under a boiler.

Ash is made up of earthy matter and other impurities which will not burn. In commercial coals its proportion may range from 4 to 25 per cent. Coals containing small percentages of ash are most valuable, not only because of their correspondingly higher heating capacity, but because there is less resistance to the free and uniform distribution of air through the bed of coal. The labor and cost of managing the fires and of handling the ashes are also correspondingly

---

ª The British thermal unit is the amount of heat required to raise the temperature of 1 pound of water 1° Fahrenheit.

less and are items to be considered in the choice of a coal. With the ordinary furnace equipment there may be a considerable loss of efficiency and capacity through a large percentage of ash. It has been found that with some kinds of equipment, as the ash increases there is a decided drop in both efficiency and capacity. In some experiments made to determine the influence of excessive amounts of ash, coal containing as high as 40 per cent would generate no steam when fired on a chain grate, and therefore the efficiency and capacity of the plant would be zero.[a] Such coal would not only be worthless, but involve a direct expense, due to the cost of handling it. Whether the result would be similar with equipment other than a chain grate has not yet been determined. However, coals so high in ash that they are unsuited to boiler furnaces can be utilized in gas producers.

The volatile part of coal as shown by the analysis may in some coals be all combustible, but it generally contains some inert matter. This varies in different coal deposits and makes it impossible to determine the heating value of the coal from the proximate analysis alone. Moreover, not all coals having the same proportion of volatile matter behave alike in the furnace. It is important to know both the chemical composition and the British thermal units in order to determine the value of one coal as compared with another for the same purpose.

Of two coals of different character, the one which contains the higher proportion of fixed carbon is most easily burned so as to give the maximum efficiency. However, if the coal containing the higher volatile matter is properly burned in a suitably designed furnace, it may be made equally efficient.

Sulphur may be present in the free state, or, as is more commonly the case, in combination with iron or other elements. Other impurities with sulphur often form a clinker which shuts out the air and increases the labor of handling the furnaces. It is possible, however, to burn coals containing up to 5 per cent of sulphur without particular difficulty from clinkers. A little steam introduced under the grate will relieve much of the trouble. Clinker may be due to other causes than sulphur, as any constituents of the ash which are easily fusible may produce it. There is need of further investigation to determine the influence of sulphur and the elements which comprise the ash on furnace fires and combustion.

The size of the coal influences the capacity of any given equipment, owing to its effect on the draft. With a poor draft fine coal can not be burned in sufficient quantities to maintain the rated capacity. If thin fires are resorted to, the efficiency is usually lowered as a result of an excessive supply of air through holes in the fire. As a

[a] Abbott, W. L., Some characteristics of coal as affecting performance with steam boilers, a paper read before the Western Society of Engineers, Chicago, Ill.

rule, when dust and very fine coal are fed into the furnace they either check the flow of air or are taken up by the draft and after being only partly burned are deposited back of the bridge wall or pass up the stack, to the annoyance of the people in the vicinity of the plant. If this dust is completely burned in passing through the furnace there is of course no loss of fuel. It has been found that coal of a uniform size is most satisfactory, as it does not pack so closely as a mixture of sizes.

In general it may be said that in any market the coal obtainable at the lowest price is the most economical, provided the furnace equipment is suitable. If the furnace is not so designed as to permit the use of the cheaper coal, it should be changed.

The results of tests tend to show that, other conditions being equal, coals of similar composition are of value in proportion to the British thermal units in the coal as received—a basis on which, indeed, all coals may be valued approximately. It should be remembered, however, that the value of a coal for any particular plant is influenced by the fact that all furnaces are not equally suitable for burning the many grades of coal. Aside from this factor, coals may be compared in terms of the British thermal units obtained for 1 cent, or on the cost per million heat units.

In the purchase of coal, then, attention should be given to the character of the furnace equipment and the load; the character of coal best suited to the plant conditions; the number of heat units obtainable for a unit price; the cost of handling the coal and ash; and the possibility of burning the coal without smoke or other objectionable features.

SPECIFICATIONS IN USE.

As the result of a letter from President Roosevelt to the national advisory board on fuels and structural materials, calling attention to the need of a uniform and efficient basis for the purchase of the Government fuel supply, the following specification was drafted by engineers in the employ of the Government and approved by this board in March, 1907:

SPECIFICATIONS AND PROPOSALS FOR SUPPLYING COAL.

United States _____

_____, 190__

PROPOSAL.

Sealed proposals will be received at this office until 2 o'clock p. m., _____, 190__, for supplying coal to the United States ___ _____ building at ___ _____ as follows:

The quantity of coal stated above is based upon the previous annual consumption, and proposals must be made upon the basis of a delivery of 10 per cent more or less than this amount, subject to the actual requirements of the service.

Proposals must be made on this form, and include all expenses incident to the delivery and stowage of the coal, which must be delivered in such quantities and at such times within the fiscal year ending June 30, 190__, as may be required.

Proposals must be accompanied by a deposit (certified check, when practicable, in favor of _____) amounting to 10 per cent of the aggregate amount of the bid submitted, as a guaranty that it is bona fide. Deposits will be returned to unsuccessful bidders immediately after award has been made, but the deposit of the successful bidder will be retained until after the coal shall have been delivered and final settlement made therefor, as security for the faithful performance of the terms of the contract, with the understanding that the whole or a part thereof may be used to liquidate the value of any deficiencies in quality or delivery that may arise under the terms of the contract.

When the amount of the contract exceeds $10,000, a bond may be executed in the sum of 25 per cent of the contract amount, and in this case the deposit or certified check submitted with the proposal will be returned after approval of the bond.

The bids will be opened in the presence of the bidders, their representatives, or such of them as may attend, at the time and place above specified.

In determining the award of the contract, consideration will be given to the quality of the coal offered by the bidder, as well as the price per ton, and should it appear to be to the best interests of the Government to award the contract for supplying coal at a price higher than that named in lower bid or bids received, the award will be so made.

The right to reject any or all bids and to waive defects is expressly reserved by the Government.

DESCRIPTION OF COAL DESIRED.[a]

Bids are desired on coal described as follows:

Coals containing more than the following percentages, based upon dry coal, will not be considered:

Ash _____ ____per cent.
Volatile matter _____ ____per cent.
Sulphur _____ ____per cent.
Dust and fine coal as delivered at point of consumption [b] _____ ____per cent.

DELIVERY.

The coal shall be delivered in such quantities and at such times as the Government may direct.

---

[a] This information will be given by the Government as may be determined by boiler and furnace equipment, operating conditions, and the local market.
[b] All coal which will pass through a ½-inch round-hole screen.

In this connection it may be stated that all the available storage capacity of the coal bunkers will be placed at the disposal of the contractor to facilitate delivery of coal under favorable conditions.

After verbal or written notice has been given to deliver coal under this contract, a further notice may be served in writing upon the contractor to make delivery of the coal so ordered within twenty-four hours after receipt of said second notice.

Should the contractor, for any reason, fail to comply with the second request, the Government will be at liberty to buy coal in the open market, and to charge against the contractor any excess in price of coal so purchased over the contract price.

SAMPLING.

Samples of the coal delivered will be taken by a representative of the Government.

In all cases where it is practicable, the coal will be sampled at the time it is being delivered to the building. In case of small deliveries, it may be necessary to take these samples from the yards or bins. The sample taken will in no case be less than the total of 100 pounds, to be selected proportionally from the lumps and fine coal, in order that it will in every respect truly represent the quantity of coal under consideration.

In order to minimize the loss in the original moisture content the gross sample will be pulverized as rapidly as possible until none of the fragments exceed one-half inch in diameter. The fine coal will then be mixed thoroughly and divided into four equal parts. Opposite quarters will be thrown out, and the remaining portions thoroughly mixed and again quartered, throwing out opposite quarters as before. This process will be continued as rapidly as possible until the final sample is reduced to such amount that all of the final sample thus obtained will be contained in the shipping can or jar and sealed air-tight.

The sample will then be forwarded to_____

If desired by the coal contractor, permission will be given to him, or his representative, to be present and witness the quartering and preparation of the final sample to be forwarded to the Government laboratories.

Immediately on receipt of the sample, it will be analyzed and tested by the Government, following the method adopted by the American Chemical Society, and using a bomb calorimeter. A copy of the result will be mailed to the contractor upon the completion thereof.

CAUSES FOR REJECTION.

A contract entered into under the terms of this specification shall not be binding if, as the result of a practical service test of reasonable duration, the coal fails to give satisfactory results owing to excessive clinkering or to a prohibitive amount of smoke.

It is understood that the coal delivered during the year will be of the same character as that specified by the contractor. It should, therefore, be supplied, as nearly as possible, from the same mine or group of mines.

Coal containing percentages of volatile matter, sulphur, and dust higher than the limits indicated on page 2 and coal containing a percentage of ash in excess of the maximum limits indicated in the following table will be subject to rejection.

In the case of coal which has been delivered and used for trial, or which has been consumed or remains on the premises at the time of the determination of its quality, payment will be made therefor at a reduced price, computed under the terms of this specification.

Occasional deliveries containing ash up to the percentage indicated in the column of "Maximum limits for ash," on page 4, may be accepted. Frequent or continued failure to maintain the standard established by the contractor, however, will be considered sufficient cause for cancellation of the contract.

PRICE AND PAYMENT.^a

Payment will be made on the basis of the price named in the proposal for the coal specified therein, corrected for variations in heating value and ash, as shown by analysis, above and below the standard established by contractor in this proposal. For example, if the coal contains 2 per cent, more or less, British thermal units than the established standard, the price will be increased or decreased 2 per cent accordingly.

The price will also be further corrected for the percentages of ash. For all coal which by analysis contains less ash than that established in this proposal a premium of 1 cent per ton for each whole per cent less ash will be paid. An increase in the ash content of 2 per cent over the standard established by contractor will be tolerated without exacting a penalty for the excess of ash. When such excess exceeds 2 per cent above the standard established, deductions will be made from price paid per ton in accordance with following table:

| Ash as established in proposal (per cent). | No deduction for limits below. | Cents per ton to be deducted. | | | | | | Maximum limits for ash. |
| | | 2 | 4 | 7 | 12 | 18 | 25 | 35 | |
		Percentages of ash in dry coal.							
5	7	7- 8	8- 9	9-10	10-11	11-12	12-13	13-14	12
6	8	8- 9	9-10	10-11	11-12	12-13	13-14	14-15	13
7	9	9-10	10-11	11-12	12-13	13-14	14-15	15-16	14
8	10	10-11	11-12	12-13	13-14	14-15	15-16	16-17	14
9	11	11-12	12-13	13-14	14-15	15-16	16-17	17-18	15
10	12	12-13	13-14	14-15	15-16	16-17	17-18	....	16
11	13	13-14	14-15	15-16	16-17	17-18	18-19	....	16
12	14	14-15	15-16	16-17	17-18	18-19	19-20	....	17
13	15	15-16	16-17	17-18	18-19	19-20	20-21	....	18
14	16	16-17	17-18	18-19	19-20	20-21	21-22	....	19
15	17	17-18	18-19	19-20	20-21	21-22	....	....	19
16	18	18-19	19-20	20-21	21-22	22-23	....	....	20
17	19	19-20	20-21	21-22	22-23	....	....	....	21
18	20	20-21	21-22	22-23	....	....	....	....	22

Proposals to receive consideration must be submitted upon this form and contain all of the information requested.

----------------------, 190__

The undersigned hereby agree to furnish to the U. S.-----------------------
building at ------------------, the coal described, in tons of 2,240 pounds

---

^a The economic value of a fuel is affected by the actual amount of combustible matter it contains, as determined by its heating value shown in British thermal units per pound of fuel, and also by other factors, among which is its ash content. The ash content not only lowers the heating value and decreases the capacity of the furnace, but also materially increases the cost of handling the coal, the labor of firing, and the cost of the removal of ashes, etc.

each and in quantity 10 per cent more or less than that stated on page 1, as may be required during the fiscal year ending June 30, 190__, in strict accordance with this specification; the coal to be delivered in such quantities and at such times as the Government may direct.

Description.	Item No.____	Item No.____	Item No.____
Commercial name			
Name of mine			
Location of mine			
Name of coal bed			
Size of coal (if coal is screened):			
Coal to pass through openings	___inches round. square.	___inches round. square.	___inches round. square.
Coal to pass over openings	___inches bar.	___inches bar.	___inches bar.
Data to establish a basis for payment.			
Per cent of ash in dry coal (method of American Chemical Society).			
British thermal units in coal as delivered			
Price per ton (2,240 pounds)			

It is important that the above information does not establish a higher standard than can be actually maintained under the terms of the contract; and in this connection it should be noted that the small samples taken from the mine are invariably of higher quality than the coal actually delivered therefrom. It is evident, therefore, that it will be to the best interests of the contractor to furnish a correct description with average values of the coal offered, as a failure to maintain the standard established by contractor will result in deductions from the contract price, and may cause a cancellation of the contract, while deliveries of a coal of higher grade than quoted will be paid for at an increased price.

Signature_____

Address_____

Name of corporation_____

Name of president_____

Name of secretary_____

Under what law (State) corporation is organized_____

As will be seen from the foregoing specification, the bidder is not required to submit a sample of his coal, but is expected to name a standard of British thermal units in the coal as it is to be delivered. This value is made the basis for purchase, because a correction is thus made for the amount of moisture in the coal. It should be noted that this value will in all cases be lower than the British thermal units in the dry coal, which is usually given in connection with the coal analysis. The percentage of ash is also specified, as it is a factor in the successful burning of the coal on the grate and as it involves an expense for removal from the premises.

If the dealer is not fairly familiar with the value of his coal, it may then be arranged to have him submit a properly selected sample with his bid, this sample to be analyzed by the Government and the results used as a standard in the contract. It is preferred, however, that the bidder use his own values.

The following circular letter was issued by the Treasury Department in connection with the Government specification, for the information of dealers desiring to bid on coals for public buildings, such as mints, custom-houses, and post-offices located in all parts of the United States:

TREASURY DEPARTMENT, OFFICE OF THE SECRETARY,
Washington, March 27, 1907.

SIR: The accompanying specifications for coal are intended to give a clear description of the coal desired by the Government, and to secure a definite statement of the quality of coal offered by the bidder, with a view to using such statement as a basis for payment in connection with the stated price per ton. The plan is not new in its essential features, as it is an extension of a system already in force in the larger United States public buildings, and is similar to that employed by a number of private consumers.

What is desired is not necessarily the cheapest or the highest grade of coal per se, but the coal which will insure the greatest net economy in plant operation. In view of these facts the description of the coal inserted by the Government on page 2 of the accompanying specifications will receive careful consideration as the boiler and furnace conditions require. It is not expected that all deliveries will be absolutely uniform or agree exactly with the standard established by the contractor, but it is necessary that all deliveries shall be within the limits set by the Government.

The limits are wide enough to permit the use of the output of any mine or group of mines provided proper care is exercised in mining and picking out slate, bone, etc. With these points in mind it is only necessary for the bidder to select coal for each proposal which will meet the description given and permit deliveries within the limits set. The standard established by the contractor should be such as to require the least correction applying to deliveries for variation in heat units and ash from values established.

It is believed that the enforcement of the provisions of the specifications will operate equitably both with respect to the Government and to the contractor, and will guarantee adequate protection to each. Many coal dealers have already signified their willingness to furnish coal on this basis, and have commended as well as indorsed the method.

The application of this system will not only enable the award of a contract to be made in an equitable manner, but will also remove many of the usual causes for dispute as to the character and quality of the coal subsequently delivered, and provide a satisfactory basis for the correction of payments for differences in quality in favor of the party in whose interest it is due.

The system of sampling, analyzing, and testing coal delivered under the Government contracts will be under the supervision of the fuel-testing division of the Geological Survey, in order to insure reliability and impartiality.

The heating value expressed in British thermal units of coal containing approximately the same percentage of ash is essentially a direct measure of the actual value to the purchaser, and for this reason the specifications provide for an adjustment of payment in direct proportion to the variation in heat units in the coal as received. As the coal is weighed when delivered and the payments are based also upon the price per ton, it is necessary to determine the heating value of the coal in the condition in which it is received, containing whatever moisture may be present at the time.

A further correction in payment will be made for variation of the ash in dry coal in order to take account of the cost of handling additional fuel and ash and its effect on the capacity of the boiler and furnace.

Respectfully,

A. F. STATTER,
Assistant Secretary.

On account of lack of information among the dealers in anthracite coal as to the heating value of the several sizes and kinds used by the Government, a number of contracts were let on the basis of the ash in dry coal. For the city of Washington these were worded in the same way as the regular specification, except the section relating to price and payment, which was as follows:

PRICE AND PAYMENT.

Payment will be made on the basis of the price named in the proposal for the coal specified, corrected for variations in ash as shown by analysis, above and below the standard established by the contractor.

For an increase or decrease up to 2 per cent in the ash content above or below the standard no correction will b· made in the price. When the variation exceeds this allowance above or below the standard, corrections will be made in the price as follows:

For furnace, egg, stove, and chestnut sizes of coal, variations from the standard percentage of ash exceeding 2 and less than 2.5 above and below will result in the deduction or addition of 15 cents per ton. For each additional one-half of 1 per cent, or fraction thereof, 3 cents more per ton will be deducted or added.

For pea coal, variations from the standard percentage of ash exceeding 2 and less than 2.5 above and below will result in the deduction or addition of 10 cents per ton. For each additional one-half of 1 per cent, or fraction thereof, 2½ cents more per ton will be deducted or added.

For buckwheat and screenings, variations from the standard percentage of ash exceeding 2 and less than 2.5 above and below will result in the deduction or addition of 8 cents per ton. For each additional one-half of 1 per cent, or fraction thereof, 2 cents more per ton will be deducted or added.

The following specification is used by a few firms in Baltimore, Md.:

SPECIFICATIONS AND INFORMATION CONCERNING SUPPLY OF COAL FOR _____

Caution.—Bidders are directed to familiarize themselves with the storage facilities and local conditions, affecting the deliveries of coal, existing at the power house.

Method of determining quantity and quality.—Weights are subject to check either by measurement or by company scales. Any deficiency will be deducted from and any excess added to the bill.

Sampling will be done by a representative of the company and contractor may have a representative present.

Sample will be taken from each delivery and kept in hermetically sealed jars.

Detail of specifications.—Coal will be semibituminous and run-of-mine. It shall be dry, well picked, and free from excessive amounts of slate, pyrites, and dirt of any kind and shall have the following composition: Moisture, not over 1 per cent; volatile carbon, not over 20 per cent; ash, not over 7 per cent; sulphur, not over 1 per cent.

ADJUSTMENTS.

Additions.—If the coal has less than 1 per cent moisture, the deficit per cent less than 1 per cent will be added to the bill. If the coal has less than 20 per cent volatile carbon, the deficit per cent less than 20 per cent will be multiplied by 2 and added to the bill. If the coal has less than 7 per cent ash, the deficit per cent less than 7 per cent will be multiplied by 3 and added to the bill.

Deductions.—If the coal contains more than 1 per cent moisture, the excess per cent above 1 per cent will be deducted from the bill. If the coal contains more than 20 per cent volatile carbon, the excess per cent above 20 per cent will be multiplied by 2 and deducted from the bill. If the coal contains more than 7 per cent ash, the excess per cent above 7 per cent will be multiplied by 3 and deducted from the bill.

The following are the essential features of the contracts on which a Chicago company is said to purchase and inspect nearly 1,000,000 tons of coal for its clients in Chicago, Indianapolis, Minneapolis, St. Louis, and other cities:

I. The company agrees to furnish and deliver to the consumer _____ _____ at such times and in such quantities as ordered by the consumer for consumption at sa'd premises during the term hereof, at the consumer's option, either or all of the kinds of coal described below; said coals to average the following assays:

Kind of coal_____			
Of size passing through screen having circular perforations in diameter_____	____ inches	____ inches	____ inches
Of size passing over a screen having circular perforations in diameter _____	____ inches	____ inches	____ inches
Per cent of moisture in coal as delivered_____			
Per cent of ash in coal as delivered_____			
British thermal units per pound of dry coal_____			
From following county_____			
From following State_____			

Coal of the above respective descriptions and specified assays (not average assays) to be hereinafter known as the contract grade of the respective kinds.

II. The consumer agrees to purchase from the company all the coal required for consumption at said premises during the term of said contract, except as set forth in Paragraph III below, and to pay the company for each ton of 2,000 pounds avoirdupois of coal delivered and accepted in accordance with all the terms of this contract at the following contract rate per ton for coal of each respective contract grade, at which rates the company will deliver the following respective numbers of British thermal units for 1 cent, the contract guaranty:

Kind of coal.	Contract rate per ton.	Contract guaranty.
_____	$_____	equal to _____ net B. t. u. for 1 cent.
_____	$_____	equal to _____ net B. t. u. for 1 cent.
_____	$_____	equal to _____ net B. t. u. for 1 cent.

Said net British thermal units for 1 cent being in each case determined as follows: Multiply the number of British thermal units per pound of dry coal by the per cent of moisture (expressed in decimals), subtract the product so found from the number of British thermal units per pound of dry coal, multiply the remainder by 2,000, and divide this product by the contract rate per ton (expressed in cents) plus one-half of the ash percentage (expressed as cents).

III. It is provided that the consumer may purchase for consumption at said premises coal other than herein contracted for, for test purposes, it being understood that the total of such coal so purchased shall not exceed 5 per cent of the total consumption during the term of this contract.

IV. It is understood that the company may deliver coal hereunder containing as high as 3 per cent more ash and as high as 3 per cent more moisture and as low as 500 fewer British thermal units per pound dry than specified above for contract grades.

V. Should any coal delivered hereunder contain more than the per cent of ash or moisture or fewer than the number of British thermal units per pound dry allowed under Paragraph IV hereof, the consumer may, at its option, either accept or reject same.

VI. All coal accepted hereunder shall be paid for monthly at a price per ton determined by taking the average of the delivered values obtained from the analyses of all the samples taken during that month, said delivered value in each case being obtained as follows: Multiply the number of British thermal units delivered per pound of dry coal by the per cent of moisture delivered (expressed in decimals), subtract the product so found from the number of British thermal units delivered per pound of dry coal, multiply the remainder by 2,000, divide this product by the contract guaranty, and from this quotient (expressed as dollars and cents) subtract one-half of the ash percentage delivered (expressed as cents).

In Cleveland, Ohio, coal is purchased for the waterworks on the basis of its heating value. The standard agreed upon is 13,624 British thermal units and was established as the result of analysis and tests made on a sample furnished by the dealer.

The following are the essential features of the specifications used by the Interborough Rapid Transit Company of New York in purchasing about 30,000 tons of coal each month for use in its plants, which are among the largest in the United States:

PRELIMINARY SPECIFICATIONS FOR BITUMINOUS COAL FOR THE INTERBOROUGH RAPID TRANSIT COMPANY.

Coal must be a good steam, caking, run-of-mine, bituminous coal free from all dirt and excessive dust, a dry sample of which will approximate the company's standard in heat value and analysis, as follows: Carbon, 71; volatile matter, 20; ash, 9; British thermal units, 14,100; sulphur, 1.50.

A small quantity of coal will be taken from each weighing hopper just before the hopper is dumped while the lighter is being unloaded. These quantities will be thrown into a receptacle provided for the purpose, and when the lighter is empty the contents of the receptacle will be thoroughly mixed, and a sample of this mixture will be taken for chemical analysis. This average sample of coal will be labeled and held for one week after the unloading of the lighter. The sample taken from the mixture for test will be analyzed as soon as possible after being taken. No other sample will be recognized.

Tests of sample taken from average sample will be made by the company's chemist under the supervision of the superintendent. Should the contractor question the results of the company's test (a copy of which will be mailed to him), the company will, if requested by the contractor within three days after copy of test has been mailed to him, forward sufficient quantity of the average sample taken from each weighing hopper to any laboratory in the city of New York which may be agreed upon by the superintendent and the contractor, and have said sample analyzed by it, and the results obtained from this second test will be considered as final and conclusive. In case the disputed values, as obtained in the company's test, shall be found by the second test to be 2 per cent or less in error, then the cost of said second test shall be borne by the contractor; but if the disputed values shall be found to be more than 2 per cent in error, then the cost of said second test shall be borne by the company.

Should there be no question raised by the contractor within the three days specified, as to the values of the first analysis, the average sample of coal will be destroyed at the end of seven days from date of discharge of coal from lighter. Should a second test be made of coal taken from any lighter as herein provided, then any penalties to be made as set forth in paragraph under "Penalties" will be based on the results as obtained from the second test.

The price to be paid by the company per ton per lighter of coal will be based on a table of heat values for excess or deficiency of its standard, but subject to deductions as given in the section under "Penalized coal," including excess of ash, volatile matter, sulphur, or dust, or less than the minimum amount required to be contained in any lighter, for coal which shows results less than the company's standard.

Premiums or deductions are based on a rate of 1 cent per ton for a variation of 50 British thermal units per pound of coal, as indicated in a table a few items of which are given below:

Table for B. t. u. values.

For coal in any lighter which is found by test to contain, per pound of dry coal, from—

15,501 and above-------------------------28 cents per ton above standard.
15,101 to 15,150, both inclusive-----------20 cents per ton above standard.
14,601 to 14,650, both inclusive-----------10 cents per ton above standard.
14,101 to 14,150, both inclusive-----------Standard.
13,601 to 13,650, both inclusive-----------10 cents per ton below standard.
13,101 to 13,150, both inclusive-----------20 cents per ton below standard.
12,101 to 12,150, both inclusive-----------40 cents per ton below standard.

No lighter of coal will be accepted which, by trial, in the judgment of the superintendent, contains an excessive amount of dry coal dust. The decision of the superintendent will be final in this respect. Coal taken from such lighter for trial will be subject to the special deduction set forth under "Penalized coal," but paid for in all other respects as herein provided.

Coal which is shown by analysis to contain less than 20 per cent of volatile matter, 9 per cent of ash, or 1.50 per cent of sulphur, will be accepted without a deduction from the bidder's price, plus or minus an amount for excess of deficiency of British thermal unit value, as herein provided. Where the analysis gives amounts for any or all elements in excess of these quantities, deductions will be made from the bidder's price in accordance with the tables of values of volatile matter, ash, and sulphur below given, plus or minus the amount for excess or deficiency of the standard British thermal unit value, in addition to any other deductions which may be made as herein provided.

Table of deductions for volatile matter.

For coal in any lighter which is found by test to contain, per pound of dry coal—

Over 20 per cent and less than 21 per cent----------------2 cents per ton.
*   *   *   *   *   *   *
Over 22.5 per cent and less than 23 per cent--------------12 cents per ton.
*   *   *   *   *   *   *
24 per cent and over--18 cents per ton.

the table is made for a difference of each one-half of 1 per cent and the reductions are at the rate of 4 cents for each 1 per cent of volatile matter.

Table of deductions for ash.

of coal in any lighter which is found by test to contain, per pound of dry coal

over 9 per cent and less than 9.5 per cent_____ 2 cents per ton.

.   *   *   *   *   *   *

over 11 per cent and less than 12   _____12 cents per ton.

.   *   *   *   *   *

. . . .   _____23 cents per ton.

. . . . difference of one-half of 1 per cent and at the . . . . increase in the ash.

the requirements of said notices from the company, or said difference may be deducted from any money then due or thereafter to become due to the contractor under the contract to be entered into.

METHODS OF SAMPLING AND TESTING.

In connection with the Survey's study of the coal deposits of the country and the best methods to prevent waste in mining and utilizing the coal supply, trained inspectors have visited nearly 300 mines in 23 States, taking two or more samples from each mine. A study of the analyses of these samples and of samples taken from cars shipped from 175 of the same mines shows that the mine sample is in most instances better than the average of the coal as shipped in cars. On the average the coal delivered contains about one-third more ash than the mine sample taken in accordance with the instructions to the miners regarding the rejection of slate and impurities. This difference is due to the failure of the miners to follow these instructions in getting out coal for shipment. The samples collected by the Government inspectors from the mines almost invariably show a higher moisture content than is usually obtained in commercial sampling, because of the precaution taken to have the sample represent the coal in the mine.[a]

Mine samples when properly taken indicate the general character of the coal and enable one to judge of its probable value for any definite purpose.

Samples taken from the cars should not be limited to a few shovelfuls of coal from the top of the car, because the heavier pieces gradually work down toward the bottom. Some samples taken at the bottom of a car have shown as much as 8 per cent more ash than the coal at the top. The moisture also may vary from top to bottom, depending on the weather. The only way to get a fair sample is to take a number of shovelfuls of coal from various points in the car, so as to procure a representative portion of the coal from top to bottom and from end to end.

Bituminous coal when exposed to the air gradually depreciates in heating value, owing to losses of volatile matter, but aside from this loss there should be the same total number of heat units in a car of coal when it reaches its destination as when it started. If rain falls on the coal it will become heavier and a greater number of pounds will be delivered, but each pound will have a correspondingly lower heat value. On the other hand, if the weather is fair and the coal dries out on the way, it will weigh less and the heating value of each pound will be correspondingly higher. In other words, under a specification such as is used by the Government, neither the dealer

---

[a] A description of the method of mine sampling is given in Bull. U. S. Geol. Survey No. 290, 1906, pp. 17–18. See also Bull. No. 316, 1907, pp. 486–517.

This table is made for a difference of each one-half of 1 per cent and the deductions are at the rate of 4 cents for each 1 per cent of volatile matter.

Table of deductions for ash.

For coal in any lighter which is found by test to contain, per pound of dry coal—

Over 9 per cent and less than 9.5 per cent_____ 2 cents per ton.

*   *   *   *   *   *   *

Over 11.5 per cent and less than 12_____12 cents per ton.

*   *   *   *   *   *   *

13.5 per cent and over_____23 cents per ton.

This table is made for each difference of one-half of 1 per cent and at the rate of 4 cents for each 1 per cent increase in the ash.

Table of deductions for sulphur.

For coal in any lighter which is found by test to contain, per pound of dry coal—

Over 1.50 per cent and less than 1.75 per cent_____ 6 cents per ton.

*   *   *   *   *   *   *

Over 2 per cent and less than 2.25 per cent_____ 10 cents per ton.

*   *   *   *   *   *   *

2.50 and over_____ 20 cents per ton.

This table is made out for each difference of one-fourth of 1 per cent and at a diminishing rate.

Should any lighter of coal delivered at the company's docks contain less than 700 tons, a deduction of 7 cents per ton will be made from the price as determined by the British thermal unit value and analysis, in addition to any other penalty provided for herein. Should any lighter of coal delivered at the company's docks be rejected by the superintendent on account of excessive amount of coal dust, then a deduction of 25 cents per ton will be made from the price as determined by the British thermal unit value and analysis, for the coal taken from said lighter, in addition to any other penalty which may be made as herein provided. Should any lighter of coal be delivered in other than self-trimming lighters as herein provided, a deduction of 7 cents per ton will be made from the price as determined by the British thermal unit value and analysis, exclusive of any other penalty which may be made as herein provided.

The contractor's bill of lading will be checked by the company's scales. Should there be a deficiency of 1 per cent or more between the bill of lading and the company's weights, then the company's weights will be taken as correct.

When the contractor has been notified by the company to deliver coal under this contract, a further notice may be given requiring the contractor to make delivery of the coal so ordered within twelve hours after the receipt of said second notice. Should the contractor, for any reason, fail to deliver the coal so ordered within twelve hours after the receipt of said second notice and in accordance with the requirements therein as to place of delivery, the company shall be at liberty to buy coal in the open market, and the contractor will make good to the company any difference there may be between the price paid by the company for said coal in open market and the price the company would have paid to the contractor had the coal been delivered by it in accordance with the requirements of said notices from the company, or said difference may be deducted from any money then due or thereafter to become due to the contractor under the contract to be entered into.

METHODS OF SAMPLING AND TESTING.

In connection with the Survey's study of the coal deposits of the country and the best methods to prevent waste in mining and utilizing the coal supply, trained inspectors have visited nearly 300 mines in 23 States, taking two or more samples from each mine. A study of the analyses of these samples and of samples taken from cars shipped from 175 of the same mines shows that the mine sample is in most instances better than the average of the coal as shipped in cars. On the average the coal delivered contains about one-third more ash than the mine sample taken in accordance with the instructions to the miners regarding the rejection of slate and impurities. This difference is due to the failure of the miners to follow these instructions in getting out coal for shipment. The samples collected by the Government inspectors from the mines almost invariably show a higher moisture content than is usually obtained in commercial sampling, because of the precaution taken to have the sample represent the coal in the mine.[a]

Mine samples when properly taken indicate the general character of the coal and enable one to judge of its probable value for any definite purpose.

Samples taken from the cars should not be limited to a few shovelfuls of coal from the top of the car, because the heavier pieces gradually work down toward the bottom. Some samples taken at the bottom of a car have shown as much as 8 per cent more ash than the coal at the top. The moisture also may vary from top to bottom, depending on the weather. The only way to get a fair sample is to take a number of shovelfuls of coal from various points in the car, so as to procure a representative portion of the coal from top to bottom and from end to end.

Bituminous coal when exposed to the air gradually depreciates in heating value, owing to losses of volatile matter, but aside from this loss there should be the same total number of heat units in a car of coal when it reaches its destination as when it started. If rain falls on the coal it will become heavier and a greater number of pounds will be delivered, but each pound will have a correspondingly lower heat value. On the other hand, if the weather is fair and the coal dries out on the way, it will weigh less and the heating value of each pound will be correspondingly higher. In other words, under a specification such as is used by the Government, neither the dealer

---

[a] A description of the method of mine sampling is given in Bull. U. S. Geol. Survey No. 290, 1906, pp. 17–18. See also Bull. No. 316, 1907, pp. 486–517.

nor the purchaser will gain or lose by change in the moisture content of the coal between the time it is weighed at the mine and the time it is weighed on delivery. The price per ton will be correspondingly lower if the coal is wet and higher if the coal is dry.

In order to determine the maximum variation in moisture in several sizes of anthracite coal the following experiments were made: The coal was soaked in water to allow it to absorb as much moisture as possible, the result representing the extreme conditions due to rains or other causes. Each sample was then weighed and allowed to dry in a room exposed to the air. When this sample ceased to lose moisture it was assumed to be air dried, which represents the condition of least moisture to be expected in a delivery of coal. The results are summarized in the following table:

Experiments to determine possible variations of moisture in anthracite coal during shipment.

	Furnace.	Pea.	Buck-wheat.
Number of samples used in experiment	13	10	12
Number of hours dried in air at ordinary room temperature	0.5 to 24	24	24
Total moisture in thoroughly wet coal_____per cent	5.12	5.74	8.44
Moisture in air-dried samples_____do	3.58	1.84	2.24
Loss of moisture_____do	.73 to 1.54	3.1 to 3.9	4.5 to 6.2
Percentage of maximum variation in moisture from wet to air-dried coal	30	68	71

The air-dried anthracite still contains from 1.8 to 3.6 per cent of moisture. Moisture in air-dried coal varies with the weather, just as it does in wood.

The moisture in air-dried bituminous coals depends on the character of the coal. It is about 1 per cent in West Virginia coal and about 7 per cent in Illinois coal. The moisture in the same Illinois coal delivered may range from 7 to 17 per cent.

Owing to these variations some method should be used to correct for the difference in moisture in coals of different character.

The following suggestions are presented for the guidance of those who wish to send samples to a laboratory for analysis:

If samples are taken at the buildings as the coal is delivered, it will usually be satisfactory to take one shovelful of coal from each third or fifth wagonload, the load being selected without the knowledge of the driver. It must be kept in mind that the main object is to obtain a portion of coal which represents as nearly as possible the entire delivery. The sample should contain about the same proportion of lump and fine coal as exists in the shipment as a whole. The practice of taking a shovelful near the bottom of the pile should be avoided, as the larger lumps of coal roll down and collect near the bottom and such a sample will not truly represent the coal.

These samples should be immediately deposited in a metal receptacle having a tight-fitting cover and provided with a first-class lock. Except when samples are being deposited or when the contents are being quartered down, this receptacle should be securely locked and the key held by a responsible employee. The receptacle should be placed in a comparatively cool location to avoid loss of moisture in the coal. When it becomes filled, or at the end of the sampling period, the contents should be emptied on a clean, dry floor, in a cone-shaped pile. The larger lumps should be broken down by a coal maul or sledge, and the pile re-formed and quartered into four equal parts, a shovel or board being used to separate the four sections. Two opposite sections should then be rejected and the remaining two again mixed, broken down, and re-formed into a pile to be quartered as before. This process should be continued until the lumps are no larger than pea size, and a quart sample is finally procured. The samples should then immediately be placed in suitable receptacles for shipping and sealed air-tight. The Geological Survey inspectors use a metal can 3 inches in diameter and 9 inches high, with a screw cap 2 inches in diameter, for making shipments to the chemical laboratory. These cans are sealed air-tight by winding adhesive electrical tape around the joint of the screw cap. Each can holds about a quart. or 2 pounds of coal.

The process of quartering down and preparing samples for shipment to the chemical laboratory for analysis should be carried on as rapidly as possible to avoid loss of moisture. The samples should be forwarded promptly and notice of shipment sent under separate cover. Receptacles should be marked plainly on the outside, and a corresponding number or description should be placed inside. A complete record of all deliveries should be kept, showing dates, names of contractor, kind of coal, total weight delivered, condition of coal (wet or dry), and any other particulars of importance.

The procedure at the chemical laboratory of the Geological Survey testing plant is described in Survey Bulletin No. 261. The samples are crushed and ground to a fine powder. and then analyzed and tested.

Persons not experienced in taking samples have a tendency to select a sample better than the average. In many cases a lump of coal is broken and shipped in a cloth sack to the laboratory. This allows the moisture to dry out; moreover, the lump selected is usually free from layers of slate and impurities and of course then represents the best coal in the lot, and shows a higher value than can be expected to hold throughout the coal delivered.

The preceding statements show that the purchaser should usually have the quality determined on the basis of coal "as received," in order to correct any excess or deficiency in the moisture content.

ANALYSES OF COALS DELIVERED TO THE GOVERNMENT.

The following tables, giving the results of tests made by the Geological Survey on coal delivered to the Government Departments, are submitted in response to numerous recent requests for information regarding the quality of coal which may be expected, and the variations in quality which may be found from month to month in coal delivered by the same dealer and presumably from the same mine or group of mines:

Average analyses of anthracite coal delivered to all Government buildings in Washington, D. C., from December 15, 1906, to April 26, 1907.[a]

	Furnace.	Egg.	Pea.	Buckwheat.
Dry coal:				
Volatile matter	2.42	3.10	3.02	2.42
Fixed carbon	87.14	86.33	80.94	79.53
Ash	10.44	10.57	16.04	18.05
	100.00	100.00	100.00	100.00
Sulphur [b]	.79	.98	.80	.68
B. t. u.	13,406	13,523	12,487	12,107
Coal as received:				
Moisture	4.08	4.16	4.81	5.09
B. t. u.	12,861	12,961	11,886	11,483

[a] Payment not based on chemical analysis.
[b] Separately determined.

Analyses of anthracite furnace coal delivered during 1907 to a Government building in Washington, D. C.[a]

	Jan. 29.	Feb. 2.	Feb. 4.	Feb. 9.	Feb. 11.	Feb. 12.	Feb. 14.	Feb. 15.
Dry coal:								
Volatile matter	2.66	2.85	3.32	2.53	3.06	2.56	2.90	2.09
Fixed carbon	86.38	86.34	87.66	86.65	85.45	86.69	86.63	86.15
Ash	10.96	10.81	9.02	10.82	11.49	10.75	10.38	11.76
	100.00	100.00	100.00	100.00	100.00	100.00	100.00	100.00
Sulphur [b]	.78	.78	.78	.78	.69	.84	.83	.95
B. t. u.	13,297	13,435	13,764	13,418	13,282	13,360	13,427	13,168
Coal as received:								
Moisture	4.13	4.19	4.67	4.41	3.83	5.83	4.90	4.33
B. t. u.	12,749	12,874	13,121	12,826	12,773	12,582	12,757	12,595

	Feb. 23.	Feb. 25.	Feb. 26.	Feb. 27.	Feb. 28.	Mar. 13.	Average.
Dry coal:							
Volatile matter	1.56	1.62	2.00	3.34	2.84	2.13	2.51
Fixed carbon	87.96	87.77	88.61	86.12	85.37	87.22	86.74
Ash	10.48	10.61	9.39	10.54	11.79	10.65	10.75
	100.00	100.00	100.00	100.00	100.00	100.00	100.00
Sulphur [b]	.93	.78	.74	.87	.86	.79	.81
B. t. u.	13,094	13,346	13,055	13,390	13,166	13,421	13,363
Coal as received:							
Moisture	3.06	3.50	3.93	4.50	5.60	2.12	4.24
B. t. u.	12,694	12,879	13,118	12,787	12,429	13,136	12,796

[a] Contract for 4,500 tons.   Payment not based on chemical analysis.
[b] Separately determined.

Analyses of anthracite egg coal delivered during 1907 to a Government building in Washington, D. C.[a]

	Jan. 29.	Jan. 31.	Feb. 5.	Feb. 9.	Feb. 11.	Feb. 13.	Feb. 14.	Feb. 15.	Feb. 21.
Dry coal:									
Volatile matter	2.96	2.83	2.57	3.90	3.17	3.53	3.05	3.69	2.86
Fixed carbon	86.29	85.58	86.27	85.39	86.76	86.04	86.90	86.36	85.72
Ash	10.75	11.59	11.16	10.71	10.07	10.43	10.05	9.95	11.42
	100.00	100.00	100.00	100.00	100.00	100.00	100.00	100.00	100.00
Sulphur [b]	.98	.92	.86	1.15	1.06	.93	1.04	1.12	.97
B. t. u	13,423	13,279	13,336	13,496	13,588	13,614	13,738	13,649	13,328
Coal as received:									
Moisture	4.18	4.34	3.38	4.01	4.11	4.17	4.73	4.51	4.23
B. t. u	12,863	12,703	12,886	12,953	13,030	13,045	13,688	13,084	12,764

	Feb. 23.	Feb. 25.	Feb. 26.	Feb. 28.	Mar. 3.	Apr. 2.	Apr. 19.	Apr. 24.	Average.
Dry coal:									
Volatile matter	3.60	3.62	3.15	3.22	2.86	3.01	2.89	3.19	3.41
Fixed carbon	87.99	86.00	86.58	86.46	85.53	87.59	86.93	86.36	86.99
Ash	9.40	10.38	10.27	10.32	11.61	9.40	10.18	10.45	9.60
	100.00	100.00	100.00	100.00	100.00	100.00	100.00	100.00	100.00
Sulphur [b]	.94	.98	.99	1.13	1.02	.99	1.08	1.01	.99
B. t. u	13,798	13,643	13,606	13,617	13,293	13,657	13,672	13,558	13,744
Coal as received:									
Moisture	4.15	3.64	4.07	4.13	3.15	4.08	3.92	4.07	4.15
B. t. u	13,174	13,291	13,054	13,055	12,874	13,100	13,136	13,006	13,174

[a] Contract for 9,000 tons. Payment not based on chemical analysis.
[b] Separately determined.

Analyses of anthracite pea coal delivered during 1906–7 to a Government building in Washington, D. C.[a]

	Dec. 22.	Jan. 9.	Jan. 24.	Jan. 28.	Jan. 31.	Feb. 5.	Feb. 6.	Feb. 7.	Feb. 11.	Feb. 12.
Dry coal:										
Volatile matter	3.38	2.58	3.73	3.07	3.33	3.39	4.06	3.30	3.77	3.23
Fixed carbon	80.51	84.17	81.34	82.16	80.22	80.80	79.37	80.57	79.95	80.09
Ash	16.11	13.25	14.93	14.77	16.45	15.81	16.57	16.13	16.28	16.68
	100.00	100.00	100.00	100.00	100.00	100.00	100.00	100.00	100.00	100.00
Sulphur [b]	.60	.67	.99	1.01	.87	.78	.78	.77	.71	.68
B. t. u	12,460	12,883	12,877	12,835	12,583	12,623	12,555	12,552	12,418	12,384
Coal as received:										
Moisture	4.35	4.62	5.24	3.40	5.62	5.86	4.27	5.50	6.04	5.78
B. t. u	11,919	12,286	12,203	12,390	11,876	11,883	12,019	11,862	11,668	11,668

	Feb. 13.	Feb. 14.	Feb. 18.	Feb. 19.	Feb. 25.	Mar.18.	Apr. 1.	Apr. 8.	Apr. 12.	Average.
Dry coal:										
Volatile matter	2.27	2.32	3.50	3.06	2.13	2.50	2.49	2.51	2.74	3.02
Fixed carbon	82.45	80.45	78.75	81.35	83.24	80.83	79.45	81.00	81.24	80.94
Ash	15.28	17.23	17.75	15.59	14.63	16.67	18.06	16.49	16.02	16.04
	100.00	100.00	100.00	100.00	100.00	100.00	100.00	100.00	100.00	100.00
Sulphur [b]	.92	.86	1.05	.89	.68	.80	.69	.64	.77	.80
B. t. u	12,300	12,317	12,160	12,464	12,627	12,410	12,088	12,284	12,479	12,487
Coal as received:										
Moisture	5.20	4.92	4.58	5.23	5.02	3.85	2.97	5.18	3.77	4.31
B. t. u	11,660	11,711	11,603	11,812	11,993	11,932	11,680	11,649	12,008	11,886

[a] Contract for 6,000 tons. Payment not based on chemical analysis.
[b] Separately determined.

CLASSIFICATION OF THE PUBLICATIONS OF THE UNITED STATES GEOLOGICAL SURVEY.

[Bulletin No. 339.]

The serial publications of the United States Geological Survey consist of (1) Annual Reports, (2) Monographs, (3) Professional Papers, (4) Bulletins, (5) Mineral Resources, (6) Water-Supply and Irrigation Papers, (7) Topographic Atlas of United States—folios and separate sheets thereof, (8) Geologic Atlas of the United States—folios thereof. The classes numbered 2, 7, and 8 are sold at cost of publication; the others are distributed free. A circular giving complete lists can be had on application.

Most of the above publications can be obtained or consulted in the following ways:

1. A limited number are delivered to the Director of the Survey, from whom they can be obtained, free of charge (except classes 2, 7, and 8), on application.

2. A certain number are delivered to Senators and Representatives in Congress for distribution.

3. Other copies are deposited with the Superintendent of Documents, Washington, D. C., from whom they can be had at practically cost.

4. Copies of all Government publications are furnished to the principal public libraries in the large cities throughout the United States, where they can be consulted by those interested.

The Professional Papers, Bulletins, and Water-Supply Papers treat of a variety of subjects, and the total number issued is large. They have therefore been classified into the following series: A, Economic geology; B, Descriptive geology; C, Systematic geology and paleontology; D, Petrography and mineralogy; E, Chemistry and physics; F, Geography; G, Miscellaneous; H, Forestry; I, Irrigation; J, Water storage; K, Pumping water; L, Quality of water; M, General hydrographic investigations; N, Water power; O, Underground waters; P, Hydrographic progress reports; Q, Fuels; R, Structural materials. This paper is the tenth in Series Q, the complete list of which follows (PP=Professional Paper; B=Bulletin):

SERIES Q, FUELS.

B 261. Preliminary report of the operations of the coal-testing plant of the United States Geological Survey at the Louisiana Purchase Exposition, St. Louis, Mo., 1904; E. W. Parker, J. A. Holmes, M. R. Campbell, committee in charge. 1905. 172 pp.

PP 48. Report on the operations of the coal-testing plant of the United States Geological Survey at the Louisiana Purchase Exposition, St. Louis, Mo., 1904; E. W. Parker, J. A. Holmes, M. R. Campbell, committee in charge. 1906. 3 parts. 1,492 pp., 13 pls.

B 290. Preliminary report on the operations of the fuel-testing plant of the United States Geological Survey at St. Louis, Mo., 1905, by J. A. Holmes. 1906. 240 pp.

B 323. Experimental work conducted in the chemical laboratory of the United States fuel-testing plant at St. Louis, Mo., January 1, 1905, to July 31, 1906, by N. W. Lord. 1907. 49 pp.

B 325. A study of four hundred steaming tests made at the fuel-testing plant, St. Louis, Mo., in 1904, 1905, and 1906, by L. P. Breckenridge. 1907. 196 pp.

B 332. Report of the United States fuel-testing plant at St. Louis, Mo., January 1, 1906, to July 1, 1907, Joseph A. Holmes in charge. 1908. —— pp.

B 333. Coal-mine accidents: their causes and prevention; a preliminary statistical report, by Clarence Hall and W. O. Snelling, with an introduction by J. A. Holmes. 1907. 21 pp.

B. 334. The burning of coal without smoke in boiler plants, a preliminary report, by D. T. Randall. 1908. 26 pp.

B 336. Washing and coking tests of coal and cupola tests of coke, by Richard Moldenke, A. W. Belden, and G. R. Delamater. 1908. —— pp.

B 339. The purchase of coal under Government and commercial specifications, on the basis of its heating value, by D. T. Randall. 1908. 27 pp.

Correspondence should be addressed to

THE DIRECTOR,

UNITED STATES GEOLOGICAL SURVEY,

WASHINGTON, D. C.

JANUARY, 1908.

O

I

DEPARTMENT OF THE INTERIOR

UNITED STATES GEOLOGICAL SURVEY

GEORGE OTIS SMITH, DIRECTOR

BULLETIN 340

CONTRIBUTIONS

TO

ECONOMIC GEOLOGY

1907

Part I.—METALS AND NONMETALS, EXCEPT FUELS

C. W. HAYES AND WALDEMAR LINDGREN

GEOLOGISTS IN CHARGE

WASHINGTON

GOVERNMENT PRINTING OFFICE

1908

CONTENTS.

3

CONTENTS.

ILLUSTRATIONS.

CONTRIBUTIONS TO ECONOMIC GEOLOGY, 1907, PART I.

C. W. HAYES AND WALDEMAR LINDGREN, *Geologists in charge.*

INTRODUCTION.

By C. W. HAYES, *Chief geologist.*

This bulletin is the sixth of a series, including Bulletins Nos. 213, 225, 260, 285, and 315, Contributions to Economic Geology for 1902, 1903, 1904, 1905, and 1906, respectively. These bulletins are prepared primarily with a view to securing prompt publication of the economic results of investigations made by the United States Geological Survey. They are designed to meet the wants of the busy man, and are so condensed that he will be able to obtain results and conclusions with a minimum expenditure of time and energy. By means of the bibliographies accompanying the several groups of papers they also serve as a guide to the economic publications, and afford a better idea of the work which the Survey as an organization is carrying on for the direct advancement of mining interests throughout the country than can readily be obtained from the more voluminous final reports.

The first two bulletins of this series included numerous papers relating to the economic geology of Alaska. In view of the rapid increase of economic work, both in Alaska and in the States, and the organization of a division of Alaskan mineral resources distinct from the division of geology, it was in 1905 considered advisable to exclude all papers relating to Alaska. These were brought together in a separate volume entitled "Report of progress of investigations of mineral resources of Alaska in 1904," Bulletin No. 259. A similar segregation of papers relating to Alaska was made in 1905 and 1906 (Bulletins Nos. 284 and 314), and will be made this year.

During 1906 a further change in the arrangement of the economic bulletin seemed desirable. The former section of iron ores and nonmetallic minerals was divided and M. R. Campbell was placed in charge of a new section devoted to the investigation of fuels. This change in Survey organization was used as a basis for a separation of the economic bulletin, based on subjects. The present bulletin is therefore restricted to the work of the Survey in 1907 in the metals, structural materials, and other nonmetals except coal. A separate bulletin will be issued later relating to Survey work on coal, lignite, and peat.

In the preparation of the present volume promptness of publication has been made secondary only to the economic utility of the material presented. The papers included are such only as have a direct economic bearing, all questions of purely scientific interest being excluded.

The papers are of two classes: (1) Preliminary discussions of the results of extended economic investigations, which will later be published by the Survey in more detailed form; (2) comparatively detailed descriptions of occurrences of economic interest, noted by geologists of the Survey in the course of their field work, but not of sufficient importance to necessitate a later and more extended description.

The papers have been grouped according to the subjects treated. At the end of each section is given a list of previous publications on that subject by this Survey. These lists will be serviceable to those who wish to ascertain what has been accomplished by the Survey in the investigation of any particular group of mineral products. They are generally confined to Survey publications, though a few titles of important papers published elsewhere by members of the Survey are included.

Material assistance in the preparation of this volume has been rendered by W. C. Phalen, and to him is largely due the promptness of its publication.

The results of the Survey work in economic geology have been published in a number of different forms, which are here briefly described:

1. *Papers and reports accompanying the Annual Report of the Director.*—Prior to 1902 many economic reports were published in the royal octavo cloth-bound volumes which accompanied the Annual Report of the Director. This form of publication for scientific papers has been discontinued and a new series, termed Professional Papers, has been substituted.

2. *Bulletins.*—The bulletins of the Survey comprise a series of paper-covered octavo volumes, each containing usually a single report or paper. These bulletins, formerly sold at nominal prices, are now distributed free of charge to those interested in the special subject discussed in any particular bulletin. This form of publication facilitates promptness of issue for economic results, and most economic reports are therefore published as bulletins. Their small size, however, precludes the use of large maps or plates, and reports containing large illustrations are therefore issued in the series of Professional Papers.

3. *Professional Papers.*—This series, paper covered, but quarto in size, is intended to include such papers as contain maps or other illustrations requiring the use of a large page. The publication of the series was commenced in 1902, and the papers are distributed in the same manner as are the bulletins.

4. *Monographs.*—This series consists of cloth-bound quarto volumes, and is designed to include exhaustive treatises on economic or other geologic subjects. Volumes of this series are sold at cost of publication.

5. *Geologic folios.*—Under the plan adopted for the preparation of a geologic map of the United States the entire area is divided into small quadrangles bounded by certain meridians and parallels, and these quadrangles, which number several thousand, are separately surveyed and mapped. The unit of survey is also the unit of publication, and the maps and descriptions of each quadrangle are issued in the form of a folio. When all the folios are completed, they will constitute a Geologic Atlas of the United States.

A folio is designated by the name of the principal town or of a prominent natural feature within the quadrangle. It contains topographic, geologic, economic, and structural maps of the quadrangle, and in some cases other illustrations, together with a general description.

Under the law copies of each folio are sent to certain public libraries and educational institutions. The remainder are sold at 25 cents each, except such as contain an unusual amount of matter, which are priced accordingly.

Circulars containing complete lists of these folios, showing the locations of the quadrangle areas they describe, their prices, etc., are issued from time to time, and may be obtained on application to the Director of the United States Geological Survey. The following list shows the folios issued since January 1, 1907, and in an advanced state of preparation, also the economic products discussed in the text of each, the products of greatest importance being printed in italic:

List of geologic folios issued since January 1, 1907, showing mineral resources described.

No.	Name of folio.	State.	Area in square miles.	Author.	Mineral products described as occurring in area of folio.
143	Nantahala	N.C.–Tenn ...	975	Keith, A	*Marble, talc,* kaolin, soapstone, mica, corundum, iron, gold.
144	Amity	Pa	228	Clapp, F. G	*Coal, oil, gas,* limestone, shale, sandstone.
145	Lancaster–Mineral Point.	Wis.–Iowa–Ill.	1,756	Grant, U. S.; Burchard, E. F.	*Zinc, lead,* cement, building stone, glass sand, clay.
146	Rogersville	Pa	229	Clapp, F.G	*Oil, gas,* coal, limestone, sandstone.

List of geologic folios issued since January 1, 1907, etc.—Continued.

No.	Name of folio.	State.	Area in square miles.	Author.	Mineral products described as occurring in area of folio.
147	Pisgah	N. C.–S. C		Keith, A	*Mica, corundum, soapstone,* talc, kaolin, gold, graphite, copper, building stone, lime, clay.
148	Joplin district	Mo.–Kans	476	Smith, W. S. T.; Siebenthal, C. E.	*Zinc, lead,* lime, barite, coal, bitumen, oil, building stone, clay, underground water.
149	Penobscot Bay	Me	857	Smith, G. O.; Bastin, E. S.; Brown, C. W.	*Granite,* limestone, building stone, copper, clay.
150	Devils Tower	Wyo	849	Darton, N. H.; O'Harra, C. C.	Coal, building stone, clay, gypsum, *underground water.*
151	Roan Mountain	Tenn.–N. C	963	Keith, A	*Iron, mica,* lime, talc, soapstone, building stone, road material, clay.
152	Patuxent	Md.–D. C	982	Shattuck, G. B.; Miller, B. L.	Clay, sand, gravel, *building stone,* marl, *diatomaceous earth, underground water.*
153	Ouray	Colo	235	Cross, W.; Howe, E.; Irving, J. D.	*Silver, gold,* building stone, lime.
154	Winslow	Ark	969	Purdue, A. H.; Adams, G. I.	Lime, clay, coal, underground water.
155	Ann Arbor	Mich	885	Russell, I. C.; Leverett, F.	Road material, building stone, clay, *cement material,* marl, peat, *underground water.*
156	Elk Point	S. Dak.–Nebr.–Iowa.	878	Todd, J. E	Clay, sand, gravel, cement material, volcanic ash, lignite, *underground water.*
157	Passaic	N. J.–N. Y	906	Darton, N. H.; Bayley, W. S; Salisbury, R. D.; Kümmel, H. B.	*Iron,* building stone, road material, *pottery clay,* brick clay, graphite, peat, underground water.
158	Rockland	Me	215	Bastin, E. S	*Lime, granite, clay,* peat, road material, gravel, underground water.

List of geologic folios in preparation.

Name of folio.	State.	Area in square miles.	Author.	Mineral products described as occurring in area of folio.
Aberdeen–Redfield	S. Dak	3,383	Todd, J. E	Lignite, clay, sand, gravel, salt, gas, *underground water.*
Accident–Grantsville	Md. - Pa. - W. Va.	460	Martin, G. C	Coal, fire clay, lime, building stone, road material.
Bellefourche	S. Dak	849	O'Harra, C C.; Darton, N. H.	Gypsum, lime, clay, building stone, bentonite, *underground water.*
El Paso	Tex	889	Richardson, G B	Tin, *clay, cement,* flux, lime, sand, gravel, *underground water.*
Franklin Furnace	N. J	226	Wolff, J. E.; Spencer, A. C.; Palache, Charles; Salisbury, R. D.; Kümmel, H. B.	*Lime, flux,* road material, graphite, *iron, zinc,* manganese, building stone.
Independence	Kans	950	Schrader, F C	*Oil, gas,* coal, building stone, road material, lime, glass sand, *cement material,* clay, underground water.

List of geologic folios in preparation—Continued.

Name of folio.	State.	Area in square miles.	Author.	Mineral products described as occurring in area of folio.
Mercersburg-Chambersburg.	Pa.	458	Stose, G. W.	*Iron*, manganese, white clay, barite, *lime*, cement material, sand, clay, building stone, marble, road material, underground water.
Philadelphia	Pa.-N. J	915	Bascom, F.; Darton, N. H.; Clark, W. B.; Kümmel, H. B.; Salisbury, R. D.	Building stone, *road material, lime*, magnesium carbonate, soapstone, iron, gravel, sand, *pottery clay, feldspar, brick clay*, marl.
Santa Cruz	Cal	950	Branner, J. C.; Newsome, J. F.; Arnold, R.	Gold, *bituminous rock, petroleum*, building stone, road material, lime, cement material, diatomaceous shale, sand, underground water.
Trenton	N. J.-Pa	912	Bascom, F.; Darton, N. H.; Clark, W. B.; Kümmel, H. B.; Salisbury, R. D.	*Pottery clay*, brick clay, *molding sand*, building sand, gravel, *marl*, building stone, road material, lime, copper, barite.
Watkins Glen-Catatonk.	N. Y	1,770	Williams, H. S.; Tarr, R. S.; Kindle, E. M.	Sand, gravel, clay, underground water.

6. *Mineral Resources.*—From 1883 to 1894, inclusive, an octavo cloth-bound volume bearing the above title was issued annually, except that the reports for the years 1883–84 and 1889–90 were included by pairs in single volumes. The first of this series was Mineral Resources of the United States, 1882; the last, Mineral Resources of the United States, 1893. In 1894 this form of publication was discontinued, in accordance with an act of Congress, and thereafter the statistical material was included in certain parts of the sixteenth, seventeenth, eighteenth, nineteenth, twentieth, and twenty-first annual reports. The separate publication of the series on mineral resources was resumed, however, in 1901, in accordance with an act of Congress, and seven volumes of the new series, Mineral Resources of the United States for 1900, for 1901, for 1902, for 1903, for 1904, for 1905, and for 1906, have been issued.

This publication contains a systematic statement of the production and value of the mineral products of the United States, a summary of new mineral resources developed, and short papers on economic geology when these are necessary to account for the new developments.

INVESTIGATIONS RELATING TO NONMETALLIC MINERAL RESOURCES.

By C. W. HAYES, *Chief Geologist.*

PUBLICATIONS ISSUED AND IN PREPARATION.

During the year 1907 the following publications, consisting wholly or in part of the results of investigations on the nonmetallic mineral resources of the United States, were issued by the Survey. Publications relating to coal, lignite, peat, etc., except folios, are not included in this list.

Bulletins:

No. 279. Economic geology of the Kittanning and Rural Valley quadrangles, Pennsylvania, by Charles Butts.

No. 296. Economic geology of the Independence quadrangle, Kansas, by F. C. Schrader and E. Haworth.

No. 300. Economic geology of the Amity quadrangle, eastern Washington County, Pennsylvania, by F. G. Clapp.

No. 304. Oil and gas fields of Greene County, Pennsylvania, by R. W. Stone and F. G. Clapp.

No. 309. The Santa Clara Valley, Puente Hills, and Los Angeles oil districts, southern California, by G. H. Eldridge and Ralph Arnold.

No. 313. The granites of Maine, by T. Nelson Dale, with an introduction by George Otis Smith.

No. 317. Preliminary report on the Santa Maria oil district, Santa Barbara County, California, by Ralph Arnold and Robert Anderson.

No. 318. Geology of oil and gas fields in Steubenville, Burgettstown, and Claysville quadrangles, Ohio, West Virginia, and Pennsylvania, by W. T. Griswold and M. J. Munn.

No. 321. Geology and oil resources of the Summerland district, Santa Barbara County, California, by Ralph Arnold.

No. 322. Geology and oil resources of the Santa Maria oil district, Santa Barbara County, California, by Ralph Arnold and Robert Anderson.

No. 324. The San Francisco earthquake and fire of April 18, 1906, and their effects on structures and structural materials, by G. K. Gilbert, R. L. Humphrey, J. S. Sewell, and Frank Soulé.

The following folios are those in which nonmetallic products of considerable importance are described. The substances printed in italics are of most importance.

Folios.

No. 143. Nantahala (North Carolina-Tennessee), by A. Keith. *Marble, talc,* kaolin, soapstone, mica, corundum, iron.

No. 144. Amity (Pennsylvania), by F. G. Clapp. *Coal, oil, gas,* limestone, shale, sandstone.

No. 145. Lancaster-Mineral Point (Wisconsin-Iowa-Illinois), by U. S. Grant and E. F. Burchard. Cement, building stone, glass sand, clay.

No. 146. Rogersville (Pennsylvania), by F. G. Clapp. *Oil, gas,* coal, limestone, sandstone.

No. 147. Pisgah (North Carolina-South Carolina), by A. Keith. *Mica, corundum, soapstone,* talc, kaolin, graphite, building stone, lime, clay.

No. 148. Joplin district (Missouri-Kansas), by W. S. T. Smith and C. E. Siebenthal. Lime, barite, coal, bitumen, oil, building stone, clay, underground water.

No. 149. Penobscot Bay (Maine), by G. O. Smith, E. S. Bastin, and C. W. Brown. *Granite,* limestone, building stone, clay.

No. 150. Devils Tower (Wyoming), by N. H. Darton and C. C. O'Harra. Coal, building stone, clay, gypsum, *underground water.*

No. 151. Roan Mountain (North Carolina-Tennessee), by A. Keith. *Iron, mica,* lime, talc, soapstone, building stone, road material, clay.

No. 152. Patuxent (Maryland-District of Columbia), by G. B. Shattuck and B. L. Miller. Clay, sand, gravel, *building stone,* marl, *diatomaceous earth, underground water.*

No. 153. Ouray (Colorado), by W. Cross and E. Howe. Building stone, lime.

No. 154. Winslow (Arkansas), by A. H. Purdue and G. I. Adams. Lime, clay, coal, underground water.

No. 155. Ann Arbor (Michigan), by I. C. Russell and F. Leverett. Road material, building stone, clay, *cement material,* marl, peat, *underground water.*

No. 156. Elk Point (South Dakota-Nebraska-Iowa), by J. E. Todd. Clay, sand, gravel, cement material, volcanic ash, lignite, *underground water.*

No. 157. Passaic (New Jersey-New York), by N. H. Darton, W. S. Bayley, R. D. Salisbury, and H. B. Kümmel. *Iron,* building stone, road material, *pottery clay,* brick clay, graphite, peat, underground water.

No. 158. Rockland (Maine), by E. S. Bastin. *Lime, granite, clay,* peat, road material, gravel, underground water.

Reports for which the field work has been completed and which are in an advanced state of preparation are the following:

Iron.

The iron ores and iron industry of the Birmingham district, Alabama, by E. C. Eckel, E. F. Burchard, and Charles Butts.

The iron ores of the Iron Springs district, Utah, by C. K. Leith and E. C. Harder. (Bulletin No. 338.)

Magnetite deposits of the Cornwall type in Pennsylvania, by A. C. Spencer.

Petroleum and natural gas.

Geology and oil resources of the Coalinga district, California, by Ralph Arnold and Robert Anderson.

Systematic comparison of the crude petroleum in the United States, by D. T. Day.

Geology of the Rangely oil field, Rio Blanco County, Colorado, by Hoyt S. Gale. (Bulletin No. 350.)

Structure of the Berea oil sand in Flushing quadrangle, Harrison, Belmont, and Guernsey counties, Ohio, by W. T. Griswold. (Bulletin No. 346.)

Oil and gas in the Sewickley, Carnegie, and Clarion quadrangles, Pennsylvania, by M. J. Munn.

Geography and geology of a portion of southwestern Wyoming, with special reference to coal and oil, by A. C. Veatch. (Professional Paper No. 56.)

Building stones, road metal, etc.

The road materials of Maine, by E. S. Bastin.[a]

Commercially important granites of Massachusetts, New Hampshire, and Rhode Island, by T. Nelson Dale. (Bulletin No. 354.)

Commercially important granites of Vermont, by T. Nelson Dale.

Cement and concrete materials.

Concrete materials produced in the Chicago district, by E. F. Burchard.

The structural materials in the District of Columbia region, by N. H. Darton.

The structural materials in the vicinity of Portland, Oreg., and Seattle, Wash., by N. H. Darton.

Portland cement mortars and their constituent materials, by R. L. Humphrey. (Bulletin No. 331. Published in March, 1908.)

Clays.

The clays of Arkansas, by J. C. Branner. (Bulletin No. 351.)

Economic geology of the Kenova quadrangle, northeastern Kentucky, and adjacent portions of Ohio and West Virginia, by W. C. Phalen. (Coal, clay, iron ore, petroleum, and natural gas, etc.) (Bulletin No. 349.)

Gypsum and magnesite.

Gypsum deposits of California, by F. L. Hess.

Magnesite deposits of California, by F. L. Hess. (Bulletin No. 355.)

Miscellaneous.

The pegmatites of Maine, by E. S. Bastin.

The occurrence of diamonds in the United States and elsewhere, by G. F. Kunz.

FIELD WORK.

As in previous years, a large part of the work on nonmetalliferous minerals during 1907 was carried on in connection with other investigations, chiefly areal surveys. By reason of the character of these deposits they can not generally be studied with advantage in advance

---

[a] To be published by the Department of Agriculture.

of detailed topographic mapping, and their relations to the areal distribution of the rock formations are such that detailed areal geologic surveys are also generally necessary.

Iron, manganese, and aluminum ores.—The investigation of iron ores of the Cornwall type in eastern Pennsylvania has been completed by A. C. Spencer, and a report is in preparation which gives the relations of these ores to the associated rocks in detail and points out the directions in which prospecting may be carried on with the expectation of locating other ore bodies in addition to those now known. The directing of exploration to the most favorable localities, whether additional ore bodies are found or not, will certainly effect great economies in future prospecting.

The study of the iron ores of Georgia has been continued by W. C. Phalen in connection with areal work on the Ellijay and adjoining quadrangles. His conclusions regarding certain of these deposits are contained in a paper in this volume.

The developments in the Lake Superior district have been followed closely by C. R. Van Hise and his assistants. A report summarizing the several monographs already published and bringing information and conclusions regarding the relations of the iron ores down to date is in an advanced state of preparation.

A systematic examination is being made of the manganese ore deposits of the United States by E. C. Harder, and a comprehensive report covering all known deposits, whether at present productive or not, will be prepared. Information regarding these deposits is in considerable demand, as it is at present necessary to import a large portion of the manganese ore used in this country.

The producing bauxite localities in Alabama, Georgia, and Tennessee were visited during the past season and information was procured looking to the preparation of a comprehensive report on the aluminum ores of the United States.

Structural materials.—During the past year the necessary field work has been carried on for a thorough revision of Bulletin No. 243, Cement Materials and Industry of the United States, by E. C. Eckel. This report has been out of print for some time, and it was found necessary, in view of the constant demand for information on this subject, to issue a new edition. It was determined, therefore, to bring the report up to date and embody the latest information available before reissue.

In cooperation with the structural-materials section of the technologic branch considerable work has been done during the past year on the geologic relations of building materials in the vicinity of several of the larger cities, particularly building stone and concrete materials. This work has been done chiefly by N. H. Darton and E. F. Burchard. Two brief reports giving some of their results appear in this volume.

The work on the New England granites by T. Nelson Dale has been continued during the year, and he has prepared a report on the granites of Rhode Island, Connecticut, Massachusetts, and New Hampshire, similar in scope to Bulletin No. 313, on the granites of Maine.

Work has also been in progress on the granites of the Southern Atlantic States by T. L. Watson and his report is well advanced. It will cover all the Atlantic States south of Pennsylvania and will be based in part on work of the State surveys, much of which was done originally by Doctor Watson.

In cooperation with the State of Arkansas, a thorough investigation is being made of the Arkansas slate deposits by A. H. Purdue. The detailed economic results of this work will be published by the State, and a brief statement of the economic results, together with the areal and structural geology, will be embodied in the Caddo Gap folio.

Much information regarding structural materials, including clays, building stone, and concrete material, has been collected by geologists engaged in various parts of the country primarily in the study of areal geology. This information will appear in the geologic folios.

Oil and gas.—Detailed surveys have been continued by M. J. Munn in the western Pennsylvania oil and gas fields, in cooperation with the State. The utility of this exact instrumental determination of the minor structural features of the oil and gas bearing rocks has been amply demonstrated by actual field tests with the maps previously published. The cost of prospecting is very greatly reduced by a reduction in the proportion of dry wells drilled and the indication of the most favorable locations for the accumulation of the hydrocarbons. At the same time it has led to important conclusions, that are of general application, as to the manner in which oil and gas accumulate in the rocks. The work, however, is very expensive and therefore has not been done extensively except in cooperating States.

The investigation of the California oil fields has been continued by Ralph Arnold, the Coalinga, McKittrick, Midway, and Sunset fields having been covered during the past season. Reports on these fields, similar in scope to those published during the year on the Santa Clara, Puente Hills, Los Angeles, Santa Maria, and Summerland fields, are in an advanced state of preparation.

In connection with the investigation of the lead and zinc deposits and the stratigraphy of northeastern Oklahoma, C. E. Siebenthal made a reconnaissance of the oil fields of that region. Although the work was not sufficiently detailed to throw much light on the occurrence of the oil, it indicated the need of work, which is being planned for the coming season.

Phosphates.—A thorough reconnaissance of the Utah-Idaho phosphate deposits was made by F. B. Weeks and the report in preparation, a brief extract of which appears in this volume, will suffice until there are adequate topographic maps of the region and the deposits are more thoroughly prospected. The work already done will serve to indicate to the prospector the general location of the phosphatic beds and their stratigraphic associations.

The phosphate deposits of Tennessee, South Carolina, and Florida were visited by F. B. Van Horn, and plans are being made for a comprehensive report on the rock phosphates of the United States.

47076—Bull. 340—08——2

INVESTIGATIONS RELATING TO DEPOSITS OF METALLIFEROUS ORES.

By WALDEMAR LINDGREN, *Geologist in charge.*

PUBLICATIONS OF THE YEAR.

During the year the following publications on subjects connected with the investigation of deposits of metalliferous ores within the United States proper have been issued by the Survey:

Bulletins:

No. 287. The Juneau gold belt, Alaska, by A. C. Spencer, and A reconnaissance of Admiralty Island, Alaska, by C. W. Wright.

No. 320. The Downtown district of Leadville, Colo., by S. F. Emmons and J. D. Irving.

Folios:

No. 143. Nantahala (North Carolina-Tennessee), by A. Keith. Contains description of gold deposits.

No. 145. Lancaster-Mineral Point (Wisconsin-Iowa-Illinois), by U. S. Grant. Contains description of zinc and lead deposits.

No. 147. Pisgah (North Carolina-South Carolina), by A. Keith. Contains description of gold and copper deposits.

No. 148. Joplin district (Missouri-Kansas), by W. S. T. Smith and C. E. Siebenthal. Contains description of the lead-zinc deposits of the Joplin region.

No. 149. Penobscot Bay (Maine), by G. O. Smith and E. S. Bastin. Contains description of copper deposits.

No. 153. Ouray (Colorado), by W. Cross, E. Howe, and J. D. Irving. Contains description of gold and silver deposits.

The following list comprises reports for which field work has been completed, but which have not yet been issued:

Economic geology of the Georgetown quadrangle, together with the Empire district, Colorado, by J. E. Spurr and G. H. Garrey, with a chapter on geology by S. H. Ball. (Professional Paper No. 63.)

Copper deposits of the Butte district, Montana, by W. H. Weed.

Economic geology of the Park City mining district, Utah, by J. M. Boutwell and L. H. Woolsey.

Geology and ore deposits of the Cœur d'Alene district, Colorado, by F. L. Ransome and F. L. Calkins. (Professional Paper No. 62.)

Geology and ore deposits of the Goldfield district, Nevada, by F. L. Ransome.

Geology and ore deposits of the Bullfrog district, Nevada, by F. L. Ransome and W. H. Emmons.

Geology and ore deposits of the Franklin Furnace quadrangle, New Jersey, by A. C. Spencer,

Resurvey of the Leadville mining district, Colorado, by S. F. Emmons and J. D. Irving.

The Tertiary auriferous gravels of the Sierra Nevada, by W. Lindgren.

Reconnaissance of the deposits of metalliferous ores of New Mexico, by W. Lindgren, L. C. Graton, and C. H. Gordon.

The copper deposits of Shasta County, Cal., by L. C. Graton.

Ore deposits of the Phillipsburg quadrangle, Montana, by W. H. Emmons.

Ore deposits of Mohave County, Ariz., by F. C. Schrader.

Ore deposits of the Dahlonega district, Georgia, by A. Keith.

FIELD WORK.

GENERAL STATEMENT.

In May, 1907, Mr. S. F. Emmons, desiring to devote his entire time to scientific work, relinquished the administrative duties of the section of metalliferous deposits. Since the organization of the Geological Survey Mr. Emmons has always occupied the position of senior geologist in the investigation of ore deposits, and for the last fifteen years he has under varying forms of administration had entire charge of this work. It is not necessary to call attention to his distinguished services nor to tell how much the science of ore deposits has gained by his brilliant work. All those who have worked under him feel deeply how much they have been helped by his guiding hand, his pregnant suggestions, and his indefatigable interest in their work.

During 1907 the funds available for the investigations of metalliferous deposits were unusually small, owing to the thorough and extended work devoted in that year to the investigation of coal deposits. The publication of the papers in hand was also greatly delayed on account of shortage in the funds devoted to printing. For this reason very few bulletins and professional papers were issued during the latter part of the year. It is confidently expected that next year will witness a great improvement both in extent of field work and number of publications issued.

As during the previous year, the supervision of the collection of metal statistics has occupied a part of the time of Messrs. Boutwell, Graton, Hess, Lindgren, and Siebenthal. This work can not be characterized better than by quoting S. F. Emmons in Bulletin No. 315, Contributions to Economic Geology, 1906, Part I:

This work comprises more than a mere compilation of figures furnished by others to show production. It involves the tracing of the metals back to their various sources and the verification, by comparison and analysis, of the necessarily varying results obtained by different lines of investigation. More than that, its object is to gather at the same time such geological data as will enable the geologists in charge of the respective branches of the work to prepare an annual review of the production and prospects of the different metals in their geological as well as their technical and commercial relations, and thus to provide data for an intelligent forecast of the progress of the industries involved and of the direction it is likely to take,

A disposition has been noted in some technical journals to criticise the diversion of geologic work in this direction. This criticism would doubtless be justified if the investigation of metalliferous deposits should permanently suffer as a consequence of this diversion of geo-logic work, but it should not be forgotten that the starting of this undertaking involved considerable labor, which will be rapidly less-ened, so far as the geologist in charge is concerned, when, after acquainting himself fully with the field, he can turn over a large part of the detailed work to other men. He will retain the supervision and the reviewing, and this connection with practical work in his branch, far from being a detriment, will enable him to obtain a much better knowledge of the resources of the nation; and, moreover, the data collected, if intelligently handled, will allow him to draw many im-portant geologic conclusions.

The following notes summarize the geologic work by members of the Survey in 1907, so far as the metalliferous deposits, except iron, are concerned:

ARIZONA.

In Mohave County, Ariz., F. C. Schrader completed in February, 1907, a three months' reconnaissance of the important deposits in the vicinity of Chloride and those in the ranges bordering Colorado River. A brief review of this investigation is contained in this bulletin.

CALIFORNIA.

In California L. C. Graton, aided by B. S. Butler, completed during the last three months of the year a detailed study of the copper depos-its of Shasta County, which have gained tremendously in importance during the last few years, and near which new reduction plants of great magnitude have recently been constructed. The field season was devoted to the study of the pyritic deposits on the western side of Sacramento River.

COLORADO.

In the northern part of Colorado H. S. Gale examined extensive deposits of gold-bearing gravels, an account of which may be found in this bulletin. This work was done in connection with the study of coals in this region.

In the central part of Colorado the writer devoted two weeks to the study of certain copper deposits, most of them of pre-Cambrian age, of which an account is also given in the following pages. The locali-ties studied are in Chaffee, Fremont, and Jefferson counties.

MONTANA.

W. H. Emmons completed in August and September the examination of the precious-metal deposits in the Phillipsburg quadrangle, Montana. On his return he spent two weeks in a reconnaissance of the gold-bearing districts of the Little Rocky Mountains. A brief report on these districts is contained in this bulletin.

IDAHO.

Little work has been done in Idaho during 1907. F. B. Weeks and V. C. Heikes spent a few days in the Fort Hall mining district, near Pocatello, and have prepared a paper on the occurrences in this region, which will be found in the following pages.

NEW MEXICO,

No geologic work was done in New Mexico in connection with metalliferous deposits. A topographic map of the Silver City quadrangle, which includes the long known gold and silver veins of Pinos Altos and the recently developed copper deposits of the Burro Mountains, has been prepared during the year.

NEVADA.

No extensive work was accomplished in Nevada during 1907. In the fall F. B. Weeks visited the gold and tungsten veins in the vicinity of Osceola, and has prepared a paper on these occurrences for the present bulletin.

Owing to ill health, F. L. Ransome was unable to visit Goldfield in order to gather the data of the latest developments in that camp for his forthcoming professional paper. A topographic map of the region extending northward from Tonopah, embracing 1° of latitude by 1° of longitude, was prepared by the topographic branch.

OREGON.

J. S. Diller and G. F. Kay investigated the gold veins and placers of the Riddles quadrangle, in southwestern Oregon, in connection with the general geologic mapping. A summary of the mining developments in that area has been prepared for the present bulletin.

MISSISSIPPI VALLEY STATES.

In the early part of the year C. E. Siebenthal completed the field work in the Wyandotte quadrangle, Oklahoma-Missouri, and studied, in connection with the general geology, the lead and zinc deposits occurring in that area.

In November F. L. Hess devoted a week to a reconnaissance of some interesting veins of antimony ore in Arkansas, a report on which is contained in this bulletin.

APPALACHIAN STATES.

In the Southern Appalachian States Arthur Keith completed the mapping of the Dahlonega special area, Georgia, which will form the basis for a report on the important gold mines of this vicinity.

D. B. Sterrett investigated the gold-bearing gravels of the Morganton quadrangle, North Carolina, in connection with the general geologic mapping.

In November, 1907, H. D. McCaskey undertook a reconnaissance of the gold deposits in Alabama, concerning which comparatively little authentic information has been available. The veins are in the main similar to those of Georgia and North Carolina. A preliminary note describing these occurrences will be found in the pages of this bulletin.

A brief reconnaissance of the molybdenum deposits of Maine was made by F. L. Hess, who gives an account of them in this bulletin.

GOLD AND SILVER.

A GEOLOGICAL ANALYSIS OF THE SILVER PRODUC-
TION OF THE UNITED STATES IN 1906.

By WALDEMAR LINDGREN.

THE PRODUCTION OF SILVER.

Previous to 1860 the production of silver in the United States was almost nominal, the yearly output, according to the tables of the Director of the Mint, reaching a maximum of only 38,500 fine ounces per annum. In 1859 the output rose to 116,000 ounces. In 1861 the figure given is 1,546,900 fine ounces, and thereafter the increase was rapid, owing to the new discoveries in Nevada and Colorado. For 1876 the figures had almost reached 30,000,000 ounces. The silver from the exceptionally rich ores near the croppings of the newly discovered deposits poured into the mints and refining works. It was the time of the bonanzas at Virginia City, Reese River, White Pine, and Eureka, in Nevada; the Silver King, in Arizona; the Lake Valley mines, in New Mexico, and the Drumlummon and Bimetallic mines, in Montana. From 1876 to 1892 the production kept on increasing, and in the latter year attained the maximum recorded, 63,500,000 fine ounces. The silver derived as a by-product from copper and lead smelting began to add its quota to that from the silver mills, and so even in the face of a declining tendency of the silver market the output of silver increased. The serious decline in the price of silver began about 1875, when the average price receded to $1.24 per ounce. For many years previous—in fact, as far back as 1833—the price had hardly ever fallen below $1.30 per ounce. From 1875 to 1892 the decline was practically continuous, and in that year of maximum production the price fell to 87 cents per ounce. The low price now began seriously to affect mining and operations had to be abandoned in many camps. Naturally the silver ores—that is, those which contained no base metals and but little gold—were

23

affected by this condition first, and the result was the decadence of the old silver milling camps where the pan-amalgamation process was the principal method of reduction. On the other hand, a large portion of the silver was now obtained as a by-product of lead and copper ores. The decline from the greatest production in 1892 covered only two years. In 1894, 49,500,000 fine ounces were mined, but the very next year the production rose to nearly 56,000,000 ounces. Since that time it has remained approximately steady, the yield in fine ounces for the last three years having been as follows: 1904, 57,-682,800; 1905, 56,101,600; 1906, 56,517,900. Meanwhile the price of silver steadily went down to a minimum of approximately 53 cents as an average for the year 1902, since which time it has again gradually risen to 67 cents per ounce as an average for 1906. The great reduction in price has found but little expression in the quantity annually recovered, and, as already emphasized, this is due to the steadying tendency of the precious metal obtained from the copper and lead smelters.

GEOLOGICAL CONDITIONS.

The silver ores of the United States are of many different types and are derived from deposits in many formations.

It is well known that no silver-bearing veins exist which are the exact counterpart of the gold-quartz veins containing native gold. Native silver is almost invariably due to oxidizing surface processes acting on primary argentiferous sulphides. One class of silver-bearing veins, of which many examples may be found in Montana, Idaho, and other States, is contained in granitic rocks or is accompanied by porphyries consolidated at considerable depth. The normal gangue is white massive quartz through which the sulphides are sparsely disseminated. Such veins are apt to be rich near the surface, where secondary sulphides and sulphantimonites have formed, but generally they are disappointing below the water level, where the primary ore is reached. Others of similar type, exemplified by certain veins in Clear Creek County, Colo., contain more abundant sulphides, among which galena generally predominates, and may be successfully worked by concentration even below the surface zone of enrichment.

Another class of silver veins cut through volcanic flows which have consolidated at or near the surface, and it can be satisfactorily proved that many of these veins themselves were formed at very moderate depths. The gangue is here also prevailingly quartzose, but usually very fine grained, chalcedonic, and drusy, and in many places it contains adularia. The quartz is characterized by primary argentite accompanied by very small amounts of lead, zinc, and copper sulphides. In dry climates oxidation and the secondary deposition of sulphantimonites have enriched the upper parts to an extraordinary degree. Illustrations of this are found in the Tonopah and Comstock ores of Nevada, the Mogollon ores of New Mexico, and the Silver City ores of Idaho.

A third class of silver deposits are found in limestones and here generally in connection with intrusive rocks—granite, diorite, monzonite, or other porphyries. Almost without exception these ores contain lead, and usually also copper and zinc, and in their primary forms the value of the base metals generally exceeds that of the silver. Both quartz and calcite appear in the gangue. Secondary silver sulphides or sulphantimonites are less common in these ores near the surface, but native silver and especially horn silver (cerargyrite) form abundantly. A mechanical enrichment by the solution of the limestone aids the ordinary concentration by oxidation, and thus the upper parts of many such deposits are extremely rich, as exemplified at Leadville and Lake Valley. Oxidized iron and manganese ores, almost free from lead but containing silver, are among the end products of the oxidation of these deposits.

Very likely it will be found that the three classes of ores mentioned are simply diverse products of the same vein-forming action, their characteristics being dependent on the depth below the surface and varying physical conditions, or on the character of the rock affected. The exposure of the veins deposited at deeper levels requires, however, long erosion, and it will usually be found that these deep-seated veins were formed at an earlier period than those breaking up through volcanic surface flows. The Leadville deposits are probably older than those of Silver Cliff or those of the Mogollon district. They were probably deposited during the early Tertiary, whereas the silver ores from the two localities last mentioned are more likely to belong to the latter part of the same period.

Our knowledge of the individual districts is as yet imperfect, but we are rapidly approaching a stage where we are justified in attempting a classification of the ores on the basis of geological occurrence. Even now it may be done in a tentative way, for the more important districts are fairly well known.

CLASSIFICATION OF ORES.

A consistent classification of ores presents great difficulties from whatever standpoint it may be undertaken. This is only natural, as there are gradations between any ores that may be selected as types. Nevertheless it is possible to arrange them in certain groups, and such a classification has been undertaken in the chapter on gold and silver in the annual volume "Mineral Resources of the United States" issued by the Geological Survey. The basis for the classification is really metallurgical. The silver product is divided according to its derivation from placers, dry or siliceous ores, copper ores, lead ores, zinc ores, and mixed ores. The small amount of silver contained in placer gold is relatively insignificant. As copper ores are designated those which contain $2\frac{1}{2}$ per cent or more of copper, equivalent to 50 pounds per short ton. The ores containing over $4\frac{1}{2}$ per cent of lead, or 90 pounds per ton, are called lead ores. The dry or siliceous ores comprise all of those which contain only small amounts of copper, lead, or zinc or which do not contain these metals at all. In the main they are siliceous ores containing only gold and silver, although necessarily included with them are small amounts of ores containing chiefly iron pyrite or else oxidized ores rich in hematite, limonite, or oxides of manganese.

In the mines report on the production of gold and silver in 1906 the division of the silver product is as follows:

Sources of silver product in United States, 1906.

	Fine ounces.
Placers	171, 058
Dry or siliceous ores	16, 792, 799
Copper ores	15, 880, 870
Lead ores	15, 328, 653
Zinc ores	98, 423
Mixed sulphide ores	·9, 090, 650
	57, 362, 453

From the standpoint of tonnage of gold and silver ores the product of the Cordilleran States, the southern Appalachians, and Alaska in 1906 is divided as follows:

Classification of gold and silver ores produced in deep mines in United States, 1906.

	Short tons.
Dry or siliceous ores	9, 230, 616
Copper ores	10, 483, 308
Lead ores	2, 270, 822
Zinc ores	68, 296
Mixed sulphide ores	1, 183, 785
	23, 236, 827

About 30 per cent of the total silver production is thus derived from dry or siliceous ores and these ores constitute about 39 per cent of the total tonnage of the deep mines.

THE SILVER–BASE METAL ORES.

It is first to be considered that there is a certain quantity of silver ore among the copper, lead, zinc, and mixed ores. If we define silver ores as those which contain not more than 3 ounces of gold per 100 ounces of silver or in which the value of the silver (at the present prices) is equal to or greater than the combined value of the other metals utilized in the ore, the following table (No. 1) can be constructed from the individual returns. In the compilation of this and the following tables I have had the assistance of Mr. James M. Hill.

TABLE 1.—*Tonnage and metallic products, by States in the United States, in 1906, of ores classed as copper, lead, zinc, and mixed ores, in which silver predominates in value.*

State.	Ore.	Silver.	Gold.	Copper.	Lead.	Zinc.
	Short tons.	Ounces.		Pounds.	Pounds.	Pounds.
Arizona	198	24,881	$174	8,844	1,221	
California	450	34,080	1,800		143,880	206,000
Colorado	178,809	3,462,654	335,785	1,564,607	16,037,819	4,868,362
Idaho	49,421	215,736	960	5,818	1,803,401	10,399
Montana	22,743	631,335	66,427	345,262	1,216,369	
Nevada	3,943	253,784	17,558	42,670	1,168,426	492
Utah	94,968	2,757,094	234,719	1,116,898	13,149,471	
	350,532	7,379,514	657,423	3,084,099	33,520,567	5,085,253

Of the great output of copper, lead, zinc, and mixed ores, approximating 14,000,000 tons, only 350,532 tons can be properly classed as silver ores. These are derived from 134 mines in seven States. Out of about 41,000,000 fine ounces of silver from such copper, lead, zinc, and mixed ores only 7,379,514 ounces, or about one-sixth, were derived from silver ores which average about 21 ounces of silver, $1.88 in gold, 8.6 pounds of copper, 94 pounds of lead, and 14 pounds of zinc per ton.

It is seen at a glance that the only two really important States, so far as these ores are concerned, are Colorado and Utah.

The product of Arizona is derived from imperfectly known districts in Gila, Mohave, and Santa Cruz counties, and in part probably represents oxidized and enriched ores. The product of California, likewise small, comes from Inyo and Orange counties.

Colorado yields about one-half of this class of ores, and they contain on an average nearly $10 in gold per 100 ounces of silver. Replacement deposits of deep-seated type in limestone of Leadville and Aspen produce 1,632,000 ounces, or about one-half of the Colorado total. A smaller part of this consists of oxidized ores. Fissure veins in various older rocks, principally deposits connected with intrusive rocks in Summit, Park, Gunnison, and Clear Creek counties, contribute 438,000 ounces. Clear Creek County alone yields 280,000 ounces mainly from sulphide ores, a small part of which were enriched by secondary sulphides. Fissure veins in volcanic flows or in close connection with them add 1,408,000 ounces to the total, and ores of this kind were confined almost exclusively to the San Juan country.

Idaho has only the small production of 215,736 ounces, which is derived from galena-tetrahedrite veins of the Wood River district and from the Gold Hunter mine in the Cœur d'Alene district. The Wood River veins cut through limestone; the Gold Hunter deposit occurs in quartzite. Both belong to the comparatively deep-seated type.

In Montana 631,335 ounces of silver are produced from ores of this type, and all of them are lead ores. Practically all these ores occur in fissure veins in or about the contacts of intrusive quartz-monzonite stocks, and a few of the deposits are contained in limestone. Silver-bow County (Butte) adds 367,000 ounces and the remainder is distributed through Beaverhead, Cascade (Neihart), Granite, Jefferson, and Lewis and Clark counties.

Few lead ores rich in silver are mined now in Nevada, although formerly large quantities were obtained, chiefly from the Eureka district. In 1906 Nevada contributed only 253,784 ounces, which were principally divided between the Lone Mountain district in Esmeralda County, the Eureka and Cortez districts in Eureka County, the Reese River and Bullion districts in Lander County, and the Hunter and Newark districts in White Pine County. So far as known most of the deposits belong to the older series associated with intrusive granitic and porphyritic rocks. Many of them, as for instance those of White Pine and Eureka, are replacement deposits in limestone; others, as for instance those of Reese River, are normal fissure veins cutting through granitic rocks.

Utah contributes 2,757,094 ounces, or almost as much as Colorado. Practically the whole output is divided between Tintic and Park City, being derived from replacement deposits in limestone associated with intrusive rocks. Tintic yields 2,200,000 ounces with copper as well as lead, and Park City furnishes the remainder.

In conclusion, it appears that of about 7,400,000 ounces derived from lead ores rich in silver 4,600,000 ounces are obtained from replacement deposits in limestone; 1,323,000 ounces from fissure veins in various rocks, standing in close connection with intrusive rocks; and finally 1,408,000 ounces from fissure veins in Tertiary lava flows.

THE SILICEOUS SILVER ORES.

GENERAL STATEMENT.

Next comes the question how much of the 9,000,000 tons of dry or siliceous ores, containing about 17,000,000 ounces of silver, is derived from ores which can be classed as silver ores. Dry or siliceous silver ores may be rather arbitrarily defined as those in which the value of the silver is equal to or greater than that of the gold, and in which copper and lead are below $2\frac{1}{2}$ per cent and $4\frac{1}{2}$ per cent, respectively. In other words, on the basis of 62 cents per ounce the ore contains at least 100 ounces of silver to 3 ounces of gold. The classification of the returns results in the following table:

TABLE 2.—*Tonnage and metallic products, by States in the United States, in 1906, of ores classed as dry or siliceous in which the proportion of gold to silver by weight is 3 : 100 or less.*

State.	Ores.	Silver.	Gold.	· Lead.	Copper.
	Short tons.	*Ounces.*		*Pounds.*	*Pounds.*
Arizona	180,160	1,147,089	$253,139	2,119,165	725
California	4,845	104,055	7,200	----------	----------
Colorado	260,891	2,438,202	88,087	3,412,963	61,601
Idaho	36,526	809,814	85,727	920,269	----------
Montana	30,399	843,434	105,789	553,188	3,600
Nevada	124,451	6,080,318	1,401,248	5,725	3,699
New Mexico	19,906	809,776	128,762	11,597	1,399
Oregon	4	111	----------	----------	----------
South Dakota	600	3,600	----------	9,000	----------
Texas	22,751	292,647	----------	----------	----------
Utah	14,866	131,051	4,684	188,987	8,457
Washington	161	6,419	20	----------	----------
	695,560	12,166,516	2,074,656	7,220,844	79,481

From this table it will be seen that the United States produced in 1906 only about 700,000 tons of dry or siliceous silver ore, containing approximately 12,000,000 ounces of silver, $2,000,000, or roughly 100,000 ounces, of gold, about 7,000,000 pounds of lead, and 80,000 pounds of copper. The silver ores clearly contain very little copper and three-fourths of this comes from Colorado. On the other hand, much of the silver accompanies the lead. Another point brought out is that the gold contained in these ores falls considerably short of the proportion established for the purposes of this table. This means, of course, that the ores in which the gold and silver are present to approximately the same value, or in which the gold is only slightly less than the silver, form only a small part of the total tonnage in the table. The proportion shown by the table is about 100 : 0.8 instead of 100 : 3, as required by the rule laid down above. The silver ore is derived from 12 States, being produced at 219 mines in 103 mining districts. The average content is about 19 ounces of silver and $1.85 in gold per ton. In the number of mines as well as in the tonnage Colorado easily leads. In the quantity of silver produced, however, Nevada occupies the first rank, and in fact contributes more than half of the total production of silver and almost three-fourths of the total production of gold. The rank of the States in silver production from these ores is as follows: Nevada, Colorado, Arizona, Montana, Idaho, New Mexico, Texas, Utah, California, Oregon, Washington, and South Dakota. The only States of importance which produce silver ore without any yield of lead or copper are California, Oregon, Washington, and Texas. In the case of Texas this is not strictly true, as the tailings from the Shafter mine contain a small amount of galena.

For further discussion it will be desirable to segregate the items given in Table 2 into three classes:

1. The ores which contain lead and copper.

2. The ores which contain no lead or copper, and in which the proportion of gold to silver ranges from 3:100 to 0.5:100 by weight.

3. The ores which contain no lead or copper and in which the proportion of gold to silver is 0.5:100 or less by weight. In other words, ores which contain $10 in gold or less per 100 ounces of silver. These ores may be called pure silver ores.

THE SILVER-LEAD-COPPER ORES.

The first class is summarized in Table 3.

TABLE 3.—*Tonnage and metallic products, by States in the United States, in 1906, of silver ores classed as dry or siliceous ores, which contain lead or copper.*

State.	Ore.	Silver.	Gold.	Lead.	Copper.
	Short tons.	*Ounces.*		*Pounds.*	*Pounds.*
Arizona	66,821	620,490	$148,617	2,119,165	725
Colorado	205,987	1,654,902	79,998	3,412,963	61,601
Idaho	19,029	91,640	99	920,260	
Montana	21,429	520,205	70,608	553,183	3,600
Nevada	271	40,832	1,361	5,725	3,609
New Mexico	1,240	21,200	115,000	11,597	1,399
South Dakota	600	3,600		9,000	
Utah	14,791	125,633	3,780	188,937	8,457
	330,117	3,078,592	419,458	7,220,844	79,452

The insignificance of copper as a constituent of silver ores is again seen. Colorado, Arizona, Montana, and Utah are here the most important States. A total of 3,078,592 ounces is derived from these ores, which average 8 ounces of silver, $1.30 in gold, and 22 pounds of lead per ton. The table in general comprises the same kind of ores as shown in Table 1, but with smaller percentages of lead. The largest part of the Arizona product is contributed by the Tombstone mine, working veins and replacements in limestone, quartzite, and shale near bodies of intrusive rocks.

In Colorado ores of this type may again be separated into three classes. Replacement deposits in limestone and quartzite, some from Leadville, but the majority from Aspen, aggregate 540,000 ounces. Fissure veins connected with intrusive rocks yield 130,000 ounces, chiefly from Clear Creek, but also from Park, Gunnison, and other counties. The fissure veins contained in volcanic surface flows yield 900,000 ounces, and these ores are derived chiefly from Mineral County (Creede), but also from San Miguel, Custer, Hinsdale, Ouray, and San Juan counties—all, with the exception of Custer, in the southwestern part of the State. For Idaho is recorded 19,000 ounces, from the replacement veins in limestone of the Wood River region.   Montana contributes 520,000 ounces, practically all from older fissure veins in Cascade, Granite, Jefferson, and Madison counties.   A small quantity, nearly 41,000 ounces, comes from the Reese River and Columbus districts in Nevada, presumably from older fissure veins.   New Mexico adds 21,000 ounces from similar deposits in Luna and Dona Ana counties.   Utah produces about 126,000 ounces from replacement deposits in limestone at Tintic and Park City.

To sum up, the silver ores containing a little lead (or copper) are derived approximately as follows:

Sources of silver-lead-copper ores in the United States, 1906.

                                                                     Ounces.
1. Replacement deposits in limestone------------------- 1, 305, 000
2. Fissure veins, connected with intrusive rocks (not
     always to be separated strictly from No. 1) ------- 712, 000
3. Fissure veins in Tertiary volcanic surface flows----- 900, 000

                                                         2, 917, 000

The small balance is from deposits of unknown character.

THE SILVER-GOLD ORES.

The second class of ores is summarized in Table 4, which shows that about 7,500,000 ounces of silver and over $1,500,000 in gold are obtained from dry or siliceous silver ores without reported lead or copper and with gold ranging from 0.5 to 3 ounces per 100 ounces of silver.   This ore comes from 8 States, being the output of 50 mines in 22 mining districts, and averages about 28 ounces of silver and $6 in gold.   Compared with the total the tonnage of these ores is small, aggregating only about 270,000 short tons.

TABLE 4.—*Tonnage and metallic products, by States in the United States, in 1906, of silver ores classed as dry or siliceous, containing no lead or copper and from 0.5 to 3 ounces of gold per 100 ounces of silver.*

State.	Ore.	Silver.	Gold.
	Short tons.	Ounces.	
Arizona--	104,549	434,697	$100,815
Colorado--	495	16,997	5,675
Idaho--	17,426	714,866	85,628
Montana---	4,638	122,111	24,428
Nevada--	124,020	6,021,284	1,399,833
New Mexico--	16,088	271,767	13,235
Utah---	35	2,818	845
Washington--	16	90	20
	267,267	7,584,630	1,629,975

The whole production of Arizona in this class is derived from fissure veins in volcanic flows, principally rhyolite or dacite, and the larger part of it comes from Cochise County.   Colorado produces

only a very small amount of these ores, about equally divided between
fissure veins in volcanic flows and those connected with intrusive rocks.
In Idaho the whole output is from veins connected with surface
flows of lavas in Owyhee County. Montana furnishes 122,000 ounces
from the Butte silver veins, in granite. By far the largest output is
contributed by Nevada, principally from Tonopah, which is responsible for nearly 5,700,000 ounces; but similar ores are also mined in
the Bullfrog and Fairview districts and from the Comstock. Smaller
amounts come from Humboldt, Elko, and Lyon counties. All of
these ores come from fissure veins or allied deposits in Tertiary volcanic flows. New Mexico yields 272,000 ounces, from veins in volcanic flows, chiefly from the Mogollon district.

To sum up, the ores of this class appear to be eminently characteristic of veins in Tertiary volcanic flows. Practically the entire output, except insignificant quantities from Colorado, Montana, Utah,
and Washington, is derived from such deposits.

THE SILVER ORES.

The third class of the dry or siliceous silver ores may be called the
pure silver ores. They consist of those which contain no reported
lead or copper and in which the proportion of gold to silver is very
small—less than 0.5 ounce of gold per 100 ounces of silver. The production of these ores by States is illustrated in Table 5.

TABLE 5.—*Tonnage and metallic products, by States in the United States, in
1906, of silver ores containing no lead or copper and less than 0.5 ounce of
gold per ton.*

State.	Ore.	Silver.	Gold.
	Short tons.	*Ounces.*	
Arizona	8,790	91,902	$3,707
California	4,845	104,055	7,200
Colorado	54,459	786,213	2,414
Idaho	70	3,308	
Montana	4,333	201,118	10,763
Nevada	160	18,202	554
New Mexico	2,578	16,809	527
Oregon	4	111	
Texas	22,751	292,647	
Utah	40	2,600	58
Washington	145	6,329	
	98,175	1,503,294	25,223

The table shows that only about 1,500,000 ounces, or one-fortieth
of the total production, are derived from these ores, which were extracted from 79 mines in 50 mining districts and average only 15
ounces in silver and 25 cents in gold per ton. Less than 100,000 tons
were mined. The ores are derived chiefly from Colorado, Texas, and
Montana. The Arizona ores are mixed and seem to be largely oxidized ores from veins connected with intrusive rocks, in the Cerbat

Range, Mohave County, also from Globe, and from mining districts in Yavapai County. The California ores are from Inyo County, also from mines near Calico, San Bernardino County, but most of them are produced near Amalie, Kern County. About 9,000 ounces are derived from volcanic flows, and the rest of the silver comes from veins probably connected with intrusive masses.

The Colorado ores form a long list. The replacement ores in limestone (here largely oxidized ores) from Leadville aggregate 114,000 ounces; those from Aspen 57,000 ounces. The mixed ores from veins genetically connected with intrusive bodies give 58,000 ounces, and veins in eruptive flows from Silver Cliff and the San Juan region aggregate 436 ounces.

Montana contributes a little over 200,000 ounces, of which by far the greater part is from Granite County and smaller amounts from Butte and from districts in Beaverhead and Jefferson counties. The product is derived entirely from veins in granitic rocks and probably genetically connected with these intrusions.

Small amounts only are derived from scattered sources in Nevada, Idaho, New Mexico, Oregon, Utah, and Washington.

From oxidized limestone replacement ores in Texas nearly 300,000 ounces are obtained; these ores contain some lead, although it is not recovered.

In conclusion, the pure silver ores are derived as follows:

Sources of pure silver ores in the United States, 1906.

	Ounces.
Replacement ores in limestone	574,000
Fissure veins in other rocks connected with intrusives	314,000
Fissure veins in volcanic Tertiary flows	466,000
Doubtful	149,000
	1,503,000

A large part of these ores consists of oxidized surface ores. Some of them are oxidized iron ores from the upper part of the deposits and really mined as flux. In earlier years this class of oxidized silver ores was far larger than it is at present.

SUMMARY AND CONCLUSIONS.

The total production of silver in 1906 according to the mines report was 57,362,455 fine ounces. Of this amount, 40,398,596 fine ounces were recovered from lead, copper, or zinc ores, and 16,792,799 fine ounces from dry or siliceous ores.

The copper, lead, or zinc ores furnished, from ores with predominating silver value, containing 3 ounces or less of gold per 100 ounces of silver and more than $4\frac{1}{2}$ per cent of lead or $2\frac{1}{2}$ per cent of copper, 7,379,514 fine ounces.

The yield from siliceous ores can be subdivided as follows:

Classification of silver produced from siliceous ores in the United States, 1906.

	Fine ounces.
From ores containing lead or copper, having 3 ounces or less of gold per 100 ounces of silver, and less than 4½ per cent of lead or 2½ per cent of copper	3, 078, 592
From ores with no recovered lead or copper; gold in the proportion of 3 ounces to one-half ounce of gold per 100 ounces of silver	7, 584, 630
From ores with no recovered lead or copper; gold in the proportion of one-half ounce or less per 100 ounces of silver	1, 503, 294
Total silver from dry or siliceous silver ores	12, 166, 516
Total silver from dry or siliceous gold ores	4, 626, 283
	16, 792, 799

The total silver from all kinds of ores with predominating silver value is as follows:

Production of silver from all kinds of silver ores in the United States, 1906, by States.

	Fine ounces.		Fine ounces.
Arizona	1, 171, 970	Oregon	111
California	138, 085	South Dakota	3, 600
Colorado	5, 900, 850	Texas	292, 647
Idaho	1, 025, 550	Utah	2, 888, 145
Montana	1, 474, 769	Washington	6, 419
Nevada	6, 334, 102		
New Mexico	309, 776		19, 546, 030

The total tonnage of these silver ores is 1,046,092 short tons.

As shown by the foregoing statement, nearly one-third of the whole silver production is derived from ores properly to be classified as silver ores, although about one-half of these probably could not be profitably mined if no other metals were present.

From a geological standpoint the ores may be classified as follows:

Geological classification of silver derived from various ores in the United States, 1906.

[Fine ounces.]

Type of deposit.	Silver ores with much lead and copper.	Silver ores with little lead or copper.	Silver-gold ores.	Silver ores.	Total.
From replacement deposits in limestone and shale genetically connected with intrusives	4,600,000	1,305,000	----------	574,000	6,479,000
From fissure veins in various rocks genetically connected with intrusives	1,323,000	712,000	314,000	----------	2,349,000
From fissure veins in Tertiary volcanic flows	1,408,000	900,000	7,500,000	466,000	10,274,000
	7,331,000	2,917,000	7,814,000	1,040,000	19,102,000

This leaves a balance from deposits of unknown character of about 444,000 ounces.

The facts cited above bring out prominently the well-known affinity of silver for lead and quite as markedly the slight degree in which copper and zinc are associated with the silver ores properly so called. They also emphasize the selective action of limestone in the precipitation of silver-lead compounds.

The relative quantities in the last table apply of course only to present conditions. During the early days of silver mining the pure silver ores of the oxidized zones greatly prevailed. Nor do the figures give the precise relation of the absolute supply of silver ores in nature. The limestone ores carrying silver and lead are sought after by the smelters. On the other hand, there is a large supply of low-grade siliceous silver ores for which at present no great demand exists. Although free-milling quartzose ores containing $5 of gold per ton may be mined and reduced economically, a corresponding grade of siliceous silver ore with sulphides would, as a rule, be unprofitable, for if wet milling processes were adopted the necessary roasting and heavy loss would render the operation too expensive, and, on the other hand, the smelter charges would probably not be less than $5 or $7 per ton on such material. Only in the pyritic smelting process and for converter lining could such ores ordinarily be used.

On further tentative generalization it seems that the copper ores, which ordinarily contain very small amounts of silver, tend to deposit under conditions of high temperature and pressure. The greater part of the lead-silver ores, which occur in deposits genetically connected with intrusive granitic or porphyritic rocks, have probably been deposited under conditions of moderate temperature and pressure. The major part of the silver-gold ores, which contain little or no lead or copper, occur in fissure veins cutting through lava flows of Tertiary age (rhyolites, dacites, or andesites). They have been deposited comparatively near the surface and under conditions of still lower temperature and pressure, though probably never much below the temperature of boiling water.

These ores in Tertiary lavas yielded in 1906 10,274,000 ounces of silver, or more than half of the total silver from silver ores. In only a small part of them, chiefly from the San Juan region, is the silver associated with notable amounts of copper and lead.

Conditions in the Cordilleran province indicate that there is some foundation for the belief that the maximum precipitation of metals in ascending thermal waters occurs in the following order: Copper, zinc, lead, silver. The precipitation of copper takes place most easily at lower depths. The precipitation of silver is most abundant near the surface. Gold is freely deposited under widely varying conditions, though most abundantly near the surface.

NOTES ON SOME GOLD DEPOSITS OF ALABAMA.

By H. D. McCaskey.

INTRODUCTION.

During a hasty reconnaissance of parts of Alabama in November, 1907, advantage was taken of visits to various gold mines, particularly those of Hog Mountain, which have been the principal producers of the precious metals in this State for several years, to obtain some notes descriptive of the deposits. Although the treatment is preliminary, this sketch has been prepared for publication with the hope that it may supply, to some extent, demands for recent information concerning gold districts of this State. For courtesies in the field the writer is indebted to so many citizens of Alabama that only general acknowledgment can be made in a paper so brief as this.

The general relief of northeastern Alabama is marked by dissected ridges which lie in a northeast-southwest direction and correspond therefore to the prevailing Appalachian trend. These watersheds confine the main streams for the greater part, although they are at intervals cut across by drainage channels. Down their slopes to the northwest and southeast flow the minor streams whose work is reducing this entire region to a peneplain.

The altitudes of this portion of Alabama are no longer great, only a small fraction of the areas of the higher hills rising above the 1,000-foot contour. The slopes are not unusually steep and the valleys are well matured.

GEOLOGY.

The southwestern portion of the Appalachian Mountain and Piedmont Plateau belts, including the gold deposits of the Southern States, is made up for the most part of metamorphic igneous and sedimentary rocks, but includes also isolated patches of slightly metamorphosed sediments and unmetamorphosed intrusives. These rocks extend into the east-central portion of Alabama, and the Piedmont Plateau disappears under the Cretaceous sediments toward the southwest. The portion of this area of metamorphic rocks lying in Alabama has roughly the shape of an equilateral triangle, with its base lying on the middle third of the Alabama-Georgia boundary line and the apex extending nearly to Calera, about 30 miles due south of Birmingham. On the northwest lie the folded Cambrian and Silurian strata, and to the southwest are the overlapping Cretaceous formations. In addition to intense folding, there has been extensive faulting apparently along northeast-southwest lines.

The area of metamorphic and crystalline rocks of Alabama, thus roughly outlined, includes the greater part or all of Cleburne, Randolph, Clay, Chambers, Tallapoosa, Coosa, and Lee counties and parts of Talladega, Chilton, and Elmore counties, and is over 4,000 square miles in extent.

The rocks within this area include the "Talladega formation," of partly metamorphosed slates and sandstones, the Alabama representative of the Ocoee group of formations, which are of Cambrian age, according to Keith; [a] a small area of later slates, probably Devonian or early Carboniferous, lying along the northwestern border of Clay County; the older gneisses and schists; and the intrusive granites and greenstones. The general relations of these were outlined some years ago,[b] and brief notes are offered here, but considerable detailed study of them remains to be done.

The Ocoee group or "Talladega formation" as described by Smith [c] embraces a series of conglomerates, quartzites, dolomites, quartz schists, and slates. The latter two are the most common rocks and include brownish, greenish, and grayish quartz and clay schists and slates, bluish graphitic slates, and magnetitic schists. Three fairly well defined belts of the Ocoee rocks, trending from northeast to southwest, and beginning next the determined Paleozoic rocks to the northwest, alternate with three roughly corresponding areas of much more highly crystalline rocks consisting chiefly of granitic gneisses. These belts may be conveniently described as the "upper," "central," and "lower slate belts." The first two of these belts are referred to and briefly defined on pages 42 and 44. The third belt, the narrowest and best defined of the three, contains the Silver Hill, Blue Hill, Gregory Hill, and other gold deposits, once famous in Alabama gold mining, but now dormant and unproductive. A number of these old mines were visited by the writer, but all the workings were found to have been abandoned for many years and the time given to them was brief. This belt is characterized in part by a conspicuous outcrop of a light-yellowish saccharoid sandstone described by Tuomey, Phil-

[a] Keith, Arthur, Geologic Atlas U. S., folio 143, U. S. Geol. Survey, 1907, p. 3.

[b] Phillips, W. B., A preliminary report on a part of the lower gold belt of Alabama: Bull. Geol. Survey Alabama, No. 3, 1892. Brewer, W. M., Smith, E. A., Hawes, G. W., Clements, J. M., and Brooks, A. H., A preliminary report on the upper gold belt of Alabama, with supplementary notes on the most important varieties of metamorphic or crystalline rocks of Alabama : Bull. Geol. Survey Alabama, No. 5, 1896.

[c] Smith, E. A., op. cit., pp. 110–115.

lips, and Brewer.[a]   This sandstone is from 20 to 80 feet thick and extends across the country to the northeast for many miles.   It is the "Devil's Backbone" of Phillips.   The sandstone is interfoliated with the bluish auriferous slates of the Silver Hill belt of Phillips and Smith, and is itself auriferous, according to Phillips.   These rocks strike N. 20°–45° E. and their planes of schistosity dip rather steeply to the southeast.   To the southeast of this belt is a broad area of crystalline rocks extending to the Georgia line.

An isolated patch of slates lying in the upper slate belt in the northern part of Clay County was found to contain fossils thought by Smith[b] to be Carboniferous, and as these rocks had previously been classified with those of the " Talladega formation " considerable doubt was thrown on the supposed Cambrian or earlier age of the " Talladega " rocks.   The problem received the attention of C. W. Hayes and David White, who concluded after field study that these slates were unconformable with the Ocoee and probably of Devonian or later age.[c]

Of the highly metamorphosed or crystalline rocks of doubtful origin the rather coarse-grained gneiss is by far the most generally distributed.   According to Smith, these gneisses are chiefly granitic in composition, but vary in the southeastern area into more basic dioritic gneisses; and in structure they may be considered means between the end terms of slightly gneissoid granites on the one hand and highly fissile mica schists on the other.   Samples of the gneiss of Pinetuckey show megascopically a medium-grained foliated rock made up of considerable quartz, somewhat less feldspar, and white and brown mica, the latter slightly in excess.   This gneiss is cut by pegmatite dikes containing much biotite in broad sheets and feldspar that is now altered, where exposed near the surface, to kaolin.   Associated with these gneisses and with the Ocoee rocks are various mica schists of equally doubtful origin.   They are chiefly fine-grained, dark-colored, rather basic schists, so altered that their composition is not readily made out without resort to thin sections.   A dark mica is common, and in the schists of the Pinetuckey mine garnets are developed.   The garnets occur also with the quartz of the vein and the garnetiferous portion of the schist next the vein carries gold and is mined with the ore.   The garnets would seem to be later than the regional metamorphism and to be genetically related to ore deposition; but as the mine was flooded at the time of the writer's visit satisfactory determination of this point could not be made.   The schists form the walls of the vein, and their planes of schistosity are parallel

---

[a] Tuomey, M., Second Bienn. Rept. Geol. Survey Alabama, 1858.   Phillips, W. B., op. cit., pp. 58, 61.   Brewer, W. M., op. cit., pp. 6, 7.

[b] Smith, E. A., Science, new ser., vol. 18, 1903, pp. 244–246.

[c] Oral communication from Dr. C. Willard Hayes.

with those of the adjoining gneiss of the country rock. The determination of the age, relation, and origin of these gneisses and associated mica schists presents many difficulties and awaits the most careful detailed work.

Of less doubtful origin is the "Hillabee green schist" described by Brewer, Smith, Clements, and Brooks,[a] and noted by the writer at Chulafinnee and Arbacoochee. This rock occurs along the northwestern border of the gneisses and as an intrusive in the Ocoee slates. It is a light-green, fine-grained, slightly foliated schist and is pyritiferous, at least in part. The slight schistosity is brought out by the appearance of the pyrite along wavy parallel planes of foliation. The rock appears to be an altered basic igneous intrusion of later age than the Ocoee slates, but earlier than the close of the regional metamorphism.

A type of acidic intrusive is present in the granite of Hog Mountain, a medium-grained holocrystalline rock made up of quartz, orthoclase, and biotite, with some muscovite and a little plagioclase. This rock, which is probably an equivalent of the granite at Villarica, in the Marietta quadrangle, occurs as a thick tabular sheet in the Ocoee slates of Hog Mountain and is exposed in the group of hills bearing this name.

ORE DEPOSITS.

The ore deposits of the mines here described belong to two structural types, that of fissure veins, as illustrated by the Hog Mountain veins, and that of lenticular bodies lying for the greater part within planes of schistosity of the inclosing rocks, as illustrated by the ore bodies of Gold Ridge, Pinetuckey, and Tallapoosa. The country rock at Hog Mountain is an intrusive granite, that of Tallapoosa and Gold Ridge consists of Ocoee slates and schists, and that of Pinetuckey is gneiss. All the ore bodies are on the border of gneisses and granites on the one hand and of the Ocoee slates on the other. At Gold Ridge and Pinetuckey a fine-grained mica schist, carrying pyrites and gold, forms the foot wall, and in this schist garnets are developed. At Pinetuckey the garnets assume both a banded structure and the form of fresh "augen" in the schist. Garnets are also found in the vein quartz at this mine, adjoining the garnetiferous schist. Thin sections of ores and rocks have as yet been studied by the writer only of samples from the Hog Mountain mines. Here the vein quartz next the granite walls contains veinlets of tourmaline and is associated with large foils of sericite. The quartz of this vein matter is of two generations, the older being a smoky blue quartz containing fluid and gas inclusions and indeterminable opaque substances and showing strong strain shadows between crossed nicols; with this

---

[a] Op. cit., pp. 84, 120, 173, 195–197.

quartz are associated some of the gold and sulphides. The younger quartz is fresh and light colored, shows few inclusions or strain shadows, and is also associated with the sulphides. The granite of the walls is somewhat altered, the feldspar being changed largely to sericite, and tourmaline and colorless garnets being present, as observed in thin sections.

The granite of Hog Mountain is intrusive in the Ocoee group. The gneiss of Pinetuckey is reported by Mr. Sam Wallace to underlie all of the Ocoee; but drill holes have shown that granite also alternates with these rocks.[a] The garnetiferous and auriferous mica schist of the foot walls of the Gold Ridge and Pinetuckey veins has not been studied in thin section and its origin is unknown. It is conformable with the Ocoee slates of the Gold Ridge and with the gneisses of the Pinetuckey mines.

The strike of all the veins conforms mainly to the general structural trend to the northeast, being in part more easterly in the Hog Mountain veins and more northerly at Pinetuckey and Gold Ridge. The dip of the veins of the lenticular type is to the southeast, ranging from 30° to 50°; that of the Hog Mountain fissure veins is from 50° to 60° NW.

The veins of the lenticular type show sheeted or banded structure, with scales of white mica developed along parting planes. The veins of Hog Mountain indicate crushing and recrystallization. The slates adjoining the Tallapoosa vein are in many places much crumpled next the vein. The granite wall rock of Hog Mountain shows strain, but has not been rendered gneissoid.

Hydrometamorphism, or weathering, has extended everywhere to water level, which is from 40 to 80 feet below the present surface. The Ocoee slates and all the schists have been completely altered by oxidation and hydration to this level, but the granite of Hog Mountain and the gneisses are but slightly changed. All the ores have been oxidized and the gold is commonly found for the most part free-milling and associated with hydrated iron oxides above water level. The vein quartz is somewhat honeycombed in the Hog Mountain ores, and rather porous in the lenticular veins, in the oxidized zone. Below water level free gold is still found in the lenticular veins, but from 60 to 80 per cent of the total gold is so closely associated with the unaltered sulphides that the ores are not amalgamable at a profit. The " blue ores " of Hog Mountain, or those of the sulphide zone, are not free-milling in any degree.

The Hog Mountain veins are fairly regular in width and values so far as explored in depth; the lens-shaped ore bodies, however,

---

[a] Nitze, H. B. C., and Wilkins, H. A. J., Gold mining in North Carolina and adjacent southern Appalachian regions: Bull. North Carolina Geol. Survey No. 10, 1897, p. 88.

pinch and swell, part and rejoin, but they also persist, both in average width and in values, as far as they have been followed, in the sulphide zone.

Data for a satisfactory discussion of the genesis of these deposits are yet incomplete. Their age is probably post-Cambrian. In all the veins the fillings are along lines of structural weakness and deposition was followed by strain, shear, and recrystallization along the same lines. All of this would seem to have taken place before the close of regional metamorphism referred to the Appalachian upheaval. Deposition occurred probably at great depth and under high heat and pressure; and that of the Hog Mountain ores at least seems to be genetically referable to igneous after effects. As stated above, the ore deposits occur along the border of the Ocoee slates and the granites and gneisses; and the latter appear to be gneissoid granites and therefore igneous rocks. It seems probable that several thousand feet of vertical extension of the veins have been removed since their deposition; and the lower limits of the ore bodies have not yet been defined.

PRODUCTION.

The production of the precious metals in Alabama has never been great for any given year, but this State has been credited with a continuous output for more than a century. The exact date when gold was first mined in Alabama is not known. It seems quite probable that the Indians found nuggets in the streams long before the advent of the white man and beat them into rude ornaments. The early Spaniards doubtless observed these ornaments and obtained some of them or learned where they came from. Nuggets are still found in Clear Creek and elsewhere in Alabama, especially after heavy rains. Evidence that the streams were worked at least a hundred years ago was observed on the banks of a small stream in the Pinetuckey district, where in extensive dumps, apparently of old placer washings, trees at least a century old have grown. However all this may be, Phillips [a] states that the earliest records seem to point to the beginning of real gold mining in Alabama in about the year 1830.

From the reports of the Director of the Mint to 1903, inclusive, and from those of the Geological Survey for 1904, 1905, and 1906, Alabama is credited with a total production of gold and silver valued at $760,470 for the one hundred and seven years beginning with 1800. This would represent an average annual production of $7,107. Only once in the twenty-four years from 1880 to 1903, inclusive, did the recorded production reach this mark; but, chiefly owing to recent

[a] Phillips, W. B., op. cit., p. 10.

activity at the Hog Mountain mines in Tallapoosa County, the production for the three years 1904 to 1906 has been as follows:

Production of gold and silver in Alabama, 1904–1906.

Year.	Gold.	Silver.	Total.
1904	$29,300	$116	$29,416
1905	41,530	208	41,738
1906	24,921	83	25,004

The production for 1907 will probably show a slight increase over that of 1906, owing to the continued activity at Hog Mountain and recent operations at the Gold Ridge mines.

DETAILED DESCRIPTIONS.

GOLD RIDGE.

Gold Ridge is in the extreme northeast corner of Randolph County, on the Randolph-Cleburne county line and about 2 miles west of the Alabama-Georgia boundary line. It is a small hamlet situated on a hill of 1,100 feet elevation bearing the same name. Considerable placer work has been done here in the past, but records giving the production are not available. Attention has recently been attracted to this place by the introduction of northern capital and by preparations for working the ores in depth.

The exposures indicate the continuation northeastward to this hill of the " Talladega " (Ocoee) beds of the central slate belt as laid down on the geologic map of Alabama.[a] This belt extends from a point 7 miles northwest of Wetumpka, where it is 2 miles wide, northeastward through Alexander and Wedowee, increasing in width to about 9 miles at the Alabama-Georgia line. On both sides of this belt occur the crystalline igneous and metamorphic rocks. Gold Ridge is on its northern border. The hill is, for the greater part at least, made up of argillaceous and siliceous slates and schists, striking from N. 10° E. to N. 40° E., and with planes of schistosity dipping from 30° to 45° S. 80° E. to S. 60° E. Certain of the clay slates and quartz schists appear to be of sedimentary origin, and the inclosing rocks of the quartz veins belong mainly to this class of rocks. The origin of the foot wall of the Eckert vein, however, which is a highly garnetiferous mica schist, and that of a quartz-magnetite schist found in association with another vein a mile to the south, is of considerable doubt. The garnets of the Eckert foot wall are much decomposed, but are shown to be of the iron-alumina variety. Many of them are dodecahedra from 2 to 3 inches in diameter. The matrix of the

---

[a] Geologic map, Geol. Survey Alabama, Eugene A. Smith, State geologist, 1894.

schist is highly micaceous. The quartz-magnetite schist is made up almost entirely of these two minerals, as shown by a megascopic examination, with light-colored micas here and there along the planes of schistosity. Many of the magnetite crystals are 0.1 inch in average diameter. A green pyritiferous chlorite-epidote schist found bordering this belt to the north and interfoliated with certain of the other schists seems undoubtedly of igneous origin and is probably the "Hillabee schist" of Brewer and Smith,[a] thin sections of which were studied by Clements and Brooks.[a]

The mine workings are situated on the east side of the hill and about 1 mile northeast of Gold Ridge. Two veins are exposed by numerous pits, trenches, and crosscuts. The upper or Black vein shows at least 10 feet of dark quartz at the best exposure, with a hanging wall of red clay resulting from weathered slates. The foot wall is not exposed, but below this vein and above the Eckert are found the red clays characteristic of certain weathered beds of the Ocoee. The foot wall of the Eckert vein is a green garnetiferous schist, much of which is slaty in appearance. The Eckert vein is from 6 to 36 inches between walls. The general strike of both veins is from N. 10° E. to N. 20° E. and the dip is from 20° to 43° S. 80° E. to S. 60° E. One dip fault, of the normal type, was observed in a drift on a narrow quartz vein similar to the Eckert and possibly a continuation of it, about a mile south of the present workings. The throw of this fault measured 4 feet. Evidences of strike faults with similar short throws and of slickensides also prove that there has been some slight fracturing and displacement of the veins.

The Black vein has not been worked and is but little developed. It is made up of rather massive quartz, stained with manganese oxides, and is reported to average in assay value from $9 to $10 across a face of 9 feet. Systematic sampling, however, has not yet been done on this vein and its real value is unknown. On the Eckert vein an inclined shaft has been sunk for about 100 feet and drifts have been turned on the 50-foot level to the north and south for about 250 feet. The vein is not yet explored below water level and the wall rocks are weathered almost beyond recognition. The pay streak of this vein is along the foot wall, which itself carries gold and for a distance of 12 to 18 inches from the quartz is mined at a profit. The vein matter is, where exposed, a rather porous quartz of schistose structure, and white mica is developed along the planes. The vein pinches and swells and thick portions of it in cross section in weathered exposures show reentrant angles simulating grooving. One "pipe," with its long axis lying in the line of strike, showed an oval cross section with corrugated border and having no connection with

---

[a] Op. cit., pp. 84, 120.

portions of the vein either above or below. A section of this "pipe" removed from the clay looked very much like a silicified tree trunk a foot or more in diameter. The ore is free-milling above water level, and sulphides of iron have yet been encountered only to a slight extent in the veins. The garnets of the foot wall are rolled and many of them are from 1 to 2 inches in diameter, giving the schist a knotted appearance.

A steam 2-stamp mill began operations on these ores toward the last part of 1907, and, according to reports, the results have been satisfactory. Work on these ores must be considered largely experimental until development underground has gone below groundwater level.

CLEAR CREEK.

Immediately to the southeast of the determined Paleozoic formations as shown on the geologic map of Alabama [a] lies a broad belt of rocks well exposed in Blue, Talladega, and Rebecca mountains, constituting what may be termed the upper slate belt, and extending from the Cretaceous sediments at Clanton northeastward through Edwardsville into Georgia. This belt is from 7 to 25 miles wide, and is made up largely of light and dark micaceous argillites, quartzites, and conglomerates, and in the Turkey Heaven Mountains, according to Smith,[b] of dark-colored graphitic and magnetitic schists. To the southeast of this belt of rocks, mapped by Smith as "Talladega," lies a continuous belt of the "Hillabee schist," noted above, bordering a broad area of acidic gneissic and granitic rocks separating the upper and central slate belts. At Chulafinnee, about 15 miles a little to the south of west of Heflin, Cleburne County, the schist is well exposed. It is there found in close association with the Ocoee auriferous slates, as it is at Arbacoochee and Clear Creek.

Between Gold Hill and Kemp Mountain, in Cleburne County, lies the well-matured valley of Clear Creek, a tributary of Tallapoosa River flowing southwestward. These are outlying hills parallel to the Horseblock-Brymer range to the north and their axes trend N. 45° E. parallel to the strike of the rock exposures. Kemp Mountain is from 1,400 to 1,500 feet in elevation, and Gold Hill is somewhat lower. The general dip of the planes of schistosity of all the rocks is about 45° S. 45° to 60° E. On the south slope of Gold Hill, near Arbacoochee, are old workings and dumps indicating the presence of the familiar lenticular quartz veins in the schists. The works have been abandoned for so long that the shafts and slopes are caved and partly filled with surface wash. Two quartz veins lying within quartz schists and dipping 45° S. 45° E. were noted in the mouth of

---

[a] Geol. Survey Alabama, 1894.   [b] Brewer and Smith, op. cit., p. 113.

an inclined shaft about one-fourth of a mile southwest of Arbacoochee. The quartz is white and shows cavities filled with limonite.
Fragments of this quartz litter the hillside down to the flood plain
of Clear Creek and may be found, together with old pits, along the
strike (N. 45° E.) through the forest for over a mile. The veins
where measured were 2 and 6 inches thick. They are reported to
have been rich in free gold.

The flood plain of Clear Creek is about half a mile wide at the
point about the same distance southwest of Arbacoochee, where placer
operations are now being carried on. The valley floor is rather over-
grown with small shrubs, and has until recently supported some
forest growth, as is attested by the large number of tree stumps remaining. Below the soil cap lies gravel to a depth of 8 to 14 feet,
with a foot or more of white clay near the bottom. The bed rock
is reported to consist of the upturned edges of the slates, striking
here to the northeast and therefore lying parallel to the average
course of the streams. The cutting of this bed rock has exposed
many narrow veins of quartz carrying scattered rich pockets of free
gold. It would seem therefore that operations here extending to the
bed rock have worked gold in place as well as alluvial gold caught
in the rock and found with stream gravel. Well-authenticated accounts have been given to the writer of rich pockets yielding $7,000
and over in gold, but it has been difficult to determine whether the
gold has been in place or not. The probabilities would seem to
favor the conclusion that the rich pockets consist mainly of alluvial
gold. Good-sized nuggets are still found along the sides of Gold
Hill after heavy rains. As might be expected, much broken vein
quartz is mixed with the Clear Creek gravel, and samples of it from
dumps of the dredge have assayed from $3 to $4 to the ton.

Placer mining on Clear Creek has long been famous in Alabama
and has undoubtedly yielded from $50,000 to $60,000 in gold from
small workings alone. In 1905 the Clear Creek Mining Company
built a small bucket dredge with a daily capacity of 600 cubic yards
for the purpose of working the gravels on a larger scale than had
been done before. Some work was done in that year but operations
were not entirely successful and the plant was closed down. Possibly
tree stumps interfered with the dredge to some extent or local strata
of white clay balled up some of the gold; certainly it was found
that some of the gold was not in alluvial form, but in comparatively
fresh broken vein matter, and it was thought that the tailing dumps
could be worked at a profit by crushing and amalgamation. Facilities for this process were therefore added to complete an extensive
plant. At the end of 1907 the Gold Ridge Mining Company was
completing preparations to sluice the gravel by the use of giants

and to elevate it by suction pipes to a series of sluice boxes of conventional type and riffles, 120 feet in length. Operations began too recently to enable the writer to judge of the success of this plan. There seems to be considerable gold still available in the valley of Clear Creek, but a close study of the economic and geologic features involved should precede final judgment on this point.

PINETUCKEY.

Between the upper slate belt in Cleburne County and the central slate belt in Randolph County lies a wedge-shaped area of acidic gneissic rocks. This area extends in a broad belt to the southwest through Ashland and Rockford and includes the granite near Alexander, which is the same as that of Hog Mountain. As mapped by Smith [a] it wedges out where the two slate belts join near the Alabama-Georgia line. Near the Cleburne-Randolph county line and about a mile to the southwest of Micaville dikes and apophyses of pegmatite occur within this area, affording considerable supplies of kaolin and of large sheets of mica. Three of these parallel dikes have been explored and they are found to trend from southwest to northeast for at least half a mile. Where they have been worked for mica and kaolin they are from 30 to 50 feet wide, and are bordered in part by zones of light-colored facies of granite free from mica or other dark minerals. Tourmaline and ilmenite were found in large crystals in some of these pegmatitic veins by Mr. Sam Wallace, who has carried on extensive exploratory work in this area.

About 2 miles south of Micaville and a quarter of a mile northwest of Pinetuckey are the historic Pinetuckey mines. Extensive lines of old trenches, pits, and tailing dumps indicate work on the surface ores extending back for many years. Considerable modern underground work has also been done and a 20-stamp mill was erected several years ago to treat the ores by the milling-amalgamation process; but owing to difficulties in obtaining a high extraction by this method and uncertainty of the best future procedure the mines and plant have been closed for several years and the mine is now flooded.

The country rock of the ore deposits is a medium-grained gray gneiss made up of quartz, feldspar, and mica. The immediate walls of the veins are of thin sheets of dark-green mica schist in which garnets are highly developed, assuming the form of " augen " and, more conspicuously, a marked banded structure. Garnets are found also at the boundary plane between the schist and gneiss. This schist of both walls is auriferous and is reported to carry from $4 to $7 to the ton in gold, partly in pyrites. The vein matter proper is hard bluish quartz with a sheeted or banded structure and contains both

---

[a] Geologic map, Geol. Survey Alabama, 1894.

free gold and auriferous sulphides. Between bands of quartz films of muscovite are developed, and garnets are found with the quartz. All these garnets would seem to be genetically related to the ore deposits.

Several veins have been worked at Pinetuckey down to water level and some very rich ore has undoubtedly been taken out. Handsome specimens showing flakes of free gold in the quartz as large as kernels of corn are easily obtainable. The vein on which most of the modern work has been done is from 6 inches to 3 feet thick and of lenticular type. On this vein have been sunk three shafts and two winzes, and from 500 to 600 feet of drifts and stopes have been worked. The vein strikes N. 10° E. to N. 30° E. and dips about 50° S. 80° E. to S. 60° E. The outcrop is traced for nearly a mile by lines of old pits and dumps. The ore shoots are somewhat irregular, but are reported to swing almost due east.

HOG MOUNTAIN.

GEOGRAPHY.

Hog Mountain is situated in the east-central part of Alabama, in the northern part of Tallapoosa County, and is about 12 miles east of Goodwater and about the same distance a little east of north of Alexander, stations on the Central of Georgia Railway.

Hog Mountain derives its name from its profile as seen from the lower country to the west. It is formed by three prominent knolls connected by saddlebacks and lying in a north-south direction. A short distance to the east, and separated from it by a rather narrow valley, lies a somewhat similar group of knolls known as Little Hog Mountain. Of Hog Mountain proper the north knoll rises slightly above 1,000 feet, the middle knoll is about 860 feet high, and the smaller knoll to the south rises slightly above the 800-foot contour.

HISTORY.

The early history of the explorations on Hog Mountain is very imperfectly known. The first definite reference to these deposits in the reports of the Geological Survey of Alabama is apparently that of W. B. Phillips,[a] who mentions them in connection with the well-defined Goldville belt lying in the slates immediately to the southeast. The deposits were first worked along the outcrops some time before the civil war, and much free gold was undoubtedly extracted by the crude methods in use at that time. More recently shafts were sunk on the south knoll, and tunnels now inaccessible were driven, to provide ore for a 10-stamp mill situated at the base of the hill on the

---

[a] Op. cit.

western side. Within the last few years the Hillabee Gold Mining
Company has carried on a large amount of exploratory work to expose the ore bodies, and of experimental work to solve the difficult
problem of the best method of treating the ores. The Hog Mountain
mines have been the principal producers of the precious metals in
Alabama during this time.

GEOLOGY.

The slates of Hog Mountain, presumably Ocoee, form the east and
west flanks and the north slope of the hill, and a fine-grained granite,
apparently intrusive, occupies the ridge of the hill and the crests of
the knolls. This granite extends southward along the ridge, and is
probably related to the great area of presumably igneous and certainly
highly metamorphosed rock bordered by the slates and extending
from southwest to northeast through Coosa and Clay counties and
wedging out in southern Cleburne County. The general dip of the
slates is here about 60° S. 60° E. and the exposed ridge of granite
cuts across these rocks in a direction almost due north and south.

Considerable detailed work will be required to make clear the
structure of Hog Mountain. The observed dips of the slate are, however, fairly persistent in the direction S. 60° E., and in two exposures
underground on the northwestern and western sides of the hill the
slate passes under the granite. Conclusive evidence of faulting was
not observed at this place. The igneous intrusive therefore seems
to have forced its way between the planes of schistosity of the slates
and to lie in this part of Hog Mountain in a great tabular mass whose
dip would roughly conform with that of the schistosity. This question of structure is of considerable economic importance, as the veins
here profitably worked are confined to the granite. As the granite
is not gneissoid it has not been folded with the slates and is therefore
later than their dynamometamorphism.

The slate is a fine-grained mica slate, dark gray in color in the
fresh specimens and weathering to reddish brown. Some quartz
and considerable white mica may be distinguished megascopically.
Planes of schistosity are brought out by weathering, but are not
readily seen in fresh samples. White mica (muscovite) is well developed along these planes. Under the microscope the rock exhibits
a finely granular texture and allotriomorphic structure. Rounded
quartz grains have been crushed and recrystallized. Feldspars are
not determinable, being probably altered to fine scales of sericite,
which occur in aggregates of irregular form. A marked development of muscovite is noted, and there is some light-brown biotite in
small flakes, apparently secondary. Magnetite grains occur in the
muscovite very plentifully and fine needles of apatite are found in
the quartz and micas. The occurrence of tourmaline in small prisms

is observed. At the granite contact and near the Blue vein considerable pyrrhotite is found in the slates. The rock seems to be a dynamometamorphosed muddy sediment that was later somewhat altered by contact metamorphism following the granite intrusion.

The granite is a normal fine-grained granite of hypidiomorphic structure, and is made up of biotite, orthoclase, and quartz. A plagioclase with albite-oligoclase characteristics occurs very rarely, and some muscovite is noted. As inclusions are found magnetite in the biotite and apatite in the orthoclase. The orthoclase shows notable alteration to sericite. Some of the crystals of this feldspar are idiomorphic and are fairly well bounded by zones showing varying extinction. The quartz is in mosaics of irregular anhedra filling spaces between the micas and the feldspars, is fairly fresh and free from inclusions, and shows few strain shadows. The muscovite is rare, but present in shreds and flakes and as sericite in minute scales and foils. Pyrrhotite occurs rather plentifully in the sample of fresh rock obtained from the vein walls and is associated with quartz and biotite. Garnet, colorless in transmitted light, occurs locally in both anhedral and euhedral forms replacing quartz and feldspar grains, and·is isotropic in part. The rock is a typical fine-grained intrusive granite showing only traces of dynamometamorphism. From a comparison of these data with unpublished notes of C. W. Hayes on the granite at Villarica, Ga., which is the youngest acidic intrusive observed by Doctor Hayes in Alabama or Georgia and in which the ore deposits of Villarica are reported to be, the rocks would seem to be remarkably alike. Mineralogically, however, the granite at Villarica contains an excess of muscovite over biotite.

ORE DEPOSITS.

The ore deposits of Hog Mountain are fissure veins in a granite that is apparently intrusive in the slates. They are confined, so far as explored, to the granite, and pinch out or fork on reaching the contact with the slates. On the north knoll the general strike of the veins is northeast and southwest. On the middle knoll, however, and on the western slopes of the saddle between these knolls the veins strike nearly east and west, although the two largest, the Barren vein and the Blue vein, swing around to the northeast on the south slopes of the north knoll and on approaching the eastern edge of exposed granite. The veins are steeply inclined and dip to the northwest. The slates are known to cover at least a portion of the flanks of the eastern area of the granite, as the Blue vein has been followed in granite for a short distance under the slates on this side of the mountain. The veins show a general tendency to decrease in width from a line about midway between the granite-slate contacts, and to taper out irregularly and gradually on approaching the slates until at the contact they pinch out. Three of the veins on the middle knoll, whose horizontal extensions, as shown by the outcrops and shallow workings, seem to be not more than from 600 to 800 feet, apparently do not reach the slates before pinching out. They range in width from 6 inches to 8 feet. The veins have in general continued of good width and values, so far as exposed from the higher surface croppings to the 100-foot level, through a vertical distance of about 300 feet. Displacement has apparently followed the fracturing of the granite to but a very slight extent. The ore shoots have so far not been definitely determined, but their pitch is apparently to the north and is therefore somewhat flatter than the dip of the veins. The greater parts of the veins seem to be ore bearing, and as the ore is almost all of comparatively low grade the limits of the ore shoots would naturally be rather poorly defined. If the intrusive granite proves to be in a broad, thick sheet, as indicated above, the economic importance of the swinging of the ore shoots away from a westerly direction is obvious.

The chief ore mineral is auriferous pyrrhotite. With this occurs a little pyrite and rarely traces of chalcopyrite. The gold obtained is associated not only with these minerals, but also with the dark-blue quartz. The gangue is quartz, of at least two generations. The older or " blue " quartz is somewhat glassy and smoky in appearance, is dark blue in color, and is fairly clouded with minute inclusions of liquids and gases and of several opaque and semiopaque substances not determinable under the highest powers of the microscope. This quartz shows abundant strain shadows and some apparent granulation and recrystallization, indicating crushing of the vein matter. The hand specimen of ore exhibits the general appearance of brecciation of the earlier quartz filling, followed by the introduction of the sulphides, in part at least, and by fresh quartz. The latter is light-colored, ordinary vein quartz, fairly free from inclusions and undulatory extinction. Near the walls of the vein large foils of muscovite and veinlets of tourmaline occur with the quartz, and the sulphides are also present in the granite itself. The orthoclase of the granite is also altered in part to sericite and includes apatite. The granite contains garnets at the contact with the slates, where fine stringers of quartz and pyrrhotite are found. The orthoclase of the slates is altered to sericite and the presence of tourmaline was noted in thin sections. Selvage is lacking.

The gold ores of Hog Mountain have in practice proved to be almost invariably of low grade. A number of assay returns were published by W. B. Phillips[a] in 1892, and four of these were from

a Op. cit., pp. 49–54.

samples collected by him at random from old dumps. These four showed gold 2.8, 0.3, 1.1, and 0.5 ounces; silver, 0.8 ounce, trace, trace, and 0.2 ounce per ton; and total values of $58.67, $6.20, $22.73, and $10.53 per ton. Mr. A. F. Hoffer had some assays made of Hog Mountain ores in 1886 and 1887, "perhaps sixty or seventy," and, according to his recollection, the values ranged from $2 to $31 and averaged about $7.50. Extensive sampling underground has returned average assay values of $9 to $10 per ton; and 99 per cent of this is in gold. Extraction, however, has so far given considerably lower returns than this. Richer ore is occasionally found, but on the other hand the values have been at times as low as $2 according to reports quoted above.

The surface or "red" ores are oxidized and are largely free-milling. The Barren vein is so called from the fact that it carried only traces of gold at the outcrop, but the vein furnishes good ore below. At the present time the "blue" or unoxidized ore is supplying much of the output and is treated with the surface ores.

The free-milling surface ores were treated in the early days by crude washing operations. Later a 10-stamp California mill was erected. A report from the St. Louis Sampling and Testing Works, submitted in 1889 as the result of a trial mill run of a small lot of Hog Mountain ore,[a] showed a saving by amalgamation of 73.7 per cent of the gold.

The present process, worked out on the ground during two years of experiments by Mr. T. H. Aldrich, jr., and Mr. A. P. Kennedy, combines the rather original feature of heating in a revolving kiln to 350° F. with coarse crushing and cyaniding. Fine crushing had been tried and found a failure. The ore is crushed between two sets of rolls to pass through ½-inch and ¼-inch screen slots, and is then introduced directly into a revolving kiln in which a temperature of 350° F. is maintained by wood firing. From the kiln the ore is trammed direct to the cyanide vats. The process is at present experimental, but has so far been fairly successful both with the usual mixture of two-thirds of "red" ore to one-third of "blue," and with the "blue" ores alone. The effect of the kiln treatment is not an oxidizing roast, as this is not desired, but seems to be the thorough shattering of the quartz, due partly perhaps to increased internal pressure of included gases, and the great increase, thereby, of porosity in the ore and of leaching by the cyanide. As complete analyses are not yet available and as minute opaque substances in the ore are yet undetermined, intelligent discussion of processes of treatment is difficult.

---

[a] Phillips, W. B., op. cit., p. 53.

THE TALLAPOOSA MINE.

Lying in the middle of the central slate belt, as outlined in the description of the Gold Ridge mines, and about 4 miles southeast of Hog Mountain, is the Tallapoosa mine. This mine is apparently along the line of strike of the Goldville belt of gold deposits described by Phillips [a] as extending from Hillabee Bridge to Goldville, in Tallapoosa County, a distance of 14 miles. The country rock is the Ocoee slate, and the ore bodies are lens-shaped quartz veins striking N. 23° E. to N. 35° E. and dipping 45° S. 67° E. to S. 55° E. One vein only has been extensively developed. This is from 6 inches to 4 feet thick and usually lies between the walls of dark-blue fine-grained mica schist or slate, but here and there crosses the planes of schistosity at slight angles. The vein is somewhat banded in structure and consists in part of alternating bands of white and dark-blue quartz with white and brown mica developed along parting planes. In places the walls of fine-grained slate contain narrow flattened " rods " of quartz lying with longer axes parallel to the strike of the schistosity and resembling the " pipes " of similar material and structure of the Gold Ridge mines. The vein swells and pinches, parts and reunites, but holds its persistent strike and dip fairly well. The slate walls are in many places much crumpled.

At the surface the wall rock is weathered to reddish-brown clay and the vein matter is usually a rather sandy and friable white or iron-stained quartz, with parting planes of white mica. The gold is free-milling and much of it occurs in large flakes or grains. With increasing depth the gold becomes finer and for the most part locked up in iron sulphides, although even below water level it is free-milling in part. The greater part of the work in this mine has been the extraction of the oxidized ores, and rich pockets have frequently been found. The average value of the ore is reported to be about $16 to the ton.

The present workings consist of an inclined shaft down 185 feet, from which drifts have been laid off on both sides. The surface ore to water level, at a vertical depth of about 40 feet, has been largely stoped out. The process of treatment has been amalgamation, stamp milling, fine crushing, and cyanidation, and a modern plant of solid construction was erected, but was run only a short time before closing down pending the company's reorganization.

---

a Phillips, W. B., op. cit., pp. 36–48.

THE MINERAL DEPOSITS OF THE CERBAT RANGE, BLACK MOUNTAINS, AND GRAND WASH CLIFFS, MOHAVE COUNTY, ARIZ.

By F. C. Schrader.

INTRODUCTION.

The field work forming the basis of this paper was a reconnaissance made by the writer under the direction of Waldemar Lindgren from October, 1906, to February, 1907. The purpose of the work was to obtain a general idea of the mineral resources of western Arizona, concerning which relatively little has hitherto been known. The writer wishes to express here his appreciation for the valuable assistance he has received from mining men throughout the field, from members of the division of chemical and physical research of the Survey, and from Mr. Lindgren, who has also made the microscopic determinations of the rocks and ores herein described.

DESCRIPTION OF THE REGION.

LOCATION.

The region containing the deposits here described, as outlined in the accompanying sketch map (fig. 1), lies in the central part of Mohave County, Ariz., on the main line of the Atchison, Topeka and Santa Fe Railway. It is bordered on the west by Nevada and California, from which it is separated by Colorado River, and on the east by the Colorado Plateau and similar highlands extending southward, from which it is separated by the Grand Wash Cliffs and their southern continuations. It extends from a point near the Big Bend of the Colorado and the mouth of the Grand Canyon on the north to Mellen and the southern extremity of the Black Mountains on the south, a distance of 90 miles. Its width is about 75 miles and its area about 7,000 square miles. Most of the region is shown on the Camp Mohave and Diamond Creek topographic sheets of the United States Geological Survey.

The principal towns are Kingman, Hackberry, Chloride, Gold
Road, and Vivian. Kingman, the county seat of Mohave County,
situated on the railroad, is the principal distributing point for
nearly all the mining districts of northwestern Arizona and adjacent
parts of Nevada. Chloride, situated about 20 miles north of Kingman and connected with it by a branch railroad, is the principal

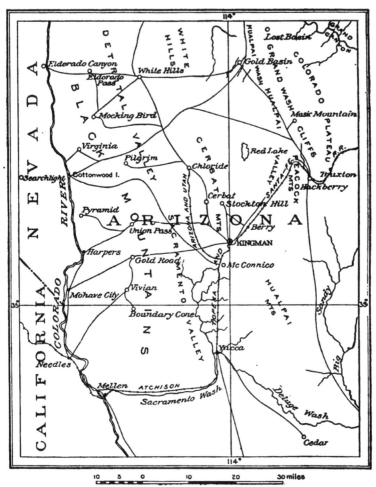

Fig. 1.—Map showing mining camps in a part of western Arizona.

trading point for the northwestern part of the region. Hackberry
is situated in the eastern part of the region, on the railroad; Gold
Road is in the western part, about 24 miles southwest of Kingman,
on the upper west slope of the Black Mountains; and Vivian is about
3 miles southwest of Gold Road.

TOPOGRAPHY AND DRAINAGE.

The principal topographic features of this region are barren desert ranges and intervening broad, plainlike, gently sloping, detritus-filled valleys, the southward extensions of the features of the Great Basin so well known in Nevada. In addition are the great trough of the Colorado on the west and the Grand Wash Cliffs on the east. There is a marked parallelism of all these features, and they trend a little north of west.

The elevation ranges from 500 feet on the southwest, at Colorado River, to 8,266 feet on Hualpai Peak, but the mountains average somewhat less than 5,000 feet in elevation, and the valleys about 2,500 feet. The general slope of the region is southwestward toward the Colorado, into which all the drainage leads.

The mountains seem in general to be due to erosion, but portions are deformational forms or fault blocks. They are more or less rugged and largely exhibit the rounded forms produced by the weathering of granite. They are composed mainly of a pre-Cambrian complex of granitoid and metamorphic rocks and are flanked or locally overlain by Tertiary or younger volcanic rocks.

The ranges, like the valleys, average 10 or 12 miles in width and their aggregate area is about equal to that of the valleys. Named in order from east to west they are the Grand Wash Cliffs, Cerbat Range, Black Mountains, and Eldorado Range. The first two are separated by Hualpai Valley, the second and third by Detrital and Sacramento valleys, and the last two by the great trough of Colorado River.

The Grand Wash Cliffs, well developed on the northeast, mark the great fault scarp between the Colorado Plateau at an elevation of 6,500 feet and the Hualpai Valley 3,000 feet below.

The Cerbat Range, which is somewhat broken above Chloride and at Kingman, consists from north to south of the White Hills, the Cerbat Mountains, and the Hualpai Mountains, as indicated on the accompanying map. The Peacock Mountains, situated between Kingman and Hackberry, are a spur or outlier of the Hualpai Mountains.

The Black Mountains lie between Detrital and Sacramento valleys on the east and the great trough of the Colorado on the west. Their western side for the most part descends from an elevation of 5,500 feet at the crest, in long, gentle, lava or gravel covered, canyon-scored, graded slopes, to the elevation of 500 or 600 feet at the river.

Rising from the great trough of the Colorado on the west to a maximum height of 6,000 feet along the eastern border of Nevada. is the Eldorado Range, which contains the Searchlight, Eldorado Canyon, and other producing camps. This range is not discussed in this paper, the descriptions being confined to the Arizona mines.

GEOLOGY.

The principal rock formations of the region are the pre-Cambrian complex, the Paleozoic sediments, and the Tertiary volcanic rocks.

PRE-CAMBRIAN ROCKS.

The pre-Cambrian complex consists essentially of coarse, in places roughly porphyritic, granite and granitoid rocks, gneisses, and schists of various kinds. It contains numerous quartz veins and lodes, in which occur the mineral deposits mined in the Cerbat Range and Grand Wash Cliffs. Its rocks have been extensively subjected to dynamic forces which have affected them in varying degrees in different regions. The dominant trend of the schistosity is about N. 30° E., with dip vertical or eastward, usually at steep angles, and the main jointing strikes north-northwest, with dip vertical or steeply inclined to the east-northeast. This latter direction is also approximately the trend of most of the fissures containing the quartz veins. These rocks are widely distributed. They largely constitute the mountain ranges and underlie the region as a whole, forming the eroded, uneven floor upon which all the other formations rest. They make up the greater portion of the Grand Wash Cliffs, whence they extend eastward beneath the Paleozoic rocks of the Colorado Plateau. They compose the greater part of the White Hills and practically the whole of the Cerbat, Hualpai, and Peacock mountains. They are also the principal rocks exposed in the northern part of the Black Mountains and are prominent in the Eldorado Range beyond Colorado River.

PALEOZOIC SEDIMENTS.

About the only Paleozoic sediments of the area are the Tonto (Cambrian?) and the Redwall (Carboniferous) formations, which form the upper part of the Grand Wash Cliffs on the east. At one time they must have covered the entire region, from which they have since been eroded.

OLDER INTRUSIVES.

The pre-Cambrian rocks are locally intruded by igneous masses and dikes which are considerably older than the Tertiary volcanic rocks next to be described, but whose age is not definitely known. The most important of these intrusives are granite and quartz syenite porphyry.

GRANITE.

The intrusive granite occurs mainly in the Cerbat Mountains, where it extends interruptedly from Stockton Hill to Chloride, being practically coextensive with the mineral belt of the mountains. The principal masses, however, occur in the Mineral Park district, where the granite forms the upper part of a prominent foothill or knob on the northwest known as·" Niggerhead." It also constitutes a considerable portion of the mountains on the opposite side of the wash to the south, and extends interruptedly throughout the greater portion of the basin eastward for a distance of 1½ miles into the slope of the range. Its structure or jointing, well shown across the northern part of the Mineral Park district, dips westward at angles of about 35°.

In a fresh specimen the granite is normally a light gray to white, medium to fine grained, granular aplitic microcline granite. It usually contains garnet and a very little biotite. The rock has a pronounced gneissoid or schistose structure, and weathers from light brown to dark, by reason of which it is not easily distinguishable in the field from its pre-Cambrian host.

QUARTZ SYENITE PORPHYRY.

The quartz syenite porphyry is a pinkish, medium to coarse grained rock, composed essentially of a quartz-orthoclase groundmass, largely orthoclase, in which the quartz occurs in granophyric growths. The principal other minerals present are biotite or chlorite derived from biotite and a small amount of magnetite, the latter occurring in small grains.

This rock occurs in two masses, one of which makes up the Hardy Mountains, an outlying group several miles in diameter, situated about 2½ miles west of Gold Road; the other occupies an area of unknown extent embracing the hills in the vicinity of the Moss mine, about 4 miles northwest of Gold Road.

TERTIARY VOLCANIC ROCKS.

GENERAL OUTLINE.

The erosion of the Paleozoic sediments from the region was followed by the eruption of a great mass of Tertiary lavas which more or less completely covered the entire area. The present bodies of these rocks, ranging from mere local outcrops to mountain masses occupying several hundred square miles and presenting sections 3,000 feet in thickness, are mere remnants of the former vast lava field which extended over the region. The rocks consist essentially of andesites, trachytes, rhyolites, dacites, and latites disposed in broad superimposed flow sheets with intercalated beds of ash, tuff, and flow breccia. They are best developed in the Black Mountains, particularly in the southern part, and contain most of the mineral deposits occurring in that range.

OLDER ANDESITE.

The oldest or basal formation of the Tertiary volcanic series, as exposed in the Vivian region, is a light-gray, purple, or pale-green medium to fine grained tufaceous andesite. It commonly contains chlorite, calcite, and biotite and is locally silicified or rhyolitic. It is at least several hundred feet thick, and occupies the border of the foothills extending from a point near the Vivian and ·Leland mines southwestward to the mesa of the Colorado, 1 or 2 miles distant. It is not known to contain mineral deposits of commercial value. West of the Leland mine it is intruded by dikes of the green chloritic andesite next to be. described.

GREEN CHLORITIC ANDESITE.

Overlying the older andesite on the northeast is the second formation of the series, an altered green andesite, locally known as " antique porphyry." It is highly porphyritic and consists essentially of innumerable phenocrysts of white feldspar about three-eighths of an inch in maximum diameter embedded in a medium to fine grained green chloritic groundmass. It is locally calcitic or pyritic. In some places it contains a little quartz and magnetite, and in others chalcedony-pyrite veinlets. In the unaltered state, however, exposures of which are few, the rock is black, with the phenocrysts fresh and glassy.

This formation extends from Vivian northward to the Mossback mine. It occupies an area of 35 or 40 square miles and has a maximum thickness of 800 feet. It is disposed in heavy beds or flows inclining gently eastward into the range.

UNDIFFERENTIATED VOLCANIC ROCKS.

Overlying the green chloritic andesite in the Gold Road district is a group of about 2,000 feet of as yet undifferentiated volcanics consisting essentially of andesites, trachytes, rhyolites, and latites. This group constitutes the bulk of the range in this locality and is important economically, as it contains the Gold Road vein and neighboring deposits. It extends from the vicinity of the Gold Road mine nearly to the summit of the range and has produced the rugged topography of that district.

RHYOLITE.

The above-described undifferentiated rocks are overlain by the fourth formation of the series, consisting of a group of rhyolites composed of tuffs, ash, flows, and breccias. In some localities this group attains a thickness of 1,000 feet or more. It extends interruptedly throughout the length of the Black Mountains, and is known as the " water rock," from springs that occur in it.

In the Cerbat Range it essentially composes the Kingman Mesa, situated between the Cerbat and Hualpai mountains, and surrounds

Kingman, where the tuffs seem in part at least to be water-laid. It also interruptedly borders a considerable portion of the western base of the Cerbat Mountains. Its dikes cut nearly all the underlying volcanic rocks as well as the pre-Cambrian complex.

YOUNGER ANDESITE.

Overlying the rhyolite and locally cutting it as dikes is the fifth formation of the series, consisting of local flows of dark-reddish andesite. It occurs in the Gold Road region and in the Cerbat Mountains northwest of Kingman. It is usually dense, but is locally vesicular and resembles the overlying basalt.

OLIVINE BASALT.

The youngest of the effusives is black olivine basalt. It occurs in sheets of two or more periods locally overlying and cutting through the older rocks, including the andesite last described and the Quaternary gravels. It is usually dense, but is locally vesicular or amygdaloidal.

MINERAL DEPOSITS.

METALLIC MINERALS.

INTRODUCTION.

The discovery of metallic mineral deposits in this region dates from early in the sixties, when rich gold ore was found at what has since been known as the Moss mine, situated about 4 miles northwest of Gold Road near the old Camp Mohave trail. A decade later the discovery of silver-gold ore in the Cerbat Range drew the attention of prospectors thither, and the rich veins of the Cerbat and Hualpai mountains were opened. The ores from these mines were so rich that large profits were returned from them, although the expense of freight and treatment ran into hundreds of dollars per ton, owing to the fact that they had to be packed long distances on burros to Colorado River, thence transported by river steamer down the Gulf of California and up the coast to San Francisco, whence they were shipped to England for treatment. This was the method of marketing ores until the advent of the railroad in 1882. During this period the region was classed as a silver camp, until the decline of silver drove prospectors back into the gold belt of the country. Later the Gold Road and Vivian mines, to which the San Francisco mining district owes its present prosperity, were found.

Up to the eighties, for want of transportation facilities and machinery, there were but few shafts more than 150 feet deep, although the production of high-grade ores ran well up into the millions.

With the advent of the railroad a new mining era began and practically the first development work below water level was inaugurated. During the last few years the industry has taken a new lease and many properties are being exploited and opened up. Old and abandoned mines are being unwatered and their dumps tested and cyanided or milled. With deeper and more systematic mining, new ore shoots are found, and the once " low-grade " ore bodies are utilized by means of the latest improved milling and value-saving apparatus. This activity, with the reduction in shipping rates to the smelters, marks an epoch in the history of mining in this region and enables the owner to handle his dump ores at a fair profit. In the early days the smelters paid for silver only, making no allowance for the gold, lead, or copper contained in the ore, but now the producer receives pay for all these metals.

GENERAL CHARACTER OF THE DEPOSITS.

The metallic deposits occurring in the ranges here described contain gold, silver, lead, copper, zinc, and tungsten, and exhibit considerable diversity in character and occurrence. They naturally fall into two very distinct groups. The first consists of quartz fissure veins containing pyrite, galena, zinc blende, and locally also arsenopyrite. The sulphides yield principally silver, but also gold. These deposits are confined chiefly to the Cerbat Range and usually occur in the pre-Cambrian rocks. It is possible that they are genetically connected with the intrusive mass of light-colored aplitic granite which breaks through the pre-Cambrian rocks in the Cerbat Mountains mainly in the Mineral Park district. They are oxidized to a depth ranging from 50 to 300 feet. The change from the oxidized to the unoxidized ore is not usually sharply defined, but occurs within a zone having a vertical range of 10 to 20 feet or more. At the present time the sulphide ores are principally utilized. In the oxidized zone cerargyrite, or horn silver, is the principal valuable mineral, but is at many places accompanied by native silver and ruby silver. The water level is found at about 400 feet below the surface.

The second group comprises the deposits of the Black Mountains. They differ markedly from those of the Cerbat Range just described in four important respects. First, they occur chiefly in the Tertiary volcanic rocks, especially in the green chloritic andesite, and are younger than the Cerbat veins. Secondly, though they occur chiefly in fissure veins, these veins seem to have originally contained a calcite gangue, which is still present in many of them. In the most valuable deposits, however, a mineralogical change has taken place by which the calcite has been replaced by quartz and adularia. Third, the values are almost exclusively gold. Fourth, the oxidation extends to a depth of 600 or 700 feet, and, as a rule, no sulphides are found. The general water level is probably about 700 or more feet below the surface.

The deposits as a whole seem to owe their origin to mineralized, principally hot-water solutions that circulated through the fissures and fractures they now occupy.

Owing to the great number of the deposits, their mode of occurrence, the " chloriding " method pursued in the past for the removal of values, and the fact that work was generally abandoned when water was encountered, the area contains a large number of small mines, which are either producing or capable of becoming productive. In the following descriptions these may be most conveniently considered by districts, of which the most important are the Chloride district in the Cerbat Range and the Gold Road district in the Black Mountains. No attempt is made to identify these areas with the legally established mining districts.

DETAILED DESCRIPTIONS.

CERBAT RANGE.

INTRODUCTION.

The principal districts in the Cerbat Range, named in order from north to south, are the Gold Basin, White Hills, Chloride, Mineral Park, Cerbat, Stockton Hill, McConnico, and Maynard. Of these the four most important are the Chloride, Mineral Park, Cerbat, and Stockton Hill districts. These are located in the middle part of the Cerbat Mountains, extending from a point just south of Stockton Hill and Cerbat to a point north of Chloride, a distance of about 12 miles, and they have certain features in common.

The rocks of this portion of the mountains, as described under " Geology," are essentially of the pre-Cambrian complex and consist of gray granite, gneissoid granite, and dark schists, including hornblende, mica, and garnet schists. They are flanked on the west in the Cerbat district by local areas of rhyolite and other Tertiary volcanic rocks. To the north, through Mineral Park and toward Chloride, they are intruded by granite masses. Furthermore, they are locally cut by dikes of pegmatite, aplitic granite, diabase, vogesite, kersantite, minette, and rhyolite.

The deposits occur in well-defined fissure veins, of which there are two sets, one striking about N. 20° or 30° W., and the other N. 70° W., usually with steep dips. Some of the deposits are also intimately associated with the vogesite, minette, and aplitic dikes.

The croppings, which are generally prominent, consist of red or dark reddish-brown iron and manganese stained quartz and altered, silicified country rock. Many of the veins are frozen to the walls;

others are separated from them by several inches of soft argillaceous or talcose gouge. The gangue is quartz and the ores are sulphides of silver, lead, copper, and zinc, generally containing gold. Silver and lead predominate in the Chloride, Mineral Park, and Stockton Hill districts, and gold and silver in the Cerbat district.

The ore deposits as seen in some mines suggest two epochs of ore deposition which have been followed by deep oxidation and sulphide enrichment. The great depth of this oxidized ore is a favorable indication for the future of the district. Many of the mines, notably in the Chloride and Mineral Park districts, which near the surface were silver mines, with increase in depth have become base-metal or lead mines, and with greater depths are becoming cupriferous. The so-called copper belt of the area extends from Mineral Park northwestward toward Chloride, a distance of several miles.

CHLORIDE DISTRICT.

General outline.—The Chloride district, the most important in the region, is situated about 20 miles north-northwest of Kingman, on the west slope of the Cerbat Mountains and the adjacent border of Sacramento Valley. It covers an irregular area about 6 miles in diameter. The mountainous portion is in part rugged, and is scored by several deep washes. Within a distance of about 2 miles the surface descends from the altitude of 6,000 feet at the crest of the range to about 4,000 feet at its base, where it meets the plain of Sacramento Valley.

Chloride, the shipping and distributing point, is favorably situated just northwest of the center of the district in the open border of the valley at 4,000 feet elevation. As a camp it dates from the early sixties and as a town from the early seventies. From that time it continued to be more or less active and in 1899 and 1900 reached its zenith, with a population of about 2,000; but for the last four or five years it has been very quiet. Several of the more important mines, however, are in operation and considerable work is being done on a score of other good properties. Lack of water has been the chief drawback, but the deeper mines show that by sinking wells plenty of water can be obtained.

Many of the veins are persistent and have an extent of nearly a mile. The ores carry principally silver and lead, with some gold and copper. They have produced several hundred thousand tons of lead and several millions of dollars in gold and silver. Exact figures are not available.

The district contains about 20 mines, located mostly in the lower slope of the mountains. Six of these mines have been opened to depths of 500 to 600 feet, and many others range from 200 to 300 feet in depth. The most important are the Tennessee, Samoa, Lucky

Boy, Towne, Pinkham, Altata, Midnight, Minnesota-Connor, Elkhart, Schuylkill, Juno, and Pay Roll, the first seven being the principal present producers. Of these the Samoa, Minnesota-Connor, and Tennessee are the most prominent.

Tennessee mine.—The Tennessee mine is located a mile east of Chloride, at the base of the mountains, its elevation being 4,050 feet. The country rock is pre-Cambrian gneiss, with granite and schist occurring near by. The gneiss is composed essentially of sericitized feldspar and crushed quartz. The mine is located on the Tennessee vein, which further north has also been opened by the Schuylkill and Elkhart mines. It is developed to the depth of 600 feet by two shafts and six levels, which aggregate about 5,000 feet of workings. It produces some water. The principal surface improvements are a well-equipped 100-ton concentrating mill and two steam hoists.

The vein dips steeply to the east. The croppings show quartz stained brown or black by iron and manganese oxides. The walls are hard, smooth, and regular and show several systems of slickensiding. In places the vein itself is fissured. The ore contains the sulphides of lead, zinc, and iron, carrying silver values and some gold and copper. Its average run of mine, omitting zinc, is about as follows: Lead, 20 to 70 per cent, concentrates 75 per cent; silver, 8 ounces, concentrates 25 ounces; gold, small amount; copper, some in deep part of mine. Of the output about one-third is high-grade shipping ore; the remainder is milled.

The mine has been productive from the surface. Thousands of tons of rich galena have been shipped to the smelter from the upper 400 feet. Here the ore shoot had a horizontal extent of about 250 feet, and was locally 15 feet in width. There is still much good ore in this section of the mine. On the 400-foot level solid galena was mined for a vein width of 21 feet and 5 inches, extending horizontally for about 40 feet. From the fourth to the fifth level there is a decrease in the value of the ore due to local increase of zinc, but from the fifth to the sixth level the ore again contains more lead. The 500-foot level contains good ore for a distance of 800 feet and the upraise from it yields much solid galena. Toward the end of the 200-foot drift north, on the 600-foot level, the vein now shows about as follows beginning on the hanging-wall side: Good ore with quartz coming in toward hanging wall, 2 feet 9 inches; milky quartz waste, 8 inches; fair-grade ore with bunches or lenses of feldspar and quartz, 7 feet. It is stated that the tailings on the dump contain much zinc blende which can be recovered by concentration. The ore is shipped to the smelter at Needles, on Colorado River, or to Deming, in southwestern New Mexico.

Samoa mine.—The Samoa mine is situated 3½ miles east of Chloride, near the crest of the range, at an elevation of about 6,000 feet.

It is developed to a depth of about 400 feet by tunnels, shafts, and drifts, aggregating over 3,000 feet of underground work. It produces some water. The principal surface improvements are two well-equipped power plants, with gasoline engines, aggregating about 90 horsepower, for operating the steam and air-compressor drills and the hoists. The country rock is principally dark medium-grained biotite granite of pre-Cambrian (?) age. It is intruded by the light aplitic granite near by. There are six veins, which strike nearly north and are either vertical or dip steeply to the east. Of these the principal producer, known as No. 3, is about 4 feet thick and its ore shoot ranges from 1 to 30 inches in width. The ore contains gold and silver, some galena, pyrite, zinc blende, and here and there a little molybdenum. As shown by the smelter return sheets from 1903 to 1906 inclusive, it averages about as follows: Gold, 1¼ ounces; silver, 15 ounces per ton; lead, 8 per cent; and zinc, 5 to 8 per cent. The total production has been about $180,000. The present rate of output is about 90 tons per month. The ore is shipped principally to the Needles smelter.

Towne mine.—The Towne mine is situated 1¼ miles southeast of Chloride, in the Sacramento Valley about one-half mile from the base of the mountains. It is developed by six shafts and drifts. It produces considerable water. The country rock is pre-Cambrian schist. A vogesite dike is associated with the vein on the foot-wall side. The vein, which is 3 to 8 feet wide, dips steeply to the north.

The gangue is quartz and the ore shoot, ranging from 3 to 18 inches in width, averages about 5 inches and favors the foot-wall or dike side of the vein. The ore contains silver, gold, copper, lead, and zinc and runs about $200 per ton mostly in gold and silver. The production from 1882 to 1906 was about $100,000.

Pinkham mine.—The Pinkham mine, perhaps the most important copper mine of the region, is located about 2 miles southeast of Chloride, near the foot of the mountains. It is developed by a 250-foot shaft and five levels containing about 1,000 feet of drift and crosscuts. It contains considerable water. The principal surface equipments are a steam hoist and two smelters, one coke and one oil, both recently installed. The country rock is pre-Cambrian granite. The vein is about 12 feet in width. It strikes N. 30° W. and is nearly vertical. The ore occurs in elongated lentils and chimneys. It is mostly chalcopyrite and bornite associated with iron sulphide, and averages about 3 per cent of copper and 18 ounces per ton of silver.

Midnight mine.—In the Midnight mine, situated near the Pinkham mine, the vein is less well defined than the Pinkham vein and contains considerable zinc. A recent carload shipment of the ore ran silver 66 ounces per ton, copper 4.5 per cent, and gold $2.50 per ton. The production under the present management is reported to be 100 tons of shipping ore, averaging $50 per ton, and 2,000 tons of milling ore, containing values of about $10 per ton.

In association with the Pinkham and Midnight deposits or cutting the pre-Cambrian rocks near by are microcline granite and diabase dikes.

General outline.—The Mineral Park district lies about 4 miles southeast of Chloride, mainly in a basin several miles in diameter, between the elevations of 4,000 and 5,000 feet. The drainage issues westward into the Sacramento Valley, mostly through Mineral Park Wash. Chloride is the principal supply point, but ore and heavy freight are hauled direct to the railroad 3 miles west.

The first locations were made in 1870, when considerable ore was soon taken out and shipped to the Selby smelter at San Francisco, at a cost for freight of $125 per ton. Production continued more or less active until 1882, since which time it has been small, only a few of the mines being worked.

The pre-Cambrian complex is here intruded by the Mineral Park mass of aplitic granite and by dikes of rhyolite, diabase, and minette. The deposits contain gold, silver, lead, and copper, which usually occur together, mostly in fissure veins or lodes, some of which are extensive. The mines, numbering about 20, are small. Few of them exceed 300 feet in depth or 1,000 feet in amount of underground work. Some of the principal ones are the Rural, Buckeye, Ark, Queen Bee, Tyler, Keystone, Fairchild, Metallic, Accident, Lady Bug, Standard, and Golden Star. The most important producers at present are the Keystone, Tyler, and Queen Bee.

Keystone mine.—The Keystone mine is situated on open ground, about one-fourth mile east of Mineral Park. It is developed mainly by a 450-foot shaft and 500 feet of drifts, mostly down to the 150-foot level, above which the greater part of the ore is worked out. The principal country rock is aplitic granite. The vein dips steeply to the north. The ore occurs mostly as lenses about 1 foot wide in a quartz gangue. It contains silver, copper, zinc, and iron, the better grade running 200 ounces per ton in silver, some of which is glance; $2\frac{1}{2}$ per cent of copper; 8 or 10 per cent of zinc; and about 12 per cent of iron. Locally it is irregularly banded toward the outside of the lenses. It is richest where the vein pinches. The ore averaging $20 or more per ton is shipped to the Humboldt smelter. None lower than this grade is handled. The output is about 20 tons per month. As most of the ore runs $12 to $15 per ton, it should be milled on the ground. The total production is stated to be $50,000.

Tyler mine.—The Tyler mine is located $2\frac{3}{4}$ miles southeast of Mineral Park, near the summit of the range, on a steep northeast slope.

It is developed by crosscut tunnels and drifts, mostly within a vertical range of 100 feet. The country rock is sheared pre-Cambrian biotite granite. The vein has a width of about 40 feet, and dips steeply southwestward into the mountain. It seems to consist mainly of an altered and replaced crushed aplitic granite or rhyolite dike. The values favor the foot-wall side of the vein, being greatest near its contact with the granite. This mine produces gold-silver-lead ore. The last carload shipment made at the time of the writer's visit averaged: Gold 3.16 ounces and silver 8 ounces per ton, and lead 17.5 per cent.

Rural and Buckeye mines.—These two mines are located in the northeastern part of the Mineral Park district, at an elevation of about 5,000 feet. They are but a few hundred feet apart and are situated on the same vein, the Rural being on the west and the Buckeye on the east side of the same gulch. The principal developments in the Rural consist of a 200-foot shaft and about 100 feet of drift, and in the Buckeye of 750 feet of drift, toward the face of which the vein is faulted off to the north by a lateral throw of about 75 feet. The Rural shaft contains water. The vein in the Rural mine dips southward at angles of about 80°, but in the Buckeye it dips to the north at angles of about 70°. It is 2 to 8 feet thick and is associated with a dike of the aplitic granite intruded into the country rock, which is pre-Cambrian schist. It locally shows a 4-inch to 20-inch ore shoot, mostly iron and copper pyrites, with streaks of arsenopyrite, black oxide of manganese, and some chert and quartz, the quartz being more prominent in the Buckeye than in the Rural. The walls are generally frozen. The ore contains silver, gold, and copper, with the values high in gold.

Golden Star mine.—The Golden Star (formerly Lone Star) mine is located about a mile northeast of Mineral Park, on open ground. It produced rich sulphides of silver, containing gold and lead, from 1870 until 1902, when the ore seems to have fallen off in grade and become base and refractory. The mine is developed principally by a shaft 300 feet in depth and two levels, with 600 feet of drift on each level. The ore is stoped down to the 100-foot level. The vein dips steeply to the south. It is 2 to 4 feet in width, and the ore is all low grade. The total production is stated to be $375,000.

Ark and San Antonio mines.—The Ark mine, located about 2 miles southwest of Mineral Park at the west base of the mountains, is developed by a 250-foot shaft and three levels, comprising about 1,300 feet of workings. It produces considerable water. The vein, which is 5 or 6 feet in width, dips steeply to the northeast. The ore is of a sulphide character and contains gold, silver, and copper. It runs about 175 ounces of silver and 3.15 ounces of gold per ton. The production is about $150,000. Adjacent to the Ark mine is the San Antonio, which has produced $75,000.

General outline.—The Cerbat district, an area about 4 miles in diameter, is situated south of the Mineral Park district, in the foothills at an elevation of 3,500 to 5,000 feet, 3 miles east of the Arizona and Utah Railroad. It has produced more than $2,000,000. It is drained principally by Cerbat Wash, which leads westward into Sacramento Valley. The mines north of this wash are gold bearing; those to the south yield silver and lead. The principal mines are the Golden Gem, Vanderbilt, Champion, Oro Plata, Paymaster, Cerbat, New London, St. Louis, Flores, and Twins, the three first named being among the most important present producers.

Golden Gem mine.—The Golden Gem mine, located on Cerbat Wash, is developed principally by a 430-foot shaft and four levels comprising 1,200 feet of drift and stopes. The stoping is on the 130-foot level, and extends 166 feet horizontally and from 62 to 81 feet vertically. This mine yields considerable water. The vein dips steeply to the northeast. It ranges from 6 to 14 feet in width, and usually carries 2 to $6\frac{1}{2}$ feet of pay ore running from $10 upward per ton. The values favor the foot wall. The gangue is quartz. The ore is gold ore and carries also silver, locally 60 ounces per ton, lead 5 to 6 per cent, antimony and zinc a trace, and some iron pyrites. The production to date is $190,000. A 40-ton mill is now turning out about $350 worth of concentrates a day from ore formerly left on the dump.

Idaho mine.—The Idaho mine adjoins the Golden Gem on the west, and the ore is similar to the Golden Gem ore. The mine has been worked in a small way since 1871, and the total production is reported to be about $200,000.

Cerbat mine.—The Cerbat mine, located about a mile northeast of the Golden Gem mine, is 200 feet in depth. The vein is 4 to 10 feet thick, and the total production is stated to be about $300,000 in gold and silver.

Paymaster mine.—In the Paymaster mine, about $1\frac{1}{4}$ miles northeast of the Golden Gem mine, the vein dips steeply to the north. The ore contains silver and gold, runs high in values, and carries much ruby silver. The production to date is said to be $200,000. Considerable water is found in this mine.

Oro Plata mine.—The Oro Plata mine, located about a mile northeast of the Paymaster mine, is 280 feet deep and is developed by about 7,000 feet of underground work. It produces considerable water. The pre-Cambrian country rock is here intruded by the aplitic granite. The ore values are chiefly in gold and sulphide of silver, with locally some lead. They run about $37 per ton. The total production is given as $500,000.

General outline.—The Stockton Hill district joins the Cerbat district on the east, being situated on the opposite slope of the mountains, about 10 miles north of Kingman. It is about 4 miles in diameter and ranges from 3,500 to 5,500 feet in elevation. It is generally rough, but the mines are all accessible by wagon roads, in the main of easy grade. The drainage issues eastward into Hualpai Valley. The principal camp is Stockton Hill, situated in the eastern part of the district. The veins in general strike northwestward. The district contains about 10 mines, of which the principal are the Banner Group, Treasure Hill, Little Chief, Cupel, Prince George, De la Fontaine, C. O. D., and '63.

Banner Group mine.—The Banner Group mine is situated near the center of the district. It is developed by more than 2,000 feet of underground work, including the " tunnel " or drift, which extends in 1,600 feet on the vein. The vein dips steeply to the northeast. It is 6 to 8 feet in width, and the ore shoot is 2 to 2½ feet thick and favors the foot-wall side. In some localities the ore consists of pure galena, but usually it contains gold, silver, zinc, iron, and copper, the gold in places amounting to several ounces per ton. The amount of zinc increases in the deeper north portion of the mine. The production is reported to be many thousand dollars in gold, silver, and lead, the zinc thus far being culled and left on the dump. The ore is shipped to the Needles smelter.

Treasure Hill mine.—The Treasure Hill mine is located in the foothills in the southeastern part of the district. It is developed by inclined shafts and drifts, and yields a large supply of good water. The veins, six in number, dip steeply to the northeast. They average about 5 feet in thickness at the surface and widen downward. They are associated with what seems to be a small stock of the aplitic granite, and the two next to it are now being worked. The ore favors the hanging wall and occurs in shoots 100 to 200 feet in extent, with intervening clay or talcose gouge and sulphides. It runs about 100 ounces of silver and $5 to $16 in gold per ton, and 7 to 10 per cent of lead. The total production is stated to be $100,000.

Cupel mine.—The Cupel mine is situated at Stockton Hill camp. It is now being reopened and an excellent 200-ton mill and plant of the Joplin type have just been installed. It is developed to a depth of 400 feet, principally by shafts, drifts, and stopes, and is said to yield about 25,000 gallons of water per day. It is located on three veins, whose general trend is northerly. The ore in general contains ruby and horn silver, together with black sulphide of silver, but in some places is rich in high-grade galena and carries about $5 per ton in gold. About 2,000 tons of ore said to run from $6 to $7 per ton lie on the dump. The production to date is reported to be about $500,000.

Prince George mine.—The Prince George mine, located about one-fourth mile southeast of the Cupel mine, is developed by a 180-foot shaft and drifts, and is said to yield about 2,000 gallons of water a day. The vein dips steeply to the north and is about 12 feet thick. The total production is about $100,000.

De la Fontaine mine.—The De la Fontaine mine, located at the west side of the district, on the crest of the range, is 400 feet deep, and comprises about 1,400 feet of drift. The vein is 7 to 10 feet in width, and dips steeply to the north. The ore runs about 35 per cent in lead and zinc, and contains some gold. Good ore bodies, 2 to 4 feet thick and of considerable extent, are blocked out in the lower 300 feet of the mine.

'63 mine.—The '63 mine, located in the southern part of the district, is 200 feet deep and is stated to have produced a total of $500,000, mostly in rich silver ore.

Little Chief mine.—The Little Chief mine, located one-fourth mile west of Stockton Hill camp, is about 100 feet deep and contains about 1,000 feet of underground work. The vein, supposed to be one of the veins of the Treasure Hill mine already described, dips steeply to the northeast. The production, amounting to many thousand dollars, is all shipping ore, averaging in silver about 350 ounces and in gold 5 to 10 ounces per ton, with 8 to 40 per cent of lead.

C. O. D. mine.—The C. O. D. mine, located 2½ miles north of Stockton Hill camp, in the upper part of C. O. D. Gulch, is developed by a shaft 400 feet deep, drifts, and stopes, on and between two main and two subordinate levels, aggregating in all about 2,500 feet of underground work. The principal surface equipments consist of a 50-ton concentrating mill and engines. The vein dips steeply northward, and is about 7 feet thick. The ore, whose principal value is in silver, runs about as follows: Silver, 160 ounces per ton; gold, 2 ounces per ton; lead, 12 per cent; with some zinc and a little copper. Except the low-grade ore, it is mostly worked out for a distance of about 400 feet on either side of the main shaft, beyond which good ore is reported. The mine closed down late in 1904 and is now full of water. The total production is reported and in part verified by smelter return sheets to be $1,300,000, that of silver alone amounting to about $1,000,000.

GOLD BASIN DISTRICT.

The Gold Basin district is situated in the eastern part of the White Hills, in the Gold Basin mining district. It extends over a hilly area about 6 miles in diameter, sloping and draining to Hualpai Wash on the east, and ranges from 2,900 to 5,000 feet in elevation. The water supply is scanty. The nearest railway station is Hackberry, 40 miles to the south.

The deposits are mainly fissure veins in the pre-Cambrian granitoid rocks. They dip southeastward or northwestward, mainly at angles of 40° to 70°. The gangue is quartz, and the metal is gold, mostly free-milling, but it is locally associated with lead or copper, copper stain being a good index of gold values.

The principal mines are the Eldorado, Excelsior, Golden Rule, Jim Blaine, Never-get-left, O. K., and Cyclopic. They are developed chiefly by shafts (some 250 feet deep), drifts, and tunnels. The production of the district is given as about $100,000, of which about $65,000 came from the Eldorado mine and $25,000 from the O. K.

WHITE HILLS DISTRICT.

The White Hills district, about 2 miles in diameter, is a part of the Indian Secret mining district. It lies 28 miles north of Chloride, in the western border of the White Hills, at an elevation of 3,000 feet. It drains westward into Detrital Valley, and the camp. White Hills, is situated in the main wash in the southern part of the district. The country rock is pre-Cambrian granitoid gneiss. It dips eastward and is flanked or overlain on the east by the Tertiary volcanic rocks, which in turn are capped by younger basalt.

The deposits are quartz veins, some of which are blanket veins. They are about 5 feet in average width, and dip to the northeast at angles of 20° to 70°. The ore is mostly silver chloride, much of it horn silver, with local values in gold.

The developments, some of which extend nearly 1,000 feet in depth, consist of inclines, shafts, and drifts, aggregating probably more than 10,000 feet. A plentiful supply of good water is usually reached at the depth of 400 to 600 feet. The district contains 12 or 15 mines, of which the most important are situated within three-fourths of a mile of the camp, but only a few of the smaller ones, located in the northern part of the district, are now worked. The camp has a well-equipped 40-stamp mill. Mineral was first discovered here in 1892 and the camp soon reached its zenith, with a population of about 1,200, but has been almost deserted since the decline in the market value of silver. The total production, which is known to be large, is reported to be about $3,000,000.

M'CONNICO DISTRICT.

The McConnico district is situated about 6 miles southwest of Kingman, east of McConnico station on the Santa Fe Railway, in the border of Sacramento Valley and the adjacent foothills and scarp of Kingman Mesa, between the elevations of 2,800 and 3,500 feet. It trends north and south and has a length of about 4 miles.

The country rock is pre-Cambrian granite, which in the valley portion of the district is covered by wash débris.

The ore deposits are contained principally in gold-bearing pegmatite dikes and shear zones, cutting the granite in a northerly direction and, in some places, associated with later basic intrusives. The principal deposits are those of the Bimetallic mine, the McKesson Group, and the Boulder Creek Group.

MAYNARD DISTRICT.

The Maynard district is an indefinite area in the Hualpai Mountains, 10 miles or more southeast of Kingman, in which strong and persistent veins, similar to those in the Cerbat Mountains, occur, mostly in pre-Cambrian red granite. Among the most important of these veins are the American Flag, Enterprise, Great Eastern Group, and Siamese Group. The ore shoots, usually rich, range from an inch to 3 feet in thickness. The ore is principally horn silver, but that of the Siamese Group is promising in copper. Water can be derived from most of the mines and pine timber suitable for mining grows near by.

Other metals reported to occur in the Hualpai Mountains are molybdenite, found in association with copper, and native quicksilver, associated with lead carbonate. Just beyond, in the Aquarius Range, about 50 miles from Kingman, is a tungsten mine which annually produces 25 tons of tungsten ore, worth $400 per ton at Kingman, whence it is shipped.

GRAND WASH CLIFFS.

The principal districts in the Grand Wash Cliffs are Music Mountain and Lost Basin.

MUSIC MOUNTAIN DISTRICT.

The Music Mountain district, about 3 miles in diameter, lies about 25 miles north of Hackberry, in the foothills of the range, between 3,000 and 4,000 feet in elevation. · The country rock is the pre-Cambrian (?) granitoid complex and it is intruded by dikes and masses, mostly basic.

The deposits occur in several quartz fissure veins, which dip steeply to the northeast. The most important is the Ellen Jane vein, about 4 feet in average width, on which the Ellen Jane mine is located. This mine is developed principally by a main shaft 200 feet deep and two levels, containing about 300 feet of drifts, and by 600 feet of adit drifts. It produces about 5 barrels of potable water per day. The ore shoot is from 4 to 6 inches thick and the ore is said to run $200 to $300 in gold per ton.

LOST BASIN DISTRICT.

The Lost Basin district lies in the northern part of the region, between Hualpai Wash on the west and Pierce Mill Canyon on the east.

The principal deposits are located about 7 miles northeast of Gold Basin, between 2,000 and 3,000 feet in elevation. They trend east and west and extend for about 6 miles. They occur principally in the pre-Cambrian rocks, in quartz fissure veins, of which there are two sets. Those on the west trend northward, with a steep easterly dip, and are principally gold bearing; those on the east trend west-northwestward and are chiefly copper bearing. The nearest water supply is Colorado River at Scanlon Ferry, 7 miles to the north.

BLACK MOUNTAINS.

INTRODUCTION.

The deposits of the Black Mountains, as stated elsewhere, differ markedly from those of the Cerbat Range, in that they occur chiefly in the Tertiary volcanic rocks. Their gangue is chiefly calcite or calcite replaced by quartz and adularia; they are deeply oxidized and, as a rule, contain no sulphides; and their values are almost exclusively gold, there being usually no base metals present.

The districts in the Black Mountains, named from north to south, are the Eldorado Pass, Gold Bug, Mocking Bird, Virginia, Pilgrim, Union Pass, Gold Road, Vivian, and Boundary Cone. Of these the most important is the Gold Road district. The two first named are in the Eldorado Pass Mining District, the next two are in the Weaver mining district, and the Gold Road and Vivian in the San Francisco mining district.

ELDORADO PASS DISTRICT.

The Eldorado Pass district, an area about 2 miles in diameter, is located west of White Hills at Eldorado Pass on the road leading to Eldorado Canyon, at an elevation between 2,500 and 3,000 feet.

The topography is one of gentle relief. The country rock is the pre-Cambrian granite, and it is intruded and locally overlain by the Tertiary volcanic rocks. The principal properties are the Burrows, Bagg, Young, and Pauly. They are in the prospect stage, but have produced some gold, the production of the Burrows being reported to be about $10,000. The metals of the three first named are gold and silver in quartz veins, and the Pauly, which seems to be at the contact between the granite and the volcanic rock, contains principally copper.

GOLD BUG DISTRICT.

The Gold Bug district is situated near the summit of the range about 3 miles south of Eldorado Pass. The only mine is the Gold

Bug mine, which is developed by several shafts and drifts to the depth of 300 feet.

The deposits consist chiefly of two main veins situated about 22 feet apart in minette or andesitic tuff similar to that at the Mocking Bird mine. (See below.) The veins dip steeply to the northeast. The ore is rich gold-bearing quartz with a little silver. It favors the hanging wall, where in part it is associated with a diorite (?) dike. The production of both shipping and milling ore has been considerable, and much ore is said to remain in the mine.

MOCKING BIRD DISTRICT.

General outline.—The Mocking Bird district lies 25 miles northwest of Chloride in a reentrant parallel side valley in the east foothills of the range, at elevations between 3,000 and 4,000 feet. The valley is bounded on the east by a spur of volcanic rocks extending in a northerly direction from the flank of the range. The district trends north and south; it has a length of about 5 miles and a width of about 2½ miles. It is hilly on the south but open on the north, merging with the Sacramento Valley, which receives its drainage. Water is scarce, but some is encountered in the mines. The principal mines are the Mocking Bird, Hall, Great West, and Pocahontas. The Mocking Bird and Hall are now producing.

Mocking Bird mine.—The Mocking Bird mine is situated in the open northern portion of the district. Its principal developments are 12 or 15 shafts, ranging from 25 to 60 feet in depth, and about 500 feet of drift. The vein lies nearly flat in a local sheet or flat-lying dike of oxidized and schistose minette or andesitic tuff. It is about 6 feet thick and consists of red and green quartz and breccia. The metal is gold with a small amount of silver. The gold occurs in a finely divided state, usually associated with hematite, of which considerable is present. The ore averages about $10 per ton. The production to date is reported to be upward of $20,000; several times this amount of ore is blocked out in the mine and valuable tailings are now on the dump.

Hall mine.—The remaining mines are situated near together in the low foothills of pre-Cambrian granite in the southern part of the district. The veins here all dip steeply to the north. The Hall mine, which works the most northerly of the veins, is developed principally by a 210-foot shaft and two levels, containing about 200 feet of drift. The vein ranges in thickness from a few inches to 2 feet, and is locally associated with diabase dikes. The gangue is quartz, some being of the honeycomb variety. Some of the ore is very rich, shows free gold, and is said to contain values of $10,000 and upward per ton. A 24-ton mill is operated at the mine.

Great West mine.—The Great West mine is developed principally by adit drifts, shafts, crosscuts, and winzes to a depth of more than 200 feet. The vein is approximately 3 feet thick and consists essentially of gold-bearing, iron-stained, oxidized quartz, reported to run from $10 to $80 per ton.

Pocahontas mine.—The Pocahontas mine is developed by a shaft and drifts to the depth of 200 feet, and a new cyanide mill has recently been built to replace the stamp-amalgamation mill formerly used. The value of the ore is in gold, which occurs mainly in white iron sulphides contained in a somewhat stained and crushed quartz gangue.

VIRGINIA DISTRICT.

The Virginia district, about 3 miles in diameter, lies 25 miles west-northwest of Chloride, near the middle of the west slope of the Black Mountains, 5 miles east of Colorado River and nearly opposite the Searchlight district, Nevada, at an elevation of about 1,500 feet.

The country rock consists of the Tertiary volcanic rocks with rhyolite and green chloritic andesite most abundant. The veins dip southwestward, usually have a calcite gangue, and locally grade into the country rock. The Red Gap vein, however, consists of more or less brecciated quartz, probably with adularia. It is similar to that in the Gold Road mine, and carries good values.

The ore of the district is mostly free-milling gold. It averages about $7 to $8 per ton, and the best values are associated with specks of hematite distributed throughout the gangue, as in much of the ore in the Union Pass and Gold Road districts.

PILGRIM DISTRICT.

The Pilgrim district, which is about 2 miles in length and trends northwestward, lies 9 miles west of Chloride, in the eastern foothills of the range, at an elevation of about 3,600 feet. The country rock is principally rhyolite and granite porphyry. The main opening is the Pilgrim mine, on the northwest. This mine is situated on a contact vein of quartz, with a little adularia, and rhyolite breccia, with trachytic rhyolite forming the hanging wall and granite porphyry the foot wall. It is developed by inclined shafts, two levels, and drifts to a depth of 360 feet, and oxidation extends to the bottom of the mine. The vein is about 20 feet thick and about half a mile in length. It dips to the west at an angle of about 30°. The ore is reported to average about $8 in gold per ton, the gold occurring free. Twelve tons of $100 ore are reported to have been shipped and about 1,000 tons of $6 to $7 ore lie on the dumps.

UNION PASS DISTRICT.

General outline.—The Union Pass district lies about 30 miles west of Kingman, on the west slope of the range. It extends from Union

Pass westward to Colorado River at Pyramid, a distance of 13 miles, and has a width of about 3 miles. It ranges in elevation from 4,000 feet on the east to 500 feet on the west. The topography is rough and mountainous on the east, with low hills, broad, open washes, and gravel-covered areas sloping toward the river on the west. The water supply is scanty except near the river. On the east it consists of a few springs or shallow wells in the mountains.

The country rock on the northwest and on the south is essentially the pre-Cambrian complex, which, in the remainder of the district, is more or less deeply buried by heavy deposits of the Tertiary volcanic rocks, consisting principally of rhyolite. On the west the deposits are fissure veins in the pre-Cambrian granite. On the east they are found chiefly along or near the contact of the rhyolite and granite; locally associated with diabase, a later intrusive; and in conjunction with one or more of these intrusives along fault planes. They occur in the form of fissure veins or lodes, blanket veins, and irregular bodies. The mineral deposits were probably formed by circulating hot solutions which accompanied and followed the invasion of the intrusives, especially the rhyolite. The most important of the deposits in the eastern part of the region trend northwestward and those in the western part northeastward; those intermediate in ·position vary between these two directions. The metal is gold. The prevailing gangue is calcite, which locally is associated with, gives way to, or is replaced by quartz and adularia.

The western part of the district has produced considerable ore, the most of which has come from the Sheep Trail, Boulevard, and Katherine mines.

Katherine mine.—The Katherine mine is situated 1¾ miles east of the river and about 450 feet above it on an open, gravel-covered slope in the pre-Cambrian granitoid gneiss. The lode is about 75 feet in width, trends N. 64° E., and to judge from outcrops in alignment, may have an extent of nearly 5 miles. In the mine it shows faulting and slickensiding.

The ore on the whole resembles that of the Gold Road mine (see below) and consists mainly of fine-grained quartz and adularia, mostly replacing calcite. The greater portion is of low grade, ranging in value from $6 to $7 per ton; but toward the north it contains a rich streak said to run $200 to $300. The ore is treated at the Sheep Trail mill, Pyramid. About 5,000 tons have been produced.

GOLD ROAD DISTRICT.

General outline.—The Gold· Road district lies about 24 miles southwest of Kingman. It extends from Meadow Creek on the east slope of the range northwestward to the Moss and Golden Star mines, about 6 miles beyond the crest. It has a length of about 10 miles and a width of about 4 miles. It lies mainly on the west slope, and ranges in elevation from 2,000 feet on the west to 4,500 feet at the top of the range. The range portion, which is about 3 miles in width, is rugged, particularly on the west, being marked by precipitous fault scarps, deep gulches and canyons. The remainder consists mainly of low rounded mountains or hills, open washes, and gravel-covered gentle slopes or mesas. On the east the drainage flows by way of Meadow Creek into Sacramento Valley and on the west through Silver Creek into Colorado River.

The principal country rock in the eastern or range portion of the district consists of Tertiary volcanic rocks; on the west appear also local areas of the pre-Cambrian complex and intrusive quartz syenite porphyry, in places covered by sheets of gravel. The deposits are chiefly gold-bearing fissure veins or lodes. They occur in the undifferentiated volcanic rocks, the green chloritic andesite, and the quartz syenite porphyry, described under "Geology," and also along certain of their contacts, where rhyolite is usually the intrusive. They number about a dozen and consist of two main types—those in which the gangue is chiefly quartz and adularia and those in which it is chiefly calcite. The former occur mostly in the undifferentiated volcanic rocks and have a general northwesterly trend; the latter occur mainly in the green chloritic andesite and trend northward. The most important of the former type is the Gold Road vein; of the latter the Pasadena and Mossback veins.

Gold Road mine.—The Gold Road mine, the most important in the district, is situated at Gold Road, on the western rugged slope of the range about 1 mile below the crest, at an elevation of about 2,900 feet. It is developed principally by a main shaft and seven levels, aggregating about 2,000 feet of drift, to the depth of 700 feet, and most of the ore has been mined for a distance of about 150 feet on either side of the shaft. The mine is located on the western part of the Gold Road vein, which extends eastward to the crest of the range, a distance of more than a mile. The vein lies mostly in a deeply cut gulch. Its croppings range in elevation from 2,800 feet on the west to 4,100 feet on the east. It strikes N. 50° W. and dips northward at angles of about 80°. It is strongest on the west, where, as developed in the mine, it is about 10 feet thick, is locally enriched by oblique stringers on the hanging-wall side, and is usually in sharp contact with well-defined walls of the country rock, consisting of andesite, trachyte, latite, and rhyolite.

The ore is nearly all milling ore and runs about $10 in gold per ton. It consists chiefly of fine-grained, light-colored, in places greenish quartz, which has a peculiar chalcedonic or drusy appearance and in which the gold is finely disseminated. Adularia is commonly intergrown with the quartz, and it is very probable that these minerals replace calcite, which is present in varying quantities. As a rule no sulphides are present, and but little limonite.

The gold is extracted on the grounds by milling and cyaniding. The ore is crushed in solution and, owing to its oxidized condition, is easily milled and treated, 25 per cent of the gold being liberated in two or three hours and 85 per cent in ten hours. The plant used consists of six Huntington mills having a total capacity of more than 200 tons per day. The present rate of output, however, is about 180 tons per day, one mill being usually held in reserve.

The principal source of the power used, amounting to about 700 horsepower, is California fuel oil, freighted by wagon from Kingman. The ore is now mined at a cost of $2.50 to $3 per ton, or mined and milled for about $5 per ton, allowing $8 ore to be handled with good profit. With the installation of cheaper power, ore of much lower grade might be worked.

The production to the end of 1906 was large, and is reported to have been considerably in excess of $1,000,000, nearly all of which was produced during 1905 and 1906. Besides the output of the Gold Road mine, nearly $1,000,000 is reported to have been produced from work done at several other points on the vein, especially on the Billy Bryan claim, about half a mile southeast of the Gold Road mine. This makes the total production of the Gold Road vein more than $2,000,000. It is stated that at the close of 1906 there were also in sight 120,000 tons of ore running about $11 per ton.

The camp has a population of about 350, of which the mine employs 180 men. It is equipped with modern improvements, is well kept, and is plentifully supplied with excellent water pumped from the east slope of the range. The mine also produces about 12 gallons of water per minute.

Pasadena mine.—The Pasadena mine is situated 1½ miles west of Gold Road, on open ground. It is developed principally by a main shaft 300 feet deep and three levels containing about 100 feet of drift. The country rock is the green chloritic andesite. It is intruded by rhyolite and basic rocks, all seemingly older than the vein. The vein or lode dips westward at angles of about 70°. It is from 10 to 140 feet in width and has a known extent of more than a mile. It consists essentially of calcite and quartz, the quartz increasing with depth. Some of it is of the Gold Road type, and this carries the best values.

In the main ore body, which ranges in width from 4 to 10 feet, the ore is reported to average about $10 per ton and is excellent cyaniding ore.

Mossback mine.—The deposit of the Mossback mine, situated about 3 miles north-northwest of Gold Road, in open country near the foot of the range, is geologically and mineralogically similar to that of the

Pasadena mine. The mine is developed to the depth of 320 feet and yields considerable water. The vein is about 23 feet wide and the main ore shoot is about 12 feet wide. About 900 tons of medium-grade milling ore, said to run from $5 to $100 per ton in gold, lie on the dump.

Moss mine.—The Moss mine lies about 4 miles northwest of Gold Road, on the southeast slope of a group of outlying hills. It is developed by a 230-foot shaft and 600 feet of tunnel and crosscuts. The vein consists mainly of calcite and a small amount of quartz. It is about 20 feet thick, and extends for more than half a mile. It is in the quartz syenite porphyry, and dips steeply southward, with fault breccia on its foot wall.

The total production is reported to be about $500,000 in gold, nearly all of which was obtained near the surface, about one-half from an excavation about 10 feet in diameter and depth. It occurred mostly as incrustations, flakes, beads, and nuggets associated with quartz and hematite, and represents a local concentration, for deeper work shows the vein to consist of a large body of low-grade ore, much of it running about $4 per ton.

Golden Star group.—The Golden Star group, including " Meals ledge," an outlying lode on the east, comprises seven or eight properties mainly in the prospect stage. It lies in open country in the northwestern part of the district, extending from a point near the Moss mine northward to Cottonwood Wash, a distance of about 2 miles.

Miller mine.—The Miller (formerly Hardy) mine is situated two miles northwest of Gold Road, on Silver Creek, on the northeast slope of the Hardy Mountains, an outlying group. It is developed by two shafts (to a depth of 289 feet) and contains 800 feet of drift, tunnels, and crosscuts. The country rock is the quartz syenite porphyry, into which rhyolite is intruded near by.

The mine is located on the north end of the Hardy vein, which dips steeply to the north. The vein is about 30 feet wide and extends about 4 miles. The Jack Pot, Homestake, and Navy Group mines are also located on or near it. It is composed mostly of quartz, probably with adularia, and contains some associated green fluor spar. The ore carries gold, but it contains more silver than that at any other mine in the district, and is stated to average on the whole about $5 per ton. In the Navy Group mine, however, 2 miles farther west, it is said to range from $8 to $30. The production to date is reported to be about $100,000.

VIVIAN DISTRICT.

General outline.—The Vivian district, an area about 4 miles in diameter, lies in the southwestern part of the region and adjoins the

Gold Road district on the south. Vivian, the principal camp, is situated west of the center of the district, about 3 miles southwest of Gold Road. The district lies in the foothills at elevations between 2,200 and 2,800 feet, and is traversed by several broad washes through which the drainage issues southwestward.

The dominant country rock is the green chloritic andesite, which, on the southwest, gives way to the underlying older andesite, the basal member of the volcanic series. So far as learned, this older andesite does not contain workable mineral deposits, and as the veins seem to practically terminate with the lower limit of the green chloritic andesite, the ability to differentiate these two formations is of vital importance to the mine operator. This ability can best be gained by studying the rocks along their zone of contact, which is well exposed crossing the ridge southwest of the Vivian mine, beyond which it ascends the wash northward to the west base of Leland Monument Mountain, the high dome-shaped portion of the ridge just west of Vivian and nearly 700 feet above it.

Leland mine.—The Leland mine is located just west of Vivian, in the upper part of Leland Monument Mountain. It is developed mainly by adit drifts and tunnels aggregating about 3,000 feet of workings, distributed within a horizontal distance of half a mile and a vertical range of 700 feet. This work was nearly all done in 1903 and 1904. The surface equipment, comprising a 40-stamp mill and railway, were installed at a cost exceeding $400,000.

The mine is located on the western part of the Leland vein, which has a known extent of about a mile and dips steeply to the south. At the mine it ranges from 5 feet in width on the east to 25 feet on the west, and is of the Gold Road type, being composed essentially of fine-grained pale-greenish quartz and adularia, locally accompanied by calcite and country-rock breccia. It contains vugs lined with quartz crystals and black oxide of manganese, and many of them carry much free gold. Throughout the mine the vein contains a large amount of $5 to $10 ore, which probably continues in depth with the green chloritic andesite. The same seems to be true of the Mitchell vein, which lies about 400 feet south of and parallel to the Leland vein, but dips steeply to the north.

In the light of the foregoing conclusion concerning the termination in depth of the veins of the district with the green chloritic andesite, the widely accepted view which holds that by sinking to a depth of 1,200 or 1,500 feet at a point midway between the Leland and Mitchell veins a large body of rich ore will be encountered at their junction should be accepted with caution, as the lower limit of the andesite may lie at too shallow a depth for the veins to meet in this formation.

The production of the mine was not learned, except that about 4,500 tons of the ore were milled, some of which averaged $15 per ton and a less amount $50 to $60 per ton.

German-American mine.—The German-American mine lies in the southern part of the district three-fourths of a mile east of Vivian. It is developed by drifts, tunnels, and shafts to a depth of about 250 feet and contains about 2,000 feet of workings, mostly at the Treadwell and 35th Parallel shafts, which are situated about 1,200 feet apart on the vein. The principal surface equipments are several steam hoists, a 10-stamp mill, and a cyanide plant.

The dominant country rock is the green chloritic andesite, which locally west of the vein on the foot-wall side is in contact with the light-gray older andesite, and the vein may lie in part on the contact of these two rocks. The vein dips steeply to the east. It extends for three-fourths of a mile and ranges from 1 to 60 feet in width. It consists essentially of calcite, some of which is replaced by quartz and adularia, with some brecciated country rock. The foot wall is usually well defined with the vein solidly frozen to it; the hanging wall is ragged. The values favor the hanging wall and are associated mostly with quartz.

The production has been 2,700 tons of ore, which averaged about $10 in gold per ton. A large dump at the 35th Parallel shaft is reported to be all good milling ore.

Tom Reed mine.—The Tom Reed (formerly the Blue Ridge) mine lies in the eastern part of the region, in Blue Ridge Wash. It is developed to a depth of nearly 200 feet by two shafts, and has three levels and drifts, aggregating about 600 feet of underground work, nearly all done in 1904 to 1906. It produces considerable water, the level of which lies at about 100 feet below the surface. The principal surface improvements are a 10-stamp mill and a gasoline hoist.

The country rock is the green chloritic andesite. The vein dips steeply northeastward. It is about 20 feet in width, with the walls usually ill-defined, and is thought to be a continuation of the Pasadena vein, in which case it has a length of about 3 miles. It is mainly of the Gold Road quartz-adularia type, with but little calcite present in the principal part of the mine, and is reported to have run $25 in gold and a little silver per ton for the first 30 feet in depth, and about $12 from that point down. As considerable gold remains in the tailings the ore should be treated by cyaniding. The production is reported to be about $120,000.

Victor-Virgin mine.—The Victor-Virgin mine is situated in the southeastern part of the district. It is developed principally by two shafts, situated 900 feet apart, to a depth of nearly 300 feet, together with drifts. Elaborate surface improvements, including the installation of a 650-horsepower plant at Needles, 16 miles distant, are now in process of construction, but hardly seem warranted for a mine of its size, in which the deposits have not yet been shown to extend to any great depth.

The dominant country rock is the green chloritic andesite, but the underlying older andesite occurs near by on the west. The vein dips steeply to the northeast. It is from 1 to 18 feet in width, but is locally associated with silicified country rock or breccia 100 or more feet wide. It is of the Gold Road type, being composed principally of quartz and adularia, with a small percentage of calcite. The ore is said to contain from $9 to $50 in gold per ton, $20 being reported as a fair average. The total production is about $500 and a considerable amount of low-grade ore lies on the dump.

Midnight mine.—The Midnight mine, located 1¼ miles northwest of Vivian, is developed by an inclined shaft, drifts, and stopes to a depth of 50 feet, and is equipped with a gasoline hoist.

The country rock is the green chloritic andesite. The vein dips westward at angles of 30° to 40°. The portion of it worked is about 3½ feet in thickness and consists principally of quartz, with some calcite, fluorite, and probably adularia. It has been worked from the surface almost to the bottom of the mine and laterally for a distance of about 120 feet, and in this extent has averaged about $18 per ton in gold, the production being considerable for so small a mine. Where now worked in the bottom of the shaft the ore is of low grade, running about $7 per ton, and consists of a mixture of partly altered soft country rock traversed by stringers and veinlets of calcite and quartz.

Vivian mine.—The Vivian mine is located about one-fourth mile below Vivian, just west of Vivian Wash. It is developed to a depth of about 270 feet by three shafts and drifts. The surface equipments, which unfortunately seem to have been prematurely installed, are new, of the best modern type, and very complete, and include a 10-stamp mill.

The country rock is the green chloritic andesite, which just west of the mine gives way to the older andesite. The vein dips steeply to the south and extends for about half a mile. It is about 3 feet in average thickness and consists essentially of calcite and a little quartz. The values are reported to occur chiefly in pockets running very high in gold, and· are richest in the quartz, which occurs in association with dark calcite. The production is small. The mill was run about forty days after its completion in 1906. The principal dumps show that the older andesite has been encountered, from which it seems probable that no workable ore will be found at greater depths.

The Boundary Cone district lies in the southwestern part of the region, just south of the Vivian district last described. It extends from Boundary Cone, a prominent landmark on the west, 3 miles eastward nearly to the crest of the range, and is 2 miles or more in width. The topography is mostly rugged, of the volcanic rock type, with the mountains rising 1,500 to 2,000 feet above the intervening washes, through which the drainage issues westward.

The country rock is mainly a dense reddish-brown or purplish andesite, locally known as " phonolite," and is underlain by an earlier tufaceous andesite, below which the pre-Cambrian granite is exposed on the extreme southwest. These rocks are all intruded by rhyolite in the form of plugs and dikes, of which Boundary Cone is a typical example.

The deposits are gold bearing and occur in fissure veins in the purple andesite and also on the contact of the intruded rhyolite and the andesite or the granite. The gangue is made up of fine-grained and usually brecciated quartz and adularia, locally containing an admixture of country rock.

The veins, five or six in number, occur in the eastern part of the district. They trend northwestward, with steep dips. The principal properties located on them are the Iowa, which has been most extensively developed, the Lazy Boy, Krause, and Highland Chief.

The Iowa mine is situated in the northeastern part of the district, at an elevation of about 2,600 feet. It is developed by a shaft 200 feet in depth, and by about 100 feet of drift and crosscuts distributed on three levels, and is equipped with a gasoline hoist. The vein dips steeply to the south. It ranges from 3 to 8 feet or more in width and consists principally of greenish brecciated quartz and a little calcite, with locally some rock breccia. It is traversed by veinlets of secondary quartz and calcite. The values run from $3 to $14 per ton, and are largely found within 3 feet of the hanging wall.

The deposits at the contact occur chiefly in the western part of the district. They are best exposed along the upper edge of the andesite collar encircling the rhyolite plug that forms the upper part of Boundary Cone. They occur mainly on the west and south sides of the cone, where they extend interruptedly for about a mile. The values average from $3 to $17 in gold, and occur in a zone of quartz and adularia 6 to 8 feet wide, chiefly on the rhyolite side of the contact. The rhyolite within 2 or 3 feet of the quartz also contains values.

NONMETALLIC MINERALS.

The principal nonmetallic minerals occurring in the region here discussed are building stone, cement rock, travertine, turquoise, and graphite.

Building stone.—The most important building stone in the region is the rhyolite tuff underlying Kingman Mesa. It occurs in abundance in heavy beds and is easily quarried in the scarps near Kingman. It is medium grained and fairly uniform in texture, and dresses well. The most important buildings in Kingman are built of it.

Cement rock.—The main cement material of the region is a fine-grained pumaceous phase of the rhyolite tuff just described. It occurs in a deposit of considerable extent west and southwest of Kingman and probably elsewhere, and is said to have been proved by experiment to be excellent for making cement. It requires no calcining. Briquets made with it are reported to have a higher tensile strength than the Vesuvian products and to stand salt-water tests with excellent results.

Travertine.—Local deposits of travertine, used as flux at the smelters, occur in Sacramento Valley near Chloride and Mineral Park.

Turquoise.—Turquoise is mined at two localities, Ithaca Peak and Turquoise Mountain, about a mile south of Mineral Park. It occurs in coarse altered granite porphyry in the form of veins and globular and irregular bodies, 1 to 8 inches in diameter, some of which are connected by stringers or mere seams and others isolated in the solid rock. The production is shipped in monthly installments to New York, where it is mostly sold in the rough for jewelry purposes.

Graphite.—Graphite, possibly of commercial value, occurs in the pre-Cambrian schists on the east slope of the Cerbat Range near the old Government trail in the first gulch north of C. O. D. Gulch, about 15 miles north of Kingman.

WATER SUPPLY.

Throughout that portion of the area which lies north of the latitude of Kingman water is scarce, the only natural source of supply besides precipitation being a few small springs, found along the base of the ranges. In the pre-Cambrian rocks, however, particularly in the Cerbat Range, as shown in the mines, ample water of good quality is usually encountered at depths of 300 to 600 feet; and in the vicinity of Kingman in the rhyolite tuff a copious supply is reached at depths of about 130 feet, as shown by numerous inexhaustible wells. In the Black Mountains also the best water-bearing formation is the rhyolite tuff, commonly known as the " water rock," extending in a belt of considerable width from Union Pass southward beyond Gold Road. It is the source of the water used at Gold Road and other localities, and, as at Kingman, the supply seems inexhaustible.

GOLD PLACER DEPOSITS NEAR LAY, ROUTT COUNTY, COLO.

By Hoyt S. Gale.

INTRODUCTION.

The existence of gold-bearing sands and gravels in a field of considerable extent in the central portion of Routt County, Colo., and adjacent parts of Wyoming, has been known for a number of years. These gold placer deposits are said to have been first discovered in 1887 in the northern part of the district, on Fourmile Gulch, and also in Dry Gulch, about 20 miles west of Baggs, Wyo., both localities near Little Snake River. Interest in the field seems to have begun about that date and to have continued more or less actively down to the present time.

The district in which these deposits occur lies west of the Elkhead Mountains, north of Yampa or Bear River, and east of the lower course of Little Snake River, along whose valley near the Colorado-Wyoming line the most extensive developments have been made. Only the southern portion of the territory thus outlined is directly concerned in the present report.

The following report is the result of observations made in the early part of July, 1907, by the author, who was at that time engaged in a study of the geology and an examination of coal lands in the fields south of this district. Especial acknowledgment is due to Mr. A. G. Wallihan, of Lay, for the locations of the claims shown on the map (Pl. I), and for assistance during the progress of the work and information concerning former prospecting in the field. The author is personally familiar only with that part of the field which lies south of the Iron Springs divide, and all statements and inferences concerning the deposits north of that line or lying in the Snake River drainage basin are based on the reports of others, interpreted from a general knowledge of the whole region.

A brief description of the territory and developments along Little Snake River prior to 1895 is given in an article by E. P. Snow.[a] In

---

[a] Fourmile placer fields of Colorado and Wyoming: Eng. and Min. Jour., vol. 60, 1895, pp. 102–105.

1901 a dredge was set up in Timber Lake Gulch, 10 miles south of Baggs, Wyo., and operated with some success for four years, but it is now lying idle. It is said that about $70,000 was cleared up during that time, but that work was suspended with the exhaustion of the richer ground. A dry washer outfit was set up in Timber Lake Gulch (T. 10 N:, R. 92 W.) during the summer of 1903. It was claimed that the machine made $20 per day on ground averaging 75 cents per yard, when the dirt was perfectly dry. The rainy season of that year prevented continuous work.[a] Recent reports state that a ditch is being constructed from Slater Creek to carry water to the Iron Springs divide, and that pipe to cover a distance of about 2 miles has been purchased and hauled into the country.[b]

In 1905 a dredge was installed in the valley of Lay Creek, on the south side of the Iron Springs divide. Since that time a small area (about 10 acres) has been worked over. Most of the delay or failure to achieve immediate results has been due, it is said, to minor difficulties in the operation of machinery and the management of the rather scanty water supply. A recent report states that the Blevins property, controlling the dredge and some 480 acres of placer filings about 7 miles north of Lay post-office, has been sold for $100,000. It is also stated that an active revival of operations in the neighboring tracts is under way, and that considerable development work is promised in this southern part of the field during the coming season.

DESCRIPTION.

The territory west of the Elkhead Mountains is a broad stretch of rolling prairie of moderate relief and monotonous topography. Geographically it may be defined as a roughly triangular area lying between Little Snake and Yampa rivers. These streams furnish the chief water supply of the district, although its eastern margin is more or less readily accessible to their tributary headwaters on the western flanks of the Elkhead Mountains. The climate, although not that of a true desert, is exceedingly dry, with but little rainfall during the summer and it is said but little snow in winter. Springs are scarce throughout the region and settlement is consequently scattering and confined chiefly to the river valleys.

The region is traversed by several well-known roads which afford access from the Union Pacific Railroad in southern Wyoming to the settlements along Yampa River in Colorado. The freighting route from Rawlins to Craig and Hayden passes through the eastern part of the district, and also furnishes the present means of communica-

---

[a] Information from Mr. John H. Marks, of Denver.
[b] Information from Messrs. L. Calvert and J. W. Cavendor, of Baggs, Wyo.

tion with the placer fields. A daily stage is run from Rawlins to Dixon and Baggs, on the Little Snake, whence connecting routes extend into Colorado. An old road that was formerly a regularly traveled stage route passes diagonally through the placer districts, running to the southwest from Baggs by way of Lay, and thence southward toward Meeker. This and the old Thornburgh wagon road were formerly the best-known routes of travel through this region, being the principal means of access to the Union Pacific Railroad from a large territory to the south in Colorado before the advent of the railroad lines on Grand River. Many minor wood and hauling wagon roads meander across the prairie hills and valleys, so that the region is fairly accessible in almost any part.

GENERAL GEOLOGY.

Precise knowledge concerning the structure and stratigraphy of the field as a whole is somewhat meager. It lies within the territory mapped by the Fortieth Parallel Survey, and is described in that Survey's reports and its geology mapped in the atlas accompanying them. The rocks are classed with the groups there denoted as Green River and Vermilion Creek[a] of the Eocene. The following description of the district is given by S. F. Emmons, the geologist who visited that region.[b]

To the north and west of Fortification Peak[c] extends a low, rolling country, covered with soft, earthy material of a prevailing red color, in which no outcrops are visible. The character of the soil, however, shows that it is probably made up of decomposed beds of the Vermillion Creek Eocene. These beds are found exposed on the western face of the Elkhead Mountains, at the baylike indentation between Mount Weltha and Navesink Peak, where they consist of coarse, red sandstones, with intercalated beds of reddish and cream-colored clays and arenaceous marls. The limits of these beds are not well defined, on account of the character of the surface in this region, but their connection can be traced, over the broad plains to the west, to characteristic outcrops, in such a manner that there can be little doubt as to the horizon to which they belong. Though they present here little difference of angle with the underlying Cretaceous beds, they are probably unconformable, as they are seen to be to the westward, and the lowest beds of the series can not therefore be definitely determined. On the little Snake River they are represented by yellow, coarse, gritty sandstones containing casts of *Melania*.

Recent investigations in territory adjoining this field, both north and south of the area under discussion, tend to corroborate the general statements of these earlier reports; but, as might naturally be expected, some details of the areal distribution of these strata were generalized or overlooked entirely in the first general maps.

---

[a] "Vermilion Creek" is equivalent in part at least to the Wasatch as used by the Hayden Survey, and the latter name, which has priority and established usage, has been adopted for these beds by the Geological Survey.

[b] King, Clarence, Rept. U. S. Geol. Explor. 40th Par., vol. 2, 1877, p. 187.

[c] Now better known as Cedar Mountain, situated about 6 miles northwest of Craig.

As described by Emmons the whole field is underlain by sedimentary strata of loosely consolidated or readily disintegrated material. The beds observed at outcrop are variable in composition, including marls of red or variously colored and banded appearance; loose coarse-grained sandstone or sandy beds, white or of darker weathered hues; and banks of more regularly bedded shale exposed here and there. At some places the varicolored beds of marl that commonly distinguish the Wasatch (" Vermilion Creek " of King) are exposed in great scars or badland washes.

Considerable areas of beds distinctly more recent than the Tertiary strata already described are present along the southern margin of the district here considered. These deposits consist of soft, friable material, made up largely of rounded quartz grains more or less consolidated by calcareous cement, with few harder consolidated strata. They are everywhere of a chalky-white appearance. They extend eastward as far as Cedar Mountain, whose summit is composed of these beds protected by an overlying cap of basalt. They correspond to the strata of the " Browns Park group " as described by Powell and others, and are markedly unconformable on all the older strata. They occur invariably in essentially horizontal position, with every appearance of having been deposited in a lake basin of late Tertiary or possibly more recent age. Although these beds at another locality are described very briefly, and with much doubt as to their age, in the Fortieth Parallel reports, they are there included with the Green River group as shown on that geologic map. They appear not to have been recognized at that time as occurring in the territory east of Little Snake River. C. A. White,[a] who afterward studied this region, described their eastward extension and regarded these beds as equivalent to the latest Eocene strata exposed south of the Uinta Mountains, although his reasons for this assumption are not clear, and now seem to have been unwarranted by the facts.

The whole region is covered to a greater or less extent with a scattered drift whose origin seems intimately connected with the source and history of the present gold deposits. This drift is described in more detail in the following paragraphs, as it forms the material of the bars or terraces from which the gold is now derived.

The only igneous rocks that are known to occur within the field are the basaltic intrusives and outflows, which are at least as recent as late Tertiary. These basalts are of very moderate extent on the open prairies, being confined chiefly to the higher summits and ridges of the Elkhead Mountains to the east. A prominent dike known as the " Rampart " extends westward from these foothills of that range, cutting the Tertiary strata a few miles beyond the Craig-

---

[a] On the geology and physiography of a portion of northwestern Colorado and adjacent parts of Utah and Wyoming: Ninth Ann. Rept. U. S. Geol. Survey, 1889, p. 691.

Rawlins stage road. Fortification Butte, or Cedar Mountain, an isolated peak on the southeast margin of the district, is capped by this basalt.

The geologic structure of the whole district is exceedingly simple. The strata occupy a broad synclinal trough or basin, over the greater part of which the beds lie approximately horizontal. At the southern margin of the basin the beds are tilted rather abruptly, rising over the Axial Basin anticline, just south of the placer field here described. This broad structural basin is considered as a southeastward continuation of the great structural depression known as the Green River Basin of Wyoming. The Bridger and Washakie basins as described by King are also subdivisions of this larger structural feature.

DISTRIBUTION OF AURIFEROUS DEPOSITS.

The map accompanying this report (Pl. I) is intended to outline the approximate extent and distribution of the more valuable ground as shown by the claims that have been filed on within the region. It can be seen at a glance that these claims are distributed along the drainage channels of the larger dry washes heading from the Iron Springs divide. A study of the ground itself, however, reveals the fact that the richest deposits lie on a bar or terrace ranging from 20 to 100 or more feet above the present creek bottom.

The Blevins dredge is now situated near the south end of the placer filings. A mile or so farther downstream Lay Creek enters a narrower valley cut in the tilted beds of harder and geologically older strata, and the upper bar is not readily distinguished below that point. The dredge stands on ground 20 or 30 feet above the level of the stream channel. Northward from this point along the main channel of Lay Creek the upper bench or bar rises more steeply than the stream grade itself. At Iron Springs the bar is about 70 feet above the creek and springs. As seen from its own level this bench appears to represent a former valley much broader than that of the modern drainage channels. Remnants of approximately the same elevation may be traced across the gulches to the flanks or tops of neighboring ridges. North of the Iron Springs bar the topography of the old valley level and of the main divide beyond is broad and open, a rolling, sagebrush-covered prairie. So far as can be gathered from reports of those who have prospected in the field, the gold-bearing gravels are distributed almost universally over all of this territory. It is stated on reliable authority that "pay" ground is found even on the highest summits of the main divide, and in fact pretty generally over all of that high ground. It also appears, from the selection of the located claims, that the most promising values are found on or near the so-called "bars," which as stated are thought to represent former drainage valleys. Although the sand and char-

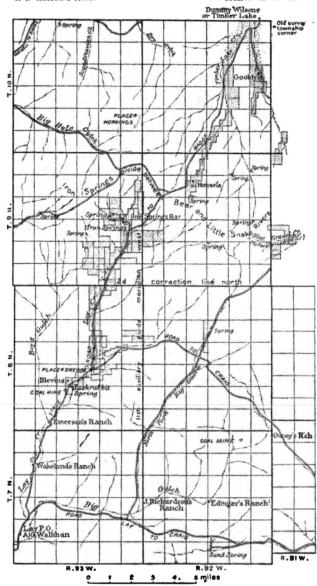

MAP SHOWING PLACER CLAIMS NEAR LAY, ROUTT COUNTY, COLO.

Dotted areas represent claims adjusted to the resurvey.

acteristic gravels are found in the bottoms of the present streams, it seems probable that these materials are only reworkings from the higher deposits and they do not appear to be as rich as the original ground itself. Colors have been found in the bed of the creek lower down on its course, as, for instance, at Lay post-office, but not of sufficient value to attract attention. It is therefore assumed that, for the most part, modern drainage channels represent a concentration of too small a portion of the gold-bearing uplands to show important values, or else that much of the gold that may have been washed into those channels has passed on downstream, and possibly has been largely carried away.

It is probable that the auriferous deposits in the Little Snake drainage basin have the same history and origin as those in the southern part of the field. This is indicated in the following paragraph, which is quoted from the report of the Fourmile placer fields previously cited:

The gravel beds are not, as is usually the case, in the various gulches, but form the mesa or upland and cover the entire country, the bed rock of the gravel being from 10 to 150 feet above the valley through which the streams flow to Snake River.

COMPOSITION OF AURIFEROUS DEPOSITS.

The beds composing the bars consist largely of loose white or light-colored quartz sand, in places containing coarser gravels and even larger, perfectly rounded pebbles and bowlders. Clear, glassy quartz grains, however, form by far the chief constituent of the sand. The gravel contains some clay which becomes evident on mixture with water, and clay also occurs in layers or beds somewhat irregularly distributed. Here and there it is more or less stained with iron, which gives it a rusty-brown or yellow color. More consolidated layers of conglomeratic material occur, evidently cemented by iron, but these are thought to be of local nature and of comparatively recent origin, having been derived from ferruginous deposits similar to the slimes now accumulating about the characteristic iron springs of the region. At some places, especially along the upper or headwater portions of the bars, and more particularly noticed near the Iron Springs divide, the material contains a variety of coarser pebbles or bowlders. A study of these pebbles shows them to have been derived from the following rocks:

Red quartzite.
White quartzite and hard sandstone.
Conglomeratic quartzite (rare).
White vein quartz.
Jet black chert.
Chert of various colors, some of it fossiliferous.

Gneissoid granitic or felsitic rock.
Crystalline granitic rocks (rare).
Pegmatite.
A feldspathic porphyry, with fine red or gray groundmass.
Silicified wood in waterworn fragments.

The character of this material is significant, denoting as it does its origin in the older formations of the Uinta and Rocky Mountain uplifts. Lower down the valleys, notably in the ground where the Blevins dredge is now situated, the coarser pebbles are more rare and the sand is very white and of a fairly uniform grain.

BED ROCK AND DISTRIBUTION OF PAY STREAKS.

No clearly defined bed rock can be traced from place to place among the prospects that show pay dirt. In places it is pointed out to be a coarse white sand, while the overlying pay dirt is composed of similar though more clayey material, the latter probably constituting the true bed rock which has served for the retention of the gold. At other places the bed rock is said to be a mottled clay of irregular pink and blue-gray patches underlying the sand or gravel pay dirt. Evidence seems to indicate that the agency which has served to retain the gold is the clay, which occurs either in the form of well-defined beds or mixed in certain layers of the gravel. The clay readily separates in water, however, and does not interfere with the recovery of the gold in washing.

To judge from the descriptions given of values and pay streaks in a number of pits and trenches visited in the vicinity of the Iron Springs bar, as well as the present shallow digging policy at the dredge, it seems a warrantable conclusion that the values are irregularly distributed in depth throughout the body of the gravel or sandbar deposits. Although it would thus not be strictly true to state that the values are uniformly distributed throughout the gravel, any project for the recovery of the gold will probably have to consider the handling of practically the whole of that material from top to bottom, inasmuch as the values seem as likely to be found at the grass roots as they are to occur at any definite horizon lower down.

CHARACTER OF THE GOLD.

The gold from this ground is said to be very pure, ranging from 885 to 935 in fineness, and bringing between $19 and $20 an ounce. It is of very fine grain, in small, well-rounded nuggets which are estimated to average about 1,000 to the cent. On magnification, the grains are seen to be well worn and of a rounded or nugget form rather than in flakes. An examination of some of this gold under the microscope showed a few copper-colored or reddish grains. The color was apparently contained only in cavities in the irregular grains and looked like an iron rust or stain. The statement has been made that a portion of the gold seems to be coated so that it is not amalgamated in the riffles of the dredge. No direct evidence of this condition could be obtained, however, and the suggestion is urged rather that the wooden riffles of the dredge are probably not well adapted to the work at that place, and may for that reason fail to hold the gold.

The final concentrates that remain with the gold are notably free from the heavy black minerals such as magnetite, ilmenite, etc. In some small samples that were obtained by panning, these dark constituents formed scarcely a third of the concentrates, the remainder being made up largely of garnet and a number of clear colorless minerals. Among the latter were noted many small, perfectly formed crystals of zircon.

ORIGIN OF THE GOLD.

The gold-bearing sands and gravels appear to have been the latest deposits laid down in the region. No direct evidence is now at hand to fix the geologic time of their deposition, but they are seen to cover the eroded surface of all other recognized formations. They seem to have preceded the cutting of the present stream channels and to have been scattered widely over an older land surface which somewhat resembled that of the present day.

The occurrence of the auriferous deposits so far removed from areas of granitic rocks in place, to which it is natural to turn as the source of such material, has been the cause of much conjecture relating to their origin. The wide distribution of the gold-bearing beds over summits, divides, and valleys alike makes the determination of the means of their transportation and deposition still more complicated. Were the deposits grouped along present or past channels of the principal rivers flowing from the Park or Gore Range, it would be easy to assume that that region was the source of the material and that these streams constituted the transporting agency.

These deposits have been discussed by White in the reports of the Hayden Survey, and they are probably related or equivalent to the scattered drift denominated the Bishop Mountain conglomerate by Powell, in the report on the Uinta Mountains, and the Wyoming conglomerate by King, Emmons, and Hague, in the reports of the Fortieth Parallel Survey. The most complete discussion is given by White, who concludes that the deposition of these beds may have been contemporaneous with that of the great northern glacial drift, and suggests that they were of glacial origin. By Powell and Emmons they are believed to have been of subaerial origin, resulting chiefly from the action of rains and streams, or, according to Emmons, representing littoral or shore deposits.

The sources from which the pebbles have been derived are not difficult to trace in a general way. The wide extent of the territory thus defined, however, leaves quite as much doubt as to the actual situation of the vein deposits from which the gold has been derived.

The greater part of the pebbles are fairly characteristic of the older formations exposed along the main uplifts of the Uinta Mountains. The red and white quartzites and conglomeratic pebbles of the same class are readily distinguished as identical in composition with the "Uinta" quartzite of Powell, which forms the core and highest portion of that range. The various chert pebbles, some fossiliferous or stained in bright colors of red and yellow, as well as the black chert, are with almost equal certainty derived from the Carboniferous limestones exposed along the flanks of the same mountain range. Here and there pebbles of the limestone itself are also found. The derivation of the red or gray feldspar porphyry is more in doubt. A few granitic rocks, such as the granite, pegmatite, and related types, seem almost as certainly to have come from the other direction, or the Park Range of the Rocky Mountain system on the east. King describes some hornblendic and metamorphic rocks at the northeast end of the Uinta Range, but it is thought that no such rocks as granite or pegmatite are known there.

The sources and general trend of all the drainage of this region would more readily explain the transportation of materials from the Rocky Mountains to the east than from the Uinta Range to the west. In fact, the gravel and bowlder deposits that cap some of the elevated mesas along Yampa River are composed almost entirely of granitic rocks of such composition that they are certainly derived from the headwater streams in the Rocky Mountains. So far as is known to the author, very little gold has ever been reported from the rocks of the Uinta Range—a fact which would also tend to support the theory that most of that metal has come from the east.

It is of interest to note in this connection that a corresponding outspread of gravel and bowlders of almost identical composition, consisting of materials apparently derived from the same sources, must have taken place near the close of Cretaceous time. Evidence of this is now afforded by an extensive and continuous bed of conglomerate, marking an unconformity in the uppermost Cretaceous strata of the region, and possibly representing the orographic disturbances which produced the adjoining mountain ranges. The outcrop of this conglomerate bed may be found in the hills south of the placer field. It crosses Lay Creek near Emerson's ranch, the bowlder bed occurring immediately above a huge white sandstone stratum which dips beneath creek level at that place. The same beds may be traced continuously eastward, passing south of Cedar Mountain and crossing Fortification Creek about 2½ miles north of Craig. To the west they extend continuously for several miles to the point where they pass beneath and are concealed by the overlying "Browns Park beds." This conglomerate has been prospected to some extent, but does not appear to have attracted much attention. It might be expected to carry values somewhat similar to those of the more recent gravel deposits, but its occurrence is limited to the narrow outcrop of that particular bed.

PROSPECTS AND DEVELOPMENT.

In 1905 a dredge was installed just above Jack Rabbit Spring, and in September of that year started operation on what is known as the Blevins property. This dredge has been worked intermittently since that date and now (August, 1907) stands near the middle of the N. ½ sec. 22, T. 8 N., R. 93 W. An area of approximately 10 acres has been worked over. The season is said to last from about April 1 until about the middle of October, after which the work is likely to be interrupted by freezing.

The water supply is rather meager, coming from Jack Rabbit Spring about one-half mile below (southwest of) the dredge. The channel of Lay Creek is normally dry and apparently no attempt has ever been made to store the winter run-off; in fact, such a project may not be feasible on account of the scant precipitation. A steam pump has been placed at the spring and the whole available water supply is raised and carried by pipe line and ditch to the working ground. When the dredge was first started much difficulty was experienced in holding the water to float the boat, but it is stated that as the dredge has advanced toward higher ground there has been a smaller seepage and less trouble from this cause. The engines at both dredge and pump are supplied with fuel from a conveniently located coal bank, which has been opened expressly for that purpose. The coal is an excellent bed nearly 12 feet thick, without seams or partings and with a good roof. It outcrops at the side of the gulch just below the spring and is readily accessible, wagons being driven directly to the face of the entry for loading. The coal, a good subbituminous grade, of lighter weight than some of the coals of the Yampa field, slacks rapidly, but burns well and gives a satisfactory heat.

The gravel bar at the dredge is low, probably 20 or 30 feet above the level of the stream channel. The pay gravel is composed almost entirely of a loose white sand with very little coarser material. Apparently no especial effort has been made to reach bed rock in the dredging ground, the digging having reached, as estimated, only 5 or 6 feet below the top of the ground. Very little appears to be known as to the nature of the underlying bed rock at that place. This is doubtless due to its rather indefinite nature, as explained on another page.

Above or north of the Blevins dredge the most important group of claims in the field south of the Iron Springs divide lies near what is known as the Iron Springs bar, in the east half of T. 9 N., R. 93 W.

The claims are located on a somewhat irregular bench or high terrace averaging 70 feet or more above the present stream channel. The bar just east of Iron Springs is about a mile long from north to south and a quarter of a mile in width, sloping southward at a grade of about 60 feet to the mile. Benches corresponding to this level appear on the neighboring ridges. The sand and gravel deposits to the north are said to rise even to the summit of the main divide. The irregular sage-covered plain affords too poor exposures for satisfactory tracing of these beds.

In sec. 14, T. 9 N., R. 93 W., a small reservoir just below the spring marks the site of some past development work. Claims here are said to have been first located by a Mr. Scrivener, who washed the narrow gulch bottom with a sluice, using the water from the spring. The claims were later taken up by a Mr. Lahr, who discovered pay gravel in the banks, and drifted in 100 feet from the bottom of the gulch, being able, it is stated, to make his living from the washings in the meantime. This drift evidently reached a depth of 30 feet or perhaps considerably more below the sand or gravel cap rock, and this fact may be taken to indicate that the gold will be found down to considerable depths in some parts of the field. The claims were later dropped and are now in other hands. An old gasoline engine used to pump the water for sluicing stands at the place. This property is said to have contained some rich pay streaks.

VALUE OF THE GROUND.

It may be assumed that the site selected for the operation of the dredge is as favorable a locality as any other in the field. Estimates made by one of the interested persons placed the value of the ground already worked over, as from 25 cents to several dollars per cubic yard. It was said on the same authority that the dredge was then standing on ground that would yield $2.40 per cubic yard.

A series of tests were conducted by Mr. Wallihan on material obtained near the Iron Springs bar, but are hardly complete enough to be considered thoroughly representative of the field. Test pans were taken from various prospects in that vicinity. The pits themselves have been dug more or less at random over the placer ground. They vary somewhat in size, at least one of them reaching a depth of 15 feet or more.

The sample taken in each case was a stricken pan containing 20 pounds of dry material. This was then panned, all of the black sands and gold being saved and the concentrates being sent to Denver for assay. The following results have been calculated from the returns of these assays, on the assumption that one cubic yard of ground will weigh $1\frac{1}{2}$ tons. A set of twelve such assays showed a range in value from 1.6 to 63.7 cents per cubic yard. The average amount of black sand was found by the same tests to be 1 ton in 368 tons of material.

It is thought, however, that the samples taken for testing are not in every case representative or an average of the whole thickness exposed in the pit. This is probably true of those samples which show the higher values, as these are very likely to have been shoveled from some particular part of the pit thought to contain the pay streaks. If this is so, allowance should be made to obtain the average grade of all the material handled.

GOLD DEPOSITS OF THE LITTLE ROCKY MOUNTAINS MONTANA.

By William H. Emmons.

INTRODUCTION.

The eastern half of Montana is in the main a nearly level country and at most places is devoid of conspicuous topographic features. A number of small buttes rise above the plains and here and there small groups of mountains relieve the monotony of the horizon. The Little Rocky Mountains form such a group, their rounded green summits contrasting in a striking manner with the featureless gray plains by which they are surrounded on every side. These mountains are in the southeast corner of Chouteau County, Mont., between Missouri and Milk rivers. The group as a whole is rudely elliptical in outline, about 10 miles in greatest length and 8 miles wide, the longer axis trending northeastward. The elevation of the surrounding plains is approximately 3,000 feet above sea level, and the highest peaks of the mountains reach elevations of about 6,600 feet.

The Little Rockies are 35 miles south of the Great Northern Railway and directly south of the Fort Belknap Indian Reservation. The principal mining camps in the mountains are Zortman, Whitcomb, and Landusky, each of which is provided with a post-office and is connected by stage with the railroad. One line of stages makes three round trips a week from Malta to Zortman. Another line makes three round trips a week from Dodson to Whitcomb and Zortman, the two stages leaving the railroad points on alternate days. A third line of stages runs three times a week from Harlem to Landusky. The trip to the mountains is made in one day from each of these points.

In 1895 Messrs. W. H. Weed and L. V. Pirsson visited the mountains to examine the mineral resources of the Fort Belknap Indian Reservation for the commissioners appointed to treat with the Indian tribes, with a view to segregating the mineral lands of that reservation. The scientific results of this visit were published the following year.[a]

---

[a] Jour. Geol., vol. 4, 1896, pp. 399–428 ; Eng. and Min. Jour., vol. 61, 1896, pp. 423–424.

At this time the mines had produced a small amount of rich ore, but very little underground development had been done, and the principal lodes now producing had not been discovered. In October, 1907, the writer, incidental to other work in Montana, made a visit to the mountains, the results of which appear herewith. He is indebted to Messrs. Weed and Pirsson, whose valuable historical and geologic data have been freely drawn upon; and to the operators and prospectors of the district, who have been uniformly courteous in facilitating his investigations.

HISTORY OF DEVELOPMENT.

The placer deposits occurring in the beds of the streams that flow southward from the Little Rocky Mountains to Missouri River had been worked intermittently but with small success for several years before the lode deposits were discovered. The lode deposits came into prominence in 1893, when gold was found in the August mine, which at that time was within the boundaries of the Fort Belknap Indian Reservation. This mine was worked quietly without the knowledge of the Indian agents, and about $32,000 was taken out in sinking a shaft 65 feet deep. The following year the Goldbug and other claims near Landusky were actively explored, the Goldbug producing a small amount of rich ore. Subsequently the mineral lands were segregated from the Indian reservation and thrown open to prospecting. As most of the ore was not of shipping grade the development of the mines was slow. About $55,000 worth of gold ore was shipped from the Alabama mine in 1899 and 1900, but aside from these shipments the production was practically nothing until the spring of 1903, when the Zortman cyanide mill was completed. This mill was built by the Alder Gulch Mining Company and was supplied with ore from the Alabama and Pole Gulch mines. It was in operation about six months of each year for four years. In October, 1907, the mill was idle and the mines of the company were under bond to the Little Rockies Exploration Company, which had a force of men doing exploration and development work.

The Ruby Gulch mill was completed in January, 1905, and began treating ore from the Independent mine. This mill was at first a 100-ton plant, but in 1907 its capacity was increased to 300 tons a day. It has been in successful operation ever since it was built, and according to the superintendent, Mr. E. E. Berry, it has produced about $600,000.

Aside from the returns from placer mining for which authentic information is not at hand, but which were probably small, the total production of the Little Rockies is about $950,000. The average annual production for the last four years, since the first cyanide mill was built, is a little less than $200,000. The returns for 1907 will probably show an increase over this amount.

The location of the various mines is shown in fig. 2.

MINING AND MILLING.

The form of the deposits and the character of the ore favor very low costs for both mining and milling. The Mint, Independent, and Pole Gulch mines are all worked by the open-cut method. The ore falls to the bottom of the hopper-shaped cavities and is drawn through chutes into tunnels driven under the deposits from 100 to 200 feet below the outcrop of the ore bodies. By this method little

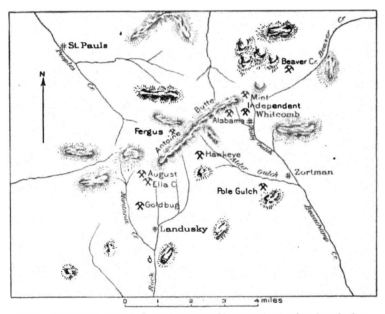

Fig. 2.—Sketch map of Little Rocky Mountains, Montana, showing location of mines.

timbering is necessary, as the walls stand fairly well; and at the Mint mine three men, one on each shift, are able to supply the mill with 300 tons of rock per day. The ore is soft and requires but little breaking for passage through the chutes, though some difficulty is experienced when the moisture content of the ore is high. After rains or melting snow, water collects in the ore pits, forming a mud with the soft clayey ore. With the present method of treatment no water is added to the ore until it reaches the cyanide tanks, and excessively wet ore is very undesirable because it clogs the ore chutes and the elevators. The water which reaches the open cuts is not surface water in the strict sense, but it probably soaks into the ground just above the cuts and seeps out into the ore chambers after a very short underground course.   There is a luxuriant growth of small pine trees surrounding the pits and this prevents surface drainage. The moisture content of the ore could probably be reduced if the country just above the pits were cleared of vegetation and if a number of ditches were dug so as to lead the surface water from points above the mine directly to the deep gulches near by, and thus prevent it from soaking into the ground and so reaching the ore pits.

The Ruby Gulch mill was the only plant in operation in 1907. This is a 300-ton cyanide plant located in Ruby Gulch, 1½ miles above Zortman.   The ore passes from a gravity tramway connected with the mine through a No. 3 Gates crusher, from which it passes over a 1-inch mesh Jeffries screen set at an angle of 45°.   From the screen the undersize goes to the tanks and the oversize is passed through a set of 14 by 24 inch rolls and from these through 12 by 14 inch rolls, of which there are two sets.   From the fine rolls the ore passes over a 1-inch mesh impact screen, the fines going to the tanks and the coarse material being returned to the same rolls.   For leaching the mill is provided with six 300-ton tanks and six 110-ton tanks.   The usual period for leaching is seven days.   Although the screens used for oxidized ore are 1-inch mesh, the angle at which they set is so steep that the material passing through is much finer.   A test on screened ore showed that 70 per cent of this material would go through a 12-mesh screen after passing through the mill.   Much of the ore is very fine as it comes from the mine.   At one time a 300-ton tank was filled with ore which had not passed through the mill at all and a satisfactory saving was made after twelve days of leaching. For the oxidized ore the saving is commonly 90 per cent, but for sulphide ore it is lower.   The strength of solution best adapted for leaching is one-twentieth of 1 per cent.   The ores assay from $3 to $21 per ton.   The average value of the ore since the mill was started is $6.55.   The usual cost of mining and milling is said to be as low as $2 per ton when conditions are favorable.   As the country is remote from a railroad, the cost of supplies is high.   Steam power is used and wood is burned for fuel, at a cost of $8 a cord delivered at the mill, which requires 8 cords per day.   The company owns coal mines and water rights on Rock Creek, and contemplates the installation of a power plant and the erection of a transmission line from the coal mines to the mill.   The lime used in the mill is burned from the Carboniferous beds which form a rim around the mountains and it is said to be of a satisfactory quality.

The Zortman mill, which is a 100-ton cyanide plant located in Alder Gulch, about 1 mile west of Zortman, treated ore from the Alabama and Pole Gulch mines.   From the Alabama mine the ore was hauled in wagons down grade about 1¾ miles to a bin just below

the mill. The ore was elevated by a short tramway to the head of the
mill, whence it passed to a Gates D crusher through a double shaking
screen, the upper plate of which was 1-inch mesh and the lower plate
⅜-inch mesh for oxidized ore and ¼-inch mesh for sulphide ore. The
ore passing over the screen was sent through coarse rolls and elevated
back upon the screen. The ore going through the 1-inch screen was
passed through finer rolls and thence was also returned to the screen;
and the ore passing through the lower screen was sent over belt conveyors into the tanks. The plant is provided with five 100-ton tanks
for leaching. The usual period for leaching was five days and the
saving is said to have been about 90 per cent for oxidized ore. The
method of treatment was essentially the same as at the Ruby Gulch
mill.

The Goldbug mill is located at Landusky. It is equipped with a 7
by 10 inch Blake crusher, 10 gravity stamps, amalgamation plates,
and two Frue vanners. It was built in 1902 to treat the lower grade
ore of the Goldbug mine, but the saving was not satisfactory and only
a few tons of ore were milled. The oxidized ores are easily cyanided,
but only a small percentage of the values can be saved by amalgamation. The quantity and value of the sulphides in the tailings from the
ore, so far as developed, are not sufficient to render mechanical concentration profitable.

GEOLOGY.

GENERAL FEATURES.

As has been shown by Weed and Pirsson, the Little Rocky Mountains form a dome-shaped uplift of sedimentary and metamorphic
rocks which has been modified by the intrusion of a thick sheet of
porphyry and by the erosion of the younger beds from the top of the
dome. In its broader features the structure is simple. As one approaches the mountains from the nearly flat Cretaceous beds which
underlie the surrounding plains he passes successively over beds of
greater age, and well toward the interior of the little mountain group
he encounters crystalline schists which are highly metamorphosed
and older than any of the bedded sedimentary rocks. The dip of the
beds is outward from the central axis of the mountains toward the
surrounding plains, at a greater angle than the average slope of the
mountains. For this reason the older beds are exposed toward
the center of the group and in general are of higher elevations than
the younger beds that overlie them.

METAMORPHIC ROCKS.

The oldest rocks are crystalline schists which are exposed in the
deep gulches in the interior of the mountains and at some places on
the higher ridges near the crest. These schists are of pre-Cambrian

age and are overlain by Cambrian quartzite, but at many places they are separated from the Cambrian by the intruding porphyry. The prevailing rock among the schists is a dark, glistening hornblende schist or amphibolite. Locally this is garnetiferous, and at some places rich in quartz and feldspar. On the road from Zortman to the Alabama mine the schists consist of thin alternating beds of different character, and at some places quartzites are included in them, showing that the series is, in part at least, of sedimentary origin. These rocks everywhere have been profoundly metamorphosed.

SEDIMENTARY ROCKS.

The Cambrian beds rest unconformably upon the metamorphic rocks or are separated from them by intruding porphyry. At the base of the Cambrian is a quartzite bed about 75 feet thick, overlain by shales and limestone, making altogether a series about 500 feet in total thickness. Above the Cambrian, with no apparent unconformity, is a succession of impure limestones, in which no fossils have been found, but which are presumably of Silurian or Devonian age. Resting upon these limestones are large massive beds of white or light-gray limestone rich in Carboniferous fossils. These rocks are more resistant to erosion than the underlying limestones and form a chain of ridges and peaks around the mountains, the continuity of which is interrupted here and there by valleys that have been cut through the limestone by numerous small streams flowing outward from the central mountain crest. This broken rim of Carboniferous rocks is a conspicuous feature of the landscape, and it has been aptly compared to the limestone girdle which encircles the Black Hills of South Dakota. On the low ridges which slope gently away from the mountains toward the plains the Jurassic limestones overlie the Carboniferous beds and these in turn are covered by the Cretaceous sandstones and shales. The Jurassic and Cretaceous formations are not known to occur within the mountain group proper, but together they cover great areas of the surrounding plains and badlands country.

PORPHYRY.

A large, thick sheet of igneous rock forms the central axis of the mountains and is the country rock for the most important ore deposits. It is composed of syenite porphyry and other closely related varieties of alkali-rich rocks. This mass is nearly circular and is about 6 miles in diameter. Some of the buttes near the outer rim of the mountains are also capped with porphyry which is separated by limestones from the main central mass. The intruding porphyry is limited so far as known to the crystalline schists and to the Cambrian beds. The horizon between the schist and the Cambrian quartzite and that at the top of the quartzite and below the Cambrian shales appear to have been planes of weakness which were especially favorable for the intruding rock.

As the structure of the mountains is that of a dome, with the dip away from the central axis, and as the porphyry is in the main a sheet intruded between the sedimentary beds, it also dips away from the center of the uplift. The highest peaks are capped with porphyry, but the crystalline schists also occur at some localities near the top of the divide and they are exposed at a large number of places in the hollows of the gulches which radiate from the center of the mountains. Small areas of both schists and limestones, some of them only a few feet in greatest diameter, are entirely surrounded by porphyry and appear to be isolated masses which were caught up in the intrusion. Other masses of limestones are probably remnants, not yet eroded, of the beds which lie upon the porphyry sheet. The porphyry appears to be thickest in the central portion of the mountains and to thin out near the margin, but its precise thickness has not been measured, as no favorable section was found. At some places it is at least 400 feet thick, and it may be thicker. The contact between the schist and porphyry shows irregularities or warpings other than the ordinary dip toward the margin of the mountains, and the distribution of the two rocks with respect to the topography as shown in fig. 3 is not such as could result if the contact were simply a tilted plane. The porphyry sheet is an intrusive rock and is younger than the rocks which inclose it. At some places it must have come up through the crystalline schists and at such places its contacts with that formation are probably steeply inclined or vertical. At the point where the porphyry rose it is likely to be found to extend downward to great depth. It is the opinion of Weed and Pirsson [a] that one of the vents through which the porphyry rose is near the Goldbug mine. It is possible that another vent is located in the vicinity of the Alabama and Mint mines, for vertical workings in the Alabama mine over 500 feet deep are said to have been entirely in porphyry. Moreover, the contacts between the porphyry and the crystalline schists, as shown on the map (fig. 3), strongly suggest crosscutting relations at this place.

The fresh unaltered porphyry is commonly light gray in color, and at some places light pink or purplish pink. The groundmass is even and fine grained and practically all the phenocrysts are feldspar. These vary in size and in some varieties of the porphyry they are half an inch long. No ferromagnesian minerals are present, though small dark specks show where biotite crystals have altered. At many places the porphyry is extensively brecciated and cemented by ma-

---

[a] Op. cit., p. 412.

terial which is of about the same composition as the fragments, but
the cement is usually less porphyritic than the fragments which it

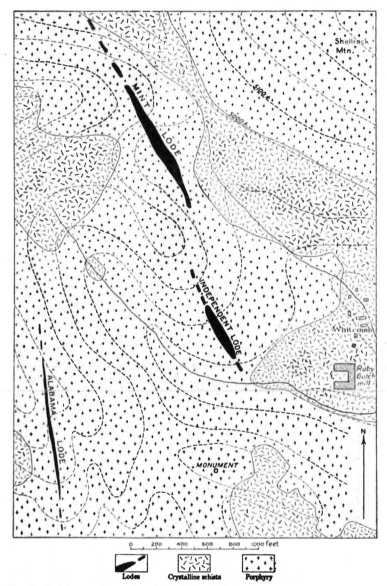

FIG. 3.—Sketch map of a portion of the Zortman mining district, Little Rocky Mountains,
Montana.

surrounds. Under the microscope the phenocrysts are seen to be
orthoclase and plagioclase, and in some varieties a few small quartz

phenocrysts are present. The largest phenocrysts are orthoclase and zonally built forms made up of orthoclase and albite. A few striated plagioclase phenocrysts with the composition of acid oligoclase are usually present. The groundmass is microcrystalline and is composed of alkali feldspar and quartz, with fine particles of magnetite. The feldspars are slightly kaolinized; even in the freshest rocks some white mica has been developed, presumably as an alteration product of biotite. More extensive alterations which have occurred in the vicinity of the ore deposits are described on page 108.

The syenite porphyry, granite-syenite porphyry, and granite porphyry which make up the central igneous mass appear to grade one into another and have not been separated in the field. On Montana Creek a short distance north of Landusky, just below the Cambrian quartzite and between it and the main porphyry mass, a dark-green porphyry, possibly a contact phase, was examined by Weed and Pirsson and proved to be tinguaite, a variety of phonolite.

ORE DEPOSITS.

GENERAL CHARACTER AND STRUCTURAL RELATIONS.

The ore deposits of the Little Rocky Mountains are (1) zones of fractured porphyry replaced and cemented by quartz and pyrite and (2) replacement deposits in limestone near intruding porphyry. The ore carries as a rule from $3 to $7 in gold and about an ounce of silver to the ton. Here and there are small bodies of high-grade shipping ore. The most important deposits are in the porphyry and these fall naturally into two groups. One of these is the Zortman group, which includes the Mint, Independent, and Alabama lodes; the other is the Landusky group, which includes the Goldbug, August, and other lodes near by. The Landusky group is about 4 miles southwest of the Zortman group. All the lodes of the Zortman group, so far as known, trend west of north; those of the Landusky group trend east of north.

There is no evidence that the movements which resulted in the formation of openings that permitted the solutions to enter the porphyry and deposit their burden produced spaces of any considerable size. These movements resulted rather in shearing and in brecciation along fissured zones, with a large number of small openings rather than a single large open space. The stresses were in part of a compressive character, for some of the country rock near the Alabama lode is noticeably sheared. The irregular width, short length, and lack of definition of some of the deposits in the Landusky group may be attributed to this cause.

The lodes in porphyry range in width from a few inches to 70 feet. They are not everywhere of a grade to pay for working, but at both the Mint and Independent cuts it was found profitable to mill the lodes to a width of more than 50 feet for 200 or 300 feet along the strike of the deposit. As the lodes are replacement deposits and grade imperceptibly into the porphyry, it is at some places a very difficult matter to determine what is ore except by assaying. This is true especially where the barren porphyry in the mineralized area is slightly stained with iron oxide.

The lodes in the Zortman district appear to be fairly persistent in length. One of them has been followed continuously for 1,200 feet along its strike, and surface workings indicate that it is longer. Nearly all the lodes are cut by fissures, commonly called "walls." These are as a rule slickensided and grooved, indicating that they

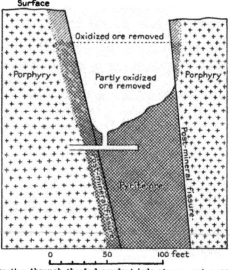

FIG. 4.—Vertical section through the Independent lode at open cut, near Whitcomb, Mont.

are planes of movement. At some places crushed quartz occurs along them, which shows that the movement was later than the deposition of the ore. They do not determine the limits of the ore body and are not walls in the strict sense of the term. At many places there is ore behind them, and some of them cut across the ore body, running at one place near the hanging-wall side and at another on the foot-wall side. Many of them split and inclose masses of ore and country rock, and some of them cross the vein at a considerable angle. Fig. 4 is a cross section at the Independent open cut, showing two such fissures crossing the ore body; followed along the strike these fissures intersect in the lode a short distance to the north of the line of the section. Some of the slickensided planes follow the lodes with singular persistence. One of them can be traced for 1,200 feet and in the entire distance does not completely cross the ore body. This parallelism of the lodes and the slickensided planes seems to indicate that the lodes are planes of weakness, and it is an illustration of the general fact that fissures have a strong tendency to follow planes along which movement has already taken place.

It is not good mining to follow these fissures or "walls" too closely, as they leave the lode at some places and strike out into the country rock, and as they do not everywhere limit the lodes. Good ore may be found beyond them or on both sides of them, and in many places frequent assaying is necessary to keep on the ore body.

The replacement deposits in limestone have not been developed sufficiently to determine their character. So far as known they all occur near the contact of the limestone with the porphyry, but they are not contact-metamorphic deposits, for the sedimentary rocks near the porhyry are not noticeably metamorphosed.

MINERALS OF THE DEPOSITS.

Gold.—Gold occurs in pyrite and in quartz and in the oxidation products of pyrite ore. It is so finely disseminated that the ores can not be successfully amalgamated. Small bodies of native gold the size of a pinhead have been found in oxidized ore from the Alabama mine, but most of the gold is not visible to the naked eye. A gold telluride occurs in several of the deposits.

Silver.—Most of the ore carries about an ounce of silver to the ton. A portion of the silver values are recovered by the cyanide process. The silver probably forms an alloy with the gold.

Quartz.—Quartz is an original constituent of the porphyry, where it occurs in the main as microcrystalline anhedra in the groundmass. Secondary quartz replaces both the phenocrysts and groundmass of the porphyry, and small veinlets of quartz, a fraction of a millimeter wide, are numerous in the porphyry near the deposits. The payable ore at some places is cut by small veinlets of milky quartz a centimeter wide, which do not carry gold.

Feldspars.—The phenocrysts of the porhyry are orthoclase and acid oligoclase and where the porphyry is not excessively altered these minerals are abundant constituents of the ore. Secondary orthoclase occurs as clear fresh crystals which have been deposited on the older altered orthoclase presumably by the vein-forming solutions.

Magnetite.—Magnetite is present as minute bodies which are primary constituents of the porphyry and as an alteration product of biotite. No magnetite was noted in the ore that replaces limestone.

Kaolin.—Kaolin is abundantly formed through the alteration of feldspar, and such alteration is not confined to the vicinity of the ore deposits.

Sericite.—Sericite, or white mica, is an abundant constituent of the ore. It occurs as minute foils replacing orthoclase and is extensively developed in the groundmass of the mineralized porphyry.

Pyrite.—Pyrite is the only sulphide which was noted in the ore and in depth is present in considerable quantity. It occurs as small irregular bodies and as well-formed crystals replacing porphyry, and in places in the Independent lode there are masses of pure pyrite of considerable size. It fills numerous small cracks which cut the replaced porphyry and commonly it coats the fragments of the crushed ore. Fibrous and globular pyrite, having the habit of marcasite but of brassy-yellow color, were noted in the Independent mine near the surface.

Fluorite.—Purple fluorite is present in all the producing mines and is a conspicuous mineral, though it forms only a small portion of the ore. It occurs as irregular bodies and as small tabular masses along planes of movement parallel to the lode and it is usually associated with the ore of high grade. It replaces both limestone and porphyry. In the porphyry it is as a rule a more or less earthy variety, and under the microscope some of it is colorless. Fine crystals of purple fluorite are present in the Pole Gulch deposit in limestone.

Calcite.—Calcite is an abundant constituent of the deposits which replace limestone. The country rock is more or less marmorized and is cut by the veinlets of white calcite, which are as a rule less than 1 inch wide. In these deposits crystals of calcite nearly half an inch in diameter are associated with quartz and fluorite. The oxidized ores of the lodes in porphyry carry a trace of calcite, but none could be found in the sulphide ore of such lodes.

Limonite.—Limonite is invariably present in the oxidized ores and some of the deposits carry a considerable quantity of this mineral. Most of it is the earthy amorphous variety, but here and there a pseudomorph after pyrite is present. Brecciated fragments of porphyry are cemented by limonite and it coats fragments of crushed ore.

Manganese oxide.—The dark-brown powder which is present in some of the ore is one of the oxides of manganese. The presence of this powder is regarded as a favorable indication of values, as it is commonly associated with the high-grade ore.

Copper carbonate.—Small bunches of green copper carbonate are present in the oxidized ore, but this mineral is rare.

REPLACEMENT PROCESSES.

The porphyry of Antoine Butte is a granite-syenite porphyry. The phenocrysts are up to 5 millimeters long and are evenly spaced through the rock. A specimen collected by Weed and Pirsson from Antoine Butte was analyzed by Dr. H. N. Stokes in the laboratory of the United States Geological Survey.[a] The theoretical composition or norm calculated from this analysis gives:

Norm of porphyry from Antoine Butte, Montana.

Quartz	0. 20
Orthoclase	. 28
Albite	. 41
Anorthite	. 05
Water and other constituents	. 06
	1. 00

The phenocrysts are made up of orthoclase and striated plagioclase—about the composition of acid oligoclase. The proportion of the plagioclase in the groundmass is very small. The action of the mineral solutions on this rock was as follows: Pyrite and quartz carrying more or less gold partly replaced both phenocrysts and groundmass; the feldspar phenocrysts were altered partly or entirely to sericite and kaolin and fresh clear orthoclase was deposited on the older feldspars. Locally the porphyry, especially where crushed, was partly replaced by fluorite; the finely powdered porphyry was almost completely replaced and this now consists of an intimate intergrowth of small square crystals of fluorite between the interstices of which are many minutely microscopic bodies of quartz.

OXIDATION OF THE ORE.

The primary ore is in the main a hard, brittle porphyry, carrying a notable amount of pyrite and secondary quartz. The fresh, unaltered porphyry is a light-colored rock with a dense, fine-grained groundmass showing practically no dark constituents. Where this has been finely crushed the powder has been completely replaced, and small angular fragments are now surrounded entirely by the quartz and pyrite. The sulphide ore is best exposed in the lower workings of the Independent lode, where at some places sulphide ore grades · into the altered porphyry without any definite dividing line between.

Near the surface the pyritiferous ore has oxidized to a variable depth. As shot down in the pits the harder ore consists mainly of porphyry, broken as a rule to small angular fragments which are covered more or less completely with a thin coating of iron oxide and cut through with thin seams of the same material. The richer portion of the ore is, however, either a reddish-brown or black material almost as fine as dust.

The lower limit of the zone of oxidation is at some places 200 feet below the surface. On the Mint claim pyrite is just beginning to appear at that depth. In the Independent open cut, oxides mixed with pyrite extend at some places to the surface and generally appear at a depth of 25 feet or less. In the Alabama mine the ore is partly oxidized at a depth of 400 feet and it contains a noticeable amount of pyrite at 100 feet. Exclusive of relatively small seams of finely powdered limonitic and manganitic ore, the gold values in the sulphides and in the oxides are said to be approximately equal.

GENESIS OF THE ORES.

The ores were deposited as sulphides and probably by ascending waters. There is a large amount of calcite in the ore of the replacement deposits in limestone, but no calcite or other carbonate could be found in the sulphide ore of the lodes in porphyry, by either chemical or microscopic examination, and there is only a trace of calcite in the oxidized ore of the lodes in porphyry. A great thickness of limestone has been eroded from above the deposits in porphyry, for residual masses of the limestone rest upon the porphyry and the porphyry mass is surrounded by a girdle of overlying limestone. If the solutions had been descending through some higher horizon now eroded, they should have passed through this limestone and would probably have dissolved lime carbonate and redeposited it with the ore. The absence of more than a trace of calcite in the oxidized ore further suggests that secondary enrichment by waters descending through the overlying limestone into the lodes or brecciated zones in porphyry was not a process of great importance.

FUTURE OF THE DISTRICT.

Owing to the great width of the lodes, their linear extent, and the depth to which oxidation has extended, the success of the camp in the immediate future does not seem to depend directly on the availability of sulphide ores. But when the oxidized ores shall have been exhausted the value and extent of the sulphide ores will be a question of first importance. Although sulphide ore of fair grade has been found in the Independent mine, this ore, because of the greater difficulty of saving the values, has not proved so attractive as the oxidized ores of the Mint claim, and the workings in sulphide ore are of small extent. The data at present available are insufficient to form a sound basis for prediction as to the extent of the sulphide ore in depth. In this connection, however, two features should be considered—(1) the character and extent of the mineralization of the crystalline schists below the porphyry; (2) the thickness of the intruding porphyry.

Concerning the character and extent of the mineralization of the crystalline schists few data are at hand. These rocks are, as already stated, of much greater age than the porphyry which contains the ore and as they were present when the ores were deposited it is quite likely that they will be found to carry sulphides where the lodes cross them. Near the head of Peoples Creek, in one of the tunnels of the Fergus Mining Company, the schists are locally replaced by pyrite. On the other hand, no important deposits have yet been discovered in the schists and they can hardly be regarded as affording a favorable ore horizon. The vertical extent of the porphyry, then, is a factor of economic importance. As already stated, the porphyry is in the main a sheet from 300 to 400 feet thick, intruded between the arched Cambrian rocks or at the base of these rocks above the crystalline schists. Necks through which the porphyry rose may extend downward through the crystalline schists and at such places the thickness of the porphyry may be considerably greater. It is possible that future developments will show this structure at the Alabama mine, where the thickness of the porphyry is known to be above the average.

On the whole the conditions on which costs depend are favorable. Owing to the isolation of the locality and the high cost of living, labor has been somewhat higher here than at most of the other camps in Montana. Power is expensive and the cost of transportation of mill supplies is heavy. Wood fuel, which has hitherto been the sole source of power, is not of the best quality and sells at a high price. Power plants can not be constructed as readily as is possible in larger mountain groups, as the streams which drain the Little Rockies are small. Good coal is said to occur on the plains to the south of the mountains and this seems to be the most available source of power. The form and character of the ore deposits themselves favor unusually low costs.

DETAILED DESCRIPTIONS.

MINT MINE.

The Mint mine is located near the head of Ruby Gulch, about 2,000 feet northwest of Whitcomb. It is owned by the Ruby Gulch Mining Company and is the most productive deposit in the Zortman district. It has yielded more than $350,000 worth of gold and silver. The lode followed northward crosses the crest of the mountains and ore has been discovered on both sides of the divide, but the main workings are on the southern slope. The lowest level is an adit driven westward about 150 feet to the lode. From the point of intersection the lode is followed southward from the adit for 200 feet and northward for about 800 feet. Several raises are turned from this adit and are used to draw ore from the open cut above, which at its highest point is about 180 feet above the level of the adit. The Carter tunnel is 175 feet higher than the main working tunnel and is driven northward along the ore body for 400 feet. What is presumably the same ore body is opened by surface workings at two or three places north of the north end of the adit.

The country rock is porphyry, which constitutes the main central axis of the mountains. The crystalline schists outcrop both to the west and to the east of the lode, only a few rods away on either side. Their position is shown in fig. 3 (p. 103). A broken line in this figure indicates that though the lode is known to be present its width is unknown at such places.

The lode is a zone of shattered, cemented, and replaced porphyry. Its average strike is N. 28° W. and it is approximately vertical. As mined it is from 20 to 70 feet wide.

The ore is highly oxidized throughout the explored portion of the lode. In a winze sunk in ore 75 feet below the level of the lowest tunnel, giving a depth of 200 feet below the surface, no sulphides whatever appear, and only small bunches of pyrite occur here and there at the north breast of the tunnel, which is about an equal distance below the surface. The ore as mined is in the main finely shattered and decomposed porphyry, covered with a thin coating of iron oxide. Some of the porphyry is highly silicified. The ore carries from $3 to $21 in gold and about 1 ounce of silver to the ton, the general average as shown by mill runs being $6.55. The richest parts of the lode are rudely tabular bodies of finely pulverized red or chocolate-brown ore, which when damp has the plasticity of clay. These bodies are approximately parallel to the lode and are from 6 inches to 6 feet wide. Generally they occur along planes of movement. Bunches of earthy fluorite occur here and there in the high-grade ore and some of them carry high values.

The open cut above the adit is 300 feet long and workable ore is said to extend beyond the cut at each end, making a shoot of workable ore which is altogether 600 feet long. North of this ore shoot is a block of ore which is of too low grade to be worked with profit, but still farther north the tunnel passes through another body of profitable ore whose dimensions have not been ascertained. The main adit is nearly everywhere driven along a slickensided plane which persistently follows the lode and which may be traced continuously for 1,000 feet. This plane of movement is east of the ore at most places and it dips to the west, hence it is regarded as the foot wall of the vein, though the ore occurs below it at many places. Another slickensided plane in the lower tunnel is crossed by slickensided slips at right angles. These slips do not displace the fissures, which run nearly parallel with the lode, nor are they displaced by them to a notable extent.

INDEPENDENT MINE.

The Independent mine, which is situated about 1,200 feet south of the Mint, is also owned by the Ruby Gulch Mining Company. The deposit closely resembles that of the Mint mine, and it is possible that the two mines are on the same lode, although the connection has not been definitely established. The ore is worked from an open cut, 225 feet long, 110 feet deep, and from 10 to 70 feet wide. An adit is driven to the bottom of this cut and another one 125 feet below it. The lower adit, driven northwestward, intersects the lode 250 feet from the portal and follows it for 300 feet. The ore is drawn through chutes to the lower tunnel and is hauled over a level tramway to the mill.

The lode is a shear zone in porphyry. Near the surface the ore is oxidized and is of the same general character as that of the Mint mine, but contains less fine material. The difference in the amount of oxidation of the two deposits is very noticeable. In the Independent open cut the pyrite extended at one place to the surface and it is generally encountered at a depth of not more than 25 feet. At 100 feet in depth the ore is essentially all pyrite. The mixed oxide and sulphide ore is said to carry about $6.50 in gold, and the unoxidized ore below the sulphide ore runs a little lower. Postmineral fissures cut the ore body, as in the Mint mine, but in the Independent they do not follow the lode so closely. Two such fissures which intersect at a large angle along their strike may be plainly seen on the surface above the open cut. There is ore on both sides of these fissures, as shown in fig. 4 (p. 105), which is a cross section of the Independent at the open cut.

ALABAMA MINE.

The Alabama mine is near the head of Ruby Gulch, about 1,500 feet west of the Independent mine. It is owned by the Alder Gulch Mining Company, and when visited in October, 1907, was under bond to the Little Rockies Exploration Company, which had a force of men engaged in exploration. The mine is said to have produced about $200,000 worth of ore, of which the richer portion was shipped to smelters and that of the lower grade was milled at the Zortman mill on Alder Gulch.

A crosscut tunnel reaches the lode some 300 feet from the portal and follows it northward for about 1,000 feet. From this tunnel a raise extends to the surface a distance of about 255 feet, and a winze is sunk near the bottom of the raise to a depth of 250 feet. Levels are turned in the lode at intervals of 50 feet. When the mine was visited the levels 150 feet below the adit were under water.

The lode is a shear zone in porphyry. The ore is a brecciated porphyry, replaced and cemented by quartz, pyrite, and other minerals. On the surface the ore is completely oxidized and at places it is partly oxidized 400 feet below the surface. A streak of rich ore composed in part of purple fluorite has furnished a large part of the shipping ore. Some of this ore shows specks of free gold, but part of the gold is a telluride and panning does not reveal its presence until after it has been roasted. The rich streak seems to have undergone more movement than the remainder of the lode and it is as a rule greatly crushed. It is from a few inches up to 2 feet wide, but it has not been found throughout the length of the ore body.

· The average strike of the lode is N. 12° W. and it dips about 87° W. Its maximum width is 30 feet. The principal ore shoot is just north of the winze, and on some of the levels is about 250 feet long. It has been stoped downward to a depth of about 450 feet. To the north of this ore shoot workable·ore has been found here and there in the adit tunnel, and at the north breast of this tunnel, which is 400 feet below the surface, the face carries 8 inches of dark sulphide ore that assays $14 in gold. Across the breast for a width of 5 feet the average value is $3, exclusive of the rich streak. In the raise from the adit to the surface, at a depth of 50 feet, the ore consists of mixed iron oxides and pyrite. The groundmass of the replaced porphyry is a black, flinty, pyritiferous ore, and large crystals of feldspar have been leached out, leaving distinct negative crystal forms, the largest of which are about half an inch long. Some of these are now filled with spongy reddish-brown silica, which carries high values in gold. Other feldspar cavities do not contain quartz but are partly filled with soft white kaolin. Postmineral fissures follow the vein closely, forming the so-called walls. Some of these are thin, knife-edge seams, grooved here and there and case-hardened on the sides. At some places the ore along them is much brecciated and locally reduced to a soft mass, cut through here and there by narrow veinlets of barren white quartz. One of the most persistent of the fissures appears to form the boundary of the lode on its east side. From the surface to the bottom of the 255-foot raise on the adit level it dips from 85° to 89° W. Near the adit level it rolls, the dip changing to 70° to 85° E. The richest ore occurs near this fissure.

On the surface above the mine, about 100 feet northwest of the top of the raise from the adit to the surface, a shaft has recently been sunk in iron-stained sheared and brecciated porphyry which is said to carry $20 to the ton in gold. This ore is apparently independent of that encountered in the older workings, but the present developments do not show the character or outlines of this ore body.

BIG CHIEF CLAIM.

The Big Chief claim is west of the Mint and north of the Alabama mine, near the crest of the main divide of the mountains. It has been prospected by several trenches, surface pits, and shallow shafts, some of which were inaccessible in 1907. A tunnel is driven westward 410 feet into the mountain, following a smooth plane of movement which strikes N. 73° W. and dips about 67° S. Eighty feet from the portal the tunnel is intersected by an incline which is also sunk on the slickensided plane.

The country rock is granite-syenite porphyry, sheeted and stained by iron oxide on fracture planes and locally silicified. The shattered porphyry is said to carry gold.

POLE GULCH MINE.

The Pole Gulch mine is 3,500 feet S. 10° E. of the Zortman mill, with which it is connected by a level tramway. The south end of the tramway is laid in an adit that is driven southward 800 feet to a point below the ore body. The country rock is limestone and all the mining has been done near the surface, mainly by the open-cut method. The caved stopes and open cuts are at most places continuous and make a large, shallow, irregular cavity about 200 feet wide and 325 feet long. The limestones encountered in the adit are grayish buff in color and dip about 15° SE. They are marmorized and to some extent replaced by quartz. When the mine was visited most of the underground workings below the ore body had caved and their relation to the structure could not be satisfactorily made out. At some places the deposits appeared to follow the bedding planes of the limestone and at other places to cut across the bedding. The ore is soft and partly decomposed limestone, locally silicified and stained with iron oxides. Bunches of earthy purple fluorite occur here and there and are regarded as indicative of good ore. The mill runs are said to have averaged from $3 to $4 a ton.

BEAVER CREEK DEPOSITS.

On Beaver Creek, 2 miles northeast of the Mint mine, there is a large body of limestone about a mile long which is nearly or entirely surrounded by porphyry. Where best exposed the limestone dips 34° NW. In several small excavations bodies of low-grade ore have been found. When the camp was visited in 1907 very little work had been done on these deposits, but they had recently been sold to the Ruby Gulch Company for $20,000 and this company was planning extensive developments in the near future. The principal deposit is near the north contact of the limestone and porphyry. It appears to strike westward and where developed is of considerable width. In the most extensive tunnel the ore body is 45 feet across and is said to average $5 in gold to the ton. What is possibly the same lode is encountered in two short tunnels about 800 and 1,400 feet to the west, where the altered country rock for 15 feet is said to carry $4 in gold to the ton. The ore consists of decomposed, iron-stained limestone, spongy brown silica, and black cherty silica. The deposits are probably not of contact-metamorphic origin, for the limestone near by does not show any marked effect of contact metamorphism.

CLAIMS OF THE FERGUS MINING COMPANY.

The Fergus Mining Company has a large group of claims near the head of Peoples Creek. They are for the most part in porphyry, but some of the workings are in the crystalline schists. The principal tunnel is driven into the hill N. 17° E. for 270 feet. At the breast a portion of the wall rock is a hard, glistening amphibole schist, carrying considerable pyrite. Higher on the hill several pits and short tunnels are driven along zones of fractured porphyry stained by iron oxide and said to carry gold.

GOLDBUG MINE.

The Goldbug mine is located about 1½ miles northwest of Landusky, on the ridge between Mill Creek and Montana Creek. It has been worked now and then since 1893 and small shipments of rich ore were made in 1904. In 1902 a 10-stamp amalgamating mill was built at Landusky to treat the lower-grade ore, but this process was apparently not successful, as only a small amount of ore was treated. Five gold-bearing lodes are said to have been discovered on the property. None of these, so far as developed, is extensive. The deposits are sheared and brecciated zones in porphyry and as a rule strike northeastward. They have been prospected at several places by short tunnels and open cuts and some stoping has been done. The stopes are short and range from a few inches to several feet in width. The ore consists of shattered porphyry and quartz cemented by quartz and limonite, and some of the porphyry is replaced by quartz. Fluorite, pyrite, and tellurium minerals occur here and there. At some places well-defined slickensided planes form the walls of the lodes.

AUGUST MINE.

The August mine is about three-fourths of a mile north of the Goldbug. It was worked secretly in the winter of 1893, while the ground was yet within the Fort Belknap Indian Reservation, and $32,000 worth of rich gold ore was taken out in sinking a shaft 65 feet deep. Since then it has produced very little ore. In 1907 it had the appearance of a mine which has long been idle. The deposit is similar to that of the Goldbug mine and the ore consists of brecciated porphyry cemented by quartz and highly stained by limonite. Northeast of the August mine a number of short tunnels have been driven on iron-stained sheeted zones of porphyry. These zones commonly show grooved, slickensided surfaces, the general trend of which is about N. 35° E.

ELLA C. CLAIM.

The Ella C. claim is situated between the Goldbug and August mines. The lode is a sheeted zone in porphyry which strikes N. 35° to 40° E. This zone is about 2 feet wide, is composed of fragments of porphyry cemented by quartz and limonite, and is said to carry $20 a ton in gold.

SUSIE CLAIM.

The Susie claim, which is a short distance southeast of the August mine, is owned by C. R. Liebert and Peter Sieh. The lode strikes northeastward and is explored in several shallow pits. It is a brecciated zone of altered porphyry that has been recemented by iron oxide and quartz, which carry gold. A slickensided fissure with hard, polished surfaces coincides approximately with the hanging wall of the lode. This fissure strikes N. 43° to 67° E. and dips about 47° SE. Below the fissure the lode is from 6 inches to 3 feet wide and is said to carry $15 in gold to the ton.

GEOLOGY AND MINERAL RESOURCES OF THE OSCEOLA MINING DISTRICT, WHITE PINE COUNTY, NEV.

By F. B. Weeks.

INTRODUCTION.

As a part of thè general reconnaissance of the Great Basin region the geology of the region was studied by J. E. Spurr [a] in the summer of 1899 and by the writer in 1900. The geologists of the Wheeler Survey [b] also published some general observations on the region. A portion of October, 1907, was spent by the writer in making a more detailed study of the geology and mineral resources of the Osceola mining district.

GENERAL DESCRIPTION OF THE SNAKE RANGE.

The Snake Range, in which is located the Osceola mining district, is one of the most prominent and extensive mountain ranges between the Wasatch and the Sierra Nevada. It extends between latitude 38° 30' and 40° 30', a distance of 135 miles parallel to and a little west of the Utah-Nevada boundary. (See fig. 5.) As an orographic feature it comprises the Deep Creek or Ibanpah Range and the connecting hills designated as Kern Mountains on the map of the Wheeler Survey. The Snake Range is about 10 miles in width. The interior portion has been eroded into sharp ridges trending in general with the range, and the east and west flanks descend in steep slopes or bold escarpments to the valleys below. Snake Valley occupies a broad depression on the east and opens into the southwest end of Great Salt Lake desert. Spring Valley, west of the Snake Range, extends from the Cedar Range on the south to the so-called Kern Mountains on the north. The difference in elevation between the valleys and the highest part of the range is about 6,000 feet. The rugged character of the range makes it a formidable barrier to east and west travel. There are only four natural passes which afford a

---

[a] Bull. U. S. Geol. Survey No 208, 1903, pp. 25–36.
[b] Rept. U. S. Geog. Surv. W. 100th Mer., vol. 3, 1875, pp. 240–242.

practicable route for wagon roads. The highest summit of the range
is Wheeler Peak (locally known as Jeff Davis Peak), which has an
elevation of 12,000 feet. In the region of the Osceola mining district
the range presents an abrupt face to the west and a long, gentle slope
to the east.

Upon a basement complex of granite and schist, only a small area
of which is exposed, there has been deposited a series of Paleozoic

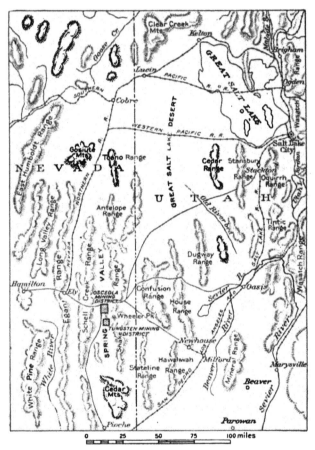

Fig. 5.—Map showing location of Osceola and Tungsten mining districts, White Pine
County, Nev.

sediments from 8,000 to 10,000 feet in thickness. As shown by their
fossils the strata were laid down in Cambrian, Ordovician, and Carboniferous time, a long interval of nondeposition occurring throughout the Silurian and Devonian periods. Some beds of unknown
age may represent the Mesozoic era, but the evidence is inconclusive.

On the lower slope of the range in certain areas there are sands and gravels which lie above the Lake Bonneville beds and which were probably laid down in late Tertiary time. At the north end of the range terraces of the Pleistocene Lake Bonneville have been noted. It seems probable that from Carboniferous time to the present the greater part of this area has been subject to erosion.

Considerable bodies of igneous materials are exposed in the northern and central portions of the range. These igneous masses intruded Carboniferous strata and so, in part, at least, are Mesozoic or younger.

The prominent structural feature of the range is a dome in the region of Wheeler Peak, which both to the north and to the south passes into anticlinal folds whose axes in general trend with that of the range. This structure has been subsequently modified by compression and faulting and by the intrusion of igneous masses, so that the sedimentary beds generally have steep dips and are in many localities separated by considerable displacements. This is especially true in the region of the Osceola mining district.

OSCEOLA MINING DISTRICT.

SITUATION AND PHYSICAL FEATURES.

This mining district is about 35 miles east of Ely, Nev., the southern terminus of the Nevada Northern Railroad. It includes the crest and western slope of the Snake Range in the vicinity of Osceola. The east-west wagon road through the district is the principal route of travel between Utah and central Nevada. Near the summit on the eastern side the road forks, a branch leading over the Sacramento Pass and descending to Spring Valley on the west.

The principal drainage lines in the mining district are Dry Gulch and Mary Ann Canyon and along them and in their alluvial fans occur the most important placer deposits. The stream beds are dry during most of the year. About one-fourth mile above Osceola, near the wagon road, are several small springs and a small stream flows from the mouth of the New Moon mine. The elevation of the district ranges from 6,000 to 9,600 feet above sea level. The region is arid, the principal precipitation being in the form of snow.

GENERAL GEOLOGY.

The distribution of the rocks in this district is shown in the sketch map forming fig. 6, and a general section is given in the following table.

General geologic section of the Osceola mining district.

No. in fig. 7.	Age.	Character.	Thickness.
			Feet.
1	Recent...........................	Gravel, coarse to fine, gold bearing...................	Up to 80
2	Upper and middle Cambrian.	Gray to white, rather pure limestones and dark-blue crystalline limestones.	1,000
3	Lower Cambrian...............	Green sandy shales, *Olenellus* zone...............	150
4	.....do...........................	White, blue, and purple quartzites, gold bearing..	2,000
5	.....do...........................	Purple argillite...........................	750
6	.....do...........................	Conglomerate...........................	100 to 160
7	Archean (?).....................	Granites and schists, with intruded granite porphyry.	....

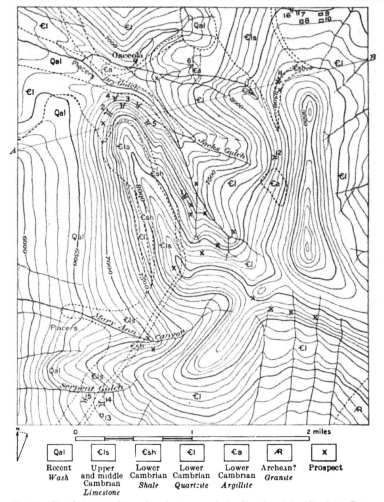

FIG. 6.—Sketch topographic and geologic map of the Osceola mining district. Contour interval 100 feet. Mines: 1, Cumberland; 2, Golden Eagle; 3, Crescent; 4, Time Check; 5, Exchange; 6, New Moon; 7, Gold Hill; 8, Gold Crown; 9, Queen; 10, King; 11, Whitney group; 12, Mulligan group; 13, Drummer shaft; 14, Mayday; 15, Serpent; 16, June. Dotted lines are boundaries of rock areas.

Fig. 7 is a cross section showing the structural relations along a line crossing Dry Gulch.

A short distance south of the mining district and near the crest of the range is an area of granite and schist overlain by a coarse conglomerate which grades into a compact argillaceous rock resembling argillite. The argillite is succeeded by a series of quartzites which pass into shales containing an *Olenellus* fauna. It appears from these observations that there is exposed here a small area of the basement complex rocks. Their structure has, however, been broken by an intrusive mass composed largely of gray and red granite porphyry which, north of the road crossing the range to Osceola, has penetrated through strata of possible Carboniferous age.[a] On the divide north of Wheeler Peak certain observations made in an area of poor rock exposures indicate that the granite porphyry cuts through the granites and schists of supposed Archean age.[b] These Archean rocks are much finer in grain and generally more basic than the intrusive rocks. By the presence of sheared zones and general

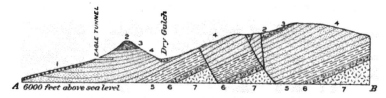

Fig. 7.—Structure section along line *A–B*, fig. 6. 1, Recent; 2, upper and middle Cambrian; 3 to 6, lower Cambrian; 7, Archean (?).

schistose structure they bear evidence of stresses and strains which were not observed in the intrusive granite porphyry.

The conglomerate which overlies the basement complex is about 100 feet thick and is formed of large subangular pebbles and bowlders derived from the older rocks. The conglomerate pebbles gradually become more rounded and smaller in size, argillaceous material forming a considerable portion of the rock, which passes into a massive bedded argillite. The argillaceous series is about 700 to 800 feet thick and is well exposed on the eastern slope of the high ridge that forms the crest of the range east of Osceola.

In this region the dip ranges from 25° to 40° NW.[c] and where the strata are cut by the intrusive porphyry they have been altered for the most part into a bluish-gray, generally schistose rock that has been called "silvery slate." This series is overlain by gray and white fine-grained quartzites. The beds have been thoroughly silicified and contain many veinlets of probably secondary quartz. They have

[a] Spurr, J. E., Bull. U. S. Geol. Survey No. 208, 1903, p. 32.
[b] Idem, pp. 26–27.
[c] Directions given in this paper are magnetic.

also been subjected to compression and numerous extensive belts of cross fracturing have been developed in which the quartzite has been broken into small angular fragments. The quartzite series has been estimated to be 2,500 feet in thickness. In the upper part of these beds occur the gold deposits and from the erosion of their outcropping edges the placer deposits in and along the sides of the gulches and in the alluvial fans have been formed.

The quartzite is succeeded by about 150 feet of green sandy and argillaceous shales. Where the outcrop of these beds crosses the south end of the ridge facing Spring Valley fossils of *Olenellus* type were found. These are the lowest beds in which fossils are known to occur, though careful search was made for them in the argillite.

Above the shale series occur dark-blue and gray limestones about 1,000 feet thick. The individual beds range from 1 to 3 feet in thickness, but in the crest of the ridge facing Spring Valley there are about 100 feet of thin-bedded and shaly blue limestones. The dark-blue limestones immediately overlie the shales and for the most part are crystalline or semicrystalline and contain numerous calcite veins. The gray limestone is comparatively pure and ranges from dark gray to white in color. Fossils have been collected at several horizons in the limestone series on the east and north slopes of the ridge facing Spring Valley and these have been determined as forms characteristic of the middle and upper Cambrian.[a]

Within the area of this mining district there are no other sedimentary rocks except the recent deposits in which occur the placers. These deposits are from a few inches to 80 feet thick. The gravel ranges from fine to coarse and contains few large bowlders.

HISTORY OF MINING DEVELOPMENT.

In 1877 work was begun on the placer deposits of Dry Gulch. A few quartz locations were made prior to that time. It is reported that 300 to 400 miners were working on the placers during 1877 to 1880 and during the latter year 400 placer and lode locations were on record. The important placer properties in Dry Gulch became the property of the Osceola Gravel Mining Company, subsequently known as the Osceola Placer Mining Company, in the early eighties. Prior to 1890 this company had constructed two ditches approximating 34 miles in length, at a cost of about $200,000. The operations of this company and of individuals continued until about 1900, when on account of light snowfall and the loss in efficiency of the ditch from leaky flumes and other causes work was discontinued.

The alluvial fan which spreads out from the mouth of Mary Ann Canyon, in the southern part of the district, is locally known as

---

[a] Fossils mentioned in this report have been determined by Mr. Charles D. Walcott.

Hogum. Here pay gravel was found several years after the discoveries in Dry Gulch and the deposits have been worked intermittently since that time.

Several attempts have been made to work the gold-quartz properties on a small scale. Three mills of 5, 10, and 20 stamps have been erected and operated, but none of them has been commercially successful. It is admitted that more than 50 per cent of the values went down the gulch with the tailings. Since field work was completed the 20-stamp mill has been partly repaired and a run of several hundred tons of ore from the Cumberland mine has been made. The results are not known.

From all accounts that have been obtained, it seems safe to estimate that the production of gold from this district approximates $2,000,000, of which about one-tenth was probably derived from the quartz mines.

MINING.

The slopes being steep, underground development is through tunnels, there being not more than half a dozen shafts in the district. In one or two mines an upper and lower tunnel have been connected by winzes. The quartzite is exceedingly hard and no timbering is required in the tunnels. In winzes and stopes a few stulls are all that is needed.

Some ore has been sacked and shipped to the smelters, but the greater part has been locally milled. Stamping and amalgamation constitute the principal method of treating the gold quartz. A small cyanide plant was constructed several years ago but was abandoned, apparently before receiving a satisfactory trial.

The Boston and Nevada Mining and Milling Company employs half a dozen men and about the same number are engaged from time to time in doing assessment work for nonresidents. The average wage for miners is $3.50 per day of eight hours.

All the mines and placers have been located by prospectors and working miners. No extensive consolidations have been made and the camp remains an aggregate of small mines and prospects on which, with the possible exception of the Osceola placers, but little outside money has been expended. The ores so far discovered have not been of high enough grade to attract lessees.

EXTENT OF PRODUCTIVE TERRITORY.

There seems to be no ground for assuming that the productive territory extends beyond the limits of the area shown on the map (fig. 6). So far developments indicate that the deposits are confined to fairly well defined zones in the quartzite. It has been thought by some that the same mineral belt extends northeastward to the recently opened

Black Horse district. It may be that the effects of the same dynamic forces which developed the lode systems in this district extended to the Black Horse district, but a slight examination of that region indicates that although the lithologic characters of the strata are in general similar, the beds at Black Horse were deposited during a later period and the ores are very different in character.

The fissures appear to be confined to the quartzite. They were not observed in the overlying shales and limestones and the underlying argillite has not been exposed in the underground workings. To the east and south of the area mapped the argillite series is well exposed and shearing with more or less movement along the bedding planes is a prominent feature, the beds being locally altered to a silvery slate.

DISTRIBUTION OF MINES.

The areas of greatest mineralization are (1) the ridge on the west and south sides of Dry Gulch, (2) the slopes of Mary Ann Canyon, and (3) the north end of the main mountain ridge. (See fig. 6.) In the Dry Gulch area are situated what is locally known as the Gold Exchange group, including the Woodman, Golden Eagle, Star, Time Check, Crescent, Exchange, and January. There also are the Butterfield and the Cumberland mines. In the region of Mary Ann Canyon are the Serpent, Mayday, Drummer, and other prospects. On the north end of the main ridge at an elevation of about 8,000 feet are the King, Queen, Gold Hill, June, and Gold Crown claims, locally known as the Summit group, and a little to the west are the Whitney and Mulligan groups.

UNDERGROUND DEVELOPMENT.

The most extensive underground workings are in the Gold Exchange group. The tunnels in these mines have a total length of about 1,000 feet, and connect with one or two shafts and several stopes and winzes. In Mary Ann Canyon several prospects have tunnels from 50 to 125 feet long. In the Summit group the tunnels range from 50 to 350 feet in length. The Whitney and Mulligan groups have about the same amount of development. The three tunnels in the Cumberland mine have a total length of approximately 1,700 feet.

STRUCTURE OF THE GOLD DEPOSITS.

There appear to be only two types of auriferous deposits in the district—(1) regular zones of fracturing or sheeting and (2) irregularly shattered masses of quartzite adjacent to these zones of fracture. In most places there is no distinct line of demarcation between the two types. There are no massive veins solidly filled with quartz such as are characteristic of many other regions.

The most characteristic structure is the sheeted zone. In this district these zones consist of narrow, nearly parallel fissures forming lodes ranging from several inches to 15 feet in width. In the wide belts, which may be considered as compound sheeted zones, there are generally one or more zones of closely spaced fissures. The sheeted zones contain some fracture planes which show slickensiding, but the displacement appears to be slight. This may be due to the brittle character of the fine-grained quartzite that forms the country rock. Beds which are broken into fine brecciated masses lie between massive beds of quartzite which apparently were not affected by the compressive stress.

Circulating waters carrying silica in solution have filled the fissures of the sheeted rock. The original openings were small, and they are as a rule completely filled. The most important exception is the Cumberland lode, in which the vein material contains many vugs lined with gold, fluorite, and other minerals.

The lodes are in places conspicuously exposed, forming bold outcrops of quartz. They are somewhat more resistant to erosion than the country rocks, but can not be followed on the surface for a very great distance. There are also lodes in the mines which do not appear at the surface. It is therefore impossible to describe in detail the lode systems, as there is a relatively small amount of underground work and the limits of the fracture zones have not been reached.

The Osceola lodes form two intersecting groups of approximately parallel fissures. In the northeastern part of the district the strike is northeast. In other parts of the area the strike varies but little from east and west, except that in the southwestern part there appears to be a northeast-southwest system of fracture zones which cut the east-west lodes. In general the lodes do not converge but maintain their direction until they can no longer be distinguished from the irregular jointing which occurs in all the rocks. The two systems seem to have formed simultaneously and they do not appear to fault each other. In Mary Ann Canyon the fissures intersect without noticeable displacement. The intersection is usually marked by an irregular broken zone, as may be seen on the outcrop and in the upper and lower tunnels of the Mayday mine.

The lodes are steeply inclined, nearly all being above 70° and many vertical. So far as the underground workings show they are fairly regular in dip. Adjacent fissures in general dip in the same direction. This is well shown in the Gold Exchange group. In many places the lodes for considerable distances are so ill defined that the dip can not be determined. It may be said in general that the east-west lodes are vertical and that the northeast-southwest lodes dip at high angles.

No systematic relation between dips, distribution of fissures, and general structure of the district has been found.

PERSISTENCE.

Very little can be said definitely regarding the persistence of the lodes in depth. The deepest underground workings are not more than 300 feet below the surface and the fissures extend to this depth. The ore shoots and sheeted zones are not necessarily coextensive, for the highly productive areas have generally proved to be moderate in extent. Detailed information concerning the length of the lodes is wanting. In the Gold Exchange group the principal lode has been fairly well traced for a distance of half a mile, the west end being cut off by erosion and the east end passing into undeveloped ground. In the Summit group not one but several fissures which appear to replace each other have been traced at irregular intervals for more than half a mile.

ORIGIN OF THE FISSURES.

The character of the stresses that fissured the strata is not easily determined. It is clear, however, that they were such as could be relieved by fracturing with only slight displacement. The hypothesis which seems to accord best with field observations is that in the readjustments, which followed the intrusion of the magma, stresses were set up that resulted in the shearing of the argillite and the fracturing of the fine-grained, brittle quartzites along vertical or highly inclined zones. Fissuring and the intrusion of the igneous magma appear to be genetically connected and were followed by ore deposition from circulating waters.

CHARACTER OF THE ORE.

The lodes of the Osceola district contain a relatively small amount of metallic or gangue minerals. Inasmuch as these minerals occur as the filling of narrow fissures or cracks in the fractured zone which usually constitutes a lode, or as the incomplete replacement of the country rock, the gangue of the ores is similar in character and composition to the rocks adjoining the fissures. Pyrite is very sparingly disseminated in grains so minute as scarcely to be distinguishable by the unaided eye. Ferruginous clays are common in the fissures. In certain lodes, particularly that in the Cumberland mine, the quartz is here and there honeycombed and contains many vugs lined with fluorite and other minerals as well as free gold. More commonly the gold occurs in flakes and also finely disseminated in quartz seams and veinlets.

So far as known gold is the only metal of commercial value in the Osceola ores. From the information available it is impracticable to estimate definitely the average gold content. Commercial assay returns show a wide range in value. Three samples taken by the writer gave assay values of $5, $32, and $77 per ton, the last representing the face of a tunnel about 4 by 6 feet. Other samples taken by the writer ranged in value from 80 cents to $4.50 per ton and represented portions of lodes not less than 3 feet in width. The return from a shipment of several tons of selected ore from the fractured country rock adjoining a fissure zone gave a value of $28 per ton. It is evident that the gold content of the lodes varies greatly, as in other known gold-bearing veins. It is not unlikely that careful prospecting will develop ore bodies of sufficient size and value to render their exploitation profitable.

OXIDATION.

The greatest depth of underground workings does not exceed 300 feet, and the sulphide zone has not been reached, so far as known. No water is found in any of the mines except in the New Moon tunnel, which crosses a fault in the argillite series. Under present climatic conditions there is very little precipitation, so that the mines are practically never wet. The district stands high above the adjacent valleys, and other conditions suggest unusual depth of ground water.

The greater number of lodes contain a considerable amount of material oxidized to a yellow or brown clay, that does not appear to be easily carried away. Oxidation, however, does not seem to have changed the composition or obliterated the structure of the lode materials to any marked degree. No evidence was obtained that there had been a secondary enrichment of the lodes from the surface downward by leaching of the ore. Such action, however, may have taken place under more humid climatic conditions, such as are believed to have existed in this region in recent geologic time.

ORIGIN OF THE ORES.

No extended discussion of the origin of the gold-bearing ores of the Osceola district can be presented here, as the examination of the mines was not made in sufficient detail to determine many questions that have an important bearing on their genesis, and it has not been possible to study the field collections prior to the preparation of this paper. From general analogy with other deposits it is considered that the ores were deposited from circulating waters within fissure zones formed by compressive stresses. If, as seems likely, the greater part of the mineralization occurred by deposition from ascending waters the silica and the fluorine in the fluor spar locally developed were derived from the originally molten magma that probably underlies the region at no great depth, in geologic terms. In an adjoining area tungsten-bearing veins in granite porphyry contain a considerable amount of fluorite. Evidence bearing on the source of the gold is inconclusive. That it was leached from the quartzite strata is not improbable, for there is some evidence that they are gold bearing. It may, on the other hand, have been separated from the intrusive magma and brought up through the fissures by magmatic waters.

DETAILED DESCRIPTION OF MINES.

GOLD EXCHANGE GROUP.

The Gold Exchange group comprises eleven lode claims, of which three are fractional. They extend from the west face of Pilot Knob Ridge around the north end, following the south and west slope of Dry Gulch. The slope is steep, but good mountain roads have been constructed to the several tunnel openings. A 20-stamp mill has been erected on the Star ground, which adjoins the Golden Eagle (No. 2 on map, fig. 6) on the west. Water from the west-side ditch has been used in operating this mill. The Star, Golden Eagle, Crescent (No. 3 on map), and Exchange (No. 5) are patented ground; the other claims are held by annual assessment work.

The underground workings on this group, except a portion of the lower tunnel on the Star ground, are in the upper part of the quartzite series. The average dip is 40° NW. and the strike is N. 10° E., but both dip and strike vary within short distances. There are many vertical or highly inclined fault planes, but the displacements observed do not exceed a few feet. The shale series overlies the quartzite and above the shales are the limestones capping the ridge. Near the mill a fault has thrown down the limestones and below these outcrops the slope is covered with débris.

The quartzite strata have been subjected to stresses resulting in two fracture zones, one having an east-west direction, the other northeast and southwest. The east-west zone is the principal one and within it occur many small displacements, the rocks showing well-marked slickensides. This zone can be traced on the surface as a succession of "blowouts." In some of the beds the fracturing extends beyond the usual limits of the lodes, but the other strata retain their massive character. The shattered beds were broken into small angular fragments.

There are two lode systems within this group, one within the Time Check and Crescent (Nos. 4 and 3 on map) and the other within the Golden Eagle (No. 2), Exchange (No. 5), and January ground. They are approximately parallel and are several hundred feet apart. The latter is apparently the more extensive and has produced the larger amount of ore.

The gold is concentrated within fissure zones of varying width and is also disseminated to a greater or less extent in the beds of finely shattered quartzite. It is not known to what extent these beds have been mineralized, as no drifts have been made in them. They evidently contain some pay ore, for in places chambers several feet in extent have been stoped. The gold is rarely visible, being very finely disseminated.

In accordance with the locally held idea that these ores are free-milling they have been treated in the ordinary stamp mill with amalgamation tables. These properties have not been worked since 1899 and at this time it is impossible to obtain definite information as to the average value of the material milled or the percentage saved. It is generally conceded, however, that there was considerable loss—possibly as much as 50 per cent of the assay values. On account of the high degree of fineness of the gold and the fact that it is not all in the free state it is believed by many that a much greater percentage of saving would result from cyanide treatment.

SUMMIT GROUP.

The Summit group of claims is situated on the crest of the range about 1 mile south of the wagon road which crosses the mountains from Osceola. Some underground work has been done on each of the claims which comprise this group.

The Gold Hill tunnel (No. 7 on map) is 309 feet in length and its direction is south. It is entirely in the quartzite strata, which strike N. 10° E. and dip 45° NW. They are generally massive bedded and contain several clay seams and fractured zones about 6 inches in width. On the eastern side of this claim there is another tunnel about 100 feet in length, having a direction S. 60° W., with a drift to the south from the face of the tunnel about 100 feet long. In these workings is exposed a broad zone of fractured and brecciated rock whose limits are not known. Considerable ore from this tunnel is said to have been milled, but no satisfactory estimate of value could be obtained.

The June tunnel (No. 16 on map) varies in direction and has a total length of 240 feet. The tunnel cuts a fault trending N. 26° E. East of the fault, in the direction of the face of the tunnel, the beds show little disturbance but contain many soft seams from which gold can be obtained by panning. To the northwest, or toward the mouth of the tunnel, the quartzite is very finely brecciated. In this fracture zone assays ranging from $8 to $15 are said to have been obtained, but no definite statement as to the width of the zone furnishing such assay values could be given. This fracture zone appears to have a general direction of N. 30° E.

The Gold Crown, Queen, and King (Nos. 8, 9, and 10 on map) are developed to a small extent. Average assays of $14 are said to have been obtained from these claims.

Within the Summit group there appear to be at least three fracture zones separated by intervals in which the quartzites are relatively undisturbed. The amount of mineralization varies greatly within these zones and extensive prospecting will be necessary to determine the distribution of the values.

WHITNEY GROUP.

The Whitney group (No. 11 on map) has been prospected by several tunnels cutting the shale and quartzite nearly at right angles to the strike. The strata are in places much broken and shattered and in others are undisturbed. Certain fault planes, indicated by slickensided surfaces, have been followed as walls in the tunnels. These fault planes dip 60° S. Considerable ore has been mined and milled from these workings, but no satisfactory statement of its value could be obtained.

MULLIGAN GROUP.

At the north end of the Mulligan group (No. 12 on map) there is an incline following what appeared to be a fault fissure nearly filled with vein quartz. The fault strikes N. 80° E. and dips 65° S. About 600 feet south of this incline is a tunnel which at the time of visit was closed, but the material on the dump showed that there must be considerable underground work in a formation of very finely crushed white quartzite. At the south end of this group is a 200-foot tunnel entirely in a crushed white quartzite.

CUMBERLAND MINE.

The Cumberland mine (No. 1 on map) has three tunnels having a common direction of S. 80° W., at vertical intervals of 100 to 200 feet. The lower tunnel is 500 feet in length and follows a fault zone in which are many small fissures showing slickensided surfaces and dipping both to the north and south at high angles. In some places a fault plane dipping steeply to the south has been followed until it became nearly horizontal in the roof and then another steeply inclined fault plane farther on in the tunnel has been used as a wall. In some places, for distances of 10 to 20 feet, the quartzite strata are unbroken and have the normal strike and dip. Many beds not showing distinct fault planes have been crushed into confused masses of small fragments. Clay seams are abundant, and some of them follow bedding planes.

The second tunnel was not examined. The third and upper tunnel is 650 feet in length. At 450 feet from the entrance is a winze 50 feet in depth and an upraise to the surface. In this upraise there is an ore shoot 3 to 4 feet in width which pitches 75° S. The greater part of this ore shoot has been worked out. Its hanging wall is a well-defined fault plane. The lower edge of the shoot is cut in the back of the third tunnel, but its pitch carries it to the south of the first and second tunnels, and no prospecting has been done to determine its extension in this direction. It is reported that most of the ore mined and milled from the Cumberland came from this upraise, but no definite information as to its average value could be obtained. The ore contains much free gold, partly in vugs with fluorite. Many beautiful specimens have been found in these ores.

OTHER PROSPECTS.

In the southwestern part of the Osceola district, in the region about Mary Ann Canyon, there has been considerable prospecting since 1900. This area is locally known as Hogum. The granite porphyry is exposed on the western edge of this area and small veins, generally of quartz, extend from it into the adjoining sedimentary strata. The derivation of the vein filling from the intrusive mass is more clearly shown here than in the other parts of the district. Although the structural features of this area have been affected by the intrusion of the igneous magma, they nevertheless are closely connected with those of other parts of the district previously described.

The Mayday claim (No. 14 on map) is developed by a tunnel 130 feet in length following the strike of the vein, which is S. 70° E. The gangue material is nearly all quartz and it pitches to the southwest, or in the direction of the granite porphyry, which is exposed about one-eighth of a mile farther south. The vein contains many gouge seams and small displacements. In the area between this vein and the granite porphyry the quartzite is fractured and broken into small angular fragments.

The Drummer claim (No. 13 on map) is developed by a shaft 18 feet in depth following an offshoot from the granite porphyry. This vein is about 4 feet wide and is formed of fine-grained granite porphyry and quartz. Its general direction is N. 30° E.

The Serpent claim (No. 15 on map) has two tunnels 50 and 80 feet in length. The entrance to the upper tunnel is in the limestone, which dips 45° NW. and strikes N. 30° E. The vein strikes N. 50° E. and dips 45° SW. The ore-bearing portion ranges in width from a 10-inch vein to a thin parting. The best ore, said to have been found where the pay streak averaged 4 inches wide, assayed $400. Coarse gold was observed on exposed faces of the vein.

The lower tunnel is 25 feet below the upper tunnel and has a direction S. 80° E. The vein is from 6 to 8 inches in width and dips 25° SW. The returns from the milling of this ore were reported as $17 to $20 per ton. Considerable lead ore is found in a parallel vein.

PLACERS.

The placer deposits of Dry Gulch range from a thin covering of the edges of the quartzite strata in the upper part of the gulch to deposits 25 to 30 feet in depth in the lower part, below which the débris spreads out into an alluvial fan. Hydraulicking and ground-sluicing methods were employed to recover the gold. The values were more or less disseminated through the gravel, the principal pay deposits being as usual near bed rock. Large nuggets were rarely found, the gold being in general very fine. There still remains a considerable area of ground to be worked, but lack of water has thus far rendered further operations impracticable.

In the southwestern part of the district, in the area locally known as Hogum, the placer deposits occur in channels buried under the material of the alluvial fan below the mouth of Mary Ann Canyon. They usually occur in a stratum overlying a so-called cement or false bed rock, of which there appear to be several at different levels. The channels are worked by sinking and drifting. The material is raised by a whim, shoveled into sluice boxes, and washed with a small quantity of water from the ditch. Here, as in every other part of the district, the gold is fine and nuggets of much size are seldom found. Frequently small potholes are encountered in the false bed rock. These have the gold concentrated around their edges, but not within them. During the summer of 1907 the Gold Bar Placer Company employed from two to four men and the operations are said to have given a satisfactory return on the investment. The pay stratum was reported to have yielded from $6 to $8 per cubic yard.

Placer mining has also been carried on east of the divide, above the town of Osceola, in Mill and Weaver creeks. This area lies to the northeast of that shown on the map and was not studied in detail. The gold is derived from the erosion of the quartzite strata, as in all other parts of the district.

GENERAL SUMMARY.

The lode systems of the Osceola district are known to be extensive. All of them carry gold, but the values are irregularly distributed along the fissure zones. Systematic and extensive prospecting must be done to determine the average value of these lodes. It seems certain that the average product of the lodes will be a low-grade ore which must be worked at a small cost and in large quantity to be profitable.

Water for milling purposes and placer mining can be obtained from the several creeks heading around Wheeler Peak, which are also available for the generation of electricity. As it will require the waters of all these creeks to fully develop the resources of the district there should be such a combination of interests as would permit the development of the water and power for the use of the various mining companies. Future development and prosperity depend on a concentration of local interests on a basis that will attract capital.

MINES OF THE RIDDLES QUADRANGLE, OREGON.[*]

By J. S. Diller and G. F. Kay.

GOLD-QUARTZ MINES OF THE RIDDLES QUADRANGLE.

By G. F. Kay.

INTRODUCTION.

The Riddles quadrangle is situated in southwestern Oregon. It is a 30-minute area and embraces parts of Douglas, Jackson, and Josephine counties. It lies immediately south of the Roseburg quadrangle, which has been described in folio 49 of the Survey.

The geology and economic resources of this area were studied, for folio publication, by Mr. J. S. Diller and myself, during the summers of 1906 and 1907. In connection with this work, the placer deposits and gold-quartz mines were examined. It is the intention in this paper to describe only the latter, the former having been studied by Mr. Diller, who describes them on pages 147-151.

My thanks are due to many persons connected with the mines for information and other kindnesses, particularly to Capt. J. S. Buck, of the Greenback; Mr. S. G. Adams, manager and secretary-treasurer of the Baby mine; Mr. John Scribner, of the Silent Friend mine; and Mr. J. H. Beeman, of Gold Hill, owner of the Lucky Bart group. I am also much indebted to Mr. Diller for many valuable suggestions.

GEOGRAPHY AND HISTORY.

The Riddles quadrangle is of moderate relief, its elevation ranging from about 600 feet to 5,274 feet above sea level. In the northern part is South Umpqua River with its large tributary, Cow Creek. The chief streams of the southern part are Grave Creek, Jumpoff Joe Creek, and Evans Creek, all tributaries of Rogue River, which crosses the quadrangle only in the southwest corner.

The Southern Pacific Railroad passes through the northwestern part of the quadrangle, reenters it about the middle of the western

---

[*] The first paper of the group on the Riddles quadrangle (Nickel deposits of Nickel Mountain, Oregon, by G. F. Kay) was published in Bull. U. S. Geol. Survey No. 315, 1907, pp. 120-127.

side, and runs southward, keeping within the quadrangle to its southern boundary, which lies between Merlin and Grants Pass. There are wagon roads along almost all the main streams, and trails run over many of the more mountainous parts. Hence the region is, in general, fairly accessible.

Much of the higher ground is covered with forest, but in many of the valleys there are small ranches. The climate is such that quartz mining can be carried on, without much inconvenience, during all seasons of the year.

. The history of gold mining in southwestern Oregon dates back for more than fifty years, the first discovery having been made about the middle of the last century. From that time to the present, this portion of the State has yielded a considerable percentage of the total gold production of Oregon. The two counties from which most of the gold has come are Josephine and Jackson, both of which lie partly within the Riddles quadrangle.

In the early days practically all of the output came from the placers. Between 1880 and 1890 there was a small but gradually increasing yield from the gold-quartz mines. During the succeeding ten years the production of the placers continued to decrease and that of the gold-quartz mines to increase. In 1905 [a] the value of the product of the placers of Josephine and Jackson counties was $165,793, whereas the value of the output of the quartz mines (including that of three quartz mines of Lane County and less than $2,000 worth of gold from copper ores) was $236,193. Of the latter amount the Greenback mine, which is in the Riddles quadrangle, contributed a large part. Since the fall of 1906, however, this mine, which was for several years the chief gold-producing mine and one of the best-equipped mines in Oregon, has been closed. Moreover, several prospects, which a few years ago were considered very promising, have proved disappointing and work on them has been suspended.

GEOLOGY.

The rocks of the Riddles quadrangle comprise both sedimentary and igneous rocks of various ages. The former belong mainly to the Mesozoic and the Tertiary, but in the southeastern part of the area are highly metamorphosed sedimentary rocks, in which, as yet, no fossils have been found, but which, owing to resemblances to fossiliferous rocks occurring farther southwest in California, are thought to be of Paleozoic age. The igneous rocks are in large part intrusive, but considerable areas show undoubted volcanic characters.

The sedimentary rocks which are thought to be Paleozoic consist chiefly of mica slates, mica schists, and micaceous quartzites. To the same system belong scattered lentils of crystalline limestone

---

[a] Mineral Resources U. S. for 1905, U. S. Geol. Survey, 1906, pp. 288–292.

found a short distance beyond the southern limits of the quadrangle. These rocks are widely separated from the Jurassic of the quadrangle by igneous rocks.

The Mesozoic rocks belong to the Jurassic and Cretaceous systems. The Jurassic sediments consist mainly of slates and sandstones with interbedded shales; conglomerates and cherts are subordinate. These rocks, particularly the sandstones, usually show quartz veining and pronounced induration. The Cretaceous beds consist chiefly of conglomerates, sandstones, and shales, and have been divided, on the basis of fossil evidence, into the Knoxville, Horsetown, and Chico formations. The Chico is now represented by but a few small remnants of the original widespread formation. The chief remnant within the area of the quadrangle is on Grave Creek about 6 miles above the small village of Placer. It is not underlain by the Horsetown or the Knoxville but by older slates and by igneous rocks. There is evidence of a slight unconformity between the Knoxville and the Horsetown. The Knoxville rocks are locally veined and indurated, but generally to a much less degree than those of the Jurassic.

The Tertiary rocks are of Eocene age. They consist of yellowish sandstones, shales, and conglomerates, the stratification being well preserved.

The evidence indicates a great unconformity between the Jurassic and the Cretaceous; a somewhat less important unconformity separates the Cretaceous from the Eocene.

The igneous rocks are of various kinds, including greenstones, peridotites, serpentines, granodiorites, dacite porphyries, and augite andesites.

The greenstones are widespread and are generally altered to such an extent as to be unsatisfactory for study. Under this name are included several kinds of rocks so related in the field that it is practically impossible to map them separately. What may be considered the normal type resembles, when fresh, a gabbro, consisting essentially of pyroxene and a lime-soda variety of feldspar. Some phases of the greenstone are dioritic, some diabasic, and some fine grained and compact, resembling basalt. Moreover, some of the rocks included as greenstone are of the nature of volcanic breccias; others show decidedly vesicular characters. All these types are, no doubt, closely related genetically, but they may vary considerably in age. All are of a more or less green color and almost everywhere they show marked evidence of extensive crushing and veining. Associated with the greenstones in the west-central part of the quadrangle are a few lens-shaped areas of rhyolite.

The peridotites consist chiefly of olivine and enstatite, the former usually predominating, but locally the pyroxene is so abundant that the rock is a pyroxenite.

The serpentines have resulted chiefly from the decomposition of the peridotites and the pyroxenites, but some areas of the serpentine are probably the result of the decomposition of basic phases of the greenstones. Much of the serpentine shows shear zones and slickensided surfaces.

The granodiorites are of granular texture and include rocks which vary considerably in composition. The more acidic approach the granites and the more basic include quartz diorites. These rocks are composed chiefly of feldspar, quartz, and hornblende or mica, or, as is more commonly the case, both hornblende and mica. The color varies, depending on the amount of dark-colored minerals present, but the prevailing color is dark gray. The feldspar is chiefly plagioclase which belongs to the acid end of the soda-lime series. It is usually present in greater amount than the quartz. The mica is generally biotite, but muscovite is also found, and in places both are present. Apatite, magnetite, and locally garnet are accessory minerals.

The dacite porphyries are thought to be closely related genetically to the granodiorites. They have a rather sparse distribution, occurring as small knoblike areas and as dikes. They are usually light colored and have as their chief constituents quartz and soda-lime feldspar, both of which minerals in much of the rock form distinct phenocrysts.

The augite andesites occur only in small dikes and have been found cutting greenstones, granodiorites, and the Horsetown formation of the Cretaceous. The dikes of this rock were found only in the eastern half of the quadrangle.

The relative ages of the igneous rocks have been fairly well worked out. The greenstones are the oldest, then come the peridotites, next the granodiorites and dacite porphyries, and finally the augite andesites. Except some of the greenstone, which is probably Paleozoic, none of the igneous rocks described are thought to be older than the Jurassic, some are younger than the Lower Cretaceous, and all, except the augite andesites, are older than the Eocene. The augite andesites are probably related to the volcanics of the Cascades, and if so related they are of Tertiary age.

THE ORE DEPOSITS.

The chief gold-quartz mines and prospects of the Riddles quadrangle are in Josephine and Jackson counties, mainly in the former, as indicated on the accompanying map (fig. 8). The gold quartz is found in small veins, veinlets, and stringers in several kinds of rock. Within Josephine County all the paying veins have been in greenstone; a few prospects but no mines have been located in serpentine.

A striking feature in connection with many of the gold-bearing veins found in the greenstones is their proximity to serpentine, but usually the veins are cut off sharply at the contact of the greenstone with the serpentine. This may indicate either that the rock from which the serpentine was derived was younger than the vein or that displacements have occurred at the contact of the greenstone and its decompo-

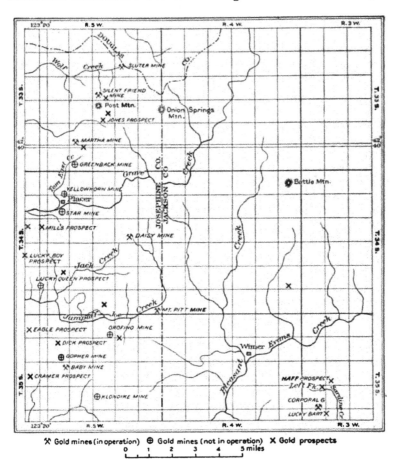

X Gold mines (in operation)   ⊕ Gold mines (not in operation)   X Gold prospects

0   1   2   3   4   5 miles

FIG. 8.—Map showing gold-quartz mines and prospects of Riddles quadrangle, Oregon.

sition product, the serpentine. In Jackson County the paying veins have been found in metamorphosed sediments, and usually they are within short distances of dikes or irregular areas of greenstone.

Quartz veinlets are also found in the granodiorites, but as yet no mine has been developed in these rocks within the area of the Riddles quadrangle. However, a few hundred yards south of the southern boundary, in the granodiorite, is the Granite Hill mine, which has been for several years an important producer of gold.

The evidence suggests that all the gold-bearing veins in the several kinds of rock are younger than the early Cretaceous and older than the Eocene. However, some of the veins in the metamorphosed sediments may be pre-Cretaceous.

The vein filling is chiefly quartz, calcite, and pyrite; here and there arsenopyrite, pyrrhotite, sphalerite, chalcopyrite, and galena are also present. The action of the mineral-bearing solutions on the country rock adjacent to the veins was such as to produce a strongly chloritized and, in places, talcose rock much of which contains carbonates or pyrite.

The gold-producing veins range in width from mere seams to veins more than 4 feet wide, but the average width of all the veins examined was less than 1 foot. In the best mine that has been found in the region, the Greenback, the average width of the vein was about 18 inches. In many places there are sheared and brecciated zones, in which occur several veinlets and stringers usually running parallel to one another, but locally irregular and running in various directions. In general, individual veins and veinlets are not continuous except for short distances, and in many localities, to further interfere with the continuity of veins, there has been a considerable amount of faulting along planes at various angles to one another. The walls of some of the veins are fairly well defined for short distances, but many of them show no distinct boundary between the vein material and the country rock.

Although gold-bearing veins and veinlets are found running in various directions, those which have been most productive trend in general east and west. For example, at the Greenback mine the vein has a direction nearly east and west; at the Martha mine, between northwest and west; at the Baby mine, northwest to nearly west; at the Corporal G mine, S. 85° W.; and the veins of the Lucky Bart group run almost west. The dips of the veins range from nearly horizontal to vertical, but usually they are at fairly high angles.

The values are found chiefly in the quartz of the veins and veinlets, but in the brecciated zones some gold is obtained in the fragments of chloritized rock which carries pyrite. The values are mostly in free gold, but the sulphides also carry gold, the amount varying considerably in different veins and in different parts of the same vein.

Some of the quartz veins which carry values are later in age than others which carry no values. This is well shown at the Baby mine, where the gold-bearing vein cuts a much wider barren vein. The barren vein appears not to have changed the strike, dip, or values of the Baby vein. In the neighborhood of the Corporal G are veins which are themselves barren, but where the younger gold-quartz veins intersect them the values are said to be enriched.

DESCRIPTIONS OF THE PRINCIPAL MINES.

Although there has been much prospecting for gold-quartz veins during the last fifteen years within the area of the Riddles quadrangle, comparatively few important discoveries have been made and some of the mines are no longer producing. Some of the most important mines are described below.

GREENBACK MINE.

The Greenback mine is situated on Tom East Creek, a branch of Grave Creek. It is worthy of note here that on the same stream, below the Greenback, is the Columbia placer mine, which is one of the most productive of southwestern Oregon.

The Greenback was discovered in 1897 by two prospectors, who lived in the vicinity of Placer on Grave Creek. They worked the deposit for about a year, treating the ore with an arrastre at Placer. They then sold the property for $30,000 to the Victor Junior Gold Mining Company, the chief owners of the stock being W. H. Brevort, of New York, and Messrs. Moffatt and Smith, of Denver. In 1902 more than 90 per cent of the stock was purchased by Mr. Brevort, and the corporation was named the Greenback Gold Mining and Milling Company. No transfer has since been made.

From the time when the property came into the possession of the Victor Junior Gold Mining Company until 1906 the development of the mine was rapid, more equipment being added each year. At first a 5-stamp mill was installed, later 5 stamps were added, and when, in 1902, Mr. Brevort became the chief owner, there were 15 stamps, besides a crusher, an air compressor, and three Wilfley tables for concentrating. Mr. Brevort's company soon began the construction of a new mill, about a quarter of a mile farther down the stream. At first 20 stamps were used in this mill. This number was increased until, in 1905, 40 stamps were being used. The new plant has three large Risdon crushers and 12 concentrating tables. There is also a cyanide plant consisting of four large tanks, with a capacity of 100 tons a day. For a while the mill was equipped with both steam and water power, but in 1905 a complete electric system was installed. The power was brought, by way of Grants Pass, from the Ray dam on Rogue River, a distance of about 30 miles. However, in the following year (August, 1906) all work at the mine was suspended and as yet it has not been resumed.

The workings of the mine are extensive, consisting chiefly of cross-cut tunnels to the vein and drifts and shafts on the vein. Much of the ore has been stoped along the whole length of the vein to a depth of about 1,000 feet from the surface. The lowest of the workings are on the twelfth level. Below the ninth level the mine is filled with water.

The country rock is greenstone, which is considerably metamorphosed, but where most free from alteration it is of the nature of a gabbro. To the east and southeast of the mine there is a considerable area of serpentine, and a short distance to the north lies the southwestern limit of a band of siliceous slates which extends for some miles to the northeast.

The Greenback vein has a direction almost east and west, and dips, in general, about 60° N. It averages about 18 inches in width, but ranges from less than 6 inches to more than 4 feet. Where it is widest there is a crushed zone of quartz stringers and country rock, forming in places a beautiful breccia; the country rock of the breccia is strongly chloritized and contains sulphides which carry values. In many places the foot and hanging walls of the vein are fairly definite, but where considerable brecciation has occurred there is no distinct boundary between the vein material and the chloritized country rock. The vein is cut off sharply to the east against serpentine and to the west by a fault. The vein between the serpentine and the fault plane has an average length of more than 500 feet and within this distance there are only minor displacements. The vein has not been picked up to the west beyond the fault plane, nor has it been found in the serpentine to the east. This latter fact tends to prove that the rock from which this serpentine was derived was younger than the vein, rather than, as is indicated in some places in the quadrangle, that the present relations are due to displacements between the greenstone and its decomposition product, the serpentine.

The vein filling consists of quartz, calcite, and pyrite, which vary in amount in different parts of the vein. The average content of the ore mined from the first and second levels was between $8 and $9 to the ton; a few assays on these levels ran above $40 to the ton. Captain Buck states that over 75 per cent of the values of the ore was free-milling. The concentrates ran about $75 to the ton and after cyaniding the ores contained less than $1 to the ton. Within the mine there is but little evidence, except near the surface, of oxidation of the ores.

A short distance to the south of the Greenback vein and running almost parallel to it is the Irish Girl vein. On this very little work has been done.

MARTHA MINE.

The Martha mine is in the SW. ¼ sec. 28, T. 33 S., R. 5 W., about 1¼ miles north of the Greenback. It was purchased by the Greenback Company in 1904 and somewhat extensively developed. The electric power of the Greenback was extended to this property, and in 1906 an aerial tramway was constructed to connect the two mines and for a few months the ore of the Martha was conveyed by the tramway to the Greenback plant and treated there. The tramway is said to have cost $20,000. The company also installed a 75-horse-power air compressor. When the Greenback was closed the company also stopped all work at the Martha.

The mine was prospected by four tunnels whose length aggregates nearly 3,000 feet. At the time the property was examined (June, 1907) it was leased by J. M. Clarke, of Golden, Oreg., who had brought in five stamps and was treating the ore which had been mined by the Greenback Company but which had not been shipped to the mill.

The country rock is greenstone. The ores resemble those of the Greenback, but do not carry as high values in gold. They are found in narrow veinlets and stringers in zones of shearing and brecciation, which have a general trend between northwest and west and which range in width from a few inches to more than 4 feet.

BABY MINE.

The Baby mine is in the northwest corner of sec. 16, T. 35 S., R. 5 W., and is owned by the Capital City Gold Mining Company. The property was located in 1897 and since that time has been extensively developed by the present and the former owners. It is now leased by R. S. Moore, of Grants Pass. During the summer of 1907 three stamps were in operation. Mr. Adams, the manager of the company, says that the mine has yielded gold to the value of more than $20,000.

There is on the property a 5-stamp mill, two boilers, a concentrating table, and a small crusher. The development consists of more than 1,500 feet of tunnels, shafts, drifts, upraises, and crosscuts.

The vein occurs in greenstone and averages about 4 feet in width, but in places a fissured zone more than 10 feet wide has within it many parallel stringers of quartz which carries gold. The vein ranges in direction from northwest to nearly west and dips to the northeast, usually at high angles, although it is in some places almost vertical and in others almost flat.

A striking feature of the mine is the prevalence of faults. These are not only numerous, but they vary considerably in direction and in amount of displacement. One of the most prominent of the fault planes runs S. 80° W.

The vein material consists of a somewhat sugary-looking quartz, some calcite, and some pyrite. The values are carried chiefly by the quartz; in many parts of the vein free gold may be seen with the unaided eye. The sulphide varies in amount in different parts of the vein, and when concentrated yields about $75 worth of gold to the ton.

SILENT FRIEND MINE.

The Silent Friend property is in the southern part of sec. 15, T. 33 S., R. 5 W., on the north slope of Post Mountain. It is owned by the original locators, John Scribner and George Henderson, both of Speaker, Oreg. They discovered the vein in 1900, worked it until 1902, then leased it for eighteen months to Joseph Dysert. From the expiration of this lease until August, 1906, no development was carried on, but from that date to the present the owners have been working the mine on a small scale. Mr. Scribner states that from the oxidized material on the surface overlying a network of small stringers he has taken gold to the value of more than $7,000.

The chief development has been by two tunnels. The lower of these is 320 feet in length and crosscuts several small stringers. The upper is 75 feet in length, with an upraise to the surface.

The country rock is greenstone, which is strongly chloritized adjacent to the veins. The chloritization is, no doubt, due to the action of the mineral-bearing solutions. The ores are found in veinlets and stringers which run in various directions, but the majority of them have a general trend between southwest and west.

The filling consists of quartz, calcite, pyrite, arsenopyrite, and, locally, chalcopyrite. Some specimens of ore, which were found to consist largely of calcite, chlorite, and arsenopyrite, showed considerable free gold visible to the unaided eye. These specimens, which were taken from the bottom of one of the drifts, appeared to represent in the mine the ore of an 18-inch brecciated zone, which could be followed for several feet.

DAISY MINE.

The Daisy mine, which is on the divide at the head of Jack Creek, is on one of six claims constituting the Oregon Mohawk gold mines, owned by G. R. Smith, of Grants Pass. It was discovered in 1890 and for a time was worked under the name of the Hammersly mine. Then the stock was acquired by Morton Lindley, of San Francisco, who later disposed of it to the present owner.

Preparations were being made during the summer of 1907 to pump the water from the mine, which had been idle for some time, and mining operations were to be resumed. Mr. Smith stated that the mine had produced gold to the value of more than $200,000.

The workings consist of an inclined shaft 175 feet in depth, from which, at a distance of 115 feet below the surface, there is a drift along the vein for 350 feet to the west and 50 feet to the east. All the ore above has been stoped. From the bottom of the shaft there is a drift running eastward on the vein for 140 feet and westward for 243 feet.

The veinlets of gold-bearing quartz carrying pyrite run about east and west and are in a chloritized greenstone. The ore-bearing zone has a width of about 3 feet.

MOUNT PITT MINE.

The following notes were obtained from J. S. Diller, who examined the Mount Pitt property.

The mine is situated in the southeast corner of sec. 36, T. 34 S., R. 5 W. It was located by H. G. Rice, of Grants Pass, the present superintendent. The property is owned by A. C. Hooper, of Portland.

The present workings consist of an entrance tunnel of 225 feet to cut the vein, a drift of 100 feet along the vein, and an upraise of 200 feet from the drift to the surface. A mill has recently been erected containing a crusher, an automatic feeder, 5 stamps, and a concentrating table.

The ore is found in small, irregular veins in sheared greenstone, the sheared zone being usually about 3 feet wide. The quartz veins are rarely well marked, the greatest width of quartz seen being 4 inches and this is not persistent for more than a yard or so. The quartz veinlets are in general parallel to the plane of shearing, but some of them are small cross gash veins nearly horizontal.

OROFINO MINE.

The Orofino property, which is located in sec. 3, T. 35 S., R. 5 W., has been closed for several months and the workings are beginning to cave. The present owners are Messrs. Monahan and Mason, of Seattle. The last work was done by B. F. Chase, of Portland, who had a lease.

C. D. Crane, of Grants Pass, stated that there had been nearly 2,000 feet of work done on this property. Fourteen carloads of ore have been shipped to smelters at Tacoma, Wash., and Ashland, Oreg. The mine had at one time considerable equipment, including a 2-stamp mill, cyanide tanks, rock crushers, boilers, and hoists, but much of this material has been sold and shipped away.

The ore occurs in veinlets and stringers in a much fractured, brecciated, and chloritized greenstone. Many of the fragments of country rock of the breccia contain considerable pyrite. The vein filling consists chiefly of quartz and calcite, and, as shown by the relations of the two, the calcite was deposited later than the quartz. Sulphides are also present in some parts of the vein in considerable amounts, but in other parts they are almost entirely absent. A large amount of ore is now lying on the dump and many sacks of ore are ready for shipment.

All the mines thus far described are associated with greenstones and the descriptions indicate that the characters of the ores of the mines and their modes of occurrence are very similar. Many other mines and prospects associated with the greenstones might be described, but they would show few new features. Some of these are now being developed; some have been extensively prospected but have never produced; others have, in the past, produced small amounts but are no longer being worked. Among such mines and prospects may be mentioned the Lucky Queen, Mill's prospect, Star mine, Olympic prospect, Spotted Fawn prospect, Blalock & Howe mine, Eagle prospect, Cramer prospect, Gopher mine, and Dick prospect, most of which are indicated on the map (fig. 8). To the north of the area shown on the map are the Gold Bluff and Levens Ledge mines, both near Canyonville.

CORPORAL G MINE.

Of the mines which are not associated with the greenstones but with metamorphosed sediments the chief are the Corporal G mine and the Lucky Bart group, which lie west of the Left Fork of Sardine Creek.

The Corporal G mine is located in the southern part of sec. 19, T. 35 S., R. 3 W. It was discovered in 1904 by J. R. McKay, who, after taking out considerable rich ore, sold it to Mrs. Nina M. Smith, of Gold Hill, the present owner. The property is now leased by J. E. Kirk.

The workings consist of three tunnels, one above another on the vein. The longest tunnel is 92 feet in length, the shortest 63 feet. The ore occurs in a small vein with fairly definite walls of micaceous quartzite and mica slate. The average width of the vein is about 7 inches; it runs S. 85° W. and dips steeply to the north. The filling consists chiefly of quartz and calcite, but pyrite, pyrrhotite, chalcopyrite, bornite, sphalerite, and galena are also present. A few of the hand specimens show free gold.

Close to the Corporal G is the Volunteer claim on which a stringer running parallel to the Corporal G was followed by a drift for 135 feet, when it pinched out. This stringer intersects a barren cross vein running about N. 30° E.; at the intersection the values in the stringer are said to have been enriched.

LUCKY BART GROUP.

The Lucky Bart group consists of eleven claims in the NW. ¼ sec. 29 and the SE. ¼ sec. 30, T. 35 S., R. 3 W. The chief claim, the Buckskin or Lucky Bart, was discovered by Joseph Cox, who sold it in 1892 for $15,000. This amount he had to share with his partner, Bart Signoretti, who had had no part in the discovery, hence the name Lucky Bart. The company which bought the property worked it for four years when one of the shareholders, J. H. Beeman, of Gold Hill, purchased the rights of his associates and became the owner. About the same time Mr. Beeman purchased adjoining claims until he had title to all the property included in the Lucky Bart group. At present mining operations are being carried on at only one of the claims, the Yours Truly. The workings on the other claims, but mainly on the Lucky Bart, are in such condition that it is unsafe to enter them. The only workings examined were those of the Yours Truly. Information with regard to the other workings of the group was obtained from J. H. Beeman and J. E. Kirk.

Ore has been mined from five veins which run in a general direction a little south of west. These veins have an average width of less than 2 feet; the country rock is metamorphosed sediment, mainly mica slates and micaceous quartzites. The general strike of these rocks in this vicinity is somewhat east of north; the dip is to the southeast and is usually at fairly high angles. The total amount of ore that has been milled exceeds 14,000 tons, which gave values ranging from $4.80 to $100 a ton of free-milling ore. The ore from the Lucky Bart claim carried an average of 3 per cent of sulphides, which ran from 4 to 8 ounces of gold to the ton and a like amount in silver. Nine tons of ore from the deepest workings of this claim were shipped to the Tacoma smelter and gave returns of $130 to the ton. Practically all the ores from the group have been treated at a mill on Sardine Creek; the sulphides were shipped to the smelters at Tacoma, Wash., and Selby, Cal.

At the Yours Truly, where work is now being done by J. E. Kirk, who has a lease on the property, the workings consist of an entrance tunnel of 75 feet to the vein, 100 feet of drifting on the vein, and a shaft of 30 feet. The country rock is mica slate. The vein has an average width of about 1 foot and runs S. 85° W. At the end of the drift there are two veinlets of 8 inches and 4 inches in width and also a small seam. Within the workings there is evidence of considerable faulting; the directions of the fault planes observed were somewhat east of north. Mr. Kirk states that the veins carry more values adjacent to the fault planes than elsewhere. The ores of the Yours Truly are highly oxidized and carry an average value of more than $30 to the ton.

CONCLUSIONS.

Of the many veins and veinlets within the Riddles quadrangle on which work has been done, comparatively few have developed into profitable mines. The chief reason is to be found in the structural features of the rocks in which the ores occur. The Paleozoic and early Mesozoic sediments, with their associated igneous rocks, were, previous to the mineralization of the region, subjected to earth movements of such a nature that no definite, continuous fissures were formed, but rather, in general, innumerable minute and irregular fractures running in various directions. Later, when the mineral-bearing solutions, which may have been connected with one or more of the igneous intrusions, passed through these rocks and deposition therefrom took place, the gold was not concentrated in definite lodes but was widely distributed through the rocks in small veins, veinlets, and stringers, few of which are continuous except for short distances. Furthermore, in those places where fairly distinct and rich veins were formed, subsequent faulting has frequently been so prevalent that it is difficult and costly to follow the values. Notwithstanding these unfavorable conditions, however, the gold-quartz veins have produced and will probably continue to produce considerable amounts of gold. But the hope of finding vein deposits which will develop into large and profitable mines is not encouraging.

The veins and veinlets have been subjected to erosion for many thousands of years, during which time an immense amount of material has been freed of its gold. Much of this gold has been deposited in the neighboring streams, from which it has been and is being mined as placer gold.

PLACER MINES OF THE RIDDLES QUADRANGLE.

By J. S. DILLER.

INTRODUCTION.

Placer mining is one of the most important industries of the Riddles quadrangle. There are 54 placer mines; 10 are in the northern half of the quadrangle in Douglas County; the remainder are in the southern half—18 in Jackson County and 26 in Josephine County. The total output of placer gold in the quadrangle up to date has been approximately $725,000. In 1906, according to the returns of Mr. Yale, of the Geological Survey, the output was $69,395, a considerable increase over that of the previous year.

The placer mines are all in gravel closely associated with the present streams. By far the greater portion of the mines are in the present stream beds or low terraces. Only a few are in gravel of the higher terraces, which rise from 100 to 400 feet above the stream.

No definite trace of ancient high-level gravels such as occur in the gold belt of the Sierra Nevada of California has yet been found in the Riddles quadrangle.

The gravels vary much in the form of the pebbles. On the higher terraces and the steeper grades of the larger streams they are generally well rounded, though some may be subangular, but in the gentler grades and especially also on the smaller lateral branches the gravel is subangular to angular. The contrast may be seen in comparing the well-rounded gravel of the Steam Beer mine on Grave Creek near Leland with the subangular gravel of the Columbia on Tom East Creek, near Placer.

The grades of the present streams range from 10 to 333 feet per mile, as made out approximately from the contour map. The major part of the placer mines are on grades not over 100 feet per mile. A smaller number are on grades between 100 and 200 feet per mile, and a few have grades greater than 200 feet per mile.

The highest terrace records are few, but if they may be depended on they seem to indicate that the grade of the streams when the gravels of the highest terrace were formed was probably lower than that of the present streams; moreover, the gravel of the highest terraces is on the whole not so coarse as that of the lower terraces and the present stream.

RIDDLES DISTRICT.

The placers of the northern part of the quadrangle are widely scattered, generally small, and for the most part unimportant. They lie on Rattlesnake, Middle, Catching, Mitchell, Jordan, Canyon and Shively creeks. The Ash mine, which covers about 3¼ acres near Mitchell Creek, is peculiar in that the material washed is chiefly slope waste of slates and sandstones. On Shively Creek extensive preparations were in progress in 1906 for mining at the forks, but later returns have not been received.

COW CREEK DISTRICT.

Starvout Creek is a tributary of Cow Creek that drains the northern slope of Green Mountain. Three of the mines on this creek, the Harrah, Booth, and Curtis Brothers, have lately been in operation, but the Mizer and O'Shea only keep up assessment work. A large tract has been covered by these placers near the present stream level, and they are reported to have been fabulously rich in the early days. The bed rock is slate except in the Curtis Brothers mine, which is near the contact of the slates, greenstone, and serpentine.

Just beyond the western limit of the Riddles quadrangle, in Cow Creek canyon, are the Victory and Gold Flat placers, or terraces, about 150 feet above the stream. A dozen miles farther down are the Cain and the Cracker Jack, on terraces over 500 feet above the creek. All these are important placers formed under conditions in strong contrast with the gentle grade of Cow Creek above Glendale.

WOLF CREEK DISTRICT.

The Wolf Creek district includes not only the three placers more or less active on the main stream above the railroad station but also the four on Coyote Creek, which joins Wolf Creek at the post-office.

In Payne's mine, near Foley Gulch, a rusty rotten gravel is well exposed. The greenstone pebbles are completely rotten; those of slate are not so thoroughly decomposed. This gravel has the aspect of great age, but this illusion is dispelled by the freshness of the dark-gray gravel upon which it rests. The mine stretches up from the creek level to the terrace nearly 100 feet above. Coyote Creek has but little fall and the Ruble elevator has been used to advantage.

Near the mouth of Bear Gulch the bed of Coyote Creek has been mined for nearly half a mile. Its richness is due to the fact that Bear Gulch drains the slope from the Martha mine and the west end of the Greenback.

GRAVE CREEK DISTRICT.

Grave Creek is not only the most important placer stream in the Riddles quadrangle, but considering its size it is one of the most important in the State. Almost a score of placers, old and new, occur along the part of its course lying in this quadrangle, and half of them are still active during the good water season.

In the vicinity of Leland the lower Lewis and the Goff mines have not been worked lately. The water of their main ditch is now used in the Columbia. A small test of one-fourth acre in 20 feet of gravel was made last winter on the Klum property. The Steam Beer, owned by H. K. Miller, has continued in full operation for a number of years and there is more ground ahead. The ditch is about 9 miles in length and supplies a head of 200 feet. The gravel terrace is 50 feet above Grave Creek, which affords excellent dumping ground. The mine exposes 25 feet of gravel, generally coarse below, and made up largely of pebbles of greenstone with scarcely any quartz. The bed rock is slate.

On Brimstone Creek the gravel mined some years ago includes much quartz that appears to come from residual material on a terrace 300 feet above the creek.

The Columbia mine, near Placer post-office, is owned by L. A. Lewis, of Portland. It is the largest placer mine of the region and is supplied with water by two ditches from Grave Creek, one giving a head of 100 feet and the other of 600 feet. The mine occupies the valley of Tom East Creek, which drains the vicinity of the celebrated Greenback mine and is advancing in that direction. The gravel ranges from 4 to 30 feet in depth and is coarsest below, with bowlders a few of which reach 3 feet in diameter. The fragments are in general subangular and almost wholly greenstone. A few are rotten, but most are solid. The gold is fine, and nuggets are rare. With three 5-inch giants nearly 6 acres are mined over annually. The grade is low and to keep the sluice clear the tailings are washed aside from the end of the sluice by a powerful side stream which piles up the gravel in a prominent heap.

Near the headwaters of Grave Creek there are a number of active placers. Most of them are on the present stream bed, which has been washed for miles, but a few are on terraces up to 150 feet above the level of the creek.

JUMPOFF JOE DISTRICT.

The lower portion of Jumpoff Joe Creek traverses an area of granodiorite and has no placers, but above the forks placers occur among the greenstone hills on both Jack Creek and the main branch. The principal mine is the Swastika, under the management of A. B. Call. It occupies a low terrace in the forks at the mouth of Jack Creek. The Swastika property is said to include a large part of Jack Creek, and prospects have been made nearly 2 miles above its mouth toward the Daisy quartz mine. The Swastika has been operated by the present company for about a year. Two 18-inch pipes were used, one with a head of 150 feet and the other about 75 feet. The sluice dump was disposed of by a strong side stream.

The gravel is from 15 to 30 feet deep and is composed of greenstone pebbles. It is coarsest below, with bowlders up to 2 feet in diameter. In many places the whole mass is rotten, so that many of the bowlders go to pieces under the stream from the giant. The bed rock of the Swastika mine and throughout the slopes of Jack Creek is greenstone.

On the main fork of Jumpoff Joe Creek there are a number of small placers near its head and a larger one 5 miles below, where Cook & Howland have stripped the shallow bed of the stream, exposing the slates for half a mile to a width of 100 to 200 feet. The slope being gentle, an elevator was used.

EVANS CREEK DISTRICT.

Pleasant Creek, a branch of Evans Creek, heads against Grave Creek and has several active placers. For over 3 miles the bed of Pleasant Creek was almost completely mined out years ago, and later efforts have been directed to the benches rising up to 100 feet. The largest amount of work has been done at Harris Gulch, where an area. of rotten gravel about 8 acres in extent has lately been removed. A smaller cut has been made in a well-marked terrace at Jamison Gulch, and farther up, between the forks, Thompson Brothers have washed off the residual material of a serpentine point 200 feet above the streams.

All the placers on Pleasant Creek except the one last mentioned are on granodiorite but near the contact with both slate and greenstone, which may be the source of the gold.

SOURCE OF THE PLACER GOLD.

The source of the placer gold is in the auriferous quartz veins, which are most abundant in the greenstones, though they occur in the slates also. The larger veins at many places are worked in quartz mines, but all the veins, both large and small, have contributed gold to the placer gravels. The greater number of placers are on slate bed rock. This does not necessarily indicate that the slates have been the chief source of the gold in the placers, but that in the process of stream erosion the slates are more readily terraced so as to preserve the gravels for mining.

PRODUCTION OF THE PLACER MINES.

Where data have been available for estimates the yield of the placers per cubic yard of gravel has ranged from 10 to 25 cents. Much of the gravel must have averaged 50 cents and in exceptional cases run as high as $1.50. To state it in another form, a number of the mines appear to have yielded from $4,000 to $6,000 per acre.

As already stated, the total production of the placers in the Riddles quadrangle has been approximately $725,000, of which considerably more than half has come from Grave Creek. A still larger proportion of the present annual output is from Grave Creek, for it has not only the greatest number of mines but includes the two largest producers. The approximate production by districts is as follows:

Total production of placer gold in Riddles quadrangle to 1907, by districts.

Riddles and Cow Creek districts	$100,000
Wolf Creek district	75,000
Grave Creek district	400,000
Jumpoff Joe district	50,000
Evans Creek district	100,000

These estimates are more likely to be below than above the truth, for little is really known of the yield of the early placers.

NOTES ON COPPER PROSPECTS OF THE RIDDLES QUADRANGLE.

By G. F. KAY.

Copper minerals have been found at several places within the Riddles quadrangle, but as yet no paying mine of copper has been developed. During the summer of 1907 work was being done on two prospects which are of sufficient interest to merit a brief description.

The Joseph Ball mine is situated in the NW. ¼ sec. 36, T. 32 S., R. 4 W., which is on the southwest slope of Cedar Springs Mountain. The elevation at the mine is about 4,250 feet. Some ore has been carried by pack train to Glendale, on the Southern Pacific Railroad, a distance of more than 20 miles. The country rock is serpentine, which has been greatly fractured and sheared, and locally, where it has been decomposed, magnesite with some strontianite is present. The ores consist of native copper, copper glance, cuprite, and the copper carbonates. They are in a faulted zone in the serpentine, which shows numerous slickensided surfaces on which are vertical striæ. Within the workings the faulted zone varies in direction and the plane of shearing is very irregular. On this plane have been found flat pieces of native copper as large as the hand; the copper glance and cuprite have also been found on this plane as nodular masses and as scattered fragments. The workings consist of an upper tunnel of 150 feet along the fault zone and a lower tunnel of 145 feet from which there is an upraise of 60 feet to the upper tunnel. At the time the mine was examined the company was preparing to sink, from the lower tunnel, a shaft on the fault plane.

The Oak mine, in the SW. ¼ sec. 4, T. 35 S., R. 5 W., was located in 1905. It is owned by the Oak Consolidated Mining and Milling Company. Copper was found on this property while a gold-quartz vein was being developed. A tunnel was being run to crosscut some quartz stringers in a fractured zone, when copper pyrites was found. The mineral occurs as small irregular masses in a fractured and chloritized greenstone. During the summer of 1907 the company was installing an air compressor, hoists, and machine drills and plans were being made to prospect the property thoroughly.

Some prospects of copper occur in greenstone near Glendale and A. D. Leroy, of Merlin, has done some work on a quartz vein carrying copper in the N. ½ sec. 8, T. 35 S., R. 6 W.

SURVEY PUBLICATIONS ON GOLD AND SILVER.

The following list includes the more important publications by the United States Geological Survey, exclusive of those on Alaska, on precious metals and mining districts. Certain mining camps, while principally copper or lead producers, yield also smaller amounts of gold and silver. Publications on such districts are listed in the bibliographies for copper and for lead and zinc. When two metals are of importance in a particular district, references may be duplicated. For names of recent geologic folios in which gold and silver deposits are mapped and described, reference should be made to the table in the "Introduction" to this volume. Gold and silver, when the most important products in the area mapped, are indicated by italics in that table.

ARNOLD, RALPH. Gold placers of the coast of Washington. In Bulletin No. 260, pp. 154–157. 1905.

BAIN, H. F. Reported gold deposits of the Wichita Mountains [Okla.]. In Bulletin No. 225, pp. 120–122. 1904.

BALL, S. H. Geological reconnaissance in southwestern Nevada and eastern California. In Bulletin No. 285, pp. 53–73. 1906. Also Bulletin No. 308. 218 pp. 1907.

BARRELL, JOSEPH. Geology of the Marysville mining district, Montana. Professional Paper No. 57. 178 pp. 1907.

—— (See also Weed, W. H., and Barrell, J.)

BECKER, G. F. Geology of the Comstock lode and the Washoe district; with atlas. Monograph III. 422 pp. 1882.

—— Gold fields of the southern Appalachians. In Sixteenth Ann. Rept., pt. 3, pp. 251–331. 1895.

—— Witwatersrand banket, with notes on other gold-bearing pudding stones. In Eighteenth Ann. Rept., pt. 5, pp. 153–184. 1897.

—— Brief memorandum on the geology of the Philippine Islands. In Twentieth Ann. Rept., pt. 2, pp. 3–7. 1900.

BOUTWELL, J. M. Economic geology of the Bingham mining district, Utah. Professional Paper No. 38, pp. 73–385. 1905.

—— Progress report on Park City mining district, Utah. In Bulletins No. 213, pp. 31–40; No. 225, pp. 141–150; No. 260, pp. 150–153.

CALKINS, F. C. (See Ransome, F. L., and Calkins, F. C.)

COLLIER, A. J. Gold-bearing river sands of northeastern Washington. In Bulletin No. 315, pp. 56–70. 1907.

CROSS, WHITMAN. General geology of the Cripple Creek district, Colorado. In Sixteenth Ann. Rept., pt. 2, pp. 13–109. 1895.

—— Geology of Silver Cliff and the Rosita Hills, Colorado. In Seventeenth Ann. Rept., pt. 2, pp. 269–403. 1896.

CROSS, WHITMAN, and SPENCER, A. C. Geology of the Rico Mountains, Colorado. In Twenty-first Ann. Rept., pt. 2, pp. 15–165. 1900.

CURTIS, J. S. Silver-lead deposits of Eureka, Nev. Monograph VII. 200 pp. 1884.

DILLER, J. S. The Bohemia mining region of western Oregon, with notes on the Blue River mining region. In Twentieth Ann. Rept., pt. 3, pp. 7–36. 1900.

———— Mineral resources of the Indian Valley region, California. In Bulletin No. 260, pp. 45–49. 1905.

———— Geology of the Taylorsville region, California. Bulletin No. —. In preparation.

ECKEL, E. C. Gold and pyrite deposits of the Dahlonega district, Georgia. In Bulletin No. 213, pp. 57–63. 1903.

EMMONS, S. F. Geology and mining industry of Leadville, Colo.; with atlas. Monograph XII. 870 pp. 1886.

———— Progress of the precious-metal industry in the United States since 1880. In Mineral Resources U. S. for 1891, pp. 46–94. 1892.

———— Economic geology of the Mercur mining district, Utah. In Sixteenth Ann Rept., pt. 2, pp. 349–369. 1895.

———— The mines of Custer County, Colo. In Seventeenth Ann. Rept., pt. 2, pp. 411–472. 1896.

———— (See also Irving, J. D., and Emmons, S. F.)

EMMONS, S. F., and IRVING, J. D. Downtown district of Leadville, Colo. Bulletin No. 320. 72 pp. 1907.

EMMONS, W. H. The Neglected mine and near-by properties, Colorado. In Bulletin No. 260, pp. 121–127. 1905.

———— Ore deposits of Bear Creek, near Silverton, Colo. In Bulletin No. 285, pp. 25–27. 1906.

———— The Granite-Bimetallic and Cable mines, Phillipsburg quadrangle, Montana. In Bulletin No. 315, pp. 31–55. 1907.

EMMONS, W. H., and GARREY, G. H. Notes on the Manhattan district, Nevada. In Bulletin No. 303, pp. 84–93. 1907.

GALE, H. S. The Hahns Peak gold field. In Bulletin No. 285, pp. 28–34. 1906.

GARREY, G. H. (See Emmons, W. H., and Garrey, G. H.; Spurr, J. E., and Garrey, G. H.)

GRATON, L. C. Reconnaissance of some gold and tin deposits of the southern Appalachians; with notes on the Dahlonega mines, by Waldemar Lindgren. Bulletin No. 293. 134 pp. 1906.

———— (See also Lindgren, W., and Graton, L. C.)

HAGUE, ARNOLD. Geology of the Eureka district, Nevada. Monograph XX. 419 pp. 1892.

HAHN, O. H. The smelting of argentiferous lead ores in the Far West. In Mineral Resources U. S. for 1882, pp. 324–345. 1883.

IRVING, J. D. Ore deposits of the northern Black Hills. In Bulletin No. 225, pp. 123–140. 1904.

———— Ore deposits of the Ouray district, Colorado. In Bulletin No. 260, pp. 50–77. 1905.

———— Ore deposits in the vicinity of Lake City, Colo. In Bulletin No. 260, pp. 78–84. 1905.

———— (See also Emmons, S. F., and Irving, J. D.)

IRVING, J. D., AND EMMONS, S. F. Economic resources of northern Black Hills. Professional Paper No. 26, pp. 53–212. 1904.

LINDGREN, WALDEMAR. The gold-silver mines of Ophir, Cal. In Fourteenth Ann. Rept., pt. 2, pp. 243–284. 1894.

LINDGREN, WALDEMAR. The gold-quartz veins of Nevada City and Grass Valley districts, California. In Seventeenth Ann. Rept., pt. 2, pp. 1–262. 1896.

―――― The mining districts of the Idaho Basin and the Boise Ridge, Idaho. In Eighteenth Ann. Rept., pt. 3, pp. 625–736. 1898.

―――― The gold and silver veins of Silver City, De Lamar, and other mining districts in Idaho. In Twentieth Ann. Rept., pt. 3, pp. 75–256. 1900.

―――― The gold belt of the Blue Mountains of Oregon. In Twenty-second Ann. Rept., pt. 2, pp. 551–776. 1902

―――― Neocene rivers of the Sierra Nevada. In Bulletin No. 213, pp. 64–65. 1903.

―――― Mineral deposits of the Bitterroot Range and the Clearwater Mountains, Montana. In Bulletin No. 213, pp. 66–70. 1903.

―――― Tests for gold and silver in shales from western Kansas. Bulletin No. 202. 21 pp. 1902.

―――― The production of gold in the United States in 1904. In Bulletin No. 260, pp. 32–38. 1905.

―――― The production of silver in the United States in 1904. In Bulletin No. 260, pp. 39–44. 1905.

―――― The Annie Laurie mine, Piute County, Utah. In Bulletin No. 285, pp. 87–90. 1906.

―――― Notes on the Dahlonega mines. In Bulletin No. 293, pp. 119–128. 1906.

LINDGREN, WALDEMAR, and GRATON, L. C. Mineral deposits of New Mexico. In Bulletin No. 285, pp. 74–86. 1906.

LINDGREN, WALDEMAR, and RANSOME, F. L. The geological resurvey of the Cripple Creek district. Bulletin No. 254. 36 pp. 1905.

―――― ―――― Geology and gold deposits of the Cripple Creek district, Colorado. Professional Paper No. 54. 516 pp. 1906.

LINDGREN, WALDEMAR, and others. Gold and silver. In Mineral Resources U. S. for 1906, pp. 111–371. 1907.

LORD, E. Comstock mining and miners. Monograph IV. 451 pp. 1883.

MacDONALD, D. F. Economic features of northern Idaho and northeastern Montana. In Bulletin No. 285, pp. 41–52. 1906.

NITZE, H. B. C. History of gold mining and metallurgy in the Southern States. In Twentieth Ann. Rept., pt. 6, pp. 111–123. 1899.

PENROSE, R. A. F., jr. Mining geology of the Cripple Creek district, Colorado. In Sixteenth Ann. Rept., pt. 2, pp. 111–209. 1895.

PIRSSON, L. V. (See Weed, W. H., and Pirsson, L. V.)

PURINGTON, C. W. Preliminary report on the mining industries of the Telluride quadrangle, Colorado. In Eighteenth Ann. Rept., pt. 3, pp. 745–850. 1898.

RANSOME, F. L. Report on the economic geology of the Silverton quadrangle, Colorado. Bulletin No. 182. 265 pp. 1901.

―――― The ore deposits of the Rico Mountains, Colorado. In Twenty-second Ann. Rept., pt. 2, pp. 229–398. 1902.

―――― Preliminary account of Goldfield, Bullfrog, and other mining districts in southern Nevada. In Bulletin No. 303, pp. 7–83. 1907.

―――― (See also Lindgren, W., and Ransome, F. L.)

RANSOME, F. L., and CALKINS, F. C. Geology and ore deposits of the Coeur d'Alene district, Idaho. Professional Paper No. 62. In press.

SCHULTZ, A. R. Gold developments in central Uinta County, Wyo., and at other points on Snake River. In Bulletin No. 315, pp. 71–88. 1907.

SMITH, G. O. Gold mining in central Washington. In Bulletin No. 213, pp. 76–80. 1903.

―――― Quartz veins in Maine and Vermont. In Bulletin No. 225, pp. 81–88. 1904.

SMITH, G. O.   (See also Tower, G. W., and Smith, G. O.)

SPENCER, A. C.   (See Cross, Whitman, and Spencer, A. C.)

SPURR, J. E.   Economic geology of the Mercur mining district, Utah.   In Sixteenth Ann. Rept., pt. 2, pp. 343–455.   1895.

———— Geology of the Aspen mining district, Colorado; with atlas. Monograph XXXI.   260 pp.   1898.

———— The ore deposits of Monte Cristo, Washington.   In Twenty-second Ann. Rept., pt. 2, pp. 777–866.   1902.

———— Ore deposits of Tonopah and neighboring districts, Nevada.   In Bulletin No. 213, pp. 81–87.   1903.

———— Preliminary report on the ore deposits of Tonopah.   In Bulletin No. 225, pp. 89–110.   1904.

———— Ore deposits of the Silver Creek quadrangle, Nevada.   In Bulletin No. 225, pp. 111–117.   1904.

———— Notes on the geology of the Goldfields district, Nevada.   In Bulletin No. 225, pp. 118–129.   1904.

———— Geology of the Tonopah mining district, Nevada.   Professional Paper No. 42.   295 pp.   1905.

———— The ores of Goldfield, Nev.   In Bulletin No. 260, pp. 132–139.   1905.

———— Development at Tonopah during 1904.   In Bulletin No. 260, pp. 140–149.   1905.

———— Ore deposits of the Silver Peak quadrangle, Nevada.   Professional Paper No. 55.   174 pp.   1906.

SPURR, J. E., and GARREY, G. H.   Preliminary report on the ore deposits of the Georgetown mining district, Colorado.   In Bulletin No. 260, pp. 99–120. 1905.

———— ———— The Idaho Springs mining district, Colorado.   In Bulletin No. 285, pp. 35–40.   1906.

———— ———— Economic geology of the Georgetown quadrangle (together with the Empire district), Colorado, with general geology by S. H. Ball.   Professional Paper No. 63.   In preparation.

TOWER, G. W., and SMITH, G. O.   Geology and mining industry of the Tintic district, Utah.   In Nineteenth Ann. Rept., pt. 3, pp. 601–767.   1899.

WEED, W. H.   Geology of the Little Belt Mountains, Montana, with notes on the mineral deposits of the Neihart, Barker, Yogo, and other districts.   In Twentieth Ann. Rept., pt. 3, pp. 271–461.   1900.

———— Gold mines of the Marysville district, Montana.   In Bulletin No. 213, pp. 88–89.   1903.

———— Notes on the gold veins near Great Falls, Md.   In Bulletin No. 260, pp. 128–131.   1905.

WEED, W. H., and BARRELL, J.   Geology and ore deposits of the Elkhorn mining district, Jefferson County, Mont.   In Twenty-second Ann. Rept., pt. 2, pp. 399–550.   1902.

WEED, W. H., and PIRSSON, L. V.   Geology of the Castle Mountain mining district, Montana.   Bulletin No. 139.   164 pp.   1896.

———— ———— Geology and mining resources of the Judith Mountains of Montana.   In Eighteenth Ann. Rept., pt. 3, pp. 446–616.   1898.

WILLIAMS, A.   Popular fallacies regarding precious-metal ore deposits.   In Fourth Ann. Rept., pp. 253–271.   1884.

WOOLSEY, L. H.   Lake Fork extension of the Silverton mining area, Colorado.   In Bulletin No. 315, pp. 26–30.   1907.

COPPER.

NOTES ON COPPER DEPOSITS IN CHAFFEE, FREMONT, AND JEFFERSON COUNTIES, COLO.

By WALDEMAR LINDGREN.

INTRODUCTION.

Colorado is not an important copper-producing State. The annual production rarely reaches 10,000,000 pounds, and was only 7,427,253 pounds in 1906. Eighteen counties contributed to this output, but only two—Lake (Leadville district) and San Juan—yielded notable amounts. The Leadville output was somewhat over 2,000,000 pounds, which was really obtained as a by-product; that from San Juan County slightly exceeded 1,500,000 pounds. The copper ores, properly so called, of Colorado comprised only 31,431 short tons, of which one-half came from San Juan County. It may be said that Colorado does not contain a single copper mine of prominence.

Of the counties mentioned in the title Chaffee produced 349,466 pounds of copper from 6,249 tons of copper ores in 1906; the yield of Fremont and Jefferson counties is negligible. Although the copper deposits in these counties are not of great commercial importance, they are in some respects unusual, and from the standpoint of genetic problems merit some discussion. Nearly all the gold- and silver-bearing deposits of Colorado, and the copper deposits as well, occur in close connection with igneous rocks of Tertiary age. They have clearly been formed during late geological epochs—chiefly during the early or late Tertiary—and their deposition closely followed the eruption of the igneous rock with which they are associated.

There is evidence, however, of a far older epoch of ore deposition, antedating, in fact, the whole of Paleozoic and Mesozoic time, and best designated as pre-Cambrian. The existence of such deposits has been shown in South Dakota,[a] Wyoming,[b] New Mexico,[c] and Arizona,[d] and it is the purpose of this paper to describe six or seven such occurrences in Chaffee, Fremont, and Jefferson counties, Colo.

Copper deposits of still another class occur as disseminations in certain sedimentary beds in New Mexico, Texas, and Arizona; also in western Colorado. These deposits are generally of low grade, but the question of their origin has given rise to much discussion, which has lately been summarized by S. F. Emmons.[e] An occurrence of this kind in Fremont County will be described, for the reason that it seems to throw light on the genesis of these concentrations of copper.

PRE-CAMBRIAN COPPER DEPOSITS IN CHAFFEE AND FREMONT COUNTIES.

TOPOGRAPHY.

The larger part of Fremont County is occupied by the Front Range and the irregular plateau to the west of it. The southwestern part of the county contains the north end of the Sangre de Cristo Range. Through the central part of it flows Arkansas River in a deep and picturesque canyon.

The western part of Chaffee County, adjoining Fremont County on the northwest, is occupied by the high snowy peaks of the Sawatch Range, and the eastern part contains a complex of moderately elevated ridges, which forms part of the general plateau south of South Park. Arkansas River traverses the county from north to south in a series of open valleys, of which that surrounding Salida is the most attractive. A short distance south of Salida the river enters the canyon, from which it emerges just above Canon City. In the western part of Chaffee County are the La Plata, Cottonwood, Alpine, Chalk Creek, Monarch, and Garfield mining districts, of whose geological features but little is known. In the southeastern part of the county near Salida are the mining camps Turret, Sedalia, and Cleora. (See fig. 9.)

---

[a] Irving, J. D., and Emmons, S. F., Economic resources of the northern Black Hills: Prof. Paper U. S. Geol. Survey No. 26, 1904.

[b] Spencer, A. C., The copper deposits of the Encampment district, Wyoming: Prof. Paper U. S. Geol. Survey No. 25, 1904.

[c] Lindgren, W., and Graton, L. C., A reconnaissance of the mineral deposits of New Mexico: Bull. U. S. Geol. Survey No. 285, 1906, p. 81.

[d] Reid, J. A., A sketch of the geology and ore deposits of the Cherry Creek district, Arizona: Economic Geology, vol. 1, No. 5, 1906, p. 417. Graton, L. C., Mineral Resources U. S. for 1906, U. S. Geol. Survey, 1907, p. 389.

[e] Copper in the red beds of the Colorado Plateau region: Bull. U. S. Geol. Survey No. 260, 1904, pp. 221–232.

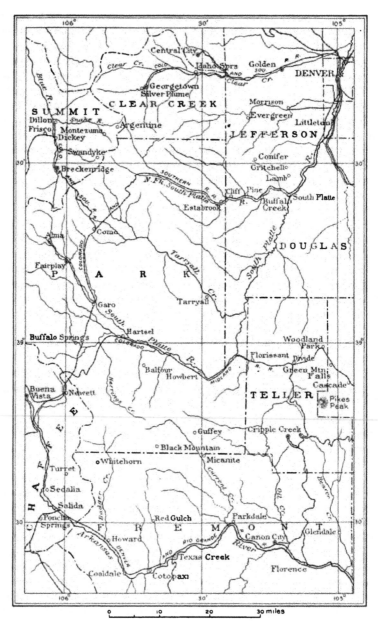

FIG. 9.—Map of central part of Colorado, showing location of copper deposits in Chaffee, Fremont, and Jefferson counties.

GEOLOGY.

Fremont County contains large areas of pre-Cambrian rocks. and their granites, gneisses, schists. and pegmatites are well exposed along the canyon of the Arkansas. In the western part of the county a belt of limestones and red beds crosses the canyon. and this belt occupies some areas north of the river in Chaffee and Fremont counties. The age of the lower beds is not well established, and the areas are outlined with imperfect accuracy on the Hayden map of Colorado. A bed of quartzite lies at the base, being succeeded by a thick series of limestones, which in turn are covered by red beds of Carboniferous or Triassic age. It is doubtful whether any Cambrian beds are present in this vicinity, but as they appear elsewhere in conformable development underneath the upper Paleozoic. the rocks underneath the great unconformity may with confidence be referred to as pre-Cambrian.

Near Cotopaxi gneissoid granites predominate, and these rocks with micaceous schists and intruded pegmatites make up the bulk of the pre-Cambrian area. At Salida, however, a series of greenstone schists have been described by Whitman Cross.[a] who states that they extend with general east-west strike from a point a few miles below Salida to a point about 5 miles north of that place. The rocks are determined as amphibolitic schists ranging into coarse-grained altered dioritic rocks (metadiorites) in the lower part of the series near Salida. Locally they are developed as chloritic. micaceous. and staurolitic schists. many of which contain garnets. They are cut by dikes of pegmatite and granite.

The pre-Cambrian complex is well exposed along the road from Salida to the small mining camp of Turret. 14 miles to the northeast. For 8 miles this road ascends in a steep gulch along which for the entire distance massive. very coarse grained dioritic rocks are exposed. in some places slightly sheared. and everywhere much jointed. Locally these rocks are epidotized. and traces of copper are seen in places. Under the microscope the structure is decidedly ophitic. and the rocks were doubtless originally diabases or gabbros. although the augite is now wholly converted into pale-green hornblende with small specks of pyrite and chalcopyrite. At an elevation of 8.500 feet. or 1.500 feet above Salida at the first divide. a block of nearly horizontal limestones covers these rocks for about 1½ miles: at the northern edge of this block the limestones are underlain by very coarse. white gneissoid granite with coarse mica. cut by pegmatite dikes with feldspar crystals a foot long. Turret. a couple of miles farther north, lies in a shallow gulch cut in white granite. containing chiefly quartz. microcline. biotite. and a little albite. From Turret another road

---

[a] On a series of peculiar schists near Salida. Colo.: Proc. Colorado Sci. Soc., vol. 4, 1891-93, pp. 286-293.

descends to the Arkansas through Browns Canyon in reddish gneissoid granite, forming conspicuous turret-like outcrops. At the river the schists become very micaceous and are injected by pegmatite dikes.

Cross believed that the schists were derived from a metadiorite, probably a surface flow. It seems more likely that we have here a large pre-Cambrian mass of gabbro or diabase, which is intrusive in sedimentary metamorphic rocks, and which in part has been made schistose by pressure.

ORE DEPOSITS.

INTRODUCTION.

The deposits in this vicinity contain chiefly copper, and the following are described or mentioned below: The Sedalia mine, the Independence and Copper King mines, and the prospects at Cleora and Cotopaxi. The Independence and Copper King mines are located near Turret, and there are at that camp two other mines, the Gold Bug and the Vivandiere, which, however, are on gold-bearing fissure veins of an entirely different type from those which form the main subject of this description. The Gold Bug mine is on a quartz-filled fissure following a dike of fine-grained porphyry. The strike is east and west, and the greatest depth attained is 500 feet. The richest ore is said to carry about 2 ounces of gold and 6 ounces of silver per ton, besides 9 per cent of copper. The Vivandiere is probably located on the same vein a short distance to the east, and is developed by a shaft 600 feet in depth. Both properties are intermittent producers. Beyond doubt they are much more recent than the copper deposits described below. It is said that similar auriferous veins are found at Whitehorn, 10 miles farther east in Fremont County.

SEDALIA MINE.

Location, history, and production.—The Sedalia mine is situated in the first foothills rising on the east side of Salida Valley, 2 miles east of Arkansas River and 4 miles north of Salida. It is said to have been discovered about twenty-four years ago and has been worked intermittently since that time. At present it is owned by the Shawmut Consolidated Copper Company. It was an active producer in 1907; the ore was shipped to various smelters, but lately it has been taken chiefly by the Canon City zinc-lead paint plant of the United States Smelting Company, which utilizes the zinc for paint and smelts the residues for copper matte. No exact data of production are available, but the Shawmut Company states that probably in all from 60,000 to 75,000 tons of ore have been shipped, containing from 5 per cent upward of copper and from $1 to $2.50

in gold and silver per ton. It is really one of the few important copper mines of Colorado. During the last two years the mine has shipped about 400 tons per month.

Developments.—The deposit is developed by three tunnels. The lowest, No. 3, is 100 feet higher than the first outcrops of schists above the sandy valley bottom and at an elevation of approximately 7.475 feet, or 475 feet above Salida. No. 2 tunnel is on the steep slope, 321 feet above No. 3, and the uppermost tunnel, No. 1, is 100 feet above No. 2. There are two levels 60 and 100 feet below No. 2. No. 3 tunnel runs in a direction a little north of east for about 1,300 feet. The total developments probably aggregate 5,000 feet.

A leaching plant was built on the sandy valley bottom 1 mile west of the mine, but as it was not successful it is now being turned into a concentrating plant for the primary sulphide ores recently found. The surface ores contain both sulphides and carbonates, and for this reason direct leaching is not applicable to them. An incline carries the ore from No. 2 to the level of the valley, whence it is transported on a tramway to the railroad spur at the concentrator.

Geology.—Seen from a distance the rocky, reddish-gray slope with scattered junipers seems to rise abruptly from the valley. The prevailing rocks are schists. Both north and south of the mine heavy dikes of light-colored pegmatite cut conspicuously through the schists, which strike almost due east and west and dip from 50° to 70° S. Cross, in the paper already cited, states that the succession consists of "fine mica schists, locally staurolitic, actinolite and chlorite schists. with garnet developed in various forms." The best exposures are found in the long No. 3 tunnel, which penetrates below the level of oxidation. For the first 1,000 feet the tunnel penetrates a hard dark-gray and blocky gneissoid rock in which the schistosity is obscured by joints striking north and south or N. 65° W., and dipping respectively 80° W. and 40° NE. This hard gneiss, which contains softer bars of schist richer in biotite, continues up to a place where the plane of the ore deposit intersects the tunnel. No ore is visible here, but the rock is softer, with some slips and copper stains. Beyond this point the rock is a mica schist with smooth planes of schistosity dipping 52° S. In this rock the crosscut continues for 300 feet, and the schist is intersected by many small dikes of white pegmatite, probably apophyses of the larger dike shown in the upper workings.

The gneiss in the first part of the tunnel is dark gray and imperfectly schistose, and consists chiefly of a fine-grained mixture of quartz and brownish-green biotite, fibrous crystals of sillimanite, abundant small grains of corundum, and pale-red garnets up to 1 centimeter or more in diameter. The garnets and the sillimanite give a pseudoporphyritic appearance to the rock; the quartz appears in allotriomorphic grains and the general structure is that intimate intergrowth characteristic of many crystalline schists which Grubenmann designates as diablastic.

The steep hillside between tunnels No. 3 and No. 2 shows the weathered forms of the gray gneissoid schist, in places containing streaks of chloritic schists with very large crystals of garnet, some of them several inches in diameter. These garnets appear near the contact of the big pegmatite dike, but as they are also noted elsewhere in the series it is doubtful whether they are the effects of contact metamorphism. Above the dike, up to the croppings of the ore deposit, the rocks are mixed and consist of amphibolite rocks, chloritic schists, dark quartz schists, and garnet schists. In the foot wall of the croppings appears fissile mica schist similar to that noted from the lowest tunnel level. Toward the east the mica schist occurs on both sides of the deposit. On the highest spur and 200 feet south of the ore body is a fine-grained aplitic dike rock, which contains bunches of amphibolitic rocks.

The principal pegmatite dike is 50 feet wide and outcrops in the vicinity of tunnel No. 2. As shown by the workings, it dips 70° N., consequently in an opposite direction to the schists. Only small stringers of pegmatite are noted in the lowest tunnel. The rock is coarse grained; the specimens examined consist of white microcline, quartz, and greenish-white muscovite, the foils of which reach 1 inch or more in diameter.

The ore deposit.—Cross, in the paper cited, referred briefly, but correctly, to the copper deposit of the Sedalia mine as " a thick bed of actinolite schist richly impregnated with copper minerals." The ore body lies conformable with the schists and is practically a flat lens, 800 feet long and at most 150 feet thick, of amphibolitic rocks of varying types. The outcrops of the ore body can be followed for many hundred feet east and west on a ridge 600 feet above the valley, and show chiefly as soft whitish rocks locally stained by copper or lead. These rocks, to a considerable extent, are made up of talc, asbestos, and partly weathered amphibolite. On the first level the ores have been followed for about 400 feet. On the second level the deposit extends for over 800 feet. There are two smaller levels 60 and 100 feet below level 2. Below that level, however, the 50-foot pegmatite dike mentioned above as dipping in a direction opposite to that of the deposit and the schists cuts off the ore body. In one place toward the east of the body the workings are said to have crossed the dike, but only low-grade and very hard ore was found beyond it. Thus practically all of the ore shipped has been taken from that part of the deposit which lies above the pegmatite dike, and tunnel No. 2 shows the character of these ores very well. They are all at least partly oxidized, and it is evident that the copper has spread through a considerable width of rock outside of its original home. In one place the mass of amphibolite is about 150 feet thick. Some of the stopes are 50 feet wide. The prevailing rocks are amphibolite of varying grain, chloritic schist, and a dark-gray quartz rock. Much low-grade copper ore, said to average 2 per cent, still remains in the workings. The ore generally consists of a mixture of limonite, malachite, cuprite, and chalcocite, with remaining unaltered grains of chalcopyrite, which evidently is the original mineral; this ore is contained in a gangue of chloritic schist or amphibolite. There is no well-defined zone of chalcocite. A little zinc blende, galena, and cerusite occur in places. The ore carries a very small amount of gold and silver. A new basic lead-copper sulphate, having the formula $PbSO_4.CuSO_4.CuO$, was found in the croppings. It appears as a dull lemon-yellow powder.

The first attempts to find the ore deposit on the level of the lowest (No. 3) tunnel were not successful, as indicated under the description of the rocks (p. 162). Recent exploration, however, resulted in the finding of a considerable body of sulphide ore 40 feet above and a little to the north of the tunnel. This ore body also lies on the contact of the biotite schist and the garnet gneiss, almost directly underneath the croppings, following the southerly dip of the schists, which seems to be a little steeper between levels 2 and 3 than it is above level 2. The ore is here about 300 feet vertically below the surface. The find had just been opened last year, and so far as could be seen from the scanty exposures, the body is about 50 feet wide, although the workable part has only one-half of that width. The rocks are fresh and their study gives a far better insight into the relations than is possible in the oxidized ores of the upper workings.

The foot wall was exposed in only one place, and is the fissile biotite schist referred to above. Next lie irregular masses of a dark-green fine-grained amphibolite composed of bluish-green and colorless amphiboles, the two colors in places combined in one prism, and of grains of labradorite, possibly also other feldspars, and some disseminated particles of magnetite. The structure is diablastic and typical of a crystalline schist. Mixed with the amphibolite are streaks and bunches of an almost black quartz schist, which besides a quartz mosaic contains slender prisms of a pale-greenish or grayish-blue amphibole with scarcely measurable extinction and a few grains of magnetite and chalcopyrite. Then follows several feet of heavy, dark zinc-blende ore, which under the microscope shows reddish-brown sphalerite and a little chalcopyrite intimately intergrown with about equal quantities of prisms of bluish-green and colorless amphibole. The bulk of the ore is massive, and contains both zinc blende and chalcopyrite, with about 10 per cent of magnetite and some pyrite; as broken it is said to contain about 20 per cent of zinc. Under the

microscope the ore shows irregular masses of reddish garnet penetrated by prisms of colorless monoclinic amphibole, which also, intergrown in the manner of the crystalline schists, makes up the bulk
of the rock. These two minerals contain grains of chalcopyrite,
pyrite, magnetite, and zinc blende in relations indicating simultaneous formation with the minerals of the crystalline schists. There are
also small grains of a dark-green spinel and a little feldspar, apparently labradorite or anorthite.

Toward the hanging-wall side the ore gradually becomes poorer
and changes into the normal garnet gneiss described under "Geology." In places it is chloritic and contains large dodecahedra
of garnet. In the garnet gneiss the same bluish-gray amphibole
referred to in a previous paragraph was observed. It is evidently
closely related to glaucophane.

The following minerals occur at the Sedalia mine: Chalcopyrite,
pyrite, sphalerite, galena, chalcocite, malachite, azurite, cerusite, a
new basic lead-copper sulphate, magnetite, spinel (zinc spinel?), corundum, quartz, garnet, hornblende, biotite, glaucophane?, staurolite,
labradorite, asbestos, talc, chlorite.

Summary and genesis.—For the origin of the Sedalia deposit the
following hypothesis is suggested:

The series of micaceous schists at the Sedalia mine is probably
of sedimentary origin. A large mass of diabase or gabbro was intruded into these rocks in pre-Cambrian time, between the Sedalia
mine and Salida; the magma contained copper, and some of the dikes
intruded into the folded sediments surrounding the igneous massif
were products of differentiation rich in copper sulphides. The Sedalia ore deposit is probably such a dike, and the ores originally
consolidated from a magma. The dike followed approximately, but
not strictly, the planes of stratification. Renewed dynamometamorphism following the intrusion accentuated the conversion of the sediments into crystalline schists, and changed peripheral parts and dikes
of the intrusion into amphibolites. The ore minerals were recrystallized and migrated in part through the wall rocks, the contacts
being made indistinct by the pressure and rearrangement of minerals.

In its present form the deposit is assuredly a product of pre-
Cambrian dynamometamorphism, and to judge from the position of
the covering of Paleozoic crust blocks near Turret the portion now
worked was at least 1,500 feet below the surface upon which the Cambrian and Carboniferous sediments were deposited. It was probably
much farther below the surface of the earth at the time of the intrusion. The intrusion and regional metamorphism of the diabase
or gabbro was followed by enormous intrusions of granites, which
seem to be barren of mineral deposits. Pressure continued after the
consolidation of these rocks, and they were made partly schistose.

The last feature in the pre-Cambrian intrusions was the pegmatite dikes. These are likewise barren and one of them has cut the Sedalia deposit in two.

Active oxidation has converted the upper 200 feet of the deposit into copper carbonates and chalcocite, mixed with residual sulphides, but even in the lowest levels, 300 feet below the surface, there are indications of incipient oxidation. The water level is at present below the lowest (No. 3) tunnel. There is no well-defined chalcocite zone. In the upper 200 feet of the deposit the zinc blende has been oxidized and almost wholly removed as soluble sulphate.

COPPER DEPOSITS AT TURRET.

A short distance west of the Turret mining camp (see p. 160) the granite becomes more micaceous and somewhat schistose. At the Independence mine, 1 mile west of Turret, at an elevation of 8,500 feet, the strike is N. 55° W. and the dip 30° NE. The schists here consist of a coarse brown " augen " gneiss; there is also some quartz-biotite-garnet schist like that at the Sedalia camp. Streaks of very coarse amphibolite embedded in this gneiss or schist contain partly oxidized chalcopyrite. The explorations have been carried down 200 feet along the dip. The mine was not entered, but the owner, Mr. P. S. Plympton, states that the width of the ore body is 30 feet, with a richer streak 5 feet wide. Molybdenite is stated to occur in this deposit. A considerable tonnage of copper ore, low in gold and silver, was hauled down to the railroad in 1907 and sold to smelters, where it is used for purposes of flux in matte concentration.

The Copper King is another deposit of very similar character situated between Turret and the Independence mine.

COPPER PROSPECTS AT CLEORA.

The Cleora district is situated 4 miles south of Salida along Arkansas River. Many prospects have been located here, but no important deposits are reported. The ore is chalcopyrite, occurring in amphibolite schist, and the occurrence is probably similar to those described above.

COTOPAXI MINE.

The little settlement of Cotopaxi is located 24 miles southeast of Salida, in the canyon of Arkansas River, at an elevation of 6,373 feet, or about 600 feet below Salida. The old Cotopaxi mine is located in a small gulch half a mile northwest of the railroad station. It has been idle for many years and the underground workings were closed in 1907, but it was at one time a considerable shipper of copper ore. The lowest tunnel has an elevation of about 6,650 feet. There are three tunnels, 40, 90, and 120 feet above the bottom of the gulch.

The prevailing rocks in this part of the canyon consist of reddish granite or gneissoid granite with imperfect schistosity. Bunches of irregular pegmatite dikes are also present. Approximately 1,500 feet above the river on the north side the Paleozoic rocks rest flat upon the eroded pre-Cambrian complex. The deposit occurs in granite gneiss, with irregular strike and dip. The strike is chiefly northeast and southwest, the dip about 45° NW. There are no extensive out-crops of the deposit, which appears to have come to the surface only at one place close to the highest tunnel. As the underground rela-tions could not be seen, the following notes are based on the character of the rocks as shown on the dump. The ore minerals are massive chalcopyrite, and dark-brown zinc blende, in intimate intergrowth. There are also a few grains of galena. The gangue consists of quartz with large amounts of biotite, reddish garnets, and dark-green amphibole. Some of the material on the dump is a soft schist com-posed only of mica and garnet with a few grains of chalcopyrite. There are also almost pegmatitic rocks consisting of quartz, labra-dorite, and dark-green zinc spinel, with a little chalcopyrite and galena. The zinc spinel (gahnite) is abundant in places, and the locality is mentioned by Dana. The deposit is probably a lenticular mass of an igneous basic ore-bearing rock, now greatly metamor-phosed and conformable to the schistosity of the gneiss.

The similarity to the Sedalia deposit is unmistakable, and there seems to be little doubt that the Cotopaxi mine should be classed with the pre-Cambrian copper deposits. If this is correct the ores lie about 1,500 feet below the eroded surface on which the Paleozoic sedi-ments were deposited. Probably they were formed at a far greater depth below the surface.

PRE-CAMBRIAN COPPER DEPOSITS IN JEFFERSON COUNTY.

LOCATION AND GEOLOGY.

Jefferson County contains a part of the Front Range just west of Denver. Between Golden, on Clear Creek, and Evergreen, on Bear Creek, are several copper prospects on some of which considerable development work has been done. The only mine which has any pro-duction to its credit is the Malachite property.

As is well known, the Front Range is in this region almost wholly built up of granite and gneissoid rocks. The geology has been briefly described and mapped in a recent paper by James Underhill.[a] Several kinds of granite and gneisses, as well as belts of diorite, are described. The conclusion of this author is that the whole series is of igneous origin, and that the rocks were possibly formed at a later

---

[a] Areal geology of lower Clear Creek: Proc. Colorado Sci. Soc., vol. 8, 1906, pp. 103–122.

period than has generally been supposed, perhaps as late as the Devonian. It is difficult to accept these results entirely. The general fact of a great unconformity at the base of the Cambrian seems fully proved throughout Colorado, and also the fact that there were no post-Cambrian epochs of dynamometamorphism in this region. The age of the gneiss complex of the Front Range, aside from some later intrusions, is pre-Cambrian beyond question. The correctness of recognizing as separate intrusions some of the granites outlined by Underhill seems at least open to discussion, and the igneous origin of some of the gneisses seems likewise doubtful. Three principal features are likely to impress the observer of the geology of this region—first, the great masses of biotite schist, compressed, folded, and contorted in the most extreme manner; second, the tremendous injection of granite and pegmatitic material to which this schist has been subjected; third, the great areas of somewhat gneissoid granite. The biotite schist bears all the earmarks of a highly metamorphosed sedimentary series, soaked in granitic and pegmatitic magmas.

Belts of amphibolite are inclosed in the schists a few miles east of Evergreen and continue in a northwesterly direction for at least a few miles. At the F. M. D. prospect the belt is at least 1,000 feet wide; the schistosity strikes due northwest and dips 75° SW. This amphibolite contains in places pegmatite dikes, showing that the rock is older than the pegmatite intrusions. In one or two places where good exposures are available dark, fine-grained diorites or diabases are seen to cut this amphibolite in sharply defined narrow dikes.

ORE DEPOSITS.

In general the copper deposits are in or near the amphibolites. At the Despatch property, 2 miles below Evergreen, the prevailing rock is granite, but there is also some schist. The shaft is reported to be 140 feet deep and two "veins" containing copper are said to have been encountered.

The F. M. D. property lies on a tributary to Bear Creek about 5 miles northwest of Evergreen, at an elevation of 6,800 feet. A vertical shaft 350 feet deep has been sunk here and three "veins" of copper-bearing ore are said to have been cut. Work is suspended and the shaft is partly filled with water. The country rock is a dark-green amphibolitic schist which contains pyrite and chalcopyrite. The rock consists of intimately intergrown prisms of ragged green hornblende and foils of biotite. In and between these minerals lies a mosaic of labradorite. Accessories are apatite and magnetite, the latter intergrown with pyrite and chalcopyrite. The rock is probably a metamorphosed diabase. Coarse-grained masses of a pale-green labradorite (?) feldspar, quartz, and biotite occur in the schist, perhaps as a dike, and contain pyrite and chalcopyrite. Fractures in this rock are coated by secondary pyrite and an unusual mineral in small aggregates of rhombohedra of brilliant luster. According to an analysis by Dr. E. C. Sullivan, this mineral is a rare zinciferous siderite containing 11.6 per cent of zinc oxide and allied to the subspecies monheimite. The pyritic deposit at the F. M. D. mine seems to trend east and west, toward the Malachite mine and obliquely to the schistosity, and it perhaps represents a dike somewhat later than the mass of the amphibolite. The continuity of the pyritic ores between the two deposits is, however, not proved. A small prospect shaft a quarter of a mile west of the Malachite, toward the F. M. D., shows amphibolite schist with streaks of chalcopyrite and pyrite containing a little garnet and epidote. Barren pegmatite dikes cut this amphibolite.

The Malachite mine is situated on a high ridge near the northerly divide of Bear Creek, 1¼ miles east of the F. M. D., at an elevation of 7,000 feet. Many years ago this mine produced a considerable amount of oxidized ores, and intermittent small shipments have been made since then. The value of the total production is estimated by the former owners to be $35,000. These ores were taken out from a shaft 150 feet in depth. A tunnel 300 feet in length has been driven in a northerly direction to tap the lower workings at the bottom of this shaft, and at the time of visit this tunnel was being reopened.

The entire length of the tunnel is in contorted biotite and amphibole schist intersected by pegmatite dikes. The deposit opened by the old shaft does not appear to be directly continued on the tunnel level, but immediately beyond the shaft the tunnel had just intersected an irregular mass of sulphides about 10 feet in width. Of the position of this ore with reference to the upper workings nothing definite can be said, as these workings were not accessible. The ore is massive and consists of coarse masses of chalcopyrite, zinc blende, and pyrrhotite, said to contain some nickel. It contains very little gold and silver. Sharply defined but small octahedra of pyrite are embedded in the pyrrhotite. The minerals occurring with the ores are augite, a pale-green feldspar in grains up to 1 centimeter in diameter, and small grains of quartz. The thin section reveals the fact that the feldspar is labradorite in broad plates; a pale-green augite is abundantly intergrown with the feldspar in gabbro structure. A few grains of apatite and titanite form the accessory minerals. The feldspar is in part sericitized and the augite is locally converted to greenish-brown hornblende. The abundant ore minerals in the sections are pyrrhotite, chalcopyrite, and dark-brown zinc blende, all three genetically equivalent and intergrown with augite and feldspar in a manner to indicate simultaneous crystallization. The sulphides include these minerals, and vice versa; the included augite is sharply defined with curved contact lines. The inevitable conclusion is that the ores form part of a gabbroitic dike rock and that they have crystallized from a molten magma.

A few of the ores are surrounded by a colorless or pale-yellow epidote and a little quartz, which probably resulted from secondary action in the consolidated rock.

The Malachite ore deposit is, then, a gabbro dike containing magmatic sulphides later than the main mass of schists, but probably earlier than the pegmatite. The proof of pre-Cambrian age is not quite as strong as at the Sedalia mine. Marked points of resemblance will easily be perceived between all these copper deposits so far described, and it seems justifiable to regard them all as differentiation products of a basic magma, in places changed and rearranged by dynamometamorphism.

The Augusta lode on Cub Creek, half a mile above Evergreen, is a very different deposit. Ore containing copper and silver was shipped from it several years ago and work has been resumed at intervals. The principal development consists of a shaft 130 feet deep. The lode is a sharply defined quartz-fluorite vein, occurring in red granite; its strike is north-northwest and its dip 70° WSW. Yellow zinc blende and chalcocite are the ore minerals. The vein shows distinct crustification, and is evidently very much younger than the deposits described above.

COPPER DEPOSITS IN THE "RED BEDS" OF FREMONT COUNTY.

LOCATION.

The Red Gulch copper district is situated in Fremont County about 9 miles due north of Cotopaxi station on the Denver and Rio Grande Railroad. The elevation at this station is 6,373 feet, and the high plateau at the Red Gulch camp has an elevation of about 8,000 feet.

GEOLOGY.

The Arkansas River canyon is here cut in pre-Cambrian gneiss and schist. On the north side of the river and about 1,500 feet above it a block of Paleozoic limestones gently inclined eastward rests upon the eroded pre-Cambrian surface. The road to the copper camp ascends steeply for a few miles in Bernard Creek, enters the limestone area, crosses eastward over into Red Gulch, the next tributary to the Arkansas from the north, and here enters the "Red Beds," which overlie the limestones and like them dip about 20° E. Following Red Gulch, the road continues northward for 3 or 4 miles in a wide, open valley, a sort of high plateau, in which the creek runs with slight fall. At the camp is a small settlement called Copperfield, and half a mile farther north another named Springfield. Immediately east of Red Gulch, and in places following the creek bed, a great north-south fault brings up the pre-Cambrian gneissoid granites above the " Red Beds," and these rocks form the whole eastern slope of the valley, which rises about 1,000 feet above the watercourse. On the western side the ridge of Paleozoic limestones is seen in the distance.

The series of sedimentary rocks exposed along Red Gulch is approximately as follows. The examination extended over only one day and was in the nature of a rapid reconnaissance.

Approximate section of " Red Beds " at Red Gulch, Colorado.

Top.	Feet.
1. Gray limestone, partly silicified	250
2. Red or green shale	20
3. Red conglomerate	20
4. Red or green shale	30
5. Conglomerate	40
6. Dark-green shales	200

The main body of Carboniferous limestones lies at an undetermined distance below No. 6. The original color of the shales and conglomerates is apparently dark green, but oxidation rapidly turns them to a deep brownish-red color. Water is scarce and evidently contains a large amount of sulphates. The age of these beds is doubtful. The limestones contain some imperfect round crinoid stems, and more careful search would probably reveal better fossils. The series probably belongs in the upper Paleozoic column.

Within a distance of a few hundred to 1,000 feet from the great fault the beds begin to dip more steeply, and at the fault they have at many places assumed a vertical position. A few smaller transverse east-west faults were noted. The only rock intrusive in the " Red Beds " is a thoroughly oxidized east-west dike, a fourth of a mile north of Springfield, which is said to contain a little silver. It has no apparent connection with the copper deposits.

COPPER DEPOSITS.

At several horizons the " Red Beds " of Red Gulch contain copper ores—chalcocite, malachite, and azurite—and the deposits are of the type made familiar by many occurrences in New Mexico, Arizona, and western Colorado. Their existence has been known for many years, but active exploration began only in 1907, and several companies are now operating in the camp. Two cars of high-grade ore and several cars of low-grade material are said to have been shipped in 1907, principally from the Red Gulch and Copper Prince mines.

The Copper Prince property lies a short distance northwest of Copperfield and covers a slope of heavy red conglomerate of well-washed pre-Cambrian rocks. Bunches of oxidized copper ores and chalcocite are found in the pits sunk in the conglomerate, apparently without exception associated with coaly material representing fossil wood. In a coaly shale bed below the conglomerate, exposed in a small shaft, chalcocite occurred more abundantly.

The Colorado Copper Company has sunk a shaft on the hill about 1,000 feet east of the creek at Copperfield and 150 feet above it; the depth attained in September, 1907, was about 250 feet. The water is kept in check by bailing. The sediments between the creek and the shaft dip eastward at moderate angles and consist of red shale with one intercalated bed of conglomerate. Near the shaft they dip at a steeper angle and the shales turn into reddish-brown limestone. At the granite contact about 100 feet east of the shaft the strata stand vertically or dip a few degrees from the vertical to the west. The fault is clearly marked, but no ore appears to follow it. Copper stains appear in various places, especially in a belt of quartzitic rock which lies near the contact, but it was not claimed that large ore bodies had been found.

The most interesting developments are at Springfield, half a mile north of Copperfield. The Red Gulch Gold and Copper Mining Company is operating at this place just east of the creek and has sunk a shaft 150 feet deep on an incline of 70° from the horizontal. Some rich chalcocite ore has been shipped. The fault lies here within a few hundred feet of the creek. On the west side are gently dipping shales and conglomerates, which near the shaft of the company gradually assume a dip of 70°. The ore follows a bed of carbonaceous shale 4 feet wide. In the foot wall lies a heavy red conglomerate; the hanging wall consists of about 50 feet of red shales with some intercalated conglomerate. The dip here becomes nearly vertical. The shales are capped by about 200 feet of gray, partly silicified limestone, which adjoins the great fault bringing up the pre-Cambrian granitic complex. The dark-gray, soft ore-bearing shale contains seams of a compact bituminous coal up to an inch or more in thickness, and in places abundantly disseminated chalcocite. The copper mineral appears to follow the coal seams very closely, and specimens show a peculiar ore of intimately intergrown chalcocite and coal. It is stated that the coal itself contains copper. Polished sections show that this is due to very minute veinlets of the copper sulphide in the coal. According to a determination by Doctor Hillebrand the coal contains some vanadium. The analysis gave 0.114 V_2O_3. The chalcocite also occurs in smooth nodules in the shale, usually 1 or 2 inches in diameter. Sections of these nodules show that the bedding planes of the shale continue through them and the inference is plain that they replace coaly shale material. The only associated mineral thus far found is barite, narrow seams of which cut both ore and coal. The pure chalcocite is said to contain at most 10 ounces of silver per ton. No ore occurs in the red shale, nor in the conglomerate or the limestone. The fault plane is likewise barren.

Half a mile north of Springfield are the Acme and Queen Princess properties. These are located on a gently dipping cupriferous dark-gray or dark-green shale, probably the lower part of No. 6 in the section (p. 171). At the Queen Princess two ore-bearing deposits, 12 feet apart, have been opened. The surface shows but little copper, but about 15 inches from the outcrops green stains and narrow seams of sooty chalcocite appear. The ore is apparently of low grade. The shale of the Red Gulch mine lies probably 200 feet above this horizon, but no copper is reported from it at this place.

CONCLUSIONS.

The copper deposits of this district are in the main similar to those in the " Red Beds " of New Mexico and Arizona and share with them certain disadvantages of exploitation, being generally of low grade and presenting difficulties of concentration, or leaching. Whether any property contains payable ore bodies can be determined only by careful sampling and close consideration of problems of transportation and reduction. As far as known to the writer the bodies of low-grade disseminated ore in the " Red Beds " have not been profitably handled anywhere, in spite of many attempts. On the other hand, the finding of rich chalcocite shipping ore in the coaly shale is a more promising feature, encouraging further exploration in favorable places.

GENESIS.

In view of the extended discussion of the genesis of the copper deposits in the " Red Beds," the data obtained in this camp are of more than passing interest. The earlier view that the copper ores were formed simultaneously with the accumulation of the strata in which they occur has lately been disputed, and theories explaining their origin by ascending thermal solutions or by precipitation from circulating surface waters have been advanced. The primary ore of the Red Gulch district is unquestionably chalcocite. Its connection with the beds containing carbon is equally unquestionable. Sections of the ore occurring in the Red Gulch mine show that the chalcocite is deposited by replacement of this coal, which of course would exercise a strong reducing action. If this theory is accepted it follows that the solution must have contained copper sulphate. The presence of barite and the known gypsiferous character of the " Red

Beds " indicate that sulphate solutions would be abundant in them. The reaction probably took place according to the following formula, in which it can be shown that chalcocite and carbon would be exchanged almost volume for volume: $4CuSO_4 + 5C + 2H_2O = 2Cu_2S + 2H_2SO_4 + 5CO_2$.

This reaction would not appear to necessitate ascending or heated waters, but could proceed at ordinary temperatures in circulating surface waters. It has been shown in preceding pages that the pre-Cambrian rocks which furnish the material for the " Red Beds " contain old copper deposits. The degradation of these pre-Cambrian deposits would distribute through the " Red Beds " an appreciable amount of copper salts, partly soluble, partly sulphides. When surface waters containing oxygen searched these beds copper would naturally be dissolved as a sulphate, and its precipitation as chalcocite would follow wherever agents of reduction, such as carbon, were available.

NOTES ON THE FORT HALL MINING DISTRICT, IDAHO.

By F. B. Weeks and V. C. Heikes.

INTRODUCTION.

The Fort Hall mining district is located in Bannock County, southern Idaho, near the city of Pocatello and along the Oregon Short Line Railroad. The present paper contains the results of a brief reconnaissance undertaken in October, 1907.

The mining district was established June 17, 1902, and its area comprises all of the ceded portion of the Fort Hall Indian Reservation lying within Bannock County. It extends approximately 26 miles north and south and 30 miles east and west, comprising an area of about 750 square miles. Pocatello lies in the northwestern part of the district.

TOPOGRAPHY AND DRAINAGE.

Portions of three mountain ranges are included within the district—the Pocatello Range in the western, the Bannock Range in the central, and the Portneuf Range in the eastern part. These ranges trend north and south and are fairly well defined. The upper and the lower Portneuf valleys and the Marsh Creek valley are the principal depressions. (See fig. 10.)

The Portneuf Range extends from the valley of Ross Fork on the north to Cache Valley on the south. On the east is a wide valley known as Portneuf Valley, and on the west it is separated from the Bannock Range by the south fork of Ross Fork, Rapid River, and a portion of Portneuf River. It is about 40 miles long and its average width is 12 to 15 miles. The principal peaks are Mount Putnam and Bonneville Peak, the latter having the greater elevation—9,200 feet. Portneuf River has cut a deep canyon across the southern part of the range nearly at right angles to its trend. South of this canyon the range is formed of several parallel ridges of less elevation, but much wider in mass than the northern part of the uplift.

The Bannock Range is bounded on the north by the valley of Ross Fork and extends southward to the headwaters of Malade Creek.

Portneuf River cuts through the range at right angles to its trend about 10 miles north of its canyon through the Portneuf Range. The northern portion of the Bannock Range is formed of the Tertiary and basalt hills south of Ross Fork, the hills west of Mount Putnam, and the ridge which faces the lower valley of the Portneuf. South of the Portneuf Canyon the range extends to the headwaters of Marsh and Malade creeks.

The Pocatello Range extends from the Snake River plains southward to the headwaters of Bannock and Malade creeks and is formed of a broad mass of hills and ridges.

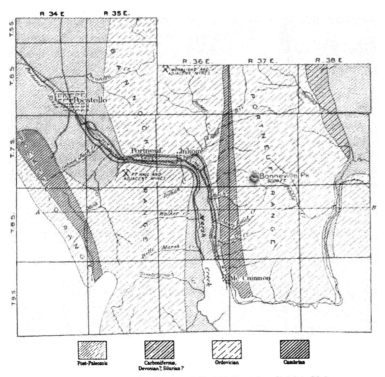

FIG. 10.—Geologic sketch map of Fort Hall mining district, Idaho.

Portneuf River and its tributaries drain the three ranges above described. This stream forms one of the principal southern branches of Snake River. Its course is very irregular. It rises in the hills north of Portneuf Valley, which form the divide separating the Great Basin from the Pacific drainage, and flows southward for about 35 miles, penetrating the eastern portion of the Portneuf Range; turns abruptly to the west and cuts directly through the range; thence flows northwestward and northward for about 10 miles, paralleling the valley of Marsh Creek, from which it is separated by a basalt flow from 1 to 2 miles wide; turns again abruptly to the west and cuts through the Bannock Range a short distance below the mouth of Marsh Creek; and thence flows northwestward into Snake River. Before reaching the Snake the Portneuf receives the waters of Ross Fork from the east and Bannock Creek from the south.

The basaltic flows through which the Portneuf has cut a channel of considerable depth and which form benches reaching back to the foot of the mountain slopes are prominent features of the valley of this stream.

GEOLOGY.

STRATIGRAPHY.

The sedimentary strata exposed in the Fort Hall mining district extend from the upper Cambrian to the lower Carboniferous. They outcrop principally along the higher portions of the ridges and have been folded into a series of anticlines and synclines. The lower slopes are usually covered by débris and the Portneuf Valley is filled by basalt flows.

CAMBRIAN AND ORDOVICIAN.

Only the upper beds of the Cambrian system are exposed in this area, in the Pocatello Range and along the western slope of the Portneuf Range. They form the northern extension of the strata exposed along the eastern side of Malade Valley, and their age has been determined by fossils collected and identified by Charles D. Walcott.[a] The strata comprise siliceous and cherty limestones and are conformably overlain by Ordovician sandstones. The Ordovician rocks cover the greater part of the district. They are mainly quartzites, conglomerates, and shales, with a series of limestones 50 to 75 feet thick in the upper part. The total thickness is approximately 3,000 feet. They represent the horizon of the Ogden quartzite of the Wasatch Range and by change in character of sediments probably include the limestones of Ordovician age, which occur above and below the Ogden quartzite in that region.

DEVONIAN, SILURIAN, AND CARBONIFEROUS.

Field work by the senior author in 1905 in the northern part of the Wasatch Range has shown that Devonian and Silurian strata occur in that region. It is possible that strata of the same age may be present on the eastern slope of the Portneuf Range beneath the known Carboniferous limestones.[b]

---

[a] Personal communication; also Sixth Ann. Rept. U. S. Geol. Survey Terr., 1873, pp. 203–204.
[b] Eleventh Ann. Rept. U. S. Geol. and Geog. Survey Terr., 1879, pp. 329–330.

Carboniferous rocks occur only in the northeastern corner of the district. They consist of limestones of Mississippian age, as shown by fossils collected by the Hayden Survey.[a]

STRUCTURE.

An anticlinal fold, formed by Cambrian and Ordovician strata, follows approximately the trend of the Pocatello Range. West and a little south of the canyon of Portneuf River through the Bannock Range this fold is broken, probably by a thrust fault. Southeast of Pocatello the valley of the Portneuf occupies a syncline which, south of the bend of the river, is broken by a fault of considerable displacement.

The Bannock Range is formed by an anticlinal fold that follows its general trend. This structure is well shown on the ridge north of the Portneuf Canyon. South of this canyon the ridge shows only the eastern side of the fold. In the tunnel of the Fort Hall copper mine the lower part of this fold is shown to be broken by a fault.

Between the Portneuf and Bannock ranges the valley of Portneuf River occupies a synclinal basin. The strata rise in a low anticline along the western slope of the Portneuf Range, exposing the Cambrian limestones in a rather narrow north-south belt extending across the district. The main ridge is an eastward-dipping monocline of Ordovician strata which merges into a low syncline in the southern part of the upper Portneuf Valley.

An east-west structure section across the district is shown in fig. 11.

IGNEOUS ROCKS.

With the exception of the basaltic flows, igneous rocks are of rare occurrence in the Fort Hall district. On the north side of the canyon of the Portneuf through the Bannock Range is exposed a grayish-green rock, which under the microscope is seen to be of igneous origin. The structure of the rock indicates that it was probably poured out through a vent or fissure as a sheet of lava. In its present outcrop it appears as more or less regularly defined beds. South of the river it is not found in any of the underground workings, so far as known, but it may, nevertheless, underlie the main ridge at no great depth. To the north and east of the river it rises on the upward pitch of the Bannock anticline and forms the surface rock over a considerable area on the western side of a second anticline that is developed in the region known locally as Moonlight.

In the hand specimen the rock is of medium grain and rather compact, but shows distinct amygdaloidal texture. It is as a rule somewhat schistose and has probably been subjected to the action of the

---

[a] Sixth Ann. Rept. U. S. Geol. Survey Terr., 1873, p. 206; Eleventh Ann. Rept., 1879, p. 563.

dynamic forces which have affected the associated sedimentary rocks. Under the microscope it appears greatly decomposed, but still preserves its character as an amygdaloidal diabase. Its most prominent characteristic is the occurrence of many elongated amygdules filled with coarse calcite and rimmed by hematite. Portions of the rock mass are very strongly schistose and stained with copper carbonates. Others, more massive, contain specks of chalcopyrite and pyrrhotite. A deep-green chlorite is the most abundant mineral, but remains of augite still show in places, as well as locally preserved triclinic feldspars of lathlike form.

ORE DEPOSITS.

OCCURRENCE.

Prospecting work has been done at several points since this portion of the Fort Hall Reservation was thrown open to the public, but the most important mining operations have been carried on by the Pocatello Gold and Copper Company, owning the Moonlight group of claims in the northern part of the district, and the Fort Hall Mining Company in the southern part. (See fig. 10.) The valuable metals of the ores are copper, silver, and gold. The occurrence of lead is limited to small quartz veins in the limestone, and a little of this metal is also associated with the copper minerals in the Fort Hall prospects. Some manganese oxide and iron ore are found in parts of the district, but no attempt has been made to ship the material.

The Moonlight property is located at the head of Rabbit Creek, about 9 miles east of Pocatello, and has been

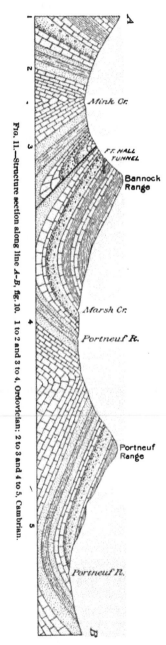

Fig. 11.—Structure section along line A–B, fig. 10. 1 to 2 and 3 to 4, Ordovician; 2 to 3 and 4 to 5, Cambrian.

worked steadily since the opening of the reservation in 1902. Soon after the opening the Pocatello Gold and Copper Mining Company (Limited) was incorporated to operate the property. During 1904 two carloads of copper ore of a good grade were shipped to the smelters. The development consists of a crosscut tunnel 830 feet long driven westward, at the end of which is an upraise opened for 90 feet for the purpose of connecting with known ore bodies. This part of the work was abandoned, however, and attention was given to taking out the ore near the surface. Two tunnels have been driven near the top of the hill, 275 feet and 75 feet long. The ore deposits occur in the conglomerate as opened in the workings near the top of the hill. The minerals, bornite and copper glance, associated with some carbonate of copper near the surface, are found in small kidneys and in fractures or fissures in the conglomerate. These have a north-south trend and dip about 40° E.

The Fort Hall Mining Company owns property located about 1½ miles west of Portneuf siding, on the Oregon Short Line, and 8 miles southeast of Pocatello. The development consists of a crosscut tunnel driven for 3,890 feet to the east through several different formations of the sedimentary rocks. These rocks have a westerly dip at the mouth, but at 1,200 feet in the dip is eastward. Near the end of this tunnel a crosscut is being driven and the sinking of a winze has been commenced.

Adjoining the Fort Hall property are claims owned by the Papoose Mining Company. The development consists of a 400-foot tunnel driven in black shale toward the west. This rock is apparently the same formation as that in which the Fort Hall Company terminates its tunnel, and the Papoose management expects to cut the Fort Hall mineral zone by extending its tunnel a few hundred feet farther. There are numerous other prospects in the district on which some development work has been done, especially on the east side of the ridge south of Portneuf River.

GEOLOGY.

The ore deposits of the Fort Hall district occur in conglomerates and shales of Ordovician age. These rocks are exposed in the anticlinal fold which forms the Bannock Range and in another fold of similar character which occupies the area between this range and the Portneuf Range, locally known as the Moonlight region. A few hundred feet beneath the strata in which the ores occur is an intercalated mass of diabase, which probably is a surface flow of Ordovician age, and which appears to have been an important factor in the ore deposition.

In the Moonlight area the ore occurs in a dark-colored conglomerate, which lies 500 to 600 feet above the shale series which contains the ore of the Fort Hall mine. The stresses that produced sharply compressed folds in the shales were relieved by fracturing and the formation of fissures in the conglomerate in which the ores were deposited.

In the vicinity of the Fort Hall mine the ore occurs in a gray calcareous shale which is No. 11 of the section in the Fort Hall tunnel, given below. This ore-bearing zone comprises approximately 125 feet of crumpled and contorted layers of shale and thin-bedded limestones. The Fort Hall tunnel shows the following section, the thickness having been measured along the tunnel level:

General section of the Fort Hall tunnel.

Feet.

1. Brown conglomerate made up of granite and quartz bowlders, ranging in size from small pebbles to bowlders 12 inches in diameter in a brown cementing material, dipping 40° W _____ 400
2. Brownish-gray quartzitic sandstone made up of subangular grains in a sericitic matrix _____ 420
3. Gray slates, thinly laminated and contorted, the ends of the beds thrust upward against a white quartz vein, which marks a plane of thrust faulting. Throughout the shales there are nearly vertical white quartz veins ranging from 1 to 3 inches in width. Small nodules of chalcopyrite and iron pyrite are disseminated in white quartz _____ 465
4. Grayish-green compact dolomitic shale, composed of angular quartz grains, sericite, and aggregates of dolomite. At the contact of the shale and conglomerate the dip is 20° E _____ 325
5. Conglomerate, fractured and brecciated. Open fissure, with much water at contact of conglomerate in 10-inch quartz vein _____ 245
6. Quartzitic sandstones dipping 40° to 54° E _____ 275
7. Brown conglomerate composed of large quartzite bowlders in a brown quartzy cementing material. Probably mineralized zone of the Moonlight area _____ 800
8. Varicolored calcareous shale _____ 50
9. Fine-grained gray sandstone, dipping 20° to 40° E _____ 300
10. White and gray siliceous and calcareous shale containing quartz seams _____ 275
11. Laminated gray shale and thin-bedded limestone in beds from 1 to 3 inches thick. Dip 26° to 40° E. *Zone of ore deposition* _____ 125
12. Compact gray to black calcareous shale _____ 210

3, 890

CHARACTER OF ORE.

In the Moonlight area the principal minerals occurring in the fissures in the conglomerate are bornite and copper glance, associated with carbonate of copper near the surface. In the southern part of the district, in the Fort Hall and adjacent mines, the most abundant mineral is chalcopyrite, occurring as veinlets in the sharply compressed folds. There is also some pyrite. A small amount of galena is contained in the limestone strata. The company reported that a general sample from the ore zone 125 feet in width along the tunnel level was tested to determine the proper methods of milling the product. A ratio of concentration of 11.05 tons of crude ore to 1 of concentrates gave the following results: Copper, 12.3 per cent; gold, 0.13 ounce per ton; silver, 4.30 ounces per ton; iron, 26.4 per cent; silica, 16.1 per cent; and lead, 0.8 per cent. The gangue consists of quartz with some calcite. A specimen consisting of chalcopyrite and quartz in about equal quantities contained, according to an analysis made by the Bureau of the Mint, 0.05 ounce of gold and 0.75 ounce of silver per ton.

The copper-bearing shale zone contains innumerable, mostly non-persistent veins and veinlets filled with quartz, calcite, and chalco-pyrite, with smaller amounts of pyrite. These veinlets are bent, corrugated, and contorted. The shale itself is a normal clay shale with sericite in minute flakes and aggregates of dolomite. Next to the seams of quartz and chalcopyrite sericite is developed more abundantly over a distance of a few millimeters. The banded limestones intercalated with shales contain seams of metasomatically developed cubes of pyrite and also some albite which has the appearance of being authigenetic.

FORM OF THE DEPOSITS.

At the Moonlight property, as stated above, the bornite and chalcopyrite occur in small masses along fractures in the Ordovician conglomerate. At the Fort Hall property the ore body appears on the surface as a broad belt of limonite-stained rock which can be traced by irregular outcrops for a considerable distance and which contains little or no copper. There are no developments of note in this surface zone, but the tunnel intersects a bed of cupriferous shale (No. 11 in above section) approximately 800 feet below the croppings (measured vertically) and in a position which indicates that it corresponds stratigraphically with them. Measured along the dip this would indicate a distance of about 1,600 feet from the outcrops to the tunnel level. The thickness of the ore-bearing bed would be about 80 feet.

From present knowledge it is doubtful whether the ore bodies shown are of sufficient value to warrant the expenditure of the large amount of capital required for their extraction and reduction. At the Fort Hall and adjacent mines deposition of minerals of economic value has taken place irregularly through a zone of sedimentary strata of considerable width and a large amount of very low grade ore must be handled. Local conditions affecting the cost of mining and milling operations should be most carefully considered in connection with plans for the mining of these ores. Some enrichment may possibly have taken place in the lower part of the zone of oxidation between the barren croppings and the poor ore of the tunnel level.

ORIGIN OF THE ORES.

Chalcopyrite ores low in gold and silver occurring in fissured zones in sedimentary rocks and not apparently associated with igneous rocks are unusual for the Cordilleran province and therefore of interest in any study of the origin and formations of ore deposits in general. The occurrence of a flow of diabase intercalated in the sediments within a comparatively short distance below the ore bodies and the presence of disseminated copper minerals in it suggest the hypothesis that the diabase is the original source from which the ore minerals in the sedimentary strata were derived.

The amygdaloid diabase probably represents an Ordovician lava flow from some deeper-seated source. The succeeding sediments were laid down upon it. After the beginning of the crustal movements by which the strata were elevated and folded into the present mountain ranges, the ores may have been leached from this diabase and deposited higher up by ascending more or less heated surface waters. Where the circulation of these waters through the interstices or along the planes of stratification or plication in the shales was arrested a deposition of minerals held in solution took place.

SURVEY PUBLICATIONS ON COPPER.

The following list includes the principal publications on copper by the United States Geological Survey or by members of its staff. In addition to the publications cited below certain of the folios listed in the "Introduction" contain discussions of the copper resources of the districts of which they treat.

BAIN, H. F., and ULRICH, E. O. The copper deposits of Missouri. In Bulletin No. 260, pp. 233–235. 1905.

BALL, S. H. Copper deposits of the Hartville uplift, Wyoming. In Bulletin No. 315, pp. 93–107. 1907.

BOUTWELL, J. M. Ore deposits of Bingham, Utah. In Bulletin No. 213, pp. 105–122. 1903.

——— Economic geology of the Bingham mining district, Utah. Professional Paper No. 38. 413 pp. 1905.

——— Ore deposits of Bingham, Utah. In Bulletin No. 260, pp. 236–241. 1905.

COLLIER, A. J. Ore deposits of the St. Joe River basin, Idaho. In Bulletin No. 285, pp. 129–139. 1906.

DILLER, J. S. Copper deposits of the Redding region, California. In Bulletin No. 213, pp. 123–132. 1903.

——— Mining and mineral resources in the Redding district in 1903. In Bulletin No. 225, pp. 169–179. 1904.

DOUGLAS, J. The metallurgy of copper. In Mineral Resources U. S. for 1882, pp. 257–280. 1883.

——— The cupola smelting of copper in Arizona. In Mineral Resources U. S. for 1883–84, pp. 397–410. 1885.

EMMONS, S. F. Geological distribution of the useful metals in the United States—Copper. Trans. Am. Inst. Min. Eng., vol. 22, p. 73. 1894.

——— Economic geology of the Butte (copper) district, Montana. Geologic Atlas U. S., folio No. 38. 1897.

——— Copper in the red beds of the Colorado Plateau region. In Bulletin No. 260, pp. 221–232. 1905.

——— The Cactus copper mine, Utah. In Bulletin No. 260, pp. 242–248. 1905.

EMMONS, W. H. The Cashin mine, Montrose County, Colo. In Bulletin No. 285, pp. 125–128. 1906.

GIGNOUX, J. E. The manufacture of bluestone at the Lyon mill, Dayton, Nev. In Mineral Resources U. S. for 1882, pp. 297–305. 1883.

184

GRATON, L. C. Copper. In Mineral Resources U. S. for 1906, pp. 373–438. 1907.

HOWE, H. M. Copper smelting. Bulletin No. 26. 107 pp. 1885.

IRVING, R. D. The copper-bearing rocks of Lake Superior. Monograph V. 464 pp. 1883.

LINDGREN, W. The copper deposits of the " Seven Devils," Idaho. In Mining and Scientific Press, vol. 78, p. 125. 1899.

——— Copper deposits at Clifton, Ariz. In Bulletin No. 213, pp. 133–140, 1903.

——— Copper deposits of Clifton-Morenci district, Arizona. Professional Paper No. 43. 375 pp. 1905.

PETERS, E. D. The roasting of copper ores and furnace products. In Mineral Resources U. S. for 1882, pp. 280–297. 1883.

——— The mines and reduction works of Butte City, Mont. In Mineral Resources U. S. for 1883–84, pp. 374–396. 1885.

PHALEN, W. C. Copper deposits near Luray, Va. In Bulletin No. 285, pp. 140–143. 1906.

PIRSSON, L. V. (See WEED, W. H., and PIRSSON, L. V.)

RANSOME, F. L. Copper deposits of Bisbee, Ariz. In Bulletin No. 213, pp. 149–157. 1903.

——— The Globe copper district, Arizona. Professional Paper No. 12. 168 pp. 1904.

——— Geology and ore deposits of the Bisbee quadrangle, Arizona. Professional Paper No. 21. 168 pp. 1904.

SPENCER, A. C. Mineral resources of the Encampment copper region, Wyoming. In Bulletin No. 213, pp. 158–162. 1903.

——— Reconnaissance examination of the copper deposits at Pearl, Colo. In Bulletin No. 213, pp. 163–169. 1903.

——— Copper deposits of the Encampment district, Wyoming. Professional Paper No. 25. 107 pp. 1904.

ULRICH, E. O. (See BAIN, H. F., and ULRICH, E. O.)

VAUGHAN, T. W. The copper mines of Santa Clara Province, Cuba. In Eng. and Min. Jour., vol. 72, pp. 814–816. 1901.

WATSON, T. L. Notes on the Seminole copper deposits of Georgia. In Bulletin No. 225, pp. 182–186.

WEED, W. H. Types of copper deposits in the southern United States. In Trans. Am. Inst. Min. Eng., vol. 30, pp. 449–504. 1901.

——— Copper mines of Las Vegas, Chihuahua, Mexico. In Trans. Am. Inst. Min. Eng., vol. 32, pp. 402–404. 1902.

——— Copper deposits near Jimenez, Chihuahua, Mexico. In Trans. Am. Inst. Min. Eng., vol. 32, pp. 404–405. 1902.

——— Copper deposits of Cananea, Sonora, Mexico. In Trans. Am. Inst. Min. Eng., vol. 32, pp. 428–435. 1902.

——— Ore deposits at Butte, Mont. In Bulletin No. 213, pp. 170–180. 1903.

——— Copper deposits of the Appalachian States. In Bulletin No. 213, pp. 181–185. 1903.

——— Copper deposits in Georgia. In Bulletin No. 225, pp. 180–181. 1904.

——— The Griggstown, N. J:, copper deposit. In Bulletin No. 225, pp. 187–189. 1904.

WEED, W. H. Notes on the copper mines of Vermont. In Bulletin No. 225, pp. 190–199. 1904.

———— The co)per production of the United States. In Bulletin No. 260, pp. 211–216. 1905.

——⌐—— The copper deposits of eastern United States. In Bulletin No. 260, pp. 217–220. 1905.

———— The copper mines of the United States in 1905. In Bulletin No. 285, pp. 93–124. 1906.

WEED, W. H., and PIRSSON, L. V. Geology of the Castle Mountain mining district, Montana. Bulletin No. 139. 164 pp. 1896.

LEAD AND ZINC.

MINERAL RESOURCES OF NORTHEASTERN OKLAHOMA.

By C. E. SIEBENTHAL.

INTRODUCTION.

The field reconnaissance on which this report is based was made during June and July, 1907, occupying in all about six weeks. The itinerary comprised two excursions—the first from Miami to Choteau, Claremore, Chelsea, Coodys Bluff, Centralia, and Chetopa; the second from Bluejacket to Coffeyville, Tulsa, Catoosa, Broken Arrow, Mounds, Haskell, Morris, Muskogee, Tahlequah, Choteau, Vinita, and Miami. The object of the reconnaissance was to determine the mutual relations of the Pennsylvanian formations of the Wyandotte, Independence, and Muskogee quadrangles. Incidentally, considerable material bearing on the stratigraphic and economic geology of the area was accumulated. This material is brought together here and summarized with all available data from other sources, in order to satisfy as best it may, until detailed surveys are made, the demand for information concerning this part of the new State.

GEOGRAPHY.

The area covered by the accompanying map (Pl. II) is 83 miles broad from east to west and 104 miles in length from north to south. It includes the northeastern portion of the Creek Nation, practically the whole of the Cherokee Nation, and all of the seven small reservations in northeastern Oklahoma, viz, the Seneca, Wyandotte, Ottawa, Shawnee, Modoc, Peoria, and Quapaw.

The topographic maps of the Five Tribes Survey were available for the region and, though somewhat out of date as regards culture, were adequate for the purposes of the reconnaissance. The region is embraced in nine 30-minute quadrangles—the Wyandotte, Vinita, Nowata, Claremore, Pryor, Siloam Springs, Tahlequah, Muskogee, and Okmulgee—also including a portion of the Nuyaka quadrangle.

The Wyandotte quadrangle has been recently surveyed topographically and geologically, and the report is now in course of preparation, a summary of the principal geologic features being incorporated herein. The Siloam Springs quadrangle was touched by this reconnaissance on the northwest and southwest corners only. The geology of the Tahlequah and Muskogee quadrangles is described in detail in the geologic folios on those areas by Joseph A. Taff, and has been summarized on the accompanying map for the purpose of showing the relation of the geologic structure in that section to the structure in the northeastern portion of the region.

In addition to this detailed geologic mapping, there have been several previous reconnaissances. Drake[a] in 1897 published a paper dealing particularly with the coal fields of Indian Territory, but embodying a map and a general description of the geology. Adams[b] in 1901 gave a preliminary account of the geology and development of the Kansas-Oklahoma oil and gas field, with an accompanying geologic map of the area. The same author[c] in 1903 correlated and described various limestone beds in the Cherokee and Osage nations and delineated their outcrop on a map. Taff[d] in 1905 described the coal measures of the Indian Territory and gave two maps, one covering approximately the area shown on the map accompanying this report; the other covering the area adjoining it on the south, showing the main coal fields of the State and based in large part on the detailed folio mapping of those fields. Taff and Shaler[e] in 1905 gave a brief description of a small oil field near Muskogee, with an account of the geologic column and the general structure.

The map accompanying the present report contains considerably more detail than others which have preceded it covering the same territory, and is designed to supplant them. In studying the distribution of the geologic formations or the structure of the west border of the Ozark uplift, it will be found helpful to use this map in connection with Pl. I of Bulletin No. 260, both having the same scale and joining along the parallel of 35° 30'.

STRATIGRAPHY.

ORDOVICIAN ROCKS.

The Ordovician rocks outcrop only in the valleys of the large streams that dissect this part of the Ozark uplift, and in general only where local elevations bring them within reach of the drainage. In the Tahlequah quadrangle they are exposed in the valleys of Barren Fork and Illinois River. They are there made up of the Burge

[a] Drake, N. F., Proc. Am. Phil. Soc., vol. 36, 1897, pp. 226–419.
[b] Adams, G. I., Bull. U. S. Geol. Survey No. 184, 1901, pp. 5–28.
[c] Bull. U. S. Geol. Survey No 211, 1903, pp. 61–65.
[d] Taff, J. A., Bull. U. S. Geol. Survey No. 260, 1905, pp. 382–401.
[e] Taff, J. A., and Shaler, M. K., Bull. U. S. Geol. Survey No. 260, 1905, pp. 441–445.

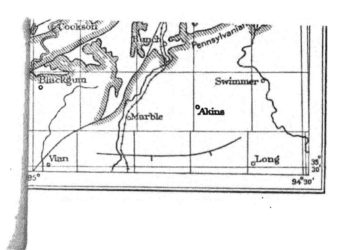

sandstone, 100 feet thick, overlain by the Tyner formation, which consists of shale with subordinate limestone and sandstone, the whole 60 to 100 feet in thickness.   What is presumably the Tyner formation is exposed in the bottom of Spring Creek 10 miles above the mouth. In the vicinity of Spavinaw dolomites and fossiliferous cherts are exposed in an area about a mile long, which has been described by Drake.[a]   According to Ulrich the fossils show the age of these rocks to be the same as that of the Jefferson City formation.   The Ordovician sediments at this locality are intruded by a dike of reddish granitic rock, 75 feet or more in width, and probably 1,800 feet in length.   Drake reports also an exposure of Ordovician rocks on Illinois River extending downstream from the Arkansas line for a distance of about 12 miles.   He states that the thickness exposed here, as at Spavinaw, is about 200 feet.   Similar dolomites and magnesian limestones with oolitic opalescent chert lenses outcrop in the Wyandotte quadrangle in the bluff of Neosho River,[b] 4 miles above the mouth of Cowskin River.   From 10 to 13 feet of the formation is shown above low water.   Where the Horse Creek anticline crosses Buffalo Creek, 3 miles above Tiff City, it brings to the surface 26 feet of the Ordovician rocks, consisting of dolomite with a little chert, in some places oolitic.   Both the Neosho River and Buffalo Creek exposures are nonfossiliferous.

SILURIAN ROCKS.

The only rocks of Silurian age in the area here discussed are found in the southern part of the Tahlequah quadrangle, and have been described by Taff in the Tahlequah folio as the St. Clair marble.

DEVONIAN ROCKS.

The Chattanooga shale is exposed in almost all the deeper stream valleys between Tahlequah and Cowskin River.   In the east bluff of Neosho River, 3 miles above the mouth of Cowskin River, the thickness is 26 feet; on Buffalo Creek it is 20 feet, at Southwest City 50 feet, at the mouth of Honey Creek 83 feet, and at Spavinaw 90 feet.   On Spring Creek 25 feet, not the full thickness, was noted. Taff reports the thickness in the northern part of the Tahlequah quadrangle as 40 feet, and in the southern part as 20 feet.

The Sylamore sandstone member, which reaches a thickness of 20 to 30 feet in the Tahlequah quadrangle, does not appear in the Spavinaw region.   On Buffalo Creek a thin bed of sandstone with a maximum thickness of 4 inches intervenes between the Ordovician rocks and the Chattanooga shale, and is tentatively referred to the Sylamore member.

[a] Proc. Am. Phil. Soc., vol. 36, 1897, p. 343.
[b] This river is mapped as the Neosho in accordance with a decision of the United States Geographic Board.   The portion below Spring River is known locally as Grand River.

CARBONIFEROUS ROCKS.

MISSISSIPPIAN SERIES.

The Boone formation is made up of an alternating series of limestones and cherts, approximately 300 to 350 feet in thickness where fully developed. It forms the surface rock over by far the largest part of the area of Mississippian rocks. At the base of the formation there is always present a limestone member, consisting at the top of a heavy ledge of coarsely crystalline encrinital limestone, or marble, which is usually 10 to 15 feet in thickness. This bed is separated by several feet of shaly limestone from a lower ledge of flaggy limestone, locally rather cherty and in many places irregularly bedded, which lies upon the Chattanooga shale. The upper ledge usually outcrops as a smooth, wall-like bluff, from which large blocks, the full thickness of the ledge, break away. It has been correlated with the St. Joe limestone member in the Tahlequah and Fayetteville folios. This limestone is normally overlain by a series of dark limestones and cherts from 50 to 80 feet thick. Above these, to the top of the formation, are lighter colored cherts and limestones, with one or more massive ledges of limestone 10 to 20 feet in thickness. The Short Creek oolite member, noted in the Joplin district, is found in the east half of the Wyandotte quadrangle wherever its horizon is exposed, though west of Spring and Neosho rivers it usually pinches out or loses its oolitic character.

The Chester group outcrops in a strip several miles wide inside the border of the Mississippian area. In the southern part the Fayetteville formation and the Wedington member are represented, but. in the Wyandotte quadrangle the group comprises the Batesville, Fayetteville (with its Wedington member), and Pitkin formations. These are persistent formations northward almost to the Kansas line, but in the area which has been subject to underground solution they occur chiefly in patches occupying solution depressions or ancient sink holes.

PENNSYLVANIAN SERIES.

Correlation of the Cherokee formation.—The Cherokee formation as defined by the Kansas geologists, includes all the various shales and sandstones in southeastern Kansas which lie between the base of the Pennsylvanian and the Fort Scott limestone. This limestone, as will be noted on the map, has been traced from a point near Chetopa, Kans., to and beyond Arkansas River. Here, in Concharty Mountain, the limestone is overlain by a bed of sandstone, the cap rock of the mountain, and underlain by a bed of coal. The sandstone strikes off southwestward, extending nearly to Baldhill. The Henryetta coal sets in a few miles south of Baldhill, in the same line of strike, and is probably the equivalent of the coal below the Fort Scott that is mined at Evans, Catoosa, and many other points to the northeast. The Henryetta coal has been traced by Taff and Shaler for many miles to the southwest, outcropping beneath a prominent escarpment of sandstone, apparently the equivalent of the Calvin sandstone of the Coalgate quadrangle. Beneath the Calvin sandstone the following formations, with the thickness indicated, outcrop successively toward the east along the parallel of 35°: Senora sandstone, 500 feet; Stuart shale, 275 feet; Thurman sandstone, 250 feet; Boggy formation, 1,200–2,000 feet; Savanna sandstone, 1,150 feet; and McAlester shale, 2,000 feet. Beneath the McAlester, in the Atoka quadrangle to the south, lie the Hartshorne sandstone, 200 feet thick, and the Atoka, 3,000 feet. This 9,000+ feet of Pennsylvanian shales and sandstones is represented at the Kansas line by a thickness of but 500 feet of Cherokee. There is a pronounced thinning of these formations both northward and southwestward from the vicinity of Canadian River, where there was apparently a basin in which the deposits were much heavier than elsewhere. As described by Taff the formations from the McAlester to the Senora inclusive decrease in thickness 50 per cent in passing from east to west across the Coalgate quadrangle. To the north a similar thinning takes place, but in addition there has been a warping of the west end of the Ozark dome, which resulted in the transgression and overlap of the later formations upon the older ones. This is shown by the pinching out of the Savanna formation just south of the area shown on the accompanying map (see Bulletin No. 260, Pl. I, p. 382) and by the discordance of the strike of the Boggy formation with that of the Fort Scott limestone and overlying formations. The base of the Boggy formation, traced northwestward, strikes Arkansas River 6 or 7 miles below Haskell. On the north side of the river the topography is gentle and the stratigraphy is partly concealed by heavy deposits of silt and sand that have been blown up from the Arkansas bottoms in the dry season by the prevailing southwest winds. As well as may be judged, however, the strike swings to the northeast in alignment with the strike of the overlying formations. This would locate the beginning of the warping of the uplift in the period between the Winslow and the Boggy and would account for the cutting out of the outcrop of the Savanna formation. The continuation of the sandstone and the shales of the Winslow beneath the overlapping formations would naturally be sought in the line of strike of the present outcrop of those rocks. Because of the probable curvature of the shore line of the uplift, the buried extension of the Winslow should be expected somewhere between Chelsea, Nowata, and Coffeyville, Kans. In this connection there arises an interesting speculation as to the relation of these buried Winslow sandstones to the oil sands of the Alluwe-Coodys Bluff field, which will be considered elsewhere, in the discussion of the oil fields.

The Cherokee formation is thus apparently the equivalent of the various formations from the Boggy to the Senora inclusive, though it may be that the upper formations overlap a portion of the Boggy, and that the Cherokee represents only the upper portion of the Boggy together with the overlying formations. The question can be settled only by detailed work in the region involved.

Fort Scott limestone.—The distribution of the Fort Scott limestone has been considered to some extent in the preceding discussion. In the Kansas reports it was at first called the Oswego limestone and it is known by that name to most drillers in the Midcontinent field. In the type locality, at Fort Scott, Kans., it consists of two beds of limestone separated by 7 feet of very dark shale. The upper limestone is from 10 to 14 feet thick, and the lower one, which is the rock used for hydraulic cement, is 4½ feet thick. Below the lower limestone there is black shale for a few feet, underlain by a bed of coal, 18 to 22 inches thick. This bed has been mined considerably in the vicinity of Fort Scott and in the Kansas reports is called the Fort Scott coal. Though in reality classed in the Cherokee formation, it has been at times more or less loosely included with the Fort Scott. As shown in the section on coal (p. 215), the bed which is extensively stripped east of Centralia lies between two limestones of the Fort Scott formation. Toward the north, however, the lower of the two limestones disappears, leaving the coal beneath all the limestone, and thus in the same position as the Fort Scott coal. This indicates a possibility that the dark shale and the coal beneath the Fort Scott formation should, in reality, be included in that formation.

In the vicinity of Coodys Bluff the Fort Scott is about 115 feet thick and consists of four limestone beds separated by shale. The limestone beds, from the top down, are respectively 10, 40, 10, and 8 feet thick and the intervening shales 5, 10, and 35 feet thick. At Sageeyah, as shown by the drill, a bed of coal comes in between the two limestones which make up the formation and outcrops in the escarpment to the east. At Catoosa, according to the drill record, the formation has a thickness of 154 feet, comprising six beds of limestone with intervening layers of shale and two beds of coal. To the south some of the beds of limestone disappear, and the intervening shales become sandy or give place to sandsone. A drill log shows that near Broken Arrow the formation has a thickness of 130 feet and contains two thin beds of limestone, one at the top and the other at the bottom. A thin bed of coal occurs just below the upper limestone and another one just below the lower limestone. The lower coal is taken to be the coal mined at Evans, 4 miles east of Broken

Arrow. The sandstone in the middle of the formation becomes more prominent south of Arkansas River, and forms the cap rock of Concharty Mountain, in the northern and eastern bluffs of which the lower limestone is exposed, underlain by coal. The extension of the formation beyond Concharty Mountain has already been considered.

Labette shale.—At the Kansas-Oklahoma line, according to the Kansas geologists, the Pawnee limestone is separated from the Fort Scott limestone by the Labette shale, which there has a thickness of only a few feet, though farther north in Kansas it is 60 feet thick, and to the south it increases to more than 200 feet. In the vicinity of Coodys Bluff and farther south a bed of massive fine-grained buff sandstone 15 to 50 feet thick is included in the upper portion of the shale.

Pawnee limestone.—The Pawnee limestone, generally known as the " big lime " by drillers, is the lower member of the formation which has been called the Oologah limestone. Its thickness ranges from 25 to 45 or 50 feet on the outcrop, but is reported to reach 80 to 100 feet in drill holes, possibly owing to confusion with some overlying or underlying bed of limestone. In many places it splits up into two or more beds separated by shale members, as exemplified in the section of the bluff east of Nowata, given on page 221, in the discussion of cement materials. Considerable white to light-buff spongy chert is included in the upper part of the limestone and forms a residual mantle over the surface. This part of the formation is especially apt to be massive and to form a perpendicular cliff from which large rectangular blocks from 10 to 25 feet in dimensions break off. The Pawnee makes a prominent escarpment along the west bluff of Big Creek and Verdigris River from the Kansas line to Catoosa and caps the high ridge from that point to Broken Arrow, beyond which it becomes thinner and difficult to trace.

Bandera shale.—The Pawnee limestone is separated from the Parsons formation above by a bed of shale, known as the Bandera formation in Kansas and there ranging from 50 to more than 140 feet in thickness. Near Coffeyville, Kans., it is shown by drill records to be 135 feet thick, and at Wimer, Okla., its thickness must be more than 100 feet. Southwestward it thins rapidly, and in the vicinity of Delaware, Nowata, Oologah, and Collinsville is but 10 or 20 feet thick. As well as can be made out from the available drill records, the shale disappears altogether in the vicinity of Owasso, and is absent from that place westward, allowing the lower limestone of the Parsons to rest directly upon the Pawnee. Between Tulsa and Catoosa, however, the shale thickens up and the two limestones are again separated by an interval of 100 feet. To the southwest the interval is as great or greater, though the correlation in that direction is not very satisfactory.

Parsons formation.—The Parsons formation is made up of two limestone beds separated by a shale member.

Drake proposed the name Oologah for the limestone outcropping in the vicinity of the town of that name. Adams [a] quoted Bennett as recognizing that there are two limestones at Oologah, the upper one of which he determined to be the equivalent of the lower limestone of the Parsons. The lower Oologah limestone, as already shown, is to be correlated with the Pawnee limestone.

The lower limestone of the Parsons has a thickness of 15 to 30 feet. It is a bluish, fine-grained crystalline limestone with considerable chert which weathers out and mantles the surface. South of the Kansas line this limestone caps the escarpment parallel to the outcrop of the Pawnee limestone along Big Creek and from 2 to 3 miles west of it. It crops out at the base of the hills on the east side of Verdigris River from Coffeyville to a point opposite Lenapah, where it crosses the river and gradually rises in the west bluffs until, in the vicinity of Nowata, it covers the tops of the hills of which the Pawnee forms the east escarpment. This relative position persists through Talala and Oologah to the vicinity of Owasso. South of Owasso the limestone spreads out along the valley of Mingo Creek to a point west of Broken Arrow, where it becomes inconspicuous. South of Arkansas River the lower limestone of the Parsons seems to be represented by the thin bed of calcareous claystone with lumps of blue limestone, weathering yellow, which crops out in the ridge 1½ miles west of Bixby and becomes more prominent in the vicinity of Duck Creek, 5 miles southwest of Bixby.

The middle member of the Parsons formation is a sandy shale 55 feet in thickness at the State line southeast of Coffeyville. It thickens to 130 feet at Nowata, and continues to increase in thickness to the south. West of Watova the Dawson coal sets in 100 feet below the upper limestone, and, as shown on the map, is traced to and beyond Mounds. In a drill hole 6 miles west of Broken Arrow the shale has a thickness below the coal of approximately 500 feet.

The upper limestone of the Parsons in Oklahoma has a thickness of 15 to 20 feet, and is a bluish, densely crystalline, clinky limestone with light blotches, which weather out as opaque white cherty lumps, in places thickly distributed through the residual clay. The formation is typically exposed in the quarry at the rock crusher 2 miles north of Lenapah. Farther south it has a broad outcrop east of the railway, extending to Delaware, beyond which it forms a low escarpment on the west side of the railway. At Nowata, where the member is 20 feet in thickness, the lower part consists of earthy encrinital limestone, the middle part is shaly, and the upper part is earthy limestone weathering into pebbly lumps. A thin limestone representing

---

[a] Adams, G. I., Bull. U. S. Geol. Survey No. 211, 1903, p. 62.

this member caps the hill 2 miles northwest of Watova. Beyond this point no more is seen of the limestone until the vicinity of Tulsa is reached. Southwest of Tulsa, according to Taff,[a] limestone which weathers yellow lies about 100 feet above the Dawson coal in the vicinity of Mounds and elsewhere, as shown on the accompanying map. This bed, in all probability, is a continuation of the upper Parsons limestone.

Drum limestone.—The Drum limestone outcrops with a thickness of 22 feet on the point of the ridge at the State line 3 miles southwest of Coffeyville, Kans., and extends westward adjacent to the State line for about 4 miles, to a point where it thins out and disappears. It does not outcrop at a corresponding elevation on the south side of Opossum Creek and was not identified elsewhere. Limestone outcrops in the valley of Opossum Creek on the headwater streams of Hickory and California creeks, but it is believed to belong to the Coffeyville formation, which lies between the Drum and the Parsons and, though in the main a sandstone and shale formation, carries some lentils of limestone.

Wilson formation.—The Wilson formation consists of the sandstones and shales which make up most of the western part of the broad ridge between Verdigris and Caney rivers. The crest of the ridge is formed by a bed of sandstone that appears as a fairly conspicuous escarpment from the Kansas line to the point where it crosses Arkansas River, 7 or 8 miles above Tulsa. This outcrop is outlined on the map (Pl. II) because it is easily traced and not because it marks the eastern limit of the formation, as do the other boundary lines here shown. The sandstone which forms the prominent escarpment west of Mounds and stretches away to the southwest occupies an analogous position above the Dawson coal and the upper limestone of the Parsons and apparently belongs to the Wilson formation, though possibly it should be correlated with some of the sandy members of the Coffeyville formation.

The Piqua limestone is the uppermost member of the Wilson formation. It outcrops at the base of several outliers of the Buxton formation on the Kansas line, due north of Wann, and also 3 miles northwest of that place. It is reported to be only 1 or 2 feet in thickness at the State line, but apparently thickens toward the south. As pointed out elsewhere, it is believed to be the limestone in use at the Portland cement plant at Dewey.

Buxton formation.—The Buxton is a shale and sandstone formation which occurs in the outliers above mentioned and forms the prominent sandstone escarpment that closely parallels the meridian of 96° from the Kansas line southward to the vicinity of Ramona and veers west of south from that locality to Arkansas River.

---

[a] Taff, J. A., Bull. U. S. Geol. Survey No. 260, 1903, p. 396.

STRUCTURE.

THE OZARK UPLIFT.

The area under consideration embraces the extreme southwestern prolongation of the Ozark dome. Along the southern limit of the area of Mississippian rocks, as shown on the map, the gentle southwestward dip of the attenuated uplift changes to a more pronounced dip to the southeast and south, and shows here and there the characteristics of a definite monoclinal fold. This feature is strengthened toward the east, in Arkansas, and, together with parallel faults having a southward downthrow, serves to limit the plateau of the Boston Mountains on the south, as pointed out by Newsom.[a]

WARPING.

As has been remarked in the section on the correlation of the Cherokee formation, the discordance of strike of the Winslow, Savanna, and Boggy with the Fort Scott and overlying formations shows a warping of the west end of the Ozark uplift somewhere in the period between the close of the Winslow deposition and the early part of the Cherokee. The southern portion was elevated, crowding the shore line far to the southwest, while the northern portion was depressed, allowing the sea to advance to the east. The northeast-southwest orientation of the shore line was well established early in Cherokee time, and this relative attitude was maintained through the remainder of the Pennsylvanian, the various formations of this age outcropping in this region in lines parallel to the Fort Scott limestone, with gentle northwesterly dips ranging from 30 feet per mile in the northern portion of the area to 50 feet in the southern portion.

FOLDING AND FAULTING.

The folds and faults of the area are so closely related that their joint discussion seems preferable.

The axis of that part of the Ozark uplift included in Oklahoma trends from northeast to southwest, and the margin of the uplift is serrated by an interesting system of parallel normal faults, which Taff has described in the Tahlequah and Muskogee folios. These faults trend parallel to the axis of the uplift, and at either end usually develop into monoclines or asymmetric anticlines and gradually die out. An inspection of the map will show the close relation of the minor drainage to the faults. Northwest of Tahlequah the parallelism of the faults is not so pronounced and the system is somewhat complicated by intersecting cross faults. These faults are

---

[a] Newsom, J. F., Am. Geologist, vol. 20, July, 1897.

of post-Winslow age, but no relation to later formations can be established.

Extending southward from Locust Grove is a prominent fault with the downthrow to the west, which brings the Chester down to a level 100 to 150 feet lower than the top of the Boone chert hills immediately east.

The next prominent structural feature toward the north is the Seneca fault, which extends from a point midway between Choteau and Pryor Creek to a point several miles northeast of Spurgeon, beyond the limits of the area shown on the accompanying map, having an almost due northeast course parallel to the main axis of the uplift. and roughly parallel to its northwest margin. This fault is double and in places multiple, letting down a long, narrow block of Boone, Chester, and overlying rocks into the Boone formation. In addition to the downthrown block, the strata for some distance, in places for a mile or two on either side, dip toward the fault. This combination has had a strong influence on the drainage, as may be seen from the map. From Seneca toward the northeast the fault closely follows the valley of Lost Creek. South of Seneca it crosses the divide to Sycamore Creek and follows down that valley to Neosho River. From the mouth of Sycamore Creek the fault cuts across the various meanders of Neosho River to a point just above the mouth of Spavinaw Creek. Southwest of this point it traverses the flat upland to and beyond Pryor Creek. Near the Neosho, where the rocks on either side are the cherty limestones of the Boone formation, it is easy to trace the down-dropped strip of Chester consisting of limestone and sandstone. Farther to the southwest, where the Chester formations become the prevailing surface rocks, the fault is difficult to follow. In the Pennsylvanian area it is quite impossible to trace the fault, but whether this is due to its absence, to the uniformity of the rocks, or to the concealment of the faulted Winslow by the overlapping Cherokee was not determined. If the last explanation is correct, as is quite likely, the fault is post-Winslow and pre-Cherokee and probably of the same age as the parallel faults of the Tahlequah-Muskogee region and as the warping to which all the faults are with little doubt genetically related. Owing to the fact that the amount of throw is variable within short distances along the fault and to the further fact that the fault line coincides so closely with the drainage lines, the Chester formations are preserved but here and there along the fault.

The intersections of this fault line with the meanders of Neosho River afford many fine cross sections of the faulted area. The width of the down-dropped block ranges from less than 200 feet to more than 1,500 feet. The fault ranges in character from a simple pair of opposed breaks with the downthrown block between them, and with the strata of the wall rock on either side dipping more or less steeply toward the faulted block, to a sort of faulted syncline, the limbs of which are made up of distributive faults with the cumulative downthrow toward the axis of the syncline. The best view of the latter phase is shown in the west bluff of the Neosho opposite the mouth of Cowskin River, where the south limb dips from 2° to 5° N., the angle increasing toward the axis, and shows four distinct dislocations, one being opposed to the other three, but leaving a resultant throw of 14 feet to the north. On the north side there is a faulted zone 55 feet wide in which there is an upthrow of 18 feet, but this is more than counterbalanced by three small faults and one with a throw of 22 feet to the south, and by the southerly dip of 2° some distance from the fault and of 11° adjacent to the fault.

The amount of displacement on either side of the block varies from place to place. In the west bluff of Neosho River 2 miles below the mouth of Horse Creek it is more than 90 feet. At the Becker mines, south of Seneca, it is from 100 to 140 feet. Between Seneca and Spurgeon it must be as much as 100 feet in many places, for it serves to bring the Chester formations down to the level of the valley, though the Boone forms the top of the hills on either side.

The Horse Creek anticline is an asymmetric fold which starts at a point on Cabin Creek, 5 miles southeast of Big Cabin station and trends east-northeastward by Cleora to the mouth of Cowskin River, where it intersects the Seneca fault. East of this point it swings a little more eastward to the vicinity of Tiff City, where it trends nearly due east for 10 miles and farther east gradually dies out. The anticline has a gently sloping northern limb and a steeper southern limb. To the south of the anticline and parallel to it is a long, low synclinal trough beyond which the strata rise again to the south, with a gentle incline. The average dip of the northern limb of the anticline is about 2°; the dip of the southern limb ranges from 5° to 18°. West of Neosho River the fold expresses itself topographically in an abrupt faultlike escarpment to the south and a low upland slope to the north. East of the Neosho the anticline is cut through on either side by many short, steep hollows, and forms the greatly dissected highland known as the Seneca Hills. In places, notably where the fold is cut through by Neosho River, the rocks lie nearly flat, but where it is crossed by Buffalo Creek and Horse Creek the dip is about 5° SE. About 2 miles west of Horse Creek Gap the dip is 18° SE. For the most part the dip of the southern limb is concealed by débris washed down from the steep slope, and can be made out only in exceptional places. It is entirely possible that for short distances along the axis west of Horse Creek the anticline may break down into small faults. Though cut across in several places by streams, this fold is nowhere breached parallel to the axis, a fact due doubtless to its monoclinal nature.

The Owasso dome lies 2 miles west of Owasso, partly in sec. 26 and partly in sec. 35 of T. 21 N., R. 13 E. One of the upper beds of the lower limestone of the Parsons formation swells up in a perfect low dome, nearly circular in outline and about three-fourths of a mile in diameter. The center of the dome has an elevation 50 feet higher than the rim, and the limestone beds conform in shape to the surface of the uplift. It is possible, however, that the quaquaversal structure extends for some distance beyond the limestone all about the dome, but is concealed owing to the lack of outcrop.

UNDERGROUND SOLUTION.

The effects of solution, which are so prominent in the Joplin mining region,[a] are here limited chiefly to a small area in the northeast corner of the territory covered by the accompanying map, occurring altogether northeast of a line drawn through Miami, Wyandotte, and Tiff City. This line is fairly parallel to the eastern margin of the Winslow formation, a parallelism which is doubtless due to their common relation to the Winslow shore line. After the close of Winslow deposition each margin retreated from the original shore line with the progress of erosion, the margin of the Winslow retreating seaward through simple erosion of its outcrop, and the margin of the area affected by underground solution retreating landward through the effacement of the shallow coast sink holes in the degradation of the surface. As a structural process, underground solution deserves consideration here chiefly because of its inseparable connection with the ore deposits of the area which it affects. In the rest of the region the contact of the Pennsylvanian with the Mississippian is that of a simple erosional unconformity; in this area it is a solution unconformity in which the results of erosion are complicated by the effects of underground solution. Advancing over the sink-hole topography of the Mississippian land, the Cherokee sea filled up the valleys and sink holes with various sediments. After the erosion of the Pennsylvanian the shales and sandstones filling the depressions were left as outliers of various shapes and sizes, forming the solution patches referred to in the description of the ore deposits.

BRECCIAS.

A basal breccia was formed where the later shale formations were deposited upon and in the residuum which mantled the surface and in part filled the caverns of the limestone land that was subject to underground solution. Such breccias, also called " mixed " or " confused " ground, are the loci of ore deposition in many places.

---

[a] See Geologic Atlas U. S., Joplin district folio (No. 148), 1907.

Locally a chert member of the Boone, with thin interstratified limestone strata, was finely shattered by warping or other tension, the more elastic limestone escaping brecciation. On being recemented in place this rock formed a sheet breccia. When the limestone was replaced later by ore and ore-bearing jasperoid, it became "sheet ground." Though important in the Joplin district, sheet ground was apparently not extensively developed in this area.

When the interbedded limestone in such a series was dissolved, but not replaced, it allowed the chert to settle irregularly, resulting in strong brecciation, in many places completely obscuring the bedding. Shale may have been carried into the openings or they may have been filled with ore or jasperoid. A horizontal tabular body of breccia, which differs in origin and in form from the sheet breccias, was thus developed. Such a formation may be called a blanket breccia. The lead and zinc ore in the Quapaw district is generally found in a breccia of this kind. In the "sheet ground" of the Joplin district the characteristic thing is the occurrence of sheets of ore and jasperoid between the ledges of chert. In the Quapaw district, however, sheets as much as a few feet in length are extremely rare, the ore being generally disseminated through the breccia. In distinction from the "sheet ground" deposits in the sheet breccias, the Quapaw ore bodies in the blanket breccias may be called "blanket ground," the term "blanket vein" comprehending both classes. Zonal breccias, after the type described as occurring in the Joplin district, have been observed in the area here considered, but are not important structurally or economically. Fault breccias were developed by the faulting which has been described as occurring around the margin of the uplift, but only in the regions affected by underground solution has siliceous cementation rendered them prominent or ore deposition rendered them important. The breccias of this kind to be considered, therefore, are limited to those associated with the Seneca fault. Typical examples occur at the mines on Sycamore Creek; at the Becker mines, southwest of Seneca; in the south bluff of Lost Creek at Seneca; and in general on each side of the Seneca fault block from Seneca northeastward to and beyond Spurgeon.

MINERAL RESOURCES.

LEAD AND ZINC DEPOSITS.

GENERAL CONDITIONS.

The workable deposits of lead and zinc ores, so far as at present known, are limited to the northeastern corner of the region here discussed, the area in which they occur being entirely within the Wyandotte quadrangle and coincident with the territory which was subject to underground solution. They are found mainly in the chert breccias in the Boone formation, but also to a considerable extent in sandstone breccias in the Chester, particularly in the new Miami district. More or less limestone and shale are included in the breccias, incorporated either at the time of their formation or since. The ores are galena ("lead") and sphalerite ("jack") with a relatively small amount of smithsonite and calamine (both known as "silicate"). The associated or gangue materials are jasperoid, which occurs as a fine-grained gray to black siliceous rock cementing the breccia, pyrite ("mundic"), calcite ("tiff"), and dolomite ("spar").

The districts where mining has been carried on, or where the indications have encouraged extensive prospecting, arranged chronologically, are the Peoria, Sycamore Creek, Quapaw, and Miami. In the following pages mines typifying the different classes of ore bodies are described somewhat fully under each of these districts, the other mines being mentioned but briefly or entirely omitted.

The ore bodies have various forms, depending on the structural features of the associated rocks. In basal, zonal, and fault breccias there are "runs" and, though rarely in this area, "circles." Sheet breccias, carrying "sheet ground" deposits, so prominent in the Joplin district, have not as yet been discovered in Oklahoma. Blanket breccias, with "blanket ground" deposits, are the main source of ore, being especially well displayed in the Quapaw district.

PEORIA DISTRICT.

General description.—The mines at Peoria were opened in 1891, on land the first lease of which is held by the Peoria Mining, Construction and Land Company, a New Jersey corporation. The most productive area adjoins the village on the northwest, and underlies the bottom and north bluff of Peoria Creek. In the creek bottom, over an area 300 feet long east and west by 100 feet wide, known as the Playhouse diggings, a solid sheet of galena was found in chert at a depth of 7 to 10 feet. This sheet, narrowing to 60 feet, extended northward for 600 feet under Monkey Hill and is reported to have been from 6 to 22 inches thick. Other shallow deposits of lead have been worked on the first and second hills west of Monkey Hill. Not much sphalerite has been mined at these places. A sheet of sphalerite from 4 to 18 inches in thickness, with a thin sheet of galena just above it, which yielded two carloads of ore, was struck about 12 feet above the level now worked for silicate.

Silicate mine.—The Silicate mine is operated by Gordon & Wilkins. The shaft is in the face of the hill just north of the creek, about 50 feet west of the edge of the Playhouse diggings, and some of the drifts extend under those old workings. The face of ore ranges from 1 to 7 feet in height, averaging 2½ feet. The drifts are carried 6 to 8 feet in height and from 10 to 12 feet in width, and have a total length of approximately 1,000 feet, covering an area less than 200 feet square. The ore occurs either in slabs or as " fish-egg silicate," in clay interbedded with red tallow clay and layers of soft, rotten chert, the whole conforming to the limestone horses and bowlders which are present here and there. In one place the walls of the run closed in, nearly pinching out the ore, which continued on through the opening in the solid limestone. The ore in the main is plainly the result of a carbonate replacement of the limestone country rock, associated with more or less underground solution, the latter in part antedating the ore deposition, giving rise to the openings through which the ore-bearing solutions passed, and in part contemporaneous with the ore deposition. The ore is concentrated on hand jigs, the quantity of fine fish-egg silicate associated with flat shapes requiring an elaborate scheme of concentration. Women and girls are employed to hand pick the screenings—probably the only instance of such employment in the Joplin region.

Chicago Syndicate Mining Company.—In 1907 the Chicago Syndicate Mining Company erected a mill over some old workings half a mile northwest of Peoria. The level worked at the mill shaft is 120 feet deep, but at another shaft on the edge of a small, oblong solution patch of shale and sandstone of Chester age, the mining was done at the 160-foot level. A considerable amount of lead was taken from this shale patch at a depth of 12 to 22 feet, the ore occurring near the base of the sandstone and shale. In the drifts now being worked the ore is sphalerite disseminated in rather coarse crystals through the bluish-gray jasperoid cement of chert breccia. In places this cement makes up one-third to one-half of the mass of the breccia, the chert bowlders and slabs being suspended in it. Considerable spar is present here and there, and where decomposition has progressed far there is much tallow clay.

Other mines.—The Poor Boys Mining Company is operating a silicate mine, and several other companies are prospecting in the immediate vicinity of Peoria.

Three miles due east of Peoria, on a tract of land belonging to S. L. Davis and adjoining the State line, in the vicinity of the Pinnick mines, there have been some recent strikes of ore. The Grimes & Williams shaft is sunk near the border of a circular solution patch, bordered by an outcrop of brecciated chert with a jasperoid matrix showing impressions of sphalerite crystals which have been leached out. Within the circle there are scattered sandstone bowlders of Chester age. The ore is sphalerite and occurs at the 90-foot level in the matrix of the chert breccia. This matrix consists of jasperoid in some places and of dolomite in others; in still others the ore itself acts as the cement. In the McKisson shaft, on the same tract, a 3-foot run of lead was struck at the 60-foot level in yellow flint ground.

SYCAMORE CREEK DISTRICT.

South of Seneca the Seneca fault cuts diagonally across the divide between Lost and Sycamore creeks, striking the latter 1½ miles west of the Missouri-Oklahoma line. At this point considerable prospecting has been done in and along the fault block, though without developing any paying bodies of ore. The first prospecting, which yielded some silicate, was done by means of a drift under the south bluff of Sycamore Creek a few feet above water level. Here the fault block is about 180 feet in width, and the throw is not sufficient to bring the sandstone of the Chester down to the present level of the surface. The limestones of the block, probably opened up more or less by the faulting, have been subjected to much solution, with the result that the space between the side faults, as exposed in the bluff, consists of a great mass of chert blocks lying topsy-turvy, the interstices being filled with tallow clay and residual clay. In the ravine just south of the bluff, near the southeast edge of the block, is a shaft from which drifts at the 75-foot and 104-foot levels extend southwestward for about 100 feet each. Thin ore was encountered in the lower drift, and better ore in the upper one. The ore consists of galena, sphalerite, and silicate. The silicate was too heavy to be separated from the sphalerite on hand jigs, and the ore could not be sold at a profitable price. Another shaft, 250 feet southwest of the deep shaft, encountered some large chunks of galena, which had to be broken up before they could be brought to the surface. Southwest of this point the displacement by the fault is greater, and at a distance of 400 feet sandstone and sandy shale of Chester age are present to a depth of 65 feet, with some galena and sphalerite in crevices in the sandstone and in the secondary limestone cement of associated chert breccia.

MINES ALONG THE SENECA FAULT.

The Sycamore Creek mines mark the southwesternmost known occurrence of ore along the Seneca fault. Northeast of this locality the fault has been prospected at short intervals to the vicinity of Spurgeon, Mo. The Becker mines (formerly the Potwin and Holmes mines) are just in Missouri, a mile south of Seneca. The ore, which is galena, sphalerite, and calamine, occurs in the chert and sandstone breccias of the southeast margin of the fault block and in the solution breccias of the adjacent Boone chert and limestone. These mines have been recently reopened, after lying idle for fifteen years. A 150-ton mill has been erected, several shafts have been sunk on shallow ore, and the original shafts have been sunk to deeper levels.

A mile northeast of Seneca are the old Huber mines, operated by the Seneca Lead and Zinc Company. The ore was found at a depth of 70 to 100 feet in broken chert ground adjacent to the northwest margin of the Seneca fault block. Breccia bowlders in the waste piles show jasperoid cement with impressions of sphalerite crystals.

The Gallemore mines are situated in the shallow saddle where the faulted area crosses a flint ridge one-half mile west of Racine. Sphalerite was struck in the shaft of the Racine Mining Company, at a depth of 105 feet. For the first 40 feet the shaft penetrated Pennsylvanian shale, in which an 18-inch bed of coal was found. This shale occurs very irregularly, not showing at all in some shafts near by, and is evidently a solution patch, the fault brecciation offering most favorable conditions for solution. Small quantities of sphalerite were taken from several shafts near by and considerable shallow lead was found in yellow flint ground adjacent to the fault on the northwest.

The Buzzard mines, on the Newman land, 2 miles northeast of Racine, have been operated more or less continuously since their discovery in 1882. They are adjacent to the fault block on its southeast side. The original discovery was of lead ground, 1½ to 2 feet in thickness, averaging 50 per cent of galena, at a depth of 8 to 10 feet, lying upon a solid sheet of Short Creek oolite and overlain by yellow clay and chert bowlders. The oolite dips slightly toward the fault and is reached at a depth of 28 feet in the shaft at present worked, at the foot of the low ridge which marks the breccia zone. Under the foot of this hill the ore consists of lead, sphalerite, and calamine, and as the fault zone is approached dips steeply downward toward it. The ore is associated with more or less chert breccia, with dark jasperoid cement. The Shaffer diggings, adjoining the Buzzard mines on the southwest, have not been worked for many years. The Hunt mines were a quarter of a mile north of the Buzzard mines, on the northwest side of the fault block.

The Baxter mines are on the fault block and along its northwest margin, in the NW. ¼ sec. 1, T. 25 N., R. 33 W. The surface of the block seems to be entirely covered with the sandstones and shales of the Chester. Much shallow lead has been taken out, at depths ranging from 2 to 30 feet, in broken sandstone and clay ground resting upon the Boone chert. Galena and sphalerite are found at deeper levels in the residual and basal breccias.

The Henderson mines join the Baxter diggings on the east and lie partly in the section just north. The mines extend from the north margin of the fault block completely across and for some distance beyond it, though most of the ore was found near the south fault zone. It consisted entirely of galena, and came from depths of 15 to 60 feet, in much the same relations as the shallow lead on the Baxter land.

The fault block has been prospected at short intervals all the way from the Henderson mines to a point half a mile northeast of Spurgeon. On the west edge of the Bucklin land at Spurgeon the sandstone outcrop marking the limits of the fault block is about 250 feet in width. Between this point and the village the sandstone and shale area widens out until it is 1,600 feet across. This widening is due mostly, however, to solution, as the fault block can be easily traced across the tract by a slight trough bounded on the northwest by the outcropping chert. The ore, consisting of galena and sphalerite, occurs in breccias—sandstone breccias in the upper levels and chert breccias in the deeper levels. Many of the sandstone fragments of Chester age in these breccias have been changed into quartzite through secondary enlargement of the sand grains. The surface of such fragments shows a glistening sheen which has given rise to the local name " glass rock."

QUAPAW DISTRICT.

General description.—The Quapaw mines extend from 1 to 4 miles east of Quapaw station, and from 5 to 7 miles south of Baxter Springs. The ore-bearing ground has been proved by drilling to extend for some distance beyond these limits on all sides, notably to the west, in the vicinity of Quapaw station, where a good strike was made in drilling the town well. The main ore deposits occur at a depth of 80 to 150 feet in blanket-ground formation, rarely in confused broken ground. Ore at shallower depths is limited chiefly to the region lying immediately south of Lincolnville, the village at the Quapaw mines. On the eastern edge of the Quapaw district the blanket ground rests upon the Short Creek oolite member of the Boone formation, which is usually penetrated by the mine sump. Though doubtless the blanket breccia forms an uninterrupted sheet throughout the area of the Quapaw mines, the oolite is not found in the western mines. This is because the bed, as noted in its description, thins out or loses its oolitic character along a north-south line which bisects the district. The ores found in the blanket ground are sphalerite and lead in about the proportion of 5 to 1. In a part of the blanket ground there has been some oxidization in the upper part and a little calamine is present. No ore is mined below the Quapaw blanket ground, although the drill has shown ore at deeper levels. In the shallow ground the ore, which is principally silicate and galena, occurs in runs and circles. In addition to the circular solution patch (with a probable circular ore deposit) on the Cherokee Lead and Zinc Mining Company's land, described below, there are several other shale and sandstone patches of the same shape. A circular shale patch at the F F F mine had much sphalerite and pyrite in the lower part of the shale, and ore continued in broken ground down to the main blanket ground at 90 feet. There is a large circular solution patch on the Red Eagle tract, in the NE. ¼ SE. ¼ sec. 31, T. 29 N., R. 24 E. The shale area is 350 feet in length north and south by 300 feet in width. A shaft near its southeastern margin shows a thickness of 70 feet of shale. The shale area is surrounded by a low rim of chert which dips away from the center on all sides. Near its contact with the shale the chert is brecciated and recemented with jasperoid from which considerable sphalerite has been leached. It seems reasonably certain that the ore here, should it occur in workable quantity, will be found in circular shape.

There are in the Quapaw district 25 steam concentrating plants, with a daily capacity of 3,300 tons, besides 7 mines operating hand-jig plants. In the typical blanket ground hand jigs can not be operated to much advantage, as the ore must be crushed very fine to free it from the rock. Those in operation are working with dirt from the shallow ground or the more oxidized portions of the blanket ground. In addition to the mines with concentrating outfits there are about 30 prospects which have shafts down to the ore. The owners of some of these prospects have planned to build mills; others are waiting for ore prices to become better.

Cherokee Lead and Zinc Mining Company.—In a field on the Cherokee Lead and Zinc Mining Company's lease just southwest of Lincolnville several wagon loads of calamine were picked up, having been plowed up and cast aside in ignorance of its value. Aproximately 50 tons of silicate with a little lead were taken from shallow open cuts less than 15 feet in depth, along the southwest margin of a circular solution patch at this place. Limestone and sandstone of Chester age cover the surface except for the circle 125 to 150 feet in diameter. The border of the circle is marked by a ring of sandstone bowlders. Inside this ring shafts and borings strike Cherokee shale with a little coal. Between the sandstone ring and the adjacent limestone there is, along the southwest margin, a strip of chert breccia with a matrix of jasperoid. This matrix has been ore bearing, as shown by the cavities from which sphalerite has been leached. The ore found in the shallow diggings lay mainly between the breccia and the limestone, but also extended in clay seams into both the limestone and the jasperoid. In the shaft just west of the circle ore was found at a depth of 35 feet in the clay and limestone bowlder filling of a solution chamber at the base of the limestone of Chester age.

A sheet of lead ranging up to 6 inches in thickness was struck at a depth of 40 feet in the shaft of the Alabama Mining Company, a hundred yards south of the locality just mentioned.

Good Luck mine.—Soft "confused ground" occurs at the Good Luck mine, as well as in some other mines in the northeastern portion of the Quapaw district. In the No. 1 shaft of the Good Luck mine the soft ground joins the blanket ground along a north-south line and the east-west drift is partly in each kind of ground. The blanket ground exposed by the drift is fractured and broken, but not recemented, and in places the bedding is entirely obscured. Clay and shale occur in the fractures and joints and between the chert beds. The blanket ground was evidently much softened by the solution which is responsible for the confused ground adjoining. The latter is the typical soft ground of chert and limestone bowlders in soapstone and yellow clay. Tiff and pink spar are present in veins and pockets in the unoxidized ground and impressions of spar and sphalerite occur in the jasperoid cement of the oxidized ground. Weathered lead occupies seams and fractures in the jasperoid. The No. 2 shaft was sunk on a drill hole showing rich lead cuttings that were found to have come from a solid chunk of galena a foot or two in thickness, which did not reach completely across the shaft. When followed to the south the lead gave out within $2\frac{1}{2}$ feet of the shaft, but the drift soon ran into good zinc ore. The ground here consists of dull chert and rotten limestone bowlders in a matrix of shale, clay, and tallow clay. Some of the bowlders are of secondary limestone, highly crystalline and very rich in ruby sphalerite in grains of the same size as those of the limestone. In some of these limestone bowlders the sphalerite constitutes from 15 to 25 per cent of the mass.

Mission mine.—The Mission mine may be chosen to exemplify the blanket-ground mines because it is typical, because its underground workings are the most extensive, and because it is the best-known mine in the district, having been the " show " mine from the beginning of the camp. The mined area extends over approximately 5 acres in the NE. ¼ sec. 1, T. 28 N., R. 23 E. The blanket ground is about 30 feet in thickness, the top of the ore being reached at about 50 feet from the surface and the limestone upon which it rests at about 80 feet. Only the upper 18 to 20 feet of this ground has been worked, though at the time the mine was visited a lower stope 10 or 12 feet in height was being taken up from the pump shaft.

In this mine, as generally in the district, there is at the top of the blanket ground 2 or 3 feet of soft ground, containing more or less black shale and yellow clay. Locally solution has progressed so far that slabs only of limestone and chert occur in the shale and clay. There is as a rule more or less spar and tiff in veins and pockets, and here and there a little bitumen. Much of the shale is slickensided. This zone is apparently not one of general movement, but one of accommodation to the slight stresses resulting from underground solution and the weight of the superincumbent strata. It seems likely that the opening at the top of the blanket breccia, into which the clay and shale have been drawn, was largely a result of the settling of the chert beds when the interbedded limestone was dissolved. The fact that the roof did not settle into this space would be explained if the solution in the blanket cherts went on by areas, portions remaining to support the roof until after the portions first dissolved and fractured had been recemented by jasperoid and rendered capable of sustaining the roof. This upper, softer sheet is in many places richer in ore than the harder ground below, but on the other hand it is here and there entirely barren.

Below the soft ground the blanket breccia consists of greatly fractured " live " blue flint, which in the more unbroken portions shows small dark spots one-eighth to one-fourth of an inch in diameter, surrounded by a lighter border. At many points the stratification of the chert is completely obscured by the brecciation, but in general it can be made out either from the chert or from the jasperoid sheets between the broken chert beds.

The brecciation of the chert ledges is uneven. In places the whole ledge is broken up; elsewhere it is broken into large bowlders with finer brecciation between them; elsewhere still the sheets are comparatively unbroken, with chert fragments suspended in the jasperoid interstrata. In many of the more unbroken portions of the chert strata traces of a former fine brecciation may be distinguished. The outlines of the fragments are faint, and the cementing material resembles a network of thin dark veins. These are not ore bearing and possibly correspond to the older sheet brecciation in the sheet-ground deposits of the Joplin district. At numerous places in the mine the chert is fractured vertically or nearly so. Some of these fractures come so close together as to constitute " sheeting " and to conceal the bedding of the chert completely. They are ordinarily not slickensided, but are usually stained dark by the circulating waters. In general they are open, containing neither ore nor jasperoid. They are probably equivalent to the sheeting in the mines of the sheet ground in the Joplin district, where also the sheeting is later than the ore deposition.

The ore in the Mission mine consists of sphalerite and lead in the proportion of 2 or 3 to 1. The latter is found in fissures and crevices in the chert, and the former is disseminated in the jasperoid as well as in the crevices and fissures. Where both the ores occur in a pocket, the galena shows a tendency to be crystallized in the upper parts and the sphalerite in the lower parts. In the mine as a whole not much differentiation can be seen, more lead occurring in the upper portion in some places and near the bottom in others.

MIAMI DISTRICT.

The Miami mines, 4 miles due north of the town, are the youngest in the State of Oklahoma, ore having been reached in the first shaft in 1907. The ore is sphalerite with some galena, and is found at depths of 90 to 130 feet, the variation being due in the main to the westward dip of the rocks. The surface formation is Cherokee shale, below which are sandstone and limestone of Chester age, resting upon the Boone chert and limestone. In certain areas the limestone of the Chester group has been carried away by solution, allowing the sandstone to settle down upon the Boone. In this settling the sandstone was more or less fractured and broken. The ore was first discovered in this sandstone breccia, though in some of the recent drilling it has been found to continue on down into the Boone chert. The main one of these old solution channels, as at present developed, crosses the Miami road near the southwest corner of sec. 6, T. 28 N., R. 23 E., and bears north-northwest. The territory has not yet been extensively drilled, and the available drill records are too few to determine the courses of any of the solution channels other than the one just mentioned, but it seems highly probable that other ramifications of the old underground drainage will be discovered. With this end in view, systematic preservation of drill records, whether they represent blank holes or not, is very desirable.

The sandstone in which the ore occurs is strongly impregnated with a heavy crude oil or bitumen. This material occurs in exactly the same fashion as the ore, in crevices in the fractured rock and between the laminæ of the sandstone. There is also in many places considerable fine-grained pyrite. Neither calcite nor dolomite has been observed in association with the ore, though they will likely be found when mining has extended downward into the Boone. The sphalerite, where it occurs in the openings between the laminæ, is very fine grained, though in the fracture openings larger masses of ore are found. The galena constitutes about one-sixth of the total amount of ore, and is found more often in the crevices than between the laminæ of the sandstone. The intimate association of the sandstone and ore requires close crushing to effect their separation, and this in turn necessitates the use of concentrating tables to save the fine ore. The bitumen must have a tendency to float off the fine-ore slimes. Experience has shown also that the bitumen collects in the jigs, and in the first cell of the cleaner jig it forms round balls with the ore. In a short time these balls completely cover the grating of the cell to a depth of 1 or 2 inches, so that it is necessary to clean off the jig bed every shift. The introduction of a jet of hot water into this cell simply delays the balling up and clogging until the next cell of the jig is reached. A single assay of the sphalerite concentrate is reported thus: Zinc, 44 per cent; lead, 4 per cent; iron, 7 per cent. Against these values penalties were assessed by the ore buyer, as follows: Moisture, 2 per cent; bitumen, 2 per cent; iron, 7 per cent. A deduction of $1 per ton of ore is made for each 1 per cent of these impurities, making a combined penalty in this case of $11 per ton.   From this showing it seems imperative that means be adopted to remove the bitumen and effect a better separation of the iron.   Some process of oil flotation, or of roasting and magnetic separation, would be likely to accomplish that end, but to roast the crushed ore before concentration would probably not be profitable.   Possibly, if the crude ore were crushed in hot water, the bitumen might be freed and mechanically skimmed or allowed to float off from the surface of a settling tank.   Roasting and magnetic separation might be employed after concentration.

The richness of the drill cuttings and of the ore itself, where it has been reached, has caused the prospecting, shaft sinking, and mill building to go on unabated.   Development was pushed rapidly in 1907.   Eight shafts were sunk to ore, many others were started, one mill was completed, and two others were begun, both of which are now finished.

A circumstance which has delayed the development of the district is that each undertaking requires the raising of 500 to 750 gallons of water per minute.   This of course has been the case in most of the new camps in the Joplin region.   With the lowering of the underground water level to the level of mining, the amount of pumping ordinarily becomes inconsiderable.   In the Miami district, however, the situation is somewhat exceptional.   The surface of the ground is low, not more than 25 feet above Tar Creek, the drainage level for the region, and the water-bearing ore stratum is capped by the unbroken Cherokee shale.   The mine water is different from the surface water in other camps, being highly charged with H_2S, like the deeper artesian waters, and the height to which it rises may represent artesian head rather than the underground water table. If this be so there may be no great diminution of the amount of water with time, but on the contrary a possible increase with the development of the underground workings.

As the water pours into the mine, the dissolved H_2S gas escapes and renders the air so noxious that the miners are unable to work a full shift, though there is much individual variation in the ability to withstand the gas.   As the gas is heavier than air, the ordinary blower at the surface will not force it out of the workings, so that an exhaust fan, with the intake at the bottom of the drift, is commonly employed in ventilating the mines.

OIL AND GAS.

DISTRIBUTION.

The areas which have yielded oil and gas are outlined on Pl. II. The most productive areas are the Alluwe-Coodys Bluff, or " shallow sand " field; the Bartlesville, or " deep sand " field; and the Glenn

" pool," near Kiefer. Between the Bartlesville and Glenn fields and in line with them, there is a series of smaller pools at such close intervals as practically to constitute one continuous field from Kiefer to the Kansas line. New but promising areas are the Delaware, Hogshooter, and Morris-Baldhill fields. Many other smaller outlying areas, generally gas bearing, are indicated on the map. Sufficient data are not at hand to discuss the depth, character, and mutual relations of the oil- and gas-bearing sands, or the origin and character of the oil.

GENERAL CONSIDERATIONS OF STRUCTURE.

Probably the most striking fact in connection with the distribution and shape of the pools, as will be noted from the map, is their grouping into belts with north-south alignment and their individual prolongation in that direction. This characteristic manifests itself with perfect independence of the topography and likewise of the general structure. The general structure of that portion of the Midcontinent field included in this report is, as noted in the preceding pages, that of a gentle monocline, dipping to the north of west. The study of a complete series of well records in connection with detailed leveling might reveal important minor features, such as low anticlines and terraces (flats or arrested anticlines), but it is difficult to see how such features could exist and be influential in the distribution of the oil and gas without showing their influence in the outcrop of the formations, unless they antedated the deposition of those formations. If the distribution of the oil and gas depends on the existence of elongated lenses of sandstone, it is likewise difficult to account for the diagonal position of these lenses with reference to surface outcrops, unless they were deposited when the land and sea relations were decidedly different from those which existed later, when the formations now at the surface were laid down. It seems a fair assumption, then, that the oil and gas reservoirs had their origin in conditions, which, whether orographic or geographic, antedated the deposition of the Cherokee and overlying formations. In the section on structure it has been shown that the warping, folding, and faulting which affected the western portion of the Ozark uplift were post-Winslow and apparently pre-Cherokee—that is to say, they were probably contemporaneous with the development of the structure influencing the distribution of the oil and gas. That they were related to that development is very probable. The orographic structure of the west end of the Ozark uplift has a very pronounced northeast-southwest trend. It is not easy to see how a north-south structure could be developed in the adjacent oil field at the same time and by the same forces which produced the structure of the uplift, or even independent of them. There remain for consideration the geographic relations of the sedimentation of pre-Cherokee time. In the section descriptive of the Cherokee formation it has been shown that there is a wide divergence in strike between the formations correlated with the Cherokee in the south and those preceding the Cherokee, and it has been pointed out that the most rational explanation of this discordance is that the west end of the Ozark uplift was warped so that the southern portion was elevated, crowding the shore line far to the southwest, while the northern was depressed, allowing the sea to spread farther to the east and to completely overlap the deposits laid down during pre-Cherokee Pennsylvanian time. The continuation of the Winslow sandstones would naturally be sought in the line of strike of the present outcrop of those rocks. Allowing for the probable curvature of the shore line of the uplift, we should expect to find the buried extension somewhere between Chelsea, Nowata, and Coffeyville, Kans. Similarly, the continuation of the Boggy formation would be expected to lie parallel to the buried Winslow, somewhat farther west, and the Savanna formation, if sedimentation in the northern portion of the area proceeded as in the southern, would be sought to the west of the buried outcrop of the Boggy. Lenses of sand with their elongation parallel to the old shore line or disconnected patches showing that alignment would naturally occur. Such sand patches, after being overlapped by the Cherokee shale, might become saturated with oil and gas forming " pools." The linear grouping of known pools into belts, the discordance of such belts with the strike of outcropping rocks, and the parallelism of these belts with the pre-Cherokee shore lines strongly suggest that they represent such sand lenses and patches laid down along those shore lines. If this be the fact, it has an important bearing on the distribution of the oil and gas sands in that the pools may in general be expected to develop greatest length in a north-south direction, thus explaining a characteristic of the known pools to which attention has already been called. Furthermore, the upper contour of the thickened lenses of sandstone determines the attitude of the succeeding deposits, developing therein a structure which may likewise influence the distribution of oil and and gas. For instance, the rapid thickening toward the west of the sandstone lentil in the Labette shale in the vicinity of Coodys Bluff might locally counterbalance the normal westerly dip, producing a depositional feature in the overlying sediments akin to a structural terrace or arrested anticline. Sandstones overlying an oil-bearing sand of pronounced lens shape might also be oil bearing over the same area.

This speculation as to the origin of the pools could be readily proved or disproved by the construction of several cross sections from a sufficient number of drill records. The records available at this writing are not adequate to affirm or deny its truth.   It is a fact that the Cherokee, or, more precisely, the Pennsylvanian below the Fort Scott limestone, thickens toward the southwest.   At the outcrop of the Cherokee on the Kansas line the formation is about 500 feet thick; at Centralia, 570 feet; at Nowata, 600 feet; at Chelsea, 600 feet; on the Verdigris River west of Chelsea, 700 feet; at Claremore, 900 feet; at Catoosa, 920 feet; and southeast of Tulsa, 700 to 850 feet. Two miles northwest of Coffeyville, Kans., the thickness of the Cherokee is but 440 feet—less than it is at the outcrop of the formation farther east.   This fact, however, is not a fatal objection to the idea of a progressive overlap, for the landward edge of a marine sedimentary deposit will ordinarily be thicker than the seaward edge, and there may thus be a continual transgression without greatly increasing the vertical section at the initial point.   In fact, it might easily happen, owing to more favorable conditions of erosion during its deposition, that the overlapping bed would be thicker at its outcrop than the combined overlying and overlapped beds at the initial point. This involves the same principle as that according to which a section at right angles to the shore line, measured across the outcropping edges of a series of deposits, sums up the maxima of thickness and thus gives an exaggerated idea of the total thickness of the series.

In corroboration of this suggestion, and quite independent of any consideration of the land and water relations of Oklahoma during early Pennsylvanian time, Haworth[a] points out that " the heavy gas-bearing sandstones of Laharpe and Iola " and " the sandstone and coal at the very base of the Cherokee and Cherryvale " imply shore conditions at those points just as strongly as do the sandstone and coal deposits at the margin of the Pennsylvanian, farther east.

COAL.

A number of beds of coal occur in the Cherokee formation and to the south in the sandstones and shales that have been correlated with the Cherokee.   These beds are for the most part thin and of no great linear extent, so far as may be judged by the openings and exposures. However, two or possibly three of these beds have been worked to supply a market more than purely local.   These, with other beds higher in the geologic column, will be briefly described.

Bluejacket coal.—A bed of coal ranging from 12 to 16 inches in thickness has been stripped in a small way in a number of places east and southeast of Bluejacket, and is penetrated by wells in that town. A higher bed occurring just beneath the sandstone cap of the hill, 3 miles southwest of town, has been considerably mined·   The area of the flat top of the hill is about 2 square miles, but the coal has been worked only along the west escarpment.   At·the Williams & Esry

---

[a] Univ. Geol. Survey Kansas, vol. 3, 1898, p. 30.

coal bank, in the NE. ¼ sec. 34, T. 27 N., R. 20 E., the coal averages 36 inches in thickness and is underlain by 2 to 6 inches of blue clay shale which rests upon 25 or 30 feet of buff sandstone. Above the coal is an 8 to 10 foot bed of irregularly bedded and curly sandstone with ironstone lenses and thin streaks of coal. This sandstone ranges from buff to red in color and is coarse grained in places and coarsely micaceous.

The preceding description would answer as well for the other coal banks on the west bluff of the hill. The coal is of good quality, and a blocky variety with rusty joint planes, locally known as " red coal," is in especial demand.

A coal bed overlain by a similar coarse-grained red sandstone has been stripped in several places on the ridge near the railroad 2 miles north of Welch station. It is very probably the Bluejacket coal. The sandstone associated with the coal does not always form a ridge, a fact which makes it difficult to trace.

In the valley of Cabin Creek, 2 miles west of Bluejacket Hill, the only coal reported is a bed 6 inches thick which can hardly represent the Bluejacket coal. Three miles farther west, near the junction of the middle and west forks of Cabin Creek, a bed of coal 18 to 24 inches thick has been opened in a number of places and stripped to a considerable amount. It is more likely that this is the continuation of the Bluejacket bed.

About 6 miles southwest of Bluejacket Hill, in sec. 13, T. 26 N., R. 19 E., coal 32 to 36 inches in thickness is reported. This is on the line of strike of the Bluejacket coal and is probably that bed.

Coarse dark-red sandstone forms a north-south escarpment 4 miles east of Chelsea, crossing the railroad at Catale and bearing northeastward along the railroad for several miles. A mile east of Catale two shafts are sunk to a bed of coal which ranges in thickness from 32 to 36 inches. This coal is about 150 feet below the sandstone, which in appearance much resembles the sandstone over the Bluejacket coal. An outlier of similar sandstone, with coal croppings reported beneath it, lies midway between Catale and Big Cabin.

In sec. 31, a mile south of Chelsea, and in the SW. ¼ sec. 1, 3 miles south of Chelsea, there are a number of old pits where a bed of coal about 16 inches thick was stripped to a considerable extent. The relation of the coal to the coarse red sandstone was not made out, but they are evidently not separated by a very great vertical interval. It is rather probable that the coal comes just under the sandstone, in that case corresponding more closely to the Bluejacket coal than the coal mined at Catale. The coarse red sandstone outcrop was followed southward to a point about 12 miles from Chelsea, but no exposures of coal were reported.

Coal near Boynton.—Coal from 12 to 16 inches thick outcrops on Pecan Creek 8 miles east of Boynton. Another bed from 14 to 22 inches thick has been opened in a number of places on the tributaries of Cloud Creek both south and east of Boynton. A bed of similar thickness has been stripped and mined considerably on Dog Creek, near Wellington, 1 and 2 miles northwest of Boynton. On practically all the tributaries of Cane Creek in the territory northwest of Boynton there are coal outcrops from 16 to 26 inches in thickness. In sec. 19, T. 14 N., R. 15 E., at the Renty bank, the coal is reported to be 4 feet thick. Considerable coal has been taken out at this opening.

Fort Scott coal.—In the Kansas reports the term Fort Scott is applied to the bed of coal in the Cherokee formation a short distance below the Fort Scott limestone. It is here taken to include as well the coal which is found interstratified with the latter formation. From the vicinity of Centralia northeastward to the Kansas line, an 18 to 22 inch bed of coal has been stripped at short intervals. On the point of the bluff east of Centralia the coal underlies a bed of limestone with numerous *Fusulina*, being separated from it by 5 feet of shale. There is another bed of limestone a short distance below the coal. This lower limestone persists about halfway to the Kansas line, to a point in the vicinity of Kinnison, where it disappears. Beyond that point the coal occurs below the limestone series, in a position analogous to that of the Fort Scott coal in Kansas. From Centralia to Catoosa, as pointed out in the section on stratigraphy, there are two beds of coal associated with the Fort Scott. They range from 1 to 2 feet in thickness and have been stripped in numerous places.

In surface wells in Catoosa coal is penetrated beneath the upper limestone of the Fort Scott formation. Three miles southeast of Catoosa, mainly in the NE. ¼ sec. 33, T. 20 N., R. 15 E., there are several large strip pits connected with the railroad at Catoosa by a tramway. The coal, which ranges from 18 to 24 inches in thickness, is of good quality and has been extensively mined. This bed is distinct from the coal mentioned as underlying the town of Catoosa, and occurs beneath a lower limestone.

At Evans, 4 miles east of Broken Arrow, there are extensive strip pits reached by a 3-mile spur from the Missouri, Kansas and Texas Railway. The coal averages from 24 to 28 inches in thickness, with a maximum of 40 inches. It is overlain by 15 to 18 feet of blue shale, which is capped by a thin limestone weathering yellow. A 3 to 5 foot bed of shelly reddish limestone outcrops in the draw near the store, and is apparently several feet beneath the coal, with fire clay in the interval.

Southwest of Evans the outcrop of the coal is traceable by scattered exposures as far as Arkansas River. No coal is known to occur beneath the limestone at Wealaka. It is present, however, in the bluffs of Concharty Mountain, and near the township corner southeast of Wealaka, where it has been opened, it has a thickness of 18 inches. From this point to Baldhill the outcrop was not followed.

Henryetta coal.—Several outcrops of coal near Baldhill, having a thickness of 16 to 18 inches, presumably represent the coal below the Fort Scott, and they almost certainly are equivalent to the Henryetta coal, as their outcrops occur within a mile or so of that bed, at nearly the same elevation, in rocks practically flat. The following notes are quoted from Taff's description of this coal at the type locality:[a]

The bed is 3 feet thick and mines in block, separating into two or three benches along stratification lines of distinct cleavage. In the southernmost strip pits east of the town a shale parting was discernible near the middle of the coal. Three and one-half miles southeast of Henryetta an outcrop of the coal shows that the shale parting has increased to 10 inches, separating the bed into two benches, the upper 12 to 15 inches, and the lower 15 to 20 inches in thickness. At the mines 2 miles north of Henryetta, 2 miles southeast of Schulter, and east of Okmulgee, the Henryetta coal maintains its thickness of 3 feet to 3 feet 4 inches and shows no appreciable changes in character except the presence of a thin shale in the openings near Okmulgee. The outcrop has been traced between these localities and to a point 6 miles east of Okmulgee.

Dawson coal.—When the upper limestone of the Parsons is traced southward from Nowata, it is found to thin out and disappear in the vicinity of Watova, but a thin bed of coal makes its appearance 100 feet or so below the horizon of the limestone and is traceable far to the southwest. At Sunday's coal bank, 3 miles northwest of Oologah, there was a great deal of coal taken out formerly, though the bank is now practically· abandoned. The thickness is reported to range from 26 to 30 inches. Three miles south of west of Oologah, near the southwest corner of sec. 30, T. 23 N., R. 15 E., there are several pits where considerable coal has been raised from a bed 30 inches thick. This bed outcrops at short intervals between that locality and Collinsville. Through this distance and at Collinsville it is from 18 to 20 inches thick. In the vicinity of the latter place it has been stripped along the valley of Middle Branch in a practically continuous line from Horsepen Creek to a point 4 miles farther south. The next exposure reported is 3 miles due west of Owasso, where two small strip pits on the west side of a small stream have opened a' coal bed 24 inches thick. The next exposures are south of the wide valley of Bird Creek. From Flat Rock Creek through Mohawk to Dawson, a distance of over 3 miles, the outcrop is very closely stripped and several shafts from 32 to 64 feet in depth reach the coal at points where it is too deep for stripping. The average thickness in this region is about 26 inches. At Scales the coal has been and is exten-

---

[a] Taff, J. A., Bull. U. S. Geol. Survey No. 260, 1905, p. 396.

sively mined in strip pits and by slopes. The thickness here averages 32 inches. The coal has been somewhat prospected in places for 2 or 3 miles southwest of Scales, beyond which it passes beneath the alluvial deposits of the Arkansas River valley.

The following notes on the occurrence of this bed southwest of Tulsa are quoted from Taff: [a]

Outcropping in association with the coal on the west and occurring nearly 90 feet above it is a thin bed of light-blue limestone which weathers to a bright yellow color. The outcrop of this limestone has been mapped from Tulsa southwestward nearly 50 miles and beyond the known occurrence of the coal. This peculiar limestone is easily recognized and its outcrop is a ready reference in locating the coal. The Dawson coal is here unusually clean for a bituminous coal. It mines in block, separating near the middle along a distinct bedding plane, and resembles very closely the Henryetta coal in physical characteristics. A very thin parting of bony coal occurs near the center, separating the bed into two benches. The shale gradually grows thicker southward, reaching about 4 inches near Mounds. It continues to increase southward beyond Mounds at the expense of the coal until the latter has decreased to 8 inches northwest of Beggs. South of Mounds the Dawson coal is not known to be of any commercial value. * * * At points 6 and 9 miles south of Red Fork the coal is 2 feet 6 inches to 3 feet 4 inches thick, and 3 feet 4 inches, respectively, where strippings have been made. Two miles northeast of Mounds the Dawson coal is 2 feet 2 inches to 2 feet 6 inches thick and contains a parting of shale near the middle.

PORTLAND CEMENT MATERIALS.

OCCURRENCE AND CHARACTER.

The formations which make up the Pennsylvanian column in the northeastern part of Oklahoma comprise an alternating series of limestones and sandstones with intervening shales, the whole dipping gently to the north of west. The limestones and sandstones, being more resistant to the effects of erosion than the shale, tend to form long gentle westward dip slopes, with steeper eastern slopes of shale, the harder rock outcropping along the crest of the ridge. Where this rock is limestone which has been protected from weathering by a thin covering of shale, the essential materials are in the most favorable position for cement-plant sites. Where, in addition, adequate supplies of fuel, especially natural gas, and good transportation facilities are available, the climax of advantageous position has been reached. It is to just such a fortunate combination of advantages that the rapid increase in the number of cement plants in the adjoining parts of Kansas is due. That a like development will overtake the eastern part of Oklahoma is not to be doubted.

In the semicircular area of the outcrop of the Mississippian rocks contiguous deposits of suitable shale and limestone will not often be

---

[a] Op. cit., pp. 396–397.

found, except along the border of the Pennsylvanian, where the pure limestones of the Chester are in association with the shales of the Pennsylvanian. Away from the border the limestone of the Mississippian is ordinarily interstratified with too much flint to be avail-. able. Near the base of the Mississippian, however, overlying the Chattanooga shale, there is a bed of encrinital, coarsely crystalline limestone, the St. Joe member, which is in general quite free from chert. No doubt the limestone and shale could be combined in suitable proportion to make Portland cement. Owing to the fact that the limestone and shale are exposed only where the body of the Mississippian rocks is cut through, they usually outcrop near the valley bottoms, as a rule in places unfavorable for economical quarrying. Furthermore, the localities of outcrop in the main have poor transportation facilities and are rather distant from fuel supplies. It is not likely, therefore, that cement plants will be located in this area under the present conditions; that is to say, without a betterment of transportation facilities and the discovery of nearer supplies of fuel. ·

Along the contact of the Mississippian and the Pennsylvanian there are many suitable locations for cement plants. The limestones of the Chester are generally heavy bedded, of good quality, and from 20 to 40 feet in thickness. The basal formations of the Pennsylvanian—the Cherokee at the north and the Winslow at the south—consist largely of shale generally suitable for Portland cement manufacture. From Wagoner northeastward the contact of the two series is in general not more than 5 or 6 miles from a railroad, and at many points closely coincides with one. The remaining factor is fuel. Good supplies of natural gas can be found within 20 or 25 miles of any point along the contact between Wagoner and Vinita. Northeast of Vinita the distance to the gas fields is greater.

In the area between the contact which has just been described and the outcrop of the Fort Scott limestone—that is, in the area underlain by the Cherokee formation and its southern equivalents—there are few if any beds of limestone of workable thickness, and consequently a lack of available sites for cement manufacture.

The map (Pl. II) shows that there are in the area lying west of the outcrop of the Fort Scott limestone a series of limestones with outcrops parallel to that of the Fort Scott and extending for varying distances to the southwest. These limestones themselves consist in places of different members separated by beds of shale. In addition there are various limestone lenses whose outcrops are not shown on the map.

The outcrop of the Fort Scott limestone in Kansas is marked by many cement manufactories. The formation is made up of several limestone and shale members, and some of these limestone beds outcrop along the streams to the west of the line marked on the map, which shows only the east edge of the formation. Below the lower limestone member is a bed of coal, which, as already noted, has been mined at short intervals throughout the length of the outcrop from the Kansas line to Arkansas River, and which, if not the exact equivalent of the Henryetta coal, is not far from it. The area of outcrop of the Fort Scott formation coincides in part with the Alluwe-Coodys Bluff, Sageeyah, and Catoosa fields, and is not far from the Morris field. The formation is, therefore, well situated as to supplies of natural gas. From Catoosa northward to a point a few miles beyond Chelsea the outcrop is but a short distance from the Frisco Railway. North of that point branch railways would be necessary for the commercial manufacture of Portland cement along the outcrop of this limestone.

The Pawnee limestone, a massive bed forming the escarpment on the west bluff of Big Creek and Verdigris River, reaching from the Kansas line southward to and beyond Catoosa, is well situated for cement manufacture from the vicinity of Nowata southward. North of this point the outcrop swings some distance to the east of the railway. The line of outcrop diagonally crosses the Alluwe-Coodys Bluff field and passes near the Sageeyah and other fields, small as at present developed, which occur at intervals from Nowata to Broken Arrow. The new cement plant planned at Nowata is to be located upon this formation.

The upper and lower limestones of the Parsons formation where they cross the Kansas line are of suitable quality and thickness to be used for Portland cement. To the south the upper limestone becomes argillaceous and probably is not available for such use south of Delaware, except from Tulsa toward the southwest. The lower limestone of the Parsons north of Delaware is for the most part too far from railway facilities to be of much value from the cement standpoint at present, but from Delaware southward to a point near Arkansas River it lies convenient to railways and to supplies of natural gas.

Between the upper limestone of the Parsons and the Wilson sandstone there are a number of lentils of limestone that outcrop in several localities between Nowata and the Kansas line. They apparently represent the Dennis or Mound Valley limestone lentils of the Coffeyville formation. The Drum limestone of Kansas thins out and disappears within half a mile south of the State line and was not seen again farther south. The limestone that is to be used at the proposed Tulsa cement plant outcrops southward to and far beyond Mounds and may be correlated with one of those lentils, though the probabilities seem to favor its being the upper limestone of the Parsons.

Limestone outcrops about Dewey and Bartlesville and in the escarpment which stretches southward along the line between the Cherokee and Osage nations—that is, the ninety-sixth meridian—and crosses Arkansas River 10 miles or so above Tulsa. To judge from its position, this limestone represents the Piqua limestone, which, thinning from 45 feet in the northern part of the Independence quadrangle, Kansas, to 2 or 3 feet at the State line, probably thickens again toward the south.

PORTLAND CEMENT PLANTS.

The Dewey Portland Cement Company has recently completed a plant at Dewey with a daily capacity of about 2,500 barrels. This plant has five rotary kilns, each 8 feet in diameter and 100 feet in length. The machinery is driven by electricity throughout, the power being furnished by gas engines using natural gas. The limestone used is probably the Piqua member, as suggested above, and the shale probably belongs to the Buxton formation. There is also considerable shaly material interbedded with the limestone, and when necessary some residual clays are used in addition to give the right constitution to the clay. On account of the variations in the quarry, no set ratio of limestone to shale can be given. The plant is situated in an excellent gas field, where the wells are noted for their large capacity, some of them yielding upward of 50,000,000 cubic feet per day. There is switch connection with both the Santa Fe and the Missouri, Kansas and Texas railways. Analyses of the limestone and shale are given in the table on page 222. The following analysis of the cement is given on the authority of the company's chemist, P. R. Chamberlin:

Analysis of cement made at Dewey plant.

Silica (SiO_2)	23.31
Alumina (Al_2O_3)	6.79
Ferric oxide (Fe_2O_3)	2.90
Lime (CaO)	63.38
Magnesia (MgO)	1.54
Sulphur trioxide (SO_3)	1.68
	99.60

The Tulsa Portland Cement Company proposes to build a plant 4 miles west of Tulsa, on the south side of Arkansas River. The limestone bed, probably the upper limestone of the Parsons, is 35 feet thick and overlies the shale at an elevation of more than 100 feet above the bottom upon which the plant is to be built, enabling the material to be handled by gravity. The capacity will be 2,500 barrels daily, with five rotary kilns, each 8 feet in diameter and 125 feet in length.

All machinery will be driven by electricity generated by gas engines. Ample supplies of gas are available in the vicinity. The plant is situated on a connecting line of the Frisco system and has ample railroad facilities. Analyses of the limestone and shale are given in the table on page 222, and show the proper ratio of limestone to shale to be 100 to 24. The preliminary tests with this ratio, made by the department of experimental engineering of Cornell University, gave a cement of this composition:

Analysis of test cement made from materials to be used in Tulsa Portland Cement Company's plant.

Silica (SiO_2)	21.22
Alumina (Al_2O_3)	11.99
Lime (CaO)	62.57
Magnesia (MgO)	1.33
Sulphur trioxide (SO_3)	1.19
	98.30

Another Portland cement plant is projected at the quarries of the Tulsa Limestone Ballast Company, $1\frac{1}{2}$ miles west of the locality just described.

The British-American Portland Cement Company is promoting a 2,500-barrel plant near Nowata. The site chosen is about 3 miles east of the town, in the valley of Verdigris River. The Pawnee limestone outcrops in the bluff, a section of which was measured along the road at the south side of sec. 22, T. 26 N., R. 16 E., about a mile north of the proposed site, as follows:

Section of west bluff of Verdigris River east of Nowata.

	Feet.
Shelly blue limestone, weathering yellow, chert scattered over surface	2–10
Bluish clay shale	35
Heavy-bedded, fine-grained gray limestone; no chert	7
Black shale	5
Blue fine-grained limestone; some interbedded shale in lower part; no chert	14
Black shale to river bottom	11

The lower two limestone beds belong to the Pawnee formation. To the southeast the strata rise gently, and a massive sandstone, 30 to 40 feet thick, is exposed about 40 feet below the lower limestone. It is proposed to pipe natural gas from the field 14 miles west of this site and after the exhaustion of the gas to use crude petroleum from the Alluwe-Coodys Bluff field, within the limits of which the plant is to be located. The analyses of the limestone and shale are given in the table (p. 222). These indicate a ratio of limestone to shale of

100 to 26. The test cement made with this ratio, as analyzed by J. Robert Moechel, has the following composition:

Analysis of test cement made from materials to be used in plant near Nowata.

Silica (SiO$_2$)	23.58
Alumina (Al$_2$O$_3$)	8.04
Ferric oxide (Fe$_2$O$_3$)	3.12
Lime (CaO)	63.40
Magnesia (MgO)	1.20
Sulphur trioxide (SO$_3$)	.45
Alkalies (Na$_2$O, K$_2$O)	Trace.
Ignition	.19
	99.98

The subjoined table includes analyses of the constituent materials of cements for the plants described in the foregoing paragraphs:

Analyses of Portland cement materials.

	Limestones.					Shales.					
	1.	2.	3.	4.	5.	6.	7.	8.	9.	10.	11.
Silica (SiO$_2$)	4.35	2.06	2.62	2.48	4.3	52.54	61.64	59.40	56.06	63.61	62.5
	1.73	1.52	1.76	.71	1.2	19.10	14.97	20.34	21.02	16.89	20.1
Iron oxide (Fe$_2$O$_3$, FeO)	1.14			.79	1.2	5.30	2.88	6.78	9.86	8.63	6.4
Lime carbonate (CaCO$_3$)	92.02	95.30	94.05	93.43	91.9	9.64	1.29	.28	.32	4.18	1.6
Magnesium carbonate (MgCO$_3$)		1.11	1.29	2.71	0.6			1.69	3.37	2.59	.7
Sulphur trioxide (SO$_3$)				.54	Tr.					.50	Tr.
Alkalies (Na$_2$O, K$_2$O)										1.08	
Loss, etc					.4	12.40	17.00				6.1
	99.24	99.99	99.72	99.66	99.6	98.98	97.78	88.49	90.65	97.06	97.4

1, 6, 7. From quarries of Dewey Portland Cement Company, Dewey. Analyses on authority of P. R. Chamberlin.
2, 3, 8, 9. From proposed site of Tulsa Portland Cement Company, near Tulsa. Analyses by department of experimental engineering of Cornell University, Ithaca, N. Y.
4, 5, 10, 11. From proposed site of British-American Portland Cement Company, near Nowata. Analyses by J. Robert Moechel, Kansas City, Mo.

BUILDING STONE.

The northeastern portion of Oklahoma is well supplied with building stones—limestone, sandstone, and granite. The following notes relate to formations and sites suitable for quarrying on a commercial scale with adequate machinery, and comprise such as came to the notice of the writer during his reconnaissance. Rock suitable for supplying small local demands can be found almost anywhere within a mile or so of the place where it is wanted.

Spavinaw dike rock.—In the vicinity of Spavinaw the Ordovician rocks are intruded by a reddish granitic dike, 75 feet or more in width, which outcrops in three places in an approximately straight line. If the outcrops are connected beneath the soil covering, the total length of the dike is about 1,800 feet. At any rate, there is in sight enough rock to supply several quarries for a long time. The stone has a pleasing warmth of color and would doubtless make a durable building stone. It is well situated for quarrying and needs only railway facilities to become a valuable deposit.

Boone limestone.—At the base of the Boone formation, just above the Chattanooga shale, the upper part of the St. Joe member usually shows from 10 to 15 feet of solid gray crystalline limestone, which, though it has a flaggy appearance on the outcropping edge, is, as a matter of fact, of massive character and breaks off in blocks the full thickness of the ledge. Where favorably situated for quarrying this rock will be found well adapted for all monumental or building purposes. It is called the St. Joe marble in the reports of the Geological Survey of Arkansas.

About 200 feet above the base of the Boone formation, 30 or 40 feet below the Short Creek oolite member, there is usually a ledge of massive limestone from 12 to 25 feet or more in thickness. This bed is free from chert and generally free from stylolitic seams, so that blocks of any desired size can be quarried from it.

The Short Creek oolite member reaches a thickness of 8 to 10 feet in the vicinity of Wyandotte, Okla., and Seneca, Mo., where it has been quarried to some extent. Where cemented firmly enough to be durable, its massive character and light color would give it value for trimmings and ornamental purposes. A perfect oolite, it is texturally a beautiful stone.

An item to be taken into account in considering for quarry purposes any of these limestones of the Boone formation is that they generally outcrop in high, steep slopes upon which the cost of stripping in a quarry would soon become prohibitive.

Chester group, limestone and sandstone.—The several limestones and sandstones of the Chester group for the most part outcrop on the flats and hilltops. At the base of the group there is a series of flaggy, sandy limestones, which, on account of ease in quarrying, have generally supplied the local demand. Near the top of this series lies a thin laminated calcareous sandstone with oolitic spherules along the partings. This stone splits into wide, thin sheets suitable for flagging, and into thicker layers that are good for building purposes. The more massive sandstones and limestones of the upper part of the Chester have also been quarried here and there.

Cherokee and related sandstones.—Massive to heavy-bedded sandstones suitable for quarrying occur at many places in the area occupied by the outcrops of the Cherokee formation and of the related formations to the south. Only the more flaggy of these sandstones have been opened up to supply local demands, for the reason that machinery is necessary to operate a quarry in the massively bedded rocks.

Four miles northwest of Vinita, in the NW. ¼ sec. 5, T. 25 N., R. 20 E., there is a quarry working a bed of massive fine-grained buff sandstone. This stone is largely used in Vinita. The same bed outcrops 2 miles northwest of Bluejacket and 2 miles southwest of Welch.

The ridge 4 miles west-southwest of Pryor Creek is capped by a 45-foot ledge of soft buff fine-grained nonmicaceous even-textured sandstone. The rock is massive and blocks as large as a house break off and roll down the hill. To all appearances it is well adapted to quarrying purposes and would furnish blocks as large as could be handled.

A very similar sandstone is exposed in the south bluff of Arkansas River midway between Wealaka and Haskell, outcroping for some distance just above the level of the river, adjacent to the railroad. It forms a bench and over considerable areas would require very moderate stripping. It is more than 30 feet thick and is a soft, massive, fine-grained sandstone suitable for quarrying by machinery.

Sandstone lentil in the Labette shale.—A soft, massive buff sandstone 15 feet in thickness outcrops in the bluff 1 mile southeast of Coodys Bluff post-office. A similar sandstone 30 to 40 feet thick is exposed in the west bluff of Verdigris River 2 miles west of the post-office. In both places it is suitable for quarrying purposes. Farther north this sandstone occurs in Centralia Mound, and to the south it outcrops in the bluff east of Talala and Oologah.

Pawnee and overlying formations.—The Pawnee limestone occurs massively in some places and is available for rough masonry, but as a rule is not suitable for general building purposes on account of its roughness and irregular texture.

Various other sandstones and limestones lying higher in the Pennsylvanian series have been used locally in the western part of the area. In the west bluff of Fourmile Creek, a mile west of Oologah, a heavy sandstone, showing large blocks on the outcrop, has been quarried more or less. It lies between the Dawson coal and the horizon of the upper limestone of the Parsons. In the vicinity of Seminole station, 6 miles north of Lenapah, a sandstone belonging to the Coffeyville formation outcrops in large blocks and would be suitable for quarry purposes. A mile southwest of Mounds a quarry has been opened in the heavy buff sandstone which forms the long escarpment west of that place.

BRICK CLAYS AND SHALES.

Surface clays for the manufacture of common brick and shales suitable for vitrified brick can be found in any part of the area occupied by the Pennsylvanian rocks. Where such deposits are within a convenient distance of transportation lines and have available a plentiful supply of natural-gas fuel, they are well situated for the extensive manufacture of such brick. It stands to reason that in a new, rapidly growing community with little timber these advantages will lead to the establishment of a thriving industry, such as characterizes the corresponding part of Kansas. Plants for the manufacture of vitrified brick are in operation at Bartlesville, Muskogee, Tulsa, and elsewhere, and numerous brick and tile plants are scattered over the area.

WATER RESOURCES.

GENERAL CONDITIONS.

The area underlain by the Mississippian rocks is abundantly supplied with water. Innumerable fine bold springs gush from the chert bluffs of the Boone formation, and all the larger and many of the smaller valleys contain perennial streams of clear pure water. Near the mouth of Salina Creek an exception to the rule occurs in the brine springs and shallow salt wells which were the site of a primitive salt industry in pioneer days.

In parts of the area underlain by Pennsylvanian rocks, however, the problem of obtaining a good water supply is perplexing. So much of the country is underlain by shale, which sheds the rainfall instead of storing it, that in a season of drouth the shallow wells and small streams go dry. At the same time Caney and Verdigris rivers, together with the smaller streams flowing through the area, are apt to have obnoxious quantities of oil floating on their surfaces. Where sandstone or limestone can be reached, however, a deep surface well will ordinarily furnish sufficient water for domestic purposes.

ARTESIAN WELLS.

It is not claimed that the following notes describe all the artesian wells in the area covered; they relate only to those which came to the notice of the writer.

In the eastern part of the area the flowing wells, with a few exceptions, obtain their supply below the Chattanooga shale. At Needmore two wells, each about 60 feet in depth, flow a gallon per minute of slightly sulphureted water, which is derived from the Boone formation. These flows depend on a slight local dip of the rocks.

At Miami there are three flowing wells—one at the waterworks, one piped to the ice plant from a point half a mile east of town, and the third 2 miles north of town. The third well has a small flow of sulphur water from a bed just below the Chattanooga shale. The well used at the ice plant is 1,680 feet in depth. and the one at the waterworks 1,000 feet. Both draw their supply from various horizons between 600 and 1,000 feet. The water is soft, with a very slight sulphur odor. The temperature is $65\frac{1}{2}°$ F. The pressure at the ice-plant well is about 12 pounds and the flow is more than 100 gallons per minute An analysis of the water is given in the table on page 227. These two wells affect each other decidedly, though half a mile apart.

Two miles east of Fairland there is a well 1,152 feet deep which is reported to have had originally a head of 16 feet and a flow of 30 to 50 gallons per minute. Through defective casing the flow has been reduced to about one-half gallon per minute. The water is slightly sulphureted. In a well 889 feet deep in the town of Fairland the water rises within 30 feet of the surface. The flow comes from sandstone 400 feet or more below the Chattanooga shale. The water is pure and sweet, as shown by the analysis given in the table on page 227.

The town well at Afton has a flow of about 4 gallons per minute of strongly sulphureted water. It comes from a depth of 650 feet. A very small flow was obtained at 600 feet, about 200 feet below the Chattanooga shale. The analysis of this water is given in the table. Two miles northwest of Afton, in the northwest corner of sec. 29, T. 26 N., R. 22 E., on J. L. Courtney's place, there is a well 788 feet in depth which at one time had a small flow, but in which the water now stands from 1 to 20 feet below the surface. The water is similar to that of the Afton well and probably comes from the same horizon. On J. M. Smith's place, in the NW. ¼ sec. 13, T. 26 N., R. 21 E., a well 1,020 feet in depth yields a fine flow of fairly strong sulphur water, which comes from a sandstone 500 feet below the Chattanooga shale. Water was also struck at shallower depths, but did not rise above the surface.

At Welch a well 1,001 feet deep draws water from a sandstone in the Ordovician rocks about 300 feet below the Chattanooga shale. The water is sulphureted and rises 4 feet above the surface of the ground.

At Bluejacket, in a well 730 feet in depth, a flow of 1 gallon per minute was struck at 660 feet and one of 7 gallons per minute at 730 feet. The bottom of the well is about 300 feet below the Chattanooga shale. The water has a fairly strong sulphur taste.

In Vinita and vicinity there are numerous artesian wells. A strong flow of slightly sulphureted water is struck just beneath the Chattanooga shale. The analysis of a typical sample of these waters is given in the table on page 227.

At Chelsea a heavy flow of slightly salty and considerably sulphureted water is struck at 794 feet in a 30-foot bed of sandstone lying just beneath the Chattanooga shale and probably corresponding to the Sylamore sandstone. The water has an artesian pressure of 35 pounds per square inch. It is used in a small bath house.

At Pryor Creek a sulphosaline water struck just below the Chattanooga shale is used at the bath house and sanatorium.

At Claremore the medicinal properties of the strongly saline sulphur water obtained below the Chattanooga shale are utilized in several sanatoriums. There are three deep wells on adjacent corners at the intersection of two streets in the northeast part of the town. The deepest one, 1,500 feet in depth, is owned by the Claremore Radium Wells Company, which operates the adjacent bath house. The water is reported to have an artesian pressure of 60 pounds to the square inch. The Radium Water Company has a well 1,150 feet in depth, and operates a bath house in conjunction with it. Analyses of the water from both these wells are given in the accompanying table. The water from the George Eaton Radium well, 1,115 feet in depth, is piped to a sanatorium and bath house in the business part of town. All these wells obtain water at about the same horizon, from 1,075 to 1,110 feet in depth. In addition to the local use, there is a considerable sale of the bottled water from the wells.

At Nowata the water from a flowing well is utilized by the Nowata Radium Sanitarium Company, which operates a bath house. The water comes from a depth of 1,315 feet, just below a 47-foot bed of black shale,.which is apparently the Chattanooga. The water prob-ably comes from the same horizon as many of the other sulphosaline waters noted above. The analysis of the water is given separately below, for the reason that the hypothetical combination of the constituents made by the chemist is not comparable with the other analyses without recalculation.

In addition to the wells described in the foregoing notes, there are numerous wells drilled for oil and gas which found flows of salt water instead.

The subjoined tables contain analyses of water from several of the artesian wells already mentioned.

Analyses of artesian waters from northeastern Oklahoma.

[Grains per United States gallon.]

	1.	2.	3.	4.	5.	6.
Iron and aluminum carbonate (Fe_2CO_3+ Al_2CO_3)	1.18					
Calcium carbonate ($CaCO_3$)	21.88	19.73	3.10	4.55	5.30	4.69
Magnesium carbonate ($MgCO_3$)	1.94	2.56	1.72	2.52	1.75	1.68
Sodium carbonate (Na_2CO_3)			13.04	1.33	1.58	
Calcium chloride ($CaCl_2$)	236.94	218.46				
Magnesium chloride ($MgCl_2$)	110.47	106.26			2.34	1.16
Sodium chloride ($NaCl$)	1,833.08	1,789.71	36.74	21.28	3.97	9.73
Magnesium sulphate ($MgSO_4$)						.65
Sodium sulphate (Na_2SO_4)			1.49	.98		
Iron and aluminum oxides ($Fe_2O_3+Al_2O_3$)			.07	.04		
Silica (SiO_2)			.60	.77		.55

1. Claremore Radium Wells Company's well, Claremore. Analysis by E. H. Keiser, Washington University, St. Louis, Mo.
2. Brown's Radium well, Claremore. Analysis by E. H. Keiser, Washington University, St. Louis, Mo.
3. Frisco Railway artesian well, Vinita. Analysis by Kennicott Water Softener Company, Chicago, Ill.
4. Deep well, Afton. Analysis by Kennicott Water Softener Company, Chicago, Ill.
5. Deep well, Fairland Analysis by department of chemistry, Oklahoma Agricultural and Mechanical College, Stillwater.
6. Deep well at ice plant, Miami. Analysis by John F. Wixford, St. Louis, Mo.

Analysis of water from Nowata Radium Sanitarium Company's well, Nowata.

[By J. Robert Moechel, Kansas City, Mo.]

	Grains per United States gallon.
Ferrous bicarbonate (Fe(HCO₃)₂)	1. 36
Calcium bicarbonate (Ca(HCO₃)₂)	524. 28
Magnesium bicarbonate (Mg(HCO₃)₂)	62. 29
Sodium bicarbonate (Na₂CO₃)	1, 111. 56
Calcium chloride (CaCl₂)	247. 47
Magnesium chloride (MgCl₂)	182. 37
Sodium chloride (NaCl)	1, 230. 78
Calcium sulphate (CaSO₄)	20. 86
Magnesium sulphate (MgSO₄)	23. 27
Silica (SiO₂)	2. 47
Alumina (Al₂O₃)	2. 78
Ammonium nitrate (NH₄NO₃)	7. 41

SURVEY PUBLICATIONS ON LEAD AND ZINC.

The following list includes the more important publications on lead and zinc published by the United States Geological Survey. In addition to the publications cited below certain of the geologic folios, especially the Joplin district folio (No. 148) and the Lancaster-Mineral Point folio (No. 145), contain discussions of the lead and zinc resources of the districts of which they treat.

ADAMS, G. I. Zinc and lead deposits of northern Arkansas. In Bulletin No. 213, pp. 187–196. 1903.

———— (See also Bain, H. F., Van Hise, C. R., and Adams, G. I.)

ADAMS, G. I., and others. Zinc and lead deposits of northern Arkansas. Professional Paper No. 24. 118 pp. 1904.

BAIN, H. F. Lead and zinc deposits of Illinois. In Bulletin No. 225, pp. 202–207. 1904.

———— Lead and zinc resources of the United States. In Bulletin No. 260, pp. 251–273. 1905.

———— A Nevada zinc deposit. In Bulletin No. 285, pp. 166–169. 1906.

———— Zinc and lead deposits of the upper Mississippi Valley. Bulletin No. 294. 155 pp.

———— (See also Van Hise, C. R., and Bain, H. F.)

BAIN, H. F., VAN HISE, C. R., and ADAMS, G. I. Preliminary report on the lead and zinc deposits of the Ozark region [Mo.-Ark.]. In Twenty-second Ann. Rept., pt. 2, pp. 23–228. 1902.

BOUTWELL, J. M. Lead. In Mineral Resources U. S. for 1906, pp. 439–457. 1907.[a]

———— Zinc. In Mineral Resources U. S. for 1906, pp. 459–489. 1907.[a]

CALKINS, F. C. (See Ransome, F. L., and Calkins, F. C.)

CLERC, F. L. The mining and metallurgy of lead and zinc in the United States. In Mineral Resources U. S. for 1882, pp. 358–386. 1883.

ELLIS, E. E. Zinc and lead mines near Dodgeville, Wis. In Bulletin No. 260, pp. 311–315. 1905.

EMMONS, S. F. Geology and mining industry of Leadville, Colo., with atlas. Monograph XII. 870 pp. 1886.

EMMONS, S. F., and IRVING, J. D. Downtown district of Leadville, Colo. Bulletin No. 320. 72 pp. 1907.

GRANT, U. S. Zinc and lead deposits of southwestern Wisconsin. In Bulletin No. 260, pp. 304–310. 1905.

HOFFMAN, H. O. Recent improvements in desilverizing lead in the United States. In Mineral Resources U. S. for 1883–84, pp. 462–473. 1885.

ILES, M. W. Lead slags. In Mineral Resources U. S. for 1883–84, pp. 440–462. 1885.

---

[a] Earlier volumes of the Mineral Resources of the United States contain discussions relating to the lead and zinc industries of the United States.

IRVING, J. D. (See Emmons, S. F., and Irving, J. D.)

KEITH, A. Recent zinc mining in East Tennessee. In Bulletin No. 225, pp. 208–213. 1904.

RANSOME, F. L. Ore deposits of the Cœur d'Alene district, Idaho. In Bulletin No. 260, pp. 274–303. 1905.

RANSOME, F. L., and CALKINS, F. C. Geology and ore deposits of the Cœur d'Alene district, Idaho. Professional Paper No. 62. In press.

SMITH, G. O. Note on a mineral prospect in Maine. In Bulletin No. 315, pp. 118–119. 1907.

SMITH, W. S. T. Lead and zinc deposits of the Joplin district, Missouri-Kansas. In Bulletin No. 213, pp. 197–204. 1903.

——— (See also Ulrich, E. O., and Smith, W. S. T.)

ULRICH, E. O., and SMITH, W. S. T. Lead, zinc, and fluorspar deposits of western Kentucky. In Bulletin No. 213, pp. 205–213. 1903. Professional Paper No. 36. 218 pp. 1905.

VAN HISE, C. R. Some principles controlling the deposition of ores. The association of lead, zinc, and iron compounds. Trans. Am. Inst. Min. Eng., vol. 30, pp. 102–109, 141–150. 1901.

——— (See also Bain, H. F., Van Hise, C. R., and Adams, G. I.)

VAN HISE, C. R., and BAIN, H. F. Lead and zinc deposits of the Mississippi Valley, U. S. A. Trans. Inst. Min. Eng. [England], vol. 23, pp. 376–434. 1902.

WINSLOW, A. The disseminated lead ores of southeastern Missouri. Bulletin No. 132. 31 pp. 1896.

WOLFF, J. E. Zinc and manganese deposits of Franklin Furnace, N. J. In Bulletin No. 213, pp. 214–217, 1903.

RARE METALS.

SOME MOLYBDENUM DEPOSITS OF MAINE, UTAH, AND CALIFORNIA.

By Frank L. Hess.

MOLYBDENUM DEPOSITS OF MAINE.

INTRODUCTION.

As Maine contains considerable areas of granitic rocks, it is not surprising to find that molybdenum, which is largely associated with granites, is here rather widely distributed. So far as known, it is found in Maine only in the form of the sulphide, molybdenite, accompanied by small amounts of its alteration product, molybdic ocher. The principal deposits now known are in Washington and Hancock counties, and these have been brought into more or less prominence through several companies which have been formed to exploit them.

WASHINGTON COUNTY.

AMERICAN MOLYBDENUM COMPANY'S PROPERTY.

Location and general features.—The American Molybdenum Company, of Boston, for several years worked upon a deposit in the town of Cooper,[a] 22 miles southwest of Calais, and spent a considerable amount of money upon buildings, machinery, and excavation. Granite, carrying molybdenite, was quarried from an open cut, and an excavation about 100 feet square and up to 10 or 12 feet deep was made.

The deposit is situated on the south slope of a gentle hill 500 or 600 feet high, overlooking the stretch of country between Lakes Cathance and Meddybemps. The rocks of the vicinity have been much scored by glaciers in an approximately northwest-southeast direction, and

---

[a] The Cooper deposits were described by George Otis Smith in " A molybdenite deposit in eastern Maine," Bull. U. S. Geol. Survey No. 260, 1905, pp. 197–199.

outcrop at many places through the thin soil. The prevailing rock is a comparatively fine-grained light-gray biotite granite, which weathers to a pinkish color. It is much broken by three prominent sets of joints. Two of these strike about northeast, one dipping 45° or less to the northwest, the other from 45° southeastward to vertical. The third set is older than the other two and runs nearly north and south with an almost vertical dip. Many of the joints of this third set have been filled with thin dikes of rather quartzose pegmatite from one-half inch to 3 inches thick. Ordinarily the most quartzose phases occur where the dike has narrowed to its smallest proportions, and at many such places molybdenite forms crusts in which the flakes lie in radial groups of considerable beauty. Between the molybdenite and the walls of the dike there is a thin layer of quartz and feldspar, the latter on the wall itself, but at some places in crystals so small as to escape casual examination. In width the molybdenite flakes commonly range between one-sixteenth inch and an inch, but may fall short of or exceed these limits. It is said that segregations occur weighing 10 or 12 pounds. In wider parts of the dikes feldspar is a conspicuous constituent, but molybdenite is more sparsely distributed and seems inclined to segregate in the middle of the dike.

Some of the thinner dikes lose their identity in that of the surrounding granite by structurally merging into it, but may reappear a few inches or a foot farther along. Others apparently stop abruptly, ending in vugs filled with crystals of quartz, feldspar, and here and there purple fluorspar, reaching one-fourth inch in thickness. Similar vugs occur in the body of some of the dikes, but they rarely carry much molybdenite in either place. A "pocket" containing a mass of molybdenite ranging from a few flakes to several pounds may occur in places at what seems to be a point just beyond the end of a dike.

The influence of the dikes extends into the granite to varying distances, in some places for 4 or 5 feet from a dike less than 1 inch thick. In such places the granite is miarolitic, that is, full of small vugs, which show well-crystallized feldspar, quartz, and occasionally purple fluorspar, giving the rock a peculiar spotted appearance. Apparently the original constituents have been dissolved by some agency and recrystallization has taken place. A little pyrite and chalcopyrite are apt to occur in both the vugs and the dikes. Here flakes of molybdenite, mostly small, few reaching one-fourth inch in diameter, and very thin, are to be found distributed through the granite. Although in such places the molybdenite flakes seem sporadic, close examination often shows minute seams running from one flake to another or to the dikes.

The joints that carry the pegmatite and the molybdenite are much less plentiful than the others. No molybdenite was seen in the northeast-southwest joints. The molybdenite has been altered to molybdic ocher in a few places, but in very small quantity. This ocher has usually been supposed to be molybdic oxide, but Waldemar T. Schaller has recently shown that it is iron molybdate in all specimens that he has examined. Besides the minerals already described, a single piece of native bismuth the size of a hazelnut was found by the writer. In Australia the occurrence of bismuth with molybdenite is frequently noted, but it is uncommon in this country.

Extent of the deposits.—A shaft said to be 50 feet deep with a drift at the bottom 200 feet long, now filled with water, is reported to have shown molybdenite throughout in as large proportion as in the quarry. A ditch 225 feet long and 4 feet deep, running southward from the quarry, showed molybdenite through its whole length, making a stretch of over 300 feet north and south in which the granite has been proved to carry the mineral. Less is known about the width at this point, but in the quarry it is at least 100 feet.

Method of treatment.—The American Molybdenum Company endeavored to extract the molybdenite by screening. The plant consisted of a 35-horsepower boiler and engine, a Sturtevant jaw crusher and roll, and four sets of special rolls, each 3 feet in diameter and 10 inches wide. The crusher was but a couple of feet above the floor, from which the material, crushed to about one-fourth inch square, was elevated to the Sturtevant roll, 18 inches in diameter by 4 inches wide, which reduced the ore to about one-eighth inch. It was then elevated to a bin at the top of the building, from which it fell to a series of two special rolls, thence was elevated to a third special roll, and run through a 34-mesh screen. The molybdenite caught on the screen was delivered to a box at the end. The material going through the screen was carried by an elevator and screw conveyor to a fourth roll, from which it fell onto a 40-mesh screen and from that to a 60-mesh screen. What went through the 60-mesh screen was elevated and sent to the tailings pile. It is readily seen that the repeated elevations of the material meant a considerable waste of power. The mill ran only six weeks, and is said to have made about a ton of concentrates, a portion of which seen by the writer was very clean. In the tailings some fine flakes of molybdenite were found, but the amount seemed small. Such a process, if mechanically perfected, might work profitably on deposits where, as in this one, the molybdenite flakes are comparatively broad, but would be wholly unsuited to deposits like many of those in Colorado and elsewhere, in which the individual flakes are of almost microscopic size.

CALAIS MOLYBDENUM MINING COMPANY'S PROPERTY.

The property of the American Molybdenum Company is adjoined on the north by that of the Calais Molybdenum Mining Company.

Less work has been done here, but the conditions of jointing and deposition appear to be similar to those on the property just described, of which this is an extension. Several test pits show the same form of molybdenum-bearing dikes and apparently the rock is of about the same richness. The molybdenite evidently still follows the north-south joints.

The width of the molybdenite-bearing granite at this place has not been determined any more definitely than on the property to the south. No molybdenite was seen in the material thrown from a prospect hole about 1,000 feet northeast of the test pits, although it was said that some had been found. None is known to exist on the north side of the road at this point, nor does any appear on the surface between the pits and the road. It would seem, therefore, that the impregnated zone is probably less than 1,000 feet wide, which is about the distance from the road to the prospect hole farthest east. Further prospecting, however, may show this estimate to be in error. The impregnated area on this property is known to extend 300 or 400 feet northward from the American Molybdenum Company's property, and may extend much farther, as molybdenite is reported from points half a mile or more to the northwest.

NEIGHBORING AREAS.

Less is known of areas lying west of those described, but a few prospect holes have shown some molybdenite. Southwest of the property of the American Molybdenum Company and south of Cooper Church, some molybdenite in small flakes and segregations has been found in a number of prospect holes. The granite here also is dotted with small vugs similar to those on the property of the American Molybdenum Company.

HANCOCK COUNTY.

CATHERINES HILL.

The known deposits of molybdenum in Hancock County lie along the north and west sides of Tunk Pond, a sheet of water about 7 or 8 miles long and a mile or more broad. Its northeast end is 8 miles west of Cherryfield, and its southwest end about 2 miles north of Tunk Pond station. At the north side of the lake a partially bare, glaciated knob, 500 or 600 feet high, known as Catherines Hill, has been prospected to some extent. The hill shows several benches with vertical faces, running northeast and southwest, from 12 to 18 feet high, due to the sheeting and jointing of the granite. The granite is light gray in color, of a prominently granular appearance, showing little or no muscovite, small amounts of biotite, and considerable variation in texture. In places the quartz grains, although only one-eighth to one-fourth inch in diameter, are the most noticeable of the minerals, while in other places the feldspars are more prominent and here and there become prophyritic, reaching three-fourths of an inch in length. The dip of the principal joints varies considerably each way from the perpendicular. The sheeting, which is practically horizontal, is also conspicuous. Several crushed zones, 2 to 4 feet wide, were noted running about north and south.

Several companies have sunk prospect holes and done some blasting along the face of a small cliff, so that the deposits have been exposed at a number of places extending over about 300 feet east and west, two-thirds of the way to the top of the hill. Molybdenite is said to be found also at the foot of the hill. It occurs in flakes, said to reach several inches in width but ordinarily much smaller, and in small bunches, generally less than three-fourths of an inch in diameter. It apparently accompanies a series of pegmatite dikes which dip in various directions but for the most part steeply. The larger number of the dikes seen are over 1 or 2 inches thick, though in places they widen to form vugs up to 4 or 5 inches across, which contain druses of quartz crystals up to 3 inches long and 2 inches thick. Some of the vugs are almost filled with quartz and feldspar crystals, with smaller amounts of biotite and molybdenite, but in general, as in the Washington County deposits, the amount of molybdenite in these enlarged portions of the dikes is small. It is, however, apt to be deposited at their ends and in the inner parts of the dikes, from which places it disseminates into the granite, here and there probably to a distance of 3 or 4 feet. As at the Cooper deposits, near the dikes the granite shows at many places a different texture, small crystals of feldspar forming rosettes, one-half inch across, around a crystal of quartz, or around pyrite, as indicated by the rust-lined cavities. It seems hardly possible that these rosettes can be due to the original structure of the granite, and they are probably the result of a readjustment which may have been caused by solutions accompanying the pegmatite. The molybdenite does not seem to occur in this recrystallized granite so much as in the less affected portion near it, which contains molybdenite flakes, thin but up to one-half inch or more wide. A peculiarity of the small pegmatite dikes is that bunches of quartz, feldspar, and mica crystals occur where the dikes apparently end. The dikes' influence, shown by a coarseness of structure, can apparently be traced in the surrounding granite for several inches. If molybdenite is present in the vugs and enlargements of the dikes, it is generally in small quantity. The dissemination in the granite would seem to indicate that the minerals were deposited from a solution, possibly a gaseous solution, as it does not seem possible that a liquid solution could have penetrated the dense granite to the distances observed.

OTHER LOCALITIES.

Halfway toward the southwest end of the lake, one-fourth mile west of Sand Cove, molybdenite has been found at several places on a flat formed of granite of the same type as that in Catherines Hill. But little work has been done on the deposits and they can not be seen to advantage. A couple of blasts were put in during the visit of the writer, but they showed nothing new, although some molybdenite was obtained at each blast. There are two principal occurrences, one of which is on the south side of a ravine cut in the granite to a depth of 20 or 30 feet. The granite is jointed vertically along a north-northwest course, which has probably determined the position of the ravine. Another series of joints cuts this set at right angles. The granite contains more biotite than that of Catherines Hill, and the porphyritic crystals of feldspar show a zonal growth, generally with the outer zone lighter in color than the inner ones. The granite is here cut by aplite dikes up to a foot or more in width. Quartzose pegmatite dikes cut both the granite and the aplite and are accompanied by molybdenite very much as in the deposits already described. One noticeable feature is that in places the dikes become very thin, some of them less than one-sixteenth inch thick, and have little influence upon the surrounding granite or aplite. Along these dikes bunches of molybdenite several inches in diameter are found. No dissemination of the molybdenite could be seen in the granite. A quarter of a mile southwest of this deposit several shots have been put in showing similar occurrences of molybdenite. Between these two places molybdenite has been found on the surface at a number of points. It is also reported to occur on the south side and elsewhere around the lake, but these localities were not visited.

CUMBERLAND COUNTY.[a]

Molybdenite has been found in the granite at a number of places in the town of Brunswick, but the deposits are of mineralogical interest only. A thin vein was exposed for some distance in digging a sewer for the town.

OXFORD COUNTY.

Molybdenite has been reported from several places in Oxford County. Three places were visited—one in the town of Greenwood, 6 miles north of Norway village; another on the Horatio N. Flagg farm in the town of Buckfield, 10 miles east of Paris; and a third at Crocker Mountain, half a mile east of Mount Mica and about 4 miles east of Paris.

At each of these places the rock is a quartz-biotite gneiss, which much resembles a crushed granite, intruded by pegmatite, also some-

---

[a] Communicated by Prof. L. A. Lee, State geologist, Brunswick, Me.

what squeezed. On the Flagg farm both gneiss and pegmatite have been crushed until the pegmatite shows merely as mashed lenses in the gneiss. The gneiss carries more or less graphite at each of these places, reaching a maximum at Crocker Mountain. The graphite is flaky and much resembles molybdenite, but specimens collected at each place and submitted to Waldemar T. Schaller, of the Geological Survey laboratory, showed no trace of molybdenum. Others report having specimens tested which showed its presence, and it would seem entirely possible that molybdenite should occur in these gneisses, considering their possible origin from a granite and later intrusion by pegmatites. Molybdenite is occasionally found in the pegmatite at Mount Mica.

ORIGIN OF THE MOLYBDENITE.

There can be but little doubt that the origin of the molybdenite is connected with the pegmatite dikes, for in the deposits described it has been seen nowhere except in connection with them. The distribution of the molybdenite through the granite adjacent to the dikes and the recrystallization of the granite seem to indicate very hot, possibly gaseous, solutions or some other form more capable of penetrating the rocks than the watery solutions as now understood. The granite is as dense and compact as most granites except where it has been altered in the vicinity of the pegmatite dikes. The molybdenite, quartz, and feldspar seem to have been nearly contemporaneous, the feldspar probably being first deposited and the quartz and molybdenite afterwards, although at times this order may have been reversed.

RICHNESS OF DEPOSITS.

The quantity of molybdenum in the deposits is very difficult of determination. Hand specimens which show a very high percentage of molybdenite may be obtained from the dikes, and specimens which also show molybdenite may be taken from certain portions of the granite. At the same time, with an equal amount of effort, a much larger number of specimens which will probably show no molybdenite may be obtained from the same area. The average assay of a hand-picked specimen is worse than useless, as it gives an apparent value very much above any possible mining values. Owing to the unequal distribution of the molybdenite in the rock it is very hard to pick a fair sample in any manner. After a sample is obtained, it is then very difficult to determine the percentage of molybdenite it contains, as the processes for separating the molybdenite are often questionable in their results. In nearly all mechanical processes loss of molybdenite is inevitable owing to the form of the mineral. It usually breaks into thin flakes, even if occurring in a mass. If a water flotation or an air-blast separation is attempted part of the flakes are usually carried away with the gangue. Sieves lose a part of the finest material and magnetic separation appears to be somewhat uncertain unless it is adapted to the individual ore in each case. In a purely chemical separation the samples analyzed are almost always very small—but a few grams at most—and thus the determination is again incompetent, for the specimen can not be quartered down with even moderate exactness owing to the segregation of the mineral in the rock and the difficulty of pulverizing the flakes of molybdenite. It seems possible that a better method might be worked out along the line of oxidizing the molybdenite by roasting, dissolving in ammonia, and determining the molybdenum from the resulting solution. In such a process large amounts of material could be used, but it would be an expensive operation.

Should such a method as that just outlined be practicable, it would give accurately the amount of molybdenite contained in a given parcel of rock, and might be applicable as a means of extraction to certain deposits where the molybdenite flakes are finer than those in the Maine deposits.

Owing to the mechanical difficulties experienced, there is likely to be a great difference between possibility and practicability of extraction of molybdenite, so that the most desirable test of a deposit is one that will show just what extraction may be expected in practice. For such results an actual mill run, by the method of work contemplated, is probably the only satisfactory mode of testing molybdenite deposits, unless they are rich enough to be worked by hand picking, in which case pockets must be close together and the masses of considerable size.

AN OCCURRENCE OF MOLYBDENITE AT CORONA, CAL.

During 1906 molybdenite was discovered in a granite quarry operated by John Fletcher, about 4½ miles northeast of Corona, Riverside County, Cal. This quarry, which was visited by the writer in the latter part of January, 1907, is located in an embayment in a group of granite hills reaching a height of about 1,500 feet above sea level, and rising about 1,000 feet above the surrounding plain. The molybdenite occurs on the west face of one of the hills, about 50 feet above its base.

The granite is cut by thin pegmatite dikes one-half inch to 2 inches wide. Molybdenite in flakes up to one-half inch across accompanies the dikes in small quantity. In one or two places small flakes may be found in the granite for a distance of 2 or 3 inches from the dikes. The only other metallic mineral found in the dike is iron pyrites in which an assay is said to have shown some copper, gold, and silver.

Molybdenite is also found on both surfaces of joints in the granite, but in such places it does not appear to penetrate the granite more than one-eighth of an inch. Along these joints no alteration of the granite beyond simple weathering is to be seen, nor have any other minerals been deposited in the cracks, unless stains of iron oxide may indicate the former presence of iron pyrite.

Molybdenite has been shown over a narrow strip for a distance of 50 feet, but the total amount is small and the deposit gives no promise of having economic value.

WULFENITE AT ALTA, UTAH.

Alta is one of the older mining camps of Utah, and for many years wulfenite (lead molybdate) has been known to occur there, but it has always been sold for its lead content, as aside from the lead it contained it has had a merely mineralogical interest until recent years. Since the rarer metals have come into more prominent use, however, the minerals of these metals have been sought for more closely and wulfenite has acquired a new importance as a source of molybdenum.

Alta is situated at the head of Little Cottonwood Creek, 16 miles east of Sandy, from which place it may be reached by stage. Its elevation is about 9,000 feet.

The Continental-Alta mine is about a mile east of Alta and 300 or 400 feet higher. It has been worked for a number of years as a silver-lead mine, but has not been a heavy producer. The country rock is a siliceous limestone, dipping 15° to 30° northeastward from a granite mass which forms the walls of the lower portion of Little Cotton-wood Canyon. There is considerable faulting, and the main fault along which most of the ore bodies are said to have been found shows in some places 30 feet of crushed limestone. Here and there the crushed fragments are rounded, apparently from solution rather than from movement, as in such places the limestone of which they are composed has taken a sugary appearance from recrystallization, and is much more friable than that in the less altered fragments. These zones probably lie in the channels of greatest circulation, either present or past.

A dike of fine-grained, almost white rhyolite 2½ feet thick is cut by the main tunnel 300 feet from the mouth. A smaller dike, 2 to 4 inches thick, has been intruded into a fault fissure close by. The effect of the dike on the limestone is apparently confined to very local metamorphism. Most of the ores mined have been oxidized—argentiferous lead carbonates and oxides, carrying small amounts of copper and gold—and with these has been mixed a small proportion of wulfenite.

The wulfenite occurs in the crushed zones, especially where the fragments are most waterworn and decayed, in delicate light-yellow translucent scales standing on edge on the fragments of limestone. In no place were they seen to penetrate the rock. Some of the crystals are covered with clay, showing that the waters are no longer depositing wulfenite in those places. The wulfenite has probably been formed by the oxidation of galena carrying molybdenite. Some powdery black oxide of manganese occurs with the wulfenite.

The ores are transported 4½ miles down the valley by an aerial tram to a concentrating plant, where the company reports having had little trouble in concentrating the wulfenite into a comparatively clean product. However, from the quantity of ore in sight at the time of the writer's visit it would seem impossible to make a profitable production of wulfenite.

The Continental-Alta property is adjoined on the south by the City Rocks, an old mine to which is credited a larger production of silver-lead ore. The ore carries more copper than that in the Continental-Alta, and much more wulfenite. The ore is all oxidized and follows a fault running N. 45° E. magnetic, dipping steeply to the northwest.

About 200 feet from the mouth of the main tunnel is an ore shoot 3 to 4 feet thick and 25 feet long, which has been followed by an upraise for 500 feet. Here also the wulfenite occurs in the crushed limestone upon much decayed fragments. Some powdery black oxide of manganese occurs in the shoot, but copper minerals are apparently absent where wulfenite occurs. Some massicot and plattnerite (yellow and black lead oxides) are found in the vein. The wulfenite occurs in the same fragile, delicate yellow scales as in the Continental-Alta, but in larger quantity. It is peculiarly beautiful in the candlelight, the hundreds of flat faces reflecting the light so that each looks like a tiny flame. It is almost impossible to transport specimens without great injury owing to the ease with which the crystals are detached.

These ore bodies are sufficiently rich to make it seem feasible to concentrate the wulfenite from the other lead minerals. However, one of the largest companies dealing in molybdenum minerals has refused to buy from this camp on account of the impurities the wulfenite contains. Though it was said that vanadium was the principal objectionable ingredient, none was found by the writer in a qualitative test of one sample. No complete analysis is at hand.

Wulfenite is said to occur in small quantity in practically all the oxidized ores of the camp, but the City Rocks mine seems to contain the greatest amount. Molybdenite is known to occur in a canyon on the south side of Little Cottonwood Canyon, but as the ground was covered with snow the locality was not visited. It is reported in small quantity from other points in the vicinity.

THE ARKANSAS ANTIMONY DEPOSITS.

By Frank L. Hess.

INTRODUCTION.

Owing to the high price reached by metallic antimony during 1906, interest was somewhat revived in the comparatively little known and recently neglected Arkansas deposits of the metal, all of which lie in the northern part of Sevier County, near Gilham, a village on the

Fig. 12.—Map showing distribution of igneous rocks in Arkansas.   (After J. Francis Williams, Arkansas Geological Survey.)

Kansas City Southern Railway. (See fig. 12.) Aside from their economic features, these deposits are of much geologic interest, as they lie in unaltered sediments and show no close connection with igneous rocks.

The following papers treating of the antimony deposits and ores of Sevier County have been published:

WILLIAMS, CHARLES P. Note on the occurrence of antimony in Arkansas: Trans. Am. Inst. Min. Eng., vol. 3, 1875, pp. 150–151.

Paper read in 1874; gives analyses of several specimens of stibnite from Sevier County, Ark.

SANTOS, J. R. Analysis of native antimony ocher from Sevier County, Ark.: Chem. News, London, vol. 36, No. 933, 1877, p. 167.

An analysis of a specimen of oxide of antimony giving a formula of $2Sb_2O_4.H_2O$. Possibly a mixture of cervantite and stibiconite.

DUNNINGTON, F. P. The minerals of a deposit of antimony ores in Sevier County, Ark.: Proc. Am. Assoc. Adv. Sci., vol. 26, 1878, pp. 181–185.

Read at the Nashville meeting, August, 1877. Describes the Antimony Bluff and Stewart mines, and gives several analyses of ores. Had not visited the mines.

WAIT, CHARLES E. The antimony deposits of Arkansas: Trans. Am. Inst. Min. Eng., vol. 8, 1880, pp. 42–52.

Gives description of Bob Wolf, Antimony Bluff, and Stewart mines, with results of a large number of assays and analyses. The best paper on the deposits.

COMSTOCK, THEO. B. A preliminary examination of the geology of western-central Arkansas: Ann. Rept. Arkansas Geol. Survey, 1888, vol. 1, pp. 136–137.

Gives general description of the geology of the Stewart, Conboy, Antimony Bluff, Valley, and Otto mines.

JENNEY, WALTER P. The lead and zinc deposits of the Mississippi Valley: Trans. Am. Inst. Min. Eng., vol. 22, 1894, pp. 206–208.

Gives a general description of the veins and ascribes an igneous origin for their derivation. His article is general, but good.

ASHLEY, GEORGE II. Geology of the Paleozoic area of Arkansas south of the novaculite region: Proc. Am. Phil. Soc., vol. 36, No. 155, 1897, pp. 306–308. Reprinted in Contributions to Biology from the Hopkins Seaside Laboratory, No. 12, Stanford Univ., 1897, pp. 306–308.

Briefly describes the antimony deposits, and gives a better idea of the distribution of the veins than the others.

BAIN, H. F. Preliminary report on the lead and zinc deposits of the Ozark region: Twenty-second Ann. Rept. U. S. Geol. Survey, p. 2, 1901, p. 133.

Describes the Bellah mine from the standpoint of zinc production.

In but one of these articles is there attempted an explanation of the origin of the veins or ores, there is but little description of the ores as such, and there is some disagreement as to the facts observed in regard to the structure of the veins and their relation to the country rock. In an endeavor to clear up some of these items and to get some idea of the amount of ore available, the writer visited the locality in November, 1907, but owing to the abandonment of the mines they were mostly filled with water and opportunities for examination were limited.

The history of the mines is not entirely clear. According to local belief, antimony was discovered in the vicinity shortly after the civil war by several hunters who set a number of rocks together to inclose a fire on which to roast venison. Some of these rocks were noticed to be peculiarly heavy, and when the fire became hot, they partially melted and gave off fumes which imparted a disagreeable taste to the meat. Portions of the rock were taken to Little Rock, where they were identified as antimony ore. Charles E. Wait[a] states that the discovery of antimony took place in the winter of 1873-74. However, work was commenced about 1873 and was carried on, whenever high prices for antimony permitted, for thirty years.

GEOLOGY.

STRATIGRAPHY AND STRUCTURE.

The rocks of the region are alternating thinly bedded sandstones and sandy or muddy shales, of Pennsylvanian and Mississippian age.[b] They are of a light-yellowish or drab color where exposed, and dark gray to black where unweathered. The rocks have been thrown into very regular parallel folds running a trifle north of east. The folds are so close that in many places the dip of the rocks approaches perpendicularity, and so regular that the strike of the rocks is sometimes used to tell direction.

The topography of the country is governed largely by the rock structure, and each fold makes its individual hill or valley, modified by the erosion of the streams. The country is rough but the relief is not great, and the hills probably nowhere reach a height of over 200 feet above the valleys. The main streams flow southward across the structural features, but the tributaries in general flow along the synclines until a weak point in the rocks allows them to cut through.

In the process of folding the shales have been compelled to slip upon the stiffer sandstone beds, and this has produced some slickensiding and crushing. The folding is supposed to be of the same age as that of the Appalachians,[c] probably post-Carboniferous.

In many places the sandstones and shales contain veins of quartz ranging in thickness from a small fraction of an inch to several feet. The main veins, wherever seen, apparently follow the trend of the structure, though locally the thinner ones form a stockwork in the fractures of the rocks. No continuous joints or faults were seen cutting the rocks across the strike, and none have been reported by others, so that veins are to be expected only along structure lines and such cracks or joints as would naturally be formed by the folding of the rocks in one direction.

---

[a] Op. cit., p. 42.
[b] Ashley, George II., op. cit., pp. 245-246.
[c] Griswold, L. S., Whetstones and novaculites of Arkansas: Ann. Rept. Arkansas Geol. Survey for 1890, 1891, p. 213.

IGNEOUS ROCKS.

The nearest known exposed igneous rock is the peridotite of Pike County, which has recently attracted much attention as furnishing a possible source of diamonds. This rock is located in secs. 21 and 28, T. 8 S., R. 25 W., between 35 and 40 miles somewhat south of east from Gilham, which is in sec. 10, T. 7 S., R. 31 W., and 25 miles from the Busby mine, the nearest point at which stibnite has been known to be found. It is probably an early Tertiary or late Cretaceous intrusion, as it cuts sandstones of Cretaceous age.[a] The next nearest igneous rock is an augitic dike in Montgomery County, somewhat farther away, and still farther east are the Magnet Cove and Fourche Mountain areas of syenitic and other rocks. These rocks are supposed to be of about the same age as the peridotite of Pike County. In other directions no igneous rocks are known for very much longer distances.

THE ANTIMONY VEINS.

BOB WOLF MINE.

The Bob Wolf property lies 3 miles east and somewhat north of Gilham. It is locally known as Wolfton, but it was first known by the former name and is so designated in the literature. It was first worked for silver under the supposition that the stibnite was galena.[b] The shaft, which is now much caved, is said by Charles E. Wait[c] to have been 60 feet deep.

The dip of the country rock, which is considerably disturbed for about 2 feet on each side, could not be measured along the vein, nor could the exact relationship between the two be determined, but in the road about 200 feet north of the vein, the rocks have a strike of N. 70° E., with a steep southerly dip. In a valley 200 or 300 feet north of the main shaft a smaller quartz vein also strikes N. 70° E.

The rock in which the main vein is located is a soft, sandy, yellow to olive "shelly"[d] shale with thin drab intercalated sandstones. Where less weathered the sandstone is a very dark gray. One-fourth mile west of the main shaft almost black shale of muddy origin is exposed. As seen in the top of the shaft and in a prospect hole a short distance east of it, the vein runs about east and west with an almost vertical dip. It varies in width up to 10 or 12 inches, and may be traced by old excavations along its course for half a mile or

---

[a] Branner, J. C., and Brackett, R. N., The peridotite of Pike County, Ark.: Am. Jour Sci., 3d ser., vol. 38, 1899, p. 55.

[b] Personal communication from William Conboy, of Gilham, Ark.

[c] Op. cit., p. 44.

[d] "Shelly" is a term much used in the oil regions to distinguish a shale which breaks in curved plates somewhat resembling shells.

more, swelling and thinning along both strike and dip. The vein is comby; that is, the quartz has formed crystals which extend from the sides toward the center and show terminal faces. It could not be seen that the solutions from which the vein was deposited had exercised any influence whatever, such as hardening or silicification, on the inclosing rock. Many pieces of shale forming nuclei from which quartz crystals radiate, yet so soft that they crumble and fall out like clay, are included in the vein. The vein is largely barren, and where stibnite occurs, it is found mostly between the combs of quartz. The blades of stibnite lie nearly parallel with the walls and so approximately at right angles to the quartz crystals. Some very small crystals of stibnite, less than one-fourth inch long, show the peculiar erratically bent forms characteristic of the mineral. Fragments of ore lying on the dump show an antimony oxide, probably cervantite (Sb_2O_4). Here and there a small amount of calcite forms thin bands in the veins. The calcite is mixed with siderite, which in places is of a uniform reddish-brown color, from oxidation.

No sphalerite, iron sulphides, or other accessory minerals were found in the pieces of ore picked up around the mine. In an analysis of a specimen from the upper part of the vein, Wait[a] found about 0.5 per cent each of arsenic and bismuth. In ore obtained near the bottom of the shaft neither of these minerals was present, but he found 0.002 per cent of cobalt. Unless the sample taken for analysis was large, so minute a quantity would be indeterminable. Under certain circumstances, in dealing with such small amounts, copper might readily be mistaken for cobalt, and as copper is noted in many veins, whereas this is the only occurrence of cobalt observed, this determination should probably be accepted with caution.

ANTIMONY BLUFF SHAFT.

The old Antimony Bluff shaft was not visited by the present writer, as he was assured that it was so badly caved that nothing could be seen of the vein. This shaft was described by Wait,[a] who states that the vein had a strike of N. 13° E. and a dip of 70° N., cutting across the stratification. What he evidently intended to say was that the main vein had a strike of 13° N. of E., for he makes the strike the same as that of the veins at the Ben Wolf and Stewart mines, and they are near the latter direction.[b] The vein contained stibnite in masses reaching 30 inches in thickness and pieces were taken out weighing 100 to 500 pounds. The ore was remarkably

---

[a] Op. cit., pp. 45–49.
[b] T. B. Comstock (loc. cit.) says that the vein dipped 79° N. 27° W.

pure. Analysis of a specimen obtained at a depth of 50 feet gave the following result:

Analysis of stibnite ore from Antimony Bluff shaft.

Stibnite_____ 99. 711
Chalcopyrite _____ . 055
Bismuthinite_____ . 005
Gangue_____ . 229
Silver _____ None.

100. 000

A minute trace of arsenic was found in a specimen examined later, and some arsenopyrite was found in needle-like crystals in "a siliceous slate" accompanying the vein at one place. There was a small vein a few inches in thickness on the north side of the shaft which contained some lead, zinc, and iron. Wait found silver up to 8 ounces per ton in the ore, but only in those specimens which contained lead.

At 60 feet in depth the ore pinched out, water came in, and the shaft was abandoned.[a]

NEW DISCOVERY SHAFT.

Half a mile or more east of the Antimony Bluff mine, a shaft said to be between 100 and 125 feet deep was sunk in 1902, and a plant consisting of a shaft house, boiler, engines, ore crusher, and blacksmith shop was put up on a vein carrying some sphalerite and galena. This property is known as the New Discovery. When visited the vein was not visible at the surface and the shaft could not be entered, but to judge from a small amount of ore in the bin the deposit is evidently somewhat different from the others seen. The ore is a brecciated sandstone cemented with small veins containing sphalerite, galena, quartz, and siderite. There are also small amounts of pyrites, which may be copper bearing. Only one piece showing stibnite was found, and it was not certainly from this shaft. David Ziedler, who worked in the shaft until it reached a depth of 40 or 50 feet, stated to the writer that he saw no stibnite in the vein down to that depth. The ore bears little resemblance to the ores from the antimony-bearing veins, and the amount of quartz present is very much less.

CONBOY SHAFT AND TUNNEL.

The old Conboy diggings are in the northeast corner of sec. 5, T. 7 S., R. 30 W., about one-half mile east of the New Discovery shaft. An inclined shaft 90 feet deep was sunk along a quartz vein out-

[a] Comstock, T. B., op. cit., p. 142.

cropping on the north side of a hill, about 100 feet above the base. The shaft follows the dip of the vein and is inclined about 45°.

The stibnite in the shaft is said to have averaged 6 inches in thickness for 50 feet from the surface, below which it became thinner until it pinched out altogether. If there was any such amount of stibnite as this in the shaft the ore shoot must have been very narrow, for the sides were not mined. Mr. Conboy stated to the writer that some of the ore was so rich in silver that it was sold for 22 cents per pound. The fragments of ore found near the shaft show a distinct banded structure, with comby quartz one-half inch thick on each side of a central band of stibnite 1 to 2 inches thick, whose crystals lay roughly parallel to the walls. Scattered through the stibnite are small crystals of quartz and some particles of siderite. Outside of the comby vein there are several thin bands of siderite with some impurities, probably stibnite. No galena was seen. A tunnel 310 feet long cut the vein about 12 feet below the bottom of the shaft. At this depth the vein had pinched to almost nothing and no ore was in sight. A drift was run 75 feet eastward along the vein and although there was no ore in the roof of the drift, it is said that in the floor 6 or 7 inches of stibnite was found at the thickest part. The drift is now caved and could not be examined.

The tunnel affords the best place to study the relations of the rocks and veins, as the walls are clean and there is no caving. The rocks strike N. 77° E., with a dip of 45° N. 13° W. The vein follows the bedding and lies between dark sandy shales and a stiffer thin bed of sandstone. Slipping between the beds, due to folding, is noticeable.

OTHER SHAFTS NEAR THE CONBOY.

About one-half mile farther east, in a general way along the strike of the vein, a couple of shafts have been sunk on a vein which may or may not be the same as that at the Conboy slope. Very little stibnite was to be seen in the fragments from the vein. There were small amounts of pyrite, sphalerite, and galena, with the usual quartz and considerable siderite. The vein seems to be not over a few inches wide.

Various other excavations have been made along this vein which show it to be parallel to the vein on which the Stewart diggings are located, the last of them being located 50 or 60 feet north of the Stewart vein.

STEWART MINE.

About the old Stewart mine there is now but little to be seen except a caved trench showing the direction the diggings followed. C. E. Wait [a] states that this antimony deposit was discovered considerably

---

[a] Op. cit., p. 49.

later than the Antimony Bluff and Bob Wolf outcrops, but according to local tradition the Stewart was probably the first discovered.[a]

It is said that at some time between 1870 and 1872 stibnite was hauled from this mine to Little Rock, over 100 miles distant, and to Fulton, on Red River, about 65 miles distant. Thence it was taken to New Orleans by boat, and from that city shipped to England.

The diggings follow a direction about N. 82° E. The dip could not be seen, but is said by Comstock[b] to be 60°–80° N. Wait[c] states that in 1877, at a depth of 32 feet, the vein was about 1 foot thick and carried a streak of stibnite 4 inches thick. He shows the remarkable purity of the stibnite by the following analysis:

Analysis of stibnite from Stewart mine.

Antimony	69.87
Sulphur	27.91
Iron	.02
Zinc	.01
Silica	2.69
Silver	None.
	100.50

He found specimens of zinkenite ($PbS.Sb_2S_3$), jamesonite ($2PbS.Sb_2S_3$), and an oxidation product of the latter, bindheimite ($PbO.Sb_2O_5.H_2O$). One specimen of jamesonite gave 0.2229 per cent of silver; 0.01 per cent or less each of copper, bismuth, and cadmium; and a trace of gold. Some galena occurred in the vein, but sphalerite is not reported, though its presence would be expected. A piece of ore picked up on the dump showed massive stibnite in which the ordinary bladed structure was wholly absent.

The total output of the mine is unknown; 25 long tons were produced in 1877,[d] and between 1886 and 1889 the yield was 50 tons of high-grade ore. Some other small amounts have also been taken out.

MAY MINE.

The May claim lies about half a mile east of the Stewart, on a vein along which a considerable number of excavations have been made and which may or may not be a continuation of the Stewart vein. It is in the NE. ¼ sec. 4, T. 30 W., R. 7 S., about 1 mile south of Antimony post-office, and about 8 miles north of east from Gilham. The excavations show that the vein strikes N. 80° E., with a southerly dip that is almost vertical. The vein follows the bedding of the country rock, which is sandy and muddy shale, a good deal shattered. There has been some horizontal movement along the vein, but no vertical movement could be noted, though the exposures are poor and it

---

[a] Personal communication from William Conboy, of Gilham, Ark.
[b] Op. cit., p. 138.
[c] Op. cit , p. 52.
[d] Wait, C. E., op. cit., p. 52.

can not be stated that there has been none. The shale is thinly bedded and is so dark a short distance below the surface that the dumps from the shafts look much like " slack " piles in the bituminous coal fields.

This claim was worked for antimony as late as 1903, and there is on it a shaft house with hoisting and air-compressing machinery in fair shape. The shaft is located almost at the foot of the south slope of a hill, about 30 feet south of the outcrop of the vein. At the time it was visited the shaft was full of water.

Mr. Paul Knod, of Gilham, one of the owners of the property, gave the following information concerning the mine:

The shaft is vertical, 8 by 9 feet, 125 feet deep, and solidly cribbed for 113 feet below the surface. At a depth of 45 feet a crosscut reached the vein at 27 feet, where only a little disseminated stibnite was carried in the quartz.

At 80 feet in depth a crosscut reached the vein in 22 feet, at which point it carried a sheet of solid stibnite 6 inches thick. A drift was run 33 feet to the east, and a stope 10 feet wide was driven 18 feet high. From 16 to 18 tons of ore was taken out of this drift and stope. At 100 feet in depth the vein was reached by a crosscut 17 feet long. A drift was run 18 feet to the east, where the ore was about 10 inches thick in the face.

The last work was done in 1903. In that year 27 tons of ore, which ran 61.5 per cent antimony and brought $57 per ton, was shipped to New York.

A small amount of bismuth and traces only of arsenic and copper are said to have been found in the ores.

There is now lying upon the dump probably between 20 and 30 tons of ore that needs some concentrating to make it marketable. The ore contains quartz, siderite, calcite, and some pyrite as impurities.

Some years before the present shaft was dug two other shafts, 300 and 600 feet west of it, were sunk simultaneously. The shaft 600 feet west reached a depth of 110 feet, and at one place had 22 inches of stibnite. The third shaft, sunk to 90 feet in depth, had a " feather-edge " of stibnite at the surface and but 5 inches at the thickest.

In 1906 plans were made by the company owning the claims to work the mines again, but before the work could be commenced the price of antimony fell too low to make it profitable.

BUSBY MINE.

The Busby mine is in the northeast corner of Sevier County and is said by Ashley [a] to have shown some stibnite, but there is no record of any production.

OTTO AND VALLEY MINES.

The Otto mine is located near the middle of sec. 20, T. 31. W., R. 7 S., about 3½ miles southwest of Gilham, in the valley of a tributary of Roaring Fork. The country rocks are similar to those described at the other mines. The mine is now caved in and nearly filled with water, so that but little can be seen. However, the vein is fairly well

---

[a] Op. cit., p. 307.

exposed at the top of the shaft, where it is split into two branches, each 10 or 12 inches thick and about 3 feet apart. They evidently join at a depth of 20 or 25 feet. The strike is about N. 72° E., and the dip approximately vertical. The rocks are somewhat disturbed at the surface, probably from weathering, so that the relation of the vein to the rocks is not certain. Half a mile farther west the rocks strike N. 80° E.

Comstock[a] visited the mine in 1887 and states that the ores contained a considerable amount of lead, and one assay gave 1.2 ounces of silver per ton. Pieces of ore picked up on the dump showed much zinc blende in small crystals. In some specimens quartz shows peculiar reentrant angles and is somewhat etched.

Ashley[b] states that at the time he visited the mine (1892) the shaft was sunk to a depth of 230 feet and was still in good ore. According to local tradition, over a thousand tons of antimony ore was produced by this mine.

The Valley shaft was situated a few hundred feet east of the Otto. It did not produce much antimony, and but little is known of its ores.

BELLAH MINE.

The Bellah mine, 6½ miles southwest of Gilham, was not visited by the writer, as it was said to be full of water. It has recently been operated as a zinc mine, and, according to Bain,[c] was worked during the war by the Confederate Government for lead. He states that a well-defined vein 3 to 8 feet wide cuts across the shales with a strike of N. 82° E., and is perpendicular at the surface, dipping 80° N. at 145 feet in depth. It was at that time (1900) developed to a depth of 160 feet. The vein shows both vertically and horizontally striated slickensides, and is made up of comby quartz containing angular fragments of the country rock. Below 115 feet zinc blende is the more important ore, with lesser amounts of galena and chalcopyrite. No mention is made of antimonial ores. Ore seen by the present writer showed a quartz comb with zinc blende spread over the ends of the quartz crystal.

OTHER VEINS.

Besides the veins on which the deposits here described are located, others which have similar trend and apparently similar relations to the country rock and belong to the same age as those in the vicinity of Gilham are reported by Ashley[b] to occur north of Gilham in Tps. 5 and 6. Some of these carry considerable chalcopyrite. Jenney[d] says that this system of veins may be followed northeastward past

---

[a] Op. cit., pp. 143–144.
[b] Op. cit., p. 308.
[c] Op. cit., p. 133.
[d] Op. cit., pp. 206–208.

Little Rock, but this may be too great a generalization. Joseph A. Taff, of the United States Geological Survey, in a personal communication, states that west of this area, in Oklahoma, there are many apparently similar but smaller quartz veins.

SUMMARY AND DEDUCTIONS.

In general the veins described are comby quartz structures following the bedding planes of the shales and sandstone in which they are found. They are slickensided, showing both horizontal and vertical striations. The country rock has not been affected by the solutions from which the quartz has been deposited, even the included fragments in the veins remaining unaltered, but forming nuclei from which the quartz crystallized.

The original minerals found in the veins are quartz, stibnite, jamesonite, zinkenite, galena, sphalerite, pyrite, chalcopyrite, siderite, and calcite. Traces are found of arsenic, bismuth, cadmium, cobalt(?), silver, and, minutely and rarely, gold. Cervantite and bindheimite occur as oxidation products of stibnite and jamesonite, respectively.

The ores have been mostly rather pure oxide and sulphide of antimony, or lead ores, in many places silver bearing, for 40 to 115 feet from the surface, below which sphalerite and other impurities begin to come in. The ores which are easily oxidizable, or those whose oxidation products are readily soluble, have been more or less completely leached from the upper portions of the veins to the depth mentioned, which probably corresponds to the lower limit of variation of the ground-water surface.

The minerals occurring in the veins are deposited upon the faces of the quartz crystals forming the combs, and are therefore younger than most of the quartz, although a certain amount of quartz has been deposited later with the metallic minerals.

There is a central area through which the veins predominantly carry stibnite; elsewhere either the other minerals preponderate or no stibnite is present. This area runs northeastward from the Otto mine to the May—a distance of about 8 miles in a direct line, and is perhaps 2 miles wide.

The ore bodies occur in thin lenticular masses whose longest dimension approaches verticality and may reach more than 100 feet. The width may be from 3 or 4 feet to 20 or even 40 feet; the thickness ranges from a " feather-edge " to $2\frac{1}{2}$ feet.

The list of minerals given above as being found in the veins at once suggests igneous origin—that is, that the solutions from which the veins were deposited had their origin in igneous magmas, or at least picked up their load of minerals from them. The rather wide distribution in Arkansas of small igneous intrusions whose general trend is about the same as that of the veins, and which, as in the vicinity of Hot Springs, even now are either directly or indirectly responsible for flowing hot waters, such as the springs for which the town is named, gives such a theory much plausibility. In the folding of the strata the upper parts of the anticlines are cracked by the tensional strains put upon them. Exactly the same thing must happen to the lowest strata turning under the synclines. Into these cracks the waters forced from below would naturally flow, but the beds in the middle and upper portions of the synclines are compressed and less fractured, so that the solutions would be turned aside to continue upward by way of bedding planes where unevenness and slipping had been caused by dissimilarity of the beds, as of shale and sandstone, and along such cracks they would travel to the surface. If the explanation that the veins were formed by waters connected with the intrusive rocks is accepted for their origin, then their age would be but little less than that of the intrusions, probably early Tertiary or late Cretaceous.

With such a deep-seated origin it is probable that the lenses of ore will alternately make and pinch out to a considerable depth, and in times of high prices for ore, such as prevailed during 1906, some of the mines might be worked at a profit. It can not be stated, however, what the vertical extent of any ore shoot may be, and below the ground-water level varying amounts of impurities consisting of zinc blende, chalcopyrite, and iron pyrite are to be expected.

ANTIMONY IN SOUTHERN UTAH.

By G. B. RICHARDSON.

INTRODUCTION.

The occurrence of antimony in southern Utah has long been known. The deposits have been worked at irregular intervals since 1880 and it is reported that more than $100.000 worth of ore has been shipped from the property on Coyote Creek, in Garfield County. This property was described by Blake[a] and a report on the general geology of the region was made by Dutton.[b]

The rise in the price of antimony in 1905 and 1906 led to renewed activity in developing the neglected deposits in the United States, and after a period of no production covering several years, 295 tons, part of which came from Utah,[c] were mined in this country in 1906. A large concentrating mill has recently been erected in the valley of Coyote Creek, and preparations have been made for active mining. The writer spent a few days at this property in September, 1907, and is indebted to Mr. Thompson Campbell, of the Utah Antimony Company, for many courtesies.

LOCATION.

Coyote Creek is a branch of the East Fork of Sevier River. The stream occupies a short, narrow valley in the northwestern part of Garfield County, in the midst of the high plateaus of Utah. The camp is about 40 miles by road southeast of Marysvale, the terminus of the San Pete and Sevier branch of the Rio Grande Western Railway. The elevation of the mines is about 7,000 feet; the Awapa Plateau to the north and the Aquarius Plateau to the south and east rise 2,000 to 3,000 feet higher.

---

[a] Blake, W. P., Mineral Resources U. S. for 1883 and 1884, U. S. Geol. Survey, 1885, pp. 643–644.

[b] Dutton, C. E., Geology of the High Plateaus of Utah, 1880.

[c] The Mineral Industry during 1906, p. 4.

OUTLINE OF GEOLOGY.

The rocks of this general region form a part of the igneous complex of south-central Utah. They consist of lava sheets, beds of tuff and volcanic conglomerate, and intrusive masses of various types. These igneous rocks cap the highest plateaus and overlie an eroded surface of Eocene strata which outcrop at lower elevations around the igneous uplands. Beneath the Tertiary rocks lie several thousand feet of Mesozoic and Paleozoic sediments, which are exposed in a series of descending benches southward from the High Plateaus to the platform in which Colorado River has cut the Grand Canyon. The plateaus are traversed by a number of normal faults of large displacement which trend in general north and south. One of these faults, together with the action of erosion, has exposed a small area of Eocene sediments in Coyote Creek valley and it is in these beds that the antimony occurs.

The valley of Coyote Creek is occupied by a variable succession of strata. At the base of the section there is 150 feet of gray conglomerate composed of rounded pebbles of quartz and quartzite up to 6 inches in diameter, in a sandy matrix. The conglomerate is overlain by a great mass of fine-textured buff and reddish sandstone, with subordinate drab and red sandy and clayey shale and thin-bedded limestone, amounting in all to several hundred feet in thickness. No fossils have been found in these rocks, but because of their lithologic resemblance to Eocene strata elsewhere in the plateau region they are provisionally referred to that period. These sediments are succeeded by about 1,000 feet of andesitic tuff and lava which cap the surrounding plateaus. The rocks lie approximately flat, though there is a general low northeastward dip. At the mouth of Coyote Canyon a fault causes the strata to dip steeply westward.

OCCURRENCE OF THE ORE.

The ore consists of stibnite and its oxidation products, which occur generally in flat-lying deposits in the sandstone and conglomerate. The chief zone of mineralization is adjacent to the contact of the conglomerate and overlying sandstone, the ore occurring most commonly in fine-textured argillaceous sandstone a few feet above the conglomerate. In many places a bed of clay shale about 5 feet thick immediately underlies the ore-bearing sandstone, and locally the upper part of the conglomerate is mineralized. The ore does not occur persistently and uniformly, though it is present most commonly at this general horizon on both sides of Coyote Creek.

In the early days of development attention was given chiefly to the lenses of ore, the " kidney " deposits. The known lenses have now all been worked out, but it is said that they ranged from several inches to 20 feet in thickness. It is reported that 55 tons of ore were removed from one of these lenses. At present no large bodies of stibnite are in sight but there is a great amount of low-grade ore.

The occurrence and character of the deposits vary in the different prospects. A common occurrence is in layer-like bodies of irregular thickness but averaging only a few inches. The "layers" are not continuous beds and they are only approximately parallel to the bedding. Many of them are undulatory and thicken and thin out irregularly. In a number of places thin bodies of ore were observed cutting across the bedding of the sandstone and connecting the more nearly horizontal deposits. The ore also commonly occurs disseminated in the sandstone, in irregular segregations.

A characteristic feature of the antimony deposits of Coyote Creek is that the ore consists only of stibnite and its oxidation products, gangue minerals being almost completely absent. Only one exception was observed, at the Emily claim, on the south side of the creek, where in a small gash vein but a few inches wide stibnite and pyrite are associated with calcite. A thin section cut along the contact of the stibnite with the country rock shows an uneven junction, the stibnite extending very irregularly into the sandstone. Locally stibnite, with well-defined crystal faces, penetrates and is partly embedded in adjacent quartz grains of the sandstone.

The stibnite occurs in a variety of forms. In the larger ore bodies it is commonly present in aggregates of prismatic crystals arranged radially or in columnar masses. In one group crystals 6 inches long were observed. It is also present in indiscriminately mixed groups of acicular crystals. Adjacent to the outcrop the stibnite is almost invariably oxidized and the steel-gray sulphide gives place to the lighter brown, yellow, and white oxidation products. A number of oxidized specimens were examined by Dr. W. T. Schaller, of the United States Geological Survey, to determine the variety of these products. He reports that they are anhydrous and easily fusible and that they are either valentinite or senarmontite—probably the former, as reported by Blake.[a] In many places the valentinite occurs in acicular crystals as a pseudomorph after stibnite. Associated with the ore and forming efflorescences on the walls of the country rock, the following minerals are locally present: Epsomite; a hydrous magnesium sulphate; a hydrous aluminum sulphate, probably alunogen; a hydrous ferrous sulphate; and gypsum.

Small quantities of arsenic minerals have been found in the valley of Coyote Creek contiguous to the antimony deposits, but, so far as the writer is aware, not immediately associated with them. On the north side of the creek, about 100 feet southeast of the stibnite pros-

[a] Loc. cit.

pect known as " Black Jack No. 2," there is a small deposit of the sulphides of arsenic in shales of Eocene age. Realgar and orpiment in irregular seams ranging in thickness from a fraction of an inch to about 6 inches and only a few inches in length occur in a blue-drab clay shale. No other vein minerals are present and the realgar and orpiment, in small crystals, are intimately associated. Other similar occurrences of small amounts of arsenic are reported in the valley of Coyote Creek.

The occurrence of the ore indicates that it is of epigenitic origin— that is, it was formed subsequently to the deposition of the rocks in which it is found, and its origin is probably connected with the adjacent igneous rocks, as suggested by Blake. The antimony may have been derived from these rocks either during their intrusion through the sediments or less probably after their eruption on the surface, the stibnite being deposited from percolating solutions in part filling existing spaces and in part by metasomatic replacement. The bed of shale which in many places immediately underlies the ore apparently arrested the solutions and determined the local concentration of the stibnite. In such places evidently the solutions were not directly ascending but moved either laterally or from above.

DEVELOPMENT.

The deposits of antimony adjacent to Coyote Creek have been worked sporadically for the last twenty-seven years. For the most part this work has been limited to the exploitation of the large lenses and little or no systematic mining has been done. There has been considerable prospecting, however, and a score or more of tunnels have been driven into the deposits at various places on both sides of the creek.

In the past work has been chiefly directed toward getting high-grade ore running between 50 and 60 per cent of antimony. The " kidney " deposits were exploited, and hand-sorted ore was shipped. One attempt was made to smelt the ore on the property, but all efforts proved that with such methods competition could not be met.

It is difficult to estimate the amount of available antimony, but in the dumps of the old prospects and in the tunnels there is a great amount of low-grade ore in sight. The problem is how to handle the material economically. Toward this end a modern concentrating mill has been erected and it is proposed to make star metal on the property.

CARNOTITE AND ASSOCIATED MINERALS IN WESTERN ROUTT COUNTY, COLO.

By Hoyt S. Gale.

INTRODUCTION.

In a short paper, published about a year ago, the author described an occurrence of carnotite at a locality on Coal Creek, in eastern Rio Blanco County, Colo.[a] During the summer of 1907, while engaged in a further reconnaissance in the northwestern part of the same State, he found opportunity to examine another occurrence of the same mineral in Routt County, in a locality about 60 miles in a direction a little north of due west from the prospects previously described. The Routt County deposits are said to have been known some time previous to the discovery of those at the Coal Creek locality. They are situated at the southern foot of Blue Mountain (known as Yampa Plateau on the early maps of the region), about 18 miles due east from the Colorado-Utah boundary. The prospects visited lie along the summit and flanks of the highest hogback, about 2 miles west of Skull Creek, which is the main east fork of Red Wash. A number of claims have been staked there and some prospecting has been done along a narrow strip of land adjacent to the sandstone hogback in which the ores occur. This strip extends from east to west through the northern tier of 40-acre tracts in sec. 35, T. 4 N., R. 101 W., of the resurvey of that part of Colorado, and the prospects are said to extend beyond these limits along the outcrop of the same group of strata. These prospects are mentioned in a recent report of the Colorado State Bureau of Mines, and analyses of some of the ores are given there.[b]

This group of claims is of interest as furnishing another instance of the occurrence of these rare minerals, and especially as Dr. Hillebrand has discovered the presence of a selenite, presumably of copper, in some of the ores collected. The deposits are also interesting

---

[a] Carnotite in Rio Blanco County, Colo.: Bull. U. S. Geol. Survey No. 315, 1907, pp. 110–117.

[b] Report for 1905–6, Denver, Colo., 1906.

on account of the remarkable simplicity of the structural and stratigraphic relations of the beds containing the ores. The Blue Mountain deposits are essentially similar to those on Coal Creek, and the description of the geology of the former in the present paper will serve by comparison as a convenient means of correcting a misapprehension derived from the tentative conclusions of the former observations. As explained in another paragraph it is now recognized that the deposits at both places occur in rocks of Jurassic age, which are therefore older than the Dakota formation.

STRUCTURE.

The ores occur in the steeply tilted ledges at the southern foot of Blue Mountain. This locality is on the southern flank of a domal flexure or uplift which has been described in the early surveys as the Midland uplift. This name is derived from that of Midland Ridge, the northern border of the same structural feature, now represented by a high and very conspicuous escarpment, partly surrounding and inclosing an interior basin eroded along a portion of the main axis of the uplift.

The principal axis of the Midland uplift lies in an east-west direction. It pitches sharply and terminates near the State line on the west and continues eastward for 30 miles or more. At the upper valley of Wolf Creek the axis bends southward, and pitching also in that direction crosses White River at the mouth of Wolf Creek, where it is lost in the flat-lying strata of the plateau ridges to the south.

STRATIGRAPHY.

The rocks in which the carnotite and associated minerals are found are of Jurassic age. The discovery of determinative fossils, notably in one of the prospect pits from which the ore itself has been obtained, is considered to have definitely established this point.[a] The rock in which the ore occurs is a coarse white sandstone, exceedingly massive and of great thickness. The deposits occur in the upper massive beds of the sandstone group. The chief distinguishing feature of this sandstone is its amazing development of cross-bedding or false-bedding structure. This character continues to the same remarkable extent as far as the formation has been traced. Measurements give the thickness of this formation near the carnotite prospects as approximately 800 feet. It is divided about midway, below the horizon of the carnotite-bearing strata, by a small group of clay beds of a brilliant red color resembling that of the much thicker series of red clay and

---

[a] The fossils were found in the principal development here described. The following species were identified by Dr. T. W. Stanton: *Trigonia quadrangularis* H. and W., *Tancredia* sp. The specimens are deposited in the National Museum.

shales below. The massive sandstone ledges are relatively more resistant to erosion than any of the adjacent formations either above
or below, and they thus commonly form the highest summits or hogbacks, the adjoining beds, where not protected by them, being eroded
to lower valley lands.

This formation was mapped and described as Triassic in age in the
work of the early Hayden Survey, probably through lack of paleontologic evidence. It is clearly the same as that described by Powell
under the name White Cliff sandstone in his "Geology of the
Uinta Mountains." From Powell's descriptions of the continuity of
outcrop of these particular beds even as far as southwestern Colorado
and adjacent parts of Utah, it seems very clear that this formation is
the same as that named La Plata by Cross in his work in that region.
It is thus of the same geologic age as the beds in which the carnotite
and associated vanadiferous minerals are described from various
localities farther south in both Utah and Colorado, and it is also
similar to these beds in lithologic character.

It is now recognized that the Coal Creek deposits are in this same
formation, the stratigraphic relation of the beds being less evident
at that place. The Jurassic sandstones were not differentiated from
the overlying Dakota in the former description.

The beds immediately underlying the white cross-bedded Jurassic
sandstone consist of a considerable thickness of red and brilliantly
colored strata, probably in large part shale, containing limestone and
sandstone layers. This thickness was found by rough measurement
to be approximately 900 feet in the vicinity of the carnotite deposits.
It is this group that forms the major part of the escarpment of the
Midland Ridge, which rises on the northern side of the Midland
Basin and is a most conspicuous and extraordinary feature of the
landscape. This great wall with its banding in vivid red and gray
may be seen from an extensive territory on the south, to and beyond
the White River valley. Below these "Red Beds" is a second group
of massive white sandstones. The "Red Beds" and underlying
sandstone probably range in age from Triassic above to Carboniferous below.

The series of beds immediately overlying the white Jurassic sandstone, which contains the carnotite deposits, is composed of a group
of variegated clays and marls with some limestone layers. A few
layers of a compact, thoroughly silicified conglomerate also occur.
The colored shales are characteristically of shades of clear pink and
green. They occupy many bare wash banks or slopes, especially
where protected by some harder beds above, but they give way so
readily to erosion that in greater part they lie in low valley areas
or badland washes. Marine Jurassic fossils occur in these beds at
several horizons near their base. The strata are very clearly the

group described by Powell as the Flaming Gorge, and probably correspond in age in their upper or fresh-water part to the formation known as Morrison east of the Rocky Mountains.

Above these varicolored shales there is another group of harder beds, consisting of conglomerate, firmly indurated sandstone, and interbedded shales. This group is commonly conceded to represent the Dakota formation, as recognized over an extensive territory in this part of the United States. The Dakota also commonly forms hogbacks. It underlies a great thickness of Upper Cretaceous and Tertiary strata.

NATURE OF THE DEPOSITS.

The best showing of the minerals was found at the summit of the highest hogback in the NW. ¼ NW. ¼ sec. 35, T. 4 N., R. 101 W. The bare rock ledge of which this ridge is composed is somewhat difficult of access and the prospects are not easily found. The ores observed were fragments and blocks thrown out upon the dump, and also the minerals themselves in place at the protected face of the deepest excavated pit.

The ores present a beautiful display of colors. The carnotite appears to constitute a relatively small percentage of the minerals found. It is in the form of a film or thin crust of powdery or amorphous material of bright canary-yellow color. The carnotite is distinguished by the chemical tests for uranium and vanadium. The ore contains also some arsenic in the quinquivalent state. The deposit as a whole occupies a brecciated zone in the rock, the minerals being concentrated in or evidently distributed from the coarser joints or more porous layers. To a less extent some of the minerals are found impregnating the more massive sandstone.

A yellow mineral closely resembling the carnotite in color and appearance was found in greater amount. This proved on testing [a] to be a vanadate of copper, containing no uranium and therefore not carnotite. These two minerals are often difficult to distinguish without chemical tests. The vanadate of copper is found as minute clusters or aggregates of folia or plates of clear crystalline appearence, scattered over the surface or jointing planes of the country rock. These folia or plates are somewhat distinct from the more amorphous carnotite substance, and are also of a darker-greenish hue. The mineral also occurs, however, as a powdery crust or impregnation, when its only distinction from the uranium-bearing mineral is its slightly darker, more greenish cast. Doctor Hillebrand considers this mineral

---

[a] All tests on the specimens described were made by Dr. W. F. Hillebrand in the chemical laboratory of the United States Geological Survey.

the same as some light-olive and yellowish-green varieties collected and described by Boutwell from deposits near Richardson, in southeastern Utah.[a]

The most conspicuous minerals present in the prospects are exhibited as bright-green stains, which penetrate the country rock to a much greater extent than either of the yellow minerals. These stains have the characteristic color of malachite, the common copper carbonate, and some of the green substance was proved on testing to be that mineral. The veins and rock containing malachite also show the blue copper carbonate, azurite, in lesser amount, usually in small rounded knots or balls. These two minerals present the beautiful contrast of green and blue so commonly known. A considerable portion of the green stain, however, especially that part found most intimately associated with the yellow copper vanadate, proved to be practically free from all carbonate material, and as it gave the reaction for sulphates and to a small extent for silicates, it is thought to be composed of those compounds of copper. It is suggested by Doctor Hillebrand that this substance is very probably the basic copper sulphate, brochantite. Without the acid test for carbonates or in a mixture of the two minerals malachite and brochantite, it seems that it would be extremely difficult to detect even under the microscope the difference between them.

A most interesting discovery of the chemical tests has shown the presence of a copper selenite associated with the green copper minerals, both sulphates and carbonates. This is, Doctor Hillebrand observes, the first occurrence noted of a selenite in the United States. It is possible that the supposed selenite may be a selenate, but as selenates have not yet been discovered in nature it seems likely that the former class is represented.

Much of the rock containing the basic copper sulphate and copper vanadate is speckled with small black blotches. These were found to contain manganese in peroxide form and also copper, perhaps as a manganite of copper, though it is not at all improbable that the other elements are present also.

In another prospect pit, about 100 feet southwest of the prospect described above, at the head of a small draw or gash eroded in the face of the bare rock ledges leading up from the south side of the ridge, the same massive sandstone rock was found stained with a green mineral, but without the yellow or blue ores. Tests of specimens of this substance showed that it was wholly unlike either of the other two green minerals described above, containing much chromium but no copper, vanadium, nor uranium. A similar sandstone

---

[a] Boutwell, J. M., Vanadium and uranium in southeastern Utah: Bull. U. S. Geol. Survey No. 260, 1905, p. 205.

is described from the Montrose and San Miguel county localities,[a] where it was thought to be possibly analogous to the vanadium mica roscoelite, as under the microscope it was found to present a chloritic appearance. The color is a bright green, very similar to the green minerals described above, with perhaps a very slightly deeper tinge of blue.

At places near both prospects the rock is stained with the more common iron deposits (oxides and hydrates), which color it rusty brown or ocher-yellow.

The deposits as a group resemble in general character the previously described occurrences of this mineral, both the Coal Creek deposits and those still farther south in Colorado and Utah. The association of silicified wood noted at the Coal Creek locality has not been found, the latter possibly being rather of an accidental nature than bearing any genetic relation to the deposition of the minerals.[b] As stated, the minerals are very evidently mere crusts or coatings, or surficial impregnations in sheared, brecciated, and jointed zones in the rock mass. These zones of brecciation evidently mark the path of the mineralized solutions from which the deposits have been derived.

The extent and practical value of the Blue Mountain deposits is not yet shown to be of much importance. Nowhere had development work been carried more than 10 or 15 feet in from the surface. At none of the prospects seen did there appear to be any great quantity of the minerals exposed by present developments. The occurrence of the minerals is of itself an interesting feature, and there is a possibility of further discoveries.

---

[a] Hillebrand, W. F., and Ransome, F. L., On carnotite and associated vanadiferous minerals in western Colorado : Am. Jour. Sci., 4th ser., vol. 10, 1900, p. 134.

[b] Cf. Boutwell, J. M., op. cit., p. 209.

TUNGSTEN DEPOSITS IN THE SNAKE RANGE, WHITE PINE COUNTY, EASTERN NEVADA.

By F. B. Weeks.

INTRODUCTION.

A brief description of the tungsten deposits in the Snake Range, eastern Nevada, was published by the writer (2)[a] in 1901, and in 1902 F. D. Smith (6) published an account of the occurrence and development of the prospects. In October, 1907, the writer made a more detailed study of the development at this locality and the character and occurrence of the ore deposition.

SITUATION.

In 1900 a mining district was formed under the name Tungsten mining district, embracing several square miles along the western slope of the Snake Range south of Wheeler Peak (locally known as Jeff Davis Peak). This range as an orographic feature begins about 25 miles south of this locality and extends northward from its southern limit about 135 miles between latitude 38° and 40°. It includes the Deep Creek or Ibanpah Range and the group of connecting hills known as "Kern Mountains." This is one of the most extensive and prominent ranges between the Wasatch and the Sierra Nevada. Its highest point, Wheeler Peak, reaches an elevation of 12,000 feet. (See fig. 5, p. 118.) In the area of the tungsten prospects the surface of the mountain slope is dissected by several wide, shallow gulches which are dry except when occupied by melting snow or storm waters. There are several small springs, but at present the water sinks in the gulch gravels.

The region is about 45 miles southeast of the nearest railroad at Ely, Nev. This road—the Nevada Northern—is 140 miles long and connects with the Southern Pacific Railroad at Cobre, Nev. The wagon road to Ely is an excellent mountain road which crosses the

---

[a] Numbers in parenthesis refer to corresponding numbers in " List of recent publications " at end of this paper.

Schell Creek Range (see map, fig. 5) over a comparatively low pass with no very steep grades. Prior to September, 1906, the outlet to the railroad was via Osceola over the Snake Range to Newhouse, Utah, a distance of 100 miles.

GEOLOGY.

The rocks of the region are granites, which may be in part the oldest rocks; Cambrian argillites, quartzites, shales, and limestones, and an intrusive granite porphyry which is younger than any of the sedimentaries. Within the Tungsten mining district the only rocks exposed are the granite porphyry and the quartzites and argillites.

The granite porphyry ranges from fine to coarse in texture and from light to dark gray and red in color. It occupies the lower part of the mountain slope and forms a portion of a considerable mass which extends to the northeast for several miles and is exposed on the eastern side of the range. There seem to be slight indications of deformation within the eruptive mass, and contact metamorphism is developed only to a limited extent. Apparently the intrusion took place since the formation of the mountain range. In general character and mode of occurrence this intrusion of granite porphyry resembles many intrusive masses in other parts of Utah and Nevada. Some of these are known to be post-Carboniferous and they may be of much more recent occurrence.

The base of the sedimentary rocks is not exposed in the Tungsten mining district. Only a small area of purplish argillite is exposed in the northwest corner of the district, overlain by 100 to 200 feet of quartzite. The quartzites are gray, blue, and purple, the gray quartzite forming the larger part of the series. The strata are cut by many quartz veinlets which are probably of secondary origin, formed during the silicification of the original sandstone. The rocks are fine grained and the alteration by silicification is very complete. In thickness the beds range from a few inches to 2 feet. The argillite is a compact purple rock in rather thick layers. In this area it is little altered, but in other parts of this region the process of metamorphism has progressed much farther and the rock has been called " silvery slate."

GEOLOGIC STRUCTURE.

The Snake Range in this region is a quaquaversal dome, having its center near Wheeler Peak. Subsequent to the uplift there was an intrusion of a considerable mass of igneous rocks that tilted the beds to a high angle in some parts of the region and displaced them in others. The steep southerly dips in Wheeler Peak and the high ridges to the south flatten to 25° in the Tungsten district. North of Wheeler Peak the fold has been broken by several northeast-southwest faults of considerable displacement, the beds having a northeast-southwest strike and dipping 45° NW.

In the area shown on the map the metamorphism and deformation which accompanied the intrusion are not so extensive as in other parts of the region.

VEINS.

GENERAL DESCRIPTION.

The veins carrying the tungsten ore are not vertical, but pitch to the northwest or southeast at varying angles, ranging from 55° to 75°, the general direction being northeast and southwest. The actual outcrop is usually limited to a few feet. From the close proximity of some of the veins it might be considered that they are branches from a main vein, but neither outcrops nor underground workings have shown this to be the case. In some places the vein splits into several narrow veins separated by the country rock. Their occurrence is irregular and from the débris it appears probable that there are veins now covered by " slide rock." In width they range from a few inches to 3 feet. The composition of the vein material is essentially quartz and hübnerite, with here and there a little fluorite, pyrite, and scheelite. The quartz is compact and contains no pores, vugs, or honeycombed areas. A few assays have been made which show the presence of gold and silver, but the amount is small and no attempt has been made to recover it. Well-defined walls are of common occurrence, but they are not persistent.

OCCURRENCE OF THE TUNGSTEN ORES.

The hübnerite occurs irregularly through the vein material. In some places there has been a concentration of the ore near the walls. Hübnerite crystals, varying in size and completely surrounding the quartz crystals, and also quartz crystals inclosing the hübnerite, are abundant. The greater part of the ore is disseminated in fine grains through the quartz or in irregular massive bodies. Where the veins pinch to a few inches in width the hübnerite occurs in thin stringers or is interlaminated with the quartz. No wolframite has yet been determined from this region. In 1901 Dr. W. F. Hillebrand made a qualitative test of two or three specimens from the principal vein which showed the ore to be hübnerite. Scheelite has been found very sparsely disseminated in zones which appear to indicate shearing. It occurs in small flakes instead of the usual granular or massive forms.

EXTENT OF MINERALIZATION.

There appears to be a general consensus of opinion among prospectors and others interested in tungsten deposits that these ore-bearing veins do not extend in depth. No workings have thus far been put down which determine this point. It may be true that some, possibly most, of the individual veins do not extend to great depths. In considering the question of depth, however, it should be remembered that in this region the intrusive mass is a part of a magma of unknown depth, which has been forced through a considerable thickness of sedimentary strata. In the area under discussion erosion has removed at least 300 feet from the upper part of the principal vein. In the light of present knowledge of veins of this kind it seems probable that there may be ore-bearing veins within the igneous mass which have not yet been exposed by erosion.

ORIGIN OF THE VEINS.

The magma which intruded the sedimentary strata probably cooled entirely beneath the surface and is now exposed by erosion as a body of granite porphyry. Before complete consolidation the magma was subjected to strains which produced cracks and fissures. These fissures, varying in width and vertical extent, were distributed irregularly through a portion of the rock, but in the main strike in a nearly uniform direction. The latest phase of consolidation consisted in the deposition of the fissure filling by magmatic waters carrying in solution silica and a small amount of certain rare metals.

MINING DEVELOPMENTS.

About 30 claims have been located within the Tungsten mining district, and at present all of them are controlled by the Tungsten Mining and Milling Company.

The principal underground workings are on the Hub claim (No. 1 on map, fig. 13). Tunnel No. 1 (fig. 13) is 225 feet in length, and the face is 125 feet below the surface, which forms the deepest working on any of the veins. At 150 feet from the mouth of the tunnel an upraise has been made to join an incline from the surface. In this tunnel nearly all the various features described under the headings " Veins " and " Occurrence of the tungsten ore " are exhibited. The vein ranges from a few inches to 3 feet in width, strikes N. 68° E., and dips 65° NW. Present developments show that this is the largest and most prominently mineralized vein in the region. Tunnel No. 2 is about 125 feet vertically above No. 1 and is 59 feet in length. This portion of the vein is split into four parts, separated by the granite porphyry. There is about 18 inches of streaky ore in the face of this tunnel. Shaft No. 1 is 37 feet in depth. Near the surface the vein is pinched, but about midway of the shaft it is about 3 feet wide. Shaft No. 2 shows the vein about 30 inches in width, with a small amount of ore. In the face of the tunnel near shaft No. 2 the vein is 24 inches wide, with ore in streaks.

Fig. 13.—Geologic and topographic sketch map of Tungsten mining district, White Pine County, Nev.

On the slope below the outcrop of this vein several tons of ore, which was reported to average about 68 per cent of tungstic acid, were picked up among the "slide rock" and shipped before underground work was begun. Grains of hübnerite are disseminated through the finer material of the slope and the bottoms of the gulches. Considerable ore has also been gathered from time to time and added to the dumps.

The development work on the Tungsten claim (No. 3 on map) consists of two tunnels and a shaft. On the Wolframite and Great Eastern claims (Nos. 4 and 5 on map) are several small trenches exposing narrow veins with ore. On the Eagle claim (No. 7 on map), just below the contact of the granite porphyry and quartzite, the vein is exposed in a trench, standing nearly vertical and striking N. 40° E. Hübnerite with a small amount of scheelite is found here. In the quartzite débris it was found that small veinlets of quartz penetrate the quartzite, a few of them carrying a little hübnerite. It is probable that this ore occurs near the contact zone. The region is said to have been thoroughly prospected and very little material of this kind has been found in the quartzite, which therefore seems unlikely to yield a deposit of commercial importance.

In the Side Issue claim (No. 2 on map), on the south side of Hübnerite Gulch, a mineralized vein is exposed in a 10-foot cut pitching 80° S. and striking N. 45° E. On the lower side of the cut the vein is 2 feet wide and it is said that from this place a piece of solid hübnerite was taken weighing 114 pounds. On the upper side of the cut the vein is split into two 6-inch veins separated by 4 feet of granite porphyry. In the bottom of the gulch these veins have pinched to a thickness of 3 inches each. The country rock is a coarse-grained, light-colored porphyry which, it is said, can be worked more easily than the rock in other parts of the district.

On the Tungstic claim (No. 9 on map) is a 4-foot vein striking N. 65° E. which shows very little ore. About 50 feet above is a 3-foot quartz vein in which no ore was seen.

In the ridge west and a little north of the Hub claim a hübnerite-bearing vein is exposed in several places. Several small veins appear to extend in a direction about N. 60° E.

The Star claim (No. 8 on map) is developed by a tunnel 32 feet long in which the vein ranges from 6 inches to 2 feet in thickness, pitching 55° SE. and striking N. 30° E. In this tunnel scheelite associated with hübnerite occurs in larger quantity than in any other known locality in the district. About 55 feet and 70 feet south of this vein are two hübnerite-bearing veins striking N. 42° E. The country rock is granite porphyry of a more pronounced reddish color than in other parts of the area. A short distance north of the tunnel a 1-foot vein striking N. 42° E. and showing considerable hübnerite is exposed in a shallow trench.

METHODS OF MINING.

The vein material is exceedingly hard and difficult to mine. Drills quickly become dulled and the rock does not shoot well. The work is all done by hand labor and tunneling is said to cost nearly $30 per running foot. At present it would appear advisable to develop the vein by open cuts at different levels with a steel-lined shoot on the surface on each side—one to care for the waste and the other for the ore. A much larger amount of material would be dislodged by each shot than when confined in a tunnel or shaft. There would be no expense for hoisting and there would always be good light for sorting. In handling the material care should be taken to save the fines, as a considerable part of the hübnerite occurs in grains disseminated through the quartz. The scheelite also is likely to be thrown away in the waste on account of its general resemblance to quartz.

On account of the large percentage of waste a considerable amount of hand sorting is necessary. After crushing, the hübnerite is easily separated from the quartz. A hand-made jig, operated by horse-power, was used and afterwards replaced by a 5-horsepower gasoline engine.

SUMMARY.

The occurrence and character of the vein material vary so much within a few feet that the depth and width of the veins and the amount of hübnerite can not be estimated. Nature has, however, done much to assist in determining the other factors which affect the commercial value of these deposits. Several springs of small flow occur at a considerable elevation above the natural location for a concentrating plant and their combined flow would be sufficient for milling purposes. Williams Creek has an estimated flow of 700 cubic feet per minute and would furnish power to generate electricity for a mill and drilling purposes. There is still sufficient timber on the higher mountain slopes to furnish mine timbers. The lower slopes are covered in spots with mountain mahogany, which makes a good domestic fuel. There are ranches in the valley which could furnish general supplies. Railroad facilities are now at a considerable distance, but surveys have been made for a railroad to connect Ely with southwestern Nevada and Salt Lake to the northeast. One of these surveyed lines crosses the Schell Creek Range into Spring Valley opposite Osceola, about 20 miles north of the Tungsten mining district.

RECENT PUBLICATIONS RELATING TO THE OCCURRENCE OF TUNGSTEN ORES IN THE UNITED STATES.

1. Trans. Am. Inst. Min. Eng., vol. 28, 1899, pp. 543–546.
2. Twenty-first Ann. Rept. U. S. Geol. Survey, pt. 6, 1901, pp. 319–320.
3. Mineral Resources U. S. for 1900, U. S. Geol. Survey, 1901, pp. 257–259.
4. Bull. South Dakota Geol. Survey No. 3, 1902.
5. Bull. South Dakota Geol. Survey No. 6, 1902.
6. Eng. and Min. Jour., vol. 73, 1902, pp. 304–305.
7. Mineral Resources U. S. for 1903, U. S. Geol. Survey, 1904, pp. 304–307.
8. Eng. and Min. Jour., vol. 78, 1904, p. 263.
9. Min. Reporter, vol. 50, 1904, p. 217.
10. Econ. Geology, vol. 2, 1907, pp. 453–463.
11. Eng. and Min. Jour., vol. 83, 1907, pp. 951–952.

NOTE ON A TUNGSTEN-BEARING VEIN NEAR RAYMOND, CAL.

By Frank L. Hess.

On the I. X. L. claim, located in the foothills of the Sierra Nevada about 12 miles north of Raymond, Madera County, Cal., a small amount of wolframite has been found in a vein which had been located for a copper property. The claim was visited by the writer in December, 1906.

The country rock is an andalusite mica schist, in which the andalusite is considerably crushed and altered. Numerous large, parallel, nearly vertical quartz veins cut the inclined schists and stand out prominently from the weathered surfaces, forming the summits of some of the hills. There are many copper stains in the rocks, and some ore is being mined from neighboring claims. Alongside the road are the ruins of an old smelter which, to judge from the remnants lying around, evidently ran on the oxidized copper ores many years ago.

No copper staining was seen upon this particular vein, which is from 4 to 16 inches wide where exposed. The vein is composed of a glassy quartz in which bunches of wolframite up to 2 or 3 pounds in weight were found. At a depth of 40 feet the vein pinched out entirely, though it could probably be picked up again by following down along the evident fault coincident with its dip. Only a small amount of wolframite, probably 200 or 300 pounds, is said to have been taken out. Pyrite is an accessory mineral.

MONAZITE DEPOSITS OF THE CAROLINAS.

By Douglas B. Sterrett.

INTRODUCTION.

Monazite has earned a prominent place in the commercial world through the rare-earth metal, thorium, which it carries as an accessory constituent. As a source of cerium and other rare-earth metals also, monazite is of great interest to chemists. In composition it is essentially an anhydrous phosphate of cerium, praseodymium, neodymium, and lanthanum in which thoria and silica are present in variable amounts. The amount of thoria in monazite ranges from less than 1 to 20 per cent or more, but its average amount in monazite obtained for commercial purposes varies between 3 and 9 per cent.

Though sometimes found in large crystals and masses of many pounds' weight, monazite for economic purposes is obtained in the form of sand, occurring in opaque to translucent and in some cases transparent grains and crystals. Monazite ranges in color mainly from light yellow to reddish yellow and brown; some of it is greenish. The freshly broken and unaltered mineral has a resinous to adamantine luster, which is especially marked on the cleavage faces. The mineral is brittle and has a hardness of 5 to 5.5. It can readily be crushed between the teeth and yields a soft grit, quite distinct from the harder minerals sometimes mistaken for it. The specific gravity ranges from 4.9 to 5.3, and is generally over 5.

The principal use made of the thoria extracted from monazite is in the manufacture of incandescent mantles for gas lighting. These mantles are made by immersing sections of a cotton gauze or netting, woven in tubular form, in a saturated solution of the salts of certain rare earths. The composition of this mixture of salts used by different manufacturers is kept secret, but it is said to contain thorium largely in excess of the other constituents. The sections of the tubes are then dried after one end has been drawn in to the form of a mantle by a platinum wire. When dry, the organic matter of the cotton is burned off and the mantle is saturated with some form of wax, which

272

holds it in shape during shipment and is readily burned off when it is set up for use.

The production of monazite in the United States for commercial purposes has, up to the present time, come entirely from North and South Carolina. The occurrence of the mineral and the development of the industry in these States have been described in reports by Henry B. C. Nitze,[a] Joseph Hyde Pratt,[b] L. C. Graton,[c] and the writer.[d]

The value of the production of monazite from the Carolinas is small compared with that of the more important minerals produced in the United States. The benefit to the region in which the monazite is mined, however, has been considerable. During the five years 1902 to 1906 there was produced in the Carolinas about 3,612,692 pounds of crude monazite, valued at $530,866. This includes a small quantity of zircon and tantalum minerals.[e] In 1906 the production was about 846,175 pounds of sand carrying 80 per cent of monazite. The value of this sand was $152,312, corresponding to a price of 18 cents per pound. During 1907 the activity in mining was not so great as in the two previous years, and the price paid for 80 or 90 per cent monazite sand was as low as 10 to 12 cents per pound.

The present paper is intended to furnish general information on monazite, including a description of the deposits in the Carolinas and of the occurrence of the mineral in them, with a discussion of their bearing on its origin. The data used were obtained during brief visits to different parts of the region during the last five years and a more detailed study of the formations in the southeastern part of the Morganton quadrangle, North Carolina, during the field seasons of 1906 and 1907.

Acknowledgment is here made for the courtesy and general information received from the various operators in the monazite field. Among these are Mr. George L. English, of the National Light and Thorium Company; Mr. W. F. Smith and Mr. M. E. Gettys, of the Carolinas Monazite Company; Mr. Hugh Stewart, formerly of the British Monazite Company; and Mr. Herman Wanke, of the German Monazite Company. Further acknowledgment is made to Mr. A. Keith for valuable criticism.

[a] Monazite and monazite deposits in North Carolina: Bull. North Carolina Geol. Survey No. 9, 1895.

[b] Monazite: Mineral Resources U. S. for 1901 to 1905, U. S. Geol. Survey, 1902 to 1906. Also Mining Industry in North Carolina, an annual publication of North Carolina Geol. Survey, 1901, 1903, 1904, and 1905.

[c] Gold and tin deposits of the southern Appalachians: Bull U. S. Geol. Survey No. 293, 1906, pp. 116–118.

[d] Monazite: Mineral Resources U. S. for 1906, U. S. Geol. Survey, 1907.

[e] Sterrett, D. B., op. cit., p. 1208.

GEOGRAPHY.

Geographically, the area in which deposits of monazite of commercial value have been found lies in the central portion of western North Carolina and in the extreme northwestern part of South Carolina.   Fig. 14 shows the area containing monazite deposits of known commercial value.   This area covers about 3,500 square miles and includes part or all of Alexander, Iredell, Caldwell, Catawba, · Burke, McDowell, Gaston, Lincoln, Cleveland, Rutherford, and Polk counties in North Carolina, and Cherokee, Laurens, Spartanburg, Greenville, Pickens, Anderson, and Oconee counties in South Carolina.   The larger towns within or near the monazite region are Statesville, Hickory, and Shelby in North Carolina, and Gaffney, Spartanburg, and Greenville in South Carolina.   The appearance

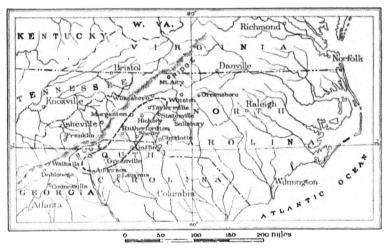

Fig. 14.—Map showing area of monazite deposits of known commercial value in southern Appalachian region.

of Alexander County, N. C., in the list of counties with valuable monazite deposits is the result of prospecting by mining companies during 1907.

PHYSIOGRAPHY.

Physiographically, North and South Carolina are divided into three parts.   These are the Coastal Plain, extending from the Atlantic Ocean northwestward for 100 to 150 miles; the Piedmont Plateau, extending from the limits of the Coastal Plain northwestward for 100 to 130 miles to the foot of the Blue Ridge; and the mountain region, extending northwestward from the Piedmont Plateau to the State lines.   The Coastal Plain and the Piedmont Plateau are prominent in both States, but only North Carolina contains a large portion of the mountain area.

The Coastal Plain is a broad, nearly flat stretch of country rising from sea level on the southeast to an elevation of a few hundred feet on the northwest, in which direction it is practically limited by the boundaries of the rock formations of which it is composed. The Piedmont Plateau is an elevated district rising from a few hundred feet above sea level on the southeast to 1,200 or 1,500 feet on the northwest. It forms a plateau much dissected by valleys from 50 to 200 or 300 feet deep, and its regularity is further disturbed by scattered mountain peaks and smaller hills rising above its general level. The features of the plateau are best observed from a prominent ridge or one of the smaller hills of the region. In the mountain region are included the Blue Ridge and its foothills, and the higher mountains to the northwest. The country in the mountain region is exceedingly rough, and the elevations range from 1,500 to over 6,500 feet.

The region in which valuable deposits of monazite have been found may be defined as a belt from 20 to 30 miles wide and over 150 miles long. (See fig. 14.) This belt lies wholly within the Piedmont Plateau and borders closely on the Blue Ridge, to whose general course it is roughly parallel.

GEOLOGY.

FORMATIONS.

The most important rocks of the monazite belt are gneisses and schists. These include the Carolina gneiss, the Roan gneiss, and gneissoid, porphyritic, and massive granites. Other rocks are pegmatite, peridotite and allied rocks, quartz diorite, and diabase.[a]

The oldest formation in the monazite region is of Archean age and is called the Carolina gneiss. It is the most extensive formation and appears in nearly every section. The composition and structure of the Carolina gneiss are varied. The most common types are mica, garnet, cyanite, and graphite gneisses and schists, or combinations of two or more of these types. These rocks range in color from light gray to dark gray, and in many places where graphite is abundant in them have a light bluish-gray to bluish-black cast. Some of the gneisses and schists are fine grained and are composed of several distinguishing minerals, as biotite, muscovite, cyanite in fine needles, and graphite, besides fine grains of quartz and other minerals; other rocks are composed of the same minerals in coarser grains or flakes. Garnet appears in both fine and coarse grained gneisses and schists and may be fairly large, even in the rocks of finer grain. The presence of much pegmatitic material is a characteristic feature of the Carolina gneiss.

---

[a] The formation names used in the description of these rocks are taken from Keith, A., Geologic Atlas U. S., folios 116 (Asheville), 124 (Mount Mitchell), and 147 (Pisgah), and others.

The Roan gneiss is the next oldest formation in the region and is also of Archean age. It consists of hornblende gneiss and schist, with here and there the less metamorphosed phase, diorite. The hornblende gneiss and schist are nearly black to dark green in color, and are composed chiefly of small interwoven and matted hornblende crystals. These hornblendic rocks grade into diorite, which is also dark colored but contains a noticeable amount of feldspar and has a granitoid texture. Bands of mica gneiss or schist are included in many both large and small masses of Roan gneiss. This formation is prominent along the northwest side of the monazite belt, throughout its length. In the central and more eastern portions, however, it is of less importance and in many places does not appear at all.

The age of many of the granites and granite gneisses has not been determined, though a part are probably Archean. In importance, granite and its different phases are second and are particularly prominent in many localities where extensive monazite deposits have been found. In composition the granite may be biotitic, muscovitic, or hornblendic; its texture may be porphyritic, massive, gneissic, or schistose. Where both porphyritic and schistose the feldspar phenocrysts generally assume an augen form, caused by crushing and elongation in the direction of shearing. Some large masses of granite gneiss have an abundant development of small red garnets. The occurrence of much quartz in veins and veinlets throughout the mass is a characteristic feature of most of the granites of this region. Some of this quartz is simply massive; at other places it has a more or less well-defined crystal form. Drusy surfaces are not uncommon on such crystals. The abundance of quartz veins is not invariably confined to the granite masses, but in numerous places extends some distance from the contact of the granite into adjacent formations.

Pegmatite is a common rock throughout the monazite region, especially in those areas where commercial deposits of monazite are found. Two principal occurrences of pegmatite are here recognized. In one it forms distinct masses or bodies with the typical composition and texture—that is, it is composed of quartz and feldspar, with or without mica and other accessory minerals, crystallized out on a large scale. The other type is a pegmatized gneiss, representing the addition of the pegmatite minerals to the gneiss, with perhaps some recrystallization of portions of the inclosing rocks. In some places secondary quartz is the principal mineral added, while feldspar appears in smaller quantities. In others the feldspar is more prominent, and is prone to assume a porphyritic form in the gneiss, producing a typical augen gneiss. Very commonly the gneisses and schists are banded with or cut at all angles by streaks of pegmatitic or granitic material. The recrystallization of the gneisses and schists, with the development of pegmatitic material or the injection of such material into the rocks, may be called pegmatization. In many places the process has gone so far that it is very difficult to distinguish pegmatized rock from granite gneiss, and especially from flow-banded and porphyritic granite gneiss. This difficulty is due partly to the fact that granite and pegmatite are composed of the same minerals and have no sharp division line between the size of their grains.

The peridotites are dark-green to greenish-black basic rocks, containing one or more of the ferromagnesian minerals olivine, pyroxene, and in places hornblende as chief constituents. So far as known the peridotites of this region are of Archean age and are apparently genetically connected with the Roan gneiss. Though composing but a very small part of the rocks of the monazite belt, the peridotites generally outcrop prominently wherever they occur, and many outcrops are marked by large rounded " nigger-head " bowlders scattered over the surface. For the most part the peridotites have altered to talcose or chloritic soapstone or to serpentine. This alteration is, in some places, only superficial, but in others whole masses have been so metamorphosed. The usual form of occurrence of the peridotites is in lens-shaped bodies parallel, or nearly so, to the schistosity of the inclosing rocks.

Quartz diorite of undetermined age is one of the less important intrusive rocks in the monazite region. It is a fine-textured rock, composed of granular quartz and feldspar with more or less hornblende, locally with garnet distributed promiscuously through it. The occurrence of quartz diorite is generally in small dikes ranging from a few inches up to several feet in thickness. The diminutive size of these dikes, however, is offset by their abundance and resistance to erosion, owing to which they leave much débris over many of their outcrops in the form of hard rounded bowlders.

Diabase is the latest intrusive rock known in the region and is probably of Triassic age. It is a dense, hard rock of dark-green to black color, composed chiefly of olivine and a feldspar rich in lime. It is rather abundant in places and the outcrop is generally marked by characteristic spheroidal " nigger-head " bowlders scattered over the surface. The diabase dikes range from a few inches up to 100 feet or so in thickness.

STRUCTURE.

The regional metamorphism, with accompanying folding and faulting, of the rocks in this area has been extreme. In many places, especially in the Carolina gneiss, it is very difficult to determine the original nature of the formations, for much of the sedimentary structure or igneous texture of the rocks has been destroyed by mashing and recrystallization. The Carolina gneiss has been intruded by rocks of later age and cut by them into irregular-shaped masses, many of which fork out into long tongues or occur as narrow streaks in the intrusives, or vice versa. There have been successive intrusions of igneous rocks of later age into the earlier formations. Thus the Carolina gneiss is cut by the Roan gneiss, and both are cut by granites of later age. Many of the granites have included blocks of the formation in which they have been intruded. In places the inclusion has been more or less absorbed by the surrounding granite, the composition of which has thereby been affected. Thus, where masses of hornblende gneiss are included in granite, the latter is generally highly hornblendic in their vicinity.

The structure of the pegmatite in the rocks of this region is extremely irregular. In some places the pegmatite occurs in the form of sheets or lenses interbedded and folded with the inclosing gneisses and schists. In other places it occurs in dikes, veins, or lenses either conformable with the inclosing rocks through part of its extent and cutting across them in other parts, or in irregular masses having no definite orientation with respect to the accompanying formations. In pegmatized rock masses the pegmatization has generally affected certain beds, which may grade into regular pegmatite in either the direction of their greatest or that of their least extension. In such rocks it is often impossible to determine the line of demarcation between the two. There is also a gradation between the pegmatized beds and ordinary gneiss.

Quartz diorite almost invariably occurs in small dikes, in places conformable with the schistosity of the country rock, though elsewhere cutting across it at all angles. The diabase dikes commonly cut across the strike and dip of all the older formations, filling a series of fissures which have a general northwest to north strike.

WEATHERING AND SOILS.

The rocks of the Piedmont Plateau have undergone such extensive weathering that good outcrops are the exception, and a thick mantle of residual soil covers much of the country. The variety of rock underlying certain soils can in many places be determined, unless decomposition has been too thorough, by studying the outcrops and graduations from such exposures into the residual soil.

The Carolina gneiss, on partial disintegration and decomposition, commonly forms a gravelly soil with a red clayey matrix. This is especially characteristic of the garnetiferous and graphite-cyanite types, which are abundant in parts of the monazite region. The pebbles are composed of small fragments of the original rock, such as tufts of cyanite impregnated with hematite or limonite, iron-stained garnets, or pieces of hematite. On more complete decomposition a fine reddish clayey soil results, with no decided characteristics. Other types of the Carolina gneiss, in which mica is an important constituent, leave a micaceous soil, much of which assumes a purplish color. Granite and its various phases, on partial disintegration and decomposition, yield light sandy soils. On more complete decomposition the granites yield soils of a light to dark reddish color, depending on the quantity of ferromagnesian minerals, as biotite or hornblende, in the original rock. The quartz grains of the granite remain as sand mixed through a clayey matrix. This quartz sand is almost everywhere to be seen at the immediate surface, from which the clays have been washed by rains. Where Carolina gneiss and granite are intimately associated, or where pegmatization has been extensive in a body of Carolina gneiss, there results a sandy soil, characteristic of granite, through which are scattered pebbles of hematite and ferruginous cyanite, characteristic of the Carolina gneiss. The relative importance of pebbles in such soils decreases as the quantity of pegmatite or of granite in the rock formations increases. These features of the soils are especially marked on the broad, flat ridges characterizing much of the Piedmont Plateau region. The roan gneiss leaves a greenish sandy soil on disintegration, and an ocher-yellow to dark reddish-brown or chocolate-colored clayey soil on decomposition. Black stains of manganese are associated with many of the soils derived from hornblendic rocks.

A clew to the nature of the rock formations in a given region is often furnished by the character of the gravels in the bottom lands and streams draining that region. Thus in this area a very light-colored gravel with much quartz débris indicates a granite or its contact or a very highly pegmatized country rock. Garnets and hematite iron ore, with which blocks of mica or cyanite gneiss are associated, indicate Carolina gneiss. Quantities of black sands in the stream gravels, containing magnetite, ilmenite, hornblende, etc., are characteristic of the Roan gneiss.

OCCURRENCE OF MONAZITE.

Up to the present time the only deposits of monazite successfully worked have been the gravel beds in streams and bottom lands, and in certain places surface soils adjoining rich gravel deposits. Prospecting and careful mill tests on monazite-bearing gneiss and schist have failed to discover deposits of a nature that could be worked extensively. The saprolite or rotted rock underlying some gravel deposits has been washed in small areas, with results reported to be favorable.

PLACERS.

Commercial deposits of monazite in gravel occur in the beds of creeks and streams and the bottom lands along them. The thickness of the gravels ranges from a foot or two, including overburden, to 6 or 8 or more feet. The distribution of the monazite in them is, as with all heavy minerals, richer near the bed rock and poorer above, grading into the overburden. In some deposits the whole bed, with the finer alluvium at the surface, is rich enough to be washed directly or sluiced down and washed. The extent and value of these deposits vary with the topography of the country and the nature of the gravels. The best deposits are more commonly associated with light-colored gravels and sands, containing considerable quartz débris and fragments of other light-colored rocks, such as pegmatite, granite, mica, and cyanite gneiss. On the other hand, the absence of much quartz and pegmatitic or granitic débris from the gravels is generally characteristic of low-grade deposits of monazite. The presence of black sands—magnetite, ilmenite, hornblende, etc.—in the gravels does not necessarily indicate a low-grade deposit, unless quartz and pegmatitic minerals are lacking also.

RESIDUAL DEPOSITS.

The surface soils on land adjoining some of the rich monazite deposits have been found to contain sufficient monazite to make sluicing down and concentrating profitable. This is the case to a depth of 3 or 4 inches or more in many residual soils that have suffered but little displacement on the surface, and to depths of several feet where the drift soil has collected on the gentle slopes below a steeper hillside. The partial concentration of monazite in the top layer of soil is caused by the washing away of the clay and other light decomposition products of the rock. The supply of monazite in the stream gravels in favorable areas is often replenished by the wash from the hillside soils during rains. This is especially true where the hills have any considerable slope and the land is cultivated. Under such conditions it is frequently profitable to work the stream gravels two or more times in a year.

The saprolite or rotted rock underlying the richer deposits of monazite is at some places sluiced down to depths of a few inches to a foot or so, along with the overlying gravels. At other places small amounts are removed and washed separately for the monazite they contain. The formations that have been found especially favorable for such work are highly pegmatized gneiss or schist. Such deposits have generally soon been lost or grown poor, probably on account of the fact that the miners have cut through the richer bed or failed to follow it in the direction of its extension. The occurrence of monazite in saprolite will be considered along with the occurrence of monazite in hard rock formations, as the former is merely an altered phase of the latter.

MONAZITE IN ROCK FORMATIONS.

Two separate companies have, at different times, undertaken to work a deposit of monazite-bearing rock about 3 miles northeast of Shelby, N. C. In each case the undertaking failed, because it was impossible to obtain sufficient ore of the high grade necessary to make the operations a success. At a number of the placer deposits ledges of rock have been found, either in the bed of the streams or near by, which contained monazite in noticeable quantity. So far the rock in which the monazite has been found in noticeable amounts is pegmatized gneiss.

It is possible at many of the mines to pan the saprolitic pegmatized gneiss under the monazite-bearing gravels almost at random and obtain monazite. The amount of the mineral obtained when the panning is done with a long-handled shovel ranges from a few grains to a teaspoonful per shovelful, according to the richness of the beds. Mr. George L. English has kindly furnished the results of a test made by him on the monazite content of the saprolite underlying the gravels at the F. K. McClurd mine, near Carpenter Knob, Cleveland County, N. C. From 30 cubic feet of saprolite 424 grams of concentrates, carrying about 40 per cent of monazite, were obtained by washing in a sluice box. This approximates closely one-third of a pound of pure monazite to a cubic yard of saprolite.

The monazite content of the rock at the deposit 3 miles northeast of Shelby, N. C., has been given a thorough test with a well-equipped mill by the British Monazite Company. The following data are given through the courtesy of Mr. Hugh Stewart, by whom the tests were made. Practically all of the rock at the mine, through a vertical height of 15 to 18 feet across the bedding, carried monazite. The quantity in different beds ranged from 0.03 per cent and less up to 1.10 per cent and more. While the mill was in operation all beds carrying 0.4 to 0.5 per cent or more were treated as ore, while lower-grade material was discarded. According to Mr. Stewart, one ore bed with a thickness of about $3\frac{1}{2}$ feet was found to average 1.10 per cent of monazite.

Most of the pegmatized gneiss bodies which are rich in monazite represent phases of the Carolina gneiss in which the original nature of the rock has been largely obliterated as a result of the addition of new minerals and the recrystallization of the original ores into pegmatitic material. The texture developed during this pegmatization is generally porphyritic, in which the feldspar phenocrysts assume somewhat of an augen form. The feldspar phenocrysts range in size from those smaller than a grain of wheat to those the size of a walnut. The porphyritic gneiss may grade into less or more highly pegmatized gneiss, and from the latter into regular pegmatite. This gradation may be between two separate beds or from one part to another of the same bed. In those beds or portions of beds where there has been little pegmatization monazite occurs sparingly. The same is true where pegmatization has been complete and but little of the original gneiss remains. It is, then, the beds of gneissic rock which are rich in secondary quartz and contain numerous small masses of feldspar throughout that carry the most monazite. In such rocks there is generally much biotite, with graphite and perhaps some muscovite and other accessory minerals, as well as abundant quartz and feldspar. The quartz occurs in layers or scattered grains throughout the rock, inclosing and replacing the other constituents. The feldspar crystals chiefly replace, though they partly displace, the other minerals of the rock. Monazite in a rock matrix almost invariably possesses crystal form, in places having brilliant faces and sharp angles.

Mostly quartz Mostly biotite, little quartz and graphite Feldspar . Muscovite Monazite

Fig. 15.—Hand specimen of monazite-bearing rock from British Monazite Company's mine, 3 miles northeast of Shelby, N. C. Three-fourths natural size.

As a typical example of rich monazite-bearing rock, that from the British Monazite Company's chosen for description. Fig. specimen of this rock and

feldspar (mostly the p
monazite, and a little z
the more or less separa
in parallel streaks, with
or grains of

smaller streaks and individual grains in a regular biotite schist. The other minerals of the section occupy various positions and show diverse relations to the minerals of these bands and to each other. The feldspar is porphyritic and occurs chiefly in individual crystals, some of which are of considerable size. A number of the feldspar phenocrysts are small bodies of pegmatite in themselves. As an example, the largest feldspar crystal shown in the section includes both quartz and muscovite. The feldspar at the lower left-hand side of this crystal also has much quartz and muscovite associated with it. As shown in the section, the feldspar phenocrysts replace the other minerals. This replacement is especially well shown by the interruption, with but little displacement, of the lower biotite band by the large crystal described above. Graphite occurs in large amounts with biotite, though it is associated with nearly every other mineral of the rock. Where present, muscovite is chiefly associated with the feldspar. Monazite seems to be indiscriminately scattered through the rock, included in or associated with all the foregoing minerals. Though generally free from inclusions it is not invariably so, and in one case a plate of graphite was observed within a monazite crystal. All the minerals observed in the rock, with the exception of zircon, have been noted as inclusions in the feldspar phenocrysts.

In microscopic sections cut from specimens from one of the ore streaks, the minerals described above were observed, together with some iron staining. The feldspar is principally orthoclase and microcline, partially kaolinized. The quartz is plainly secondary, and occurs in bands or streaks of grains parallel with the schistosity of the rock. In some places the quartz has been deposited in the fractures or between the grains of other minerals; in others it replaces or includes fragments of such minerals as biotite and graphite.

Gas cavities and inclusions of very fine acicular needles, probably rutile, are abundant in the quartz. Biotite occurs in interwoven laths and crystals roughly parallel to the banding of the rock. The pleochroism of the biotite is light yellow-brown to greenish brown or dark purplish red. Graphite occurs as plates and laths, in general lying parallel to the banding of the rock. Some of it is interbanded and even interleaved with biotite; elsewhere the plates are turned across the foliation. In one section a lath of graphite was observed inclosed in quartz which filled a fracture across the foliation of a biotite crystal. Monazite occurs in contact with the various minerals of the sections, though it is more commonly surrounded by or included in grains of biotite and quartz. The position of the monazite in the biotite indicates replacement, and the biotite foliæ are not displaced around the crystals. In the microscopic sections sufficient feldspar was not observed to determine its relation to the other minerals.

The rock has been so thoroughly recrystallized that it is difficult to give the relative order of formation of the minerals. Biotite, if not still in its original condition, was probably the first mineral to form during recrystallization. Part of the graphite was probably contemporaneous with the biotite. Some, however, was introduced later and formed at the same time with the quartz. The small amount of muscovite in the rock was probably next to form, followed closely by quartz. From the small amount of feldspar in the microscopic sections, it was not possible to state its relative period of formation. From the hand specimen, however, shown in fig. 15, it is evident that the feldspar was introduced later than the quartz, or possibly contemporaneously with part of it.

ORIGIN OF MONAZITE.

Monazite has been observed in pegmatite, pegmatized gneisses and schists, and granite gneiss. The occurrence of monazite in pegmatite is that of an accessory original constituent, with the crystal form more or less well developed. But few occurrences in granites have been observed by the writer, and those were in highly gneissic porphyritic granite. The occurrence in pegmatized gneisses and schists indicates either a gathering together of the proper elements from the original rock and their formation into monazite during recrystallization, or the introduction of the proper elements from external sources, along with the materials causing pegmatization. It is probable that pegmatization in which much quartz with but little feldspar has formed represents a phase of recrystallization, in which the quartz may either, in part or wholly, have come from the original rock itself or may have been added by solutions passing through the formations. In either case the materials do not represent the work of active magmatic solutions or magmas such as might give rise to regular pegmatite bodies. In those recrystallized or pegmatized rocks where the feldspathic component of pegmatite is not plentiful, monazite occurs but sparingly. On the other hand, monazite is found more abundantly in rock formations in which feldspar plays a prominent part. The common proximity of this form of pegmatization to granite masses gives evidence of its formation through magmatic agencies. Such pegmatized gneisses are probably the result of active magmatic solutions passing through the rock, both aiding in recrystallization of the original constituents, and depositing the materials held in solution when conditions of temperature or agents of precipitation were favorable. As evidence in favor of the association of monazite with the agencies that produce pegmatite may be cited the occurrence of large crystals of that mineral in the pegmatite worked for mica in Mitchell County, N. C.

The monazite of rock formations has, then, probably been derived from aqueo-igneous solutions such as give rise to certain forms of pegmatite and have in these cases affected large masses of rock.

SUMMARY.

The commercial value of monazite is due to the presence in the mineral of a small percentage of thorium. This element forms the basis for the manufacture of various forms of incandescent gas lights. The value of the production of monazite in the United States is small compared to that of other important minerals. Monazite deposits of commercial value have been found within an area of about 3,500 square miles, lying wholly in the Piedmont Plateau region of North and South Carolina. The principal rocks of this region are mica, garnet, cyanite, graphite, hornblende and granite gneisses and schists, massive granite, pegmatite, peridotite, quartz diorite, and diabase. The structure of the rock formations is complex and in many localities metamorphism has been so extensive that the original nature of the rocks can not be determined. The rocks are in many places concealed by a heavy mantle of residual soil, but their character can often be learned by a study of these soils.

The only deposits of monazite that have been extensively and successfully worked are placers. These deposits are richest in regions where granitic rocks and pegmatized gneisses and schists abound. Residual surface soils and monazite-bearing saprolite are in some places sluiced down from small areas and concentrated. The best-known occurrence of monazite in a rock matrix is in porphyritic pegmatized gneiss. In ordinary gneiss and in highly pegmatized gneiss, in which the pegmatite is so abundant that but little of the original rock remains, monazite occurs sparingly. In beds where pegmatization is prominent but not extreme monazite occurs more plentifully. Monazite in pegmatized gneiss is thought to be derived from aqueo-igneous solutions passing through the rock and depositing and recrystallizing portions of it into the minerals of pegmatite.

MINERALS OF THE RARE-EARTH METALS AT BARINGER HILL, LLANO COUNTY, TEX.

By Frank L. Hess.

GENERAL DESCRIPTION OF THE DEPOSIT.

Baringer Hill is located about 100 miles northwest of Austin, Tex., on the west bank of Colorado River, near the western edge of the Burnet quadrangle as mapped by the United States Geological Survey. It is 12 miles north of Kingsland, the nearest railroad point, 16 miles west of Burnet, and 22 miles northeast of the town of Llano. It is a low mound rising above the flood plain of the Colorado, and formed by the resistance to erosion of a pegmatite dike intruded in a porphyritic granite.

Few if any other deposits in the world, and certainly no other in America, outside of the monazite localities, have yielded such amounts of the rare-earth metal minerals as Baringer Hill.

The writer visited this region in the latter part of February, 1907, fortunately at a time when Mr. William E. Hidden, who has been largely instrumental in making this locality famous through his contributions to mineralogical literature on the rare minerals found here, was conducting mining operations.

The hill is named for John Baringer, who discovered in it large amounts of gadolinite about 1887. No one in the neighborhood knew what the mineral was and specimens were sent to a number of places before it was identified. A piece fell into the hands of Mr. Hidden, who at once looked up the deposit and afterwards obtained possession of the property. Meanwhile Mr. Baringer had taken out a quantity of gadolinite estimated at 800 to 1,200 pounds, which was largely picked up and carried off by persons in the neighborhood as curiosities. Some of the choicer pieces, showing crystal form, found their way into various museums. The property is now controlled by the Nernst Lamp Company, of Pittsburg, Pa., and is worked by that concern for yttria minerals. Since its acquirement by this company a considerable amount of work has been done on the deposit, consisting mostly of open cuts around the edge of the pegmatite, reaching a depth of 30 or 40 feet. A large block, 30 feet in height and more in diameter, consisting mostly of quartz, is left standing in the middle.

In general the "Llano region," in the heart of which Baringer Hill is located, is an island of pre-Cambrian rocks intruded by plutonics and surrounded by an irregular zone of Cambrian and other Paleozoics, including some that are possibly Devonian and some Carboniferous rocks. The inner portion includes parts of Burnet, Llano, and Mason counties, and is situated at almost the geographic center of Texas. The history of this island has been considerably discussed, and views differ as to whether it was an island during the deposition of the Cretaceous, by which the area is almost entirely surrounded, or whether it has been exposed by denudation of the later rocks. The coal measures extend to the north from the region, giving some evidence of an area of high land previous to the deposition of the Cretaceous. Personally, the writer is inclined to agree with the view that the region has been denuded, although his investigations have been but superficial. The plutonics are granitoid rocks of many textures, and differ considerably in composition. Large areas are composed of the rather coarse red granite, the principal outcrops of which occur near Marble Falls and from which the State capitol of Texas was built. Peculiar dikes of a chocolate-brown granite near Llano contain blue quartz.[a]

Other dikes containing this blue quartz are of a reddish color. In many localities the granite is very porphyritic, containing feldspars from 1 inch to 2 inches in longer diameter. There are also gray and fine-grained red granites, and in some places they have taken a gneissoid form. The granites are, at least in part, intrusive in crystalline schists and gneisses of uncertain origin, which are here and there graphitic and contain interbedded strata of crystalline limestone. There are some later dikes of diabasic character, which are comparatively fresh. Southwest of Llano are areas of serpentine and other basic rocks.

In many places the granites are cut by pegmatite dikes, ranging in width from a few inches to 60 feet, which show a much greater percentage of quartz than of feldspar and other constituents, and afford beautiful illustrations of the most acidic phase of pegmatites. In a 6-inch dike there may be but a few feldspar crystals from 1 inch to 3 inches long fringing the edges of the dike; in other dikes, or in other portions of the same dike, gradations from pure quartz to almost pure feldspar may be observed.

Baringer Hill is formed by such a dike on a huge scale. It is a small mound which, before mining was begun, rose perhaps 40 feet

---

[a] Described by Joseph P. Iddings, Quartz-feldspar porphyry (graniphyro-liparose-alaskose) from Llano, Tex.: Jour. Geol., vol. 12, 1904, pp. 225–331.

above a surrounding flat, was about 100 feet wide, and from 200 to 250 feet long. The longer axis runs east and west and is nearly at right angles to the course of the Colorado River at this point. The country rock is a coarse porphyritic granite with feldspar phenocrysts about 1 inch long. This granite seems to weather and erode rather easily, and the river has cut a flood plain perhaps one-fourth of a mile wide at this point, while the dike, owing to its greater hardness and freshness, has better withstood the erosion. The pegmatite, an unsymmetrical body with irregular walls, is intruded into the granite in what seems to be a pipe or short dike.

At the edges of the intrusion is a graphic granite of peculiar beauty and definite structure, being more like the text-book illustrations than the usual graphic granite found in the field. The altered band is from 1 foot to 5 or 6 feet thick, and apparently surrounds the pegmatite. No segregation of the feldspar or quartz in particular parts of the dike can be noted, except that the feldspar may possibly be more inclined to occupy the sides of the intrusion. As far as shown it occupies most of the western and southern sides, and the quartz occupies the center and much of the eastern side.

One quartz mass is more than 40 feet across. The quartz has distinct white bands, from one-eighth to one-half inch wide, which seem to be due to a movement akin to flowage and are similar to those found in many pegmatitic masses in other portions of the country. The white banding is due to small liquid inclusions, many of them containing bubbles which either do not move from change of inclination of the fragment containing them, or do so but slowly. The cavities are minute, largely of irregular, angular shapes, suggesting at first glance particles of broken minerals, and occur in straight or broken lines that probably follow fine cracks which were later cemented. Groups of these cracks, with their inclusions, form the bands, which seem to lie approximately parallel to the walls of the dike or at such angles with them as might easily be formed by the flowage of the material into the space it occupied in the granite. The condition of the quartz seems to show that the pegmatite, after being forced into the granite, partly cooled and solidified and then made another small movement, or a series of slight movements, at which time the minute fractures were formed in the quartz and the magmatic fluids were forced into them, but as the mass was not yet totally solidified the cracks were effectually healed and the fluid was inclosed. Such movements may be supposed to have been consequent on the readjustment of the mass on cooling. Between the fracture bands the quartz is glassy and clear. At one place a vug was found large enough for a man to enter, lined with "smoky" quartz crystals reaching 1,000 pounds or more in weight. This would seem to indicate that the pegmatite had been intruded in a pasty or semifluid condition and that the vugs represent the spaces occupied by segregated water that was squeezed from the magma as the minerals took their final solidified form.

The feldspar is an intergrowth of microcline and albite, of a brownish flesh color, beautifully fresh, and occurs (1) in large masses reaching over 30 feet in diameter, and (2) as huge crystals, many of which, though they rarely show terminal planes, have one or more sharply defined edges, especially where partially surrounded by quartz. An edge 34 inches long was measured on one crystal thus embedded. A smaller crystal was seen which was about a foot long, weighed perhaps 20 pounds, and showed fine terminations and twinning planes.

A large amount of feldspar has been mined and thrown on the dump, and it is possible that in time the dump material may be utilized, either for its potassium content, as a fertilizer, or for pottery making.

Large crystals of fluorspar, measuring a foot along the edge, occur in the quartz, but this mineral does not form any considerable percentage of the mass. The fluorspar ranges from almost colorless to violet so dark that it is practically opaque. Where found alone in the quartz it was, so far as observed, of lighter color than where found with dark-colored minerals. Mr. Hidden informed the writer that it sometimes becomes luminous at the temperature of a living room.

Ilmenite occurs in radiating bunches of sheets or blades ranging from 1 inch to 10 or 11 inches in width and from one-sixteenth to one-fourth of an inch in thickness. In cross section the ilmenite looks like the ribs of a fan, with the outer ends from one-fourth to three-fourths of an inch apart. Similar aggregations take different angles, and numbers of such groups are found lying close together. With them occurs biotite mica in like bunches, the sheets of which are said to reach 3 feet in width by an inch in thickness. The mica is reported by Mr. Hidden to contain cæsium and rubidium, and to be close to lepidomelane in constitution. Small flakes of lithia mica reaching half an inch in diameter are found, generally along cracks in the quartz. No muscovite was seen, but it is said to be found occasionally. Compared with the mass the total amount of mica is very small.

THE RARE-EARTH MINERALS.

The greatest interest in the dike centers in the accessory minerals, particularly in the occurrence of the rare-earth metal minerals, which, as stated, probably have never been found at any other place in such large masses and in such quantities as in this locality. So far the excavations are comparatively shallow, and such minerals as are found are more or less weathered. Many show their crystalline form, but owing to alteration the crystals are now imperfect.

Allanite, a variable silicate of calcium, iron, aluminum, and the cerium metals (cerium, praseodymium, neodymium, and lanthanum), and in smaller amount those of the yttrium group, occurs in large masses, one of which weighed 300 pounds and was embedded in purple fluorspar. It is a dense black mineral with a fine luster, and a hardness of about 6. Around the edges and along cracks it shows alteration to a brown substance having a hardness of about 5.5. The percentage of yttria ordinarily occurring in allanite is small and rarely exceeds 2½ per cent.

Cyrtolite is rather common in the dike in peculiarly fine, polysynthetic groupings with curved faces. It is brown on the surface, with a darker or nearly black interior, and is evidently a mixture of substances. It carries a considerable amount of zirconia and some yttria, and is supposed by Mr. Hidden to be an alteration product of zircon. If it is such a derivative, the original mineral was probably much more complicated than ordinary zircon. It makes a fair radiograph, which also gives evidence of its nonhomogeneity.

Fergusonite, a variable columbate of the yttrium group and other of the rare-earth metals, occurs in four varieties, so different as to be almost distinct minerals. The difference between them is due to oxidation and hydration. No anhydrous varieties are found. It is found in crystalline form surrounded by decomposition zones. Bunches of irregular crystals have been broken out, weighing over 65 pounds. It is generally a mixture of minerals, as may be easily seen on a smooth surface, from the different colors. The difference in composition is strikingly shown in a radiograph, the variations being marked by difference in radiation. According to the two analyses by Hidden and Mackintosh,[a] the fergusonite obtained here carries from 31.36 to 42.33 per cent of yttria and accompanying rare-earth metals, and 42.79 to 46.27 per cent of columbium dioxide. The two analyses give 1.54 per cent and 7.05 per cent of uranium oxides. These are probably very irregularly distributed through the material, as shown both by the mineral itself and especially by its radiographs, which are of striking beauty.

Gadolinite, a silicate of beryllium, iron, and yttrium, is the most important of the minerals found here. It contains about 42 per cent of the yttrium oxides, with a molecular weight of 260, and occurs in crystals and masses of irregular shape up to 200 pounds in weight. The outer portion of the mineral and that adjacent to the cracks is altered to dense brick-red material, but the mineral

---

[a] Hidden, W. E., and Mackintosh, J. B., Yttria and thoria minerals from Llano County, Tex.: Am. Jour. Sci., 3d ser., vol. 38, 1889, pp. 483–484. The minerals of this locality have been well described by these writers in a number of papers.

itself is of a fine, glassy black, with a smooth conchoidal fracture. Thin splinters are bottle-green in color. It has a specific gravity of a little over 4.2, and a hardness of 6.5 to 7. A specimen collected makes no impression on a photographic plate with fifty hours' exposure.

Polycrase, a columbate and titanate of yttrium, erbium, cerium, and uranium, occurs in grains, small masses, and plates, the last associated with ilmenite in such a manner as to suggest the probability of replacement. It normally contains between 20 and 30 per cent of yttrium oxide, but is in too small amount to be commercially important. It is very radioactive, and quickly affects a photographic plate.

Other rare-earth metal minerals found in the dike are yttrialite, rowlandite, nivenite, gummite of several varieties, thorogummite, mackintoshite,·and tengerite. These minerals are apt to occur in any part of the dike, either in the quartz or the feldspar, but have so far been found mostly along the outer portions. A peculiarity of their occurrence is that they are found in bunches from which, if in quartz, radial cracks extend in every direction, and by following such cracks the minerals are found. An illustration of such an occurrence was published by William E. Hidden in 1905.[a] The cause of these " stars," as they have been called by Mr. Hidden, is not clear, but the thought suggests itself that the rare-earth metal minerals may have crystallized first from the magma, and the solidifying quartz, being unable otherwise to accommodate itself to the incompressible nucleus, cracked in this manner.

Mr. Hidden stated that in mining ore of the largest pockets the faces and hands of himself and his assistant were affected as if by sunburn, and, as in sunburn, the covered flesh was not irritated. He suggested radioactivity as the cause, and inasmuch as the minerals under consideration are radioactive, the explanation seems plausible.[b]

The following was given by Mr. Hidden in a personal communication as a complete list of the minerals found in Baringer Hill:

Minerals found in Baringer Hill, Llano County, Tex.

SILICATES.

Albite; }
Microcline; } occur as intergrowths making up the mass of the feldspar.

Allanite; a variable silicate of calcium, iron, the cerium metals, and less amounts of the yttrium group, in masses weighing up to 300 pounds, embedded in purple fluor spar.

Biotite; close to lepidomelane.

Cyrtolite; hydrated silicate of zirconium, yttrium, and cerium. Radioactive, abundant.

---

[a] Some results of late mineral research in Llano County, Tex.: Am. Jour. Sci., 4th ser., vol. 19, 1905, p. 432.

[b] Mr. Hidden has described this incident in the article referred to above.

Gadolinite; a silicate of beryllium, iron, and yttrium in masses weighing up to 200 pounds.

Lithia mica; apparently a later deposition in cracks in quartz. Small flakes one-half inch or less across.

Orthoclase; not abundant.

Yttrialite; an anhydrous silicate of thoria, yttrium, and cerium earths. Contains about 30 per cent silica, 46 per cent yttria, 10 to 12 per cent thoria, and 5 to 6 per cent ceria. Does not occur in large quantity.

Rowlandite; practically a hydrated yttrium silicate. Contains 5 per cent fluorine.

COLUMBATES.

Fergusonite; four varieties, due to oxidation and hydration. Neither is anhydrous. Purest, 5.65 specific gravity. So different as to be almost distinct minerals. Crystals surrounded by decomposition zones.

Polycrase; columbate and titanate of yttrium, erbium, cerium, and uranium. Contains about 25 per cent of yttria.

OXIDES.

Hematite; specular, small quantity.

Magnetite; without metallic acids or rare earths.

Ilmenite; iron-titanium oxide in beautiful crystals, as well as plates up to 8 or 9 inches broad.

Rutile; titanium oxide, in prismatic and reticulated forms one-fourth inch thick.

Quartz; large masses and crystals of white quartz and "smoky" crystals up to 1,000 pounds in weight. Amethysts of gem quality reach 1 inch by one-half inch.

URANATES.

Mackintoshite; 3 parts thorite to 1 part uraninite; contains 13 per cent silica and a small amount of yttria. Radioactive; several times more so than its alteration product.

Thorogummite; formed from mackintoshite by addition of H_2O and alteration of UO_2 to UO_3.

Nivenite; a uranate of uranium, thorium, yttrium, and lead. Contains 10 per cent of lead. The most soluble uranate yet discovered; soluble in 5 per cent solution of SO_3. Prints well and gives great detail. Occurs in cubes and masses. (See Dana's System of Mineralogy, p. 889, for two analyses.) Alters to gummite.

Gummite; several varieties.

PHOSPHATE.

Autunite; hydrous phosphate of uranium and calcium; secondary, not analyzed.

CARBONATES.

Tengerite; carbonate of yttrium and beryllium. Generally globular, but occurs also as crystals up to one-sixteenth inch in length singly and as little nests. May be a mixture of beryllium and yttrium carbonates.

Lanthanite; carbonate of lanthanum, containing also cerium, praseodymium, and calcium. In incrustations on allanite.

SULPHIDES.

Chalcopyrite; iron-copper sulphide, massive, in small amount.

Pyrite; iron sulphide, cubic and octahedral.

Sphalerite; zinc sulphide; the purest fergusonite contains some zinc.

Molybdenite; molybdenum sulphide in scales 5 inches wide, which form masses weighing up to $10\frac{1}{2}$ pounds. Alters to powellite.

MOLYBDATE.

Powellite; calcium molybdate, in white crusts lining cavities where MbS has been. Sugary white radiating or plumose crystals, one-fourth to three-fourths inch long. Locally greenish.

It is interesting to note that among the numerous minerals in this dike no tourmaline, zircon, beryl, monazite, cassiterite, garnet, or tungsten minerals have been found. Cassiterite has been reported from the neighborhood, but its occurrence is extremely doubtful.

With the exception of the alteration products and probably of the lithia mica, which, as noted, occurs along cracks in the quartz, all the minerals are believed to be original constituents of the dike.

The possibility of finding dikes having a like variety of minerals at once suggests itself, and much prospecting has been done for them. A few specimens of the rare-earth metal minerals have been found at other places in the neighborhood, but only a few, and in small quantity. However, similar dikes occur, as already stated, and these have not all been thoroughly investigated. It is to be remembered that these minerals form but a small fraction of 1 per cent of the mass, and it might easily happen that comparatively large amounts could exist in a dike and not be exposed at the outcrop. They are minerals which are altered to softer products by exposure, and would thus be easily removed by erosion and weathering. The cracks surrounding nuclei of the minerals should be useful in prospecting.

ECONOMIC VALUE.

The economic interest in the rare-earth metal minerals centers in their incandescence on being heated, and owing to this property they have been much sought. Thoria, beryllia, yttria, and zirconia show it in the greatest degree. It was found, however, that thoria and beryllia, which form the bulk of the incandescent oxides used in gas mantles are too easily volatilized to be used in an electric glower, such as that of the Nernst lamp. Yttria and zirconia, however, will stand the necessary high temperature. Up to the discovery of this deposit it was practically impossible to get sufficient yttria-bearing minerals to manufacture the lamps, but fergusonite and gadolinite, with lesser amounts of cyrtolite, are found here in large enough quantity to meet the requirements. The zirconia is obtained from zircon brought from other localities.

In the manufacture of the glowers for the Nernst lamp, a paste consisting of 25 per cent of yttria and 75 per cent of zirconia is squirted into strips of the proper thickness, baked, and cut into the required lengths. When cold the mixture is nonconducting, but after being heated it becomes a conductor and gives a brilliant light.

The needs of the Nernst Lamp Company, which owns the deposit, require only the occasional working of the mine. After enough yttria minerals are obtained to supply its wants for a few months ahead the mine is closed. But a few hundred pounds per year are extracted.

By Arthur J. Collier.

INTRODUCTION.

One of the mineral discoveries reported during the year 1907 which has attracted considerable attention is that of tin ore at Silver Hill, southeast of Spokane, Wash. Prospects of silver-bearing galena had been known at this place, and a search for metalliferous minerals and also for coal had been carried on here for several years. The tin-bearing mineral, cassiterite, was identified as such by Richard Marsh, of Spokane, in the summer of 1906, but prospecting for tin was not commenced before March, 1907. By the 1st of June several carloads of selected ore had been mined and piled on the dump. The first authentic report of this discovery published outside of the local newspapers was by A. R. Whitman,[a] of Spokane, June 1, 1907.

As the locality is a new one for tin ore, none having been previously reported from the State of Washington, the writer spent several days early in the season examining the prospects, and again visited the region in October to note the developments made during the summer. The owners, Messrs. Charles Robbins and Richard Marsh, of Spokane, provided every facility for this examination and cheerfully supplied all of the information resulting from their explorations. Mr. A. R. Whitman also offered the results of his observations and accompanied the writer on one of his visits to the field. The present report is based on these limited investigations, and is necessarily incomplete, although many of the facts relating to the occurrence of the ore have been ascertained.

GEOGRAPHY.

Spokane, the most important city of eastern Washington, is situated about 18 miles from the Idaho State line and 90 miles south of the Canadian boundary, on the lines of the Northern Pacific,

---

[a] A tin deposit near Spokane: Min. and Sci. Press, vol. 94, June 1, 1907, p. 697; vol. 95, July 13, 1907, p. 49.

Great Northern, and Union Pacific railroad systems. It is a center for extensive agricultural, lumbering, and mining interests. Vast expanses of rich wheat lands in the Palouse and Big Bend countries lie to the south and west. Great forests of valuable timber lie within 100 miles to the east, and the mines of the Cœur d'Alene district, the chief producers of lead-silver ores in the United States, are within 100 miles to the southeast. Spokane River has a fall of several hundred feet and furnishes the surrounding country with

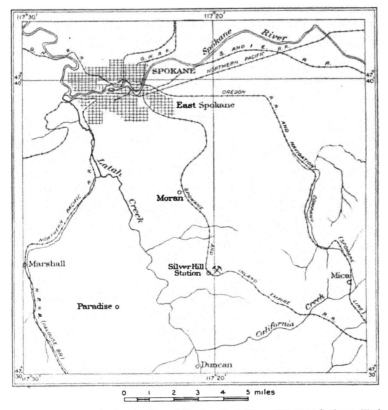

FIG 16.—Index map showing location of tin deposit at Silver Hill, near Spokane, Wash.

water power. Several electric interurban railroads radiating from Spokane are operated with power from the falls.

The tin deposits (see fig. 16) are situated on one of these roads within half an hour's ride from the center of the city. The location is an ideal one for mining, milling, and shipping ore, provided it is found in sufficient quantities, the only element lacking being coal. No deposits of coal suitable for smelting tin ores are known within 150 miles.

TOPOGRAPHY.

The area covered by the index map (fig. 16) has a total relief of 1,900 feet, the lowest point along Spokane River being less than 1,800 feet above sea level, and the elevation of the highest point in a mountain between the two railroads southeast of the city being 3,700 feet. It presents three notable topographic features—a broad, flat valley 1,900 to 2,000 feet above the sea, a level plateau 2,300 to 2,400 feet above the sea, and the mountainous area already noted. The plateau is a constructional feature and approximately represents the surface of the Yakima basalt,[a] which extends westward and southward for several hundred miles underlying the wheat lands of the Big Bend and Palouse countries. It is well represented at Moran Prairie, west of the Inland Empire Railroad between Spokane and Silver Hill. Latah Creek occupies a narrow canyon cut out of the plateau west of the prairie, and the slopes of the mountainous area rise more or less abruptly from its eastern margin, which for several miles is marked by the line of the Inland Empire road. The topography of the mountainous area is not remarkably rugged, and although the slopes are steep in many places they are usually covered with soil. Spokane Valley, several miles wide and 300 feet lower than the plateau, extends east and west across the north end of the area mapped. At the falls in the center of the city the river plunges into a gorge cut below the valley floor.

The history of this topography is about as follows:[b] Previous to the outpouring of the basalts, the region south and west of Spokane was one of well-developed drainage, with mountains, hills, and valleys produced by erosion. Then occurred one of the most remarkable volcanic outbursts that the earth has known. The old topography for hundreds of miles was submerged in seas of lava and basalt, which welled up from fissures in the earth's crust. Only the higher mountains, like that southeast of Spokane, projected above the basalt in islands and promontories. The basalt surface was later elevated to its present position, and the rivers and creeks flowing across it have eroded valleys and canyons like those of Spokane River and Latah Creek. This subsequent drainage was more or less modified in some places by the ice invasion during the glacial period, but there is no evidence of ice action in the vicinity of Silver Hill.

GEOLOGY.

GENERAL CONDITIONS.

The geology of the Spokane region has not been examined in detail, but the rocks of the vicinity fall into three groups whose relations are known in a general way. The rocks which seem to be the oldest in the region comprise a complex mass of gneisses and schists with many igneous intrusions of various kinds. These rocks form the mountains southeast of Spokane and contain the tin deposits. They probably extend westward under the basalts for an indefinite distance. South of Spokane such rocks crop out through the basalts at intervals for 40 or 50 miles, and they are almost continuously exposed around the northern edge of the basalt area to the north-central part of the State. Deposits of tungsten ore have been reported from them near Deertrail, Stevens County, 40 miles northwest of Spokane, and also near Loomis, Okanogan County, 130 miles distant in the same direction. East of the crystalline schists and gneisses there is an extensive area of only slightly metamorphosed slates, quartzites, and limestones which have been studied in great detail in the Cœur d'Alene mining district of Idaho[a] and are known to be of Algonkian age. The contact relations of the metamorphic rocks near Spokane with the Algonkian rocks have not been definitely determined, although the former apparently underlie the latter. This fact, together with their highly metamorphosed condition, suggests strongly that the gneisses and schists at Spokane are older than the Cœur d'Alene rocks and are, therefore, Archean in age, and this opinion is held by the writer, although he is aware that Carboniferous fossils have been obtained from what appear to be rocks of the same complex in the north-central part of the State.

The basalts are of Miocene age and overlap the crystalline schists in the vicinity of Spokane, their eastern limit coinciding approximately with that of Moran Prairie. For several miles between Spokane and Silver Hill the Inland Empire Railroad follows this contact.

LOCAL GEOLOGY.

The tin deposits occur in the metamorphic rocks regarded as Archean. These rocks are very much shattered and here include biotite gneisses, dark-colored quartzites, and mica schists, in many places graphitic and spotted with large crystals of andalusite, as well as numerous intrusive bodies of granite, pegmatite, aplite, quartz, and a more basic rock somewhat resembling basalt. Exposures in the

---

[a] Ransome, F. L., Ore deposits of the Cœur d'Alene district: Bull. U. S. Geol. Survey No. 260, 1904, pp. 274–303. Ransome, F. L., and Calkins, F. C., Geology and ore deposits of the Cœur d'Alene district, Idaho: Prof. Paper U. S. Geol. Survey No. 62 (in press).

railroad cuts indicate that the structure is very complex, and on the map (fig. 17) the schists and quartzites are not differentiated from the gneisses, owing to the meager evidence afforded by the outcrops.

Dikes or veins of pegmatite, aplite, and quartz believed to be aqueo-igneous intrusions in the metamorphic rocks make the most conspicu-ous outcrops. They will be described in connection with the ore de-posits. The larger masses of such rock usually have their longer dimensions parallel with the bedding or schistosity, but some of the smaller veins cut across it. Near their contacts with the surrounding rocks these veins and dikes usually consist of the ordinary pegmatite and aplite minerals—quartz, orthoclase, and muscovite, with tourma-line and in some places apatite as accessory constituents The larger

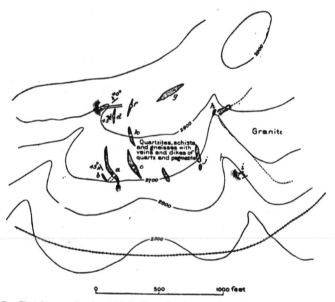

Fig. 17.—Sketch map showing distribution of outcrops of granite, pegmatite, and quartz at Silver Hill, near Spokane, Wash. See text for explanation of reference letters.

masses away from the contacts vary greatly in composition. Some of them contain cassiterite associated with more or less sillimanite and andalusite; others consist of nearly pure quartz.

Granite occurs in stocks or bosses, one of which is shown on the map (fig. 17). It is a moderately coarse-grained rock which shows no evidence of shearing, and consists essentially of quartz, biotite, and both orthoclase and plagioclase feldspar, having approximately the composition of monzonite. Tungsten ores, wolframite and scheel-ite, have been found at the contact of this granite with the sedimen-tary rocks, and the granite is regarded as a possible magma from which the tin ores emanated.

The metamorphic rocks exposed in the railroad cuts include some irregular masses, too small to show on the map, of a reddish-brown igneous rock resembling basalt, and what is thought to be a sill of the same rock cuts one of the bodies of tin ore. In some places the rock contains small veins of pegmatite and is evidently older than the latest granitic intrusions. Although the specimens obtained of this rock are more or less weathered, its mineral constituents have been partly determined. It consists essentially of plagioclase feldspar, augite, and biotite, and is tentatively regarded as kersantite.

The distribution of the rocks, as indicated by their outcrops, is shown on the sketch map (fig. 17), the area covered being about 20 acres. Pegmatites and aplites of varying composition are in many places intimately associated with masses of nearly pure granular quartz and are not differentiated on the map from the quartz veins. Tin ores have been found in these intrusives at the four points marked a, f, h, and j. The principal developments for tin ore consist of an open cut called the "west cut" and a shaft and drift at the point marked a. A deep shaft at the point f and a tunnel at e are old excavations made on an outcrop of quartz in a search for lead-silver ores. The tunnel at e crosscuts the strike of the metamorphic rocks and exposes a thickness of about 100 feet of sedimentary rocks, mostly quartzite, cut by many small veins of pegmatite and aplite. A tunnel at the point marked i was excavated several years ago in a search for coal, the rock discovered being a very black, smutty graphitic schist, spotted with large phenocrysts of andalusite. An open cut, called the "east cut," near the point h exposes a contact of granite with quartzite along which tungsten ores have been found.

Veins of granular quartz containing some silver-bearing galena are exposed at the points marked b and d, and several outcrops of similar quartz in which no metallic minerals have been found are located near the point marked g.

THE TIN ORE.

CHARACTER.

The cassiterite found at Silver Hill is nearly black and without definite crystal outlines. It is distributed through a nearly white fine-grained rock characterized by slight tinges of pink, in grains from the size of a pin head to several inches in diameter. An analysis of the crushed rock from which the cassiterite had been removed by panning gave the following result:

Analysis of tin-bearing rock at Silver Hill, Washington.

[Richard Marsh, analyst.]

SiO₂	74. 02
Fe	2. 24
Mg	1. 22
CaO	2. 28
Al₂O₃	17. 05
	96. 81

When examined with a lens, the peculiar appearance of the gangue is found to be due largely to a fine fibrous mineral in radiating aggregates. All masses of this rock that have been examined shade off into bodies of nearly pure sugary granular quartz, much of it yellowish from iron stain. Where the tin-bearing rock joins the quartz the two interlock along the contact, producing a texture resembling that of pegmatite. This texture is much more evident in a large outcrop marked *c* on the sketch map (fig. 17), in which no cassiterite has been found. Characteristic pegmatite consisting of coarse-grained quartz, orthoclase, muscovite, and black tourmaline occurs at the contact of the tin ore with the hanging wall at the west cut (*a*, fig. 17).

The mineral characteristics of the tin-bearing rock were determined microscopically in thin sections. It consists essentially of quartz, orthoclase feldspar, sillimanite in slender radiating crystals, and a highly refractive, faintly pleochroic mineral without definite outlines, which has been determined as andalusite. There are also patches of sericite and kaolin. The quartz contains minute fluid and gaseous inclusions arranged in parallel lines along many of which fractures have been developed. The feldspars are slightly clouded in the thin sections examined, but present no evidence of decomposition other than weathering. Where feldspar is in contact with andalusite or sillimanite its boundaries are distinct, and here and there sillimanite fibers are included in the quartz. Neither fluorite nor lithia mica has been identified in the tin ores or the inclosing rocks. The cassiterite as seen in the thin section presents no evidence of being due to secondary deposition. Except for the andalusite the mineral characteristics of the ore-bearing rock are similar to those of a rather fine-grained pegmatite.

Sillimanite has often been found as a constituent of granite and pegmatite, but andalusite has seldom, if ever, been reported as a constituent of pegmatite, and in no instance known to the writer has it been found in association with tin ore, though it has been reported as a constituent of granite masses near their contacts. The inclusion of andalusite in this pegmatite mass is, therefore, apparently a unique occurrence. Many of the inclosing schists contain andalusite as an essential constituent, and it is probable that the andalusite of the ore bodies has in some way been derived from them, for there is no evidence of its having been produced by the decomposition of the feldspars. Examination of a number of thin sections of the granular quartz associated with these ore bodies shows that the rock consists of rather large interlocking grains of quartz with only scattered foils of muscovite mica. The quartz grains are marked by parallel lines of inclusions and fracture planes, like the quartz of the tin ore, and the texture also seems to be similar. These masses of quartz are regarded by the writer as more siliceous portions of the pegmatite bodies, due to a different phase of the same aqueo-igneous action. Pegmatites and aplites are regarded as products of granite intrusions in which the more siliceous minerals are concentrated. When granite masses gradually cool and solidify the dark-colored minerals crystallize more readily than the siliceous minerals such as quartz and orthoclase, and in the last stages of consolidation the portions remaining liquid contain an excess of silica, together with a large percentage of the water contained in the whole of the original magma. The material is practically a solution of siliceous minerals in superheated water, and if it escapes into fissures in the surrounding rocks it forms pegmatite dikes near the sources of emanation and may deposit quartz veins farther away.

The cassiterite at Silver Hill is apparently an original constituent of the pegmatite. In this respect the Spokane tin ore resembles that of North Carolina [a] and the Black Hills. Silver-bearing galena, wherever it has been found, is confined to the more siliceous veins. Wolframite and scheelite also seem to be associated with the masses of quartz near their contacts with the granite or pegmatites. In the open cut along the contact of granite and quartzite at the point marked h (fig. 17), and also in the main shaft at the point marked a, these ores form nodules up to 2 inches in diameter in masses of nearly pure quartz. Scheelite is more common than wolframite, and was probably deposited first. Some of the nodules of scheelite are surrounded by a thin crust or rim of wolframite. Mr. A. R. Whitman has inferred from this that the scheelite was first deposited and afterwards partly altered to wolframite by solutions containing iron and manganese. The alternative hypothesis, that the wolframite was originally deposited as such around the scheelite nodules, is equally possible, however, and equally well supported by the evidence at hand.

DEVELOPMENTS AND FORM OF ORE BODIES.

Tin ore of the type described has been found at four localities in this area. At the point f (fig. 17) a large bowlder of such ore was found on the surface near the old shaft. At the point h a bowlder

---

[a] Graton, L. C., Gold and tin deposits of the southern Appalachians: Bull. U. S. Geol. Survey No. 293, 1906, p. 82.

weighing about 500 pounds and containing approximately 10 per cent of cassiterite was found, but it was not traced to its bed-rock source, although considerable excavating has been done. At the point *j* there is an extensive outcrop of andalusite-bearing pegmatite like that containing the ore, but only a small amount of cassiterite has been found in a few small fragments on the surface.

The principal workings and the largest amount of tin ore found are at what is known as the " west cut " (*a*, fig. 17). This ore body was first developed by an open cut 150 feet long, in which a mass of pegmatite and quartz was uncovered. The tin-bearing rock dips to the southwest at an angle of about 45°, and lies between well-defined walls. The hanging wall is a biotite gneiss or gneissoid granite with the foliation parallel to the ore body. The foot wall consists of quartzite and black andalusite schist, but owing to the fact that there is a casing of more or less altered rock along the contact of the ore with the foot wall, the attitude of the bedding or schistosity has not been definitely determined. A sill of the basaltic rock, provisionally determined as kersantite, divides the vein into an upper and a lower portion. This sill ends abruptly against the hanging wall, but

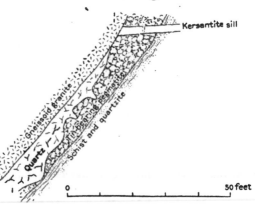

FIG. 18.—Section of ore body at Silver Hill, near Spokane, Wash.

probably extends for some distance into the foot wall. It dips toward the northwest at an angle of about 5°, and reaches the floor at the end of the cut. The relations of the different rocks exposed are shown in the section (fig. 18). A small amount of cassiterite was found very much disseminated through the pegmatite above the kersantite sill. Below this sill the pegmatite for 20 or 30 feet contained sufficient cassiterite to make it a commercial ore. The mineral was most abundant in the central part of the pegmatite mass. The maximum thickness of ore found at any point was probably not less than 10 feet. From the point where the richest ore was found a shaft has been sunk to a depth of 125 feet on an incline of approximately 45°. In depth the pegmatite is displaced by barren quartz on the hanging wall, as is shown in the section (fig. 18). Some cassiterite was found to a depth of 50 feet, although it gradually decreases in amount below the bottom of the cut.   Below the 100-foot level the walls of the ore body come together, and for some distance the vein or dike is not well defined, although large nodules of tungsten ore have been found.   On the 100-foot level a drift has been run to the north for a distance of 100 feet.   This drift is partly in barren quartz and partly in the pink pegmatite regarded as tin-bearing rock. About 35 feet from the shaft this rock contained a notable amount of cassiterite, and part of it was rich enough to be regarded as tin ore. At the end of the drift a second sill, probably of kersantite, was encountered dipping to the southeast.   These excavations have gone far enough to show that the tin-bearing intrusion is of irregular form, that the tin is not uniformly distributed through it, and that it may be confined to an ore shoot pitching to the northwest.

VALUE OF THE ORE.

In the course of these excavations the tin ore has been carefully selected from the barren rock and piled on the dump, where, at the present time, there is probably from 100 to 200 tons.   The quantity was roughly estimated at 125 tons by the writer.   It is exceedingly difficult to make a close estimate of the value of this ore, as it consists of pieces varying in size up to 100 pounds or more, in which the cassiterite is unevenly distributed in grains from the size of a pin head to several pounds in weight.   Mr. Richard Marsh has estimated the cassiterite or black tin contained at 6 per cent.   Mr. A. R. Whitman estimates the metallic tin at 3 per cent.   From an inspection of the dump, the writer is of the opinion that the former estimate may be more nearly correct.   The cassiterite is reported by both Mr. Marsh and Mr. Whitman to be remarkably free from impurities, and it is believed that a concentrate containing 70 per cent metallic tin can be obtained.   Scheelite and wolframite in small quantities have been found at a number of places in this vicinity, but they are not closely associated with the cassiterite in the lodes, and are not found in the concentrates.   At present prices (February, 1908), metallic tin is worth about 28 cents per pound, and tungsten ores are worth twice as much per unit[a] as the tin ore.   The tailings resulting from milling this ore make white sand of good quality, for which builders in Spokane are reported to be willing to pay as much as 75 cents a ton.

---

[a] The unit referred to is 1 per cent of metal contained in 2,000 pounds of ore.   For the fluctuations in the price of tin and the tungsten minerals, see Mineral Resources for 1906, U. S. Geol. Survey, 1907.   This report can be obtained on application to the Director of the Survey.

PROSPECTING FOR PLACER TIN.

The mountain mass from which Silver Hill is a spur owes its present contour to erosion, and the tin ores from the portions thus removed should be concentrated in stream gravels at no very great distance; the quantity of such placer tin, if it could be determined, would be an index to the quantity of tin in the lodes. No tin-bearing gravels have yet been discovered, but very little prospecting for such deposits has been done. As has been noted, most of the erosion of the mountain mass was accomplished before the outpouring of the basalts which filled all the valleys and submerged the lower hills. It is to be expected, therefore, that the greater part of the tin-bearing gravels, if such exist, are buried below the basalts and may never be discovered.

CONCLUSIONS.

Inasmuch as cassiterite is a mineral not affected by processes of secondary or surface enrichment, and in the present instance it is an original constituent of the igneous rocks in which it occurs, it is reasonable to expect that the quantity of ore exposed in outcrop will approximate that to be found at lower levels. The ore is contained in rather irregular aqueo-igneous veins or dikes, several of which outcrop in the area under examination. It has been found in four such outcrops, one of which yielded approximately 125 tons of ore. The others produced smaller amounts, but may not have been thoroughly prospected. No bodies of tungsten ore large enough to be of economic value have been discovered up to the present time.

Veins and dikes of pegmatite are not uncommon elsewhere in the metamorphic rocks of this region, which are continuously exposed for several miles to the east and outcrop at intervals for at least 40 miles to the south. No cassiterite has yet been found in these rocks except at Silver Hill, but such discoveries are to be expected.

The developments at Silver Hill indicate that the tin ore is to be found in detached masses whose relations to each other can not yet be forecast, and the economic value of the deposit will depend to a considerable extent on the amount of excavation necessary to locate other ore bodies. This can be determined only by experience involving a further outlay of capital and possibly requiring several years' time, but the discoveries already made are of sufficient value to warrant such investigations.

SURVEY PUBLICATIONS ON ANTIMONY, CHROMIUM, NICKEL, PLATINUM, QUICKSILVER, TIN, TUNGSTEN, URANIUM, VANADIUM, ETC.

The principal publications by the United States Geological Survey on the rarer metals are the following:

BECKER, G. F. Geology of the quicksilver deposits of the Pacific slope, with atlas. Monograph XIII. 486 pp. 1888.

——— Quicksilver ore deposits. In Mineral Resources U. S. for 1892, pp. 139–168. 1893.

BLAKE, W. P. Nickel; its ores, distribution, and metallurgy. In Mineral Resources U. S. for 1882, pp. 399–420. 1883.

——— Tin ores and deposits. In Mineral Resources U. S. for 1883–84, pp. 592–640. 1885.

BOUTWELL, J. M. Quicksilver. In Mineral Resources U. S. for 1906, pp. 491–499. 1907.

CHRISTY, S. B. Quicksilver reduction at New Almaden [Cal.]. In Mineral Resources U. S. for 1883–84, pp. 503–536. 1885.

COLLIER, A. J. Chromite or chromic iron ore. In Mineral Resources U. S. for 1906, pp. 541–542. 1907.

DAY, D. T. Platinum. In Mineral Resources U. S. for 1906, pp. 551–562. 1907.

——— and RICHARDS, R. H. Investigations of black sands from placer mines. In Bulletin No. 285, pp. 150–164. 1906.

EMMONS, S. F. Platinum in copper ores in Wyoming. In Bulletin No. 213, pp. 94–97. 1903.

GALE, H. S. Carnotite in Rio Blanco County, Colorado. In Bulletin No. 315, pp. 110–117. 1907.

GLENN, W. Chromic iron. In Seventeenth Ann. Rept., pt. 3, pp. 261–273. 1896.

GRATON, L. C. The Carolina tin belt. In Bulletin No. 260, pp. 188–195. 1905.

——— Reconnaissance of some gold and tin deposits in the southern Appalachians. Bulletin No. 293. 134 pp. 1906.

——— (See also Hess, F. L., and Graton, L. C.)

HESS, F. L. Antimony. In Mineral Resources U. S. for 1906, pp. 511–516. 1907.

——— Bismuth. In Mineral Resources U. S. for 1906, p. 517. 1907.

——— Nickel, cobalt, tungsten, vanadium, molybdenum, titanium, uranium, and tantalum. In Mineral Resources U. S. for 1906, pp. 519–540. 1907.

——— Tin. In Mineral Resources U. S. for 1906, pp. 543–549. 1907.

——— Arsenic. In Mineral Resources U. S. for 1906, pp. 1055–1058. 1907.

——— Selenium. In Mineral Resources U. S. for 1906, p. 1271. 1907.

——— and GRATON, L. C. The occurrence and distribution of tin. In Bulletin No. 260, pp. 161–187. 1905.

HILLEBRAND, W. F., and RANSOME, F. L. On carnotite and associated vanadiferous minerals in western Colorado. In Bulletin No. 262, pp. 9–31. 1905.

HOBBS, W. H. The old tungsten mine at Trumbull, Conn. In Twenty-second Ann. Rept., pt. 2, pp. 7–22. 1902.

—— Tungsten mining at Trumbull, Conn. In Bulletin No. 213, p. 98. 1903.

KAY, G. F. Nickel deposits of Nickel Mountain, Oregon. In Bulletin No. 315, pp. 120–127. 1907.

KEMP, J. F. Geological relations and distribution of platinum and associated metals. Bulletin No. 193. 95 pp. 1902.

PACKARD, R. L. Genesis of nickel ores. In Mineral Resources U. S. for 1892, pp. 170–177. 1893.

RANSOME, F. L. (See Hillebrand, W. F., and Ransome, F. L.)

RICHARDS, R. H. (See Day, D. T., and Richards, R. H.)

RICHARDSON, G. B. Tin in the Franklin Mountains, Texas. In Bulletin No. 285, pp. 146–149. 1906.

ROLKER, C. M. The production of tin in various parts of the world. In Sixteenth Ann. Rept., pt. 3, pp. 458–538. 1895.

ULKE, T. Occurrence of tin ore in North Carolina and Virginia. In Mineral Resources U. S. for 1893, pp. 178–182. 1894.

WEED, W. H. The El Paso tin deposits [Texas]. Bulletin No. 178. 6 pp. 1901.

—— Tin deposits at El Paso, Tex. In Bulletin No. 213, pp. 99–102. 1903.

WEEKS, F. B. An occurrence of tungsten ore in eastern Nevada. In Twenty-first Ann. Rept., pt. 6, pp. 319–320. 1901.

—— Tungsten ore in eastern Nevada. In Bulletin No. 213, p. 103. 1903.

IRON AND MANGANESE.

AN ESTIMATE OF THE TONNAGE OF AVAILABLE CLINTON IRON ORE IN THE BIRMINGHAM DISTRICT, ALABAMA.

By Ernest F. Burchard.

INTRODUCTION.

In a previous paper [a] it was stated that a forthcoming more detailed report on the iron ores of the Birmingham district would contain an estimate of the red-ore reserves in the district. This detailed report has been completed, and it is expected that it will be published some time after July 1, 1908. The above-mentioned estimate of ore reserves has been prepared by the writer and included in the text of the detailed report, but for the sake of more prompt publication it is given here, necessarily, however, without the mass of data, comprising measurements of thickness and extent of ore seams, and the chemical analyses and other experimental results on which the calculations are based.

By the Birmingham district is meant the area from which the furnaces at Birmingham, Ensley, and Bessemer derive their iron ores, and it is practically coextensive with Birmingham Valley, the heart of the Alabama red-ore field. This valley extends from the vicinity of Springville, on the northeast, beyond Vance on the southwest, and from the Warrior coal field, or Sand Mountain, on the northwest, to the Cahaba coal field, or Shades Mountain, on the southeast. To the southwest the inclosing ridges pass below unconsolidated Cretaceous and Tertiary clays and sands, so that the iron-bearing rocks are deeply buried. Birmingham Valley therefore has a length of nearly 75 miles, an average width of more than 6 miles, and an area of 450 to 500 square miles.

The red ores occur in the Clinton (Rockwood) formation, which consists of shale, sandstone, iron ore, and a little ferruginous limestone. This formation extends in a northeast-southwest direction on both sides of the valley, dipping away from it on each side, and there are a few small areas or strips within the valley, principally in its

---

[a] Burchard, E. F., The Clinton or red ores of the Birmingham district, Alabama: Bull. U. S. Geol. Survey No. 315, 1907, p. 150.

southwestern portion, but only in Red Mountain has the ore been found of sufficient thickness and purity to be worked on an important scale. The geologic relations of the rocks and ores have been described in the paper previously mentioned.[a]

DIVISIONS OF THE DISTRICT.

Owing to the considerable extent of Birmingham Valley, to the distribution of the ore beds along the margins and at the ends of the valley, and to the variation in the character of the ore from place to place, the district has been divided, for convenience of description in the complete paper, into seven parts. The order of the divisions from A to G represents in a general way their commercial importance, based on quality of ore, quantity of ore, structure of ore beds, accessibility, and distance from smelters. It should be understood, however, that this outline of divisions is not intended as a definite estimation or appraisal of relative values. Such facts as were obtained in the field study of the district will be presented in the later report, so that interested persons may draw their own conclusions therefrom.

Division A includes that part of Red Mountain which extends from Morrow Gap, in sec. 32, T. 16 S., R. 1 W., southwestward to Sparks Gap, in sec. 32, T. 19 S., R. 4 W., a distance of about 26 miles. This is the only portion of the district considered in the present paper, and its outline is shown in fig. 19.

All but two of the productive mines of the district are in this strip of Red Mountain. In all there were 30 workings in operation in 1906, including ·slopes, open cuts, and combination mines. These mines are served by the Birmingham Mineral Division of the Louisville and Nashville Railroad, which is built along the slope of the ridge 100 to 350 feet below the summit. The railroad runs first on one side of the mountain, then on the other, threading its way back and forth through several natural passageways, such as Sadlers Gap, Lone Pine Gap, Walker Gap, and Readers Gap. From Readers Gap it runs into Bessemer, with a spur extending southwestward along the ridge to the Potter slopes. Some mines, especially those in that part of the mountain where the railroad passes along the west side, are so situated that their tipples can be built directly on a siding. Others, facing the east, have built spurs reaching back into lateral ravines. Through Red Gap, between Irondale and Gate City, five railroads enter Birmingham from the east and north; at Graces Gap the Louisville and Nashville Railroad passes southward across the Cahaba coal field to the Gulf, and at Sparks Gap the Southern Railway finds an outlet southeastward. This portion of the district is therefore well supplied with transportation lines and consequently its development has been facilitated.

---

a Burchard, E. F., op. cit., pp. 132-146.

In July, 1906, the deepest slope in Red Mountain was reported to be more than 1,800 feet long. Three other slopes have been driven for nearly 1,800 feet each, and there were twelve slopes between 900 and 1,500 feet long. All the slopes 900 feet or more in length are in the strip of mountain southwest of Birmingham. The newer mines at the extremities of the district have slopes ranging between 260

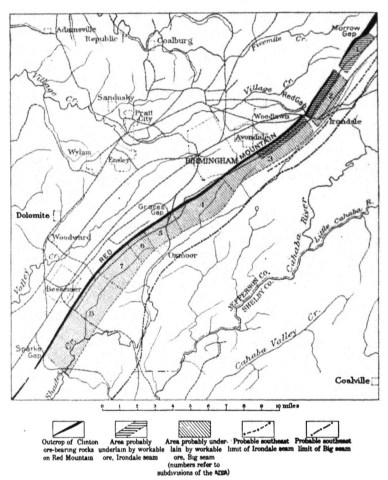

| Outcrop of Clinton ore-bearing rocks on Red Mountain | Area probably underlain by workable ore, Irondale seam | Area probably underlain by workable ore, Big seam (numbers refer to subdivisions of the area) | Probable southeast limit of Irondale seam | Probable southeast limit of Big seam |

FIG. 19.—Sketch map of main portion of Birmingham iron-ore district, Alabama, showing subdivisions on which are based estimates of iron ore reserves.

and 800 feet in length. The deepest slope goes down on beds whose average dip is about 22°, so that its present depth is about 650 feet below the level of the valley at a point directly above the bottom of the slope. Projected at the same angle to a point directly below Little Shades Creek the slope would have a length of about 6,400 feet and a depth below the creek of 1,800 feet. It is not known whether the ore extends with an unchanged dip and thickness to this depth. Drill records obtained farther south in Shades Valley indicate that the ore beds with their associated strata flatten out and locally rise toward the surface. The surface rocks in the valley indicate irregularities in structure, including faulting, which would naturally be shared by the beds below.

No great deterioration in either quality or thickness of the hard ore in the direction of the dip has yet been disclosed by the deeper slopes—an encouraging fact in so far as it can be used as a measure of the ore ahead of shorter slopes. At one of the larger mines, centrally located, systematic analyses have been made of the ore at intervals of a few feet from the outcrop to the bottom of the slope and throughout the extent of each entry to the right and left of the slope. The composition of the ore has been found to vary appreciably from place to place and the degree of variation is likely to be as great within a few yards as it is between remote parts of the mine, but the average run of the hard ore in the mine is remarkably regular. The facts brought out by this series of analyses show that the content of metallic iron increases about 1 per cent for each 1,000 feet below the upper limit of the hard ore, that the lime (CaO) decreases about 1 per cent in the same distance, and that the silica content increases a trifle. Slightly different facts are shown, however, by a series of analyses of ore from a mine also on the Big seam, northeast of Birmingham, and distant about 18 miles from the mine just mentioned. Here the lime is increasing slightly with the depth, while the insoluble material as well as the iron is decreasing slowly. This change can be accounted for, in all probability, by the fact that the iron ore here is still being mined from the zone of transition from soft to hard ore and that the completely hard ore has not yet been reached.

Studies by members of the Alabama Geological Survey extending over many years have shown that the Clinton (Rockwood) formation tends to thin out and become sandier toward the southeast. There is no reason why this change should not be shared proportionately by the inclosed ore beds, and it is believed that the drill records just referred to indicate that such is the case. However, the complete drill records available from the valley east of Red Mountain are so few that reliable conclusions can be based on them only regarding that part of the ore basin which lies in the southern third of the district. Ore can, perhaps, be expected to underlie the valley southeast of Red Mountain, probably as far as Shades Mountain. The width of Shades Valley is a rough indication of the relative extent of the Red Mountain ore toward the southeast, and the width of the valley is sensibly greater southwest than it is northeast of Readers Gap.

ORIGIN OF THE CLINTON ORE AND ITS BEARING ON ORE SUPPLY.[a]

The answer to the question as to how the ores were formed has a very practical bearing on the extent and quantity of unexploited ore. Several theories have been advanced to explain its formation and three of these have received attention from persons who have considered the subject. Briefly the processes may be outlined as follows:

1. Original deposition: The ores were formed at the same time as the rocks with which they are associated.

2. Residual concentration: The ore beds represent the weathered outcrops of ferruginous limestones, from which the lime and other soluble matter have been leached.

3. Replacement: The ores have been formed by the replacement of beds of limestone by iron-bearing waters, and are therefore of much later origin than their inclosing rocks.

If the ore beds are due to replacement or to surface decay of limestone beds they can be expected to decrease in value regularly and at a fairly rapid rate with distance from the outcrop, until the beds consist entirely of limestone. Moreover, this condition should be encountered within distances less than the lengths of some present mine slopes.

If, however, the ores originated with their inclosing rocks, no regular decrease in richness is to be expected as the beds are exploited deeper beyond the limit of soft or leached ore. Areas of low-grade ore or even barren rock may be struck, but such areas are the result of original deposition, and a mine slope may pass onward through such a patch of lean ore or rock into ore of good grade. Finally, the ore bed may be expected to thin and disappear or to split and become shaly, and in this way to so deteriorate as to become unprofitable to work, but unless structural complications render it unworkable the ore should continue down the dip of Red Mountain well toward Shades Mountain.

It may be stated that all the new facts observed in the course of the work in the Birmingham district are in accordance with the hypothesis that the ore is the result of original deposition of ferruginous sediments. The transition, vertically, between sandstone and ore or between shale and ore is as sharp as that between coal and its inclosing rocks. The variation in composition of an ore bed from place to place is not unlike the local changes in composition and character of a coal bed. The lenslike form of the beds is common to both coal and ore. Finally, as the lens thins, whether of coal or of ore, it tends to become shaly and siliceous rather than calcareous. That the ore is due entirely to the replacement of limestone seems

---

[a] This subject is discussed much more fully by Mr. E. C. Eckel in the complete report.

hardly possible when it is considered that instead of a marked decrease in percentage of iron and an increase in that of lime, with depth, until the bed becomes a limestone, very little tendency toward that condition has been noted. The lime in the bed is perhaps an accessory deposit, as is the silica. The term "depth" in this connection may be subject to misconception, for the sediments were deposited in a horizontal position, or nearly so, and their present attitudes are the result of subsequent foldings. The depth to which the beds now extend is therefore incidental, and in no way affects their character beyond the soft-ore limit. Indeed, the best criterion for judging the character of the unexploited ore beds in the direction of their dip, or in the basin southeast of Red Mountain, is the strike section of the same beds that has been afforded by the mine workings. From northwest to southeast there are likely to occur changes similar in nature to those that are known to take place from northeast to southwest, although the changes will probably be found to be more abrupt, for the reason that the former direction is toward the shore of the water body in which the sediments were deposited, whereas the latter is parallel to this shore. Keeping all these possibilities in mind and using such data as are suggested below, the geologist or the engineer should be able to make a fairly close estimate of the tonnage of the red-ore reserves in the district, or in any portion of it.

METHOD OF MAKING ESTIMATES.

First, the area should be divided into parts in somewhat the manner outlined on page 309. Then each division should be subdivided again and again until areal units are obtained in which the cubical contents of the ore can be calculated with not more than 10 per cent of error. The percentage of recoverable ore should enter into the calculations, as well as the specific gravity of hard ore carrying not more than the average percentage of metallic iron.

THE ORE RESERVES IN DIVISION A.

It has been stated that red ore might be expected to underlie the valley southeast of Red Mountain probably as far as Shades Mountain, and there are indications that it extends still farther, as a thin ore seam is brought to the surface by faults east of Cahaba River. It is hardly probable, however, that the ore continues with workable thickness beyond the line of Shades Mountain, and it is not likely that it continues workable that far toward the southeast, to judge from the thinness of the seam where it is faulted up, and from drill records in Shades Valley. The lenslike character of the ore beds and the thinning and other changes in the beds that take place more abruptly at right angles to the ancient shore line than parallel to it make it reasonable to assume that at a certain distance from the outcrop the ore bed will naturally become so thin as to be negligible. Structural conditions indicate that this line, which contains what may be called the "vanishing point" of the ore, lies below Shades Mountain, the border of the heavy cover of coal measures to the southeast. As there are several seams of ore, this maximum distance naturally applies to the largest and most persistent bed, viz, the Big seam. The other, smaller seams, such as the Ida and the Irondale, probably would not continue so far, to judge from their extent and the relations exhibited along their strike. If there is a vanishing point, or a point beyond which the ore continues only a few inches in thickness, it will not be practicable to mine the ore as far as this point, and the limit to which it will pay to drive slopes will be determined by the minimum thickness at which the ore can be mined with profit. In using these factors, some of which are to a certain extent hypothetical, as a basis for estimating the tonnage of ore still in the ground in Division A, it is also necessary to assume that there is a fairly regular decrease in thickness of the seams from their outcrop to the vanishing point, and that therefore they form long, wedge-shaped bodies, the thick end of the wedge lying along the outcrop on Red Mountain, and the thin end, somewhat less regular in outline, lying below the crest of Shades Mountain, with the limit of workability following a northeast-southwest line intermediate between the two extremes.

An estimate of the ore reserves in Division A has been made in connection with the study of this subject, but the fact is here emphasized that while many more details have been considered than there is space to enumerate here or necessity for describing at present, the estimate must be regarded as only approximate. The tonnage of ore that should be contained under the assumed conditions, first in the Irondale seam, from Morrow Gap to Clifton Gap, and second in the Big seam, from Morrow Gap to Sparks Gap, has been computed. From the sum of these estimated quantities is subtracted the total tonnage of red ore that has been produced in Alabama from 1880 to 1907, inclusive. In making this estimate, Division A is subdivided into eight parts, in two of which the Irondale seam is considered of sufficient importance to be regarded as a source of future ore supplies. These eight units of area (see fig. 19), whose ore-bearing strata outcrop along Red Mountain, are as follows: (1) From Morrow Gap to and including the Olivia mine (Irondale seam); (2) Bald Eagle to Clifton Gap (Irondale seam); (3) Bald Eagle to Lone Pine Gap (Big seam, upper bench); (4) Lone Pine Gap to Graces Gap (Big seam, upper bench); (5) Graces Gap to a point beyond Ishkoodo (Big seam, upper bench); (6) Ishkoodo to Tennessee Coal, Iron and Railroad Company's slope No. 10 (Big seam, upper bench); (7) Tennessee Coal, Iron and Railroad Company's slope No. 10 to middle of Woodward Iron Company's property (Big seam, upper bench); (8) middle of Woodward Iron Company's property to Sparks Gap (Big seam, upper bench). The estimate is considered to be conservative for the following reasons: (1) No account has been taken of any possible available ore except that in Red Mountain; (2) no ore seams besides the upper bench of the Big seam and the Irondale seam have been considered; (3) only such portions of the outcrop of these seams have been considered as are known to be workable, and wherever the seams are faulted out or badly broken up, such portions are not included in the area on which estimates are based; (4) the percentage of recoverable ore has apparently been placed low enough to be on the safe side; (5) conservative figures have been used as representing the average workable thicknesses at the outcrop and the minimum workable thickness, as under favorable conditions the former may be considerably greater and the latter may be less; (6) the percentage of the metallic iron used as a factor in determining the specific gravity of the hard ore has been taken with a view to the possible reduction rather than increase of iron content with depth; (7) in deducting the tonnage of red ore already produced the total red ore for the State has been taken, which is greater than that produced by the Birmingham district, and consequently in excess of that produced by this area, the main portion of the Birmingham district. In regard to this last factor it should be stated that the excess is not great, however, for the Birmingham district has produced almost 90 per cent of the red ore of the State, and Red Mountain between Morrow Gap and Sparks Gap has produced between 97 and 98 per cent of the red ore of the district.

In obtaining the specific gravity of the hard ore in relation to its content of metallic iron, use has been made of the laboratory determinations of Mr. R. T. Pittman, chief chemist of the Sloss-Sheffield Steel and Iron Company at Birmingham. The experiments consisted of grinding lumps of ore down to cubes 1 inch on an edge, determining the specific gravity of each by displacement of water, and afterwards analyzing the ore thus treated. The results of certain of these tests and analyses are as follows:

Specific gravity tests and analyses of calcareous hematite.a

Sample No.	Weight in air of 1 cubic foot of ore (pounds).	Specific gravity.	Analyses.		
			Fe.	Insoluble.	CaO.
1	213.47	3.42	36.25	13.80	17.98
2	215.97	3.46	37.05	12.40	18.14
3	219.23	3.50	37.60	11.42	17.43
4	220.71	3.53	38.05	10.60	17.52
Average	217.35	3.48	37.24	12.05	17.78

a Experiments by R. T. Pittman, Birmingham, Ala.

If we assume, then, that the ore in a certain seam within a given area forms a fairly regular prism, the base and altitude of which may be measured, and that the minable ore of this seam constitutes a truncated portion of this prism, the cubic contents of this truncated prism of minable ore may be calculated conveniently by substituting in a formula the values of the average thickness, length, and width of the truncated prism of ore. From this result (in cubic feet) may be deduced in the same operation the tonnage of ore of a definite grade by use of the factors, percentage of recoverable ore and specific gravity, based on the average percentage of metallic iron in the hard ore. Multiplying by 62.5, the weight in pounds of a cubic foot of water, will give the pounds of ore, which can then be reduced to long tons by dividing by 2240.

Therefore, to establish a general formula for calculating the ore content for a given ore seam in a given area, let—

L = Length of outcrop.

V = Average distance of "vanishing point" from outcrop.

T = Average thickness of ore seam at outcrop.

t = Minimum thickness to which ore may be worked.

D = Distance from outcrop at which thickness of ore seam becomes t, or maximum distance practicable to drive slopes.

R = Per cent of recoverable ore.

C = Average per cent metallic iron in hard ore.

G = Specific gravity of ore based on value of C.

Then to obtain the value of D in terms of the known quantities T, t, and V, $T:V::t:V-D$, whence $D = \dfrac{T\,V - t\,V}{T}$ and the total tonnage is $\dfrac{\frac{1}{2}(T+t) \times L \times D \times R \times G \times 62.5}{2240}$. On applying this formula to the area included in Division A, we obtain the result given in the following table:

Estimated ore reserves in main portion of Birmingham district.

Subdivision.[a]	L	V	D	T	t	R	C	G	Total ore.
	Feet.	*Feet.*	*Feet.*	*Feet.*	*Feet.*	*Per cent.*	*Per cent.*		*Long tons.*
1...................	11,000	8,000	2,720	4.54	3	80	32	3	7,896,804
2...................	32,000	6,800	2,000	4.24	3	80	35	3.27	16,910,570
3...................	37,000	10,600	5,000	8.5	3.5	80	35	3.27	100,846,671
4...................	16,500	12,000	7,300	9	3.5	80	33	3.08	51,755,855
5...................	10,000	12,000	6,750	8	3.5	80	34	3.17	27,445,981
6...................	6,000	13,000	7,167	7.8	3.5	80	36	3.36	18,222,097
7...................	12,000	16,800	9,656	8.23	3.5	80	36.6	3.42	51,835,132
8...................	29,500	24,000	9,500	8.32	5	60	36.8	3.44	107,488,381
Grand total.......									382,401,491
Production 1880 to 1907, inclusive..									43,683,445
Total red-ore reserves in main portion of Birmingham district....................									338,718,046

a 1. Irondale seam, Morrow Gap to point beyond Olivia mine.
  2. Irondale seam, Bald Eagle to Clifton Gap.
  3. Big seam, upper bench, Bald Eagle to Lone Pine Gap.
  4. Big seam, upper bench, Lone Pine Gap to Graces Gap.
  5. Big seam, upper bench, Graces Gap to point beyond Ishkoodo.
  6. Big seam, upper bench, Ishkoodo to Tennessee Company's mine No. 10.
  7. Big seam, upper bench, Tennessee Company's mine No. 10 to middle of Woodward property.
  8. Big seam, upper bench, Middle of Woodward property to Sparks Gap.

It is frankly admitted that the magnitude of the figures obtained by this estimate is rather surprising. When it is considered that the present annual production of red ore in Alabama is not greatly in excess of 3,000,000 long tons, and that this production has not increased rapidly in recent years and does not promise to increase rapidly in the near future, the results of the estimate indicate that the iron-ore reserves in this district will last for seventy-five to one hundred years longer at the present rate of output. If the estimate of the writer, 340,000,000 long tons of red ore in the Birmingham district workable under present conditions, is compared with the estimate of 1,000,000,000 long tons of red ore in reserve in the State of Alabama, recently published by E. C. Eckel,[a] it would appear that the present estimate is fairly conservative, when it is recalled that the Birmingham district probably contains 90 per cent of the workable red ore of the State. In explanation of Mr. Eckel's apparently higher estimate it should be stated that much ore at present unworkable has been included therein.

It should be repeated, in conclusion, that the present estimate is based on the belief that the ores are the result of original deposition, that they occur in the form of regular lens-shaped bodies, that their content of metallic iron does not greatly diminish from the point where the hard ores are first encountered in the mine slopes to the point where the minimum workable thickness is reached, and finally, that the structure remains fairly constant as indicated in the foregoing discussions. This last element, it should be remembered, is one of the most uncertain, and can be rendered more certain only by thorough and systematic prospecting with the drill between Red Mountain and Shades Mountain. Unexpected structural complications and "horses" of barren rock may greatly reduce the quantity of workable ore counted on in this estimate. On the other hand, in the less favorably regarded divisions of the district which are described in detail in the forthcoming paper, there are large reserves of ore which have not been included at all in this estimate of ore tonnage available in the Birmingham district.

---

[a] Production of iron ores and iron products in 1906: Mineral Resources U. S. for 1906, U. S. Geol. Survey, 1907, p. 79.

THREE DEPOSITS OF IRON ORE IN CUBA.

By Arthur C. Spencer.

INTRODUCTION.

The Iron Age for August 15, 1907 (vol. 80, pp. 421–426), contained a description of a large deposit of iron ore in the Mayari district, Cuba, which has been under development by the Spanish-American Iron Company since January, 1904. The engineers of this company believe that the deposit contains more than 500,000,000 tons of ore, carrying above 40 per cent of iron, and it is pointed out that this amount adds 5 per cent to the world's reserve of iron ore, as estimated in 1905 by the Swedish geologist Törnebohm. Though the statistical importance of the deposit is considerably decreased by the latest estimate of the iron-ore reserves of the United States, indicating that our home supply is at least 10,000,000,000 tons,[a] the industrial importance of this new source of iron ore is in no manner affected.

It is stated in the article referred to above that explorations since 1898 have revealed many iron-ore deposits in various parts of the island. "Deposits of a few tons were numerous and those of a few hundred thousand tons were perhaps three in number." Data presented in the following pages indicate that the word "thousand" in the sentence quoted should read "million," so that, to the uninformed reader, the paragraph of which the sentence quoted forms a part is misleading in regard to the prospective importance of Cuba's medium-grade iron-ore deposits.

The Mayari ores are distinct, both in kind and in occurrence, from the well-known iron ores occurring in the Sierra Maestra near the south coast of Oriente Province, or Santiago de Cuba as it was formerly called. The Cuban ores which have been mined up to the present time are hard hematites with an admixture of magnetite, containing rather high sulphur and a small amount of copper. They occur as large and small irregular masses associated with a variety of

---

[a] Eckel, E. C., Advance chapter from Mineral Resources U. S. for 1906, U. S. Geol. Survey, 1907.

embedding rocks, which include hornblende and epidote schist, marble, diorite, and porphyry.[a]

The ores of the Mayari type are essentially hydrous brown iron ores, coming under the general head of limonite. They occur in blanket form as a surficial mantle covering massive serpentine and related rocks. Average analyses show a small percentage of chromium, rather high alumina, very low sulphur, and phosphorus below the Bessemer limit. Iron, ranging from 30 to 50 per cent, is usually above 40 per cent.

The three ore fields to be described pass under the names Moa, Mayari, and Cubitas. The first two are in Oriente Province, near the north coast, and the third is in Camaguey Province (formerly Puerto Principe), midway between Camaguey City and the north coast of the island. (See fig. 20.)

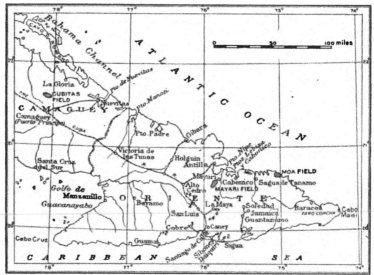

FIG. 20.—Sketch map of Camaguey and Oriente provinces, Cuba. 1, Iron mines in operation; 2, abandoned iron mines; 3, copper mines; 4, manganese deposits.

Until the announcement of the results of explorations carried on by the Spanish-American Iron Company appeared, the only available account of the occurrence and character of the Cuban iron-ore deposits of the Mayari type was a short note contributed by the present writer to a report on the mineral resources of Cuba, which was published in

a Kimball, J. P., Geological relations and genesis of the specular iron ores of Santiago de Cuba: Am. Jour. Sci., 3d ser., vol. 28, 1884, pp. 416–429; The iron-ore range of the Santiago district of Cuba: Trans. Am. Inst. Min. Eng., vol. 13, 1885, pp. 613–634.

Graham, T. H., Sigua iron mines, Cuba: Iron Age, vol. 41, p. 140.

Chisholm, F. F., Iron-ore beds in the province of Santiago, Cuba: Proc. Colorado Sci. Soc., vol. 3, 1891, pp. 259–263.

Cox, J. S., Eng. and Min. Jour., vol. 16, pp. 745–758.

Wedding, H., Stahl u. Eisen, vol. 12, p. 545; Iron Age, vol. 49, p. 607.

Spencer, A. C., Eng. and Min. Jour., vol. 72, pp. 633–634.

the annual report of the military governor of Cuba for the year 1901.[a]
The following paragraphs are quoted from this report:

Occupying the general region between Nipe Bay and Moa Bay and somewhat back from the northern coast there is a region reaching a general elevation of from 1,500 to 2,000 feet, and occupied by serpentines and other igneous rocks. Upon the top of this sierra there are many large areas which are practically level, and these are always covered by a thick mantle of red clay which contains a large proportion of iron ore in the form of spherical pellets. Locally this material entirely replaces the clay, and the separate particles are cemented together by ferruginous materials, making a spongy mass of brown iron ore. Similar occurrences of shot and massive ore were noted upon the tops of certain hills lying to the north of the city of Puerto Principe, and following the general trend of the Sierra Cubitas. The rock in this vicinity is also serpentine, and the ores have identical characteristics with those of the region mentioned above. Analyses were made from samples of these residual ores collected near Rio Seco along the trail between Mayari and San Luis.

	1.	2.
Moisture...		0.56
Iron...	52.00	54.69
Manganese...	.364	.594
Phosphorus..	.0368	.0189
Silica..	2.62	2.51
Chromium..	Trace.	Present.
Titanium..	.25	....

1. Iron ore from Sierra Nipe near trail crossing of Rio Naranjo, about 10 miles from Mayari, Santiago Province.
2. Iron ore from Sierra Nipe near Rio Seco, Santiago Province.

These residual ores are locally known as "tierra de perdigones," or "moco de herrero,' signifying shot soil and blacksmith's waste, either of which terms is a very apt designation. Rodriguez Ferrer is authority for the statement that hydrated oxide of iron in the form of pellets in the soil occurs at various points in the island. The following localities are mentioned: Province of Pinar del Rio, between Consolacion del Sur and Candelaria; Matanzas Province, in the Sierra Morena, between Cardenas and Sagua la Grande; Loma Iman, near the city of Puerto Principe; and Monte Libano north of Guantanamo, Santiago Province. The amount of these ores in various parts of the island is certainly very large, and it seems not improbable that they may eventually find a market in the United States in cases where they are situated near a sufficient supply of running water for washing them free from the clay with which they are mixed.

MAYARI DISTRICT.[b]

The Mayari iron-ore fields are situated south of Nipe Bay in the northern part of Oriente, so named because it is the easternmost province of Cuba. The ore deposit is a blanket formation, extending as a practically unbroken mantle over a gently rolling elevated plateau, roughly10 miles long and 4 miles wide; or, more accurately, about 27,870 acres in extent. Except for a few groups of hard-wood trees in moist situations, the ore field is covered by pine forest, averaging about 40 trees of medium size to the acre. The direct distance to the shipping point on Nipe Bay is about 12 miles, but the distance over the transportation route, including two inclines and

a Report on a geological reconnaissance of Cuba made under the direction of Gen. Leonard Wood, military governor, by C. Willard Hayes, T. Wayland Vaughan, and Arthur C. Spencer.
b Abstract of article in Iron Age, August 15, 1907.

a railroad, is somewhat more than 15 miles. The seaward edge of the plateau is about 1,600 feet above tide, from which elevation there is a gradual rise to 2,200 and 2,300 feet. Two peaks of 2,600 and 3,200 feet elevation are stated to have no iron on their slopes, from which it may be inferred that ore does occur upon the crests. The mantle of ore extends out onto the ridges between the streams which drain the edges of the plateau, and the deposit stops only where abrupt declivities begin.

The ore, which is chiefly earthy, is dark red near the surface and yellow at greater depths. In addition to this earthy ore, which forms the great bulk of the deposit, from an inch to a foot of gravelly ore composed of particles like bird shot occurs at the surface of the ground, and locally near the sources of streams similar ore particles cemented together form large lumps and flat-lying layers amounting in the aggregate to several million tons. The yellow clayey ore rests upon serpentine rock which is partly decomposed and locally soft enough to be penetrated by augers. The ore is said to be mainly a limonite, though some analyses indicate the presence of hematite as well.[a]

The deposit was explored in part by pits, but was sampled principally by means of 2-inch augers, borings being spaced 100, 300, 500, and 1,000 meters apart. During one year the average depth of borings was nearly 17 feet and the maximum depth attained was 51 feet. In all 3,030 borings were made and more than 15,000 analyses. A fair average depth of the ore over 18,525 acres is 15 feet, which, at 20 cubic feet to the ton, gives 605,000,000 tons. It is thought that this figure may be exceeded when every acre of the deposit has been examined, and it is regarded as certain that not less than 500,000,000 tons of ore is accessible for profitable mining. With the average given and the amount of ore at 605,000,000 tons, the tonnage per acre is 34,159. About 5 per cent of the borings were in material carrying below 27 per cent of iron and correspondingly high in silica or alumina, or both. An idea of the general run of the ore is presented by the table here given:

Composite analysis of iron ore from Mayari district.

[Dried at 212° F.]

	Number of samples.	Per cent.
Fe	918	46.03
SiO_2	918	5.50
Al_2O_3	889	10.33
Cr	889	1.73
P	889	.015

[a] Some of the ore pellets are attracted by a pocket magnet suggesting that magnetite is also present. My notes of 1901 indicate that the accumulations of shot ore at the surface have been washed out of the underlying deposit by the action of rains, and that to a depth of several feet great numbers of similar pellets are disseminated through the dark-red clayey matrix.—A. C. S.

In 67 samples the hygroscopic water averaged 31.63 per cent and in 37 samples the combined water averaged 13.62 per cent. The large amount of hygroscopic water and high alumina combine to give a clay-like consistency to the ore, such that in shafts which have stood two years every pick mark is still visible.

The uniformity of the deposit is shown by a table indicating that 94 per cent of the material is workable ore:

Percentages of iron in Mayari iron ore.

	Per cent.
10 to 20 per cent of iron	4
20 to 30 per cent of iron	2
30 to 40 per cent of iron	6
40 to 43 per cent of iron	6
Above 43 per cent of iron	82
	100

The occurrence and nature of the ore are favorable for steam-shovel mining, but because of its sticky character it has been necessary to design a special form of car to insure ready dumping. The presence of 45 per cent of moisture also makes drying indispensable to avoid import duty and freight charges on so much water, and the ore will have to be clinkered to make it physically suitable for use in the blast furnace.

The company plans the production of 2,500 tons of dry ore per day, but will install a plant capable of handling a very much larger tonnage. No description of the drying plant or details of mining methods are given. As Cuba is deficient in mineral fuel, it will be necessary to import coal for the calciners.

MOA DISTRICT.

The Moa iron-ore fields are contiguous to the protected deep-water harbor of Moa Bay, on the north coast of Oriente Province, about 50 miles east of Nipe Bay. The deposit is very much like the one at Mayari, but its situation makes it more easily accessible for mining and ocean transportation.

The Moa district lies upon the northern and seaward slope of the group of mountains which, with several other groups, enter into the make-up of a general range extending from the Mayari district south of Nipe Bay to the east end of Cuba and there merging with the mountains which border the south coast of Oriente Province. The general summit elevations of these northern mountains may be estimated at 2,000 to 2,500 or perhaps 3,000 feet. In the Moa country the highest summits are roughly about 2,500 feet, and it is probable that the northward-flowing streams of the district all rise within 10 to 15 miles of the coast. Looking toward the Moa Mountains from the sea one notes a gradual rise of the land from the shores of Moa Bay, and sees that the valleys of the larger streams are shallow and narrow at their mouths and that they gradually widen and deepen upstream. A traverse of the region bears out this first impression, and between each river broad, smooth surfaces are found to rise from the water's edge toward the interior with an average grade of 250 feet to the mile. These interstream areas extend from 5 to 8 miles back from the coast, and upon them the ore deposits occur in the form of a surface mantle.

The district is well watered and at the same time mostly well drained. Pine forest covers the greater part of the ore ground, the trees being more thickly set and on the average somewhat larger than in the Mayari district. Hard-wood forest and thick jungle is encountered everywhere along the watercourses and in moist patches on the upland surface.

The whole district appears to be underlain by serpentine, for no other rock is found either on the slopes of the river valleys or in the stream gravels, part of which must have been derived from the distant mountains. The country is dissected by river courses in such a manner as to exhibit the relations of the iron ore and the bed rock in many places. The typical occurrence is as follows: At the surface there is several feet of red clay containing from 50 to 75 per cent by weight of brown iron ore in the form of round pellets from a minute size up to that of a buckshot or a cherry. Beneath the shot ore is a more or less continuous layer of spongy brown ore which is usually seen to be composed of similar round pellets bound together by a ferruginous cement. Below the solid layer lies yellow clay, in places containing scales and irregular concretions of brown ore. It will be noted that the above description corresponds in a general way with that of the Mayari deposit. The thickness of the different layers named varies from place to place and locally any one, any pair, or all of the layers may be missing.

The mantle of ore is a prominent feature within an area of about 60 square miles, being practically continuous, except where it has been cut out by erosion along the stream valleys. It can hardly be doubted that the deposit was formerly not only actually continuous within the present field of its occurrence, but also of considerably greater extent.

The 60 square miles referred to has been fully explored, and all of it has been denounced or claimed by different parties. It is roughly estimated that 60 per cent of the area taken up, or about 36 square miles, will afford ore of minable grade and quantity. The ratio of barren ground to ore ground becomes less and less as one proceeds from the mountains toward the coast, the most nearly continuous deposits occurring in a strip of country 2 or 3 miles wide adjacent to the bay, where the river valleys are both shallow and narrow.

A considerable part of the field has been systematically prospected by the Spanish-American Iron Company, but other owners have not yet adequately explored their holdings. Though the writer's examination of the field was too cursory to form the basis of a trustworthy estimate of tonnage, it is thought that the amount of ore per unit area is likely to exceed that of the Mayari field, provided the yellow clay at Moa shows the same high iron content as that at Mayari. The thickness of the shot ore was seen to range from a few inches up to 20 feet; that of the layer ore up to 12 feet; and the yellow clay is locally more than 50 feet thick.

Taking the area of workable ore as 36 square miles, and, as at Mayari, taking the average depth as only 15 feet and allowing 20 cubic feet of material per ton of dried ore, we get 752,000,000 tons for a first approximation of the available tonnage of iron ore in the Moa field. The shot and layer ore from Moa, when dried at 212° F., give practically the same analysis as the Mayari ore. (See composite analysis, p. 321.)

CUBITAS DISTRICT.

The Cubitas iron-ore fields are situated from 12 to 15 miles north of Camaguey City, in the province of Camaguey. The port of Nuevitas, on the north coast, lies about 40 miles northeast of the district, which is as yet without transportation facilities. Within an area measuring roughly 10 miles east and west and 4 miles north and south, there are several flat-topped mesas rising 300 to 400 feet above the general level of an almost featureless plain which extends for many miles in all directions except toward the north. Two or three miles north of the ore fields there is a narrow range of rugged limestone mountains, known as the Sierra Cubitas.

The ore deposits are all surface mantles covering the plateau-like mesas. Shot ore in the red clay matrix and brown spongy layer ore are exposed in many places, and as at Moa where both are present the layer ore is invariably beneath the shot ore. Observations made in several separated localities indicate that the two varieties of ore are together at least 10 feet thick over much of the ore ground. It is probable that yellow clay exists beneath the brown ore, though no exposures of this material were noted by the writer during his short examination of the deposits. The observations on which the present notes are based were so casual that no trustworthy estimate can be given of the areal extent of the deposits. It is thought, however, that there must be at least 6,000 acres of the ore ground and that at least 150,000,000 tons of ore exists within the field.

In their physical character and occurrence the Cubitas ores are practically identical with those of the Moa and Mayari districts. No analyses of the ores can be given at the present time, but there is every reason to believe that the material will show the same chemical features as the ores from Moa and Mayari.

The bed rock of the district, being serpentine, is identical with that underlying the two other ore fields here described, but in the Cubitas field there is practically no pine or other timber, the principal growth being palmetto, bracken ferns, and coarse grasses of little or no value for grazing. Serpentine occurs also over much of the surrounding country, and in places other crystalline rocks are found. Outside of the serpentine country, but within the general region in which the Cubitas deposits are found, there are several bed-rock deposits of hard ore, consisting of hematite and magnetite and showing above 60 per cent of iron. Several bodies of rich chromite ore are also known.

ORIGIN OF THE DEPOSITS.

The yellow ferruginous clays and associated brown iron ores of the Moa, Mayari, and Cubitas districts are considered as residual materials, derived from the serpentine rocks upon which they rest, through the process of surficial weathering and decomposition. Strong evidence that the serpentine has contributed the material of the deposits lies in the fact that all the ores contain chromium, which is an element known to be characteristically associated in occurrence with certain serpentine and related rocks, and which may therefore be assumed to be present in small amounts in the serpentine that underlies these particular accumulations of iron ore. The case of these ores is closely similar to that of certain small deposits of brown iron ore embedded in brownish, earthy material, which occur on Staten Island, New York, and which have been described and commented upon by Dr. T. S. Hunt,[a] as follows:

This material rests immediately upon the serpentine rock of the region into which it graduates, and from the subaerial decay of which it has evidently been derived; the lower portion of the earthy matrix still preserves the peculiar jointed structure of the underlying serpentine.

* * * This limonite which is now mined to a considerable extent, contains, as several analyses have shown, from 1 to 2 per cent of chromic oxide, which is also known to be present in small amounts in the serpentine. An impure argillaceous specimen, containing 59.63 of ferric oxide, yielded the writer 2.81 of chromic oxide in a condition readily soluble in chlorhydric acid.

* * * It is, I think, evident that the decay of the serpentine, and the concentration, in the residuum, of its iron in the form of limonite, was a process anterior to the glacial erosion, and that the ore banks are areas of the decayed material which has escaped this action.

The Clealum iron ores of Washington, as described by Smith and Willis,[b] show a general similarity in composition to the Cuban ores, but contain nickel as well as chromium. They occur in the form of lenses, from a foot or two up to 30 feet in thickness, on an old eroded surface of an extensive formation of serpentine, and lie at or in the

---

[a] Mineral Physiology and Physiography, 1886, p. 268.
[b] Trans. Am. Inst. Min. Eng., vol. 30, 1901, pp. 356–366.

base of the Swauk sandstone.   The ore, which contains both hematite
and magnetite, is partly of oolitic nature, and is thought to have been
deposited originally in the condition of ferrous carbonate or hydrous
sesquioxide of iron.   It contains from 1.9 to 5.2 per cent of Cr_2O_3
and from 0.20 to 0.68 per cent of NiO.   A sample of serpentine analyzed gave 0.47 and 0.10 per cent, respectively, of Cr_2O_3 and NiO, so
that the ore is regarded as the result of concentration from the serpentine.   It is, however, thought to be not strictly a residual deposit
accumulated in place, but instead a sedimentary deposit of which the
materials were furnished mainly by the serpentine.

That the Cuban ores here under discussion were derived from the
serpentine rocks with which they are associated can be accepted
without reservation, and there is good reason to believe that they
were formed by decomposition of the rock in place rather than that
the materials of which they are composed were brought into their
present positions either as washed-in sediments or as dissolved salts.
Evidence for this last conclusion is seen in the facts that the ore
material contains no grains of sand and that the deposits are spread
over extensive areas which are almost completely lacking in local
topographic relief, except such as can be seen to have originated
since the accumulation of the ores.   It is thought that the surfaces
of low relief upon which the ore mantles occur must be remnants of
formerly more extensive plains, which have been uplifted and warped
by mountain-building forces and largely destroyed by subsequent
erosion.   At Moa the destruction of the ore ground is evidently even
now progressing at a geologically rapid rate.   Theoretically, plains
of this sort are the natural end products of subaerial erosion continued through very long periods undisturbed by mountain-building
forces, and the thick accumulation of strictly residual materials is the
natural and perhaps necessary accompaniment of the later stages of
the process of planation.   Peneplains (surfaces of extremely low relief reduced by subaerial decomposition, decay, and solution of the
rocks over a wide region) are recognized as existing over a large part of
the Atlantic and Gulf slopes of the United States.   An extensive and
now considerably dissected peneplain which is seen in the Piedmont
Plateau of our Eastern States is known to have been formed during
late Tertiary time and to have become partly buried by sedimentary
materials that are now grouped together and described as the Lafayette formation.

Orange-colored and red sands and gravels characterize this formation in many districts and though no direct means of correlation are
at present known, it seems very likely that those uplifted and warped
peneplains of Cuba which are covered by deep soils and clays may
eventually be found to correspond in age with the pre-Lafayette, or,
as it is more commonly called, the Tertiary peneplain of the eastern
United States.

Though the general conditions under which the Cuban iron ores accumulated were probably those which have been outlined, the actual mode of origin is a problem that demands a large amount of detailed chemical work and a comparison with similar problems presented in other regions. Consideration of the fact that the red and brown materials, consisting of shot ore in an earthy matrix and of porous layer ore, occur in the surficial part of the residual masses leads to the idea that they are the products of secondary weathering of the yellow clays which are thought to be the direct residuals from the decomposition of the serpentine. What may be termed the primary change, from serpentine to ocherous clay, involved the depletion of silica, magnesia, and lime and the concentration of alumina, iron oxide, and chromic oxide; the secondary change to brown and red materials seems to have consisted mainly in partial dehydration and deoxidation. These last effects, though probably slight, are evidenced by the marked difference of color, by the slightly higher iron content of shot ore occurring on the immediate surface, by the reported presence of some hematite in addition to the hydrated oxides that constitute the bulk of the iron ore, and by the presence of magnetite, which is indicated by the magnetic property of a portion of the shot ore.

MINING CONCESSIONS.

The Republic of Cuba as a former colony of Spain has inherited mining laws based on the principle that the ownership of all deposits of metallic minerals is inseparable from the State. In the case of iron ores, concessions of any desired size may be acquired from the Government by the procedure known as denouncement. The expense of denouncement, attendant surveys, and title of concession amounts to $2.86 per hectare, equivalent to 2.47 acres. The law provides for an annual tax of $2.50 per hectare, nonpayment of which is the only cause of forfeiture. Collection of this charge has been suspended since 1901, so at the present time mining rights once secured by a small expenditure may be held without further cost.

Where the surface rights to land containing mineral deposits are held by private interests arrangements must be made to satisfy all damages which may ensue, but in the case of wild or forest lands owned by the Government the concessionary apparently acquires full surface rights with the title of the concession. The feature last mentioned is of great importance both at Moa and at Mayari, where large tracts which have been held by the Church of Rome became Government lands when the Republic was established.

With the exception of a few small claims at Moa, all the ore lands in the Moa and Mayari districts have been denounced within the last four years, so that in these districts titles are likely to be clear. In the

Cubitas district most of the denouncements were made many years ago, and in some cases, at least, titles to ownership are more or less clouded.

METALLURGY.

The high alumina of the Mayari and similar ores produces an unusual slag, demanding careful and intelligent operation. The chromium which is present goes into the pig iron and most of it must be removed in the manufacture of steel. Extended experimentation by the engineers of the Pennsylvania Steel Company has resulted in the discovery of a method by which chromium can be sufficiently eliminated in the Bessemer converter, and the effect of the portion of chromium remaining on the quality of the product has been carefully studied. The following account of the process which has been developed for the removal of chromium appeared in the Engineering and Mining Journal October 7, 1905:

An invention calculated to increase the value of chromiferous iron ores has lately been patented by H. H. Campbell, of Steelton, Pa. (U. S. No. 795193.) Using iron ores containing from 1 per cent to 5 per cent chromium, he succeeded in producing a steel containing only 0.08 per cent Cr. It is well known that, if a steel has a content of Cr in excess of a certain small proportion, it is unfit for use for most purposes in engineering works; and also that it has not been practical to make use of iron rich in Cr as a starting metal for the manufacture of steel on a profitable scale because of the inability to effect the economical removal of the Cr. Mr. Campbell's method is to first treat the chromium iron in a basic Bessemer converter, producing a basic slag, and then to oxidize the Cr by prolonging the blow beyond the usual period, which causes the Cr to enter the slag. The charge is then drawn from the converter into a ladle (having a device for drawing the metal from beneath the slag), and then stopping or controlling the flow of slag. The subsequent treatment depends on the final use for which the steel is intended. For low-carbon steel, where the dechromized metal does not contain much oxygen, it is incorporated with ferromanganese to obtain the usual reaction; where the dechromized metal contains considerable oxygen, it is charged in a second converter having a siliceous lining with an addition of unblown molten iron (free from Cr and having a higher carbon content) and finally recarburizing.

The removal of chromium from pig iron in the open-hearth process of steel making has been described by A. W. Richards in a recent paper entitled "A method of producing high-class steel from pig iron containing chromium, nickel, and cobalt,"[a] from which the following extract is taken:

Doctor Massenez, the inventor of the new process, conceived the idea of working in a basic or neutral lined open-hearth furnace provided with slag notches, thinking that the chromium might be removed by forming successive voluminous quantities of oxidizing slags and, as the chromium oxide was formed, removing these slags through the slag notches before they became too thick.

Into an open-hearth furnace is charged 3 tons each of lime and basic slag, the latter produced by melting hematite pig iron on a basic hearth, and over this is added 3 tons of hematite iron ore. The mixture is heated until the ore becomes pasty. Two ladles

a Jour. Iron and Steel Inst., No. 1, 1907, pp. 114–120. United States patents Nos. 754154 and 772164.

containing 10 to 11 tons each of molten chrome nickel iron are then added. Thick and foaming slag is gradually formed, containing a portion of the chromium as chromic oxide, and this at the end of forty-five minutes to one hour is run out at the slag notches. A second slag is then formed from the materials remaining on the bottom of the furnace hearth, which do not all melt at first, and by further additions of basic slag, fluor spar, and lime. This slag after melting contains an additional quantity of chromic oxide, and is also removed, but owing to its viscous character it can only be made to pass the slag notches by mechanical assistance. This operation of forming slag is repeated according to the percentage of chromium in the iron. When working with an iron containing 4 per cent of chromium it is generally necessary to make and run out four such slags, after which practically all of the chromium is removed. When this point is reached the metal is worked like an ordinary open-hearth charge. The resulting steel contains about 0.3 per cent of chromium, as it derives some of this during the decarburizing period in the open hearth, being reduced from the chromic oxide in the slag.

The boil commences in from nine to ten hours after the furnace is charged, and only lasts an hour before the steel is ready for tapping.

The slag contains from 5 to 10 per cent of chrome oxide.

The steels produced by Mr. Richards contain from 0.52 to 0.69 per cent of carbon, 0.016 to 0.032 per cent of phosphorus, and 0.12 to 0.51 per cent of chromium. The iron used was made from ore of the following analysis, the source of the ore not being stated:

Fe (as received)...................................	45. 69
Mn...	. 232
Cr...	2. 35
Ni } Co }	. 90
SiO$_2$..	12. 32
Al$_2$O$_3$..	9. 25
CaO...	. 55
MgO...	2. 14
S...	. 029
P...	. 02

The pig iron contains—

Cr...	4. 00
Ni } Co }	1. 75
Si...	. 40

And the steel made from it contains—

Ni...			1. 50
Co...			. 25
Cr...	.12	to	. 51
P...	.016	to	. 032
C...	. 52	to	. 69

IRON ORES NEAR ELLIJAY, GA.

By W. C. Phalen.

INTRODUCTION.

The Ellijay quadrangle of the United States Geological Survey lies mainly in the northern part of Georgia, but includes narrow strips of Tennessee and North Carolina along its northern border. The area is entirely within the mountain division of the Appalachian province.

The existence of iron ores in and near this area has long been known, but it is only recently that systematic effort has been made to prospect them thoroughly with a view to their commercial exploitation. It is the purpose of this paper to present briefly the latest information regarding these ores, gathered in the field season of 1907.

GEOLOGY.

The rocks in the area are for the most part southwesterly continuations of belts which have been carefully studied by Arthur Keith in the Murphy and Nantahala quadrangles on the north and northeast and are in part of sedimentary and in part of igneous origin. Nearly all have been metamorphosed, some to such an extent that all traces of their original nature have disappeared, a condition which adds greatly to the difficulty of their study. According to Keith, they range in age from the Archean into the Cambrian, those of the latter system constituting the Ocoee group. Named in order from the top of the geologic column represented they comprise the Murphy marble, Valleytown formation, Brasstown schist, Tusquitee quartzite, Nantahala slate, and Great Smoky conglomerate, all probably of Cambrian age, and granites and gneisses belonging in the Archean. It is probable that the nomenclature given above may be changed in part when the Ellijay folio is published, as it is quite certain that some of the lithologic units used in mapping the Nantahala quadrangle either merge with others or die out completely in passing southward from North Carolina into Georgia. Most of these rocks, except the Murphy marble, are cut by dikes of igneous origin, many of which are quartz diorite, though much of the intruded rock is more basic than this.

All the rocks of this region have been folded and faulted, as a result of which they have been greatly compressed and metamorphosed. Faulting has aided in the formation of certain ore deposits to be subsequently described.

330

THE IRON ORES.

GENERAL OUTLINE.

Iron ore occurs at many localities in and near the Ellijay quadrangle, but in comparatively few places is it present in sufficient quantity to repay systematic development. Most of the ore belongs in the class of hydrous ferric oxides, the so-called "brown ore," but the available analyses given in this paper are so incomplete that little can be told about the original content of water in the ore, and thus it is uncertain exactly which hydrous oxide is present—turgite, goethite, or limonite. From the fact that the ores are entirely amorphous it is clear that they are more recent than the rocks with which they are associated and are also later than the periods of deformation during which the neighboring rocks have been affected. The deposits therefore are not original but secondary in nature.

The occurrences visited and studied are located as follows:

(1) Ore Knob, a small and somewhat isolated peak between Whitepath and Little Turniptown creeks, just east of the Atlanta, Knoxville and Northern division of the Louisville and Nashville Railroad, and nearly 4 miles northeast of Ellijay.

(2) About a mile east of Ellijay and just north of Cartecay River.

(3) About 6 miles southwest of Ellijay, just outside of the Ellijay quadrangle, near the mouth of Talona Creek and 1½ miles north of Talona station.

DEPOSITS AT ORE KNOB.

Occurrence.—At Ore Knob, iron ore has been exposed by several open cuts and pits, most of which are in a quartzitic sandstone or quartzite, though one test pit from which considerable ore has been removed has a foot wall of lustrous blue schist, characteristic of the Valleytown formation as exposed along the road between Blue Ridge and Ellijay. Several shipments of ore are reported to have been made from the property, one of 29 cars having been specified. All the ore seen is situated well above drainage level.

As mentioned above, the rock with which the bulk of the ore is associated appears highly siliceous. Along the public road south of Ore Knob, silvery mica schist, probably of the Valleytown formation, outcrops in several places, east of which were seen ore and sandy débris. A short distance to the east, in the valley of Little Turniptown Creek, the typical blue Nantahala slate was observed. All the observations made in the locality indicate dips of 40° or more, generally to the southeast. The strike of the rocks is usually east of north, but in places swings around nearly to north, and as nearly as could be determined the trend of the ore belt conforms closely to that of the adjacent rocks. Thus the ore occurs in a belt close to the contact of the Valleytown formation and Nantahala slate. As the beds are disposed the Nantahala overlies the Valleytown, being thrust over on it from the east. The ore apparently lies along the line of the thrust fault.

Character and origin of ore.—This conclusion as to the location of the ore with respect to the fault is in accord with observations on the character of the ore itself, some of which is a quartzite breccia cemented by iron oxide. Much of the ore is highly siliceous, being in some places a mere film of iron oxide on a quartzitic nucleus. Some of it is black, suggesting an admixture of manganese oxide; but the bulk of the material is the common limonite or brown ore. Though the ore is siliceous in part, much of it is very pure, and the study of that already uncovered indicates that with careful culling good ore exists in large enough quantity to make the occurrence an inviting one for future development.

A faulted zone is one along which waters might either descend or ascend. The ores here discussed appear to have been leached from the surrounding rocks and deposited by descending waters, for there are no sulphides associated with the ore, as would probably be the case if it were due to ascending waters. If this hypothesis as to the origin of the ores be correct, the deposits will be limited in depth but may have considerable linear extent. Such being the case, careful search along the western edge of the blue-slate belt, where these rocks are thrust upon the Valleytown formation and Murphy marble, may reveal the presence of other valuable deposits.

DEPOSITS NORTH OF CARTECAY RIVER.

Occurrence.—The deposits a mile east of Ellijay are located just north of Cartecay River, on the west side of Randall Branch. The hill in which the ore occurs rises about 300 feet above the bed of the river. The old workings are now inaccessible, but according to reports the ore occurs in at least two veins, a main vein and another of less width, which are exposed in a tunnel 200 feet long, driven from a point near the east base of the hill. In the tunnel a winze was reported nearly 50 feet deep, wholly in ore. A surface cut has also been made at one point, exposing an ore body said to be 40 feet thick, though the width shown in the tunnel is much less than this, or about 25 feet. A few open cuts and an old shaft said to be 70 feet deep have also been sunk on the ore. On the south side of the hill at the roadside another tunnel has evidently been started. The ore can be traced intermittently for several hundred feet along the hill to the north, and the underground and surface observations point to a fairly large body above drainage level. The rocks in the immediate locality, among which the blue Nantahala slate predominates, have variable strikes, ranging from N. 15° E. to N. 44° E., and the strike of the main ore body is about N.

30° E., with a dip of 50° SE. The ore body is thus conformable to the inclosing slates. The country rock immediately east is blue slate, regarded as characteristic Nantahala. Farther south, in the town of East Ellijay, the same slate outcrops along the roadside, striking about 30° east of north, and dipping at high angles to the southeast. At the bridge over Cartecay River the blue slates also outcrop. Some of the rock in the valley of Randall Branch is grayish mica schist, not entirely characteristic of the Nantahala, and it is possible that here, as at Ore Knob, the ore is situated close to a fault contact.

Character and origin of ore.—The ore itself is the usual brown ore or limonite. Some of it is rather siliceous, but the available analyses indicate a high-grade ore, as shown below:

Analyses of iron ore east of Ellijay, Ga.

	1.	2.	3.	4.
Metallic iron	51.48	54.61	55.71	50.88
Silica and insoluble	8.29	....	4.65	9.28
Phosphorus	1.06	.917	.226	1.90
Sulphur	Trace.	....	.148	.01
Moisture at 212° F	....	....	....	1.12

1 From an average sample taken across the big vein in the main tunnel, by Hall Brothers, mining engineers, Atlanta, Ga. Analysis made in the N. P. Pratt laboratory, Atlanta, Ga.
2. Ore from vein in main tunnel. John M. McCandless, chemist, Atlanta, Ga.
3. Ore from outcrop. Hodge & Evans, chemists, Anniston, Ala.
4. From sample taken entirely across the face of vein and collected by S. W. McCallie. Analysis made in the N. P. Pratt laboratory, Atlanta, Ga.

These analyses show an ore with a fair content of metallic iron, high in phosphorus, and, with the exception of the outcrop sample analyzed by Hodge & Evans (No. 3), low in sulphur.

Like the occurrence at Ore Knob, the ores east of Ellijay appear to be due to surface or shallow underground waters which have leached the iron from the garnet, staurolite, pyrite, and other iron-bearing minerals contained in surrounding rocks and deposited it in its present position. The planes of schistosity which exist in the slates of this region, especially if the rock is at all calcareous, would facilitate the movement of such solutions. Along a fault which, it is quite possible, exists here, the movement of such waters would probably be more pronounced, and in such a position the ore most likely would occur in largest amount.

DEPOSITS NEAR TALONA.

Occurrence and development.—Near the point where Talona Creek crosses the Louisville and Nashville Railroad, about 1½ miles north of Talona station, a wide vein of iron ore has been exposed in an open cut on the west side of the valley near the valley floor. The Murphy marble underlies the valley here and extends upward a short distance on the hillside to the west. The ore occurs between the marble on the east and a satiny schistose phase of the Valleytown formation, which in the immediate vicinity is a dull reddish brown or blue, on the west.   The attitude of the schist could be readily determined; not so easily, however, that of the marble, owing to the difficulty of identifying true bedding planes.   The schist strikes between 20° and 25° east of north and dips from 40° to 50° SE.   In the open cut already excavated the ore appears as a practically solid ledge. To judge from ore in place to the west, the lead is nearly 50 feet thick, although it may not be all solid ore.   It can be traced for a long distance to the north and not so far southward.   At the time of visit several hundred tons of ore had been mined and placed on the stock pile since the beginning of operations in August, 1907, and the North Georgia Marble Company, which controls the property, was at that time building a spur track to the railroad.

Character and origin of ore.—The ore is limonite and does not differ materially from that above described.   Its origin, however, is distinctly different in that it is apparently not along a fault plane.   The Murphy marble, as has been mentioned, underlies the valley floor and the schist to the west is probably in the Valleytown formation, lying normally below the marble.   Such a contact between insoluble schists and a marble offers a natural channel for the descent of superficial waters, and it is quite possible for the iron oxides leached from the ferruginous minerals in the adjacent formations to concentrate along such a contact and to be subsequently oxidized.   The ore body will probably be found to extend only to moderate depths, but the quantity of ore above ground-water level will be large owing to the linear extent of the deposit.

Two analyses of this ore, kindly furnished to the writer by Mr. H. A. Field, of Ellijay, are as follows:

Analyses of Talona iron ore.

	1.	2.
Metallic iron	49.00	51.80
Silica	16.50	12.04
Manganese		.30
Phosphorus	.37	.67

1. Childers & Hunter, Knoxville, Tenn., analysts.
2. Analysis made by the Virginia Coal and Coke Company.

These analyses show the Talona ore to be of a fairly good grade. Phosphorus is high, but if it does not run any higher on further development it is not objectionable.

SURVEY PUBLICATIONS ON IRON AND MANGANESE ORES.

A number of the principal papers on iron and manganese ores published by the United States Geological Survey or by members of its staff are listed below. In addition to these papers, several geologic folios contain descriptions of iron ore deposits of more or less importance. When iron is an important resource in the particular area covered by a folio, it is listed in italics in the tables (PP. 9–11).

BALL, S. H. The Hartville iron ore range, Wyoming. In Bulletin No. 315, pp. 190–205. 1907.

—— Titaniferous iron ores of Iron Mountain, Wyoming. In Bulletin No. 315, pp. 206–212. 1907.

BARNES, P. The present technical condition of the steel industry of the United States. Bulletin No. 25. 85 pp. 1885.

BAYLEY, W. S. The Menominee iron-bearing district of Michigan. Monograph XLVI. 513 pp. 1904.

—— (See also Clements, J. M., Smyth, H. L., Bayley, W. S., and Van Hise, C. R.; also Van Hise, C. R., Bayley, W. S., and Smyth, H. L.)

BIRKINBINE, J. American blast-furnace progress. In Mineral Resources U. S. for 1883–84, pp. 290–311. 1885.

—— The iron ores east of the Mississippi River. In Mineral Resources U. S. for 1886, pp. 39–98. 1887.

—— The production of iron ores in various parts of the world. In Sixteenth Ann. Rept., pt. 3, pp. 21–218. 1894.

—— Iron ores. In Nineteenth Ann. Rept., pt. 6, pp. 23–63. 1898.

—— Manganese ores. In Nineteenth Ann. Rept., pt. 6, pp. 91–125. 1898.

BOUTWELL, J. M. Iron ores in the Uinta Mountains, Utah. In Bulletin No. 225, pp. 221–228. 1904.

BURCHARD, E. F. The iron ores of the Brookwood district, Alabama. In Bulletin No. 260, pp. 321–334. 1905.

—— The Clinton or red ores of the Birmingham district. In Bulletin No. 315, pp. 130–151. 1907.

—— The brown ores of the Russellville district, Alabama. In Bulletin No. 315, pp. 152–160. 1907.

CHISOLM, F. F. Iron in the Rocky Mountain division. In Mineral Resources U. S. for 1883–84, pp. 281–286. 1885.

CLEMENTS, J. M. The Vermilion iron-bearing district of Minnesota. Monograph XLV. 463 pp. 1903.

CLEMENTS, J. M., SMYTH, H. L., BAYLEY, W. S., and VAN HISE, C. R. The Crystal Falls iron-bearing district of Michigan. Monograph XXXVI. 512 pp. 1899.

DILLER, J. S. Iron ores of the Redding quadrangle, California. In Bulletin No. 213, pp. 219–220. 1903.

DILLER, J. S. So-called iron ore near Portland, Oreg. In Bulletin No. 260, pp. 343–347. 1905.

ECKEL, E. C. Utilization of iron and steel slags. In Bulletin No. 213, pp. 221–231. 1903.

——— Iron ores of the United States. In Bulletin No. 260, pp. 317–320. 1905.

——— Limonite deposits of eastern New York and western New England. In Bulletin No. 260, pp. 335–342. 1905.

——— Iron ores of northeastern Texas. In Bulletin No. 260, pp. 348–354. 1905.

——— The Clinton hematite. In Eng. and Min. Jour., vol. 79, pp. 897–898. 1905.

——— The iron industry of Texas, present and prospective. In Iron Age, vol. 76, pp. 478–479. 1905.

——— The Clinton or red ores of northern Alabama. In Bulletin No. 285, pp. 172–179. 1906.

——— The Oriskany and Clinton iron ores of Virginia. In Bulletin No. 285, pp. 183–189. 1906.

——— Iron ores, pig iron, and steel. In Mineral Resources U. S. for 1906, pp. 67–102. 1907.

——— Manganese ores. In Mineral Resources U. S. for 1906, pp. 103–109. 1907.

——— (See also Hayes, C. W., and Eckel, E. C.)

HARDER, E. C. (See Leith, C. K., and Harder, E. C.)

HAYES, C. W. Geological relations of the iron ores in the Cartersville district, Georgia. In Trans. Am. Inst. Min. Eng., vol. 30, pp. 403–419. 1901.

——— Manganese ores of the Cartersville district, Georgia. In Bulletin No. 213, p. 232. 1903.

HAYES, C. W., and ECKEL, E. C. Iron ores of the Cartersville district, Georgia. In Bulletin No. 213, pp. 233–242. 1903.

HOLDEN, R. J. The brown ores of the New River-Cripple Creek district, Virginia. In Bulletin No. 285, pp. 190–193. 1906.

IRVING. R. D., and VAN HISE, C. R. The Penokee iron-bearing series of Michigan and Wisconsin. Monograph XIX. 534 pp. 1892.

KEITH, A. Iron-ore deposits of the Cranberry district, North Carolina-Tennessee. In Bulletin No. 213, pp. 243–246. 1903.

KEMP, J. F. The titaniferous iron ores of the Adirondacks [N. Y.]. In Nineteenth Ann. Rept., pt. 3, pp. 377–422. 1899.

KINDLE, E. M. The iron ores of Bath County, Ky. In Bulletin No. 285, pp. 180–182. 1906.

LEITH, C. K. The Mesabi iron-bearing district of Minnesota. Monograph XLIII. 316 pp. 1903.

——— Geologic work in the Lake Superior iron district during 1902. In Bulletin No. 213, pp. 247–250. 1903.

——— The Lake Superior mining region during 1903. In Bulletin No. 225, pp. 215–220. 1904.

——— Iron ores in southern Utah. In Bulletin No. 225, pp. 229–237. 1904.

——— Genesis of the Lake Superior iron ores. In Economic Geology, vol. 1, pp. 47–66. 1905.

——— Iron ores of the western United States and British Columbia. In Bulletin No. 285, pp. 194–200. 1906.

LEITH, C. K., and HARDER, E. C. The iron ores of the Iron Springs district, southern Utah. Bulletin No. 338. In press.

SMITH, E. A. The iron ores of Alabama in their geological relations. In Mineral Resources U. S. for 1882, pp. 149–161. 1883.

SMITH, GEO. O., and WILLIS, B. The Clealum iron ores Washington. In Trans Am. Inst. Min. Eng., vol. 30, pp. 356–366. 1901.

SMITH, P. S.   The gray iron ores of Talladega County, Alabama.   In Bull. 315, pp. 161-184. 1907.

SMYTH, H. L.   (See Clements, J. M., Smyth, H. L., Bayley, W. S., and Van Hise, C. R.; also Van Hise, C. R., Bayley, W. S., and Smith, H. L.)

SPENCER, A. C.   The iron ores of Santiago, Cuba.   In Eng. and Min. Jour., vol. 72, pp. 633-634. 1901.

——— Manganese deposits of Santiago, Cuba.   In Bulletin No. 213, pp. 251-255 1903.

——— Magnetite deposits of the Cornwall type in Berks and Lebanon counties, Pa. In Bulletin No. 315, pp. 185-189. 1907.

SWANK, J. M.   The American iron industry from its beginning in 1619 to 1886   In Mineral Resources U. S. for 1886, pp. 23-38. 1887.

——— Iron and steel and allied industries in all countries.   In Sixteenth Ann. Rept., pt. 3, pp. 219-250. 1894.

VAN HISE, C. R.   The iron-ore deposits of the Lake Superior region.   In Twenty-first Ann. Rept., pt. 3, pp. 305-434. 1901.

——— (See also Clements, J. M., Smyth, H. L., Bayley, W. S., and Van Hise, C. R.; Irving, J. D., and Van Hise, C. R.)

VAN HISE, C. R., BAYLEY, W. S., and SMYTH, H. L.   The Marquette iron-bearing district of Michigan, with atlas.   Monograph XXVIII. 608 pp. 1897.

WEEKS, J. D.   Manganese.   In Mineral Resources U. S. for 1885, pp. 303-356. 1886.

——— Manganese.   In Mineral Resources U. S. for 1887, pp. 144-167. 1888.

——— Manganese.   In Mineral Resources U. S. for 1892, pp. 169-226. 1893.

WILLIS, B.   (See Smith, G. O., and Willis, B.)

WOLFF, J. E.   Zinc and manganese deposits of Franklin Furnace, N. J.   In Bulletin No. 213, pp. 214-217. 1903.

YALE, C. G.   Iron on the Pacific coast.   In Mineral Resources U. S. for 1883-84, pp. 286-290. 1885.

ALUMINUM ORES.

SURVEY PUBLICATIONS ON ALUMINUM ORES—BAUXITE, CRYOLITE, ETC.

The following reports published by the Survey or by members of its staff contain data on the occurrence of aluminum ores and on the metallurgy and uses of aluminum:

BURCHARD, E. F. Bauxite and aluminum. In Mineral Resources U. S. for 1906, pp. 501–510. 1907.

CANBY, H. S. The cryolite of Greenland. In Nineteenth Ann. Rept., pt. 6, pp. 615–617. 1898.

HAYES, C. W. Bauxite. In Mineral Resources U. S. for 1893, pp. 159–167. 1894,

——— The geological relations of the southern Appalachian bauxite deposits. In Trans. Am. Inst. Min. Eng., vol. 24, pp. 243–254. 1895.

——— Bauxite. In Sixteenth Ann. Rept., pt. 3, pp. 547–597. 1895.

——— The Arkansas bauxite deposits. In Twenty-first Ann. Rept., pt. 3, pp. 435–472. 1901.

——— Bauxite in Rome quadrangle, Georgia-Alabama. Geologic Atlas U. S. folio No. 78, U. S. Geol. Survey, 1902, p. 6.

——— The Gila River alum deposits. In Bulletin No. 315, pp. 215–223. 1907.

HUNT, A. E. In Mineral Resources U. S. for 1892, pp. 227–254. 1893.

PACKARD, R. L. Aluminum and bauxite. In Mineral Resources U. S. for 1891, pp. 147–163. 1892.

——— Aluminum. In Sixteenth Ann. Rept. U. S. Geol. Survey, pt. 3, pp. 539–546. 1895.

SCHNATTERBECK, C. C. Aluminum and bauxite [in 1904]. In Mineral Resources U. S. for 1904, pp. 285–294. 1905.

SPURR, J. E. Alum deposits near Silver Peak, Esmeralda County, Nev. In Bulletin No. 225, pp. 501–502. 1904.

STRUTHERS, J. Aluminum and bauxite [in 1903]. In Mineral Resources U. S. for 1903, pp. 265–280. 1904.

PETROLEUM AND NATURAL GAS.

THE MINER RANCH OIL FIELD, CONTRA COSTA COUNTY, CAL.[a]

By RALPH ARNOLD.

Location.—The Miner ranch oil field is located on Lauterwasser Creek, one of the branches of San Pablo Creek, near De Laveaga, Contra Costa County, Cal. It is about 8 miles north-northeast of Oakland and is reached by road from that city and also from the several towns south of San Pablo Bay and west of Mount Diablo. The topography in the vicinity is characterized by moderately steep-sided canyons and rounded hills, some of which attain an elevation of over 1,500 feet above the adjacent valleys. The elevation at the wells, which are in the hills immediately south of Lauterwasser Creek, is between 500 and 700 feet above sea level.

HISTORY OF PETROLEUM PROSPECTING IN CONTRA COSTA COUNTY.[b]

Contra Costa County was one of the first counties in California in which petroleum was discovered, its presence being known as far back as 1864, when prospect wells were drilled 1½ miles south of the Empire coal mine. The following is a summary of the attempted petroleum devolopments in the county to date:

1864. J. W. Cruikshank, about 1½ miles south of the Empire coal mine. Several experimental wells, one 300 feet deep; green oil of high specific gravity; pumped about 15 barrels.

1865. Adams Petroleum Company, on Coates estate, south of Empire coal mine. Several shallow wells, from which some oil was obtained.

a The writer wishes to acknowledge his indebtedness to Dr. J. C. Merriam, of Berkeley, and Mr. W. E. Holbrook, president of the American Oil and Refinery Company, of San Francisco, for courtesies extended and assistance rendered during the examination of this field.

b The information contained in this section is derived largely from reports by W. L. Watts in Thirteenth Rept. California State Mineralogist, 1896, pp. 570-571, and Bull. No. 19 California State Mining Bureau, 1900, pp. 156-157.

1889. Chandler well, on Miner ranch, south bank of Lauterwasser Creek.   One well, 200 feet deep, yielded small quantity of heavy oil and water.

1895. Cumming well, on Miner ranch, one-fourth mile east of Chandler well.   Penetrates 20 feet of petroliferous shale and then 280 feet of sandstone; yielded traces of petroleum.

1896. Sonntag well, on Allen ranch, one-half mile east of the Cumming well.   100 feet deep, in light-colored sandstone.

1899. J. W. Laymance, on Old Tar ranch, 2 miles east of San Pablo.   170-foot well, encountered seepages of oil.

1900. Mount Diablo Oil Company, on Old Tar ranch.   Several wells drilled many years ago; drilling operations begun again in 1900, but failed to get oil.

American Oil and Refinery Company.   (See body of this report.)

Contra Costa Oil and Petroleum Company, on Coates estate, 1½ miles south of Empire coal mine.   One or more wells.

Grand Pacific Oil Company, on Hodges ranch, 1 mile east of Lafayette.   One or more wells.

Tide Water Oil Development Company, Coates estate, 1 mile south of the well site of the Contra Costa Oil and Petroleum Company; ceased operations in 1904.

Sobrante Oil and Investment Company, on Castro tract, a little over 3 miles northeasterly from San Pablo.   One or more wells, but no production; abandoned.   ·

San Pablo Oil Company, on Mulford ranch, 1 mile northeast of San Pablo.   One well 670 feet deep; traces of oil and considerable gas; now abandoned.

Point Richmond Oil Company, on Mulford ranch, 3 miles northeast of San Pablo.   Two 100-foot wells, drilled near seepage, but got no production.

Flood ranch, 1½ miles south of Miner ranch.   Old well with traces of oil on water.

National Paraffin Company, 1¼ miles northeast of Lafayette.   One well, 1,694 feet deep; no production.

Near the corner of secs. 9, 10, 15, and 16, T. 1 N., R. 1 E., Mount Diablo meridian.   An old well is said to have shown traces of oil.

Geology and structure.—The principal structural feature in the region of the Miner ranch is a southeastward-plunging anticline which crosses Lauterwasser Creek just east of the ranch house. Along the axis of this fold the following formations are exposed in order, beginning a mile or so northwest of the field and extending to its southeast edge: Tejon (Eocene) sandstone; 200± feet of brown sandstone, possibly Vaqueros (lower Miocene); 800± feet of Monterey (middle Miocene), 300 feet shale and 500± feet sandstone, in which are intercalated minor quantities of soft shale; and, finally, the feebly coherent fresh-water conglomerates, sandstones, and shales of the Orindan (Pliocene) formation of Lawson and Palache.[a] The petroleum deposits are in the shales and sandstones of the Miocene.

The sandstones below and above the Monterey shale are practically alike, being brown to gray in color, medium grained, and largely quartzitic. The upper sandstone is locally fossiliferous, although

---

a Bull. Dept. Geology, Univ. California, vol. 2, 1902, p. 371.

the state of preservation of the fossils usually precludes anything but a rough identification. The Monterey shale is fairly hard and is dark colored in fresh exposures, but weathers to a much lighter color. It contains many of the yellow and gray calcareous concretions so characteristic of the Monterey at most places in the Coast Range. The shale is largely organic in origin, foraminifers and diatoms being found in it abundantly. The organic remains in the shale are believed to be the source of the oil. The shale is considerably contorted, especially near the axis of the fold, dips of 42° S. 86° E. and 40° S. 70° E. being recorded within a short distance of each other in the bed of Lauterwasser Creek, northeast of the Miner ranch house. At the same locality the shale is exceedingly petroliferous, yielding a very prominent scum of light oil when the rock in the stream bed is disturbed with a pick. The dips in the overlying sandstone are not as easily obtainable as those in the shale, but it is thought that the dip on the northeast flank of the fold grows gradually less toward the northeast.

Wells.—Eight wells have been drilled in the Miner ranch field, none of which have so far been successful. They range in depth from about 570 feet to more than 2,750 feet. All lie on the northeast flank of the Miner ranch anticline, and all start in the sandstone above the Monterey shale, penetrating at first the upper sands and intercalated soft shales, and the deeper ones eventually reaching the Monterey shale. All the wells have shown more or less gas; in fact, the abundance of the gas is one of the characteristics of this field. It is said that the gas pressure was responsible for the collapse of the casing in at least two of the wells. One well is said to have encountered a pocket of oil at a depth of about 1,300 feet, which flowed 300 barrels of oil in nine hours. This is the only well in the field that has actually produced with the exception of the Flood well, 1½ miles to the south, which is reported as having yielded five barrels of 29° oil. The oil and gas apparently occur in pockets or lenses and no well-defined oil sand or petroliferous zone has yet been proved to be present.

Characteristics of the oil and gas.—The most interesting item in connection with the oil from the Miner ranch field is its relatively light gravity, said to be about 29°, as compared with the oils from the other fields in the State. As none of the wells were producing oil at the time of the writer's visit (September, 1907), it has been impossible to get samples of the oil for analysis in time to be included in this report. The gas from the Miner ranch field is noteworthy because of its relatively high marsh-gas content as compared with many of the gases from the eastern fields, especially those of

Kansas. Two analyses[a] of the gas taken at Miner ranch, one in 1904 and the other in 1907, are as follows:

Analyses of natural gas from Miner ranch, Contra Costa County, Cal.

1.		2.	
Marsh gas (CH_4)..............per cent..	93.0	Marsh gas (CH_4)..............per cent..	92.3
Hydrogen (H_2)...................do....	3.3	Hydrogen (H_2)...................do....	7.5
Carbonic acid gas (CO_2)..........do....	.4	Loss and unestimated...........do....	0.2
Nitrogen (N) and residuum.......do....	3.3		
	100.0		100.0
Specific gravity (air=1), calculated......	0.558	Calorific value..British thermal units..	1,059.3
Calorific value, calculated, British thermal units...........................	1,009		
Combustible.................per cent..	96.3		
Noncombustible.................do....	3.7		

1. By California Gas and Electric Co., San Francisco, 1904.
2. By A. Auchie Cunningham, San Francisco, 1907.

Conclusions concerning future development.—The Miner ranch oil field seems to have been pretty well prospected with the drill, and as no productive wells have so far been brought in and no well-defined oil sands or zones discovered, it appears reasonable to suppose that future development will fail to disclose any important deposits of oil. There is no question that considerable quantities of oil are present in the Monterey shale and adjacent beds, not only here but in other parts of the county, as is clearly indicated by the prospect holes and surface evidence; but there are also many reasons for believing that this oil is so uniformly disseminated in the shales and sands, with the possible exception of local and relatively unimportant pockets, as to preclude its withdrawal in commercial quantities through wells. The development of gas in the territory is another matter, but as the oil and gas have the same origin and are influenced similarly by the same conditions, it is believed that no large bodies of gas are contained in the formations of the region. Another item that must be considered in drawing conclusions concerning the future of this field is that the organic Monterey shale, which is believed to be the source of both the oil and gas, is here but about 300 feet thick, an amount entirely inadequate under the most favorable conditions for supplying large quantities of hydrocarbons. The structural conditions, on the other hand, are in general favorable for the accumulation of the oil and gas were they present in sufficient quantities to pay for exploitation.

---

[a] These analyses have been kindly furnished by Mr. W. E. Holbrook, president of the American Oil and Refinery Company, which at present holds the Miner ranch property.

PETROLEUM IN SOUTHERN UTAH.

By G. B. RICHARDSON.

INTRODUCTION.

The recent discovery of petroleum near Virgin City, Utah, has caused much local excitement and attracted considerable capital. Outside of newspaper items, however, very little has been published concerning the oil, and the following note has been prepared to help supply the demand for information. The writer's personal knowledge of the field is limited to the information obtained during a day's visit to Virgin City shortly after the announcement of the discovery.

LOCATION.

Virgin City is situated on Virgin River in Washington County, in the southwest corner of Utah, and is distant about 90 miles by road from Lund, the nearest station on the San Pedro, Los Angeles and Salt Lake Railroad. The new oil field is in the Plateau province, near the eastern boundary of the Basin Ranges. The country rises northeastward from an elevation of about 3,250 feet at Virgin City to over 10,000 feet on the crest of the plateau 30 miles distant. The ascent is accomplished by successive benches which rise steplike one above another. This region is drained by Virgin River and its tributaries, which for many miles flow through steep narrow canyons among some of the grandest scenery on the continent. The area here considered is included in the region covered by C. E. Dutton's report on "The Geology of the High Plateaus of Utah," published in 1880.

OUTLINE OF GEOLOGY.

STRATIGRAPHY.

This portion of the Plateau province is underlain by almost flat-lying strata which range in age from Carboniferous to Eocene. The several formations are distinctly marked lithologically and are characteristically colored so that they can be readily distinguished.

343

They outcrop in broad belts extending in a general east-west direction, the harder rocks forming escarpments and the softer ones the intervening stretches. The oldest formation in the Virgin City region is a thick, massive gray limestone of upper Carboniferous age which underlies the broad plateau between Virgin River and the Grand Canyon of the Colorado. Above this limestone there is a mass of red beds of variable thickness, in the vicinity of Virgin City approximating 3,000 feet. These are in the main soft, thin-bedded rocks, chiefly argillaceous and calcareous shales, with some beds of sandstone and limestone. The group of red rocks is separated into two distinct parts by a formation composed of gray sandstone and conglomerate, which in the area considered is less than 100 feet thick, though in Arizona it is reported to be much thicker. Occurring between softer rocks, this siliceous formation is prominent and commonly constitutes a broad bench capping the underlying beds in a scarp, while the upper softer rocks have been eroded from the platfrom and form the slope of the next succeeding escarpment. The rocks beneath the prominent sandstone and conglomerate are probably of Permian age, and the conglomerate, with the overlying red beds, is considered to be Triassic. The lower red beds are the oil-bearing rocks.

Above the soft red beds there is a great development of sandstone which in this region is about 2,500 feet thick. The lower part of this sandstone is characteristically dark red; the upper part is peculiarly cross-bedded and is of a prevailing light color. This great mass of sandstone is the most conspicuous geologic feature of the region. It forms prominent cliffs which can be followed for many miles and through which deep canyons have been cut.

The sandstone is succeeded by about 1,200 feet of generally soft varicolored beds, including reddish and green shales, white limestone, and gypsum of Jurassic age. These rocks are commonly eroded into badland topography. They are overlain by about 3,000 feet of buff and gray sandstones and shales which contain workable beds of coal and are of Upper Cretaceous age. Above these rocks are varicolored shale, sandstone, and limestone of Eocene age which outcrop in the Pink Cliffs and cap the summit of the high plateaus.

Although the rocks of the plateau region in the vicinity of Virgin City are prevailingly sedimentary, there are small areas covered by basaltic lavas of post-Eocene age.

STRUCTURE.

The strata in general dip northeastward at a low angle, averaging possibly between 1° and 2°. The continuity of the beds is broken, however, by a number of faults trending in general north and south, some of which have displacements of 1,000 feet or more. One zone of dislocation extends approximately along Hurricane Cliff, through which Virgin River cuts its way about 7 miles below Virgin City. This zone of fracture has been traced from the Grand Canyon to Virgin River and northward along the western base of the plateau in the vicinity of the Mormon settlements of Torquerville, Belleview, Kanarra, and Cedar City. Along portions of this zone, especially between Cedar City and Kanarra, the strata are much disturbed and are steeply tilted. Another conspicuous line of disturbance extends along the headwaters of Virgin River in what is known as Long Valley.

OCCURRENCE OF PETROLEUM.

The Carboniferous limestone outcrops a few miles west of Virgin City and the town is immediately underlain by the Permian (?) red beds, in which Virgin River has cut a relatively broad valley. The overlying sandstone-conglomerate formation marks a prominent bench north, east, and south of the town, beyond which, to the north and east, the upper red beds slope up to the base of the escarpment made by the massive red sandstone.

Oil seeps have long been known in the vicinity of Virgin City. One of them occurs close to the river about 1½ miles west of the town, and it is reported that the existence of this seep was the cause of sinking the discovery well in the summer of 1907. This well is located in the flood plain of North Creek, a tributary of Virgin River, about 2 miles north of Virgin City. The boring was started in the lower red beds and apparently did not pass through them, though it must have stopped not far from the bottom of the formation, near the Carboniferous limestone. A complete record of the drill hole was not kept. Oil was struck on July 13, 1907, at 566 feet below the surface and the well was sunk to 610 feet. The oil is reported to stand in the well 300 feet below the surface, thus being under pressure sufficient to cause it to rise 266 feet. A few hundred barrels are said to have been pumped when work was stopped by a flood on July 27. This stage of development was reached when the writer visited the field a few days later.

The following statement of conditions at Virgin City in January, 1908, is extracted from a letter from Mr. Thomas Downey, of the Paraffin Virgin Oil Company. Fifteen oil rigs were then in the field, but only four were being operated. Seven wells had been sunk supposedly to the oil horizon and some oil was found in each, but the amounts were not given by Mr. Downey. He states, however, that none are as good as the discovery well, which is reported to produce about 10 barrels in twenty-four hours. Claims have been staked far and wide, but oil has not yet been reported outside of the immediate vicinity of Virgin City.

The occurrence of petroleum in red beds is unusual. Such beds in general are believed to have accumulated in bodies of water in which there was little life, for the presence of much organic matter would tend to reduce the ferric salts of the pigment to more somber-colored compounds. If barren conditions existed in this area during the deposition of the red beds, the source of the petroleum probably must be sought in the decomposition of organic matter in the underlying Carboniferous limestone.

QUALITY OF THE OIL.

A small sample was collected by the writer from an open vat in which the oil had been exposed to the weather for a week or more. This sample was examined by David T. Day, who reports that it has a specific gravity of 0.9225, equivalent to 22° Baumé, and that it contains some paraffin, a large percentage of asphalt, and apparently considerable sulphur, including hydrogen sulphide. A larger sample, received by Dr. Day, was analyzed by him with the following results:

Chemical examination of crude petroleum from Virgin City, Utah.

Color, black.
Odor, hydrogen sulphide.
Specific gravity, 0.918 = 22.5° Baumé.
Results of distillation:
    Sample began to boil at 60° Centigrade.
    Distillate obtained—   Per cent.
        Below 150° C.—gasoline and naphtha.................... 2.1
        Between 150° and 300° C.—illuminating oil (specific
          gravity, 0.784)....................................... 19.5
        Residue (specific gravity, 0.9475)........................ 78.4

Examination of the gasoline and illuminating oils obtained above showed both to be principally saturated hydrocarbons, probably chiefly of the paraffin series. Examination of the residue showed it to contain 49.7 per cent of asphalt and 29.4 per cent of paraffin wax, the remainder consisting of heavy oils and resinous material. From the above it is evident that though a satisfactory illuminating oil can be obtained from this Utah crude petroleum, the yield is comparatively small and the petroleum is better suited to use as a fuel oil. This rendered the determination of the sulphur advisable, and by Carius's method the result was 0.45 per cent. Much of this is in the form of hydrogen sulphide, easily separated by steaming, hence the oil is preferable for fuel purposes to Texas oil. The percentage of sulphur obtained is lower than that found by other analysts, the difference being probably due to the fact that this sample was taken from a barrel which had been standing a month or more since taken from the well.

FUTURE OF THE FIELD.

The encouraging news that petroleum of a fair grade was found in promising quantity in the first well is offset by the fact that six others have been sunk without encountering oil in paying amounts. Yet, considering the present scanty knowledge of the conditions, little can be predicted concerning the future of this field. Whether oil exists here in profitable amount can be determined only by the drill. To judge from what is known of the geology, the general conditions are not unpromising, although there are unfavorable complications. In many oil-bearing areas an anticlinal structure has prevented the escape of petroleum stored in the rocks, but the strata here are not folded; moreover, the Virgin field is traversed by profound faults that possibly provided means of escape for oil that may have been present. The thickness of the oil-bearing stratum, which appears to be a layer of sand in the lower red beds, has not been reported and whether or not it is persistent over a wide area is undetermined. However, the stratigraphy of the lower red beds is known to be varied and it is probable that the oil-bearing rocks occur as lenses rather than as persistent beds. If the petroleum has accumulated in lenses of porous sandstone the surrounding relatively impervious shale would tend to prevent its escape, so that under the circumstances this possible mode of occurrence of petroleum in the Virgin field is fortunate rather than otherwise. But, on the other hand, such hypothetical reservoirs can not be predicted by surface indications and an unusually large element of chance confronts the prospector.

In prospecting in the possible eastward continuation of this field, the outcrop of the massive red sandstone that lies above the red shale will serve as a valuable aid. It would be futile to attempt to strike the Virgin City oil horizon in wells situated above this formation, because of the great thickness of the rocks that would have to be penetrated. The sandstone-conglomerate formation that separates the upper and lower red beds is also an important horizon marker in following the oil-bearing rocks. It should be borne in mind that the Virgin City oil occurs in the red beds beneath this siliceous formation, which usually is conspicuous.

GAS FIELDS OF THE BIGHORN BASIN, WYOMING.

By Chester W. Washburne.

INTRODUCTION.

A strong flow of natural gas has recently been obtained in a well near Gray Bull, Wyo. In view of the wide distribution of the strata that contain the gas at Gray Bull, and the signs of gas at other places, suggesting the further extent of the gas-producing region, a description of the field seems desirable.

The field work on which the present paper is based was done in the summer of 1907 in the course of an investigation of the coal deposits and general geology, made under the direction of C. A. Fisher. Miss A. Pishel and Homer P. Little acted as field assistants. For well records and similar data, the writer is indebted to Henry Sherard and Philip Minor, of Basin, Wyo., and to L. A. Corey, of Bridger, Mont.

The only previous mention of oil or gas within the field studied by the writer is in the report by C. A. Fisher on the geology and water resources of the Bighorn basin.[a] Fisher mentions the escape of gas from alluvial sands near Byron, Wyo., and from a dug well 3 miles east of Basin, Wyo.

The Bonanza field, close to the southeast corner of the area studied by the writer (see Pl. III), has been described by several geologists. In 1888 L. D. Ricketts[b] described an oil spring in sec. 23, T. 49 N., R. 91 W., and gave an analysis of the oil. Later W. C. Knight[c] also mentioned the same spring. Knight visited the Bonanza oil field in 1903 and found two abandoned wells, one of which is reported to have struck some oil, the other only artesian water. The wells, according to Knight, were poorly located. Fisher[d] describes the oil springs near Bonanza and gives an analysis of the oil.

---

a Prof. Paper U. S. Geol. Survey No. 53, p. 59.
b Rept. Territorial Geologist of Wyoming, 1888, pp. 39–40.
c Bull. Wyoming Exp. Sta. No. 14, 1893, p. 11.
d Loc. cit.

348

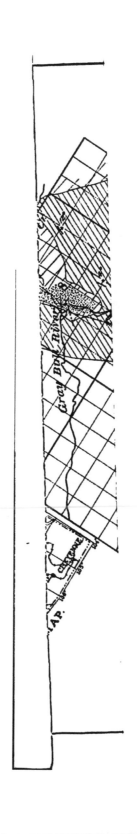

GEOLOGIC CONDITIONS.

STRATIGRAPHY.

GENERAL OUTLINE.

A knowledge of the stratigraphic column is most important to the driller.   This column is as follows:

Stratigraphic column in Bighorn basin, Wyoming.a

System or series.	Group.	Formation.	Thickness (feet).	Characteristics.
Tertiary.	Lower Eocene.	Wasatch.	500+	Bright-colored terrestrial clays.
		Fort Union.	1,000 to 2,000	Dark-colored shale, with coal, and massive sandstone.
Upper Cretaceous.	Montana.	Laramie.	150 to 700	Massive sandstone with subordinate shale, coal bearing.
			150	Dark marine shale.
			300 to 400	Variegated terrestrial clays and soft sandstone.
			400 to 500	Massive fresh- and brackish- water sandstones and dark shale.
			150 to 225	Massive fresh- and brackish- water sandstones, separated by carbonaceous shale, usually coal bearing.
		Colorado.	4,400	Dark shale with one or two conspicuous sandstones, not divisible in this field, though more than 1,500 feet of the lower part of the group is known to be equivalent to the Benton shale.
(?) Lower Cretaceous.		Cloverly.	0 to 275	Bright-colored terrestrial clays, with massive sandstones at the top and bottom.
(?) Jurassic.		Morrison.	250 to 350	Bright, variegated terrestrial clays and soft sandstone.
Jurassic.		Sundance.	225	Marine limestone, shale, and sandstone.
(?)		Chugwater.	600 to 800	"Red Beds," bright-red sandstone.
Pennsylvanian.		Embar.	215	Marine limestone with red shale at base.
		Tensleep.	85	Sandstone.
		Amsden.	90	Red shale and purplish sandstone, with a little limestone.
Mississippian.		Madison.	1,000	Massive-bedded limestone.

a For the recognition of the formations in the field the writer is indebted to the guidance of C. A. Fisher.   The subdivisions of the Montana group are correlated with similar subdivisions proposed by Stanton and Hatcher, in the Judith River region of northern Montana.   (Stanton, T. W., and Hatcher, J. B., Geology and paleontology of the Judith River beds: Bull. U. S. Geol. Survey No. 257, 1905.)   This correlation also is made possible through the work of Fisher, who has traced the formations southward from the type locality into Wyoming and who will soon publish his conclusions on the subject in one of the scientific journals.

To the prospector for gas the essential part of this stratigraphic column is the Colorado formation, the lower part of which is summarized in the following table:

Generalized section of lower part of Colorado formation in Bighorn basin, Wyoming.

	Feet.
Sandstone, thin bedded. Referred to as sandstone B. Absent in the northern part of the district	20–45
Shales with a few thin beds of sandstone containing a little gas in the Torchlight dome, and both gas and oil in the well 3 miles northeast of Basin	275
Sandstone, massive yellow. Referred to as sandstone A	65–100
Shales, black	185
Shale, gray, sandy, hard, dense, siliceous, a very persistent bed and a conspicuous ridge maker at the top of the Mowry shale	3–5
Shales, gray and black, with many fossil fish scales; contain numerous layers of hard flinty shale, and one 3-foot bed of bentonite; the Mowry shale	200
Shales, dark bluish and black, with a few beds of volcanic ash and white clay (bentonite) in the upper part	250
Shales, black, carbonaceous, in many places oily, locally containing one or more lenses of sandstone in the lower 100 feet	300
Sandstone, thin beds 3 to 18 inches thick, weathering brown, which are separated by partings of black shale 1 to 12 inches thick; the "rusty beds"	20–100

In most sections the total thickness of this part of the formation is about 1,350 feet.

THE GAS HORIZON.

Some uncertainty arises from the lack of precise knowledge of the gas horizon. There can be no doubt that the gas sand is close to the base of the black Colorado shales, but it can not yet be determined whether it is a sandstone in the lower part of the shales, the "rusty beds" of thin-bedded sandstone at the base of the shales, or the underlying Cloverly sandstone. This range of uncertainty amounts to about 150 feet. In the writer's opinion, based on the log of the Gray Bull well, the gas is obtained either from the "rusty beds" at the base of the Colorado, or from some sandstone in the shales less than 100 feet above the "rusty beds," with the probabilities strong in favor of the latter position. Coarse, hard, porous sandstones in the latter position have been observed at a number of localities, but they are lenticular and in many places absent. The "rusty beds" are a constant feature of the base of the marine Cretaceous. Seemingly they are as a group a true basal sandstone, resting upon a rather smooth surface of erosion. Beneath this erosional surface at some localities is a heavy sandstone, probably the lower sandstone of the Cloverly formation; but at most places the Cloverly sandstone is absent and the "rusty beds" rest upon maroon, pink, or bright-green shales which are regarded as part of the Morrison formation, though they may belong to the Cloverly. There can be no doubt as to the lenticular nature of the Cloverly sandstone and its absence over most of the area. The field evidence indicates that the sandstone was removed by erosion before the deposition of the overlying marine strata of the Upper Cretaceous.

If the only gas horizon is that of the Cloverly sandstone, prospecting in the region will be most uncertain because of the limited distribution of that bed, as a well might be drilled where the structure was favorable for the accumulation of gas, yet not obtain gas because of the absence of the Cloverly sandstone. This condition is illustrated diagrammatically in fig. 21. A is an anticline in which the lenticular sandstone near the bottom of the Colorado, the "rusty beds," and the Cloverly sandstone are all three present. B is an anticline in which the "rusty beds" are the only possible gas-bearing strata, the Cloverly sandstone having been removed by pre-Colorado erosion. In the absence of the Cloverly, the "rusty beds" could

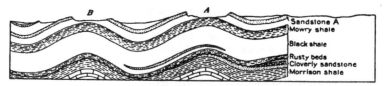

FIG. 21.—Hypothetical section showing gas-bearing strata present in anticline A and absent in B.

probably serve as a gas reservoir. Uncertainty prevails also as to the distribution of the lenses of sandstone in the black basal Colorado shales. These lenses may or may not be present at any locality where a well might be located.

SOURCES OF THE OIL AND GAS.

The source of the oil and gas of the Colorado sandstones is probably the black shales at the base of the Colorado. These shales are oily in most exposures throughout the Rocky Mountain region. The oil of the "Red Beds" and of the Madison limestone, mentioned on page 361, is probably derived from the latter rock. Both of the sources here noted as possible are marine formations.

STRUCTURE.

The Bighorn basin gas fields are on the slopes of the anticline of the Bighorn Mountains. The dips of the region are in general westward at angles of 5° to 20°, but this westward slope is interrupted by numerous small anticlines. It is in these minor folds that gas may be looked for. They are isolated from each other and rise abruptly from the surrounding areas. Most of them have the form of slightly elongate domes with broad crests and gently inclined ends. The ground plans of the anticlines, as shown by the maps, are elliptical, with the major axis of each ellipse parallel to the adjacent front of the Bighorn Mountains.

The anticlines of importance as possible sources of gas are situated about 15 to 20 miles from the Bighorn Mountain front. Their crests at the surface are near the sandstones about 1,000 feet above the base of the Colorado, referred to as sandstones A and B, or in the underlying Mowry shale or the black carbonaceous shale at the base of the Colorado. Anticlines whose tops are in strata over 1,500 feet above the base of the Colorado need not be considered as possible sources of gas, for many years, because of the great depth of the gas horizon. When the field has been proved these anticlines may become sources of gas, if the value of the product should warrant such deep drilling. Likewise, those anticlines whose crest is in rocks close to the base of the Colorado shale are not to be considered as sources of gas, because of insufficient cover. The remaining anticlines which may be considered as possible gas reservoirs are nine in number, as follows:

Silvertip anticline, 22 miles south of Bridger, Mont.
Frannie anticline, 1 mile west of Frannie, Wyo.
Sage Creek anticline, 2 miles south of Frannie, Wyo.
Garland anticline, 5 miles northeast of Garland, Wyo.
Byron anticline, 3 miles north of Byron, Wyo.
Ionia anticline, 8 miles northeast of Lovell, Wyo.
Alkali anticline, 10 miles south of Lovell, Wyo.
Peay Hill anticline, 2 miles southwest of Gray Bull, Wyo.
Torchlight anticline, 3 miles east of Basin, Wyo.

DETAILED DESCRIPTIONS.

PEAY HILL ANTICLINE.

Along Bighorn River near Gray Bull, Wyo., there is a small, low anticline about 2½ miles long and 2 miles wide. On account of its broad, flat form it is known to the oil drillers as the "Peay Hill dome." A gas well has been drilled in the central part of this dome on the east bank of Bighorn River, in the northwest corner of sec. 21, T. 52 N., R. 93 W. This well was drilled for the Peay Hill Oil and Development Company by Philip Minor and Henry Sherard, to whom the writer is indebted for the following record:

Log of the Peay Hill Oil and Development Company's gas well, 1½ miles southeast of Gray Bull, Wyo.

	Thickness.	Depth.[a]
	Feet.	*Feet.*
Shale..	17	17
Shale, hard, dark, probably the ridge-making layer of the Mowry....................	3-4	20
Shale, dark colored...	300	320
Clay, soft, white, probably bentonite, causing the well to cave in badly...........	3	323
Shale, dark...	127	450
Clay, white, sandy...	40-50	495
Shale, black..	155	650
Shale, thin layer, hard...	1(?)	651
Shale, black..	49	700
Rock, thin layer, very hard..	1(?)	701
Shale, black..	49	750
Rock, very hard...	7	757
Shale, black..	38	795
Rock, hard...	5	800
Sandstone, containing gas under high pressure (base not reached).................	1	801

a "Depth" in all the logs in this paper refers to the base of the respective strata.

The top of the well is 165 feet below the base of a massive sandstone, 65 to 75 feet thick, which forms the top of the bluffs above Gray Bull. This sandstone will be referred to as sandstone A. It is conspicuous in nearly all its outcrops, and hence a very important horizon marker in the Colorado formation. The Gray Bull gas horizon is 950 to 965 feet below the bottom of this sandstone.

The diameter of this well is $5\frac{1}{8}$ inches and its depth is 801 feet. The well is cased down for 500 feet. It has maintained a steady roaring flame over 50 feet high almost from the time it was drilled, July 14, 1907, to the date of this writing, January, 1908. It is reported that .the initial height of the flame was about 70 feet, but this can not be verified; certainly the gas well is one of remarkable volume and through it large quantities of gas have escaped from this reservoir. Some idea of the pressure of this well may be obtained from the fact that the escaping gas is sufficient to lift a large log chain composed of $\frac{1}{2}$-inch iron. No instrumental measurements of pressure have been made, but a rough calculation based on the size of stones which the gas would eject shows that the pressure in September, 1907, was over 600 pounds to the square inch. Whether or not this pressure is diminishing can not be determined now, for according to the most reliable information no change has been detected up to the present time. One of the strangest features of this occurrence is the fact that the well is drilled within a few feet of a normal fault, yet obtains gas under high pressure. The throw of the fault is about 24 feet. One-fourth mile north of the well is another fault the throw of which is 60 feet. Henry Sherard reports a 10-foot fault in the sandstone 250 yards south of the last-mentioned fault.

From a study of the nearest available geologic section, made about 10 miles northwest of Gray Bull, it would seem highly probable that the gas in the Gray Bull well is obtained from a lenticular, nonpersistent sandstone about 100 or 150 feet above the base of the Colorado formation, in a series of black shales. Such a sandstone outcrops on the wagon road from Shell to Lovell, about 6 miles northeast of the gas well. This sandstone is coarse grained and of loose, porous texture. It is about 10 or 15 feet thick and from its porosity it would seem to be well suited to hold gas. When examined in the outcrop, however, no odor of oil or gas could be detected in this rock. Another possible position of the gas horizon is in the "rusty beds" about 50 or 100 feet below this sandstone Moreover, it is thought by some that the gas is obtained from the Cloverly sandstone, which is the next underlying member of the stratigraphic column. The maximum stratigraphic range of uncertainty is about 150 feet. The only way the Gray Bull gas horizon can be definitely located at present is by reference to the first sandstone above the Mowry shale. This sandstone is exposed on the river bluffs opposite Gray Bull, and may be designated sandstone A. The gas horizon is 950 to 965 feet below its bottom.

The drilling of the Peay Hill well near Gray Bull is very important for an understanding of the economic geology of the region. It is the first well to obtain a flow of gas in commercial quantities, and it furnishes an important lesson for the drilling of wells in the future by indicating the caving nature of the soft shales encountered in drilling and the need of strong casing. On account of the lack of proper casing in this well, it has been found impossible to close it and to prevent the escape of gas in valuable quantities. Unless some means are found to stop this flow of gas very soon, an appreciable diminution of pressure and ultimate exhaustion of the well may be expected.

GARLAND ANTICLINE.

One of the sharpest and most pronounced anticlines of the region is known as the Garland anticline. This fold is about 7 miles long, extending from Polecat Creek, near the crossing of the Cody branch of the Chicago, Burlington and Quincy Railroad, to Shoshone River, 1 mile above Byron, Wyo. Three wells have been drilled in this anticline about 2 miles west of Byron for the purpose of obtaining oil, but the boring did not penetrate to the base of the Colorado. The wells were drilled by the Montana and Wyoming Oil Company, of Billings, Mont. The depth of the wells is about 900 feet. The well first drilled furnished sufficient gas to run the engine during the drilling of the second and third wells. Moreover, gas has been observed by Fisher[a] escaping from alluvial sands overlying this anticline. The wells do not reach the horizon of the beds that yield the gas near Gray Bull, and it is not known whether or not the anticline contains commercial quantities of gas.

The Gray Bull gas horizon at the base of the Colorado would be found at a depth of about 1,500 feet beneath the highest point of the axis of the Garland anticline. Gas might occur in commercial quantities at a higher horizon if sandstone were encountered of sufficient thickness to serve as a reservoir. This may be inferred from the occurrence of gas in the wells near Byron at various horizons in the lower part of the Colorado shale. A satisfactory test of the gas field can be made only by drilling entirely through the Colorado into the underlying sandstone.

The three wells mentioned above, which were drilled about 100 yards apart, have all furnished small quantities of oil. Accurate logs of the wells could not be obtained from the officers of the company, and hence the oil horizon can not be located closely. Descriptions furnished by the drillers, however, leave no doubt that the oil is obtained from a thin sandstone, not over 3 feet thick, in the upper

---

a Fisher, C. A., Prof. Paper U. S. Geol. Survey No. 53, 1906, p. 59.

part of the black basal Colorado shales. This "oil sand" is overlain by 3 or 4 inches of fine-grained limestone. ·So far as the writer knows, this is the only limestone ever found in the Colorado formation in the Bighorn basin. If the limestone outcrops at the surface, it has never been observed. The quality of the oil is very high, as shown by the following report by David T. Day on oil from well No. 1:

The oil is light red by transmitted light, with brilliant green fluorescence. It contains no water. The odor is almost like that of Pennsylvania oil, and apparently the oil contains no sulphur, therefore no determination was made. Specific gravity at 15° C., compared with water at 4° C., is 0.8315. The specimen in the small bottle submitted by you shows 0.816.[a] Distillation of the sample gave the following results: Initial boiling point 77° C.

	Per cent.
Naphtha (specific gravity 0.722)	14
Illuminating oil (specific gravity 0.761)	28
Light lubricating oil	17.5
Residue suitable for cylinder oil	36
Loss	4.5

Such an oil as this would make very satisfactory oil, if transportation facilities were afforded, for by properly adjusting the distillation method a larger percentage of illuminating oil could be obtained. This is shown by the low specific gravity of the distillates.

The quantity of oil that could be obtained from these wells is not known to the writer. They are kept tightly closed, and no tanks or other means of storing the oil have been prepared. Until the wells are opened and their flow is measured for a period of several days, their capacity must remain unknown. In the absence of such tests the general inference is that these wells have not yielded oil in commercial quantities.

TORCHLIGHT ANTICLINE.

The Torchlight anticline, or "dome," as it is known to the prospectors, is about 3 miles east of Basin, Wyo. The dome is small, being about 1 mile long and one-half mile wide, and forms part of a much larger anticline extending a mile or more to the northwest. The dips of the larger anticline are so gentle that it seems doubtful whether its structure is sufficiently pronounced to favor the accumulation of gas. The small dome, however, must be considered a favorable structural feature. The crest of this little dome is in black shales, probably between sandstones A and B. (See section, p. 350.) If such is the case, the gas horizon would be found at a depth of 1,150 or 1,200 feet. This figure is uncertain because there is doubt as to the correlation of the sandstone surrounding the gas field, here designated sandstone B. It is thought to be the same as a sandstone occurring in many places 275 feet above sandstone A.

---

a This specimen is from the well of the Union Gas and Oil Company, 3 miles northeast of Basin, Wyo.

The Torchlight Drilling and Mining Association (Limited), of Basin, Wyo., has two wells on this anticline. Both wells are 2 inches in diameter. Well No. 1 is 106 feet deep. It obtains gas from sandy shale between depths of 15 and 85 feet, with a very strong flow of gas from a thin sandstone at a depth of 60 feet. This sandstone carries water highly charged with bitter salts. The log of well No. 2 is given below. The principal gas horizon of this well is a 30-foot thin-bedded sandstone encountered at a depth of 192 feet. The sandstone is saturated with gas and contains some oil. The pressure of the gas is between 35 and 55 pounds per square inch, so that ordinarily the water in the well, standing 65 feet from the surface, holds the gas quiet, but when the water is pumped out to a depth of 110 feet there is a good flow of gas. Part of this gas is used to supply a 4-horsepower engine and two stoves at the driller's camp.

The soil on the Torchlight dome is in places impregnated with bituminous matter which has escaped from the underlying rocks. An open pit, 8 feet deep, is said to give off a strong odor of ammonia, probably a product of organic decomposition.

Log of well No. 2, Torchlight Drilling and Mining Association, 3 miles east of Basin, Wyo.

	Thick- ness.	Depth.
	Feet.	*Feet.*
Shale, sandy..	125	125
Sandstone A, containing "sulphur water, with some soda and iron," traces of gas.	35	160
Shale, with some sandstone...	32	192
Sandstone, thin bedded, containing gas and a little oil.............................	30	222

OTHER FAVORABLE LOCALITIES.

SILVERTIP ANTICLINE.

The Silvertip anticline is on the State line between Montana and Wyoming, about 20 miles northwest of Garland, Wyo. The heart of the anticline is a depression known locally as Elk Basin. The boundaries of this basin correspond approximately with the boundaries of the gas field which might be developed by very deep drilling. As thus limited, the gas field would be about 3 miles long and less than a mile wide. The crest of the anticline is in the upper part of the Colorado formation, and the depth to the rock which is gas bearing at Gray Bull is about 3,000 feet. This great depth precludes the possibility of an early development of the field, if indeed it should be exploited at all. But the structure is so favorable for the accumulation of gas at this point that the possibility of the field needs to be pointed out. The anticline is cut by many cross faults trending at right angles to the strike. The throw of some of these faults is over

200 feet, but on account of the depth of the gas horizon, it is believed that they would not interfere with the accumulation of gas.

FRANNIE ANTICLINE.

One mile west of Frannie, Wyo., there is a low anticline in sandstone A, a stratum which has been used as a horizon marker throughout the extent of the field studied. The depth of the Gray Bull gas horizon beneath the crest of this anticline is about 1,000 or 1,100 feet, but it is of course doubtful whether or not the bed is actually gas bearing at this place. The anticline is crossed by a fault trending northeastward, with a downthrow of about 50 feet on the southeast side. The most favorable location for a prospect well on this anticline would be near the southeast corner of the SW. ¼ SW. ¼ sec. 24, T. 58 N., R. 98 W.

SAGE CREEK ANTICLINE.

Another anticline about 2 miles south of Frannie station is here designated the Sage Creek anticline, because of its nearness to Sage Creek. It is possibly connected with the Frannie anticline, which lies 3 miles farther northwest. It is a low dome of sandstone 200 yards west of Sage Creek, on the south line of sec. 6, T. 57 N., R. 97 W. The anticline is poorly exposed, being partly covered by drifting sand, but the fact that sandstone A outcrops in flat beds at this point is sufficient to indicate the presence of an anticline. The area within which it would be practicable to reach the Gray Bull gas horizon by drilling, in this anticline, is very small. The most favorable location for a well would be about 200 yards northeast of the south quarter corner of sec. 6. The gas horizon at this place would be found at a depth of about 1,100 feet.

BYRON ANTICLINE.

About 3 miles northeast of Byron, Wyo., there is an anticline on which a well is being drilled by the Montana and Wyoming Oil Company, for the purpose of obtaining oil or gas. This anticline is favorably situated in the geologic column for the accumulation of gas, and it may be found by prospecting that the Gray Bull gas horizon is not too deep. The writer has not visited this locality and hence has no reliable data as to the depth of the gas horizon, but it will probably be deeper here than in the Gray Bull anticline.

IONIA ANTICLINE.

On the north bank of Shoshone River about 8 miles northeast of Lovell, Wyo., there is a low anticline exposing the lower Colorado sandstones. These sandstones make a low ridge about the margin of the anticline within which the Mowry shale outcrops in an anticlinal basin. The structure is complicated by the presence of several normal faults. The throw of these faults probably does not exceed 200 feet, but it is not definitely known, and it is therefore impossible to say whether or not they interfere with the accumulation of gas. Probably, however, there is no such interference even if the faults cut the gas-bearing strata, because similar faults are known in the Gray Bull anticline close to a productive gas well. Moreover, there are extensively faulted gas fields in California, Roumania, and other parts of the world. The depth of the Gray Bull gas horizon beneath the crest of this anticline would probably be about 800 feet, but as in all other localities not yet tested by adequate drilling, it is not known that the beds of this horizon actually contain gas in the Ionia anticline.

ALKALI ANTICLINE.

About 10 miles southeast of Lovell, Wyo., there is a long, narrow anticline in the Colorado shale. The northwest end of the anticline is in the SW. ¼ sec. 28, T. 55 N., R. 95 W. From this point the anticline extends over 6 miles southeastward to a point within 14 miles of Gray Bull. In most of the anticline the basal Colorado shales are exposed and eroded probably nearly to their base. Hence the anticline may not be a good gas reservoir on account of insufficient cover. If gas occurs in the top of the Cloverly sandstone or in the sandstones of the "rusty beds," as thought by some, it would be found at very shallow depths in the greater part of the anticline. Not having seen the anticline, the writer is unable to estimate the actual depth of the Gray Bull gas horizon. However, it is probable that if found at all, the gas would be at a depth of less than 200 or 300 feet. This is true especially of the central part of the anticline. In case it seems likely that the cover of shale is sufficient to protect the gas reservoir, the most favorable locality for a well would be near the west quarter corner of sec. 11, T. 54 N., R. 95 W.

The northern end of this anticline includes a subordinate dome of the Mowry shale. Within this part of the anticline, the gas stratum would be covered by 500 or 600 feet of shales, and for this reason the locality would be favorable for the accumulation of gas. The crest of the anticline, which may be considered the most favorable point for the location of a gas well, is in the SE. ¼ SW. ¼ sec. 28, T. 55 N., R. 95 W.

A POSSIBLE DOME NEAR BASIN.

A well is being drilled for gas by Henry Sherard, of the Union Gas and Oil Company, about 3 miles northeast of Basin, Wyo., near the northwest corner of sec. 11, T. 51 N., R. 93 W. The structure at this locality is that of a very gentle anticlinal flexure on the flanks of a larger anticline that extends from the Peay Hill dome near Gray Bull to the Torchlight dome 3 miles east of Basin. The structure can not be thoroughly worked out because of the lack of exposures, but from the tracing of a thin sandstone along the hillside it would seem probable that the well is located on the side of a small minor anticline of domical form. The rocks at the well are not exposed, but from the depth of sandstone B in the well, it is probable either that there is a fault between the hillside and the well, or that the concealed rocks dip westward at an angle of over 10°.

The possible dome on which this well is located is so small and gentle that it must be regarded as a structure of doubtful favorability for gas. Such a faint structure may indicate a more pronounced dome and a good gas reservoir below, or the structure may die out downward within a short distance and no gas reservoir be present. In the latter case, the location of the well, which is on the limb of the larger anticline, would be better for oil or water than for gas, should either exist in commercial quantities. The depth of the Gray Bull gas horizon at this well is between 1,450 and 1,500 feet.

Log of the Union Gas and Oil Company's Well No. 1, 3 miles northeast of Basin, Wyo.

	Thickness.	Depth.
	Feet.	*Feet.*
Shale, yellowish..	6	6
Shale, gray	111	117
Sandstone, light gray, probably sandstone B	43	160
Shale, light gray	37	197
Shale, black with thin beds of brown fine-grained sandstone containing some gas and oil	22	219
Shale, light gray, with a little grit; small amount of gas and oil	27	246
Shale, sandy, with some gas	79	325
Shale, dark, with some very hard layers	75	400
Shale, gray	41	441
Sandstone, with gas and water; doubtless sandstone A	64	505
Shale, sandy, with a little water	42	547
Shale, dark gray	157	704

DOUBTFUL ANTICLINES.

At a few places along the eastern border of the Bighorn basin there are indications of anticlinal folds whose precise form is unknown. These anticlines were not studied in the field and it is not possible at this time to say whether or not they may be possible gas fields.

In the NW. ¼ sec. 14, T. 7 S., R. 23 E., about 4 miles south-southeast of Bridger, Mont., there is a small dome or partial dome of this sort in sandstone A. Only the western and northern sides of this dome were seen and it is possible that the structure is not complete, as no eastward or southward dips have been determined. However, the locality is worthy of investigation by any prospector who may desire to bore for gas in that locality.

In the flats north of Cowley and Lovell, Wyo., there may be small anticlines that were not discovered in the field. At the time the geology of that neighborhood was studied, the economic importance of anticlines was not realized, and no special pains were taken to find them. They would be difficult to detect in the flats because there is no outcrop of any rock except the Colorado shale, in which faint structural features are hard to see.

The shale hills east of Basin, in the western part of T. 51 N., R. 92 W., and the northeastern part of T. 51 N., R. 93 W., have a broad anticlinal structure. Gas might accumulate in this anticline, but the depth to the Gray Bull gas horizon would be more than 2,000 feet.

Outside of the region embraced in this report, there are doubtless a great many anticlines just as suitable for the accumulation of gas. These have not been tested as yet, and it is not wise to predict that gas will or will not be found in them until such tests have been made. The map of the Bighorn basin made by C. A. Fisher [a] shows that favorable structural conditions are present in the southern and western margins of the basin. One of the anticlines shown on this map is about 6 miles southwest of the Morrison ranch, in T. 44 N., Rs. 97–98 W.; another is near Sunshine, in T. 47 N., R. 101 W.; another near Fourbear, in T. 48 N., R. 103 W.; another near Pitchfork ranch, in T. 48 N., R. 102 W.; another in the southern part of Oregon basin, about 10 miles southeast of Cody, Wyo.; and another on Shoshone River at the mouth of Cottonwood Creek, 1 mile east of Cody. Fisher [b] mentions a well drilled for oil on the Cottonwood Creek anticline, but does not state the depth of the well, and hence it is uncertain whether or not it reached the Gray Bull gas horizon at the base of the Colorado.

E. G. Woodruff, of the United States Geological Survey, informs the writer that the lower part of the Colorado shale is exposed in an anticline in upper Buffalo basin and in another anticline on Grass Creek in the southwest part of the Bighorn basin. These anticlines are indicated on Fisher's map by patches of so-called "Pierre shale" in T. 47 N., R. 100 W., and T. 46 N., R. 98 W.

W. C. Knight [c] has described an anticline in Colorado shale near the head of Cottonwood Creek, 15 miles east of Worland. By digging out a mud spring on this anticline, in or near sec. 29, T. 47 N., R. 90 W., Knight was able to obtain some light-green oil which was analyzed by E. E. Slosson.[d]

It is, of course, not possible to predict the occurrence of gas in anticlines so far away from those which have actually been tested. The writer merely desires to call attention to the existence at these places of favorable structural features in the same strata that bear gas near Basin, Gray Bull, and Byron. At present there is no reason to think

a Prof. Paper U. S. Geol. Survey No. 53, 1906 Pl. III

b Op cit., p. 59.

c The Bonanza, Cottonwood, and Douglas oil fields: Bull. No. 6, Petroleum series, Univ. Wyoming School of Mines, July 1903, pp. 14–17.

d Knight, W. C., op. cit., p. 27.

that gas will be found on the west side of the Bighorn basin, though the possibility must be admitted.

It should be kept in mind that at the present time only one well has been drilled down to the principal gas horizon. This well, which is 2 miles south of Gray Bull, yields a strong flow of gas, and inasmuch as the strata containing the gas here remain apparently unchanged in character throughout the region and as gas escapes from these strata at widely separated places, an extension of the gas field may well be inferred. Certainly the existence of favorable structure in the same rocks close to the Gray Bull well warrants the drilling of test wells. It should be noted that the writer makes no prediction that gas will or will not be found in any of the anticlines.

INDICATIONS OF OTHER POSSIBLE GAS AND OIL HORIZONS.

The Madison limestone in Sheep Canyon, 15 miles north of Basin, shows many signs of oil. The rock is black in places from the contained carbonaceous material and has a distinct oily smell when freshly broken. It is reported that there was a small oil spring in this canyon before the railroad was built but that the spring has been covered by the railroad grade. The occurrence of asphaltum or other solid hydrocarbons in many small cavities and fissures in the Madison limestone is further proof of the existence of oil in that rock.

Considering the fact that oil has been found near Lander, Wyo., in rocks of Carboniferous age, it would seem not unlikely that it might also be found here. However, the wells which have been drilled near Bonanza, about 25 miles southeast of Basin, where the structure is favorable for the accumulation of oil and gas, were unsuccessful. These wells are described in Fisher's report on the Bighorn basin.[a]

W. W. Peay, of Basin, reports a bed of asphalt mixed with sand and rock on the head of Alkali Creek, one-half to 1 mile north of the wagon road between Hyattville and Shell. Mr. Peay states that in T. 52 N., R. 90 W., the following land has been filed on for the purpose of quarrying the asphalt: SE. ¼ sec. 29, SW. ¼ sec. 28, NE. ¼ sec. 32, NW. ¼ sec. 33. The asphalt occurs in "blanket" form, as a bed 10 feet thick, lying on a hilly surface of limestone or sandstone. About half of the bed is reported to be asphalt, and the remainder to be sand and rock which the asphalt binds together in a firm mass. The bed is overlain by 10 feet or more of incoherent rock fragments and soil. Apparently the asphalt has exuded from the Pennsylvanian rocks at this place and cemented the base of the rock mantle. The occurrence of so large a deposit of asphalt is an excellent argument for the presence of oil in the upper Paleozoic strata.

---

a Fisher, C. A., Prof. Paper U. S. Geol. Survey No. 53, 1906, p. 59.

The strata next overlying the Pennsylvanian rocks are the "Red Beds," which also contain signs of oil. On an anticline 7 miles southeast of Bridger, Mont., according to C. A. Fisher, the "Red Beds" contain a stratum of coarse-grained greenish sandstone about 10 feet thick highly impregnated with oil.

UTILIZATION OF THE GAS.

So far, no use has been made of the gas escaping from the Gray Bull gas well. Except in the city of Basin, which in 1907 had about 1,000 inhabitants, there will be little demand for gas for lighting purposes until the population of the district increases considerably. An important use might be in connection with sugar factories, which are greatly needed to care for the sugar beets grown on the irrigated lands. At present the nearest sugar factory is at Billings, to which the beets are shipped 195 miles by rail at a cost of 5 cents per hundredweight. As coal is abundant and would be cheap, if mined on a large scale, the gas can not compete with it except very close to the wells. Fortunately the productive well near Gray Bull is less than 2,000 feet from the Chicago, Burlington and Quincy Railroad, beside which there is a most excellent location for a sugar-beet factory or other industrial plant. The well being drilled by the Union Gas and Oil Company near Basin and some of the other possible gas fields described in this paper, are also close to good factory sites on the railroad.

SUGGESTIONS TO PROSPECTORS.

The conditions on the east side of the Bighorn basin are favorable for the prospector. Rock exposures are good and abundant; the structure is simple; the topography is of gentle relief; there is no timber except on the river bottom. The Chicago, Burlington and Quincy Railroad traverses the field for its entire length, and in the summer heavy machinery can easily be hauled from the railroad to almost any point in the basin over the wagon roads and plains.

Drilling is not expensive. It is reported on good authority that the cost of drilling the present gas wells near Basin and Gray Bull is less than $1 a foot, for a well 800 feet deep. To this should be added about $1 a foot for casing. The softness of the Colorado shale makes this low cost of drilling possible; unfortunately, it necessitates casing the greater part of every well. Some beds of white clay, probably bentonite, are especially troublesome. When the water of a well reaches this clay it softens and flows slowly into the well. There is much caving also in all parts of the Colorado shale.

In searching for possible gas fields in this region the important strata to observe are the sandstones in the lower part of the Colorado formation referred to in this paper as sandstones A and B (see section on p. 350), and the immediately underlying Mowry shale. The latter are hard siliceous and calcareous shales, containing many fossil fish scales and forming conspicuous ridges. Both the sandstones and the Mowry shale have prominent outcrops and hence they are most serviceable to the prospector in his search for anticlines. Anticlines in the Colorado shales probably exist, but none have been found, possibly on account of the poor exposure of these shales when they have low dips. The map (Pl. III) shows all the favorable anticlines within its borders that are indicated by outcrops of sandstones A and B.

THE LABARGE OIL FIELD, CENTRAL UINTA COUNTY, WYO.

By Alfred R. Schultz.

INTRODUCTION.

This paper is a brief statement of some observations made by the writer during the summers of 1905 and 1906 while examining the coal fields in southern and central Uinta County, Wyo. It is the purpose to give a short description of the occurrence of oil in Uinta County and to point out the probable geologic relations of the oil-bearing beds furnishing the oil recently discovered east of Labarge Ridge to the oil-bearing shale that gives rise to the oil springs and wells in southern Uinta County.

HISTORICAL SKETCH.

The occurrence of oil in southwestern Wyoming has been known for nearly three-fourths of a century. Many of the early trappers and fur traders, who built Fort Bonneville and the trading post at Fort Bridger, knew the location of the oil springs in this region and visited them in their annual trapping tours. The first published account of oil in southwestern Wyoming was the result of an examination made by the Mormons in 1847 on their pioneer journey across the great plains. For a brief historical sketch of the discovery of the oil springs on Hilliard Flat, the Carter oil spring, and those in the Fossil syncline near Fossil, Wyo., the reader is referred to the preliminary report [a] on coal and oil in southern Uinta County.

A few miles southwest of the Carter oil spring, in sec. 7, T. 14 N., R. 118 W., oil was found by the Oregon Short Line Railroad in 1900 and 1902, while constructing the Aspen tunnel, and a considerable oil seepage was encountered along the fault plane about 1,600 feet from the west portal of the tunnel. The oil springs along the east front of Absaroka Ridge north of Kemmerer were probably referred to in Lander's report of 1859,[b] where he makes the general statement that

---

[a] Veatch, A. C., Bull. U. S. Geol. Survey No. 285, 1906, pp. 342–344.

[b] Lander, F. W., Preliminary report upon explorations west of South Pass for a suitable locality for the Fort Kearney, South Pass, and Honey Lake wagon route; 35th Cong., 2d sess., Senate Doc No. 36, vol. 10, p. 33.

364

Pre-Cretaceous   Cretaceous   Tertiary

MAP SHOWING THE LABARGE OIL FIELD, CENTRAL UINTA COUNTY, WYO.

in the mountains along the divide in latitude 42° north there are "beds of coal, iron, and slate and a spring of peculiar mineral oil which by chemical process may be made suitable for lubricating machinery." No further description of the spring is given and the exact location is not known. The later geologic reports do not mention oil springs north of the Fossil locality. A brief description of the oil springs in southern Uinta County is given in the Wyoming reports by W. C. Knight and E. E. Slosson.[a] For a full discussion of the oil discovery in the vicinity of Spring Valley and of the developments in southern Uinta County from 1900 to 1905, the reader is referred to the reports of A. C. Veatch.[b] Since 1905 prospecting and development work have continued in the region about Spring Valley. The Pittsburg-Salt Lake Oil Company has filed proof of labor on most of its property and the people interested in oil are holding their locations. The Pittsburg-Salt Lake Oil Company discontinued drilling about December 15, 1907, and will commence drilling again in the spring. The International Consolidated Oil Company is putting down a couple of wells and has been working all winter. Two other companies expect to begin work soon and the outlook is very promising for a great deal of development work the coming summer. During the last three months of 1907 the Pittsburg-Salt Lake Oil Company shipped seven cars of refined oil and two cars of gasoline.

North of Kemmerer no prospecting for oil was carried on during the oil excitement in southern Uinta County. Oil discoveries have been reported, however, at various times from several localities along the east front of Absaroka Ridge and from Green River basin east of Meridian Ridge and Thompson Plateau. Considerable excitement was caused during the summer of 1907 by the discovery of oil east of Labarge Ridge in T. 27 N., R. 113 W. Numerous placer claims were soon staked out over the country between Labarge Ridge and Green River. Plans were outlined to prospect this region by churn- and diamond-drill borings during the coming season. While visiting the Labarge Ridge locality in September, 1907, the writer had an opportunity to examine this field hurriedly and collect a sample of oil from one of the shallow prospect wells shown on the accompanying map (Pl. IV).

LOCATION AND TOPOGRAPHY.

The Labarge oil field lies along the east base of Labarge Ridge and extends from Labarge Creek northward to the vicinity of South Piney Creek in T. 28 N., R. 113 W. (See Pl. IV.) The greater portion of the area forms a plain sloping gently eastward toward Green River.

---

[a] Bull. No. 3, Petroleum series, School of Mines, Univ. Wyoming, 1899.
[b] Bull. U. S. Geol. Survey No. 285, 1906, pp. 342–353; Prof. Paper U. S. Geol. Survey No. 56, 1907, pp. 139–162.

Tertiary topography, with its characteristic mesas and highly colored escarpments, is prominent in the western half of the area and along part of Green River.   On the west the area is bounded by Labarge Ridge, which forms a prominent range 500 to 1,500 feet higher than the adjacent country and attains an elevation of 9,200 feet at several points along its crest.   The topographic features of this range, which is composed of Carboniferous, Devonian, and Cambrian rocks, afford a marked contrast to those of the Tertiary beds east of the range.   For a brief description of the surface features of the region west of Labarge Ridge, the reader is referred to a preliminary report[a] on the coal in central Uinta County.

GEOLOGIC SUCCESSION.

The succession of the Tertiary and Cretaceous rocks in this general region, together with their economic importance, is given in the accompanying table.

Only a part of the geologic section is exposed in the Labarge Ridge locality.   The beds composing that ridge consist of "Upper Cambrian," Devonian, and Carboniferous rocks, and east of the ridge are small exposures of Adaville and Hilliard beds, covered throughout the greater part of the region by the nearly horizontal Tertiary strata.

Generalized section of Tertiary and Cretaceous rocks in central Uinta County, Wyo.

System.	Group.	Formation.	Thickness (feet).	Characteristics.	Economic value.
Tertiary.	Green River.	Green River.[b]	200	Thin-bedded shales, sandstones, and limestones, for the most part light colored.	
	Wasatch.	Knight.	c 500	Beds of red and yellow sandy clays interlaminated with white, gray, and yellow sandstones. Local areas of concretionary limestone.	East of Labarge Ridge yields oil which has probably risen from underlying Cretaceous beds.
		Unconformity.			
		Almy.	500	Red and yellowish-white conglomerates, sandstones, and sandy clays.	
	Upper Laramie.	Evanston.[c]	9,500	Gray and yellow shales and clays, with gray and yellow sandstone beds, containing several minor coal beds, none of which are developed. Same age as the Almy coals near Evanston, Wyo.	**Coal bearing.** Several minor coal beds of workable thickness have been observed. None have been prospected or developed. Coal similar to the Evanston and Almy coals of southwestern Uinta County.
		Unconformity.			

[a] Schultz, A. R., Bull. U. S. Geol Survey No 316, 1907, p 212.
[b] Estimate of Clarence King of maximum thickness in Green River basin southeast of this region is 2,000 feet; only a portion of the beds occur in this area.
[c] Upper limit not seen.

Generalized section of Tertiary and Cretaceous rocks in central Uinta County, Wyo.—Con.

System.	Group.	Formation.	Thickness (feet).	Characteristics.	Economic value.
Cretaceous.	Lower Laramie. / Montana.	Adaville.	2,800	Gray, yellow, and brown clays and shales with irregularly bedded brown and white sandstones and numerous beds of coal. The lower beds of this formation contain plants and invertebrate remains that are referred to the uppermost Montana; the upper beds contain lower Laramie leaves.	Prolifically coal bearing throughout. A few prospect pits only are opened in this area. At Sayle's mine a 180-foot tunnel has been opened in a 6-foot bed and considerable coal mined for local use. Several other mines supply coal for ranch use.
	Colorado. — Niobrara.	Hilliard.	3,000	Gray and black sandy shales and shaly sandstones that weather readily and afford few exposures. Usually a region of low relief.	
	Colorado. — Benton.	Frontier.	2,400 to 3,800	Alternating beds of gray and yellow clays, shales, and sandstones containing numerous beds of coal. Forms pronounced ridges or hogbacks in southern part of area east of Absaroka Ridge. Near top of formation is a pronounced bed of coarse sandstone, locally conglomeratic, containing numerous large oysters. This is the Oyster Ridge sandstone. Farther north this formation loses its characteristic hogback topography.	Prolifically coal bearing throughout the area. Farther south the Kemmerer, Willow Creek, Carter, and Spring Valley coals have been developed. The Kemmerer coals are extensively mined at Frontier, Diamondville, Oakley, Glencoe, and Cumberland. Within this area only Wright's mine and a few prospect pits have been opened, the coal being supplied to ranchers. Contains good building stone.
		Aspen.	1,200 to 1,800	Gray and black shales, shaly sandstone, and beds of compact gray sandstone and bluish limestone containing fish scales; commonly weathers silver-gray and shows little white specks in some of the sandstones.	Oil bearing. Contains oil developed in wells northeast of Spring Valley, probable source of oil in Hilliard, Carter, and Fossil oil springs and in springs north of Kemmerer and in vicinity of Labarge Ridge.
	Bear River.	Bear River.[a]	800 to 1,500	Black shale, shaly sandstone, and shaly limestone with abundant invertebrate fossils. Several thin beds of coal and bituminous shale.	Coal bearing. Coal beds so far as noted are too thin and impure to be of any value. Oil bearing. Oil in this formation southeast of Spring Valley in two wells.

[a] The Bear River beds are underlain by marine Jurassic. The beds appear to be conformable. Other evidence seems to indicate that an unconformity exists between the Bear River and Jurassic beds.

STRUCTURE.

The east base of Labarge Ridge, or the eastern boundary of the Paleozoic rocks, marks the location of an overthrust fault, which here brings "Upper Cambrian" beds in contact with Montana shales and sandstone. East of the fault lies the axis of a low anticline. This axis was observed in the cretaceous beds east of Labarge Ridge in T. 28 N., R. 113 W. The dips on both sides of the anticline are from 20° to 35°. Within a short distance to the south all traces of the east limb of the anticline in the Cretaceous beds are lost beneath the Tertiary beds, which here dip toward Green River at approxi-

mately 5°.   The beds along the west limb of the anticline are exposed
at several localities and dip at 20° to 45°, N. 70° W.   The southward
extension of the anticlinal crest may be represented by the low arch
seen in the Tertiary beds in the southern portion of T. 27 N., R. 113 W.

OCCURRENCE AND ORIGIN OF THE OIL.

The oil-bearing shale of southern Uinta County does not outcrop
in the Labarge Ridge locality.   The oil, however, in this field is
believed to come from the same horizon as in southern Uinta County,
namely, that of the Aspen (Benton) shale.   None of the natural oil
springs in southern Uinta County occur along the outcrop of the
shale that supplies the oil in the Spring Valley wells.   So far as field
observations have been made, no trace of oil was seen anywhere
along the outcrop of ·the Aspen shale.   The springs are all in the
region of profound disturbance along the Absaroka fault and its
associated secondary faults.   The oil springs of Hilliard Flat, the
Carter oil spring, and the seepage near the west end of the Aspen
tunnel are located along a secondary fault, but those on Twin Creek
lie along the line of the main fault.   The oil observed at all these
springs probably represents leakage from the oil-bearing shale along
the fault line, having been forced up through the water which has
penetrated to this shale along the fault contact.

North of Hams Fork oil indications have been observed at several
places along the east base of the Absaroka and Salt River ranges,
near the fault line.   Oil was observed on the water along some of
the streams tributary to Fontenelle Creek.   In Pomeroy Basin oil
indications were observed on the water in a number of marshes, in
quaking asp groves and low depressions.   It is reported that about
12 miles north of Kemmerer, along Mammoth Hollow, there is a spot
where gas makes its escape and on a damp morning can readily be
detected by its rank odor.   Indian tradition has it that many years
ago there used to be near this same locality an oil spring from which
oil flowed.   No trace of this spring was seen during the course of the
writer's work.   However, owing to the heavy covering of talus and
timber in this vicinity, as well as to the numerous springs that rise
on the mountain sides or flow from snowbanks near the crest of the
range, traces of oil are not so readily seen or recognized as in the
southern part of the county.   In the northern part of the field examined in 1906, about 2½ miles west of Snake River, along the north
line of T. 39 N., R. 116 W., oil was observed on the water and in
footprint depressions.   The oil seen here has a distinct odor and a
greasy feel.

Farther north in Idaho, east of the Pierres Hole (Big Hole) Mountains, W. E. McDonald observed strong surface indications of the
presence of oil in the vicinity of David Breckenridge's ranch and

later vigorously prosecuted the work of boring for oil on this ranch. So far as the writer was able to learn, Mr. McDonald struck a 10-foot bed of coal at a depth of 650 feet, but found nothing in the way of oil that indicated values.

Although no prospecting, drilling, or development work has been done north of Kemmerer on the oil-bearing Aspen and Bear River shales to determine whether they contain as much oil as the beds at Spring Valley, their occurrence throughout this region, east of the Absaroka and Salt River ranges and east of the Wyoming Range, is certain. They extend from Spring Valley in southern Uinta County to Snake River in northern Uinta County, and it is not improbable that they contain oil throughout this area in much the same abundance as has been found in the southern part of it. The approximate distribution of these oil-bearing shales can be inferred on consulting the map accompanying the preliminary report on the coal fields in a portion of central Uinta County,[a] as they occupy a narrow belt along the east side of the areas mapped as containing Frontier coals. No oil is found along the outcrop of these beds so far as observed. This fact, however, does not prove that oil is not present, as the oil near the surface may all have escaped or settled toward the synclinal axis, so that it is not noticeable on the surface.

In the vicinity of Spring Valley,[b] the only locality where the Aspen shale has been developed, the oil is found in sandy layers in a black shale near the base of the formation. Failure to obtain oil in this locality has been recorded in three types of wells—(1) those not deep enough to reach the oil-bearing beds; (2) those which on account of irregularities of the sandy layers in the Aspen (Benton) shale fail to produce oil, although oil is present in adjacent wells; (3) those located on the outcrop of the shale, particularly near the lower or eastern edge, where the bed is less than 500 feet thick. Although in general no oil is found along the outcrop of the oil-bearing shale, the amount increases down the dip. The conditions of the oil problem in the Spring Valley locality as well as in much of the territory north of that place, can best be set forth by the following statement:[c]

The oil-bearing beds are entirely dry when the oil is pumped out of the wells; no water follows. Water occurs in the overlying Wasatch beds and in the sandstones of the Frontier formation, and is also reported in a sandstone several hundred feet below the main oil sands, as in the Jager well and the Consolidated Oil Company well. The occurrence of large quantities of water in the Bettys well and the Baker well has been regarded by some as affecting the oil situation, but the water-bearing beds here are in no way connected with the oil-bearing strata. The anticlinal theory, according to which oil accumulates by floating upon water on the flanks or crests of

[a] Schultz, A. R., Bull. U. S. Geol. Survey No. 316, 1902, p. 212.
[b] Veatch, A. C., Bull. U. S. Geol. Survey No. 285, 1906, pp. 342-353; Prof. Paper U. S. Geol. Survey No. 56, 1907, pp. 143-144.
[c] Veatch, A. C., Geography and geology of southwestern Wyoming: Prof. Paper U. S. Geol. Survey No. 56, 1907, p. 158.

anticlines, does not seem to apply to this field, for one of the essential factors in the theory—the water in the oil-bearing sand—is not present. The absence of water in the oil-bearing sands, together with the fact that springs do not occur along the outcrops of the beds and the irregularity shown in the position of the oil-bearing sands in adjoining wells, suggests that the oil has been formed from the shale in which it is found and that the oil-bearing shales represent local sandy layers more or less perfectly surrounded by shale in which the oil has accumulated. This is the case also in the Boulder and Florence fields, although at those localities the shales are geologically younger. In the absence of water, oil tends to move down the dip and, so far as the continuity of the porous beds will allow, to collect in the troughs of the synclines. This is apparently the case in this field, and the position of this syncline and the depth of the oil-bearing shale at its lowest point then become matters of considerable economic importance.

Because of the rising and pitching of the Lazeart syncline, the oil-bearing shale in the synclinal trough lies at various depths below the surface along the axis. In a part of the region the depth of the oil-bearing shale along this axis is practically prohibitive to development work. However, the soft character of the beds suggests that the pressure of the superincumbent rocks may be great enough to practically close the pore space, so that the maximum accumulation of oil may be found at some point on the limb of the syncline, between the axis and the outcrop. In the vicinity of the fault, where the oil beds are at great depth, the oil leakage along the fault contact may be partly cut off for the same reason and the oil may be stored on the limb of the syncline. If the above-outlined conditions are true, prospecting in much of this field should be restricted to the shallow portions of the synclinal basin and to the region between the axis of the syncline and the outcrop of the oil-bearing shale on the west flank of the Meridian anticline, as the depth of the oil-bearing shale along part of the axis of the syncline is practically prohibitive to profitable development.

One of the most favorable localities for oil prospecting is in the vicinity of Wright's ranch, in T. 23 N., R. 116 W., where the oil-bearing shale lies from 2,500 to 4,000 feet below the surface along the center of the syncline. Almost as favorable a locality is that along Fontenelle Creek, near the north end of the Lazeart syncline. The depth of the oil-bearing shale in the center of the syncline is such that wells could be readily sunk, and test holes in this region are likely to yield results. Similar results may be expected in the synclinal trough that crosses Little Greys and Snake rivers in the northern part of the field.

In southern Uinta County the well of the Pittsburg-Salt Lake Company, in sec. 10, T. 14 N., R. 118 W., developed an oil-bearing bed in the lower part of the Bear River formation.[a] The oil is black and more in the nature of a lubricating oil than that of the Aspen shale.

---

[a] Veatch, A. C., Geography and geology of southwestern Wyoming: Prof. Paper U. S. Geol. Survey No. 56, 1907, p. 159.

Although the Bear River formation extends throughout this area, lying conformably below the Aspen shale in a narrow belt along its east side, nothing further was learned about these oil-bearing beds. At no point within this field have wells been drilled to test the oil-bearing properties of either the Aspen or the Bear River shale.

East of Labarge Ridge no outcrop of Aspen shale was seen. The oil springs observed in this locality resemble those in the Fossil region in that the strata around the springs belong to the Tertiary. The Wasatch beds here lie in a gentle anticline, with dips of about 5°, and may represent the southern continuation of the anticline above mentioned at the north end of Labarge Ridge. It is believed that the oil comes from the Aspen shale, which here lies from 2,000 to 4,000 feet below the surface. The oil is probably forced up through the water which has penetrated to the oil-bearing shale along the fault east of Labarge Ridge, and escapes at various points along the fault line. Part of the oil may be collected along the low anticlinal crest in the fairly waterlogged beds of the Wasatch and makes its escape into the valley where the oil prospects are located.

Whether the oil-bearing sands in this locality are dry like those at Spring Valley, or whether there is sufficient water present to float the oil and make it accumulate on the crests of anticlines, can not be determined until drill holes are put down to the oil horizons. The success of much of the future prospecting depends on this factor, the correct determination of which becomes therefore a matter of considerable economic importance. If the anticlinal theory of oil accumulation applies here drilling should be done along the anticlinal axis east of Labarge Ridge; on the other hand, if the rocks are dry the oil has collected in the troughs of synclines and prospecting should not be carried on along the anticlinal crest. If the synclinal occurrence prevails here the conditions in the Labarge oil field are manifestly more unfavorable for the accumulation of oil in commercial quantities.

QUALITY OF THE OIL.

All of the oil obtained from springs in southern Uinta County is a dark, heavy oil which may have been derived from the Aspen shale oils by the evaporation of the more volatile portions. Slosson[a] gives the gravity of the Carter oil as 21.5° Baumé and that of the Fossil or Twin Creek oil as 19.7° Baumé. The gravity of the Fossil oil is given by the Union Pacific Railroad[b] as 26.75° Baumé. The results of analyses of the Spring Valley petroleum and a sample collected from a shallow well about 3 feet square and 6 feet deep, sunk near the center of a drain about 4 miles east of Labarge Ridge, are given below. The

---

[a] Slosson, E. E., Bull. No. 3, Petroleum series, School of Mines, Univ. Wyoming, 1899, p. 31.
[b] Mineral Resources U. S. for 1885, U. S. Geol. Survey, 1886, p. 154.

latter sample was taken from an open pit which contained more than a foot of dark-colored, heavy oil. Considerable oil was taken from this pit during the summer by various persons who visited the region. The oil was also used by the ranchers in Green River basin as machine oil and proved highly satisfactory. It did not rise to the surface, but appeared to drain into the soil that filled the valley. At several points in the valley oil was encountered by sinking shallow wells a few feet into this soil.

Tests of oil from well of Pittsburg-Salt Lake Oil Company in sec. 22, T. 15 N., R. 118 W., 1 mile north of Spring Valley.

[By C. F. Mabery, Cleveland, Ohio, 1906.]

Temperature (°C.) at which gas was given off on distillation.	Percentage.	Gravity (°Baumé).	Nature of product.
50–150	21.3	65	Gasoline.
150–305	39.7	44	Burning oil.
305–350	16.4	36	Gas oil.
350–380	15.4	37	Oil partly cracked.

Residue, 7.2. Specific gravity, 0.81, =44° B. The oil begins to crack at 350°; of course this product is really gas oil. The distillates at 305°–350°, 350°–380°, and the residue contain much paraffin. These oils become solid when cooled in tap water with paraffin, so the yield is large. We refined some of the burning oil, not, however, with reference to flash or complete absence of color; it refines very easily, and gives a very fine grade of burning oil. Of course the proportions of products will be somewhat different on a refining scale (1,000 barrels)—probably larger, rather than smaller, than is given on the small scale. This petroleum is different from any of the numerous specimens that I have previously examined from Wyoming. A large amount of very light gasoline can be separated by strong cooling. With respect to the large proportion of gasoline and of burning oil, also of paraffin, this petroleum is one of the most valuable that I have ever examined. It is a nonsulphur oil; percentage of sulphur, 0.03.

Test of oil from shallow pit east of Labarge Ridge, Green River basin, in sec. 34, T. 27 N., R. 113 W.

[By Dr. David T. Day, United States Geological Survey, January 8, 1908.]

Temperature (°C.) at which gas was given off on distillation.	Percentage.	Specific gravity.	Nature of product.
Below 150	Trace.		
150–300	34	0.891	Suitable for burning.

Specific gravity of the original oil, 0.9435=18.75° Baumé. The oil was collected from seepage into a shallow well. It had evidently suffered oxidation, as shown by the considerable amount of resins contained. These resins made it difficult to completely separate water from the oil. The distillation was, therefore, slow and somewhat unsatisfactory. There is no indication of sulphur in the oil, no quantitative test being obtained by oxidation, and there is no odor of sulphur.

The specific gravity of the oil suitable for burning was so high that this portion was treated with sulphuric acid to determine whether the oil consisted of hydrocarbons of the paraffin (Pennsylvania) series. The amount absorbed by sulphuric acid was not abnormally large, and left a pleasant-smelling, refined product. The examination of the residue not distilling below 300° was extremely interesting. In addition to an oil soluble in cold alcohol, probably plain paraffin hydrocarbons, it gave a considerable amount soluble in boiling alcohol, which should have consisted entirely of paraffin wax, but did consist to a large extent of resins entirely absorbed by strong sulphuric acid, and giving evidence of being terpenes. The portion insoluble in boiling absolute alcohol, which should consist ordinarily of asphalt, gave, instead of the usual hard black asphalt, a soft sticky material characteristic of the transition stage of resins into asphalt.

While the oil has suffered too much oxidation to be interesting from the refiner's standpoint, it is extremely interesting scientifically, on account of the effects of the oxidation, showing, as given above, the intermediate stage between oil and ordinary hard asphalt.

SURVEY PUBLICATIONS ON PETROLEUM AND NATURAL GAS.

The following list includes the more important papers relative to oil and gas published by the United States Geological Survey or by members of its staff. Certain of the geologic folios contain references to oil, gas, and asphaltum; when these commodities are of importance in a particular area, they are listed in italics (PP. 9–11).

ADAMS, G. I. Oil and gas fields of the western interior and northern Texas coal measures and of the Upper Cretaceous and Tertiary of the western Gulf coast. In Bulletin No. 184, pp. 1–64. 1901.

ADAMS, G. I., HAWORTH, E., and CRANE, W. R. Economic geology of the Iola quadrangle, Kansas. Bulletin No. 238. 83 pp. 1904.

ANDERSON, R. (See Arnold, R., and Anderson, R.)

ARNOLD, R. The Salt Lake oil field, near Los Angeles, Cal. In Bulletin No. 285, pp. 357–361. 1906.

—— Geology and oil resources of the Summerland district, Santa Barbara County, Cal. Bulletin No. 321. 67 pp. 1907.

—— (See also Eldridge, G. H., and Arnold, R.)

ARNOLD, R., and ANDERSON, R. Preliminary report on the Santa Maria oil district, Santa Barbara County, Cal. Bulletin No. 317. 69 pp. 1907.

—— —— Geology and oil resources of the Santa Maria oil district, Santa Barbara County, Cal. Bulletin No. 322. 124 pp. 1907.

BOUTWELL, J. M. Oil and asphalt prospects in Salt Lake basin, Utah. In Bulletin No. 260, pp. 468–479. 1905.

CLAPP, F. G. The Nineveh and Gordon oil sands in western Greene County, Pa. In Bulletin No. 285, pp. 362–366. 1906.

—— (See also Stone, R. W., and Clapp, F. G.)

CRANE, W. R. (See Adams, G. I., Haworth, E., and Crane, W. R.)

ELDRIDGE, G. H. The Florence oil field, Colorado. In Trans. Am. Inst. Min. Eng., vol. 20, pp. 442–462. 1892.

—— The petroleum fields of California. In Bulletin No. 213, pp. 306, 321. 1903.

ELDRIDGE, G. H., and ARNOLD, R. The Santa Clara Valley, Puente Hills, and Los Angeles oil districts, southern California. Bulletin No. 309. 266 pp. 1907.

FENNEMAN, N. M. The Boulder, Colo., oil field. In Bulletin No. 213, pp. 322–332. 1903.

—— Structure of the Boulder oil field, Colorado, with records for the year 1903. In Bulletin No. 225, pp. 383–391. 1904.

—— The Florence, Colo., oil field. In Bulletin No. 260. pp. 436–440. 1905.

—— Oil fields of the Texas-Louisiana Gulf coast. In Bulletin No. 260, pp. 459–467. 1905.

—— Oil fields of the Texas-Louisiana Gulf coastal plain. Bulletin No. 282. 146 pp. 1906.

FULLER, M. L. The Gaines oil field in northern Pennsylvania. In Twenty-second Ann. Rept., pt. 3, pp. 573–627. 1902.

FULLER, M. L. Asphalt, oil, and gas in southwestern Indiana. In Bulletin No. 213, pp. 333-335. 1903.

—— The Hyner gas pool, Clinton County, Pa. In Bulletin No. 225, pp. 392-395. 1904.

GRISWOLD, W. T. The Berea grit oil sand in the Cadiz quadrangle, Ohio. Bulletin No. 198. 43 pp. 1902.

—— Structural work during 1901-2 in the eastern Ohio oil fields. In Bulletin No. 213, pp. 336-344. 1903.

—— Petroleum. In Mineral Resources U. S. for 1906, pp. 827-896. 1907.

—— Structure of the Berea oil sand in the Flushing quadrangle, Ohio. In preparation.

GRISWOLD, W. T., and MUNN, M. J. Geology of oil and gas fields in Steubenville, Burgettstown, and Claysville quadrangles, Ohio, West Virginia, and Pennsylvania. Bulletin No. 318. 196 pp. 1907.

HAWORTH, E. (See Adams, G. I., Haworth, E., and Crane, W. R.; also Schrader, F. C., and Haworth, E.)

HAYES, C. W. Oil fields of the Texas–Louisiana Gulf coastal plain. In Bulletin No. 213, pp. 345-352. 1903.

HAYES, C. W., and KENNEDY, W. Oil fields of the Texas–Louisiana Gulf coastal plain. Bulletin No. 212. 174 pp. 1903.

HILL, B. Natural gas. In Mineral Resources U. S. for 1906, pp. 811-826. 1907.

KENNEDY, W. (See Hayes, C. W., and Kennedy, W.)

KINDLE, E. M. Salt and other resources of the Watkins Glen quadrangle, New York. In Bulletin No. 260, pp. 567-572. 1905.

McGEE, W J. Origin, constitution, and distribution of rock gas and related bitumens. In Eleventh Ann. Rept., pt. 1, pp. 589-616. 1891.

—— (See also Phinney, A. J.)

MUNN, M. J. (See Griswold, W. T., and Munn, M. J.)

OLIPHANT, F. H. Petroleum. In Nineteenth Ann. Rept., pt. 6, pp. 1-166. 1898.

—— Petroleum. In Mineral Resources U. S. for 1903, pp. 635-718. 1904. Idem for 1904, pp. 675-759. 1905.

—— Natural gas. In Mineral Resources U. S. for 1903, pp. 719-743. 1904. Idem for 1904, pp. 761-788. 1905.

ORTON, E. The Trenton limestone as a source of petroleum and inflammable gas in Ohio and Indiana. In Eighth Ann. Rept., pt. 2, pp. 475-662. 1889.

PHINNEY, A. J. The natural gas field of Indiana, with an introduction by W J McGee on rock gas and related bitumens. In Eleventh Ann. Rept., pt. 1, pp. 579-742. 1891.

RICHARDSON, G. B. Natural gas near Salt Lake City, Utah. In Bulletin No. 260, pp. 480-483. 1905.

—— Salt, gypsum, and petroleum in trans-Pecos Texas. In Bulletin No. 260, pp. 573-585. 1905.

SCHRADER, F. C., and HAWORTH, E. Oil and gas of the Independence quadrangle, Kansas. In Bulletin No. 260, pp. 442-458. 1905.

SHALER, M. K. (See Taff, J. A., and Shaler, M. K.)

STONE, R. W. Oil and gas fields of eastern Greene County, Pa. In Bulletin No. 225, pp. 396-412. 1904.

—— Mineral resources of the Elders Ridge quadrangle, Pennsylvania. Bulletin No. 256. 86 pp. 1905.

STONE, R. W., and CLAPP, F. G. Oil and gas fields of Greene County, Pa. Bulletin No. 304. 110 pp. 1907.

TAFF, J. A., and SHALER, M. K. Notes on the geology of the Muscogee oil fields, Indian Territory. In Bulletin No. 260, pp. 441-445. 1905.

WEEKS, J. D. Natural gas in 1894. In Sixteenth Ann. Rept., pt. 4, pp. 405-429. 1895.

WILLIS. BAILEY. Oil of the northern Rocky Mountains. In Eng. and Min. Jour., vol. 72, pp. 782-784. 1901.

ASPHALT.

SURVEY PUBLICATIONS ON ASPHALT.

The following list comprises the more important papers relative to asphalt published by the United States Geological Survey or by members of its staff:

BOUTWELL, J. M. Oil and asphalt prospects in Salt Lake basin, Utah. In Bulletin No. 260, pp. 468–479. 1905.

DAY, W. C. The coal and pitch coal of the Newport mine, Oregon. In Nineteenth Ann. Rept., pt. 3, pp. 370–376. 1899.

ELDRIDGE, G. H. The uintaite (gilsonite) deposits of Utah. In Seventeenth Ann. Rept., pt. 1, pp. 909–949. 1896.

—— The asphalt and bituminous rock deposits of the United States. In Twenty-second Ann. Rept., pt. 1, pp. 209–452. 1901.

—— Origin and distribution of asphalt and bituminous-rock deposits in the United States. In Bulletin No. 213, pp. 296–305. 1903.

FULLER, M. L. Asphalt, oil, and gas in southwestern Indiana. In Bulletin No. 213, pp. 333–335. 1903.

HAYES, C. W. Asphalt deposits of Pike County, Ark. In Bulletin No. 213, pp. 353–355. 1903.

HILGARD, E. W. The asphaltum deposits of California. In Mineral Resources U. S. for 1883–84, pp. 938–948. 1885.

HOVEY, E. O. Asphaltum and bituminous rock. In Mineral Resources U. S. for 1903, pp. 745–754. 1904. Idem for 1904, pp. 789–799. 1905.

McGEE, W J. Origin, constitution, and distribution of rock gas and related bitumens. In Eleventh Ann. Rept., pt. 1, pp. 589–616. 1891.

RICHARDSON, C. Asphaltum. In Mineral Resources U. S. for 1893, pp. 626–669. 1894.

SMITH, C. D. (See Taff, J. A., and Smith, C. D.)

TAFF, J. A. Albertite-like asphalt in the Choctaw Nation, Indian Territory. Am. Jour. Sci. 4th ser., vol. 8, pp. 219–224. 1899.

—— Description of the unleased segregated asphalt lands in the Chickasaw Nation, Indian Territory. U. S. Dept. Interior, Circular No. 6. 14 pp. 1904.

—— Asphalt and bituminous rock. In Mineral Resources U. S. for 1906, pp. 1131–1137. 1907.

TAFF, J. A., and SMITH, C. D. Ozokerite deposits in Utah. In Bulletin No. 285, pp. 369–372. 1906.

VAUGHAN, T. W. The asphalt deposits of western Texas. In Eighteenth Ann. Rept., pt. 5, pp. 930–935. 1897.

BUILDING STONES.

MARBLE OF WHITE PINE COUNTY, NEV., NEAR GANDY, UTAH.

By N. H. DARTON.

Introduction.—In September, 1907, in compliance with a request from Hon. J. K. Taylor, Supervising Architect of the Treasury, I made an examination of a marble deposit in eastern Nevada. The purpose was to determine its amount, conditions of occurrence, and commercial prospects and to collect samples for analysis and physical tests by the technologic branch of the Geological Survey.

Occurrence.—It was found that the marble is a member of a series of metamorphic pre-Cambrian rocks of the Snake Range. The strata are uplifted in a broad anticline and deeply incised by canyons emptying into Snake Valley. The exposures are all in the first canyon south of some warm springs and begin about 5 miles west by south of Gandy post-office, or 4 miles west of the Nevada-Utah State line. They extend up the canyon for 2 miles, constituting the greater part of the walls of the main canyon and of several of its branches. The marble member is about 150 feet thick and it is included between metamorphic schists. The underlying schists appear in a low arch near the "camp" in the canyon. They are about 40 feet thick and lie upon white quartzite of which only the top is exposed. The overlying schists are in turn overlain by Cambrian limestones of dark-blue color, which are prominent in the adjoining higher slopes and ridges. The marble constitutes the walls of the canyon for nearly 2 miles, but it pitches downward at both ends of the exposure.

Marble member.—The marble is a completely metamorphosed or recrystallized limestone, partly gray in color and partly white. A great variety of tints of various colors appear, but the larger part of the deposit is dark bluish gray, banded or mottled with light gray or white. Some beds show regular alternations of white and gray marble in thin layers which are usually wavy or contorted. A thick deposit of white marble occurs near the upper part of the member. This white marble is in very thick, massive beds in the western portion of the canyon, where its thickness is about 35 feet. Eastward it outcrops along the canyon walls at various elevations and near the east end of the canyon, where it passes beneath the surface, it is 30 feet thick, but not massively bedded. In an exposure on slopes a short distance south of the canyon a low dome brings the white marble to the surface in an area of a few acres. Here portions of the rock are pink, in part in general tone and in part in mottlings. The extent of the pink marble is not revealed. The white marble in the upper part of the canyon is uniformly white, with a very slight but pleasing tinge of cream. Very little pure white rock was observed. The gray marble is the predominant variety, and although it varies in the proportion of white and gray bandings and mottlings, large bodies of it are of a uniform general tint. Some of the faces present from 75 to 100 feet of this rock. A few of the beds are separated by thin partings of mica, which considerably diminish the strength of the marble, but much of it does not part readily along its bedding planes.

Structure.—The marble deposits as exposed lie mainly in a wide, low arch with nearly flat, slightly undulating top, as shown in fig. 22. This

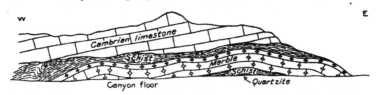

FIG. 22.—Section of marble in canyon west of Gandy, Utah. Scale: 1 inch=1 mile, approximately

illustration shows that the marble is but very little disturbed. The beds are free from noticeable faults, but they are traversed by various joint planes, mostly far apart, so that they will facilitate quarrying. It is believed that the minor joint planes which occur in some of the outcrops will disappear as the surface material is removed. Most of the cliffs show large bodies of unbroken marble.

Character.—Analyses of the marble made in the St. Louis laboratory are given at the end of this report. So far as observed the rock carries no pyrites or other metallic minerals which would disfigure it on weathering. The natural weathered surfaces indicate that it would weather satisfactorily, so far as could be judged from experience in other regions. The crystalline structure of the rock is so complete that it polishes beautifully, and this character appears to extend through the entire deposit.

Quarry conditions.—The marble deposits are favorably situated for quarrying, as they lie nearly level for a long distance and are presented in sloping canyon walls. The sketch map (fig. 23) shows the general topographic conditions. The amount of marble is great, and there appears to be no reason why quarries should not be successful if they are properly opened and operated. So far this marble has not been developed, except at a few points where samples of the surface material have been wedged off.

Shipment.—The canyon has a good road, over which the marble can be easily brought on a gentle down grade to the main wagon road in the adjoining Snake Valley.   Thence, however, the distance to railroads ranges from 65 to 80 miles, over mountains or along valleys, where the expense of haulage is great.

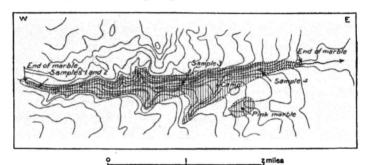

FIG. 23.—Sketch map of canyon west of Gandy, Utah   Contour Interval, 100 feet. Ruled area represents marble.

Crushing strength.—Four samples of the marble were collected, but all were either from outcropping ledges or within a few inches of the surface; they do not, therefore, indicate the strength which the quarry stone may be expected to have.   Samples No. 1 and No. 2 were white marble from the west end of the canyon.   No. 3 is banded gray marble and No. 4 is white marble obtained near the east end of the exposures. The blocks were 4-inch cubes, and they were tested in the structural-materials laboratories of the Geological Survey at St. Louis, under supervision of Mr. R. L. Humphrey.

Physical tests of marble from White Pine County, Nev.

Cube No.	Specific gravity.	Weight per cubic foot (pounds).	Crushing strength (pounds per cubic foot).	Absorption (per cent).
1	2.663	166	16,780	0.50
2	2.656	165.6	16,995	.57
3	2.696	168.1	20,145	.18
4	2.705	168.4	18,355	.16

Composition.—The subjoined analyses were made of the four samples in the St. Louis laboratory.   The materials were air dried.

Analyses of marble from White Pine County, Nev.

	1.	2.	3.	4.
Laboratory No.	262	263	264	265
Silica (SiO_2)	.60	.40	.28	.50
Alumina (Al_2O_3)	.33	.26	.23	.26
Ferric oxide (Fe_2O_3)	.22	.24	.34	.34
Manganese oxide (MnO)	.02	Trace.	.08	Trace.
Lime (CaO)	48.58	53.69	49.56	53.69
Magnesia (MgO)	5.66	1.43	5.00	1.35
Sulphuric anhydride (SO_3)	.10	.08	.07	.07
Water at 100° C	.00	.00	.01	.00
Carbon dioxide (CO_2)	44.39	43.78	44.50	43.68
Ignition loss	.05	.02	.03	.05
	99.95	99.90	99.97	99.94

SURVEY PUBLICATIONS ON BUILDING STONE AND ROAD METAL.

The following list comprises the more important publications on building stone and road metal by the United States Geological Survey. The annual volumes on Mineral Resources of the United States contain not only statistics of stone production but occasional discussions of available stone resources in various parts of the country. Many of the Survey's geologic folios also contain notes on stone resources that may be of local importance.

ALDEN, W. C. The stone industry in the vicinity of Chicago, Ill. In Bulletin No. 213, pp. 357–360. 1903.

BAIN, H. F. Notes on Iowa building stones. In Sixteenth Ann. Rept., pt. 4, pp. 500–503. 1895.

DALE, T. N. The slate belt of eastern New York and western Vermont. In Nineteenth Ann. Rept., pt. 3, pp. 153–200. 1899.

—— The slate industry of Slatington, Pa., and Martinsburg, W. Va. In Bulletin No. 213, pp. 361–364. 1903.

—— Notes on Arkansas roofing slates. In Bulletin No. 225, pp. 414–416. 1904.

—— Slate investigations during 1904. In Bulletin No. 260, pp. 486–488. 1905.

—— Note on a new variety of Maine slate. In Bulletin No. 285, pp. 449–450. 1906.

—— Recent work on New England granites. In Bulletin No. 315, pp. 356–359. 1907.

—— The granites of Maine. Bulletin No. 313, pp. 69. 1907.

DALE, T. N., and others. Slate deposits and slate industry of the United States. In Bulletin No. 275. 1906.

DILLER, J. S. Limestone of the Redding district, California. In Bulletin No. 213, p. 365. 1903.

ECKEL, E. C. Slate deposits of California and Utah. In Bulletin No. 225, pp. 417–422. 1904.

HILLEBRAND, W. F. Chemical notes on the composition of the roofing slates of eastern New York and western Vermont. In Nineteenth Ann. Rept., pt., 3, pp. 301–305. 1899.

HOPKINS, T. C. The sandstones of western Indiana. In Seventeenth Ann. Rept., pt. 3, pp. 780–787. 1896.

—— Brownstones of Pennsylvania. In Eighteenth Ann. Rept., pt. 5, pp. 1025–1043. 1897.

HOPKINS, T. C., and SIEBENTHAL, C. E. The Bedford oolitic limestone of Indiana. In Eighteenth Ann. Rept., pt. 5, pp. 1050–1057. 1897.

KEITH, A. Tennessee marbles. In Bulletin No. 213, pp. 366–370. 1903.

RIES, H. The limestone quarries of eastern New York, western Vermont, Massachusetts, and Connecticut. In Seventeenth Ann. Rept., pt. 3, pp. 795–811. 1896.

SHALER, N. S.   Preliminary report on the geology of the common roads of the United States.   In Fifteenth Ann. Rept., pp. 259–306.   1895.

——— The geology of the road-building stones of Massachusetts, with some consideration of similar materials from other parts of the United States.   In Sixteenth Ann. Rept., pt. 2, pp. 277–341.   1895.

SIEBENTHAL, C. E.   The Bedford oolitic limestone [Indiana].   In Nineteenth Ann. Rept., pt. 6, pp. 292–296.   1898.   [See also HOPKINS, T. C., and SIEBENTHAL, C. E.]

SMITH, G. O.   The granite industry of the Penobscot Bay district, Maine.   In Bulletin No. 260, pp. 489–492.   1905.

CEMENT AND CONCRETE MATERIALS.

CONCRETE MATERIALS PRODUCED IN THE CHICAGO DISTRICT.

By Ernest F. Burchard.

INTRODUCTION.

In connection with laboratory studies of the structural materials of the United States at the structural-materials laboratories of the United States Geological Survey in St. Louis, the writer spent several weeks in Chicago and vicinity, in the summer of 1906, obtaining representative samples of concrete materials. The location, extent, and geologic relations of the deposits sampled were noted, so as to supplement the experimental data obtained, and a general familiarity with the processes of preparation of material was gained. When the more important laboratory work on the concrete materials of this district shall have been completed a separate bulletin on the subject will probably be published. The present paper consists mainly of abstracts from the text of the proposed bulletin.

The term Chicago district as used in this paper is applied to the area in northeastern Illinois and southeastern Wisconsin in which concrete materials are produced principally for the Chicago market. The main portion of the district is bounded rather definitely on the east by the Illinois-Indiana State line; on the south by an east-west line passing about 7 miles south of Joliet; on the west by the west line of Kane and Kendall counties, and on the north by an east-west line passing just north of Lake Geneva, Wis. The area thus embraced is a quadrilateral 80 miles from north to south and 55 miles from east to west. About 500 square miles of this quadrilateral lies in Lake Michigan, so that there remains as land area about 3,900 square miles. (See fig. 24.) Concrete materials used principally in the Chicago market are pro-

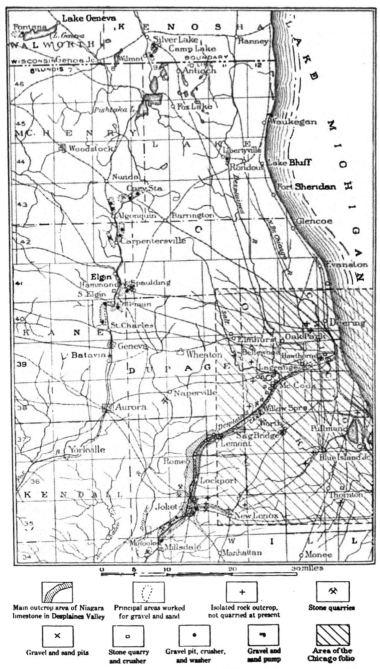

FIG. 24.—Map of main portion of district from which Chicago derives concrete materials.

duced also at three places beyond the area thus outlined, viz, Kankakee, Ill., and Beloit and Janesville, Wis. In a city having more than 2,000,000 inhabitants it is to be expected that a great deal of structural material such as dimension stone, granite blocks, clay for bricks, tile, and terra cotta, as well as limestone, sand, and gravel for concrete purposes should be brought from considerable distances. The significant fact has been brought out, however, during a brief study of the field, that Chicago and her environs, included within the area described above, produce practically all the concrete material that is used locally, besides nearly all the ordinary dimension stone and common clays. The granite, marbles, and finer grades of dimension stones, flagging, curbing, and fire clays are brought from various outside States. The granite areas near Green Lake, Wis., the limestone area near Bedford, Ind., the sandstone area near Berea, Ohio, and the clay mines of Illinois, Indiana, and Ohio, are all important contributors to Chicago construction work.

The importance to Chicago of a near-by and adequate supply of raw materials for concrete purposes is very great, especially as concrete construction in its various forms is at present making more rapid advances than any other type of such work.

CONSTITUENT MATERIALS.

VARIETIES.

The concrete materials produced in the Chicago district consist, in the order of quantities produced, of crushed magnesian limestone; sand; gravel, part of which is crushed; and Portland and natural cements. The crushed stone constituent closely approximates dolomite in composition, and is derived from the Niagara formation, which underlies the entire district, either outcropping at the surface or lying below thicknesses of glacial drift ranging from a few feet to 125 feet. The sand and gravel are derived from three types of material—(a) glacial drift and outwash from the drift sheet, (b) shore deposits of the present Lake Michigan, and (c) deposits on old beaches of the former extended glacial lake. Cements, though manufactured at Chicago, are not strictly of local materials, the limestone that enters into their composition being brought from Fairmount, a point in east-central Illinois about 100 miles south of the district as defined in this paper.

VALUE.

It has proved difficult to ascertain exactly the value of the various materials produced in the Chicago district that are used in concrete work, for the reason that in making returns producers are not always able to state definitely the uses to which the whole of their output of broken stone is put. However, if we consider as concrete material all the crushed stone produced, except that sold for flux and for lime burning, the approximate value of this material produced in the district in 1906 was nearly $2,000,000. The value of the sand produced during the same period was $205,500, and that of the gravel was $198,034. The total value of these concrete materials was therefore a little less than $2,500,000. Returns for 1907 are not yet available, but it is likely that the figures for that year will not exceed those of 1906, as general building operations were greatly curtailed during 1907. The extensive construction work at the new town of Gary, Ind., being built by the United States Steel Corporation, probably offset in part at least the general inactivity in Chicago, the stone for Gary being almost wholly derived from the Chicago district.

DETAILED DESCRIPTION OF MATERIALS.

NIAGARA LIMESTONE.

CHARACTER AND DISTRIBUTION.

The Niagara limestone of the Silurian system underlies all but the southwest corner of the district. It consists mainly of highly magnesian limestone, but contains some shale near the base. Under probably nine-tenths of the area the rock is covered by glacial drift and recent soil and alluvium, the total thickness of which, in places, is as great as 125 feet, although generally it ranges between 30 and 80 feet. The exposures of Niagara rock are mostly in the southeastern quarter of the district as defined in this paper, and they are due (a) to irregularities in the bed-rock surface, (b) to stream erosion, or (c) to a combination of these two causes. The preglacial surface or bedrock topography was undulating as compared with the present flat plain upon which Chicago stands, and the ancient hills of limestone are consequently buried by a less thickness of drift than the valleys— in fact, in several places these limestone hills reach the present surface.

Within the city limits of Chicago there are 10 or more places where the limestone either is exposed in a small area or else has been found to be so thinly covered by drift that stripping and quarrying are practicable. West, southwest, and south of the city there are 25 or more small, isolated exposures, at most of which quarrying is now or has been carried on.

The main outcrop area in the district extends along the valley of Desplaines River from Sag Bridge to a point 10 miles below Joliet. Here the rock forms the valley floor, overlain in places by a few feet of alluvium or by outwash sand and gravel, and locally it rises 30 to 50 feet in the bluffs. A few exposures occur also along other streams within the district, such as Salt Creek near Lagrange, Dupage River near Naperville, and Fox River at Batavia, St. Charles, and South Elgin. (See map, fig. 24.)

The total thickness of the Niagara formation in the district ranges from 250 to more than 400 feet, and it is probable that the original thickness was greater than this, because there was opportunity for preglacial erosion of beds lying above the present surface.

The character of the rock at the various outcrops and quarries within the area covered by the Chicago geologic folio is described in that folio by William C. Alden.[a] Since the folio was published important new quarries have been opened within that area, particularly at Gary, Lagrange, and McCook.

Just southeast of Lagrange, on the northeast side of the Chicago Junction Railway, are the quarries of the Federal Stone Company and the Lagrange Stone Company, both of which have been opened within the last two or three years, and are about 20 feet deep. The strata do not outcrop at this place, although they approach within a foot or two of the surface, where the cover is thinnest. The surface of the rock is uneven, and a short distance to the east and northeast the cover becomes too thick for stripping. The rock has generally a slight dip to the southeast, but in places dips as steep as 20° were noted. The top rock is rather thin bedded, and generally is oxidized to a buff color 2 to 10 feet below the surface, but is fairly white below the oxidized zone. The composition of the rock is shown by analyses 4 and 5 on page 394.

A new quarry was being opened near McCook in the summer of 1907 by the United States Crushed Stone Company. The stripping is thin and when removed discloses beds that are much fractured and weathered to a light-buff color the full depth of the opening, about 15 feet. Work is being pushed at this quarry, the excavation being facilitated by use of a steam shovel. Clay pockets are encountered in places in the limestone. The product is crushed and sold at present mainly for fluxing material. The average analysis (No. 7, p. 394) indicates the composition of the rock, and illustrates the fact that, although the material may be uniformly of a buff color, rather than white, it is quite as free from impurities as the unoxidized beds found at greater depths.

From the analysis which was submitted by the stone producers the rock is seen to be a fairly pure magnesian limestone, closely approximating the composition of dolomite, and therefore highly desirable for fluxing purposes. On account of its buff color its value should not be less for concrete material, except where a very light-colored stone is required for exposed construction. After the quarry reaches greater depth the lighter colored stone will be found.

---

[a] Description of the Chicago district: Geologic Atlas U. S., folio 81, U. S. Geol. Survey, 1902.

At Gary, southwest of McCook, is the new quarry of Dolese & Shepard. In August, 1907, this opening comprised about 15 acres and showed a section about as follows:

Section of Niagara limestone at Dolese & Shepard quarry, Gary, Ill.

Soil and drift.................................... 6 inches to 4 feet.
Magnesian limestone, buff colored, slightly stained and
 weathered into thin strata............................ 4 to 10 feet.
Magnesian limestone, light grayish blue, ranging from fine
 grained and dense to fairly porous; the beds are thicker
 than the surface rock, reaching a thickness in places of 15
 inches; a few clay seams are present where the bedding
 planes are irregular and in joint planes; fossils are abundant.. 27 to 30 feet.

The rock lies almost horizontal, and is cut by two sets of joints nearly at right angles to each other and extending northeast-southwest and northwest-southeast. In the section are three or four bands of light pinkish-gray porous rock, 7 to 12 inches thick, that can be traced halfway round the opening or farther, and such rock is found to make the most excellent lime. Crushed stone, rubble, and flux are the principal products here, and lime is soon to be burned.

The composition of the rock at Gary is shown by analysis No. 8, page 394, which represents an average of twenty-seven analyses, one being made each week throughout the last half of 1903, a total of 811 carloads having been sampled.

The following section represents a quarry at Lemont:

Section at quarry of Western Stone Company, Lemont.

	Ft.	in.
Soil...	1	6
Magnesian limestone, thin bedded, cherty...................	4	
Magnesian limestone, in beds about 1 foot thick; contains some chert nodules................................	3	
Magnesian limestone, similar to above, but in beds 10 inches thick.....................................	1	8
Magnesian limestone, massive, very cherty, in two beds of equal thickness................................	2	5
Magnesian limestone, sparingly cherty...................	1	9
Magnesian limestone, cherty, with "hackly" fracture, in four thin beds................................	1	3
Magnesian limestone, gray, fine grained, chert free, massive bed.	2	5
Magnesian limestone, gray, fine grained, chert free, in two beds, 1 foot 5 inches thick.........................	2	10
Magnesian limestone, fine grained, chert free, in one bed, called "Washington ledge"..........................	1	3
Magnesian limestone, gray, fine grained, chert free, in two beds, 3 and 7 inches thick.........................		10
Magnesian limestone, gray, fine grained, chert free, one massive bed...............................		3–11
Water level of quarry.		

The cherty beds at this quarry can not be used for dimension stone, but they make good road material and ballast.

On the north side of Desplaines Valley, about three-fourths of a mile northeast of Lemont, the Niagara limestone rises 35 to 50 feet in the bluff at the quarry of the Young Stone Company. Here the following section is exposed:

Generalized section at quarry of Young Stone Company, Lemont.

	Ft.
Soil	½
Gravel	3–4
Magnesian limestone, thin bedded, weathered, and fractured	5
Magnesian limestone, even, medium bedded, fine grained, buff, noncherty	6
Magnesian limestone, even, medium bedded, cherty, gray	7
Magnesian limestone, even, heavy bedded, cherty, light colored	2
Magnesian limestone, thin to medium bedded, fine grained, non-cherty	6
Magnesian limestone, heavy bedded, fine grained, noncherty, light colored	10

Rubble, dimension, and crushed stone are quarried from these beds.

Down the river from Lemont to a point about 10 miles below Joliet, the Desplaines Valley is cut in the rock so that exposures are numerous, both along the valley sides and in quarries, excavated in the valley bottom. The sanitary and ship canal of Chicago, extending nearly to Joliet, has been cut in the rock along this part of its course and consequently a large quantity of broken rock is available here. This material is being gradually utilized for riprap, ballast, filling, and crushed stone. At Lemont broken stone from the spoil banks of the canal is being loaded on barges and carried to Chicago, where it is used in lake front improvement work at Lincoln Park. Two miles below Lemont the Western Stone Company operates two crushers which are converting the rock of the spoil bank into concrete material, and the product is shipped to Chicago via the canal. The weathering of the rock where it has been piled for ten years has not been great. The material is mainly hard and gritty, but the surface rock is, of course, oxidized on the outside. Some portions of the spoil bank naturally furnish rock that is preferable to that in other localities, depending on the texture and the amount of chert and of clay present.

For about 3 miles below Lemont the valley sides are lined with abandoned quarries, where excellent dimension stone was obtained in the days before concrete construction was extensively employed. The rock suitable for dimension stone, known to the trade as "Athens marble," is found in its best development at and near Lemont, although good beds of it are found as far south as Joliet. Quarrymen have applied the term "tame stone" to rock that is fine grained, smooth textured, even bedded, and noncherty, and such rock makes the best dimension stone. They have likewise applied the term "wild rock" to rock that is irregularly bedded, breaks with a rough fracture, and contains argillaceous material or chert or both. Such rock often makes very desirable crushed stone, and although it had to be discarded before the era of concrete, it is now as valuable for crushing purposes as the "tame stone," and by some producers is held to be preferable, for some of it is found to yield on crushing a more nearly cubical fragment than the "tame stone," which tends to crack into thin chips when crushed.

Within the city of Joliet, and for 2 or 3 miles north and south from its center, the quarrying industry is active, about 15 important openings having been noted in September, 1907. On the west bluff of Desplaines Valley, in the SE. ¼ sec. 33, Lockport Township, the quarry of the Commercial Stone Company shows the following section:

Section at quarry of Commercial Stone Company, near Joliet.

	Feet.
Soil	1–2
Gravel	4–6
Magnesian limestone, buff colored, weathered, thin bedded, and cherty	5–9
Magnesian limestone, light gray, even grained, in medium to thick beds, with a few cherty strata near bottom	37±

These beds dip 2° to 3° NW., and are cut by two very prominent sets of joints. One of these sets extends N. 40° E. and the joint planes are vertical, clean cut, or enlarged by solution, and are spaced at intervals of 46 to 50 feet. The other set of joints extends practically at right angles to the first set, but the planes are less regular and persistent, and they pitch steeply to the southeast. Water descending from the gravel above the limestone has opened numerous large channels through the rock, and many of these are filled with clay when opened in quarrying. Along one joint plane so much rock has been removed by solution that the upper beds have caved down into the opening. Rubble is the principal product of the quarry at present.

On the east side of the valley, about one-fourth mile south of the north line of Joliet Township, a quarry and crushing plant is operated by the State Penitentiary. The quarry is excavated below the level of the valley bottom, and shows the following section:

Section at State Penitentiary quarry, Joliet.

	Feet.
Black soil and limestone débris	1–2
Argillaceous limestone, thin bedded and flaggy, somewhat stained to buff or light brown color	3
Magnesian limestone in fine grained, medium-thick, even beds	5
Magnesian limestone, rough grained, irregularly bedded, in medium-thick beds	3
Magnesian limestone, hard, in thin to medium-thick strata, irregularly bedded, with rough fracture and films of blue, hard, claylike material distributed through the mass. The color of the rock is light pink	6

The rock obtained here is used for road making throughout the State.

The Western Stone Company operates a large quarry near South Richards street, in the southern part of Joliet. In this and neighboring quarries the Niagara limestone is exposed for more than one-half mile along the Michigan Central and Elgin, Joliet and Eastern tracks to depths ranging from 15 to 50 feet. The following section shows the general character of the upper part of the rock and its cover at this place:

General section at quarry of Western Stone Company, Joliet.

	Feet.
Soil, gravel, peat, and calcareous clay, with minute shells	2–20
Magnesian limestone, thin bedded, flaggy, and weathered to yellow or buff color on top; the rock is even bedded and fine grained (tame stone). Lower beds become lighter colored and reach thicknesses of 2 feet	17
Magnesian and argillaceous gray limestone (wild rock) in rough-surfaced, irregular beds, 1 to 3 feet thick, mostly chert-free	10–15
Beds similar to above, but thicker bedded, and containing a little chert in small nodules, as well as considerable bluish-green argillaceous materials on the pitted surfaces of the beds	12–15

The "tame stone" is used for dimension stone, flagstone, and curbs. The "wild rock" is entirely crushed. It is very hard, and crushes into well-shaped lumps.

The beds are jointed, the planes trending nearly due northwest-southeast and northeast-southwest. The first-mentioned joints appear usually to be inclined to the northeast, but those of the second set are more commonly vertical. There has been rather general though slight slipping of the strata on the northwest-southeast joint planes. The displacement reaches 2 inches in a number of places, and the downthrow is toward the direction of inclination of the joint plane, or usually toward the northeast. Where the hade, or inclination, is in the opposite direction the downthrow is there found to be in the direction toward which the plane is inclined, or, in other words, the miniature fault is everywhere a normal one. In working the rock advantage is taken of this general drop on the northeast side of the joint planes, as it is possible thus to pry loose and move slabs and blocks with greater facility than where there is no offsetting in the beds.

As nearly all the quarries at Joliet are comparatively shallow, few, if any, additional facts would be brought out by further descriptions. In general it is shown that cherty beds usually outcrop in the river bluffs, and that below these cherty beds there are alternations of noncherty "tame" and "wild" rock, and in places beds that are sparingly cherty at 25 feet or more below the level of the flood plain.

Southwestward down Desplaines Valley to the mouth of Rock Run the normal magnesian-limestone character of the Niagara remains constant, although exposures below the south line of Joliet Township are fewer because of the presence of gravel terraces in the valley. In the vicinity of Rock Run, however, and extending southeastward to the vicinity of Millsdale, is a bed of shale very similar to the Maquoketa shale that lies below the Niagara formation. About 11 feet of this shale is exposed at the pit of the Millsdale Pressed Brick Company on the edge of the valley one-half mile east of Millsdale station. To the southwest of and stratigraphically below this shale lies a coarse-grained, roughly weathering fossiliferous limestone. It is cherty and in places contains large numbers of calcite nodules. This rock is exposed in the wagon road near Desplaines River south of Millsdale, below the Atchison, Topeka and Santa Fe Railway culvert 1 mile southwest of Millsdale; in Rock Run just below the bridge of the Chicago, Rock Island and Pacific Railway; along Dupage River above the Rock Island Railway bridge, and at other places in the vicinity. Fossils collected from the exposure on Rock Run, where the relations of the limestone to the shale are very clear, and from the Millsdale locality, were submitted to Dr. Stuart Weller of the Illinois Geological Survey, and were pronounced by him to be Niagara forms. The shale did not yield any fossils where examined. The limestone below the shale is very dissimilar to Niagara limestone. It bears some resemblance to the Galena limestone, but unless further detailed studies demonstrate the contrary, the rock must be considered as belonging to the Niagara formation, on the paleontologic evidence furnished by Doctor Weller.

From an economic standpoint this limestone below the shale bed can not be regarded as of present importance to the concrete industry for the following reasons: (a) Its texture is not sufficiently uniform, as it contains a mixture of calcite, magnesian limestone, and chert; and (b) its outcrop area is too remote from markets to enable it to compete with the better Niagara limestone, which occurs in practically inexhaustible quantities in more advantageous situations. Therefore the survey of the Desplaines Valley for limestone concrete material available to the Chicago market was terminated with Millsdale as its southwestern limit.

The thickness of the Niagara limestone in a city well at Ottawa street and Crowley avenue, Joliet, was reported by the city engineer to be 220 feet. Below this the record showed a bed of shale 140 feet thick (Maquoketa), and next below was 225 feet of limestone (Galena). There is thus at Joliet no record of a thin bed of shale toward the base of the Niagara.

On Fox River the Niagara rock was observed to outcrop at about six places, and inasmuch as no especial search was made for outcrops in the gravel district there are doubtless others. Two of these outcrops are on the east and west sides of Fox River about 1 mile north of the center of Batavia, and quarrying in a small way for local use has evidently been carried on here. Another outcrop was observed west of the river, in the northern part of St. Charles.

At South Elgin, on the west side of Fox River, the Niagara limestone comes to the surface of the valley bottom and is being exploited at the quarry of Magnus & Hagel. The rock occurs in thin beds with irregular, horizontal, wavy bedding planes usually coated with thin seams of bluish-green clay. The rock breaks with irregular rough fracture and is rather cherty. At 20 feet below the top there are 3 feet of beds in which the chert nodules are large and almost predominate in the strata. The material is typically a "wild rock" and is highly magnesian. The top 5 to 10 feet of beds are weathered and stained to a buff color. The rock is sparingly fossiliferous and in places contains crystals of dolomite and pyrite.

On West Branch of Dupage River, three-fourths of a mile southwest of the railroad station at Naperville, is a small area of Niagara limestone exposed by this stream. The rock has been quarried extensively here in former years, but the workings are now abandoned and the pits are filled with water. The cover that was stripped ranged in thickness from 4 to 15 feet, principally of drift, and the cuts were from 40 to 70 feet deep. The quarries formerly furnished bridge stone, dimension stone, rubble, and crushed stone. The rock was apparently used largely in the construction of the older buildings at Naperville. Much of the rock obtained was massive bedded and even grained, and some was evidently cherty.

CHEMICAL COMPOSITION.

An important use to which the Niagara magnesian limestone is put in the vicinity of Chicago is as a flux in iron and steel making. Vast quantities of Lake Superior ore are smelted and the iron is converted into steel at the works of the Illinois Steel Company at South Chicago and Joliet. At Indiana Harbor, Ind., the plant of the Inland Steel Company has commenced operations, and the United States Steel Corporation is erecting works of such magnitude near the lake shore in northwestern Indiana that the construction of the new town of Gary has been begun. As the Lake ore contains a very low percentage of lime and magnesia, good fluxing stone is very much in demand and many working analyses of the Niagara rock are available.   A few of these are given below:

Analyses of Niagara limestone.

No.	SiO₂.	Al₂O₃.	Fe₂O₃.	MnO.	CaCO₃ᵃ	MgCO₃.ᵃ	Na₂O.	K₂O.	SO₂.	P.	S.	H₂O.	Authority.
1...	1.12	0.91	0.83	......	54.73	42.79				0.005	0.04	......	Illinois Steel Co., Chicago.
2...	1.23	.55	.37	0.03	54.04	42.96	0.19	0.14	Tr.			0.29	U. S. Geol. Survey structural-materials laboratories.
3...	27.27	5.63	1.62	.02	33.50	27.95	.02	2.94	Tr.			.26	
4...	.40	......	.40	......	59.40	39.80				Tr.	.04		Mariner & Hoskins, Chicago.
5...	.70	......	.90	......	53.41	45.22							Featherstone Foundry Co., Chicago.
6...	1.04	.86	.80	......	54.82	43.13							Illinois Steel Co., Chicago.
7...	.28	.31		......	55.38	43.93							Inland Steel Co., Indiana Harbor, Ind.
8...	1.10	.93	.86	......	54.68	42.84							Illinois Steel Co., Chicago.
9...	17.30	1.33	.96	......	36.00	41.00			ᵃ			1.00	J. V. Q. Blaney.
10..	1.99	.62	1.15	......	53.73	42.13				.014			
11..	1.90	.64	2.08	......	52.61	41.84				.012	.054		

ᵃ The lime and magnesia are here given in terms of the carbonate in order more readily to show how closely the composition of the rock approaches that of dolomite (CaCO₃=54.35 per cent; MgCO₃=45.65 per cent).

1. From Dolese & Shepard, Hawthorne quarry. Average of 18 analyses made on 878 carloads of stone in 1904.
2. From basal beds at quarry of Brownell Improvement Co., Thornton. Rock is burned for lime.
3. From beds near middle of face of same quarry at Thornton. Rock can not be burned for lime.
4 and 5. Samples averaged from top 20-foot face at quarry of Federal Stone Co., Lagrange.
6. From Dolese & Shepard, McCook quarry. Average of 36 analyses made on 803 carloads of stone in 1903.
7. From United States Crushed Stone Co., McCook. Average of 6 analyses
8. From Dolese & Shepard, Gary (Ill.) quarry. Average of 27 analyses made on 811 carloads of stone in 1903.
9. "Athens marble" from Lemont. Analysis furnished by Western Stone Co.
10 and 11. Averaged analyses from quarry of Joliet Flux Stone Co., Romeo.

GENERAL METHODS OF PREPARATION OF CRUSHED STONE.

As nearly all the stone quarries in the Chicago district are in the form of pits excavated below the surrounding surface, the problems that have to be met are those peculiar to this type of quarry, and therefore the same fundamental principles are very generally observed. Methods vary considerably, however, throughout the district, according to the size of the quarry, its stage of development, the character of the rock, and the uses for which it is intended. In the initial stage the rock must be stripped of its overlying cover of soil and glacial débris. This is usually done by means of scrapers, but in the case of deposits 5 or 6 feet thick a steam shovel may be advantageously employed, particularly where the same shovel can be used for further work in handling the broken upper courses of rock. A thickness of 6 feet of cover is considered to be about the maximum limit profitable to strip at present. Drilling and blasting are universally employed to break up the rock, but here again occurs a wide diversity in practice. Both steam and compressed air are used in drilling, the latter preferably on long lines. The depth drilled ranges from 3½ feet to 24 feet. The charges also vary considerably in number, character, and strength. Most quarrymen use dynamite, although a few prefer black powder. The thoroughness with which the stone is broken up in blasting contributes toward the economical operation of a quarry. At some quarries large quantities of rock are shot out in huge blocks and these require reblasting and also a great deal of subsequent breaking with sledges to reduce the stone to a suitable size for the crusher. At one quarry, operated by an expert powder man, holes are drilled 2 feet apart, 4 to 10 feet from the face, in two rows, and set "staggering." At times as many as 100 shots are fired at once, and as a result very little reblasting is found necessary.

At practically all the quarries equipped with crushers, the crushers are situated at the surface, above the quarry, so that the rock has to be raised to them. At the greater number of quarries the broken rock is loaded by hand and hauled in automatically dumping cars up an incline by cable. At a few of the larger but shallow quarries the rock is loaded by steam shovels. Rock for lime burning or for rubble is generally selected and loaded by hand. In very deep pits platform elevators are in use, so built that they carry one or two loaded cars at a time. The character and capacity of the tram cars vary according to the general character of the equipment of the quarry, cars of wood or steel that hold 2 to 3 yards of broken stone being used. In crushing the stone several types of equipment are employed, but each aims to break the stone and to separate it into definite sizes by dry screening. For concrete purposes the stone should be as free from dust as it is possible to make it, and therefore plants which pay especial attention to the screening end of the process produce the best grade of concrete material.

One of the largest and most efficient plants in the district consists of two mills, one equipped with a No. 8 and the other with a No. 7½ Gates gyratory crusher, besides two No. 5 crushers each. The capacity of the two sets of crushers is respectively about 200 tons and 170 tons per hour, giving an average daily output of about 3,000 tons. The rock is put through rotary cylindrical steel screens, that give the following sizes: "Screenings," less than one-fourth inch; "roofing," one-fourth to one-half inch; "concrete," one-half to 1 inch; fine medium, 1 inch to 1½ inches; "medium," 1½ to 2½ inches; macadam, 2½ to 5 inches; and fluxing stone, 5 to 7 inches, the last size being rejected by the coarsest screen. Crushed stone is screened dry as contrasted with the washing process to which gravel is subjected when crushed and screened. As a rule the broken stone comes from the quarry with little or no foreign material, and whenever a clay seam or pocket is encountered it is cheaper to extract that material in the quarry than to remove it by washing in the mill. Another large, newly built plant is equipped with one No. 8, two No. 5, and two No. 3 crushers and four screens. The reported product of this mill at the start was 1,600 to 3,000 cubic yards[a] per day. One plant, equipped with one No. 7½ and two No. 4 McCully crushers, is reported to average 700 to 800 cubic yards per day of stone in five grades ranging from seven-eighths inch to 2 inches, besides screenings. Still another system of crushers in use is the Austin. At a plant equipped with one Austin No. 7, one Austin No. 4, one Gates No. 3 crusher and two screens, the capacity per day is reported to be 300 yards.

AVAILABLE LIMESTONE.

The reserves of Niagara limestone in the Chicago district suitable for crushing into concrete material are practically inexhaustible. The supply in those city quarries that are hemmed in by streets and buildings is of course limited because city values will prevent areal enlargement of the pits, and they must be sunk deeper and deeper until they reach the limit of depth beyond which it is impracticable to raise rock, or until they reach the underlying shale. It is thought that the deepest quarries still have more than 100 feet of stone below their lowest levels, so that their continuation is mainly a question of costs, and in such quarries slightly increased costs of working are offset by central location and consequent decrease of cost of delivery to consumers.

In the discussion of the Niagara formation the distribution of available material has been outlined in connection with the description of working quarries. The main areas are shown on the map (fig. 24). The Desplaines Valley will probably always continue to furnish the greater supply of crushed stone, although there is room for much more excavation at Stony Island, Blue Island, Thornton, Lagrange, Naperville, and at points on Fox River.

Sanitary and ship canal spoil bank.—The broken stone piled along the rock-cut portion of the sanitary and ship canal constitutes an important stock of material that is available without having to be quarried. Tests of this material made by the Chicago city engineering department show that although the rock tested was necessarily taken from the outside, or weathered portion of the spoil bank, the character still remains good, and it must reasonably be expected that on the inside of the pile also it should be sound. (See page 389.) From Willow Springs to Lockport, a distance of 15 miles, the channel is cut through rock. It is 160 feet wide at the bottom and 162 feet wide at water line, and the depth in this section averages 35 feet. The grade of the channel is 3¼ inches to the mile. The walls in the rock cut, having been cut by channeling machines, are smooth and perpendicular, with offsets. The total amount of solid rock that has been excavated

---

[a] The cubic yard is regarded as equivalent to 1¼ short tons.

is estimated by the engineers of the sanitary district to aggregate 12,912,000 cubic yards. When broken up by blasting and piled in miniature mountain masses along the borders of the channel, the cubic contents of the material was largely increased. After nearly eight years of construction work, water was turned into the canal January 2, 1900, and for several years afterwards these mountains of stone piled along the right of way were regarded simply as an incumbrance. Recently it has been planned by the sanitary district board of trustees to turn this incumbrance into an asset by selling the broken rock to parties who will erect crushers and convert it into stone for concrete, paving, etc.

The board has estimated that there are about 20,000,000 cubic yards of stone in these piles—material enough to construct concrete docks from the mouth of Chicago River throughout the length of the canal, Desplaines and Illinois rivers to St. Louis, following the course of the proposed inland deep waterway, or else the material could be used to construct a chain of concrete factory buildings and warehouses from Robey street, Chicago, where the canal begins, to Joliet, 40 miles inland. The price basis on which the rock is to be disposed of by the sanitary district is 10½ cents a yard and a portion of the net profits.[a] A beginning has already been made toward utilizing this spoil-bank stone. As mentioned on page 389, the Western Stone Company is operating two crushers near the county line west of Lemont, and east of Lemont the broken stone is being removed from the bank by steam shovel and shipped, without crushing, via barges on the canal to the lake front at Lincoln Park, Chicago, where it is used for riprap. This rock compares favorably in quality with freshly quarried limestone.

Importance to proposed deep waterway.—All the available rock, both in the spoil bank and in place in the Desplaines Valley, is adjacent to rail and water transportation facilities and can be cheaply handled. These facts, in connection with the almost limitless reserves of high-quality stone, are not only of importance in assuring to Chicago a plentiful supply of stone for crushing, but they have an important bearing on the economical construction of the proposed deep waterway from the Lakes to the Gulf. It must be remembered that for most of its length Illinois River, along which much concrete work would be necessary, flows through the coal-measure area, cutting into soft shale and sandstone and exposing few limestone beds thick enough to quarry until the area of Mississippian rocks is reached, near its mouth. Therefore supplies of crushed stone for concrete work would have to be obtained at the extremities of this inland waterway, and the Chicago end may be said to be well prepared to furnish the larger share of the needed material.

---

[a] From an industrial pamphlet issued by the sanitary district of Chicago, 1907.

GLACIAL SAND AND GRAVEL (OUTWASH AND MORAINE MATERIALS).

CHARACTER AND DISTRIBUTION.

Another source of concrete materials in the Chicago district may be found in the sand and gravel of glacial origin, derived mainly from the drift of Wisconsin age. The deposits here considered lie mainly within the morainal areas, but the character of many of the deposits is that of outwash material—that is, nearly clay-free, stratified gravel and sand, as distinguished from the morainal material, which is composed of clay, bowlders, and sand mingled in a confused mass. While the moraines of the Wisconsin drift sheet in northeastern Illinois and southeastern Wisconsin were being formed, there were streams of water issuing from the ice sheet and escaping to Mississippi River by way of Rock River and the tributaries of Illinois River, the Fox, Dupage, Desplaines, and Kankakee. These streams became overburdened with sand, gravel, and silt derived from the glaciers, and as a result filled up their beds and valley bottoms to a greater or less extent. In some places they spread out the detritus in terraces, or subsequently cut a new channel through the filled-up valley, leaving residual terraces on the valley sides.

The principal deposits of this type which are of economic importance in the Chicago district are situated, as shown on the map (fig. 24), along Fox River between Camp Lake, Wis., and St. Charles, Ill.; on Desplaines River at Libertyville and at and below Joliet, and on Long Run, Spring Creek, and Hickory Creek, small eastern tributaries of Desplaines River near Joliet. Beyond the area of the map, in the valley of Rock River, outwash deposits are exploited for the Chicago market at Janesville, Wis., and in Winnebago County, Ill., 1 mile south of Beloit, Wis. Besides these outwash deposits there is a thick deposit of morainal gravel being worked at Fontana, at the west end of Lake Geneva, Wis.

The important Fox River sand and gravel deposits in Illinois are near Cary, Algonquin, Carpentersville, Elgin, and St. Charles, and the general characteristics of the deposits being worked may be indicated by descriptions of a few typical workings. At Cary the deposits form a terrace on the north side of the river both east and west of the Chicago and Northwestern Railway, and are worked by the railway and the Lake Shore Sand Company. East of the railway the Lake Shore Sand Company has opened a face nearly one-half mile long. The present workings are at the northwest end of the face and disclose a bank about 40 feet high. The material ranges from fine quartz sand to bowlders, a few of which are 18 inches in diameter or larger. The bank is reported to average 75 per cent sand and 25 per cent gravel, including everything larger than torpedo sand. The upper 25 feet of the bank carries more gravel than that below, and in the middle third is found the coarsest gravel. There are a few ledges of partly consolidated gravel conglomerate, and locally near the base of the cut is 4 to 6 feet of sand that has been indurated by a dark ferruginous cement, forming a sandstone. Such hardened crusts of sand and gravel are termed by the quarrymen "hardpan" and this material has to be discarded, as the pit is worked by steam shovel and the hardpan ledges can not be cut by the shovel nor economically broken by blasting. West of the Chicago and Northwestern Railway the Lake Shore Sand Company is working a pit about 75 feet deep, below which water and quicksand are encountered. The material here runs irregularly as to its content of sand and gravel, but will probably yield a higher percentage of sand than the bank east of the railway. The middle third (vertically) will probably yield 75 per cent of sand, the upper part a little less, and the lower part a great deal more. The character of the material varies greatly from place to place, lenses or pockets of sand and gravel occurring without apparent system. For instance, on the southeast side of the present pit there is a bed of fine sand, extending 30 to 40 feet above the bottom, whereas on the opposite side of the pit alternate layers of gravel and sand extend down within 10 or 15 feet of the base. At this cut there is apparently no "hardpan" present, a fact which also illustrates the variability of the local deposits.

For 2½ miles north of Algonquin sand and gravel are found on the sides of the small valley through which the Chicago and Northwestern Railway passes. Northward toward Crystal Lake the deposits of sand and gravel are reported to grow thinner. The deposits at present worked form the shoulder or border of the upland lying between this small valley and Fox River to the east. At the bank of the Ætna Sand and Gravel Company, about 2 miles north of Algonquin, the heaviest deposit of sand and gravel is about 50 feet thick, with 2 to 4 feet of soil above and reddish clay below. This clay floor is about 25 feet above the creek bed. The banks worked here are from 20 to 40 feet thick, and they yield on an average about one-third gravel and two-thirds sand. The gravel runs rather small and contains only a few bowlders, which are found at the base of the deposit. At the top of the deposit and following the contour of its surface is a bed 5 to 10 feet thick containing an equal if not greater quantity of gravel than sand. Below this the gravel and sand are interstratified in layers from a few inches to 4 or 5 feet thick, and also are mixed together. Cross-bedding is seen at many places in the section and some beds having this structure are so firmly consolidated by a calcareous cement as to form hard conglomerate or hard sandstone. In places this material has assumed tubular or "pipy" shapes. Such material softens with exposure but does not disintegrate entirely. The finest sand is nearly all made up of quartz and other crystalline rock, but gives some effervescence in acid. The coarse sand effervesces more freely, showing a large proportion of calcareous material. A carload of 1-inch gravel showed nearly 20 per cent (roughly estimated) of crystalline pebbles, the remainder being mainly dolomite with some chert.

On the east side of Fox River, 1½ miles below Algonquin, is the pit of the Richardson Sand Company. The bank worked here is in the top 50 feet of the range of hills that rise 150 feet above Fox River at this place. A general section of the material exposed in the cut is as follows:

General section at Richardson Sand Company's pit near Algonquin.

	Feet.
Soil, dark brown	1–4
Gravel and bowlders, very coarse, in places 1½ to 2 feet in diameter	10–12
Gravel, medium sized, with a small proportion of coarse material; partly cemented to hard conglomerate	9–12
Gravel, medium sized and streaks of cross-bedded sand	8–14
Sand, fine grained to torpedo, in cross-bedded lenses with pockets of gravel	12–15
Clay.	

This deposit contains an unusual proportion of coarse material, some of the bowlders being angular slabs of Niagara dolomite so thick as to show more than one stratum. Many of the large bowlders are of crystalline rock. The above section can not be regarded as persistent, however, for the variation in the character of material from point to point is very abrupt. The yield of sand and gravel is about equal in quantity, although there is a larger proportion of sand than gravel in the bank. This is due to the fact that part of the sand is too fine to be caught by present methods of separation and is consequently washed away with clay and silt to the settling pond. When stripped, the surface of this gravel is almost level.

From Carpentersville to Algonquin on the east side of Fox River deposits of sand and gravel are found in places, but not continuously. For much of the distance the clay which underlies the gravel rises high and lies nearly parallel to the contour of the hills, so that the gravel is too thin to be profitably worked. On the west side of Fox River valley there are also high bluffs, largely of clay, on top of which sand and gravel occur, but the deposits have not yet been worked because of lack of transportation facilities and irregularity in thickness of the material, and because the present demand is supplied from deposits more advantageously situated.

South of Elgin there are sand and gravel deposits worked on the west side of Fox River near Coleman and 1 mile north of St. Charles, and on the east side of the river at Hammond and 1½ miles east of Coleman.

Near Coleman, between the Illinois Central Railroad and Fox River, gravel deposits are worked by the Richardson Sand Company. At this pit the working face is 15 to 30 feet thick, although the clay which underlies the deposits has so uneven a surface that the gravels thin in places to 6 or 8 feet. Overlying the sand and gravel is 4 to 5 feet of fine-grained silt. The material being worked yields about 3 · parts of sand to 1 part of gravel. The gravel is mainly small in size, and the sand is rather coarse, mostly a torpedo grade. The 1-inch to 1½-inch gravel appears to contain 10 to 15 per cent of crystalline material, and the finer gravel a still higher percentage. At the base of the deposit are many rather large bowlders of dolomite and granite, 2 to 3 feet in diameter. The sand and gravel instead of occurring in separate strata as in the region near Algonquin, are rather uniformly mixed together, and no "hardpan" or consolidated conglomerate was noted.

East of the river and south of the Chicago, Milwaukee and St. Paul Railway at Hammond is the pit and plant of the Chicago Gravel Company. The sand-gravel deposit lies upon a clay floor that is slightly uneven. The deposit reaches a total thickness of 27 feet in places and the stripping averages about 2 feet. A few bowlders 2 to 3 feet in diameter occur at the base, but in the bank the gravel is unusually uniform in size, rarely running into large cobblestones. The sand is a good sharp torpedo, not very fine. It contains a small proportion of lime, reported to be about 2 per cent. The proportion of sand to gravel in the bank is said to average about 55 to 45. No conglomerate nor "hardpan" was noted, but lenses of clay were found to occur in the bank. One of these noted at the time of visit was 6 feet thick in a bank 22 feet in length. Most of this clay, fine-grained and silt-like material, can be kept out of the product by a skillfully manipulated steam shovel, although some of it is certain to be loaded with the sand and gravel and it can not be eliminated entirely in the washing.

One mile north of St. Charles, on the west side of Fox River, is the pit of the American Sand and Gravel Company. The deposits here reach a thickness of 35 feet. Below the sand and gravel water is encountered in quicksand, before the underlying clay is reached. The base of the deposit is therefore low, not far above the level of Fox River and of the creek to the south of the pit. Gravel and sand in about equal parts appear to constitute the bulk of the material. The gravel ranges from small to coarse sizes, and some cobblestones go to the crusher in nearly every yard of material excavated.

Notes on the gravel pits along Fox River would not be complete without mention of two points just north of the State line in Kenosha County, Wis. At Capp Lake are some abandoned pits owned by the Wisconsin Central Railway. The deposits here are reported to have been thoroughly prospected, but to have proved not to be of promising thickness nor cleanness. Below 2 to 3 feet of soil there lies about 5 feet of fairly good gravel in the higher parts of the bank. Below this there are alternate seams of clay and quicksand containing heavy bowlders. Washing and crushing would therefore be involved to too great an extent for practical purposes.

Near Wilmot, Wis., is a pit the output from which is taken by the American Sand and Gravel Company. The pit is a straight cut into a terrace of Fox River and shows the following section:

Section of gravel pit near Wilmot, Wis.

	Feet.
1. Soil	1–2
2. Gravel and sand. The gravel is clean and contains about 15 per cent of crystalline rock, the remainder being dolomite. About 15 per cent of the gravel runs larger than 2 inches	8–14
3. Quicksand and silt, very fine-grained material containing about 60 per cent of quartz, the balance being clay minerals	14–15
4. Gravel, similar to the upper gravel bed (No. 2)	20
5. Sand, fine grained, to quicksand. In the aggregate this bed is coarser than bed No. 3, contains a higher percentage of silica and less clay	15
6. Quicksand and water.	

The beds of quicksand inclosed in the gravel vary in thickness and do not conform in contour to the present surface of the terrace. In general they appear to dip toward the northeast and to pinch out in various directions as if lens-shaped. The material is reported to be composed of the various grades in about the following proportions:

Proportions of sand and gravel in pit at Wilmot, Wis.

	Per cent.
Concrete sizes, one-half inch to 1½ inches, of which about 20 per cent is crushed gravel	20
Roofing gravel, one-eighth inch to one-half inch, total	20
Sand, fine grained, with a small proportion of torpedo size. It is mostly quartz and fairly free from quicksand and clay	60
	100

On the west side of Desplaines River at Libertyville is the sand and gravel pit of the Lake Shore Sand Company. The deposits seem to be mainly west of the river in this vicinity and are comparatively thin. This deposit is the thickest in the locality, and it ranges from 5 or 6 to 25 feet in thickness above water level. Test wells are reported by the operators of the pit to show 20 feet of gravel below water level. The water level varies 1 to 2 feet during the year, and the cut is deeper or shallower accordingly. A section made at the southeast end of the cut, where material was being obtained October 3, 1907, is as follows:

Section of sand and gravel bank at Libertyville.

	Feet.
Soil	1–2
Fine sand, loam, and a little gravel	0–4
Clay lens, saucer-shaped in profile	0–1
Gravel and torpedo sand in alternate beds, 1½ to 2 feet thick, cross-bedding common. The gravel is mostly smaller than 4 inches. The proportion of sand to gravel ranges from 2 to 1 to 1 to 1, but will average close to 1.5 to 1	16–20
Gravel, sand, and water (reported 20 feet to clay).	

The usual sizes of sand and gravel are produced here and an additional product worthy of note is the unwashed run-of-bank sand and gravel, including all material smaller than 1½ inches that is used in road making. It is stated that the loam present exerts a cementing action that makes the material of value as a bond when laid in alternate layers with crushed stone in macadamizing roads. Unsuccessful efforts have been made to pump the sand and gravel that lie below water level, the result being that the pumps were soon choked by the gravel. It is proposed to attempt at some future time the dredging of these deposits. It would seem worth while to utilize these submerged materials, as the visible supply of gravel above water level is diminishing rapidly in this locality.

Farther down the Desplaines Valley deposits of sand, gravel, and bowlders are scattered at irregular intervals and many of these are worked from time to time in a small way for local purposes. One such deposit is about a mile north of Willow Springs. The main deposits, those that are at present affording material sufficient for the operation of crushing plants, are at and below Joliet. Hickory and Spring creeks have built up deposits of gravel in their valleys and in the Desplaines Valley near the junction of the two creeks.

On East Washington street, Joliet, the Chicago Gravel Company operates a pit and crusher. The deposit varies greatly from place to place.

The gravel and sand are cross-bedded in places. The clay content of the gravel averages about 20 per cent. The gravel consists mostly of dolomite, with a few crystalline pebbles. In places the material is hardened by calcareous cement to a conglomerate. The sand is fine to coarse grained and of dark color. It contains comparatively a high percentage of limestone and dolomite grains and of clay, with relatively a low proportion of silica.

The clay seams thin out to the north and the south. Characteristics of the bank on the east side of the pit are that no beds of fine sand appear in the section and that the gravels are more even bedded than elsewhere in the pit.

Another plant of the Chicago Gravel Company is on the east side of the valley about 1½ miles above Millsdale, adjacent to the Santa Fe and Chicago and Alton railways. The deposit worked here is in the form of a terrace or bar in the Desplaines Valley and consists of material ranging from sand and loam to bowlders 2 feet in diameter. The bottom of the cut is in gravel, but reaches ground-water level, which is practically at the level of the water in the river and fluctuates with it. The deposit is about 10 feet thick on one side of the cut and 20 feet thick on the other. The gravel and bowlders are composed principally of hard dolomite, but about 5 per cent of crystalline material is present. The loam and sand are highly calcareous. The sand and gravel deposits in this vicinity and southward to the mouth of the Dupage are extensive, and thus far have been only very slightly utilized.

Certain important deposits of sand and gravel which, although at considerable distance from Chicago, are so directly connected by railroads with the city that they are worked to advantage should be mentioned in these notes. Such localities are in southeastern Wisconsin, at Fontana, Janesville, and Beloit.

At Fontana, at the west end of Lake Geneva, the Lake Geneva Gravel and Sand Company is exploiting a thick gravel bank. The deposit is part of the Darien moraine of the Delavan lobe of the Lake Michigan glacier, according to Alden.[a] The maximum thickness of the cut is about 90 feet. Two or three feet of soil is stripped from the top by means of scrapers. A general section shown by the cut is as follows:

General section of gravel pit at Fontana, Wis.

	Feet.
Soil with a few large bowlders at base	2–4
Clayey, loamy, fine sand, of brownish color, containing a little gravel	6–7
Coarse, cobblestone gravel	20
Sand, thin ledge	1–3
Gravel, rather coarse, with some sand, mainly concealed by talus to bottom of cut	30–60

The gravel runs unusually large, as compared with the Fox River and Rock River deposits. In the larger gravel there is a fairly large proportion, perhaps 15 per cent, of crystalline rocks, many of which are dark colored. The remainder of the gravel is mostly dolomite and limestone, largely of Niagara age. The proportion of gravel to sand is reported by the operators of the pit to run about 3 to 2, and in places a still higher proportion of gravel is found. At the west end of the pit there is considerable firmly cemented conglomerate.

---

[a] Alden, W. C., The Delavan lobe of the Lake Michigan glacier: Prof. Paper U. S. Geol. Survey No. 34, 1905, Pls. IV, V, and X.

There are apparently similar deposits still undeveloped in many of the hills at the west end of Lake Geneva, although none are so easily accessible as the bank just described.

Near Janesville, Wis., the outwash deposits of Rock River valley and tributary valleys are worked for sand and gravel. On South Main street, about 1 mile southeast of the middle of Janesville, a sand and gravel bank is exploited for the manufacture of sand-lime brick, cement shingles, and concrete blocks and posts. The face of the bank is about 25 feet in height. The upper 8 to 10 feet carries sand and gravel in the proportion of about 5 to 3, but below this the ratio increases to about 10 to 1. The gravel is small, few of the pebbles exceeding 3 or 4 inches in diameter. The material is very clean, and the sand is rather fine and composed almost entirely of quartz.

On the line of the Chicago, Milwaukee and St. Paul Railway, about 2½ miles east of Janesville, is a sand and gravel bank worked by the Knickerbocker Ice Company. The face of the bank is 50 to 70 feet in height. The material consists of small, clean gravel and clean quartz sand, much of which is of rather fine grain. The upper half of the bank is reported to carry sand and gravel in about equal quantities, but in the lower part sand predominates in the ratio of about 5 to 2. The sand occurs in beds of fine to torpedo size and in beds with gravel; and at the bottom is a sand bed probably 25 feet thick, only 12 feet of which is utilized, as the material is a little too fine for torpedo size. The normal stripping is 2 to 4 feet, but in ravines that cut down into the deposit it will run as thick as 10 feet. The gravel rarely runs larger than 3 or 4 inches and yields concrete gravel containing 50 per cent or more of crushed rock. The product goes mainly to Chicago markets.

About 1 mile south of Beloit, Wis., in Winnebago County, Ill., is situated the sand and gravel bank of the Attwood-Davis Company. This bank is on the east side of the Chicago and Northwestern Railway main line and is in the Rock River valley. The cut extends about one-third of a mile from north to south and is about 35 feet in height. The gravel is overlain by 1½ to 2 feet of black soil at the north, but to the south and east there is a bed of fine sand, 6 to 12 feet thick between the gravel and the top soil. This bed of sand forms a low ridge and also fills a shallow depression in the surface of the deposit. It is troublesome, as the sand is too fine for torpedo size and does not contain sufficient clay, except in small pockets, to make a molding sand. Below this, gravel and gravelly sand alternate in layers 2 to 3 feet thick. The gravel ranges in size from small to medium, 4-inch pebbles being about the largest. About 15 per cent of foreign crystalline material is present in the gravel. The cut is worked to the level of the underflow in the valley, but sand and gravel are reported to extend at least 50 feet farther down, as determined by a well point. The average run of the bank, as reported by the superintendent of the pit, is about 3 parts of sand to 2 of gravel. About 40 per cent of the concrete sizes produced consist of crushed gravel.

Besides the sand and gravel pits here noted, there are many small pits scattered here and there in the suburbs of Chicago worked by pick and shovel, with wagon haulage, to supply local needs. Many of these pits are in the extinct beaches of Lake Michigan, several miles from the present shore line. The location of these old beaches is shown in the areal geology maps of the Chicago geologic folio.

GENERAL METHODS OF PREPARATION OF SAND AND GRAVEL.

The preparation of cleaned sand and gravel begins with its excavation from pit or bank, and involves moving from pit to mill, screening to separate the sand and smaller gravel, crushing to reduce the small bowlders and gravel larger than 1½ or 2 inches in diameter, and washing to free the material of silt, clay, organic matter, and resultant discoloration. The method of handling the material depends somewhat on local conditions.

On reaching the crushing plant the gravel is screened under a stream of water. A set of screens usually comprises screens having some or all of the following sizes of perforations: 2-inch, 1½-inch, 1-inch, ¾-inch, ½-inch, and ¼-inch. They are of both rotating and stationary types.

After the gravel and sand have been sorted by screening, crushing, and washing, into the required sizes, the material is stored in bins which are readily emptied through spouts by gravity into cars on a convenient siding. By discharging two or more bins at once into the same car, and by regulating the rate of flow of sand and differently sized gravel, a mixture containing these materials in almost any desired proportion can be obtained, as, for instance, a mixture that will be suitable for concrete on the addition of the required quantity of Portland cement. During the winter months, when freezing interferes with washing operations, dry screens are used, when needed, at several plants in the district.

At the majority of plants in the district materials are separated into sizes about as follows: Torpedo sand (grains that pass ⅜-inch sieve), roofing gravel (passing ½-inch but not ⅜-inch), and concrete gravel (passing 1½-inch but not ½-inch). There is some variation from these sizes, of course, and larger sizes than 1½-inch are produced. The proportion of gravel of concrete size, which is sharp and angular as a result of crushing, depends on the coarseness of the deposit. Where the percentage of gravel in the bank is high and a large proportion of it is more than 1½ inches in diameter, the proportion of crushed stone in the product is of course relatively high, and has been known to reach 60 per cent.

An interesting use to which the coarser gravel is put is as a flux in iron melting at Carpentersville, Ill., and this is possible because of the large percentage of dolomite pebbles in the gravel.

An important factor in the sand and gravel business is an adequate supply of water. Some plants are situated so close to Fox or Desplaines River that they may obtain water by pumping directly from the stream. Others reach an underflow at the base of the pit, or the base of the pit may be determined by water-saturated sand and gravel, and in such places an abundance of water may be obtained by driving pipes a few feet into the water-bearing gravel and pumping therefrom. Less advantageously situated with respect to water supply are those banks that are remote from a stream or high on the valley rims, but usually in this well-watered country sufficient water may be caught in reservoirs so constructed as to receive the run-off from some gully or wet-weather stream, or such a reservoir may be partly supplied by pumping or by utilizing the flow of a small spring. Where conservation of water is necessary, settling basins must be constructed, and space must be provided for them. Water thus used over and over again can be kept fairly clean, but is hardly as desirable as a copious supply obtained from wells or from a clear, flowing stream such as Fox River.

AVAILABLE SAND AND GRAVEL.

In the foregoing portion of the text suggestions have been given as to possible extensions of workings along Fox and Desplaines rivers. In review it may be said that in the valley of Fox River, from the southern part of Kenosha County, Wis., to Geneva, Ill., and perhaps farther south, there are many unworked deposits of sand and gravel. A large part of the moraines and outwash deposits left by the melting of the glaciers in southeastern Wisconsin is made up of sand and gravel. The character of these extensive deposits has been discussed in detail by William C. Alden.[a] Their distribution is shown on the maps accompanying his report. There are vast amounts of sand and gravel yet to be utilized in the tracts indicated. These deposits, especially the moraines, vary greatly in character, however, from point to point, and much of the material is not now readily accessible for transportation. In the Desplaines Valley the best deposits are found between Joliet and the mouth of Desplaines River. Proved but undeveloped deposits occur in the areas shown on the map forming fig. 24 (p. 384), but owing to its necessarily small scale it has been impossible to show locations in the desired detail. On account of the irregularities in deposition which are common to glacial material, more particularly to morainal deposits than to outwash gravels, although somewhat characteristic of the latter, it is essential that thorough prospecting be

---

a The Delavan lobe of the Lake Michigan glacier: Prof. Paper U. S. Geol. Survey No. 34, 1905.

done before arrangements are begun to work a pit or bank on a large scale. A common method of prospecting a tract is to sink a number of test wells 3 to 5 feet in diameter and as deep as desired in order to determine the thickness of cover, proportion of gravel to sand, size and character of gravel, whether or not any clay or "hardpan" is present, whether or not the materials are mixed or stratified, at what depth water is encountered, total thickness of deposit, and all such factors as have a bearing on the economical development of the deposit. In many places such test wells have to be curbed by planks to prevent the loose wall material from caving in, and often it is impossible to remove the planks from the well, so strong is the compression exerted by the deposit.

LAKE SHORE DEPOSITS.

AVAILABLE MATERIAL.

Fine-grained sand occurs in inexhaustible quantity on the present beach of Lake Michigan, and in places there is more or less coarse sand and gravel mixed with it. As a source of supply for concrete material, however, these deposits are not now of great importance, for the following reasons: The sand is mostly of finer grain than torpedo sand, which is most desirable; the material requires special methods for the separation of sand and gravel; the deposits are constantly shifting with shore currents; and the occupancy of the lake front by docks, railroads, parks, boulevards, and private grounds has made much of the beach unavailable or too valuable to be exploited for sand and gravel.

At the south end of the lake, in Indiana, sand dunes have furnished much of the filling used in track elevation, and this area, together with a few others temporarily worked south and north of Chicago, is still furnishing supplies of sand for local use, chiefly for lime mortars and plaster.

METHOD OF OBTAINING AND PREPARATION.

At Waukegan sand and gravel are obtained in a unique manner from the beach deposit. The Waukegan Sand and Gravel Company was operating in August, 1907, a sand pump or "sand sucker" in a shallow lagoon between the Ludington Salt Company's docks and a spit occupied by the Elgin, Joliet and Eastern Railroad. The outfit consists of a centrifugal pump having a 6-inch intake and a 7-inch outlet pipe. The pump is driven by a 20-horsepower engine, and the whole apparatus is floated on a covered barge. Water, sand, and gravel as large as $3\frac{1}{2}$-inch are together pumped from the bottom of the lagoon, and are discharged through a pipe of variable length into screens and thence into cars. It is possible to pump material from as great a depth as 20 feet, and to carry the delivery pipe to cars at least 600 feet distant, provided a slight fall is given the pipe. Gravel larger than 3½-inch is excluded by a screen over the mouth of the intake. Occasionally the gravel that passes into the pump clogs it and makes trouble. It is reported that the capacity of such a plant is about 10 to 35 yards per day of ten hours.

The character of the deposits worked near Waukegan varies from place to place and also from season to season. Some deposits have been found to yield only 2 to 4 per cent of gravel, whereas others have yielded 33½ per cent. The material being raised at Waukegan was clean and of good quality. The gravel was composed principally of dolomite, granite, dark crystalline pebbles, quartz, and chert. This material is used locally for the most part, although it is occasionally bought by Chicago dealers when an extra-clean gravel is required.

TESTS OF MATERIALS.

In a separate bulletin which is in preparation describing in greater detail the sources and character of the concrete materials produced in the Chicago district, it is expected to publish the results of official tests made on these materials at the structural-materials laboratories of the Survey at St. Louis.

In the office of the city engineer of Chicago there is available an instructive set of results of tests, mostly of the Niagara limestone, made by the testing division of the bureau of engineering of that city. These tests were made on a uniform basis, and afford valuable data (a) for comparison of the various samples of rock with each other; (b) for comparison of broken stone taken from the spoil bank of the sanitary and ship canal with freshly quarried material; (c) for comparison of gravel concrete with concrete containing crushed stone; and (d) for general information as to the strength and wearing power of the limestone.

The sampling and testing of the materials were carried on under the immediate supervision of P. C. McArdle, city engineer of tests. Samples were taken from 31 localities, 14 of which were along the spoil bank of the sanitary and ship canal between Willow Springs and Lockport, and the remainder were from the various quarries delivering crushed stone and gravel to the Chicago market.

An analysis of the results of these tests shows that the average compressive strength on 6-inch concrete cubes of three spoil-bank samples of limestone is 66,444 pounds (1,846 pounds per square inch), whereas the average strength of the 17 quarry samples of limestone is 64,684 pounds (1,763 pounds per square inch). In the compressive test on limestone blocks 1 inch by 1 inch by 1½ inches the average strength with 14 spoil-bank samples is 11,834 pounds; that with the 17 quarry samples is 12,397 pounds. Similarly, in abrasion tests, the average loss of weight in 14 spoil-bank samples is 21 per cent; in the 17 quarry samples, 19.57 per cent.

Tests were also made on two cubes of crushed-gravel concrete, in one of which the gravel was of large size and in the other of small size. The results with the large gravel concrete were among the highest of all the tests made; the other cube showed a good average result.

The opinion of the city engineer of tests, based on these results and on several years of field experience in canal construction work, is that the rock taken from the spoil bank is in general as good as that taken from any of the quarries of the district. Rotten stone, however, may be found in almost any quarry, as well as in the spoil bank, and this is particularly true of the quarries in the Lemont district and of the spoil bank along sections 12 and 13, but this rotten stone can be readily detected by the observer.

LITERATURE AND MAPS.

There is a long list of papers dealing with subjects mainly of purely scientific interest in connection with this area, but few of them have practical value in relation to the subject of concrete materials. In the following papers will be found useful data regarding the character and distribution of the limestone, sand, and gravel in the vicinity of Chicago:

ALDEN, WILLIAM C. Description of the Chicago district: Geologic Atlas U. S , folio 81, U. S. Geol. Survey, 1902.
—— The Delavan lobe of the Lake Michigan glacier: Prof. Paper U. S. Geol. Survey No. 34, 1905.
LEVERETT, FRANK. The water resources of Illinois: Seventeenth Ann. Rept. U. S. Geol. Survey, pt. 2, 1896, pp. 695–849.
—— The Pleistocene features and deposits of the Chicago area: Bull. Chicago Acad. Sci. No. 2, Geol. and Nat. Hist. Survey, 1897.
—— The Illinois glacial lobe: Mon. U. S. Geol. Survey, vol. 38, 1899.

The available United States Geological Survey topographic maps of portions of northeastern Illinois and southeastern Wisconsin include the following quadrangles, the first four of which compose the area described in the Chicago folio: Chicago, Riverside, Calumet, Desplaines, Wheaton, Joliet, Wilmington, Morris, Highwood, Waukegan, Racine, Silver Lake, Lake Geneva, Delavan, Shopiere, and Janesville.

PORTLAND CEMENT MATERIALS NEAR EL PASO, TEX.

By G. B. RICHARDSON.

INTRODUCTION.

The considerable cost of Portland cement at El Paso, Tex., owing to its distance from the nearest plant, and the fact that this rapidly growing city is the commercial center of a large area, cause the local presence of the raw materials for making cement to be a matter of importance. The object of this paper is to call attention to large deposits of lime and clay materials in the vicinity of El Paso.

GENERAL GEOLOGY.

The geology of the El Paso region has already been outlined by the writer[a] and for the present purpose the following sketch will suffice. The city of El Paso is situated in the Rio Grande valley, at the mouth of a narrow gap which the river has cut through highlands in passing from the Mesilla valley to the Hueco Bolson. The Franklin Mountains lie east of the gap and the Cerro de Muleros west of it. The Franklin Mountains are composed of sedimentary and igneous rocks which range in age from Cambrian to Cretaceous. The strata dip steeply westward and the mountains as a whole have the general characteristics of a Basin Range block, but they differ from the type by being complexly faulted internally. The Cerro de Muleros is a laccolithic mountain with a porphyry core flanked by Cretaceous sediments. This mountain also has been much faulted, especially contiguous to the pass through which the Rio Grande flows. The bolson deposits consist of gravel, sand, and clay, and similar materials also compose the flood plain of the river. Limestones are abundantly developed in the Franklin Mountains and both limestone and shale are present in the Cerro de Muleros and in outlying hills between the two mountains.

---

[a] Richardson, G. B., Reconnaissance in Trans-Pecos Texas: Bull. No. 9, Univ. Texas Mineral Survey, 1904.

CLAY MATERIALS.

The clay materials can be classed as bolson clay, flood-plain clay, and shale. The bolson clays are extremely irregular in their occurrence. They are locally exposed in the terraces above the river and numerous lenses of clay have been found in the wells which have been sunk in the Hueco Bolson. As yet none of these deposits have been developed.

Flood-plain clay occurs in several localities in the Rio Grande valley near El Paso. The material, derived from rocks that outcrop higher up in the drainage area of the river, has been brought down in suspension by the stream and deposited on the flood plain. In this manner deposits of clay intercalated with sand and gravel have accumulated, the mode of origin causing the deposits to be of irregular extent and composition. The beds range in thickness from a few inches to many feet, and in character from a rather pure clay to one containing large admixtures of sand. More or less organic matter also is usually present. The analysis of clay from Whites Spur, about 10 miles north of El Paso (P. 413), shows the composition of what is perhaps a typical sample of flood-plain clay. This clay is manufactured into common wire-cut brick at several plants in the valley—at Vinton and Whites Spur, above El Paso, and at others below the city. The product is of a fairly good grade, and several million bricks from this source are made yearly. Adobe bricks, made of sun-dried flood-plain clays, are manufactured extensively by the Mexican inhabitants of the Rio Grande valley and are used in the construction of their picturesque buildings.

The deposits of shale are more important for cement making than the flood-plain clays because of their uniform texture and general freedom from coarse particles. The shale is a blue-gray clay shale of Lower Cretaceous age and occurs interbedded with sandstone and limestone on the flanks of the Cerro de Muleros. It is well exposed in the pass along the west bank of the Rio Grande and also occurs in small areas east of the river. The composition of four samples of this shale is shown by the accompanying analyses. The figures indicate a considerable variation, silica ranging from 49.08 to 75.15 per cent, alumina from 10.90 to 20.71 per cent, and lime from 0.66 to 13.56 per cent. The analyses show that the shale is well adapted for making cement, with the exception of No. 3, which contains too much silica and relatively too much aluminum and iron for an ideal Portland cement clay. Because of the variability in composition indicated by the analyses, more tests are desirable to determine the extent of the different grades.

Analyses of shale, clay, and limestone from the vicinity of El Paso.

[Fusion of air-dried material. Analyst, P. H. Bates, U. S. Geological Survey fuels and structural materials testing laboratory.]

	1.	2.	3.	4.	5.	6.
SiO₂	49.08	55.54	75.15	58.73	64.22	3.22
Al₂O₃	10.90	15.72	13.76	20.71	14.02	.78
Fe₂O₃	7.74	6.96	2.35	4.67	2.16	.28
FeO					1.25	
MnO	.11	.13	.04	.05		.31
CaO	13.56	4.88	.66	2.05	4.01	52.36
MgO	1.36	2.43	.45	1.71	1.84	1.01
SO₃	.22	.28	.45	.44	.10	.12
Na₂O	.20	.51	.15	.05	1.04	.00
K₂O	1.26	1.64	.96	1.70	2.19	.11
Water at 100°	1.59	2.63	.58	.80	2.30	.16
CO₂					1.10	
Ignition loss	14.37	9.25	5.48	8.91	5.74	41.74
	100.39	99.97	99.93	99.82	99.97	100.09

1. Shale one-fourth mile south of Courchesne quarry.
2. Shale one-fourth mile north of Courchesne quarry.
3. Shale from El Paso Brick Company's property.
4. Shale from El Paso Brick Company's property.
5. Flood-plain clay from Whites Spur, 10 miles above El Paso.
6. Limestone from Courchesne quarry.

Bricks of excellent quality are made from this shale, three grades being manufactured—pressed brick, common wire-cut brick, and fire brick. Many thousands of the first two grades are made daily, but at the present time only small quantities of fire brick are manufactured, their chief use being in the brick kilns. An analysis of fire clay is given under No. 3 in the table. It shows a small content of fluxing impurities, although the high percentage of silica, 75.15 per cent, indicates only moderate refractoriness.

LIMESTONE.

The limestones of the El Paso region aggregate more than 5,000 feet in thickness and are separable into five formations, based on their ages, as follows: Lower Ordovician, Upper Ordovician, Silurian, Carboniferous, and Cretaceous. Without fossil evidence the different limestones can not always be recognized, although each has physical properties peculiar to itself. They are all massive and are in the main gray in color, but some are whitish and others are almost black. Some are more crystalline than others and they contain variable amounts of chert. A characteristic difference is their content of magnesia, as shown by the following analyses:

Lime and magnesia in limestones from the vicinity of El Paso.

	Lower Ordovician.	Upper Ordovician.	Silurian.	Carboniferous.	Cretaceous.
CaO	32.12	30.82	28.77	53.52	52.36
MgO	16.00	18.01	18.56	.58	1.01

The three older formations contain abundant magnesia, in quantities to constitute the rocks almost a true dolomite, but the magnesia content of the younger limestones is very small.   On account of the high magnesia in the older limestones they are unfit for cement making, but those of Carboniferous and Cretaceous age are well adapted for this purpose.

The distribution of these limestones in general is distinct.   The older formations outcrop along the crest and form the "backbone" of the Franklin Mountains.   The Carboniferous limestone lies along the northwestern slope of this range and is particularly well developed adjacent to the Texas-New Mexico boundary line.   The Cretaceous limestone outcrops along the flanks of the Cerro de Muleros and occurs also in the gorge above El Paso on both sides of the river.   The greater accessibility of the Cretaceous limestone and its occurrence near the shale make it probable that this will be first used, in preference to that of Carboniferous age.

Both the magnesian and nonmagnesian limestones are burned for lime in the vicinity of El Paso.   For this purpose the Ordovician limestones are quarried at the south end of the Franklin Range and the Cretaceous limestone at the pass above the city.   Large quantities of the Cretaceous limestone are also quarried and crushed for use as furnace flux by the smelter in the valley 4 miles above El Paso. The rock is also extensively used for foundations and for road-making macadam.

SURVEY PUBLICATIONS ON CEMENT AND CEMENT AND CONCRETE MATERIALS.

The following list includes the principal publications on cement materials by the United States Geological Survey, or by members of its staff. Besides the publications cited, the technologic branch of the Survey has in preparation numerous bulletins dealing with the results of tests on concrete beams and the constituent materials of concrete, etc.

ADAMS, G. I., and others. Economic geology of the Iola quadrangle, Kansas. Bulletin No. 238. 80 pp. 1904.

BALL, S. H. Portland cement materials in eastern Wyoming. In Bulletin No. 315, pp. 232–244. 1907.

BASSLER, R. S. Cement materials of the Valley of Virginia. In Bulletin No. 260, pp. 531–544. 1905.

BURCHARD, E. F. Portland cement materials near Dubuque, Iowa. In Bulletin No. 315, pp. 225–231. 1907.

BUTTS, C. Sand-lime brickmaking near Birmingham, Ala. In Bulletin No. 315, pp. 256–258. 1907.

CATLETT, C. Cement resources of the Valley of Virginia. In Bulletin No. 225, pp. 457–461. 1904.

CLAPP, F. G. Limestones of southwestern Pennsylvania. Bulletin No. 249. 52 pp. 1905.

CRIDER, A. F. Cement resources of northeast Mississippi. In Bulletin No. 260, pp. 510–521. 1905.

——— (See also Eckel, E. C., and Crider, A. F.)

CUMMINGS, U. American rock cement. A series of annual articles on natural cements, appearing in the volumes of the Mineral Resources U. S. previous to that for 1901.

DURYEE, E. Cement investigations in Arizona. In Bulletin No. 213, pp. 372–380. 1903.

ECKEL, E. C. Slag cement in Alabama. In Mineral Resources U. S. for 1900, pp. 747–748. 1901.

——— The manufacture of slag cement. In Mineral Industry, vol. 10, pp. 84–95. 1902.

——— The classification of the crystalline cements. In Am. Geologist, vol. 29, pp. 146–154. 1902.

——— Portland cement manufacturing. In Municipal Engineering, vol. 24, pp. 335–336; vol. 25, pp. 1–3, 75–76, 147–150, 227–230, 405–406. 1903.

——— The materials and manufacture of Portland cement. In Senate Doc. No. 19, 58th Cong., 1st sess., pp. 2–11. 1903.

——— Cement-rock deposits of the Lehigh district. In Bulletin No. 225, pp. 448–450. 1904.

ECKEL, E. C. Cement materials and cement industries of the United States. Bulletin No. 243. 395 pp. 1905.

——— The American cement industry. In Bulletin No. 260, pp. 496–505. 1905.

——— Portland cement resources of New York. In Bulletin No. 260, pp. 522–530. 1905.

——— Cement resources of the Cumberland Gap district, Tennessee-Virginia. In Bulletin 285, pp. 374–376. 1906.

——— Advances in cement technology, 1906. In Mineral Resources U. S. for 1906, pp. 897–905. 1907.

——— Lime and sand-lime brick. In Mineral Resources U. S. for 1906, pp. 985–991. 1907.

ECKEL, E. C., and CRIDER, A. F. Geology and cement resources of the Tombigbee River district, Mississippi-Alabama. Senate Doc. No. 165, 58th Cong., 3d sess. 21 pp. 1905.

HUMPHREY, R. L. The effects of the San Francisco earthquake and fire on various structures and structural materials. In Bulletin No. 324, pp. 14–61. 1907.

——— Organization, equipment, and operation of the structural-materials testing laboratories at St. Louis, Mo. Bulletin No. 329. In press.

——— (in charge). Portland cement mortars and their constituent materials: Results of tests, 1905 to 1907. Bulletin No. 331.

KIMBALL, L. L. Cement. A series of annual articles on the cement industry and the production of cement in the United States. In Mineral Resources U. S. for 1901, 1902, 1903, 1904, and 1905.

LANDES, H. Cement resources of Washington. In Bulletin No. 285, pp. 377–383. 1906.

NEWBERRY, S. B. Portland cement. A series of annual articles on Portland cements, appearing in the various volumes of the Mineral Resources U. S. previous to that for 1901.

RUSSELL, I. C. The Portland cement industry in Michigan. In Twenty-second Ann. Rept., pt. 3, pp. 620–686. 1902.

SEWELL, J. S. The effects of the San Francisco earthquake on buildings, engineering structures, and structural materials. In Bulletin No. 324, pp. 62–130. 1907.

SMITH, E. A. The Portland cement materials of central and southern Alabama. In Senate Doc. No. 19, 58th Cong., 1st sess., pp. 12–23. 1903.

——— Cement resources of Alabama. In Bulletin No. 225, pp. 424–447. 1904.

TAFF, J. A. Chalk of southwestern Arkansas, with notes on its adaptability to the manufacture of hydraulic cements. In Twenty-second Ann. Rept., pt. 3, pp. 687–742. 1902.

CLAYS.

CLAYS IN THE KOOTENAI FORMATION NEAR BELT, MONT.

By Cassius A. Fisher.

INTRODUCTION.

Clays of different varieties are more or less abundant throughout the Kootenai formation in the vicinity of Belt, Mont., and along Otter Creek, where this product has been prospected and mined to some extent. Clay deposits of commercial value are found in this formation at many different places throughout the Great Falls region, but at no other locality, so far as they have yet been observed, are they of as good quality, apparently, as those near Belt. These clays are locally known as "flint" and "plastic" clays. The former term, however, is not here used in a strictly technical sense, being applied to a light-tan-colored, highly siliceous rock, unlike the typical flint clays of Pennsylvania. The latter term is used to designate a fine-grained slate-colored plastic clay of good quality. The so-called flint clay was formerly used to some extent in the manufacture of brick, and the plastic clay is now shipped to Anaconda, Mont., where it is burned into refractory products used in the large smelters at that place. With the completion of the new Billings and Northern Railroad, which passes near some of the best deposits, an excellent opportunity will be afforded for renewed activity and increased development of the clay resources of this district.

LOCATION AND EXTENT.

The area in which the best clays have been observed is situated in the eastern part of Cascade County, which is located near the center of Montana. It comprises about 145 square miles, including in its eastern part the plains region lying between the Little Belt and Highwood mountains and in its western part Belt Creek valley and a small portion of the adjoining plains. (See Pl. V.) Belt, a small coal-mining town, and Armington are located in the extreme northwest corner of the area.   The Billings and Northern Railroad extends diagonally across it, and the Neihart branch of the Great Northern Railway leaves the above-mentioned road at Armington and extends up Belt Creek to the town of Neihart.

GEOLOGIC OCCURRENCE OF CLAY.

The rocks that outcrop in the district here described range from Carboniferous to Cretaceous in age.   They comprise the Quadrant, Ellis, Morrison, and Kootenai formations, and a portion of the Colorado shale.   Over a great part of the area, however, the Kootenai formation occupies the surface, and in this formation at several different horizons clay and shale deposits of commercial value are found.   The Kootenai formation has a thickness of about 450 feet, and consists mainly of sandstones, sandy shales, and clays occurring in alternate succession.   In the lower part sandstones predominate and are massive in character, but higher in the formation the proportion of sandstone decreases and the beds consist largely of red shales and clays, with here and there a thin layer of sandstone or a bed of limestone.   The formation rests with apparent conformity upon the variegated sandy shales and sandstones of the Morrison, and is overlain by the somber-colored sandstones and shales of the Colorado. A generalized section of the Kootenai formation of this district is given below:

Generalized section of the Kootenai formation (Lower Cretaceous) in the vicinity of Belt, Mont.

	Feet.
Colorado formation (Upper Cretaceous).	
Shale, red, sandy, and clay, with few thin sandstone layers...	195
Shale, red, sandy, capped by sandstone.......................	20
Limestone..	5
Shale, red, sandy, capped by sandstone.......................	20
Shale, red, sandy at top....................................	30
Clay and shale, the former light tan color and the latter red, sandy, with an occasional thin sandstone layer (clay formerly mined)...	50
Shale, red, containing lenses of impure limestone capped by gray limestone..	30
Sandstone and blue clay in alternating layers...............	6
Sandstone, gray, compact....................................	1
Clay, slate colored, fine grained, homogeneous; mined at present...	4½
Sandstone, gray, massive....................................	20
Clay, bluish, sandy...	6
Coal..	4½
Sandstone and sandy shale...................................	60
Morrison formation (Jurassic).	
	452

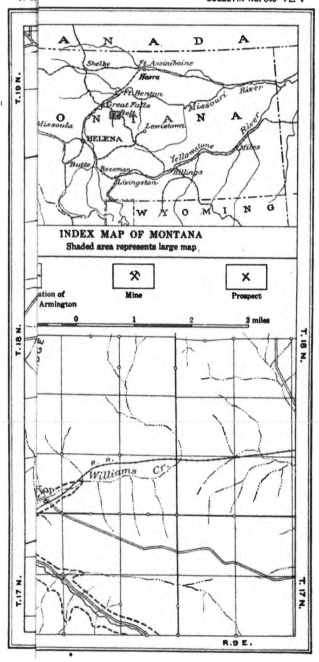

INDEX MAP OF MONTANA
Shaded area represents large map

The foregoing section is compiled from a number of detailed measurements made in various parts of the field, and as individual beds vary materially from place to place, it can not be regarded as typical of any one locality. It is introduced here mainly to show the succession of the beds in the Kootenai formation and the relative position of the clays here described, also their relation to the Kootenai coal. The red shales in the upper part of the section are in many places associated with clays which are good brickmaking material.

STRUCTURE.

The rocks in this region lie nearly horizontal, dipping at small angles (3° or 4°) to the northeast, away from the Little Belt Mountains and toward the plains. Although they are nearly horizontal, a close examination of the beds along the sides of the canyons shows that they are in reality gently folded into a series of low anticlines and shallow synclines, most of which are not perceptible to the casual observer. No large faults occur within this district, but minor faults are not uncommon, especially in the vicinity of Belt. The throw of these faults is slight, ranging from 5 to 15 feet, and their presence is therefore difficult to detect on the surface. They are usually first encountered by miners working the coal beds, and some of them have presented considerable difficulty to coal-mining operations. In Armington Coulee, about half a mile above the mouth, a short distance east of the Anaconda Copper Mining Company's clay pit now being worked, there is a sharp fold in the rocks which may possibly be more or less fractured along the axis of the fold that here trends northward toward Belt Butte. Exposures at this place were inadequate for determination on this point, but it is a structural condition which should be seriously considered in any extensive development of the clay deposits of Armington Coulee.

DETAILED DESCRIPTIONS OF CLAYS.

ARMINGTON COULEE.

On the north side of Armington Coulee, a small tributary of Belt Creek entering from the east just above the town of Armington, there is a clay mine owned by the Anaconda Copper Mining Company, of Anaconda, Mont. The clay deposit worked at this mine has a thickness of 4 feet 6 inches and occurs about 90 feet above the base of the Kootenai formation and 26 feet above the Kootenai coal horizon. The massive gray sandstone which overlies the coal in this general region has a variable thickness, ranging from about 20 to 80 feet, and in Armington Coulee its thickness is near the minimum.

A section at the Anaconda Copper Mining Company's mine, showing the position of the clay with respect to the coal, is given below:

Section at Anaconda Copper Mining Company's clay mine, Armington, Mont.

	Feet.
Sandstone and slate-colored clay occurring in alternating layers...	6
Sandstone, gray, compact..	1
Clay, light gray to slate colored, fine grained, homogeneous; deposit worked.....	4½
Sandstone, gray, massive, weathering tan..........................	20
Clay, slate colored, sandy..	6
Coal..	4½
	42

The clay at this mine is light gray to slate colored, fine grained, and uniform in texture. It has a greasy feel and a subconchoidal fracture, and in places indistinct laminations can be seen. Although the material contains all the minerals ordinarily found in clay of this character, the only one which could be detected with the naked eye or a hand lens is pyrite, in crystals most of which are cubical in shape, although pyritohedral forms are also present. Throughout the thickness of the deposit the clay is homogeneous in character and notably free from sandy lenses. The workings extend back more than 150 feet from the mouth of the entry and in this distance the clay maintains a uniform thickness. Samples of this clay were shipped to St. Louis, where ultimate analysis was made in the structural-materials laboratory of the United States Geological Survey. The result of this analysis is given below:

Analysis of clay worked by Anaconda Copper Mining Company at Armington, Mont.

Silica...	55.38
Alumina...	30.86
Ferrous oxide...	.86
Lime...	.40
Magnesia...	1.03
Sulphuric anhydride...	.16
Ferric sulphide...	2.23
Alkalies $\begin{cases} Na_2O \\ K_2O \end{cases}$	.26 / 1.04
Water at 100° C...	.84
Ignition loss...	6.86
	99.92

The analysis shows that this clay has an average percentage of silica, not high enough to cause it to be sandy, and that alumina, the refractory ingredient, is moderately high, so that the clay compares favorably in this respect with plastic clays of Woodbury, N. J., and St. Louis, Mo. Its fluxing constituents, lime, magnesia, and the alkalies, and also the iron are low.

In addition to the above analysis the writer obtained from the reduction works of the company at Anaconda an analysis of an average sample of this clay as utilized, which is here given:

Average analysis of fire clay from Armington, Mont.

Silica	53.70
Alumina	27.20
Ferric oxide	5.00
Lime	Trace.
Magnesia	Trace.
Alkalies	2.90
Sulphuric anhydride	.30
Ignition loss	10.85

The mine has been in operation at this place for about two years. The main entry extends more than 150 feet back from the outcrop, with side entries in either direction. The clay is mined out by hand at the bottom of the bed and then blasted down from above. It is hauled in wagons to the station and shipped to Anaconda, where it is manufactured into refractory products for utilization in the smelters at that place. The output of the mine at present is not large, and only a few men are employed.

About 45 feet above this clay bed stratigraphically there is a light-tan to yellowish sandy clay or impure sandstone which was formerly mined on the south side of the coulee for the manufacture of brick. The material is hard, gritty to the touch, and breaks with an irregular fracture. Although in general it is light yellow in color, there are bands of limonite running through it, and along all joint planes a thin film of limonite occurs. Ruins of the abandoned brick plant can be seen on the south side of Armington Coulee, nearly opposite the Anaconda Copper Mining Company's mine. Clay for this plant was obtained from the bluffs near by, also about one-fourth mile farther up the coulee, on the east side of the Belt-Lewistown stage road.

An analysis made several years ago has been furnished by the Anaconda Copper Mining Company, and is given below:

Analysis of clay formerly mined in Armington Coulee.

Silica	77.6
Ferric oxide	2.3
Alumina	12.2
Lime	.2

The location of this abandoned clay mine in Armington Coulee is shown on the map (Pl. V.)

WILLIAMS CREEK.

On either side of Williams Creek near Spion Kop there is a workable clay bed in the Kootenai formation, a short distance above the coal. The deposits have been prospected at a number of places on Lewis Larson's ranch, which is located near the mouth of the creek. The clay at this place is believed to occur at about the same horizon as that found in Armington Coulee, but an exact correlation with that deposit can not be made, owing to the fact that the clay and shale, also the sandstone members, of the Kootenai formation are very lenticular and individual beds thicken or thin within a very short distance. Sections of the clay on both sides of Williams Creek, measured by A. J. Hazlewood, are given below:

Section of clay deposit on north side of Williams Creek, near Spion Kop, Mont.

	Ft.	In.
Sandstone, gray, impure, soft......................................	3	
Clay, gray, very sandy at base....................................	3	8
Clay, light gray to slate colored, somewhat sandy, hard...........	5	6
Beds concealed.		

Section of clay deposit on south side of Williams Creek, near Spion Kop, Mont.

	Ft.	In.
Clay, light gray..	4	
Sandstone, gray, soft..	2	6
Clay, light bluish gray...	5	
Beds concealed.		

It was impossible to obtain a continuous section of the beds between the clay and the coal at this place, but it is reasonably certain that the stratigraphic interval between the two deposits does not exceed 40 feet. The massive gray sandstone overlying the coal is present here, but it is not a conspicuous feature and exposures were not sufficiently good for an exact measurement of its thickness. In a railroad cut on the north side of Williams Creek near the wagon road, a bed of gray clay about 6 feet thick was observed. The material appears to be of an inferior quality and occurs in a bed of variable thickness. It is underlain by a sandstone believed to be the one immediately overlying the coal. The correlation of this clay with the workable beds described in the foregoing sections is not certain, but it is probably at about the same geologic horizon.

About half a mile southwest of the locality just described, on the north side of Otter Creek, another railroad cut exposes a deposit of slate-colored clay about 8 feet thick, which contains a 1½-foot layer of red clay near the middle and calcareous concretionary bands near the base and in the upper part. The workable clay in this bed thickens and thins within short distances, so that the deposits can not be regarded as of commercial importance. This lack of persistence in character and thickness in the clay deposits of this region makes correlation of individual beds often uncertain.

OTHER FAVORABLE LOCALITIES.

Although the clay at the horizon of the bed worked at the Anaconda Copper Mining Company's mine has been prospected at only a few places in this general region, clay-bearing sediments of this part of the Kootenai formation are exposed in the bluffs bordering Otter and Belt creeks and their tributaries. The accompanying map (Pl. V) shows the approximate position along these streams of the clay worked at Armington. Clays of equal commercial importance may possibly be found at different horizons a short distance above or below the one described, or in fact throughout a zone 150 feet thick overlying the coal. It is probable that the clay at Armington is not a persistent deposit, but that it dovetails in with deposits above and below of similar character and equal value. The fact that the zone in which the above clays are found occupies a position in the bluffs about 100 to 200 feet above the valley along both Otter and Belt creeks, where lines of transportation have already been constructed, makes the conditions favorable for exploitation.

CLAYS OUTSIDE THE AREA DESCRIBED.

About one-eighth mile west of the Boston and Montana smelters at Great Falls a sandy clay or argillaceous sandstone, mined by Coombs & King, occurs on the north bluffs of Missouri River. The material has a greenish-gray color, is very sandy, and occurs in a bed 30 feet thick near the top of the Kootenai formation. It is used for the manufacture of ordinary brick, also as a lining for converters at the smelters. The clay is satisfactory for these purposes, but it would probably never be mined so extensively were it not for its convenient location to the smelters and the city of Great Falls. It is mined continuously and a considerable proportion of the deposit is utilized.

There is also a sandy clay of some commercial value in the upper half of the Kootenai formation at Fields, a railroad siding 3½ miles east of Great Falls. A large amount of this clay was formerly used in the Boston and Montana smelters, probably before the Coombs & King clay bed was exploited. This material is at about the same horizon as that near the smelters, 4 miles farther north, but exact correlation of these beds can not be made. The clay at Fields is about 42 feet thick, light gray to slate color above, and reddish below. It is very siliceous and in portions resembles a fine-grained impure sandstone. Close examination of a hand specimen shows a large number of minute pyrite crystals and small rounded sand grains embedded in the clay. In addition to its use in the Great Falls smelters this clay was also manufactured into fire brick by the Hosford Fire Brick Company, located at Fields, but this enterprise proved not to be practicable, owing to the high percentage of iron in the clay. The deposit has not been worked for several years.

SURVEY PUBLICATIONS ON CLAYS, FULLER'S EARTH, ETC.

In addition to the papers named below, some of the publications listed under the heading "Cement" contain references to clays. Certain of the geologic folios also contain references to clays, fuller's earth, etc; when these materials are of importance in a particular area, they are printed in italics in the list of folios (pp. 9–11).

ASHLEY, G. H. Notes on clays and shales in central Pennsylvania. In Bulletin No. 285, pp. 442–444. 1906.

BASTIN, E. S. Clays of the Penobscot Bay region, Maine. In Bulletin No. 285, pp. 428–431. 1906.

BRANNER, J. C. Bibliography of clays and the ceramic arts. Bulletin No. 143. 114 pp. 1896.

—— The clays of Arkansas. Bulletin No. 351. In preparation.

BUTTS, C. Clays of the Birmingham district, Alabama. In Bulletin No. 315, pp. 291–295. 1907.

CRIDER, A. F. Clays of western Kentucky and Tennessee. In Bulletin No. 285, pp. 417–427. 1906.

ECKEL, E. C. Stoneware and brick clays of western Tennessee and northwestern Mississippi. In Bulletin No. 213, pp. 382–391. 1903.

—— Clays of Garland County, Ark. In Bulletin No. 285, pp. 407–411. 1906.

FENNEMAN, N. M. Clay resources of the St. Louis district, Missouri. In Bulletin No. 315, pp. 315–321. 1907.

FISHER, C. A. The bentonite deposits of Wyoming. In Bulletin No. 260, pp. 559–563. 1905.

FULLER, M. L. Clays of Cape Cod, Massachusetts. In Bulletin No. 285, pp. 432–441. 1906.

GARDNER, J. H. (See Shaler, M. K., and Gardner, J. H.)

HAWORTH, E. (See Schrader, F. C., and Haworth, E.)

HILL, R. T. Clay materials of the United States. In Mineral Resources U. S. for 1891, pp. 474–528. 1892.

—— Clay materials of the United States. In Mineral Resources U. S. for 1892, pp. 712–738. 1893.

LANDES, H. The clay deposits of Washington. In Bulletin No. 260, pp. 550–558. 1905.

LINES, E. F. Clays and shales of the Clarion quadrangle, Clarion County, Pa. In Bulletin No. 315, pp. 335–343. 1907.

MARBUT, C. F. (See Shaler, N. S., Woodworth, J. B., and Marbut, C. F.)

MARTIN, LAWRENCE. (See Phalen, W. C., and Martin, Lawrence.)

MIDDLETON, J. Clay-working industries. In Mineral Resources U. S. for 1906, pp. 933–983. 1907.

PHALEN, W. C. Clay resources of northeastern Kentucky. In Bulletin No. 285, pp. 412–416. 1906.

424

PHALEN, W. C., and MARTIN, LAWRENCE. Clays and shales of southwestern Cambria County, Pa. In Bulletin No. 315, pp. 344–354. 1907.

PORTER, J. T. Properties and tests of fuller's earth. In Bulletin No. 315, pp. 268–290. 1907.

RIES, H. Technology of the clay industry. In Sixteenth Ann. Rept., pt. 4, pp. 523–575. 1895.

———— The pottery industry of the United States. In Seventeenth Ann Rept., pt. 3, pp. 842–880. 1896.

———— The clays of the United States east of the Mississippi River. Professional Paper No. 11. 298 pp. 1903.

SCHRADER, F. C., and HAWORTH, E. Clay industries of the Independence quadrangle, Kansas. In Bulletin No. 260, pp. 546–549. 1905.

SHALER, M. K., and GARDNER, J. H. Clay deposits of the western part of the Durango-Gallup coal field of Colorado and New Mexico. In Bulletin No. 315, pp. 296–302. 1907.

SHALER, N. S., WOODWORTH, J. B., and MARBUT, C. F. The glacial brick clays of Rhode Island and southeastern Massachusetts. In Seventeenth Ann. Rept., pt. 1, pp. 957–1004. 1896.

SIEBENTHAL, C. E. Bentonite of the Laramie basin, Wyoming. In Bulletin No. 285, pp. 445–447. 1906.

STOSE, G. W. White clays of South Mountain, Pennsylvania. In Bulletin No. 315, pp. 322–334. 1907.

VAUGHAN, T. W. Fuller's earth of southwestern Georgia and Florida. In Mineral Resources U. S. for 1901, pp. 922–934. 1902.

———— Fuller's earth deposits of Florida and Georgia. In Bulletin No. 213, pp. 392–399. 1903.

VEATCH, O. Kaolins and fire clays of central Georgia. In Bulletin No. 315, pp. 303–314. 1907.

WILBER, F. A. Clays of the United States. In Mineral Resources U. S. for 1882, pp. 465–475. 1883.

———— Clays of the United States. In Mineral Resources U. S. for 1883–84, pp. 676–711. 1885.

WOODWORTH, J. B. (See Shaler, N. S., Woodworth, J. B., and Marbut, C. F.)

WOOLSEY, L. H. Clays of the Ohio Valley in Pennsylvania. In Bulletin No. 225, pp. 463–480. 1904

LIME AND MAGNESITE.

SURVEY PUBLICATIONS ON LIME AND MAGNESITE.

In addition to the papers listed below, which deal principally with lime, magnesite, etc., further references on limestones will be found in the lists given under the heads "Cement" and "Building stone."

BASTIN, E. S. The lime industry of Knox County, Me. In Bulletin No. 285, pp. 393–400. 1906.

BUTTS, C. Limestone and dolomite in the Birmingham district, Alabama. In Bulletin No. 315, pp. 247–255. 1907.

HESS, F. L. Some magnesite deposits of California. In Bulletin No. 285, pp. 385–392. 1906.

RIES, H. The limestone quarries of eastern New York, western Vermont, Massachusetts, and Connecticut. In Seventeenth Ann. Rept., pt. 3, pp. 795–811. 1896.

STOSE, G. W. Pure limestone in Berkeley County, W. Va. In Bulletin No. 225, pp. 516–517. 1904.

YALE, C. G. Magnesite deposits in California. In Mineral Resources U. S. for 1903, pp. 1131–1135. 1904.

——— Magnesite. In Mineral Resources U. S. for 1906, pp. 1145–1147. 1907.

GYPSUM AND PLASTERS.

SURVEY PUBLICATIONS ON GYPSUM AND PLASTERS.

The more important publications of the United States Geological Survey on gypsum and plasters are included in the following list:

ADAMS, G. I., and others. Gypsum deposits of the United States. Bulletin No. 223. 123 pp. 1904.

BOUTWELL, J. M. Rock gypsum at Nephi, Utah. In Bulletin No. 225, pp. 483–487. 1904.

BURCHARD, E. F. Gypsum and gypsum products. In Mineral Resources U. S. for 1906, pp. 1069–1078. 1907.

ECKEL, E. C. Salt and gypsum deposits of southwestern Virginia. In Bulletin No. 213, pp. 406–416. 1903.

——— Gypsum and gypsum products. In Mineral Resources U. S. for 1905, pp. 1105–1115. 1906.

ORTON, E. Gypsum or land plaster in Ohio. In Mineral Resources U. S. for 1887, pp. 506–601. 1888.

RICHARDSON, G. B. Salt, gypsum, and petroleum in trans-Pecos Texas. In Bulletin No. 260, pp. 573–585. 1905.

SHALER, M. K. Gypsum in northwestern New Mexico. In Bulletin No. 315, pp. 260–265. 1907.

SIEBENTHAL, C. E. Gypsum of the Uncompahgre region, Colorado. In Bulletin No. 285, pp. 401–403 1906.

——— Gypsum deposits of the Laramie district, Wyoming. In Bulletin No. 285, pp. 404–405. 1906.

GLASS SAND, ETC.

SURVEY PUBLICATIONS ON GLASS SAND AND GLASS-MAKING MATERIALS.

The list below includes the important publications of the United States Geological Survey on glass sand and glass-making materials:

BURCHARD, E. F. Requirements of sand and limestone for glass making. In Bulletin No. 285, pp. 452–458. 1906.

———— Glass sand of the middle Mississippi basin. In Bulletin No. 285, pp. 459–472. 1906.

———— Glass-sand industry of Indiana, Kentucky, and Ohio. In Bulletin No. 315, pp. 361–376. 1907.

———— Notes on glass sands from various localities, mainly undeveloped. In Bulletin No. 315, pp. 377–382. 1907.

———— Glass sand, sand, and gravel. In Mineral Resources U. S. for 1906, pp. 993–1000. 1907.

CAMPBELL, M. R. Description of the Brownsville-Connellsville quadrangles, Pennsylvania. Geologic Atlas U. S., folio 94, p. 19. 1903.

COONS, A. T. Glass sand. In Mineral Resources U. S. for 1902, pp. 1007–1015. 1904.

STOSE, G. W. Glass-sand industry in eastern West Virginia. In Bulletin No. 285, pp. 473–475. 1906.

WEEKS, J. D. Glass materials. In Mineral Resources U. S. for 1883–1884, pp. 958–973. 1885.

———— Glass materials. In Mineral Resources U. S. for 1885, pp. 544–555. 1886.

ABRASIVE MATERIALS.

TRIPOLI DEPOSITS NEAR SENECA, MO.

By C. E. Siebenthal and R. D. Mesler.

INTRODUCTION.

A very light, porous variety of decomposed siliceous rock, resembling the weathered chert popularly known in the region as " cotton rock," is rather extensively quarried in the vicinity of Seneca, Mo., and marketed under the commercial name tripoli stone. Production began in 1888 with the manufacture of scouring bricks and tripoli powder or flour. The product came into competition with similar articles made from tripoli (kieselguhr, kieselmehl, bergmehl) and tripoli slates (kieselschiefer)—infusorial deposits—and, though of entirely different origin, resembled those articles so much both in appearance and use that it was sold under the same name.

LOCATION.

As indicated on the sketch map (fig. 25), about 2 square miles of land in the neighborhood of Seneca and Racine is owned in fee or under lease by the various companies interested in the tripoli industry. Tripoli is likewise known from the vicinity of Fairland, Wyandotte, and Grove, Okla., and Neosho, Mo. There are doubtless other deposits in the territory between these localities, as well as elsewhere over the area underlain by the cherts and limestones of the Boone formation, though probably not all are suitable for use as filters or scouring powder.

GEOLOGY.

The tripoli deposits occur in the Boone formation, which in this region consists of a series of alternating limestones and cherts, with an average thickness of probably 350 feet. The only stratum which can be readily recognized is a bed of oolitic limestone 6 to 9 feet in thickness, which in the region to the north has been called the Short Creek member of the Boone formation. The deposits of tripoli are mostly found above the horizon of this limestone, though some are below it, in particular those south of Grove.

Some of the deposits occur in the steeper bluffs of the hills, but in such locations, by reason of the fact that tripoli is formed by weathering processes, they are not likely to be of workable extent. Most of the deposits now being exploited are on the tops of the hills, owing to the greater economy of operation and the greater likelihood of extensive deposits in such a situation. The bodies of tripoli range from 4 to 12 feet or more in thickness, and are overlain by chert, gravel,

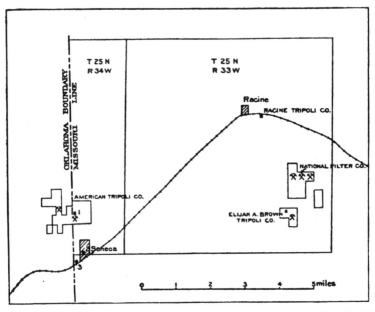

FIG 25.—Sketch map showing location of tripoli deposits near Seneca, Mo. 1, 2, 3, Plants of American Tripoli Company.

and red clay, such as are common to the region. Some of the deposits rest upon impure spotted or discolored tripoli; in others the bed rock is a stratum of solid chert. In the spotted tripoli below the regular quarry bed in the American Tripoli Company's quarry in Oklahoma, the dark spots occur in a horizontal band and strongly suggest the dark material that is deposited in the pits of stylolites.

In the quarries north of Seneca the tripoli is massive, with scarcely a trace of stratification, but divided into irregular blocks by vertical, horizontal, and variously inclined and curved seams or joints. Chert occurs in lenses or more commonly in " balls " through the body of the tripoli itself, locally in such quantity as to force the abandonment of the quarry. Chert is also present in a streak of varying thickness on both sides of the slickensides which occur here and there in the tripoli. Such a cherty " slickenside " may form the limiting wall of a quarry. In one of the quarries there are several pronounced and characteristic slickensides in the north wall, each with its cherty faces. These extend downward for 2 or 3 feet and then die out, leaving no trace in the tripoli below. Another one, a foot in length from top to bottom, dies out at both ends. It is conceivable, of course, though not probable, that the slickensiding beyond that now evident was destroyed in the alteration to tripoli. Otherwise the conditions here seem to indicate that slickensides require very little displacement for their production.

In the quarries south of Racine the tripoli bed is itself massive and shows no lamination, but is overlain by horizontally banded and flaggy rotten cherts. Similar material occurs below the tripoli bed where the underlying beds were seen. In an abandoned portion of the quarry of the National Filter Company there is a sink hole which permits a view of the underlying formation and shows to some extent how effective the leaching surface waters have been in this region. A section here is as follows:

Section at National Filter Company's quarry near Racine, Mo.

	Thickness.	Depth.
	Feet.	*Feet.*
Soil, waterworn gravel, and waxy red clay	3	3
White and red streaked shelly tripoli	2	5
White tripoli	5	10
Banded red and white soft chert	3	13
Red and white rotten chert	5	18
Soft white rotten chert rock resembling tripoli	8	26

The tripoli at this place is very free from the irregular and curved jointing which was noted in the Seneca district, there being a few vertical crevices bearing due northwest and one seam of waxy red clay 2 inches wide reaching to the bottom of the quarry.

CHARACTER.

Tripoli is a light, even-textured, minutely porous rock, crumbling easily to the touch in the green state, but preserving its form very well when air dried. Owing to its extreme porosity, it is highly absorbent. This porosity is thought to be due, as shown later, to the solution of the limy portion of a calcareous chert, the siliceous skeleton of chert grains being left behind. According to Hovey,[a] the

---

[a] Sci. Am. Suppl., July 28, 1894, p. 15487. There are numerous references to the occurrence and nature of the Seneca tripoli deposits in various text-books, State reports, and the statistical annuals, but they are all apparently based on the observations of Dr. Hovey, as reported in the above citation.

tripoli grains " are very minute, by far the most of them being not over 0.01 mm. (0.0004 inch) in diameter, though occasional grains measure 0.03 mm. across. These particles are double refracting and are probably chalcedony."

In the quarries the more open seams are filled with red clay, which in many places stains the contiguous rock to a uniform brownish or reddish tint. Here and there occurs a beautiful regular banding perfectly independent of the original bedding planes, but showing more or less relation to the jointing and resembling the banded iron stains sometimes observed in weathered sandstone. The color thus varies from white to cream and light red. No traces of diatoms, radiolaria, or other fossils have been observed in the deposits. They either were absent in the original rock or have been removed in the process of the alteration which produced the tripoli.

The composition of the tripoli in these deposits is shown in the following analyses:

Analyses of tripoli from deposits near Seneca, Mo.

	1.	2.	3.
Silica (SIO₂)	98.28	98.100	98.10
Alumina (Al₂O₃)	.17	.240	.24
Iron oxide (FeO and Fe₂O₃)	.53	.270	.27
Lime (CaO)	Trace.	.184	.33
Potash (K₂O)	.17		
Soda (Na₂O)	.27	.230	.23
Ignition	.50	1.160	1.17
Organic matter		.008	
	99.92	100.192	100.34

1. By R. N. Brackett. Rept. Arkansas Geol. Survey for 1892, vol. 5, p. 267.
2. By W. H. Seamon. Sci. Am. Suppl., July 28, 1894, p. 15487.
3. Missouri Geol. Survey, vol. 7, 1894, p. 731.

ORIGIN.

The commonly accepted explanation of the origin of the tripoli is that it results from the decomposition of chert. The analyses show that it is practically pure silica. In the quarries lenses and solid or hollow balls of more or less decomposed chert are found, and slickensided crevices are usually faced with solid undecomposed chert. In none of the quarries or natural exposures is it possible to trace the tripoli laterally into unaltered rock. The massive, jointed character of the rock, the absence of bedding, and the lack of fossils would indicate that a profound alteration must have taken place if the tripoli is derived from the ordinary fossiliferous cherts of the region, which occur in beds rarely over 1 or 2 feet thick. "Cotton rock," the ordinary weathered and decomposed surface rock, is quite as fossiliferous as the unaltered chert from which it is derived, though some of it is almost as light as tripoli. It is difficult to see how further weathering alone could completely obliterate the bedding and fossils. On the whole, it is practically inconceivable that the usual chert of this region could weather into such an even-tinted, homogeneous rock as the tripoli actually is.

Locally there are beds of massive white to gray dull-looking rock ranging up to 10 or 15 feet in thickness. This rock is fine granular and unfossiliferous, and breaks with the subconchoidal fracture of a dense, fine-grained limestone. Immersed in acid it effervesces freely, but preserves its original volume after treatment, showing it to be a siliceous limestone or highly calcareous chert. The siliceous residue, after being washed and dried, is a porous granular rock closely resembling a poor grade of tripoli. It seems most likely that the tripoli deposits were formed by the leaching of the lime from some such bed as this. If so, this would explain the massive character and the localization of the tripoli, as well as the absence of fossils. Such a rock would be subject to the same concretionary processes as other limestones, and chert lenses and nodules, once formed, would show a much greater resistance to solution or leaching than the body of the rock, and apparently would remain much as they are now found in the tripoli. The cherty slickensides likewise afford corroboration of this view. Thus the original calcareous chert adjacent to the slickensides might readily be silicified by water entering along the crevice, and this cherty seam would persist when the body of the rock was altered to tripoli. On the other hand, if we assume that the original rock was chert, it seems unlikely that the area which would be entirely unaffected by the decomposition and weathering would lie along openings that permitted free access to the circulating water.

QUARRYING.

In quarrying these deposits, after the mantle of 2 or 3 feet or more of clay, gravel, and residual chert is stripped from the tripoli, vertical channels 12 inches wide are cut to the bottom of the deposit, or to such depth as is desired. These channels are easily made with a light pick of ordinary shape. Where the rock is much cut up by fissures and clay seams, the channels are cut along the most prominent of these joints, to lose as little as possible of the dimension stone. Two such channels, several feet apart, are run into the quarry face as far as it is desired to loosen the stone. A 2-inch hole is then drilled between the ends of the channels, filled with unslacked lime, and tamped. By absorption of quarry sap the lime is slacked, swells, and lifts the stone, the steadily increasing pressure having a tendency to loosen up the blocks along the already existing joints rather than to make new fractures. The shape and size of the blocks thus obtained depend on the number and attitude of the joints. The larger blocks of good quality are sent directly to the filter shop. Spalls and pieces unsuitable for filters are sent to the dry sheds, to be later ground into tripoli flour. When rock for grinding only is desired, that is to say, when it is too much jointed or for some other reason is unsuitable for filter stones, powder is used instead of lime in raising the rock, as it gives blocks of smaller size and saves some hand breaking before crushing.

Where the rock is not so closely beset with joints and fractures, narrow 2-inch cross channels are cut the length of the handle with a narrow-eyed pick, the eye being no wider than the cutting edge of the pick. In this way pieces of regular dimensions are obtained. Blocks 2 by 2 by 5 feet are as large as are ordinarily desired.

It seems that channeling machines, such as are used in quarrying massive limestones, might be advantageously employed in those quarries where there is not so much jointing. Special care would need to be taken to keep the channel clear of the muck formed of the comminuted tripoli.

MANUFACTURE.

FILTER STONES.

The rough blocks from the quarry are taken direct to the mill, where they are sawed up into blocks of the proper dimensions to be turned into filter stones of various shapes and sizes. In practice it was found that the " set " of an ordinary circular saw soon wore down to the thickness of the body of the saw, when it would cramp. With a saw set with diamond teeth, the setting of the diamonds soon wore loose under the abrasion of the stone. A wire saw was likewise tried without success. A special form of saw, which is thoroughly satisfactory, consists of a simple disk in the margin of which, at intervals of 2 or 3 inches, slots slightly narrowing toward the bottom are cut to a depth of three-eighths of an inch. . A steel " tooth," five-sixteenths of an inch in width, about five-eighths of an inch in length, and just thick enough to stick tightly, is driven in each slot. This saw cuts the stone rapidly and does not clog up. The teeth are occasionally dislodged, and at intervals those missing have to be supplied. The stone as sawed is usually so full of sap that water oozes out just in front of the saw. The sawed blocks are air dried before the next step, which in the case of cylindrical and tubular filter stones is to make them round. For this purpose a vertical sandpaper disk is employed. The ends of the block are squared to the proper length, and the block is placed lengthwise between two centers, which allow it to revolve on a vertical axis parallel to the face of the disk. The operator presses the corner edges successively against the sandpaper, revolving the block with a motion opposite to that of the disk, until it is roughly cylindrical. Then the block is allowed to rotate freely on its axis, the motion being communicated by the disk, and the diagonal attrition rapidly produces a perfect cylinder. Filter stones of special shapes are produced on regular turning lathes and special boring machines. Defective blocks and trimmings, as well as lathe dust picked up by the dust collector, go to the tripoli flour mill.

TRIPOLI FLOUR.

Spalls and small or waste blocks from the quarry, together with waste from the filter mill, after having been thoroughly dried for two or three weeks, are crushed, ground on buhrs, and bolted by machinery of simple flouring-mill type. Two grades are marketed, depending on the degree of fineness. The grade O. G. (once ground) will pass through a No. 60 wire mesh, and the grade D. G. (double ground) passes through a No. 140 mesh or a No. 14 silk bolting cloth. Three colors of the flour are made—" white," " cream," and " rose." These colors are obtained by hand sorting the blocks in the dry sheds, those blocks with the most iron stain making the " rose " flour, and the mixed and spotted blocks making the cream colored. The bolted product is sacked or barreled and shipped just as ordinary flour.

PRODUCTION.

According to published reports from various sources, the production has increased from less than 200 tons of tripoli flour in 1888 to 1,000 tons in 1893, 1,375 tons in 1894, and 4,000 to 5,000 tons in recent years. Figures are not available for the tonnage of rough tripoli blocks worked up into filter stones in previous years, but for 1906 it was approximately 600 tons. The present price of tripoli flour f. o. b. cars at Seneca is from $6 to $7 per ton. It is impossible to put a unit price on the finished filter stone, the price of the individual pieces varying with the size and the amount of work done on each. The value of the combined production of tripoli flour and filter stones was, in 1893, $25,000; in 1894, $35,500; in 1905, $50,000; in 1906, approximately $60,000. According to report, about 40 per cent of the flour is exported to foreign countries.

USES.

Tripoli stone has a moderate sale for blotter blocks and scouring bricks, but the important use is for filter purposes. The local mills turn out the filter stones in shapes and quantities to suit the manufacturers of filters. The size ranges from the ordinary house filter to single filters with a capacity of 400 gallons per hour, or batteries of such filters with any desired capacity. For such purposes ocular evidence proves that tripoli stone will remove much of the matter mechanically suspended. As pathogenic germs infesting waters are largely attached to such suspended matter, it follows that they would be likewise removed. Bacteriological examination of water before and after its passage through tripoli filters, it is claimed, demonstrates the sterilizing efficiency of the stone.

Tripoli flour is used as an abrasive, for general polishing, burnishing, and buffing. It is used also as an ingredient of various scouring soaps. Unsuccessful attempts have been made to mold filter stones from tripoli flour to which a binder has been added, but in all experiments so far the binding agent has fatally impaired the porosity of the filter.

Formerly diatomaceous earth (kieselguhr) was used as the absorbent base in the manufacture of dynamite. In recent years this inert base has been superseded by a compound of sodium nitrate, wood pulp, marble dust, and various other substances, which has the advantage of entering into the combustion of the explosive. A rough estimate of the additional force thus gained places it at 5 per cent. The cost of diatomaceous earth is from $25 to $30 or more per ton, depending on the shipping distance. Wood pulp costs $30 per ton, or more, and is constantly advancing in price. The price of tripoli flour, finely bolted, sacked, and hauled to the railway, as shown above, is between $6 and $7 per ton. Merely ground, unbolted, hauled in bulk from the mill, the flour could be laid down at the two powder plants in the vicinity of Joplin for approximately half that sum. It is believed that tripoli flour might be substituted for the wood pulp, either in whole or in part, without materially impairing the explosive value of the compound, at a possible saving of $3 or $4 per ton in the cost of the powder. If experience should show a serious impairment, it is believed that this could be remedied by the addition of more nitroglycerine and that there would still be a notable saving. As the Joplin plants have a combined daily production of about 30 tons of dynamite, it seems that the point is one worthy of consideration.

FIRMS ENGAGED IN THE INDUSTRY.

The American Tripoli Company is the largest as well as the pioneer company in the industry, having erected a grinding mill in 1887, to which was soon added machinery for the manufacture of filter stones. This grinding mill, variously enlarged and remodeled, is in use today, and has a daily capacity of 15 tons of tripoli flour. It is situated on the highland just east of the Missouri-Oklahoma State line, a mile north of Seneca, and the quarries are near by on each side of the line. In 1905 the Seneca Filter Works, established in 1894, were absorbed by this company, which thenceforward carried on the filter business at the new plant. In 1907 the company built in Seneca, adjacent to the Frisco Railroad, a new and much larger mill, though of the same type as the original mill. The new mill is equipped with two crushers, five runs of vertical buhrs, 20 sieve reels, flour packers, etc., and has a capacity of 30 tons per ten-hour day, power being furnished by a 150-horsepower boiler and a 100-horse-power engine. In this plant the attempt is made by the installation of large dust collectors and by hidden journals to lessen as far as possible the friction and wear due to flying tripoli dust. It is planned to double the capacity of the filter works and consolidate them with the mill. The company also owns 80 acres of land with tripoli quarries in the Racine district just south of the National Filter Company's quarry.

Filter works at Racine, Mo., have been operated in connection with the tripoli quarries 2 miles south of that place for a number of years. The National Tripoline Company built a tripoli mill at Kirkwood, Mo., in 1904, to operate the same quarry, but it was shut down after a few months. A reorganization under the name Racine Tripoli Company was effected in 1907. Water power was secured just south of Racine station, and it is planned to start a tripoli flour mill shortly.

The Elijah A. Brown Tripoli Company during 1907 completed a small new mill 3 miles due south of Racine. It is equipped with a crusher, run of buhrstones, sieve reel, saw, sandpaper disk, and boring machine. Though small it is complete and will turn out both filter stones and tripoli flour. The company owns 120 acres of land on which are good deposits of fine-grained tripoli.

The National Filter Company owns 80 acres in the Racine tripoli district. It quarries out the tripoli stone in large blocks, 2 by 2 by 4 or 5 feet in size, which are shipped to the firm's manufacturing plant in Chicago, Ill. Only the drying sheds and quarry machinery are located at the quarries. Owing to the practical absence of joints and clay seams, no blasting is done at this quarry, the rock being wholly quarried out with hand picks.

C. C. Martin & Son own a quarry in the Racine district, and make odd lots of filter stones to order on an ordinary turning lathe.

Other tracts of land near the quarries above described are known to be underlain by tripoli stone of good quality, but the quarries mentioned comprise all that are now in operation.

SURVEY PUBLICATIONS ON ABRASIVE MATERIALS, QUARTZ, FELDSPAR, ETC.

The following list includes a number of papers, published by the United States Geological Survey or by members of its staff, dealing with various abrasive materials:

ANDERSON, ROBERT. (See Arnold, Ralph, and Anderson, Robert.)

ARNOLD, RALPH, and ANDERSON, ROBERT. Diatomaceous deposits of northern Santa Barbara County, Cal. In Bulletin No. 315, pp. 438–447. 1907.

BASTIN, E. S. Feldspar and quartz deposits of Maine. In Bulletin No. 315, pp. 383–393. 1907.

——— Feldspar and quartz deposits of southeastern New York. In Bulletin No. 315, pp. 394–399. 1907.

——— Quartz (flint) and feldspar. In Mineral Resources U. S. for 1906, pp. 1253–1270. 1907.

CHATARD, T. M. Corundum and emery. In Mineral Resources U. S. for 1883–84, pp. 714–720. 1885.

ECKEL, E. C. The emery deposits of Westchester County, N. Y. In Mineral Industry, vol. 9, pp. 15–17. 1901.

FULLER, M. L. Crushed quartz and its source. In Stone, vol. 18, pp. 1–4. 1898.

GOLDING, W. Flint and feldspar. In Seventeenth Ann. Rept., pt. 3, pp. 838–841. 1896.

HIDDEN, W. E. The discovery of emeralds and hiddenite in North Carolina. In Mineral Resources U. S. for 1882, pp. 500–503. 1883.

HOLMES, J. A. Corundum deposits of the southern Appalachian region. In Seventeenth Ann. Rept., pt. 3, pp. 935–943. 1896.

JENKS, C. N. The manufacture and use of corundum. In Seventeenth Ann. Rept., pt. 3, pp. 943–947. 1896.

PARKER, E. W. Abrasive materials. In Nineteenth Ann. Rept., pt. 6, pp. 515–533. 1898.

PRATT, J. H. The occurrence and distribution of corundum in the United States. Bulletin No. 180. 98 pp. 1901.

——— Corundum and its occurrence and distribution in the United States. Bulletin No. 269. 175 pp. 1905.

RABORG, W. A. Buhrstones. In Mineral Resources U. S. for 1886, pp. 581–582. 1887.

——— Grindstones. In Mineral Resources U. S. for 1886, pp. 582–585. 1887.

——— Corundum. In Mineral Resources U. S. for 1886, pp. 585–586. 1887.

READ, M. C. Berea grit. In Mineral Resources U. S. for 1882, pp. 478–479. 1883.

STERRETT, D. B. Abrasive materials. In Mineral Resources U. S. for 1906, pp. 1043–1054. 1907.

TURNER, G. M. Novaculite. In Mineral Resources U. S. for 1885, pp. 433–436. 1886.

——— Novaculites and other whetstones. In Mineral Resources U. S. for 1886, pp. 589–594. 1887.

WOOLSEY, L. H. Volcanic ash near Durango, Colo. In Bulletin No. 285, pp. 476–479. 1906.

MINERAL PAINT.

SURVEY PUBLICATIONS ON MINERAL PAINT.

BURCHARD, E. F. Southern red hematite as an ingredient of metallic paint. In Bulletin No. 315, pp. 430–434. 1907.

ECKEL, E. C. The mineral paint ores of Lehigh Gap, Pennsylvania. In Bulletin No. 315, pp. 435–437. 1907.

—— Metallic paints of the Lehigh Gap district, Pennsylvania. In Mineral Resources U. S. for 1906, pp. 1120–1122. 1907.

—— (See also Hayes, C. W., and Eckel, E. C.)

HAYES, C. W., and ECKEL, E. C. Occurrence and development of ocher deposits in the Cartersville district, Georgia. In Bulletin No. 213, pp. 427–432. 1903.

PHOSPHATES.

PHOSPHATE DEPOSITS IN THE WESTERN UNITED STATES.

By F. B. Weeks.

INTRODUCTION.

The field work on the western phosphate deposits in 1907 occupied about six weeks in August and September. It has been considered advisable that a brief statement of progress be prepared for publication, to be followed later by a detailed report to be published as a separate bulletin. The present report will, therefore, be limited to a general description of the field developments in 1907 and the various conditions which affect the industry.[a]

THE PHOSPHATE SERIES.

GENERAL CHARACTERISTICS.

The phosphate-bearing series ranges from 60 to over 100 feet in thickness. The main phosphate bed, which is the one of commercial importance under present conditions, usually occurs at the base of the series and is from 5 to 6 feet thick. It is almost entirely oolitic in structure, the small, black, well-rounded grains being readily distinguishable in the hand specimen. There is very little matrix material, and it effervesces slightly with hydrochloric acid. The lower part of the bed is hard and blocky; its upper part is softer and more shaly. The material has a bituminous odor which in early days was taken to indicate the presence of oil or coal, and considerable prospecting was done to find these materials. This bed averages high in its content of P_2O_5 and the whole bed is mined and shipped.

Above the main phosphate bed there are alternating layers of phosphate, limestone, and shale. Some of the phosphatic beds, a few inches thick, show a high percentage of P_2O_5, but they can not

---

[a] For a preliminary paper on these deposits, including considerable general information in regard to them, see Bull. U. S. Geol. Survey No. 315, 1907, pp. 449–462.

be mined separately, and at present there is no practicable method of separating the valuable material from the waste. In the upper and middle portions of the series there are also large lime nodules associated with shaly material. The upper part of the phosphate series is shaly in structure and contains a much larger proportion of impurities.

The general character of the series is very persistent, and, as a rule, the rocks are easily recognized where exposed. A prominent exception occurs at the Hot Springs locality, in Idaho, on the eastern side and near the north end of Bear Lake. Here a considerable part of the lower portion of the series has become so completely silicified and the beds have been so displaced by faulting that it is difficult to recognize their original character.

GEOGRAPHICAL DISTRIBUTION.

The phosphate series has been found in nearly every mountain range from central Utah to eastern-central Idaho and in western Wyoming. The beds are in many localities tilted at high angles, and the area of outcrop is then correspondingly small. Where the dip is low the series usually forms a considerable portion of a mountain slope and is covered by soil.

UTAH.

A few miles northeast of Thistle Junction, Utah County, Utah, a bed of phosphate 12 to 18 inches thick has been found. In the vicinity of Midway, on the east side of the Wasatch Range, and in one of the canyons east of Salt Lake City, a bed of about the same thickness and character occurs. These beds are not commercially valuable. The phosphate series, having a thickness of 60 to 90 feet, is exposed in Weber Canyon, and the side gulches from 1½ to 3 miles west of Devils Slide station, on the Union Pacific Railroad. From this point northward the beds become of economic importance, their present value depending on their accessibility. The next known locality toward the north is about 16 miles west of Woodruff. The series is extensively exposed on the western slopes of the Crawford Mountains, along the Utah-Wyoming boundary. During the last year a new discovery has been reported in the high ridge east of Bear Lake about 10 miles northeast of Laketown.

WYOMING.

The northern extension of the phosphate beds exposed in the Crawford Mountains has been found in the low isolated hills 3 miles west of Sage, Wyo. The series also occurs several miles northeast of this railroad station along the slopes of Rock Creek. Two miles east of Cokeville the beds are exposed on the north side of Smith Fork and they follow the trend of the Sublette Range northward for a distance of 25 miles. To the north the extension of the same belt lies on both sides of the Salt River valley.

IDAHO.

In Idaho the most southerly outcrop of the phosphate beds occurs near the Hot Springs, on the eastern side of Bear Lake near its north end. The beds are extensively exposed in the Preuss Range east of Montpelier, and follow the trend of this range northward until they pass beneath the lavas of the Snake River plain. Reports have been received that the beds occur on the western slope of the Bear River - Range in the vicinity of Paris.

GEOLOGIC OCCURRENCE.

The general character and sequence of the Paleozoic sedimentary strata of this region are given in the following section:

Section of Paleozoic strata in southeastern Idaho, southwestern Wyoming, and northeastern Utah.

Carboniferous:
    Mostly light-colored limestones; *phosphate beds near base.*
    Series of red, white, and green quartzites and sandstones.
    Massive blue and gray limestones.
Devonian:
    Limestone, where present.
Silurian:
    Thin-bedded limestone, where present.
Ordovician:
    White and green quartzites.
    Light-colored, generally thick-bedded limestone.
Cambrian:
    Thin-bedded blue and gray limestone.
    Quartzites, mainly white in some areas, purple in others.

The phosphate series, as shown in the above section, occurs in the lower part of the upper division of the Carboniferous strata. The underlying sandstones and quartzites are usually exposed and the phosphate series lies approximately 200 to 400 feet above it. The lithologic character of the series renders it very susceptible to erosion and its outcrop is usually concealed by soil or slide material. A careful study of the overlying and underlying strata and the occurrence of phosphate float will generally indicate the position of the phosphate beds.

DEVELOPMENTS IN 1907.

At the time the previous report was prepared, about the close of the year 1906, an average of 2 carloads of phosphate per day was shipped from Montpelier, Idaho, by the San Francisco Chemical Company. It was found that the margin of profit after paying the freight charges was too small to warrant the continuation of operations and the work was discontinued. The period of shipments extended from September, 1906, to March, 1907. It is doubtful if there is another locality where mining and transportation can be carried on at less cost than at Montpelier. The result indicates that the successful exploitation of the phosphate deposits depends in large measure on the cost of transportation from the field to the consumer. The operations of the Bradley Brothers near Sage and of the Union Phosphate Company near Cokeville, Wyo., corroborate the above statement.

Development during the summer of 1907 was for the most part limited to the annual assessment work. In the greater number of localities this consisted of digging shallow trenches to expose the phosphate series where covered. Considerable work was done in the Crawford Mountains for the purpose of securing a patent to the ground. In the Preuss Range in Idaho the phosphate field was shown to extend northward to Blackfoot River.

CONDITIONS AFFECTING THE INDUSTRY.

In considering the question of the commercial value of the western phosphate beds three important factors should be borne in mind—(1) other sources of supply that will be brought into competition, (2) the local and physical conditions which determine the cost of production, and (3) markets.

COMPETING FIELDS.

The production of the South Carolina phosphate field has steadily declined for several years and it seems probable that this decline will continue.

The increase in production from the Tennessee field has been considerable, but has been due to added facilities rather than to new discoveries or extension of the phosphate-producing area. At the present time the life of this field can not be estimated, but it will continue to be an important factor in the phosphate industry for a number of years.

The extent of the Arkansas phosphate field is not definitely known. The production to the present time has been small. It apparently contains an important bed of phosphate which in the future may come into competition with the product of the western field.

The production from the Florida phosphate field has steadily increased until it now amounts to 1,300.000 tons per annum. It appears to be the consensus of opinion that this field can not be greatly extended. The present rate of production may be continued or even increased, but this will be due to added facilities rather than to new discoveries.

LOCAL CONDITIONS.

Among local conditions which affect the industry as a commercial enterprise are the topography and geologic structure of the region.

Where variations in elevation are considerable so that the material can be handled by machinery run by gravity, the cost of mining is much less than in a region where it must be elevated. The continuity of the beds and their angle of dip materially affect the cost of handling the phosphate and the amount of timbering that must be done. The most favorable situation, and one which is of rare occurrence, is found where the beds dip with the surface slope and the strata overlying the phosphate have been removed by erosion. In such a locality the work is quarrying rather than mining.

At present most of the main phosphate bed, 5 to 6 feet in thickness, is being worked, and this constitutes a small part of the phosphate series. The remainder consists of interbedded thin layers of limestone, shale, and phosphate, no part of which can be mined separately because the waste or low-grade material would reduce the average content of P_2O_5 below a paying basis. The most pressing need, which if successfully met will increase the possible production of this field to an enormous extent, is a process which will separate the thin phosphatic layers from the associated lime and shale and also concentrate the low-grade material. The possibilities of production from this field can hardly be realized until an attempt is made to estimate the amount of phosphate rock in a given area. To this must be added the acid, which forms about one-half the bulk of the treated rock.

Accessibility to the railroad is another important factor affecting cost of operations. Wagon roads in a mountainous region are expensive in first cost and in subsequent maintenance. Heavy snowfalls are liable to interfere with transportation. In some localities it will be possible to develop water power to generate electricity for operating tramways and electric railways to transfer the material to the steam roads.

In any mining enterprise where the bulk of material to be handled is large and its relative value small, local and physical conditions frequently determine whether it can be made profitable. A careful study of these factors in some of the recent attempts to work these phosphate beds would have shown that under present conditions or those which are likely to exist for a number of years the enterprise could not be made a financial success.

MARKETS.

At the present time the raw phosphate rock must be shipped by rail to the Pacific coast and there manufactured into a commercial fertilizer. The home market is a small but growing one and is confined mainly to California. This fertilizer when shipped abroad comes into competition with the product of foreign phosphate fields. With the completion of the Isthmian Canal it will hardly be possible for this product to compete successfully with the Florida phosphate, at least while the present large output from that State is maintained. In order to market the product of the western phosphate field successfully existing conditions must be materially changed, and this seems possible only by a considerable reduction in the cost of rail transportation.

This field embraces the largest area of known phosphate beds in the world, and at some future time it will doubtless furnish a large part of the world's production of commercial fertilizer. The development of intensive farming as the result of the reclamation of arid lands in the West will afford an increasing home market.

MINERAL LOCATIONS UNDER THE PRESENT MINING LAWS.

The mining laws of the United States do not adequately provide for locating beds of phosphate or other beds of economic value having a similar occurrence and origin. The phosphate beds must be entered as placer or lode locations, but they do not properly belong in either class. It may therefore be desirable to consider briefly the characteristics of lode and placer deposits.

A lode is formed by the deposition and concentration of metallic substances from mineral-bearing solutions circulating in the crevices of a rock mass. It is therefore a process taking place subsequent to the formation of the material with which it is associated or of which it forms a part. A placer is formed of material which results from disintegration and erosion of a rock surface and which by the aid of gravity and running water is removed to a lower level and spread out as a covering of the underlying material. The formation of a lode is largely the result of chemical and mechanical action, whereas the formation of a placer takes place, for the most part, by mechanical action and in a measure resembles the formation of a sedimentary stratum. A lode varies in width but is generally confined to so-called "walls." A placer may and usually does have a larger areal extent, but this is limited by the carrying power of the running water by which it was formed.

The western phosphate beds were probably deposited on the ocean bottom as a part of the sediments which had been brought down from a land surface subjected to erosion during a long period of geologic time and were in part also the result of chemical precipitation in the ocean waters during the same time. They therefore constitute a part of the sedimentary strata of the earth's crust. By warping and folding of the crust the strata have become land. They are therefore bedded deposits covering a wide extent of territory, and they differ materially in origin and formation from either lode or placer deposits.

If the phosphate beds were located as lodes it would be possible, according to the present interpretation of the law, to follow these beds for long distances and therefore defeat the implied purpose of the law. If located as placers the limits of a claim would be determined by the areal extent of the surface, which the law defines. It would appear, therefore, that to comply with the spirit of existing mining law the phosphate beds should be located as placers.

In actual practice locations have been made both as lodes and placers, and in some instances both forms of locations have been made on the same ground. To avoid useless expenditure of time and money and legal controversies in the future it seems desirable that a decision should be made that would determine the form of entry for which patent to the ground could be obtained. One precedent has been made by granting patent to phosphate ground as a placer, but it was specifically stated in the decision that this applied only to the patent in question, and it can not therefore be considered as an established precedent.

SURVEY PUBLICATIONS ON PHOSPHATES AND OTHER MINERAL FERTILIZERS.

The following papers relating to phosphates, gypsum (land plaster), and other mineral materials used as fertilizers have been published by the United States Geological Survey or by members of its staff. Further references will be found under the head of " Gypsum."

ADAMS, G. I., and others. Gypsum deposits in the United States. Bulletin No. 223. 127 pp. 1904.

DARTON, N. H. Notes on the geology of the Florida phosphates. In Am. Jour. Sci., 3d ser., vol. 41, pp. 102–105. 1891.

ECKEL, E. C. Recently discovered extension of Tennessee white-phosphate field. In Mineral Resources U. S. for 1900, pp. 812–813. 1901.

——— Utilization of iron and steel slags. In Bulletin No. 213, pp. 221–231. 1903.

——— The white phosphates of Decatur County, Tenn. In Bulletin No. 213, pp. 424–425. 1903.

ELDRIDGE, G. H. A preliminary sketch of the phosphates of Florida. In Trans. Am. Inst. Min. Eng., vol. 21, pp. 196–231. 1893.

FERRIER, W. F. (See Weeks, F. B., and Ferrier, W. F.)

HAYES, C. W. The Tennessee phosphates. In Sixteenth Ann. Rept., pt. 4, pp. 610–630. 1895.

——— The Tennessee phosphates. In Seventeenth Ann. Rept., pt. 2, pp. 1–38. 1896.

——— The white phosphates of Tennessee. In Trans. Am. Inst. Min. Eng., vol. 25, pp. 19–28. 1896.

——— A brief reconnaissance of the Tennessee phosphate field. In Twentieth Ann. Rept., pt. 6, pp. 633–638. 1899.

——— The geological relations of the Tennessee brown phosphates. In Science, vol. 12, p. 1005. 1900.

——— Tennessee white phosphate. In Twenty-first Ann. Rept., pt. 3, pp. 473–485. 1901.

——— Origin and extent of the Tennessee white phosphates. In Bulletin No. 213, pp. 418–423. 1903.

IHLSENG, M. C. A phosphate prospect in Pennsylvania. In Seventeenth Ann. Rept., pt. 3, pp. 955–957. 1896.

MEMMINGER, C. G. Commercial development of the Tennessee phosphates. In Sixteenth Ann. Rept., pt. 4, pp. 631–635. 1895.

MOSES, O. A. The phosphate deposits of South Carolina. In Mineral Resources U. S. for 1882, pp. 504–521. 1883.

ORTON, E. Gypsum or land plaster in Ohio. In Mineral Resources U. S. for 1887, pp. 596–601. 1888,

PENROSE, R. A. F. Nature and origin of deposits of phosphate of lime. Bulletin No. 46. 143 pp. 1888.

PURDUE, A. H. Developed phosphate deposits of northern Arkansas. In Bulletin No. 315, pp. 463–473. 1907.

STOSE, G. W. Phosphorus ore at Mount Holly Springs, Pennsylvania. In Bulletin No. 315, pp. 474–483. 1907.

——— Phosphorus. In Mineral Resources U. S. for 1906, pp. 1084–1090. 1907.

STUBBS, W. C. Phosphates of Alabama. In Mineral Resources U. S. for 1883–84, pp. 794–803. 1885.

WEEKS, F. B., and FERRIER, W. F. Phosphate deposits in western United States. In Bulletin No. 315, pp. 449–462. 1907.

WILBER, F. A. Greensand marls in the United States. In Mineral Resources U. S. for 1882, pp. 522–526. 1883.

SALINES.

SURVEY PUBLICATIONS ON SALINES, INCLUDING SALT, BORAX AND SODA.

The more important publications of the United States Geological Survey on the natural lime, sodium, and potassium salts included in this group are the following:

CAMPBELL, M. R. Reconnaissance of the borax deposits of Death Valley and Mohave Desert. Bulletin No. 200. 23 pp. 1902.

—— Borax deposits of eastern California. In Bulletin No. 213, pp. 401–405. 1903.

CHATARD, T. M. Salt-making processes in the United States. In Seventh Ann. Rept., pp. 491–535. 1888.

DARTON, N. H. Zuñi salt deposits, New Mexico. In Bulletin No. 260, pp. 565–566. 1905.

DAY, W. C. Potassium salts. In Mineral Resources U. S. for 1887, pp. 625–650. 1888.

—— Sodium salts. In Mineral Resources U. S. for 1887, pp. 651–658. 1888.

ECKEL, E. C. Salt and gypsum deposits of southwestern Virginia. In Bulletin No. 213, pp. 406–416. 1903.

—— Salt industry of Utah and California. In Bulletin No. 225, pp. 488–495. 1904.

HILGARD, E. W. The salines of Louisiana. In Mineral Resources U. S. for 1882, pp. 554–565. 1883.

KINDLE, E. M. Salt resources of the Watkins Glen district, New York. In Bulletin No. 260, pp. 567–572. 1905.

PACKARD, R. L. Natural sodium salts. In Mineral Resources U. S. for 1893, pp. 725–738. 1894.

RICHARDSON, G. B. Salt, gypsum, and petroleum in trans-Pecos Texas. In Bulletin No. 260, pp. 573–585. 1905.

YALE, C. G. Borax. In Mineral Resources U. S. for 1883–1884, pp. 494–506. 1902.

—— Borax. In Mineral Resources U. S. for 1906, pp. 1058–1062. 1907.

450

SULPHUR AND PYRITE.

SULPHUR DEPOSITS AT CODY, WYO.

By E. G. WOODRUFF.

INTRODUCTION.

The paper here presented is the result of an investigation made during the field season of 1907. The hot sulphur springs and travertine deposits near Cody were described by George H. Eldridge,[a] of the United States Geological Survey, in his report on northwest Wyoming. A more recent account of these deposits is given by C. A. Fisher,[b] also of the Survey, in his description of the Cody Hot Springs, published in a report on the Bighorn Basin. Since the mines have been opened L. W. Trumbull,[c] professor of geology and mining at the State University of Wyoming, has described the occurrence of the sulphur and explained the method used to reduce the ore.

LOCATION AND EXTENT.

The deposits here described are located about 3 miles west of Cody, along the base of Cedar Mountain, on the south side of Shoshone River, in secs. 3 and 10, T. 52 N., R. 102 W. Though geologic conditions are favorable for the formation of sulphur both north and south of the river, the deposits considered workable at the present time are confined to a belt 2 miles long and less than one-fourth mile wide, extending southeastward from the Hot Springs along the foot of the mountains to Sulphur Creek. All the mines now producing sulphur are included in a small area a few acres in extent, located near the north end of the mineralized zone south of Shoshone River. The known area of sulphur-bearing rocks is shown on the accompanying map (Pl. VI).

---

[a] A geological reconnaissance in northwest Wyoming: Bull. U. S. Geol. Survey No. 119, 1894, p. 67.

[b] Geology and water resources of the Bighorn Basin, Wyoming: Prof. Paper U. S. Geol. Survey No. 53, 1906, p. 61.

[c] Sulphur mining and refining: Mines and Minerals, February, 1907, p. 314.

SURFACE FEATURES.

The area treated in this paper includes a part of the eastern slope of Cedar Mountain and a narrow belt of the adjoining plains. Shoshone River crosses the north end and a branch of Sulphur Creek drains the southern part. Cedar Mountain is the southern extension of the Rattlesnake–Cedar Mountain anticline, which is structurally a spur of the Absaroka Range. In this region the anticline pitches sharply to the south, with steep or vertical dips on the west and gentle dips on the east of the axis. That part of the anticline with which this report is concerned presents a uniformly dipping slope somewhat trenched by watercourses. The plains adjoining the bottom of this dip slope constitute part of a great gravel apron which slopes gently eastward from the base of the mountains. Shoshone River crosses both the anticline and the plains, flowing through the former in a deep canyon and through the latter in a gorge 300 to 400 feet wide and 150 to 200 feet deep. There are, in the area here described, portions of two terraces which form part of a series extending along Shoshone River and rising by successive elevations from the stream channel to the level of the plains. The gravel plains are covered near the mountains by travertine terraces, upon the surface of which there occur small cones and circular depressions characteristic of hot-spring deposits. (See Pl. VI.)

Such springs are now active in Shoshone Canyon, where the water issues from a number of vents in crevices in the upper members of the Carboniferous limestone. Several of these springs have their outlets above the water level, while others discharge beneath the river, and are visible only at periods of low water. The waters, which issue at a temperature of 98° F., contain large quantities of hydrogen sulphide and carbon dioxide, and hold in solution compounds of calcium, magnesium, iron, aluminum, lithium, sodium, potassium, chlorine, and some organic matter. The water is clear and emits so strong an odor of sulphur that it may be detected in the canyon 2 miles downstream from the springs.

GEOLOGIC RELATIONS.

The Paleozoic rocks which arch over the summit of the Rattlesnake–Cedar Mountain anticline have been lifted 3,000 feet above the surrounding plains and 8,000 feet above sea level. In the walls of Shoshone Canyon there are exposed massive limestones constituting the Madison and Bighorn formations, which weather into bold, castellated cliffs, overlain by soft red shales of the Amsden formation and resting upon dull-green shales of Cambrian age. Below the Cambrian shales is a mass of granite several hundred feet of which is ex-

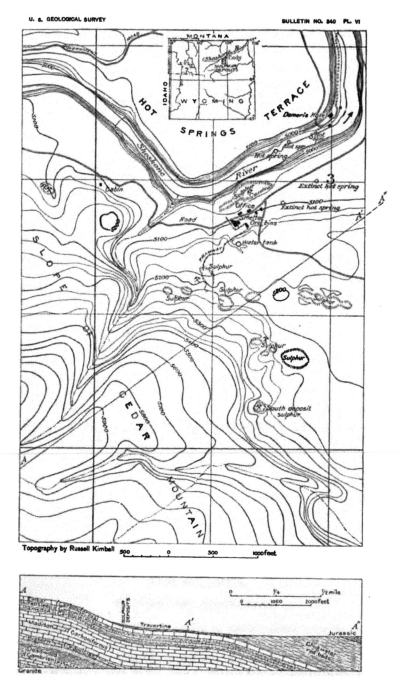

MAP SHOWING SULPHUR DEPOSITS NEAR CODY, WYO., WITH SECTION OF STRATA
ACROSS SULPHUR-BEARING AREA.

posed and which weathers into rugged, precipitous cliffs and is cut by large intrusive dikes of quartz-bearing diabase.

Around the base of the mountains occur the steeply dipping beds of softer Mesozoic rocks, across the upturned ends of which Quaternary gravels have been laid down in the form of terraces. Within and above these gravels heavy layers of travertine have been deposited throughout a narrow zone bordering the base of the uplift.

That portion of the travertine flat which lies north of Shoshone River has a gently sloping surface upon which appear small, low cones of extinct hot springs and shallow depressions such as occur in regions where thermal waters flow over a plain. South of the river the surface is less regular, but it exhibits the same general features that are found on the north.

Geysers may have existed in the region prior to the period of present hot-spring activity and built these cones, which show imperfect tubes extending down into their interior. The travertine is composed in part of crystals of selenite, a variety of gypsum, which seems to be an alteration product of the travertine, due to the action of sulphur waters. Sulphur deposits occur along the upper edge of the travertine flats, where they lie upon the limestone which forms the mountain slope; in fact, the richest sulphur deposits are found in the limestone just under the terrace material, at the upper edge of the flats, where formerly hot springs must have been most active. Some sulphur is also present in the sandstone which underlies the limestone, and other deposits are in the travertine. In one such deposit on the slope from the lowermost broad terrace to a lower, narrow terrace on the edge of the present gorge, the travertine covers a steep scarp and forms the cement of a coarse gravel conglomerate, the pebbles being derived from the terrace gravels. This is intersected by numerous irregular veins of sulphur, which here and there expand into pockets of considerable size. In addition to the sulphur associated with the limestone noted above, that contained in the travertine and sandstone has been tested to a considerable extent, but it is found that the limestone contains the only workable deposits. The cross section at the bottom of Pl. VI shows the relations of the travertine and sulphur to the underlying formations.

THE SULPHUR.

NATURE AND OCCURRENCE.

The sulphur, which occurs native in small yellow crystals and in gray streaks in the rocks, is now being mined and smelted at the plant of the Bighorn Sulphur Company, near the mouth of Shoshone Canyon, on the south side of the river. Where the sulphur is mined it is found in irregular beds in limestone and travertine associated with fine white crystalline aggregates, filling cavities 2 to 8 inches in diameter, and disseminated through the limestone, where it has been deposited by sulphur-bearing gases permeating small crevices in the rock. The cavities just mentioned seem to be portions of subterranean channels through which the hot sulphur-bearing waters flowed, and on the walls of which the sulphur was gradually deposited until the chamber was partly or entirely filled. No regular arrangement of the cavities can be discovered, though they seem to be in groups at places where the waters found free passage. In the areas between the groups of cavities only a small amount of sulphur is found, but in the enriched pockets the amount of sulphur reaches 30 to 50 per cent of the rock and becomes commercially important. Laterally a deposit may be rich at one point and barren 10 feet away. The depth of the mineralization is not known, as mining and prospecting have not been carried below 20 feet, but it seems improbable that rich pockets of sulphur will be found far below the surface. As previously stated, a small amount of sulphur is found in the sandstones and travertine, but the quantity in these rocks is too small to pay for mining.

GENESIS OF THE ORE.

That the sulphur was deposited by hot springs is indicated by the presence of sulphur in the waters now issuing from the Hot Springs in Shoshone Canyon, and also by the evidence that ore is found in channels such as water forms when it flows through openings in limestone. The water probably comes from the Rattlesnake–Cedar Mountain anticline, but may have a more distant source. As the source of the water is not known, the cause of the heat remains undiscovered. There is no evidence of solfataric origin and the indications are that only a small amount of the heat is derived from chemical reactions, because but little of the material produced by chemical changes below is brought to the surface. It is known that intensive activity must necessarily take place to maintain the water at a temperature above 98° F., the temperature found in these springs. It is supposed, therefore, that the heat is derived from slowly cooling magmatic bodies, probably intrusions, and increased slightly by heat from chemical reactions.

CHEMICAL CONDITIONS.

The same chemical changes which produce the increased heat are thought to free the sulphur and permit it to come to the surface. As the waters approach the surface the sulphur compounds are cooled and oxidized and the mineral is deposited. Both of these processes, oxidation and loss of heat, can best take place near the surface; hence it is expected that the best deposits will be shallow. It is of interest also to note that the ore is associated chiefly with calcareous deposits and not with siliceous rocks, though the solutions must have traversed sandstone beds in their passage from below. This fact is believed to indicate that the limestone is essential to the precipitation of the sulphur from the solutions. Prof. Chase Palmer, of the United States Geological Survey, has made a special study of the chemical changes supposed to be involved in the deposition of sulphur, and suggests the following reactions:

It has long been known that a deposit of sulphur may be caused by direct oxidation of hydrogen sulphide in water through the agency of atmospheric oxygen alone: $2H_2S+O_2=2H_2O+2S$, to which reaction the formation of sulphur deposits has sometimes been attributed.

Winogradsky[a] has recently shown, however, that certain bacteria, which are found only in sulphur waters, utilize hydrogen sulphide, and by oxidizing it store the sulphur product in their cells. The sulphur does not long remain thus in storage, but is soon oxidized further by the cells to sulphuric acid, which is neutralized by the bicarbonates usually present in such waters, forming sulphates: $H_2SO_4+H_2Ca(CO_3)_2=CaSO_4+2CO_2+2H_2O$. The presence of these bacteria probably accounts for the deposition of only a part of the sulphur formed by hot-spring activity.

Most of the sulphur deposited from natural sulphur waters is probably the result of several changes in which various compounds are involved. One of these various reactions was noted by Béchamp,[b] who found that insoluble calcium carbonate is readily attacked when suspended in water charged with hydrogen sulphide, in which case two soluble calcium compounds are formed, viz, calcium hydrosulphide and calcium bicarbonate: $2CaCO_3+2H_2S=Ca(SH)_2+H_2Ca(CO_3)_2$.

The importance of the limestone in the deposition of the sulphur from hot waters is shown by some comparative experiments recently made by myself on the solvent action of sulphur waters on calcium carbonate, which indicate that the presence of sodium chloride increases materially the solvent power of a hydrogen sulphide water on calcium carbonate.

It is also known that the soluble polysulphides are decomposed by acids, reproducing hydrogen sulphide, and causing copious deposits of sulphur. Even carbon dioxide is capable of decomposing the polysulphides in this manner: $CaS_2+2CO_2+2H_2O=H_2Ca(CO_3)_2+H_2S+S$. The thiosulphates which often accompany the sulphides in natural waters are also similarly attacked by carbon dioxide, yielding a deposit of sulphur: $Na_2S_2O_3+CO_2+H_2O=NaHCO_3+NaHSO_3+S$.

All the essential conditions mentioned above are operative in the area here described. Hot springs are active, limestone is abundant, and the hot waters hold H_2S, CO_2, and sodium and potassium salts in solution. It seems very probable, therefore, that the methods of deposition indicated by Professor Palmer are at least the active if not the essential reactions by which the sulphur was deposited.

---

[a] Lafar, Franz, Technische Mykologie, 1897 edition.
[b] Annales de chimie et de physique, 4th ser., vol. 16, 1869, p. 202.

MINING, SMELTING, AND MARKETING.

Mining is carried on by open-pit quarry methods, in which promising places are located, small drill holes are put down, and the rock is blasted with powder. The rock is then sorted by hand and all ore estimated to contain over 30 per cent of sulphur (the yearly average of the ore smelted is 35 per cent) is taken to the smelter by wagon or tram. At the smelter the ore is placed in bins, from which it is discharged into small steel cars with perforated sides, each holding about 1¼ tons of ore. A string of three cars is then run into a large cylindrical retort, the door closed, and steam admitted at 65 pounds pressure for an hour and three-quarters. The sulphur is melted and flows to the bottom of the retort, from which it escapes through a trap into bins, where it is allowed to cool. When the sulphur has been melted from the rocks the cars containing the gangue are removed from the retort, other cars admitted, and the process repeated. This process is not considered highly efficient, as only about two-thirds of the sulphur which the rock contains is melted out; the remainder, being contained in the gangue, is thrown on the refuse dump. After the sulphur is cooled it is ground in an 8-inch Blake crusher and pulverized in a rotary grinder to a powder apparently equal in fineness to flowers of sulphur. The sulphur powder is sacked and taken to Cody, 3 miles distant, for shipment.

PRODUCTION.

The sulphur refining and milling plant was built in 1906, and during the first year of operation 850 tons of sulphur were produced from 2,833 tons of ore. Of this amount 350 tons were sent to Omaha, Nebr., the chief distributing point, and the remainder was used in compounding sheep-dipping preparations in Wyoming and adjacent States. The market price is $35 per ton at Cody.

SURVEY PUBLICATIONS ON SULPHUR AND PYRITE.

The list below includes the important publications of the United States Geological Survey on sulphur and pyrite:

ADAMS, G. I. The Rabbit Hole sulphur mines, near Humboldt House, Nev. In Bulletin No. 225, pp. 497–500. 1904.

DAVIS, H. J. Pyrites. In Mineral Resources U. S. for 1885, pp. 501–517. 1886.

ECKEL, E. C. Gold and pyrite deposits of the Dahlonega district, Georgia. In Bulletin No. 213, pp. 57–63. 1903.

——— Pyrite deposits of the eastern Adirondacks, N. Y. In Bulletin No. 260, pp. 587–588. 1905.

LEE, W. T. The Cove Creek sulphur beds, Utah. In Bulletin No. 315, pp. 485–489. 1907.

MARTIN, W. Pyrites. In Mineral Resources U. S. for 1883–84, pp. 877–905. 1886.

RICHARDSON, G. B. Native sulphur in El Paso County, Tex. In Bulletin No. 260, pp. 589–592. 1905.

ROTHWELL, R. P. Pyrites. In Mineral Resources U. S. for 1886, pp. 650–675. 1887.

SPURR, J. E. Alum deposit near Silver Peak, Esmeralda County, Nev. In Bulletin No. 225, pp. 501–502. 1904.

MISCELLANEOUS NONMETALLIC PRODUCTS.

A COMMERCIAL OCCURRENCE OF BARITE NEAR CARTERSVILLE, GA.

By C. W. HAYES and W. C. PHALEN.

INTRODUCTION.

The original source of barium is to be found in the silicates of the metal contained in the igneous rocks, from which it is derived as the carbonate during the ordinary processes of weathering. Though commonly regarded as one of the less common constituents in the earth's crust, F. W. Clarke[a] has shown that its oxide is the thirteenth in order of abundance, an average of 617 analyses indicating the presence of 0.11 per cent.

One of its commonly occurring forms is the sulphate ($BaSO_4$) or the mineral barite. This substance is more commonly found in mineral veins than in sedimentary rocks. It has also been observed as a sintery or stalactitic deposit and as the cementing substance in sandstone. All of its occurrences indicate that it is a mineral of aqueous origin and has resulted either from direct deposition in water or as a precipitate when waters of certain composition mingled.

Barite is known to occur at several localities in the vicinity of Cartersville, Bartow County, Ga., and at one place between $2\frac{1}{2}$ and 3 miles southeast of the town it is mined on a fairly large scale by the Nulsen, Klein & Krausse Manufacturing Company, of Lynchburg, Va., and St. Louis, Mo. Although originally deposited from water its present position is in a residual mantle of clay and gravel.

GEOLOGY.

The geology of the Cartersville district has been worked out by the senior writer. The district is in the southeastern half of Bartow County, and the outcropping rocks are the older crystalline schists,

---

[a] Bull. U. S. Geol. Survey No. 228, p. 17. Clarke's more recent figure, contained in an unpublished manuscript and based on an average of 678 analyses, remains the same, namely, 0.11 per cent BaO.

gneisses, and granites, and the partially or completely metamorphosed sedimentary rocks of the Ocoee group, including gneisses, schists, slates, and conglomerates, which cover the surface in the southeastern half of the area, and a series of unaltered sediments consisting of quartzite, limestone, shale, and dolomite, all Paleozoic, in the northwestern half. The former rocks make up the more mountainous parts of the area. The Paleozoic rocks outcrop in the valley region to the west. In descending order and as at present differentiated, they are as follows:

Knox dolomite.
Conasauga and Rome formations.
Beaver limestone.
Weisner quartzite.

All these formations except the Knox dolomite belong in the middle or lower Cambrian, and it is probable that the lower portion of the Knox should also be classed with the Cambrian. Of these formations, only the Weisner quartzite and Beaver limestone are of interest in this connection, for it is in these rocks, or rather associated with them, that the deposits of barite are found.

The Beaver limestone rarely outcrops at the surface. It occupies valleys parallel with the ridges formed by the adjacent harder formations and is generally covered by a heavy residual deposit of clay and wash from the higher land on either side. It contains a large amount of clayey impurities, and its weathered outcrops appear as a clay shale.

The Weisner quartzite forms a fairly continuous belt about 19 miles long, extending from the northeastern nearly to the southern edge of the area. Its greatest width, about 3 miles, is nearly due east of Cartersville, and the average width of the belt is between 1 and 2 miles. It consists chiefly of fine-grained vitreous quartzite, although it contains some beds of fine conglomerate and siliceous shale. The thickness of the formation is probably between 2,000 and 3,000 feet, and may be considerably more; but it can not be accurately determined because of the intense folding which it has undergone and the absence of satisfactory exposures. In addition to the folding, it is doubtless intersected by numerous faults, the evidence of which is seen in its crushed and brecciated condition at many points. The mechanical and chemical conditions resulting from the deformation of the quartzite beds are responsible for the deposition of certain of the iron ore and ocher deposits of the region which are contained in it, and also originally of the barite deposits.

OCCURRENCE OF THE BARITE.

The barite is so intimately associated with certain of the iron ores of this region, particularly with the ocher, that conclusions regarding the origin of the latter are of importance in this connection. The iron oxide [a] forming the yellow ocher deposits in the Cartersville district is a direct replacement of the silica in the Weisner quartzite. It appears that the faulting of the region, by fracturing the rocks, afforded favorable conditions for the percolation of surface waters to great depths, and that as the faulting was doubtless accompanied by the development of considerable heat, the region was probably characterized by numerous thermal springs. As is well known, silica, under favorable conditions, especially under great pressure and at high temperature, becomes one of the readily soluble rock constituents.

With reference to the barite occurring at the mine of the Peruvian Ocher Company, at the wooden bridge over Etowah River, where perhaps the best opportunity exists to study the past action of the thermal waters and the development of the iron ocher and barite, the following statement is quoted from the paper cited above:

Numerous open passages and cavities penetrating the quartzite and the bodies of ocher are met in mining. The smaller cavities are generally lined with a crust of small quartz crystals, while the larger ones frequently contain beautiful crystals of barite, which were probably deposited after the conditions favorable for the solution of silica and the deposition of ocher had passed. Groups of acicular crystals of this mineral, several inches in length, are not uncommon. It also occurs in white granular veins. The barite is called "flowers of ocher" by the miners. It remains in the residual soil which covers the quartzite outcrops and affords the best means of tracing the ocher deposits. It is found at numerous points on the low quartzite ridge north and south of the Etowah River. and prospecting at these points has never failed to reveal more or less extensive deposits of ocher.

The south end of the quartzite belt above described has a distinct anticlinal structure, as shown in the gap where Etowah River cuts across it. It is on the eastern side of this anticline that the principal barite deposits occur. The quartzite dips eastward at about 45° and the overlying shaly Beaver limestone, although not well exposed, probably underlies the valley occupied by the railroad between the quartzite and the crystalline rocks to the east. At various places along the eastern side of this anticlinal quartzite ridge deposits of limonite have been worked for many years. South of Etowah River these deposits occur well up on the side of the ridge, upon the slope immediately below the outcrop of the quartzite. The limonite is seen to rest directly upon the quartzite, and large exposures of the latter are found in the old ore workings. The limonite appears to have a definite relation to the quartzite and forms a nearly continuous belt of irregular deposits parallel to its outcrop. Parallel to this belt and adjacent to it down the slope is another belt occupied by the deposits of barite. The extent of the barite is not yet proved, and

[a] Hayes, C. W., Geological relations of the iron ores in the Cartersville district, Georgia: Trans. Am. Inst. Min. Eng., vol. 30, 1900, pp. 403–419.

it probably does not extend in workable amount throughout the length of the quartzite ridge, though it is known to be present in variable amount northward to Etowah River and for some distance beyond.

While this belt has not been sufficiently prospected to prove the continuity of the deposits, their thickness is indicated by a tunnel which has been driven into the side of the hill about 300 feet north of the present workings. This tunnel penetrates 120 feet of the barite-bearing clay and then enters the iron ore. (See fig. 26.)

The accompanying sketch shows the relations of the limonite and the barite to the underlying formations, so far as these relations can be determined from the present development. One striking feature is the sharp separation between the two minerals; each appears to occupy a definite horizon, and so far as observed there is no intermingling. While both in their present condition are residual, they

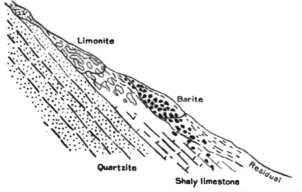

Fig. 26.—Sketch section showing relations of barite and limonite to underlying formations near Cartersville, Ga.

appear originally to have replaced distinct beds in the shaly limestone overlying the quartzite, the iron being deposited immediately adjacent to the quartzite, and the barite, for the most part, at least, in higher beds. It is highly probable that deposition of the barite and limonite took place, in part, at least, simultaneously. It is quite certain that gravity has been largely instrumental in concentrating the former into a workable deposit.

The barite deposit, as shown in the present workings, has a thickness of about 50 feet normal to the slope of the surface. It is intermingled with residual material, chiefly red, brown, and yellow clay, with some fragments of quartzite. The barite itself makes up about one-third of the material removed in mining. It consists of irregular or slightly rounded bowlders ranging from a few ounces up to several hundred pounds in weight. It is for the most part of a massive, compactly granular structure, and of a pure-white or faint-bluish color. As it occurs in the face of the open pit its presence might not be suspected owing to films of ferruginous clay which cover the nodules. The iron stain is confined almost entirely to the surface of the bowlders.

The deposit is worked in an open pit. The ore and accompanying clay are loosened by blasting and shoveled into small cars, which dump into a steeply inclined sluiceway with a stream of water. This carries the ore to a log washer at the base of the hill and at the same time frees it from much of the associated clay. After passing through the washer the ore is separated into three grades by hand picking. The highest grade contains some iron oxide, but merely as a thin film on the outside, which may be removed readily by acid. The inferior varieties contain more or less iron oxide disseminated throughout their mass, making its complete separation expensive and interfering with its utilization for certain purposes.

Work has been in progress about six months, and during this time more than 1,000 tons of barite, valued at approximately $4,000, have been removed. The deposits will in all probability prove fairly extensive, and the future prospect of the industry seems good.

GRAPHITE DEPOSITS NEAR CARTERSVILLE, GA.

By C. W. HAYES and W. C. PHALEN.

Graphite is a mineral the demand for which in the United States at present exceeds the supply.[a] A new occurrence, therefore, or the possible extension of an occurrence already known, merits attention. It is quite possible that the utilization of the deposits of graphitic slate in northwestern Georgia may furnish an abundant supply of certain grades of this mineral.

The extreme northwestern portion of Georgia is occupied by sedimentary formations of Paleozoic age—limestones, sandstones, and shales in great variety. To the east and south of these are metamorphic and crystalline rocks whose age is not definitely determined, but which are probably older than any of the determined Paleozoic formations. The contact between these two classes of rocks passes southward from the Tennessee line to the vicinity of Cartersville, and thence westward to the Alabama line a little south of Rock Mart. It intersects in Georgia the following counties: Murray, Gordon, Bartow, Polk, and Tallapoosa. To the east and south of this contact and forming a nearly continuous belt from a few hundred yards to several miles in width occurs a graphitic talcose slate. Similar but less continuous belts also occur to the east within the metamorphic area. This slate, particularly the first-described belt, has been prospected at numerous places for its graphite, and is worked on a considerable scale at two points in the vicinity of Cartersville. The slate was originally, in all probability, a carbonaceous clay shale. The metamorphism to which it has been subjected has converted the carbonaceous material, probably derived originally from plants, into graphite and the clay into talcose minerals of somewhat indefinite composition and mineralogical character. While the color of the rock is generally black, some portions of it contain little if any graphitic carbon and are light bluish gray in color. Selected samples of the rock would probably run as high as 12 or 15 per cent of graphitic carbon, although the content of carbon in the rock as a whole is much lower. The material going to the mill of the American Chemical Mining Company,

---

[a] Smith, George Otis, Mineral Resources U. S. for 1905, U. S. Geol. Survey, 1906, p. 1265.

in the vicinity of Emerson, Ga., immediately north of the Western and Atlantic Railroad, averages about 4 per cent.   In this mine, however, and for the purpose for which the product is used, a high percentage of carbon is not essential.   All classes of the slate are, therefore, mined together, both that containing a high percentage of carbon and the bluish bands, which are practically free from carbonaceous matter.   At the mine of the Cherokee Chemical Company, about 2 miles south of Emerson, the slate is more uniformly graphitic, and the product runs about 9 per cent of carbon.   The slate here contains some stringers of vein quartz with traces of gold and also shows the presence of considerable pyrite, with here and there a distinct copper stain.   Oxidation of the pyrite gives rise to an abundant efflorescence of alum and other sulphates.

The rock is mined by open pits and can be delivered to the mill at a very slight cost.   Both at the points of present production and elsewhere along the belt of slate above described the supply of raw material is practically unlimited.

The graphite appears to be chiefly if not entirely of the amorphous variety.   Examined under the microscope no crystalline outlines could be detected, the graphitic material occurring in irregular fragments, with jagged edges, associated chiefly with quartz.   In some of the material collected from the pit of the Cherokee Chemical Company there is a suggestion of crystalline outlines.   When the material is fused with alkaline carbonate and caustic alkalies the carbonaceous matter appears to be wholly oxidized and to go completely into solution, whereas high-grade flake graphite, at least, is supposed to resist such treatment.[a]   Boiled with nitric acid (specific gravity 1.4) there is apparently no decomposition; at least the solutions were not decolorized as is the case when anthracite, bituminous coal, and charcoal from various sources are similarly treated.[b]   It would seem, therefore, that although the material can not be classed as graphite in the sense that Wittstein has determined the chemical behavior of the mineral, it nevertheless has some of the chemical properties of graphite, and for all practical purposes may be so considered.

Even with the very finest grinding practicable in commercial practice it may be questioned whether it will be found possible to effect a separation from the impurities, in so far as they are flaky in character, such as mica and talc.   The specific gravity of graphite ranges from 2.1 to 2.5; of talc, 2.7 to 2.8; and of muscovite (mica), 2.76 to 3— all too close to attempt perfect separation.   As a lubricant, the presence of the talc can not be regarded as deleterious, though care would have to be exercised to remove even the last traces of quartz and

---

a Wittstein's method for the determination of graphite, Lunge, Chemische-technische Untersuch-ungsmethoden, 4th ed., vol. 2, p. 785.

b Classen, Ausgewählte Methoden der analytischen Chemie, p. 637.

other gritty material when used for such a purpose. There should, however, be no difficulty in removing the quartz, pyrite, and similar minerals when the purpose to which it is to be applied requires it.

The principal use to which the Georgia product is put is as a filler in fertilizer. It is claimed that when mixed with potash and ammonium salts and the other ingredients which enter into fertilizers it prevents the absorption of moisture and consequent caking of the mass. This facilitates grinding, and the fertilizer retains its powdery state after being ground and may thus be more evenly and economically applied to the land. However this may be, it is quite certain that there is nothing in the graphite itself that enriches the land.

Some of the graphite is also used in paint, and practical tests on iron work, particularly where subjected to high temperatures, are reported to have been highly satisfactory. Its use in stove polish, in electrotyper's powder, and for foundry facings was not reported, though it should prove entirely suitable for any or all of these applications, where low-grade material only is required.

The great extent of the deposits of raw material, their convenience to lines of transportation, and the cheapness with which the graphite can be mined and prepared for use should insure these deposits a large development.

47076—Bull. 340—08——30

MEERSCHAUM IN NEW MEXICO.

By Douglas B. Sterrett.

INTRODUCTION.

Meerschaum has been extensively used for over a century in the manufacture of pipes and other articles for the use of smokers, as cigar holders, mouthpieces, etc. The principal source of supply has been, for many years, the deposits in the plains of Eskishehr, in Anatolia, Asia Minor, about 120 miles southeast of Constantinople. Deposits of the mineral are also reported to occur in Greece, on the island of Eubœa; in Moravia, Austria, near Hrubschitz; in Spain near Vallecas, Madrid, and Toledo; and in Morocco. The meerschaum deposits of Eskishehr have been briefly described by J. Lawrence Smith [a] as occurring in a valley filled with drift material from the surrounding mountains that has been consolidated by lime. The meerschaum is scattered through the drift in rounded nodular masses, with pebbles and fragments of magnesian and hornblende rocks.

The mineral sepiolite, or meerschaum, as it is commonly called, is a hydrous silicate of magnesia with the probable composition $H_4Mg_2Si_3O_{10}$ or $2H_2O + 2MgO + 3SiO_2$.[b] Many analyses of meerschaum only roughly approximate the formula given above. This is probably due largely to the uncertain relation of the water to the other elements of the mineral. A large amount of hygroscopic moisture is driven off below 110°, sometimes amounting to nearly half of the total lost at a red heat.

Pure meerschaum is a white, porous mineral, with a specific gravity of about 2. In much of it, however, the porosity is so great that blocks of the mineral will readily float on water. This property, along with its snow-white color, gives rise to the name, meerschaum, from the German for sea foam. In a similar way the French often call it "écume de mer." It absorbs water strongly and becomes somewhat plastic, but returns to its original condition on drying.

---

[a] Am. Jour. Sci., 2d ser., vol. 7, 1849, p. 285.
[b] Dana, J. D., System of Mineralogy, 6th ed., p. 680.

When saturated it will not, of course, float on water. The hardness of meerschaum is from 2 to 2.5.. It is very tough, breaking with a conchoidal to earthy fracture. Some forms have a leathery or fibrous texture, and in these the toughness is very pronounced. The luster is dull and earthy, somewhat like that of plaster.

The ease with which meerschaum can be carved, its whiteness, and the fine polish it takes with wax render it especially suitable for elaborate carving and artistic treatment in the manufacture of pipes. Meerschaum pipes are prized for the rich cream-brown or brown color which the bowl assumes after being smoked a while. This color is caused by the mixture of the nicotine from the tobacco with the wax used in polishing the pipe, permeating through the mineral. As long as there is absorbed wax in the meerschaum the color of a pipe will grow darker and nearly black with continued smoking. It is, therefore, necessary to "fix the color" of the pipe when the proper shade is obtained. Though the principle employed is the removal of the wax and boiling in linseed oil to harden the mineral and render it less porous, there are trade secrets in the process which the writer is not at liberty to divulge.

The manufacture of meerschaum, together with clay, amber, horn, wood, metals, etc., into pipes and similar articles is a thriving industry in parts of Germany and Austria. The headquarters of the industry in Germany is at the town of Ruhla, in the Thuringian Forest. According to Consul George N. Ifft,[a] of Annaberg, there are between 3,000 and 4,000 workmen employed at this industry, which was started in 1767. It is said that the supply of meerschaum is becoming low and that the manufacturers experience great difficulty in obtaining the necessary material to keep their factories going. This scarcity is said to be caused partly by failure of the mines in Asia Minor to meet the demands of the trade and partly because American and English agents have gained control of the Asia Minor production. Consul U. J. Ledoux, of Prague, reports similar difficulties in the Austrian meerschaum industry.

The treatment that meerschaum receives before reaching the consumer is varied. At Eskishehr the crude mineral is mined by systematic pits and galleries.[b] The nodular masses are first roughly scraped to remove the earthy matrix; then dried, scraped again, and polished with wax. The roughly polished nodules, in almost every conceivable peculiarity of form, are then shipped to the manufacturers. Pipe bowls are first turned out on lathes or carved by hand. The bowls are then smoothed down with glass paper and Dutch rushes, and after being boiled in wax, spermaceti, or stearin, are carefully polished with bone ash or chalk.

---

[a] Daily Cons. Rept., April 25, 1907.   [b] Encyclopædia Britannica, vol. 15, p. 825.

Artificial and imitation meerschaum are also manufactured for the trade. Artificial meerschaum is made by consolidating waste chips and fragments by pressure. Imitation meerschaum is sometimes prepared by treating hardened plaster of Paris with wax and coloring with gamboge and other suitable materials. Many of these imitations are nearly perfect.

NEW MEXICO DEPOSITS OF MEERSCHAUM.

LOCATION.

Two deposits of meerschaum have been located in Grant County, N. Mex. Both are in the upper Gila River valley. One, taken up by the Meerschaum Company of America, lies in the Alunogen mining district, about 23 miles in an air line east of north from Silver City, near Sapillo Creek; the other deposit has been taken up by the Dorsey Meerschaum Company, and lies in the Juniper mining district, about 12 miles northwest of Silver City, in the canyon of Bear Creek.

The Dorsey meerschaum mine, consisting of six claims, was visited by the writer in October, 1907, when the information here given was obtained. The deposit lies in the bottom and walls of the canyon of Bear Creek. This canyon has steep cliffs about 100 feet high at the base and other cliffs above rising to a height of several hundred feet.

GEOLOGY.

The rock forming the canyon walls is chiefly light and dark gray to brownish-gray limestone, with some sandstone strata included. The limestone contains a large amount of cherty matter in certain beds, both in nodules and small bands up to an inch or two in thickness, lying parallel with the bedding. In some places these smaller cherty bands are very numerous and give the limestone a banded or ribbon structure. The limestone shows the first stages of crystallization and under the microscope numerous small calcite crystals can be seen scattered through the finer homogeneous rock. Such rock contains a notable amount of magnesia, but as it effervesces in cold dilute hydrochloric acid it is not a true dolomite. A sample from a prominent bed of brownish-gray sandstone included in the limestone was found to consist of well-rounded quartz grains with the interstices almost entirely filled with calcite as a cementing material. The slightly brownish color of the stone is due to small patches of ferric oxide inclosed in the cementing material. It is said that granite outcrops in the canyon about half a mile up the creek from the meerschaum deposit. Bowlders of a coarse porphyritic granite (or possibly monzonite) with large red feldspar phenocrysts were seen in the creek débris, which contained also bowlders of a dark traplike rock resembling diabase.

The structure of the formations is not simple and in the brief examination of the locality was not worked out. At a point about a quarter of a mile below the deposit where the canyon is narrow, the limestone beds have a strike west of north, with a dip ranging from nearly vertical to 60° E. Farther up the creek the dip becomes less until it is nearly flat above a prominent sandstone bed, 15 to 20 feet thick, among the meerschaum deposits. Farther above the sandstone ledge the strike of the beds is northerly and nearly parallel with the course of the creek. The dip seems to be low and in opposite directions on each side of the canyon, which is therefore the axis of an anticlinal fold that apparently pitches slightly to the north. The limestone has been badly broken up by large fractures, with some faulting, and smaller joints. Several prominent fractures filled with vein material were observed in the bottom of the canyon, where the beds have a low dip. These veins had a northerly strike with a nearly vertical dip. There were many other fractures and veins, however, some of them prominent and others merely joints and seams, cutting across the direction of these larger breaks.

The age of the rocks has not been definitely determined. Two fragments of brachiopod shells, from a large angular limestone bowlder in the bottom of the canyon, were identified as strophomenoids by Mr. E. M. Kindle, and the rock is probably Ordovician in age.

OCCURRENCE.

The meerschaum occurs in veins, lenses, seams, and balls in the limestone. All but the balls are fillings of fractures and joints, which do not seem to be confined to any definite direction. The veins are filled with chert, quartz, calcite, clay, and meerschaum. Chert is the most important gangue mineral and occurs in the veins with meerschaum in bands, lenses, and nodules. Both the crystallized quartz and calcite were observed in small veins, in which also there was a small amount of meerschaum. The largest vein seen contained considerable reddish clay with chert and meerschaum.

The mineral occurs in two different forms—(a) in nodules of irregular shape and (b) somewhat massive, with a finer and more compact texture than the nodular form. Some of the veins and seams are filled with massive meerschaum having practically the same texture as that in the nodules, though not in nodular form. In other veins there is both compact massive and nodular meerschaum, generally embedded in red clay.

The nodules range from less than an inch up to several inches in diameter and are of all shapes, with small, rounded knots and bumps protruding from the surface, which is generally coated or stained with the inclosing clay. The nodules are exceedingly tough and have to be vigorously beaten with a hammer before they will break.

The fracture of this kind of meerschaum is very uneven and the texture is fibrous, or rather leathery and porous. The color is pure white except where iron stains have worked in from the red-clay matrix. Small fragments from the nodules sometimes float for a while when dropped into water, though the greater part sink. Some of the meerschaum that was not light enough to float became so after it had been heated. After absorbing water this meerschaum, like that from other localities, becomes somewhat mushy and has a soapy feeling.

The massive meerschaum is finer grained, less leathery, and heavier. It is very tough, however, and some pieces break with a conchoidal fracture. Small fragments floated on water a minute or two after heating. This variety does not absorb water so rapidly as the nodular varieties.

The occurrence of meerschaum in balls was observed chiefly in one layer of limestone 5 or 6 feet thick. The balls ranged up to 2 or 3 inches in diameter and were irregularly distributed through the limestone. In some places they were plainly connected with one another either by merging or by veins; in other places they were apparently unconnected with other bodies of meerschaum. These balls, so far as observed, did not contain meerschaum of commercial value, but were composed partly of calcite and another fine-textured white mineral. Some whole balls were composed of these minerals; others contained a core or breccia of chert or dark limestone.

CHEMICAL PROPERTIES.

A chemical analysis made by George Steiger, of this Survey, on selected mineral showed the composition of the meerschaum from the Dorsey mine to be approximately that called for by the chemical formula previously given ($2H_2O + 2MgO + 3SiO_2$). In the following table are given (1) the results of Mr. Steiger's analysis, (2) the theoretical composition, and (3) Mr. Schaller's analysis of material from the deposit of the Meerschaum Company of America on Sapillo Creek:

Analyses of meerschaum.

	1.	2.	3.
SiO_2	57.10	60.8	60.97
Al_2O_3	.58	....	9.71
Fe_2O_3	Trace.	....	
MgO	27.16	27.1	10.00
CaO	.17	....	.22
CO_2	.32	....	
Water	14.78	12.1	19.14
	100.11	100.0	100.04

The water determination in analysis 1 was made on material which had been ground in an agate mortar for several hours. As compared with the theoretical composition of meerschaum the water is too high by 2.68 per cent. Experimental work by Mr. Steiger revealed the fact that during the process of grinding for eighteen hours meerschaum absorbed 2.58 per cent of water. This test was made on other material than that used for the analysis, with the following results:

Per cent.

Original water in fragmental meerschaum	11.75
After six hours' grinding	13.48
After twelve hours' grinding	13.49
After eighteen hours' grinding	14.33

The percentage of water given in column 1 is therefore doubtless too high, as the material probably absorbed water during grinding. A rough determination made on fragments of the same sample also showed, however, more water than is required by the theoretical composition.

The nature of the water in meerschaum is very uncertain. Total water determinations were made at a high heat with calculations for loss of CO_2. A special test made by Mr. Steiger on the ground meerschaum as used for analysis 1 gave the following results:

Water lost by meerschaum at various temperatures.

Temperature (° C.).	Loss of water (per cent).
60	3.49
105	3.17
150	1.07
275	2.40
Total at 275°	10.13

From this test it will be seen that considerable percentage of the water is probably hygroscopic, but just where the line between hygroscopic water and water of composition should be drawn it is impossible to say.

A specimen from the deposit owned by the Meerschaum Company of America was sent to the Survey by Ledoux & Co. of New York. An analysis of this material (No. 3 in the table), made by Waldemar T. Schaller, showed the presence of considerable alumina, which was also found by other chemists, as indicated by analysis published in the report of the company. Though the analysis shows that the specimen was not true meerschaum, the character of the two minerals is very much the same. In fact, the properties possessed by this mineral are so like those of true meerschaum that a company has undertaken to develop the deposit.[a]

---

[a] Collins, A. F., Mining meerschaum in New Mexico: Mining World, June 1, 1907, p. 688.

PHYSICAL PROPERTIES.

The meerschaum from the Dorsey mine is very similar in appearance to that from Asia Minor as it appears on the market ready for carving. The Asia Minor meerschaum is a little lighter and more spongy than the surface material from the Dorsey deposit, owing in part, probably, to the fact that the former has been dried before shipping.

The results of tests made by pipe makers on two different pieces of the nodular meerschaum from the Dorsey mine were not highly satisfactory, though probably as good as could be expected from surface material. Small pipes were turned down on a lathe and after treating with wax were polished. Besides the iron stains in certain lines through the meerschaum the wax gave a mottled appearance to the surface by rendering one portion a little more translucent than others. The statement of each pipe maker was that the material was heavier than usual for meerschaum. One of them, Mr. S. Heyman, of Baltimore, reported it to be much harder than any other meerschaum he had ever worked, and mentioned a fine sand or grit in the sample he tested. The sandpaper used in smoothing the pipe down was also worn out more rapidly than is usual with meerschaum. Mr. Heyman turned out a small, thin pipe which was somewhat translucent and mottled. The mottling was apparently caused by fibrous white tufts in a more translucent matrix. These tufts appeared to compose the grit mentioned above, since they remained as little lumps above the surface of the pipe after polishing. These lumps were found to be very tough when cut with a knife, and doubtless caused the sandpaper to wear away rapidly when the pipe was being smoothed down, by dragging the sand grains away from the paper. When tested for coloring qualities the pipe readily absorbed the wax and nicotine, which appeared on the outside of the bowl in its characteristic color.

The specimen of meerschaum from which Mr. E. Butzen, of Chicago, turned out a pipe was evidently more compact and smoother grained. The bowl received a fair polish, though it was badly mottled by iron stains.

Another piece of meerschaum carved into a rough pipe and sent to the Survey by Mr. W. P. Dorsey, of Silver City, N. Mex., appeared to be pure white in color and of an even texture. On boiling in paraffin, however, mottling became prominent, showing irregular seams of rather translucent material in white tufted fibrous meerschaum.

The value of the New Mexico meerschaum has not yet been proved. At the Dorsey mine numerous outcrops of seams and veins of various sizes have been located. In chemical composition the material corresponds closely to meerschaum, though the physical properties of material from the outcrop have not as yet proved of good grade. Whether valuable meerschaum will be found below the surface can be learned only by opening some veins to a depth where surface movement and weathering have not affected the mineral. It is probable that the fibrous tufts described above are natural to the nodular meerschaum and will not disappear with depth. The more compact massive material should, however, be found free of stains and more sound below the surface.

As to the origin of this meerschaum in the limestone formation, the deposits were not sufficiently studied to permit the expression of an opinion.

SURVEY PUBLICATIONS ON MISCELLANEOUS NONME-TALLIC PRODUCTS, INCLUDING MICA, GRAPHITE, FLUORSPAR, ASBESTOS, AND BARITE.

The following list includes a number of papers, published by the United States Geological Survey or by members of its staff, dealing with various nonmetallic mineral products:

BAIN, H. F. Fluorspar deposits of southern Illinois. In Bulletin No. 225, pp. 505–511. 1904.

BALL, S. H. Mica in the Hartville uplift, Wyoming. In Bulletin No. 315, pp. 423–425. 1907.

—— Graphite in the Haystack Hills, Laramie County, Wyo. In Bulletin No. 315, pp. 426–428. 1907.

BREWER, W. M. Occurrences of graphite in the South. In Seventeenth Ann. Rept., pt. 3, pp. 1008–1010. 1896.

BURCHARD, E. F. Barite. In Mineral Resources U. S. for 1906, pp. 1109–1113. 1907.

—— Fluorspar. In Mineral Resources U. S. for 1906, pp. 1063–1066. 1907.

DILLER, J. S. Asbestos. In Mineral Resources U. S. for 1906, pp. 1123–1129. 1907.

EMMONS, S. F. Fluorspar deposits of southern Illinois. In Trans. Am. Inst. Min. Eng., vol. 21, pp. 31–53. 1893.

FULLER, M. L. The occurrence and uses of mica. In Stone, vol. 19, pp. 530–532. 1899.

HOLMES, J. A. Mica deposits in the United States. In Twentieth Ann. Rept., pt. 6, pp. 691–707. 1899.

KEITH, A. Talc deposits of North Carolina. In Bulletin No. 213, pp. 433–438. 1903.

KEMP, J. F. Notes on the occurrence of asbestos in Lamoille and Orleans counties, Vt. In Mineral Resources U. S. for 1900, pp. 862–866. 1901.

—— Graphite in the eastern Adirondacks. In Bulletin No. 225, pp. 512–514. 1904.

PHILLIPS, W. B. Mica mining in North Carolina. In Mineral Resources U. S. for 1887, pp. 661–671. 1888.

SMITH, G. O. Graphite in Maine. In Bulletin No. 285, pp. 480–483. 1906.

—— Graphite. In Mineral Resources U. S. for 1906, pp. 1139–1143. 1907.

SMITH, W. S. T. (See Ulrich, E. O., and Smith, W. S. T.)

STERRETT, D. B. Mica deposits of western North Carolina. In Bulletin No. 315, pp. 400–422. 1907.

—— Mica. In Mineral Resources U. S. for 1906, pp. 1149–1163. 1907.

STOSE, G. W. Barite in southern Pennsylvania. In Bulletin No. 225, pp. 515–517. 1904.

ULRICH, E. O., and SMITH, W. S. T. Lead, zinc, and fluorspar deposits of western Kentucky. In Bulletin No. 213, pp. 205–213. 1903.

INDEX.

O

Lightning Source UK Ltd.
Milton Keynes UK
UKHW031135290119
336362UK00009B/266/P

9 781528 505383